Lecture Notes in Computer Science   3634

Commenced Publication in 1973
Founding and Former Series Editors:
Gerhard Goos, Juris Hartmanis, and Jan van Leeuwen

T0223708

Luke Ong (Ed.)

Computer Science Logic

19th International Workshop, CSL 2005
14th Annual Conference of the EACSL
Oxford, UK, August 22-25, 2005
Proceedings

 Springer

Volume Editor

Luke Ong
Oxford University Computing Laboratory
Wolfson Building, Parks Road
Oxford OX1 3QD, United Kingdom
E-mail: lo@comlab.ox.ac.uk

Library of Congress Control Number: 2005930337

CR Subject Classification (1998): F.4.1, F.4, I.2.3-4, F.3

ISSN   0302-9743
ISBN-10   3-540-28231-9 Springer Berlin Heidelberg New York
ISBN-13   978-3-540-28231-0 Springer Berlin Heidelberg New York

Springer is a part of Springer Science+Business Media

springeronline.com

© Springer-Verlag Berlin Heidelberg 2005
Printed in Germany

Typesetting: Camera-ready by author, data conversion by Olgun Computergrafik
Printed on acid-free paper   SPIN: 11538363   06/3142   5 4 3 2 1 0

Preface

The Annual Conference of the European Association for Computer Science Logic (EACSL), CSL 2005, was held at the University of Oxford on 22–25 August 2005. The conference series started as a programme of International Workshops on Computer Science Logic, and then in its 6th meeting became the Annual Conference of the EACSL. This conference was the 19th meeting and 14th EACSL conference; it was organized by the Computing Laboratory at the University of Oxford.

The CSL 2005 Programme Committee considered 108 submissions from 25 countries during a two-week electronic discussion; each paper was refereed by at least three reviewers. The Committee selected 33 papers for presentation at the conference and publication in these proceedings.

The Programme Committee invited lectures from Matthias Baaz, Ulrich Berger, Maarten Marx and Anatol Slissenko; the papers provided by the invited speakers appear at the front of this volume.

Instituted in 2005, the Ackermann Award is the EACSL Outstanding Dissertation Award for Logic in Computer Science. The award winners for the inaugural year, Mikołaj Bojańczyk, Konstantin Korovin and Nathan Segerlind, were invited to present their work at the conference. Citations for the awards, abstracts of the theses, and biographical sketches of the award winners are at the end of the proceedings.

I thank the Programme Committee and all the referees for their work in reviewing the papers; Jolie de Miranda for her sterling work as editorial assistant; the other members of the local organizing team (William Blum, Matthew Hague, Andrzej Murawski and Sam Sanjabi), as well as many other Computing Laboratory colleagues who helped in various ways, for arranging the event itself; the organizers of CSL 2004; and Andrei Voronkov, whose EasyChair system greatly facilitated the handling of submissions and reviews.

Finally, I thank Merton College and the Computing Laboratory, which provided both financial support and much time from their staff.

June 2005   Luke Ong

Programme Committee

Albert Atserias *UPC Barcelona*
David Basin *ETH Zürich*
Martin Escardo *U. Birmingham*
Zoltan Esik *U. Szeged*
Martin Grohe *Humboldt U.*
Ryu Hasegawa *U. Tokyo*
Martin Hofmann *LMU München*
Ulrich Kohlenbach *T.U. Darmstadt*
Orna Kupferman *Hebrew U. Jerusalem*
Paul-Andre Mellies *U. Paris 7*

Aart Middeldorp *U. Innsbruck*
Dale Miller *E. Polytechnique*
Damian Niwinski *U. Warsaw*
Peter O'Hearn *Queen Mary U.*
Luke Ong *U. Oxford (Chair)*
Alexander Rabinovich *U. Tel Aviv*
Thomas Schwentick *Philipps U.*
Alex Simpson *U. Edinburgh*
Nicolai Vorobjov *U. Bath*
Andrei Voronkov *U. Manchester*

Additional Referees

Klaus Aehlig
Yoji Akama
Arnold Beckmann
Lev Beklemishev
Michael Benedikt
Ulrich Berger
Stefan Blom
Mikolaj Bojanczyk
Udi Boker
Julian Bradfield
Olivier Brunet
Krishnendu Chatterjee
James Cheney
Jacek Chrzaszcz
Bob Coecke
Giovanni Conforti
Bruno Courcelle
Olivier Danvy
Nachum Dershowitz
Mariangiola Dezani
Roy Dickhoff
Kevin Donnelly
Thomas Eiter
Zoltán Fülöp
Nicola Galesi
Lilia Georgieva
Erich Grädel
Esfandiar Haghverdi
Masahito Hasegawa
Yoram Hirshfeld

Hendrik Jan Hoogeboom
Dominic Hughes
Michael Huth
Andrew Ireland
Florent Jacquemard
Jan Johannsen
Neil Jones
Felix Klaedtke
Hans-Joachim Klein
Bartek Klin
Konstantin Korovin
Margarita Korovina
Viktor Kuncak
Ugo dal Lago
Francois Lamarche
Martin Lange
Stephane Lengrand
Giacomo Lenzi
L. Leustean
Paul Levy
Leonid Libkin
Ralph Loader
Markus Lohrey
Hans-Wolfgang Loidl
Michael Macher
Patrick Maier
Harry Mairson
Janos Makowsky
Sebastian Maneth
Claude Marché

Jerzy Marcinkowski
Jean-Yves Marion
Maarten Marx
Ralph Matthes
Richard McKinley
Ichiro Mitsuhashi
Georg Moser
Misao Nagayama
Hans-Juergen Ohlbach
Vincent van Oostrom
Martin Otto
Leszek Pacholski
Dirk Pattinson
Reinhard Pichler
Axel Polleres
Shaz Qadeer
Femke van Raamsdonk
Michael Rathjen
Eike Ritter
Tatiana Rybina
Hiroyuki Sato
Alexis Saurin
Andrea Schalk
M. Schmidt-Schauss
Aleksy Schubert
Nicole Schweikardt
Helmut Seidl
Hiroyuki Seki
Masaru Shirahata
Dieter Spreen

Christoph Sprenger
Kristian Stovring
Thomas Strahm
Thomas Streicher
Hayo Thielecke
Cesare Tinelli
Alwen Tiu

Sándor Vágvölgyi
Luca Vigano
Marina De Vos
Johannes Waldmann
Daria
  Walukiewicz-Chrzaszcz
Volker Weber

Andreas Weiermann
C.P. Wirth
James Worrell
Hongwei Xi
Byoung-Tak Zhang

Local Organizing Committee

William Blum
Matthew Hague
Jolie de Miranda
Andrzej Murwaski
Luke Ong
Sam Sanjabi

Table of Contents

Linear Logic and Ludics

Constraints

Finite Models, Decidability and Complexity

Verification and Model Checking

Constructive Reasoning and Computational Mathematics

Implicit Computational Complexity and Rewriting

Appendices

XML Navigation and Tarski's Relation Algebras

Maarten Marx

Informatics Institute, Universiteit van Amsterdam
The Netherlands

Navigation is at the core of most XML processing tasks. The W3C endorsed navigation language XPath is part of XPointer (for creating links between elements in (different) XML documents), XSLT (for transforming XML documents) and XQuery (for, indeed, querying XML documents). Navigation in an XML document tree is the task of moving from a given node to another node by following a path specified by a certain formula. Hence formulas in navigation languages denote paths, or stated otherwise binary relations between nodes. Binary relations can be expressed in XPath or with first or second order formulas in two free variables. The problem with all of these formalisms is that they are not compositional in the sense that each subexpression also specifies a binary relation. This makes a mathematical study of these languages complicated because one has to deal with objects of different sorts. Fortunately there exists an algebraic formalism which is created solely to study binary relations. This formalism goes back to logic pioneers as de Morgan, Peirce and Schröder and has been formalized by Tarski as *relation algebras* [7]. (Cf., [5] for a monograph on this topic, and [8] for a database oriented introduction). A relation algebra is a boolean algebra with three additional operations. In its natural representation each element in the domain of the algebra denotes a binary relation. The three extra operations are a constant denoting the identity relation, a unary conversion operation, and a binary operation denoting the composition of two relations. The elements in the algebra denote *first order definable* relations. Later Tarski and Ng added the Kleene star as an additional operator, denoting the transitive reflexive closure of a relation [6].

We will show that the formalism of relation algebras is very well suited for defining navigation paths in XML documents. One of its attractive features is that it does not contain variables, a feature shared by XPath 1.0 and the regular path expressions of [1]. The connection between relation algebras and XPath was first made in [4].

The aim of this talk is to show that relation algebras (possibly expanded with the Kleene star) can serve as a unifying framework in which many of the proposed navigation languages can be embedded. Examples of these embeddings are

1. Every Core XPath definable path is definable using composition, union and the counterdomain operator $\sim$ with semantics $\sim R = \{(x, x) \mid \text{not } \exists y : xRy\}$.
2. Every first order definable path is definable by a relation algebraic expression.
3. Every first order definable path is definable by a positive relation algebraic expression which may use the Kleene star.

L. Ong (Ed.): CSL 2005, LNCS 3634, pp. 1–2, 2005.
© Springer-Verlag Berlin Heidelberg 2005

4. The paths definable by tree walk automata and certain tree walk automata with pebbles can be characterized by natural fragments of relation algebras with the Kleene star.

All these results hold restricted to the class of finite unranked sibling ordered trees. The main open problem is the expressive power of relation algebras expanded with the Kleene star, interpreted on this class of models. Is this formalism equally expressive as binary first order logic with transitive closure of binary formulas? Whether the latter is equivalent to binary monadic second order logic is also open [2, 3]. So in particular we do not know whether each regular tree language can be defined in relation algebras with the Kleene star.

References

1. S. Abiteboul, P. Buneman, and D. Suciu. *Data on the web*. Morgan Kaufman, 2000.
2. J. Engelfriet and H. Hoogeboom. Tree-walking pebble automata. In *Jewels are Forever, Contributions on Theoretical Computer Science in Honor of Arto Salomaa*, pages 72–83. Springer-Verlag, 1999.
3. J. Engelfriet and H. Hoogeboom. Automata with nested pebbles capture first-order logic with transitive closure. Technical Report 05-02, LIACS, 2005.
4. J. Hidders. Satisfiability of XPath expressions. In *Proceedings DBPL*, number 2921 in LNCS, pages 21–36, 2003.
5. R. Hirsch and I. Hodkinson. *Relation algebras by games*. Number 147 in Studies in Logic and the Foundations of Mathematics. North-Holland,, 2002.
6. K. Ng. *Relation Algebras with Transitive Closure*. PhD thesis, University of California, Berkeley, 1984.
7. A. Tarski. On the calculus of relations. *Journal of Symbolic Logic*, 6:73–89, 1941.
8. J. Van den Bussche. Applications of Alfred Tarski's ideas in database theory. *Lecture Notes in Computer Science*, 2142:20–37, 2001.

Verification in Predicate Logic with Time: Algorithmic Questions

Anatol Slissenko[1,2,*]

[1] Laboratory for Algorithmics, Complexity and Logic, University Paris-1, France
[2] Dept. of Informatics, University Paris-12
61 Av. du Gén. de Gaulle, 94010, Créteil, France
slissenko@univ-paris12.fr

Abstract. We discuss the verification of timed systems within predicate logics with explicit time and arithmetical operations. The main problem is to find efficient algorithms to treat practical problems. One way is to find practically decidable classes that englobe this or that class of practical problems. This is our main goal, where we concentrate on one approach that permits to arrive at a kind of small model property. We will also touch the question of extension of these results to probabilistic systems that will be presented in more detail elsewhere.

1 Introduction

Even not so long ago testing was the main, uncontestable practical method of program validation [14, 16]. Though the foundations of testing pose interesting theoretical problems, the field of theoretical footing of the testing process remains, in some way, secondary with respect to the development of foundations of verification, that is regrettable. We mention this question, that is out of the scope of the present text, because it is clearly related to the concept of 'small model property' whose particular realization will be discussed below.

Nowadays the verification, one of the activities aimed at software validation, is gaining ground. Verification based on model checking becomes more and more widespread. Logic based verification, though less represented in conference activities, is of growing importance. The both approaches have their success stories. Their advantages and disadvantages are well known.

Verification presumes that, given a requirements specification Φ_{Req} and a program specification Φ_{Prg}, we have to prove that Φ_{Prg} verifies Φ_{Req}. And "to prove" involves some logic.

Suppose that a sufficiently expressible logic is at our disposal. Denote by Φ_{Runs} a formula representing the runs (executions) of our program. We may suppose that we have some notion of time (here we speak about 'physical' time) represented by a linearly ordered set $\mathbb{T}$, and every run is a mapping from $\mathbb{T}$ to an interpretation of the vocabulary of our program. "Formula Φ_{Runs} represents runs" means that every model of this formula is a run of the program,

---

* Member of St Petersburg Institute for Informatics, Russian Academy of Sciences, St-Petersburg, Russia

and inversely. Requirements usually consist of two parts (see, e.g., [16]), namely, requirements on functioning (e.g., safety, liveness) and a description of the environment where our program works – this part of requirements can be also viewed as a part of the description of the set of runs. We will denote a formula representing the environment by Φ_{Env} and a formula describing the demands on functioning by Φ_{Func}. Formula Φ_{Env} may describe properties that restrict the types of input signals or may relate the input and output signals to some functions introduced for the user convenience.

The most essential part of the verification problem can be formulated as proving the formula $\mathfrak{F} =_{df} ((\Phi_{Runs} \wedge \Phi_{Env}) \rightarrow \Phi_{Func})$. But this is not all. An important property to prove, without which proving Φ may become trivial, is that the program has a run for each input. If a program has no runs at all then any property of its runs is true. For a control program, the existence of runs means that for any input signal that satisfies Φ_{Env}, the program computes an output. This is a second-order property, as it is formulated in terms of second order quantifiers over input/output signals, and each signal in this context is a function of time. The verification literature rarely mentions this problem.

The formula Φ_{Runs} is relatively easy to get, even automatically, from a program specification written in a language with a rigorous semantics. As for Φ_{Env} and Φ_{Func}, it is harder and usually demands some abstractions. The more powerful is our formal language of verification, the easier is the task.

Having arrived at $\mathfrak{F}$ we can start the verification using proof search procedures or other ones, e.g., decidability or quantifier elimination algorithms. This is a logic based approach. Model checking approach usually demands to construct simplified abstractions of the set of runs and that of requirements. It may be laborious. The advantage is in many practical automatic tools that can be applied thereafter. But the results are much more relative than in the case of logical proof, even a found error may happen to be an error in the chosen abstraction.

The methodological disadvantage of the existing model-checking approaches is in enormous number of logics to (partially!) model the requirements and in a big amount of formalisms to model the programs.

Since a longtime one can see a trend of convergence of the both approaches, and we discuss one way to go in this direction.

We are interested in the verification of timed systems whose specification involves arithmetical operations and parameters. The parameters may represent abstract time constants, the number of processes etc. These just mentioned two types of parameters are often used in high level specifications.

We will mainly speak about continuous time, though many of our methods work also for discrete time. Continuous time will be represented by non negative reals, and discrete time – by natural numbers. For some systems, like real time controllers, some protocols, continuous time is more intuitive. Notice that in our reasoning about programs we always use time, and often continuous one. The unpleasant feature of algorithms with continuous time is that it is not so easy to find a mathematically precise definition of their semantics. For the present there is no 'universal' semantics that works in all situations relevant to practical verification. A detailed discussion of algorithms with continuous time can be found in the special issue of *Fundamenta Informaticae*, 2004, vol. 69.

Moreover, algorithmics of arithmetics over continuous time is simpler that that of discrete time. The known worst-case complexity bound for the theory of real addition is exponentially better than the one for the theory of integer addition (Presburger arithmetic). For the theory of real addition and multiplication (Tarski algebra) these bounds, that are the same as for the theory of real addition, are even 'infinitely' better than those for the theory of integer addition and multiplication (formal arithmetics) that is undecidable, and even not enumerable.

The approach that will be presented embeds all the specifications into a special type of predicate logic that will be called FOTL (*First Order Timed Logic*). Such a logic takes a theory that represents necessary mathematical functions and that has 'good algorithmic properties', and extends it with abstract functions needed to describe our systems. 'Good algorithmic properties' means decidability, quantifier elimination or simply practically efficient algorithms to deal with it. We will mainly speak about theories that have quantifier elimination algorithms.

FOTL turns to profit the quantifier elimination for the theories we use. This procedure provides a quantifier-free description of counter-models (of a given complexity) when the verification formula is not true – a property highly appreciated in the verification domain, as counter-models help to identify errors. Moreover, if the verification formula contains parameters for reals, this procedure returns a description of the scope of parameters for which the formula is true, or the 'forbidden parameters' for which it is false.

On the basis of FOTL we describe classes of verification problems that are not only decidable, but have the mentioned property of a quantifier-free description of counter-models. These classes are described in terms of what we call finiteness properties: *finite refutability* and *finite satisfiability*. The properties related to the functioning of a program, like safety or liveness, are usually *finitely refutable*: if there is a counter-model for such a property then the contradiction is concentrated on a small piece of this counter-model. For example, consider a distributed algorithm with N processes, and a property $R(t,p)$ that says that at moment t a particular event (R-event) occurs in p. We can express that "an R-event cannot be absent in the same process for a duration greater than d" by the formula

$$\forall p \, \neg \, \exists t \, \exists t' \, \left(\, (t' - t) > d \wedge \forall \tau \in [t, t') \, \neg R(\tau, p) \, \right). \tag{1}$$

This formula (1) is an example of Φ_{Func}. If the property (1) is false then there is a process p_0 and 2 time instants t_0 and t_1 such that

$$\left(\, (t_1 - t_0) > d \wedge \forall \tau \in [t_0, t_1) \, \neg R(\tau, p_0) \, \right). \tag{2}$$

So whatever be the behavior of processes different from p_0 or whatever be the behavior of p_0 at other time instants, the property will remain false. Hence, the 'core' of the counter-model is concentrated on a piece of interpretation of complexity $O(1)$.

A more complicated finiteness property concerns the behavior of programs. It is called finite satisfiability. Its simpler version [8] looks as follows. Take a run and some finite partial sub-run in it. *Finite satisfiability* means that we can extend this partial sub-run to a total finite run with a controlled augmentation of complexity.

In general this property is false even for rather simple timed systems, for example for timed automata [1] as shown in [7]. However, for practical systems we often have this property or the more general one [6]. This more general finite satisfiability property deals with runs that have a finite description involving infinitely many time intervals. It says that if we take a run and some finite partial sub-run in it, then we can extend this partial sub-run to a run consisting of ultimately repetitive pieces with a controlled augmentation of complexity. We will describe this property in section 4.

Combining both properties, namely finite refutability and finite satisfiability, we define a decidable class of implications ($\Phi \to \Psi$), where Φ is finitely satisfiable and Ψ is finitely refutable with a fixed complexity. The verification formulas from these decidable classes can be efficiently reduced to quantifier-free formulas that describe all counter-models of a given complexity, and we know that if a counter model exists then there is a counter-model of this complexity. This is our finite model property, though it is better to speak about bounded complexity model property, as the models we consider are not finite.

FOTL permits to describe rather directly (see [3, 4, 8]) the runs of basic timed Gurevich Abstract State Machines (ASM) [12] (we used this type of ASMs in[8]). Such a basic ASM consists of one external loop inside which one executes in parallel **If-Then**-operators whose **Then**-part is a list of assignments executed again in parallel. The parallelism is synchronous. The runs of timed parallel while-programs can be also represented in FOTL without complications.

The decidability algorithm for the simpler class (without infinite models of bounded complexity) was implemented and showed encouraging results [2], [3].

A shortcoming of the approach is that the finiteness properties are undecidable in general [7]. Finite refutability is a typical property of safety, and for safety it is usually quite evident, but it is less evident for liveness. Finite satisfiability may be hard to prove even for practical systems. For example, for usual abstract specifications of practical cryptographic protocols it is a hard open question.

One can use our approach along the lines of bounded model-checking. Recall that the basic idea of bounded model checking is to check the requirements for runs whose length is bounded by some integer k. It is feasible if the set of these runs of bounded length is of reasonable size, or if we can use some symbolic representation of these runs. In some cases we know that if there is a counter-model run, i.e., a run that does not satisfy the requirements, then there exists such a run whose length is bounded by a constant known a priory. This constant is called a completeness threshold [10, 17]. Otherwise, we can increase k until we can process the runs of length k in a feasible way, and stop when a counter-model is found or the checking becomes unfeasible. In the latter case we have some partial verification. Our analogue of such procedure is to increase the complexity of models to consider and to check whether there exists a counter-model of the chosen complexity.

Bounded model checking is being developed first of all as a practical tool which accelerates symbolic model checking. Completeness threshold was estimated in cases when the verification is a priory decidable, and the found bounds are very high. In our setting we deal with logics for which the verification problem is undecidable in general, and that are much more expressive than the logics used in model checking. Our notion of bounded model is also much more general. So when we proceed, like in practical bounded model-checking, by increasing the complexity of models to try, we seek counter-models in a much larger class. For the concrete problems that we studied, the complexity bounds on models to consider are very small, and thus, the search for such models is feasible. Moreover, these concrete problems are out of the scope of model-checking.

The structure of text below, that mainly presents results of my colleagues and myself cited above, is as follows. Section 2 briefly describes decidable theories of arithmetical operations that underlay our FOTLs. We mention also a particular theory that is conjectured to be decidable and that features a way to look for theories adequate to the applications. Section 3 gives an example of FOTLs that are well supported by decidability algorithms, even better to say, by counter-model constructing algorithms. These algorithms are based on finiteness properties described in section 4. In Conclusion we discuss some open problems and draw attention to logics with probability.

2 Decidable Theories with Arithmetical Operations

We briefly summarize some known decidable theories that are apt to the presented approach and give an example of a theory whose conjectured decidability would put the verification of parametric clock synchronization in a decidable class.

One of the simplest theories that deal with arithmetics and have quantifier elimination is the theory of real addition. It may be defined as a theory with real addition, unary multiplications by rational constants, arithmetical order relations and rational numbers as constants. With this set of constants only rational numbers are representable. The known quantifier elimination procedures have worst-case complexity (roughly) of the form $L^{n^{O(\alpha)}}$, where L is the length of formula, n is the number of variables, and α is the number of quantifier alternations. In practice, with good simplification algorithms, some implemented procedures work well.

The worst-case lower bound is exponential. I repeat my argument that all known proofs of lower bounds (absolute or relative, like hardness) are irrelevant to computational practice as they concern diagonal algorithms that never appear in practice. Thus, these bounds say something useful about the theory under consideration but not about instances that we are interested in.

We can add to the theory of real addition linearly independent constants, for example, a finite number of infinitesimals, and we will have the same algorithmic properties. Adding real constants whose linear dependence is unknown may pose problems.

The theory of integer addition (Presburger arithmetic) has similar algorithmic properties, but its known complexity is one exponent tower higher.

The theory of mixed real/integer addition has both real addition and integer addition, and rational (or natural) constants. If to add integer part to the vocabulary this theory has a quantifier elimination [19] whose complexity is the same as for Presburger arithmetic.

This theory can be combined with finite automata representation of numbers to resolve constraints [9]. We do not go into details of this kind of theories as we have not yet studied how to use them in our framework.

The theory of real addition and multiplication (Tarski algebra) has the same theoretical complexity of quantifier elimination as the theory of real addition. If the constants are rational numbers then one can represent algebraic numbers. One can add transcendental constants and represent the corresponding rational functions. Practical complexity of quantifier elimination is better for Collin's cylindrical decomposition than for the algorithms that are the best from theoretical viewpoint (remark that the complexity of cylindrical decomposition is exponent tower higher than that of the theoretically best algorithms).

Now we describe a quantifier-free theory that is sufficient to represent the verification of a clock synchronization algorithm. We mean the Mahaney–Schneider protocol as it is described in [18]. This theory is conjectured to be decidable. We do not need to go into details (see [15]), and just describe some important points. The protocol deals with N processes $\mathbb{P} =_{df} \{1, 2, \ldots, N\}$, each one having its clock. At the beginning the clocks are δ-synchronized for some δ. There are delays in communications that are much smaller than δ.

The processes start to synchronize the clocks not simultaneously but with some shifts in time implied by these non deterministic delays. They exchange messages that arrive again with some non deterministic shifts of time. However, for a process p only $N + 1$ time instants are important: the arrival of starting signal and the instants of receiving clock values from other processes. Some processes are Byzantine, their number is $B < \frac{N}{3}$. To calculate its clock update a correct process filters the received data evaluating the cardinality of some finite sets (subsets of $\mathbb{P}$) and calculating at the end the mean value of N individual updates calculated before.

Notice that N is a parameter, and that δ, the delays, etc. are also parameters. If N is concrete, then we can use our decidability algorithms (because of arithmetics even this case is out of the reach of model checking). However it is more interesting to resolve the parametric case. The decidability of the following quantifier-free theory (modulo minor technical details) of parametric addition would suffice for it.

Atomic formulas of the theory are constructed in the following way:

• Inequalities of the form $\eta \cdot N + \xi \cdot B + a_1 \cdot \alpha_1 + \cdots + a_k \cdot \alpha_k \, \omega \, 0$, where η and ξ are real abstract constants or rational numbers; a_i are rational constants; α_i are expressions of the form $f(p)$ or $f(p, q)$ constructed from a symbol of real valued function f, and of abstract constants p, q for processes from $\mathbb{P}$; ω is any usual order relation. Below we will refer to such inequalities as *basic inequalities*, and each basic inequality is a *formula*. The left part of the inequality above is a *simple sum*.

- An expression $\#\{p : L(p)\}$, where $L(p)$ is a basic inequality, and $\#$ means the cardinality, is a *cardinality term*. A cardinality term can be used to construct *formulas* of the form
$\#\{p : L(p)\}\omega(a \cdot N + b \cdot B + c)$, where a, b and c are rational constants, and ω is an order relation, as above.
- An expression $\sum(p, L(p), \theta(p))$, where p and $L(p)$ are as above, and $\theta(p)$ is a simple sum of the form $a_1 \cdot \alpha_1 + \cdots + a_k \cdot \alpha_k$ (here we use the same notations as above). This expression means a sum of $\theta(p)$ over p from the set $\{p : L(p)\}$. Such a sum is a *parametric sum* and can be used to construct a formula with the help of an order relation over reals: either we compare two parametric sums, or a parametric sum and a simple sum.

Formulas are constructed from atomic formulas with the help of propositional connectors. The set of all these formulas constitutes our theory.

In a less studied domain of verification of probabilistic systems we can also find some particular restricted theories with exponential functions that may be decidable.

3   First Order Timed Logic (FOTL)

The starting idea of FOTL is to choose a decidable theory to treat arithmetics or other concrete mathematical functions, and then to extend it by abstract functions of time that are needed to specify the problems under consideration. In some way, the theory must be minimal to be sufficient for a good expressivity. For concreteness we take, as such an underlying theory of arithmetical operations, the theory of mixed real/integer addition with rational constants and unary multiplications by rational numbers. This theory is known to have quantifier elimination [19] if one extends it with the floor function $\lfloor \ \rfloor$.

Syntax and Semantics of FOTL

The vocabulary W of a FOTL consists of a set of *sorts*, a set of *function symbols* and a set of *predicate symbols*. A set of variables is attributed to each sort, these sets are disjoint. The sorts for numbers and Boolean values are sorts by default, as well as the corresponding constants (see below).

If a finite sort has a fixed cardinality it can be considered as pre-interpreted because it is defined modulo notations for its elements. Interesting finite sorts are those whose cardinality is not concrete, say, given by an abstract natural constant (or not given at all). For example, the set of processes in a distributed algorithm. It is often convenient, without loss of generality, to interpret such a sort as an initial segment of natural numbers.

FOTL Syntax

A FOTL syntax is defined by a vocabulary composed of:

Pre-interpreted sorts: $\mathcal{R}$ (reals), $\mathcal{Z}$ (integers), $\mathcal{N}$ (natural numbers), $\mathcal{T}$ (time, i.e., non negative reals), *Bool* (Boolean values), $Nil = \{nil\}$ (a sort to represent the 'undefined', included in all other sorts), and of a finite number of finite sorts of concrete cardinality

Abstract sorts: finite number of symbols maybe supplied with abstract natural constants that denote their respective cardinalities (strictly speaking, these constants must be declared below among the functions of the vocabulary).

Pre-interpreted functions:
    – *Constants*: **true**, **false**, *nil*, and integers $\mathbb{Z}$ (each of type $\to \mathcal{Z}$) and rational numbers $\mathbb{Q}$ (each of type $\to \mathcal{R}$).
    – *Arithmetical operations and relations*: $+, -, =, <, \leq$ over reals and integers.
    – *Boolean operations*: $\wedge, \vee, \neg$.
Abstract functions and predicates: function symbols of type $\mathcal{T} \times \mathcal{X} \to \mathcal{S}$ or $\mathcal{X} \to \mathcal{S}$ or $\to \mathcal{S}$, where $\mathcal{X}$ is a direct product of finite sorts and $\mathcal{S}$ is an arbitrary sort (recall that $\mathcal{T}$ is time).

Semantics of FOTL

A priori, we impose no constraints on the admissible interpretations. Thus, the notions of interpretation, model, satisfiability and validity are treated as in first order predicate logic modulo the pre-interpreted part of the vocabulary.

    Remark that an interpretation of a function f of type $\mathcal{T} \times \mathcal{X} \to \mathcal{S}$ describes a family of temporal processes with values in the interpretation of $\mathcal{S}$ parameterized by the elements of the interpretation of $\mathcal{X}$.

    Clearly, even a FOTL based on the theory of real addition with two unary predicates is undecidable (this follows from [13]).

4   Finiteness Properties and Decidability

Here we introduce specific classes of interpretations of a finite complexity. These interpretations play a key role in our decidability algorithm.

    From the point of view of an algorithm that we wish to verify all functions are piecewise constant. However, their 'physical' interpretation may be of other nature. For example, to represent a piece of linear function $a \cdot t + b$ on an interval σ, we give two values a and b for the coefficients and two values σ^- and σ^+ for the end of σ. And these values remain constant up to the instant when the algorithm calculates the next linear piece of this function. But the 'physical' interpretation of this function, that may be used in guards of the algorithm, is not constant – however, it is described as a term of the vocabulary, for example, it may appear in a guard as a term $(a \cdot CT + b)$, where CT is Current Time (the value of 'physical time').

    A *U-FOTL* is a FOTL extended in the following way. For every abstract function f of type $\mathcal{T} \times \mathcal{X} \to \mathcal{S}$ there is associated a finite set $\mathcal{U}_f$ of terms with values of type $\mathcal{S}$ constructed only from variables and pre-interpreted functions.

The vocabulary of FOTL does not give many possibilities to construct a term $U_f \in \mathcal{U}_f$. We will consider the following types of terms: first, those of the form z with z being a variable for an abstract sort, if $\mathcal{S}$ is an abstract sort, and second, the terms of the form $a_0\tau + a_1\lambda + z$, where $a_0, a_1 \in \mathbb{Q}$ and τ, λ and z are real variables whose role is defined as follows: τ is a time variable, λ is the left end of the interval on which we consider our function, and z is a real parameter.

Below a U-FOTL is supposed to be fixed. For technical simplicity we assume that the types $\mathcal{T} \times \mathcal{X} \to \mathcal{S}$ of functions contain only one abstract sort $\mathcal{X}$, not a direct product (direct products can be treated as in [8]) We will write U_f of real type also as $U_f(\tau, \lambda, z)$ to make the parameters explicit. We say that f_x is U_f-defined on an interval ζ by $z_0 \in \mathbb{R}$, if for $t \in \zeta$, the value $f_x(t)$ is defined as $f_x(t) = U_f(t, \zeta^-, z_0)$.

A *partition* of $\mathcal{T}$ is a sequence $\pi = (\zeta_i)_{i \in \overline{N}}$ of non empty disjoint intervals such that:
- $\overline{N}$ is a prefix of $\mathbb{N}$,
- $\bigcup_{i \in \overline{N}} \zeta_i = \mathcal{T}$,
- $\zeta_i^+ = \zeta_{i+1}^-$ for $0 \le i \le |\overline{N}| - 1$,
- $\zeta_0^- = 0$, $\zeta_k^+ = \infty$ if $\overline{N}$ is finite and k is its last element.

Repetitive Interpretations

We define the interpretations that will be used in the description of our decidable class of formulas of FOTL. Below we use the following abbreviations:
PI for *partial interpretation*; *FPI* for *finite partial interpretation*.

For an abstract function f of type $\mathcal{T} \times \mathcal{X} \to \mathcal{S}$ and an interpretation $\mathcal{X}^*$ of $\mathcal{X}$, a *(finite) partial interpretation* $f_{x^*}^*$ of f_{x^*}, where $x^* \in \mathcal{X}^*$, is given by
- a (finite) set of disjoint intervals
and for each interval by
- a term $U_f \in \mathcal{U}_f$ and by a value of z to be put into U_f to define f_{x^*} on this interval.

This set of intervals is called the *support* of the (F)PI $f_{x^*}^*$.

A FPI has *complexity* c if the number of intervals in its support is c. In the context of several complexity parameters, that will be introduced later, we will call this complexity *interval complexity*.

A *(finite) partial interpretation* of $f : \mathcal{T} \times \mathcal{X} \to \mathcal{S}$ is a subset $\mathcal{Y}^*$ of an interpretation $\mathcal{X}^*$ of $\mathcal{X}$ and a collection of (F)PIs, one for each f_{y^*}, $y^* \in \mathcal{Y}^*$.

A *(finite) partial interpretation* of vocabulary V is a collection of (F)PIs, one for each abstract function of V.

A PI $\mathcal{M}'$ of a function f_{x^*} is an *extension* of a PI $\mathcal{M}$ of f_{x^*} if every interval of $\mathcal{M}$ is contained in an interval of $\mathcal{M}'$, and the restriction of $\mathcal{M}'$ on intervals of $\mathcal{M}$ gives $\mathcal{M}$. In a similar way we define an *extension* of a PI of a vocabulary.

Now we go to more general finitely definable interpretations.

An interpretation $\mathcal{M}$ of f_{x^*} is *ultimately repetitive of complexity c and period h* if it is a finite interpretation with complexity c or it is a concatenation of a finite interpretation of complexity c on some interval, say $[0, h_0)$, followed by an interpretation of the following 'almost periodic' structure:

   – any interval $I_i = [h_0 + i \cdot h, h_0 + (i+1) \cdot h)$, $i \geq 0$, is partitioned into c consecutive intervals $\zeta_{i,j}$, $0 \leq j \leq (c-1)$ such that $|\zeta_{i,j}| = |\zeta_{i+1,j}|$ (that means that the partition has a periodic structure starting from h_0)

   – moreover, on each $\zeta_{i,j}$ the function f_{x^*} is defined by a $U_{f,j}(t, \zeta_{i,j}^-, z_j)$, where $U_{f,j} \in \mathcal{U}_f$ and z_j do not depend on i.

The intervals $\zeta_{i,j}$ are called *period defining intervals* and I_i is called *defining interval* of this ultimately repetitive interpretation.

Our *main* notion concerning interpretations of finite complexity is that of a chain of repetitive interpretations.

A finite prefix of an ultimately repetitive interpretation of a f_{x^*} is *exact* if its end coincides with the end of one of its defining intervals I_i. Its *complexity* is defined similar to the complexity of ultimately repetitive interpretations (in fact, this complexity is the maximum of the interval complexity of the prefix and of the interval complexity of the period.)

We say that an interpretation of a f_{x^*} is a *chain of ultimately repetitive interpretations with complexity* (L, c) if it is a concatenation of at most $(L-1)$ finite exact prefixes of repetitive interpretations and of one infinite ultimately repetitive interpretation, each of complexity c. We will sometimes refer to L as to *chain complexity*.

Equivalence

To reduce the complexity of interpretations in spite of a possibly large amount of elements in abstract sorts we introduce a notion of equivalence of interpretations, and on this basis we will generalize the complexity measures for PI of individual f_{x^*}. Given an interpretation of the vocabulary, such an equivalence is defined over elements of the interpretation of abstract sorts for each f.

Without loss of generality, an abstract sort $\mathcal{X}$ is interpreted as an initial segment $\mathcal{X}^*$ on natural numbers.

In Definitions that follow, $\mathcal{X}^*$ stands for an interpretation of a sort $\mathcal{X}$.

An equivalence E over $\mathcal{Y}^* \subset \mathcal{X}^*$ is *interval-wise* if its classes of equivalence are intervals. An equivalence E over $\mathcal{Y}^*$ is f-*compatible* if for any two elements $u^*, v^* \in \mathcal{Y}^*$ the equivalence $u^* E v^*$ implies that the functions $f_{u^*}^*$ and $f_{v^*}^*$ are equal.

Complexity of Partial Interpretations

A PI of f over $\mathcal{Y}^* \subset \mathcal{X}^*$ is a FPI of *complexity* (m, c) if there is an interval-wise equivalence E on $\mathcal{Y}^*$ with at most m classes which is f-compatible, and such that each $f_{y^*}^*$, $y^* \in \mathcal{Y}^*$, has complexity c (without loss of generality we assume that the partition of time, the terms from $\mathcal{U}_f$ and parameters z that define $f_{y^*}^*$ are the same for all y^* of the same equivalence class).

A FPI of V of *complexity* (m, c) is a collection of FPIs with complexity (m, c), one for each abstract function. A FPI of complexity (m, c) will be also called a (m, c)-PI.

The parameter m is called *equivalence complexity*.

An interpretation of f over $\mathcal{X}^*$ is *ultimately repetitive with complexity (m, c)* if there is an interval-wise equivalence E on $\mathcal{X}^*$ with at most m classes which is f-compatible, and such that for each class, all $f_{x^*}^*$ with x^* in this class are ultimately repetitive with complexity c.

An interpretation of f over $\mathcal{X}^*$ is *a chain of ultimately repetitive interpretations with complexity (m, L, c)* if there is an interval-wise equivalence E on $\mathcal{X}^*$ with at most m classes, which is f-compatible and such that for any class, all $f_{x^*}^*$ with x^* in the class are chains of ultimately repetitive interpretations with complexity (L, c).

An interpretation of V of complexity (m, L, c) is a collection of interpretations with complexity (m, L, c), one for each abstract function.

We introduce classes of interpretations used below, in particular, the class used in our decidability result.

Notations

- Below $\mathcal{K}$ is a complexity of the form (m, c), and $\mathcal{L}$ is a complexity of the form (m, L, c).
- For a class $\mathcal{C}$ of interpretations we denote by $\mathcal{C}(\kappa)$ the set of interpretations in the class $\mathcal{C}$ with complexity κ, where κ has the form defined for this class of interpretations.
- $\mathcal{UR}$ is the class of ultimately repetitive interpretations.
- $\mathcal{UR}^*$ is the class of chains of ultimately repetitive interpretations.
- $\mathcal{UR}^*(\mathcal{L}, \Lambda)$, where $\Lambda \subset \mathbb{Q}_{>0}$, is the set of interpretations from $\mathcal{UR}^*$ with complexity $\mathcal{L}$ whose period lengths are from Λ. (Recall that for a given ultimately repetitive interpretation $f_{x^*}^*$, the period length is fixed, so the set Λ specifies possible period lengths for interpretation of different functions f_x.)
- $\mathcal{UR}^*(\Lambda)$ is the union of all $\mathcal{UR}^*(\mathcal{L}, \Lambda)$ over $\mathcal{L}$.   □

Finite Refutability and Finite Satisfiability

Recall that we fixed some FOTL so when we speak about a formula then, by default, we mean a formula of this FOTL.

A formula F is *$\mathcal{K}$-refutable* if for every its counter-model $\mathcal{M}$ there exists a $\mathcal{K}$-FPI $\mathcal{M}'$ such that $\mathcal{M}$ is an extension of $\mathcal{M}'$, and any extension of $\mathcal{M}'$ to a total interpretation is a counter-model of F.

Finite satisfiability, defined just below, is a notion that is, in some way, dual to finite refutability. It represents the following property. If in a *model* we take any piece of a given complexity (imagine that this piece is defined on some number of separated intervals) then we can fill the gaps between these defined parts to get a total model whose complexity is bounded as a function of the complexity of the given initial piece. This bounding function is the augmentation function used below. The main point is with what kind of interpretations we will fill the gaps.

By α we will denote a total computable function transforming a complexity value of the form (m, c) into a complexity value of the form (m, c), when we speak about class $\mathcal{UR}$, or into a complexity value of the form (m, L, c), when we speak about class $\mathcal{UR}^*$. Such a function will serve as an *augmentation* function in the notion of finite satisfiability below.

A formula F is $(\mathcal{C}, \mathcal{K})$-*satisfiable with augmentation* α if for every $\mathcal{K}$-FPI $\mathcal{M}$ extendable to a model of F there is an extension $\mathcal{M}'$ of $\mathcal{M}$ from $\mathcal{C}(\alpha(\mathcal{K}))$ that is a model of F.

A formula F is $\mathcal{C}$-*satisfiable with augmentation* α if for every $\mathcal{K}$, for every $\mathcal{K}$-FPI $\mathcal{M}$ extendable to a model of F, there is an extension $\mathcal{M}'$ of $\mathcal{M}$ from $\mathcal{C}(\alpha(\mathcal{K}))$ that is a model of F.

The finiteness properties introduced above permit to describe our class of formulas, such that the validity of closed ones is decidable, and for any formula we can effectively describe its counter-models of a given complexity as a quantifier-free formula in a theory with 'good' algorithmic properties. The class is motivated by the verification problem, that is why it consists of implications that tacitly refer to the structure of verification formulas explained in Introduction.

Class $VERIF(\Lambda, \mathcal{K}, \alpha)$ of FOTL-Formulas

- $\mathcal{C}_h =_{df} \mathcal{UR}^*(h \cdot \Lambda)$, where h is a real number, Λ is a *finite* set of rational numbers and $h \cdot \Lambda$ is the set of reals of the form $h \cdot \lambda$ with $\lambda \in \Lambda$.
- $VERIF_h(\Lambda, \mathcal{K}, \alpha)$ is the class of FOTL-formulas of the form $(\Phi \rightarrow \Psi)$, where formula Ψ is $\mathcal{K}$-refutable and Φ is $(\mathcal{C}_h, \mathcal{K})$-satisfiable with augmentation α.
- $VERIF(\Lambda, \mathcal{K}, \alpha) = \bigcup_{h \in \mathbb{R}_{>0}} VERIF_h(\Lambda, \mathcal{K}, \alpha)$.

Notice that our description of $VERIF(\Lambda, \mathcal{K}, \alpha)$ admits not closed formulas in the class.

Decidability and Quantifier-Free Description of Counter-Models

Theorem 1 *Given a complexity $\mathcal{K}$, a computable augmentation function α and a finite set of positive rational numbers $\Lambda \subset \mathbb{Q}_{>0}$, the validity of (closed) formulas from $VERIF(\Lambda, \mathcal{K}, \alpha)$ is decidable. Moreover, for any formula of this class, its counter-models of complexity $\alpha(\mathcal{K})$ can be described by a quantifier-free formula.*

For the class $VERIF$ that uses ultimately repetitive models the quantifier-free formulas may contain $\lfloor\ \rfloor$. For the class $VERIF$ that uses only finite models and is based on the theory of real addition, the quantifier-free formulas are formulas of the theory of real addition that are much easier to deal with. The worst case complexity of our algorithms is the same as that of the underlying theory of arithmetical operations.

Theorem 2 below gives precisions on the role of h in this description (h is a parameter of $VERIF_h(\Lambda, \mathcal{K}, \alpha)$)

Theorem 2 *Given a FOTL-formula F and a complexity $\mathcal{L}$, one can construct a quantifier-free formula that describes all h and all repetitive models (we mean chains of ultimately repetitive interpretations) of F of complexity $\mathcal{L}$ in $\mathcal{C}_h$.*

Corollary 1 *The existence of h for which there is a model of complexity $\mathcal{L}$ in $\mathcal{C}_h$ for a formula F, or the existence of a model of F of complexity $\mathcal{L}$ for a concrete h, is decidable.*

Conclusion

Several questions important for practical application of the presented methods remain open.

How to describe formulas corresponding to practical problems of verification and what is their complexity?

A related question is to find sufficient syntactical conditions on programs that ensure finite satisfiability.

What about decidability of the second order properties that were mentioned in Introduction and that are formulated in a theory that is clearly undecidable in general?

What are other practically relevant finiteness properties?

As for theoretical questions, one question seems to be of a growing importance: verification of probabilistic systems. For many protocols and especially for the security properties, the models are probabilistic. The argument that we rarely know the probabilities is not a real objection, as we can try different plausible distributions if we have good algorithms to deal with them. On the basis of the results of verification for these various distributions and our experience, we can make conclusions about practical value of the system that we analyze. It is even better to treat probability distributions as parameters and to try to find a description of these parameters for which the property we are interested in, is true.

Logics with probabilities that are decidable (see [11] that gives an excellent presentation of the subject) are not sufficiently powerful; the same is true for predicate logics. But for logic of probability it is much harder to find decidable classes of practical value. Even decidable model checking is not so easy to find (e.g., see [5]).

Continuous time that is quite intuitive also in the probabilistic framework poses a problem of semantics from the very beginning because quantification over non countable domain may give non measurable sets, and because arithmetical operations over stochastic processes are problematic. Thus, we have to define the syntax of theories more carefully, maybe without composition and iteration all the used constructors. To ensure measurability it is better to avoid, for example, formulas like $\mathbf{P}\{\forall t\, \varphi(t)\} > p$, where $\mathbf{P}$ stands for probability. Imagine that $\forall t\, \varphi(t)$ is a safety property. On the other hand, can we be satisfied with proving $\forall t\, \mathbf{P}\{\varphi(t)\} > p$ that is a property different from the first one?

In order to be able to define finiteness properties we have to consider finitely definable probability spaces. This is the case in applications of the probability theory. Usually, a 'practical' probability space is either finite or countable or is a sub-interval of reals or a mixture of the previous ones. But for stochastic processes we have a product $R^{\mathbb{T}}$ of simple spaces R over time $\mathbb{T}$, and the probability measure is over this $R^{\mathbb{T}}$.

However all these difficulties seem to be surmountable, at least, in some cases of practical interest. So how to define finite refutability for a formula with probability?

References

1. R. Alur and D. Dill. A theory of timed automata. *Theoretical Computer Science*, 126:183–235, 1994.
2. D. Beauquier, T. Crolard, and E. Prokofieva. Automatic verification of real time systems: A case study. In *Third Workshop on Automated Verification of Critical Systems (AVoCS'2003)*, pages 98–108. University of Southampton, 2003.
3. D. Beauquier, T. Crolard, and E. Prokofieva. Automatic parametric verification of a root contention protocol based on abstract state machines and first order timed logic. In K. Jensen and A. Podelski, editors, *Tools and Algorithms for the Construction and Analysis of Systems: 10th International Conference, TACAS 2004, Barcelona, Spain, March 29 – April 2, 2004. Lect. Notes in Comput. Sci., vol. 2988*, pages 372–387. Springer-Verlag Heidelberg, 2004.
4. D. Beauquier, T. Crolard, and A. Slissenko. A predicate logic framework for mechanical verification of real-time Gurevich Abstract State Machines: A case study with PVS. Technical Report 00–25, University Paris 12, Department of Informatics, 2000. Available at http://www.univ-paris12.fr/lacl/.
5. D. Beauquier, A. Rabinovich, and A. Slissenko. A logic of probability with decidable model-checking. *Journal of Logic and Computation*. 24 pages. To appear.
6. D. Beauquier and A. Slissenko. Periodicity based decidable classes in a first order timed logic. *Annals of Pure and Applied Logic*. 38 pages. To appear.
7. D. Beauquier and A. Slissenko. Decidable verification for reducible timed automata specified in a first order logic with time. *Theoretical Computer Science*, 275(1–2):347–388, March 2002.
8. D. Beauquier and A. Slissenko. A first order logic for specification of timed algorithms: Basic properties and a decidable class. *Annals of Pure and Applied Logic*, 113(1–3):13–52, 2002.
9. B. Boigelot and P. Wolper. Representing arithmetic constraints with finite automata: An overview. *Lecture Notes in Computer Science*, 2401:1–19, 2002.
10. E. Clarke, D. Kroening, J. Ouaknine, and O. Strichman. Completeness and complexity of bounded model checking. In Levi G. Steffen, B., editor, *Proceedings of the 5th International Conference on Verification, Model Checking, and Abstract Interpretation (VMCAI'2004), Venice, Italy, January 11-13, 2004*, volume 2937 of *Lecture Notes in Computer Science*, pages 85–96. Springer-Verlag Heidelberg, 2004.
11. R. Fagin and J. Halpern. Reasoning about knowledge and probability. *J. of the Assoc. Comput. Mach.*, 41(2):340–367, 1994.
12. Y. Gurevich and J. Huggins. The railroad crossing problem: an experiment with instantaneous actions and immediate reactions. In H. K. Buening, editor, *Computer Science Logics, Selected papers from CSL'95*, pages 266–290. Springer-Verlag, 1996. Lect. Notes in Comput. Sci., vol. 1092.
13. J. Halpern. Presburger arithmetic with unary predicates is π_1^1-complete. *J. of symbolic Logic*, 56:637–642, 1991.
14. J. Sanders and E. Curran. *Software Quality*. Addison-Wesley, 1994.

15. A. Slissenko. A logic framework for verification of timed algorithms. *Fundamenta Informaticae*, 69:1–39, 2004.
16. I. Sommerville. *Software Engineering*. Addison-Wesley, 4th edition, 1992.
17. M. Sorea. Bounded model checking for timed automata. *Electronic Notes in Theoretical Computer Science*, 68(5), 2002.
    http://www.elsevier.com/locate/entcs/volume68.html.
18. G. Tel, editor. *Introduction to Distributed Algorithms*. Cambridge University Press, 1994.
19. V. Weispfenning. Mixed real-integer linear quantifier elimination. In *Proc. of the 1999 Int. Symp. on Symbolic and Algebraic Computations (ISSAC'99)*, pages 129–136. ACM Press, 1999.

Note on Formal Analogical Reasoning in the Juridical Context

Matthias Baaz

Technische Universität Wien, A-1040 Vienna, Austria
baaz@logic.at

Abstract. This note describes a formal rule for analogical reasoning in the legal context. The rule derives first order sentences from partial decision descriptions. The construction follows the principle, that the acceptance of an incomplete argument induces the acceptance of the logically weakest assumptions, which complete it.

> "The common law is tolerant of much illogicality. especially on the surface, but no system of law can be workable if it has not got logic at the root of it." (Lord Devlin in *Hedley Byrne and Co. Ltd. v. Heller and Partners Ltd. (1964)*)

1   Introduction

4000 Years ago, mathematical arguments were given by examples. When Sumerian and Babylonian mathematicians wanted to present a general statement, they provided examples such that scholars were able to grasp the principle by calculating these particular cases (Gericke [6]).

4000 years ago, legal reasoning developed along the same lines, but for different reasons. In mathematics, the general principle relates to the one-for-all validity of the single case, in legal systems, to the stability of the system, or in more traditional terms: to justice.

Although this fact is nowadays generally overlooked, mathematics and law have remained connected throughout the history of human civilizations, with law being for the majority of the time, the methodologically more developed part. When the notions of argument and proof were established in Greek logic, they used expressions from law. The Greek notion of proof is the ancestor of modern mathematics.

Nowadays, both of the renowned sciences of mathematics and jurisprudence look at each other with suspicion. Some law schools disclaim even the admissibility of general principles in judgments (Holme's Maxim), relating this to a

L. Ong (Ed.): CSL 2005, LNCS 3634, pp. 18–26, 2005.

reductive argument of Kant[1], which would apply to the general principles in mathematics as well if valid. Mathematicians on the other hand reject the possibility of formally correct reasoning in law, since they are unable to explain the effectiveness, both at the theoretical and practical level, of the interplay of deductions and actions in juridical procedures.

In this note we try to establish a bridge between mathematical logic and jurisprudence by developing a proof theoretic concept of analogical reasoning, which we consider as the most fundamental deduction principle of juridical logic.

2   Analogical Reasoning in Mathematics

Analogical reasoning occurs frequently in mathematics (cf. Kreisel [11]) but is rarely documented. A notable exception is Eulers computation of the sum

$$\sum_{n=1}^{\infty} \frac{1}{n^2} = \frac{\pi^2}{6} \tag{1}$$

that made him famous [5], cf. Polya [13, page 17 ff.]. The problem of computing this sum, which was readily seen to be convergent but for which nobody could guess the value, was posed by Jacques Bernoulli.

Let us consider Eulers reasoning. Consider the polynomial of even degree

$$b_0 - b_1 x^2 + b_2 x^4 - \ldots + (-1)^n b_n x^{2n}. \tag{2}$$

If it has the $2n$ roots $\pm\beta_1, \ldots \pm \beta_n \neq 0$ then (2) can be written as

$$b_0 \left(1 - \frac{x^2}{\beta_1^2}\right) \left(1 - \frac{x^2}{\beta_2^2}\right) \ldots \left(1 - \frac{x^2}{\beta_n^2}\right). \tag{3}$$

By comparing coefficients in (2) and (3) one obtains that

$$b_1 = b_0 \left(\frac{1}{\beta_1^2} + \frac{1}{\beta_2^2} + \ldots + \frac{1}{\beta_n^2}\right). \tag{4}$$

Next Euler considers the Taylor series

$$\frac{\sin x}{x} = \sum_{n=0}^{\infty} (-1)^n \frac{x^{2n}}{(2n+1)!} \tag{5}$$

---

[1] "General logic contains and can contain no rules for judgment … If it sought to give general instructions how we are to subsume under these rules, that is, to distinguish whether something does or does not come under them, that could only be by means of another rule. This, in turn, for the very reason that it is a rule, again demands guidance from judgment, And thus it appears that, though understanding is capable of being instructed, and of being equipped with rules, judgment is a peculiar talent which can be practiced only, and cannot be taught." [7]

which has as roots $\pm\pi, \pm 2\pi, \pm 3\pi, \ldots$ Now by way of analogy Euler *assumes* that the infinite degree polynomial (5) behaves in the same way as the finite polynomial (2). Hence in analogy to (3) he obtains

$$\frac{\sin x}{x} = \left(1 - \frac{x^2}{\pi^2}\right)\left(1 - \frac{x^2}{4\pi^2}\right)\left(1 - \frac{x^2}{9\pi^2}\right)\cdots \tag{6}$$

and in analogy to (4) he obtains

$$\frac{1}{3!} = \left(\frac{1}{\pi^2} + \frac{1}{4\pi^2} + \frac{1}{9\pi^2} + \cdots\right). \tag{7}$$

which immediately gives the expression (1). This solution of the problem caused much amazement and astonishment at the time. The "leap of faith" used by Euler to arrive at the solution was later rigorously justified, but it is important to realize the role that reasoning by analogy played in finding the initial solution. Using analogy Euler found a proof *scheme* that would lead to the desired conclusion[2]. This proof scheme suggested to him the *preconditions* he had to prove to formally justify his solution.

The structure of Eulers argument is the following.

(a) $(2) = (3)$ (mathematically derivable)

(b) $(2) = (3) \supset (4)$ (mathematically derivable)

(c) $(2) = (3) \supset (5){=}(6)$ (analogical hypothesis)

(d) $(5) = (6) \supset (4)$ (modus ponens)

(e) $((2) = (3) \supset (4)) \supset ((5) = (6) \supset (7))$ (analogical hypothesis)

(f) $(5) = (6) \supset (7)$ (modus ponens)

(g) (7) (modus ponens)

(h) $(7) \supset (1)$ (mathematically derivable)

(i) (1) (modus ponens)

To transform Eulers argument in a rigid proof in the sense of mathematics it is sufficient to verify (c) and (e). On the other hand, if one is determined to uphold this argument one has at least to uphold the weakest preconditions that verify (c) and (e). This provides a connection to juridical reasoning.

3   The Derivation of Weakest Preconditions from Proofs

The main logical problem of legal reasoning lies in the conflict of the following:

---

[2] In mathematical terms the scheme can be formulated as follows. To calculate $\sum_{v \in \Gamma} \frac{1}{v^2}$ search a function f such that $f(x) = \sum_{i=0}^{\infty} c_i x^i$, $\Gamma = \{x | f(x) = 0, x > 0\}$, and $f(x) = 0 \Leftrightarrow x \in \Gamma \wedge -x \in \Gamma$. Then $\sum_{v \in \Gamma} \frac{1}{v^2} = \frac{c_1}{c_0}$.

(i) Arguments should be demonstrably sound.

(ii) Decisions have to be achieved within a priori limited time and space.

The solution is provided by minimalist systems such as English Common Law and maximalist systems such as continental legal systems. In minimalist systems, completeness is achieved by the admitted generation of legal norms from juridical decisions (stare decis), which logically represent preconditions of the decisions (ratio decidendi) in the sense of incomplete reasoning. In maximalist systems extensive interpretations treat the inherent incompleteness of the system. The system obtains stability by the application of teleological interpretations, which restrict the derivable conclusions in conflicting situations.

Let us consider how the ratio decidendi is established according to the English doctrine of precedent[3] (Wambaugh's test):

> "First frame carefully the supposed proposition of law. Let him then insert in the proposition a word reversing its meaning. Let him then inquire whether, if the court had conceived this new proposition to be good, and had had it in mind, the decision would have been the same. If the answer be affirmative, then, however excellent the original proposition may be, the case is not a precedent for that proposition, but if the answer be negative the case is a precedent for the original proposition and possibly for the other propositions also. In short, when a case turns only on one point the proposition or doctrine of the case, the reason for the decision, the *ratio decidendi*, must be a general rule without which the case must have been decided otherwise." [4, p52]

In a mathematical sense the ratio decidendi of a decision is the weakest reasonable precondition completing the otherwise incomplete argument. We will formally specify the notion of weakest precondition as follows. Let a proof system be given by schematic axioms and rules.

A *partial proof skeleton with respect to* T, T a set of formulas, is a rooted tree whose vertices are labelled by the inference rules. Further, the order of the given vertices is marked on the tree. Some of the initial nodes are designated by axiom schemes or formulas from T. If all initial nodes are designated in this way the partial skeleton is called *total*. The information which the skeleton does not contain are the terms and variables used in quantifier rules. Every proof determines uniquely its partial skeleton with respect to T, but we do not require a skeleton to be determined by some proof.

$A_1 \ldots A_n$ are *preconditions* with respect to the partial proof skeleton S, T, and end formula E, if the assignment of $A_1 \ldots A_n$ to the non-designated initial nodes of S can be extended to a proof of the end formula E.

To calculate (weakest) preconditions is however not sufficient to represent analogy in legal reasoning. Decisions by judges should be general in the sense that

---

[3] In continental e.g. German legal practice analogical reasoning occurs less explicit. However all specific rules of German legal reasoning such as argumentum a simile, argumentum e contrario, argumentum a fortiori, argumentum ad absurdum [9], are easily reducible to analogical reasoning.

they are independent[4] of the concrete persons etc. involved. This is completely (but in practice tacitly) specified before the decision is taken. We represent this constraint by a (possible empty) set of constants Γ.

$W_1 \ldots W_n$ are *weakest preconditions* with respect to the partial proof skeleton S, T, end formula E, and Γ, if they are preconditions in the above sense and whenever $\{W_1 \ldots W_n\}\sigma, \Gamma \vdash \Delta$ then $\{A_1 \ldots A_n\}\sigma, \Gamma \vdash \Delta$, for any Γ, Δ in the original language of the derivation, and any choice of preconditions $A_1 \ldots A_n$. This includes all substitutions σ for the free variables in E and the constants in Γ (considered as variables).

We need therefore a proof system that allows for the calculation of weakest preconditions as a formal basis of analogical reasoning: The weakest precondition will allow for the derivation of new information in subsequent decisions and therefore enforce stare decis.

4   LK Is Not Suitable for Calculating Weakest Preconditions

A first proof theoretic approach is to consider LK as theoretical basis. We define partial proof skeletons with respect to T, T a set of sequents, preconditions, weakest preconditions, as above, only that for the exchange rule the label contains also the number of the pair to which it should be applied.

In this section we show that LK in the full first order language with cuts does not allow for the calculation of weakest preconditions.

Theorem 1. *Let L be a language containing a unary function symbol s, a constant 0, and a binary function symbol. There is a partial skeleton S and a given end sequent such that it is undecidable whether a sequent may serve as precondition and where no weakest precondition exists.*

Proof. By Orevkov [12] and Krajicek and Pudlak [10] we have that for every recursively enumerable set $X \subseteq \omega$ there exists a sequent $\Pi \to \Gamma, P(a)$ and a total skeleton S' such that $n \in X$ iff $\Pi \to \Gamma, P(s^n(0))$ has an LK proof with skeleton S'. The argument however uses the concrete form of $\Pi \to \Gamma, P(a)$. We therefore extend the proof skeleton S' as in Figure 1. This guarantees that $P(s^n(0))$ is doubled and the form is preserved in the end sequent $\Pi \to \Gamma, \exists x A(x)$. Preconditions are consequently $\{P(s^n(0)) \to \mid n \in X\}$. The property of being a precondition is therefore undecidable if X is undecidable, and there are in general no weakest preconditions.

5   A Model Calculus for Analogical Reasoning

We consequently restrict to propositional LK with schematic sets of sequents T [5] in the definition of partial proof skeleton, preconditions, and weakest pre-

---

[4] Independence can be considered as a proof theoretic interpretation of justice.

[5] Alternative approaches in full first order languages can be based on LK without the cut rule [10], [3], or on LK with blockwise inferences of quantifiers with the cut rule [2].

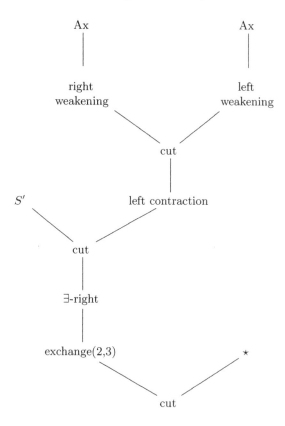

Fig. 1. Proof skeleton S'

conditions. A derivation from T is therefore a derivation from instances of the schemata in T. We construct the weakest preconditions with respect to partial proof skeleton S, T, end sequent E, and Γ as follows.

1. *(Reconstruction step)* Reconstruct the propositional matrix of the derivation using the end sequent and the partial proof skeleton. When passing applications of the cut rule represent the cut formula by a new propositional variable in the premises. This reconstruction is possible as otherwise there would be no proof with this partial proof skeleton.
2. *(Unification step)* Unify both sides of the logical axiom sequents and unify initial sequents with the appropriate schemata from T if the partial proof skeleton assigns them in this way. (The schemata from T have to be chosen variable disjoint.) Apply the most general unifier to the propositional matrix.
3. *(Introduction of Skolem terms)* Extract the initial sequents corresponding to the initial nodes in the partial proof skeleton which are not assigned logical axioms or schemata in T. Replace all constants in $c_1, \ldots, c_n$ in Γ by new variables $a_1, \ldots, a_n$ Replace all first order variables y different from $\{a_1, \ldots, a_n\}$ which do not occur in the end sequent by Skolem

terms $f_y(a_1, \ldots, a_n, x_1, \ldots, x_m)$, where $x_1, \ldots, x_m$ are the free variables in the end sequent. Replace propositional variables X by Skolem predicates $F_X(a_1, \ldots, a_n, x_1, \ldots, x_m)$, where $x_1, \ldots, x_m$ are the free variables in the end sequent. Here f_y is a new function symbol and F_X is a new predicate symbol.

Proposition 1. *The construction above is adequate.*

Proof. The extracted initial sequents are obviously preconditions. To switch from a derivation from the extracted initial sequents to a derivation from arbitrarily chosen preconditions replace the Skolem terms and Skolem predicates under σ everywhere in the derivation by adequate terms and formulas.

We consequently define the analogical reasoning with respect to a proof P, T, and independence conditions Γ.

First, read the partial proof skeleton S and the end sequent E from P. Then calculate weakest preconditions $W_1, \ldots, W_n$ with respect to Γ. Then

$$(S, T, E, \Gamma) \vdash W_i$$

for all i by *analogical reasoning*.

Example 1. Brown vs. Zürich Insurance (1977). The plaintiff claimed compensation for a car damage which was denied given the established principle that a car is not insured if it is not in roadworthy condition and the fact that the plaintiff's car had bald tires. The formalization of this decision might look as follows.

pc	plaintiff's car
bt	bald tires
rw	roadworthy
$I(x)$	x is insured
$COND(x, y)$	x is in condition y

$$\frac{\qquad \qquad \dfrac{COND(pc, bt) \rightarrow \neg COND(pc, rw) \qquad \neg COND(pc, rw) \rightarrow \neg I(pc)}{COND(pc, bt) \rightarrow \neg I(pc)}}{\rightarrow \neg I(pc)}$$
$$\rightarrow COND(pc, bt)$$

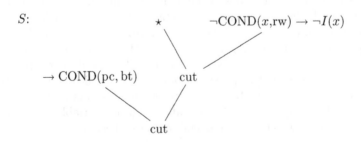

$$T = \{ \rightarrow \text{COND}(\text{pc}, \text{bt}), \neg\text{COND}(x,\text{rw}) \rightarrow \neg I(x) \}$$
$$E = \neg I(\text{pc})$$
$$\Gamma = \{\text{pc}\}.$$

$$\cfrac{\rightarrow X \qquad \cfrac{X \rightarrow Y \qquad Y \rightarrow \neg I(\text{pc})}{X \rightarrow \neg I(\text{pc})}}{\rightarrow \neg I(\text{pc})}$$

$$\sigma = \{ X \Rightarrow \text{COND}(\text{pc}, \text{bt}), Y \Rightarrow \neg\text{COND}(\text{pc}, \text{rw}), x \Rightarrow \text{pc} \}$$

$$(S, T, E, \Gamma) \vdash \text{COND}(x,\text{bt}) \rightarrow \neg\text{COND}(x,\text{rw})$$

6   Conclusion

The rule for analogical reasoning as described in this note suggests the development of a calculus based on formulas and partial decision descriptions. The question remains why not to represent the partial decision descriptions by the generalized weakest preconditions and to omit the partial decision descriptions altogether. The reason is, that with respect to the hierarchy of courts and to the development of the legal apparatus in time[6] regulations and decision have to be canceled to keep the apparatus consistent. This might affect other partial decision descriptions, because accepted statements might become free preconditions, and therefore generalized weakest preconditions have to be calculated again. (This continuous adaption of the calculus can be considered as a mathematical interpretation of Kelsen's Stufenbau der Rechtsordnung, cf. Kelsen [8] und Baaz und Quirchmayr [1].)

References

1. M. Baaz and G. Quirchmayr, *Logic-based models of analogical reasoning*, In *Expert Systems with Applications*, volume 4, pages 369–378. Pergamon Press, 1992.
2. M. Baaz and G. Salzer, *Semi-unification and generalizations of a particularly simple form*, In L. Pacholski and J. Tiuryn, editors, *Proc. 8th Workshop CSL'94*, LNCS 933, pages 106–120. Springer, 1995.
3. M. Baaz and R. Zach, *Generalizing theorems in real closed fields*, Ann. Pure Appl. Logic 75 (1995) 3–23.
4. R. Cross and J. W. Harris. *Precedent in English law*, 4th edition, Claredon Law Series, Oxford University Press 1991.
5. L. Euler, *Opera Omnia*, ser. 1, vol. 14, 73–86, 138–155, 177–186.
6. H. Gericke, *Mathematik in Antike und Orient. Mathematik im Abendland*, Wiesbaden: Fourier Verlag, 1992.
7. I. Kant, *Critique of Pure Reason*, trans. N. Kemp Smith, St Martins Press, 1929, A133/B172.

---

[6] Cf. in German law: Lex posterior derogat legem priorem.

8. H. Kelsen, *Reine Rechtslehre*, Verlag Österreich, 2000 (reprinted from 2nd edition from 1960)

9. U. Klug. *Juristische Logik*, 4th edition, Springer, 1982.

10. J. Krajicek and P. Pudlak, *The number of proof lines and the size of proofs in first order logic*, Arch. Math. Log. 27 (1988) 69–84.

11. G. Kreisel, *On analogies in contemporary mathematics*, 1978 UNESCO lecture, in: Hahn and Sinacèur (eds.), Penser avec Aristote, Erês 1991, 399–408.

12. V. P. Orevkov, *Reconstruction of the proof from its scheme*, (Russian abstract), 8th Sov. Conf. Math. Log. Novosibirsk 1984, p133.

13. G. Polya, *Induction and analogy in mathematics*, Vol. I of Mathematics and plausible reasoning, Princeton University Press, 1954.

An Abstract Strong Normalization Theorem

Ulrich Berger

University of Wales Swansea
u.berger@swan.ac.uk

Abstract. We prove a strong normalization theorem for abstract term rewriting systems based on domain-theoretic models. The theorem applies to extensions of Gödel's system T by various forms of recursion related to bar recursion for which strong normalization was hitherto unknown.

1 Introduction

In his seminal paper [1] Plotkin introduced a domain-theoretic method for proving termination of higher-order rewrite systems. His Adequacy Theorem says that if a closed PCF term of ground type does not denote $\bot$, then its call-by-name reduction sequence terminates. Similar domain-theoretic methods were developed in [2] and [3] to prove strong normalization for simply and polymorphically typed rewrite-systems.

In this paper we isolate what one could call the essence of these methods. It turns out that the common idea of these methods, namely the simultaneous approximation of the operational and the denotational semantics, is not tied to a particular typing discipline or λ-calculus. The argument works for very abstract notions of term, reduction and denotational semantics which can be characterized by simple functoriality and naturality conditions.

Given such an abstract term system T and a domain-theoretic model of it we introduce the notion of a *rewrite structure*, which is a triple $\mathcal{R} = (\mathcal{C}, \rightarrow, \alpha)$ consisting of a set of constants $\mathcal{C}$, an operational semantics of the constants given by an abstract reduction relation $\rightarrow$ on $\mathsf{T}(\mathcal{C})$, and a denotational semantics given by an interpretation α of the constants in the model. We formulate conditions expressing that the two semantics nicely fit together. *Monotonicity* says that reducing a term can only increase its denotational semantics, while *strong normalization* says that if a term does not denote $\bot$, then it is strongly normalizing with respect to $\rightarrow$.

Furthermore we define what it means for a rewrite structure $\mathcal{R}_\omega$ to *approximate* a rewrite structure $\mathcal{R}$. Typically, $\mathcal{R}_\omega$ will have for every constant c of $\mathcal{R}$ a sequence of constants c_n corresponding to the finite stages of the (usually) recursive definition of c. Theorem 1 says that through approximation strong normalization is transferred from $\mathcal{R}_\omega$ to $\mathcal{R}$. Because the proof of strong normalization of $\mathcal{R}_\omega$ is usually easy (recursion is replaced by finite iteration), or, more precisely, can be easily inferred from strong normalization of the underlying typed λ-calculus, we obtain a method for proving strong normalization for $\mathcal{R}$.

L. Ong (Ed.): CSL 2005, LNCS 3634, pp. 27–35, 2005.

As an application we show that our method can be used to prove strong normalization for various extensions of Gödel's system T which were used in [4–8] to give computational interpretations of classical analysis in finite types.

The paper is organized as follows. Sections 2 and 3 introduce the general method and prove the main abstract result (Theorem 1). The abstract definitions are accompanied by examples in which the method is applied to a simply typed λ-calculus with an (unspecified) higher-order rewrite system based on pattern matching. In Section 4 we choose a concrete example of a higher-order rewrite system and prove its strong normalization using a (technically) simple totality argument. Section 5 gives an informal summary of the method in form of a 'recipe' for further applications and compares it with other approaches.

2 Abstract Term Systems and Their Domain-Theoretic Models

In this section we introduce an abstract notion of a 'term over a set of constants' and its domain-theoretic models. The approach is extremely loose, in particular no initiality assumptions are made. We just collect the minimal set of conditions sufficient for our purpose.

Definition 1. Let Const be a subcategory of the category CountSet of countable sets. An object $\mathcal{C}$ of Const is called a *system of constants*, a Const-morphism $\theta: \mathcal{C} \to \mathcal{C}'$ is called a *constant substitution*. A *term system* is a (covariant) functor

$$\mathsf{T}: \mathsf{Const} \to \mathsf{CountSet}.$$

For every system of constants $\mathcal{C}$ the set $\mathsf{T}(\mathcal{C})$ is called the *set of terms with constants in* $\mathcal{C}$. If $\theta: \mathcal{C} \to \mathcal{C}'$ is a constant substitution and $M \in \mathsf{T}(\mathcal{C})$, then we write $M\theta$ for $\mathsf{T}(\theta)(M)$ and call $M\theta$ the result of applying θ to M.

Note that the functoriality of T means that $M\mathrm{id} = M$, where $\mathrm{id}: \mathcal{C} \to \mathcal{C}$ is the identity, and $M(\theta \circ \theta') = (M\theta')\theta$.

Example 1. Consider simple types generated from the base types boole and nat by the formation of function types, $\rho \to \sigma$. Let an object $\mathcal{C}$ of Const be a countable set of typed constants c^ρ, and a morphism $\theta: \mathcal{C} \to \mathcal{C}'$ a type respecting constant substitution. Terms (and their types) over $\mathcal{C}$ are built from typed variables, x^ρ, and constants, $c^\rho \in \mathcal{C}$, by the formation of constructor terms, 0^{nat}, $\mathsf{S}(M^{\mathsf{nat}})^{\mathsf{nat}}$, $\#\mathsf{t}^{\mathsf{boole}}$, $\#\mathsf{f}^{\mathsf{boole}}$, definition by cases, (if P^{boole} then M^ρ else $N^\rho)^\rho$), abstraction, $(\lambda x^\rho M^\sigma)^{\rho \to \sigma}$, and application, $(M^{\rho \to \sigma} N^\rho)^\sigma$. We let $\mathsf{T}(\mathcal{C})$ be the set of closed terms over $\mathcal{C}$. For a constant substitution $\theta: \mathcal{C} \to \mathcal{C}'$ we let $\mathsf{T}(\theta)(M)$ be the result of replacing each constant c^ρ in M by $\theta(c^\rho)$. Clearly this defines a term system, i.e. a functor $\mathsf{T}: \mathsf{Const} \to \mathsf{CountSet}$.

Definition 2. A *model* of a term system T consists of a Scott domain D [9–11] together with a family of continuous functions

$$\mathsf{val}_\mathcal{C}: D^\mathcal{C} \to D^{\mathsf{T}(\mathcal{C})} \qquad (\mathcal{C} \in \mathsf{Const})$$

which is 'natural in $\mathcal{C}$'. More precisely, val is a natural transformation between the contravariant functors $D^{\cdot}, D^{\mathsf{T}(\cdot)}: \mathsf{Const} \to \mathsf{DOM}$ where DOM is the category of Scott domains and continuous functions (which has countable products). If $\alpha \in D^{\mathcal{C}}$ and $M \in \mathsf{T}(\mathcal{C})$, then we write $[M]_{\mathcal{C}}\alpha$, or just $[M]\alpha$, for $\mathsf{val}_{\mathcal{C}}(\alpha)(M)$ and call this the *value of M under the constant interpretation α*.

Note that the naturality condition for val means that for all $M \in \mathsf{T}(\mathcal{C})$, $\theta: \mathcal{C} \to \mathcal{C}'$, and $\alpha \in D^{\mathcal{C}'}$ we have

$$[M]_{\mathcal{C}}(\alpha \circ \theta) = [M\theta]_{\mathcal{C}'}\alpha$$

which is the usual substitution lemma (restricted to constant substitutions) in denotational semantics. Note also that by continuity the function $[M]: D^{\mathcal{C}} \to D$ is monotone, i.e. $\alpha \sqsubseteq \beta$ implies $[M]\alpha \sqsubseteq [M]\beta$, and

$$[M] \bigsqcup_{n \in \mathbb{N}} \alpha_n = \bigsqcup_{n \in \mathbb{N}} [M]\alpha_n$$

for every increasing sequence of constant interpretations $\alpha_n \in D^{\mathcal{C}}$.

Example 2. We define a model (D, val) for the term system T of Example 1 as follows. For every type ρ we define a Scott domain D_ρ by $D_{\mathsf{boole}} = \{\bot, \#\mathsf{t}, \#\mathsf{f}\}$, $D_{\mathsf{nat}} = \{\bot, 0, 1, 2, \ldots\}$, the flat domains of booleans and natural numbers, $D_{\rho \to \sigma} = D_\rho \to D_\sigma$, the domain of continuous functions from D_ρ to D_σ. For every constant interpretation α assigning to each $c^\rho \in \mathcal{C}$ some $\alpha(c) \in D_\rho$, and every variable environment η assigning to each variable x^ρ some $\eta(x) \in D_\rho$ we define the strict semantics, $[M]\alpha\eta \in D_\rho$, as follows.

$$[x]\alpha\eta = \eta(x)$$
$$[c]\alpha\eta = \alpha(c)$$
$$([\lambda x\, M]\alpha\eta)(a) = [M]\alpha\eta_x^a$$
$$[MN]\alpha\eta = \begin{cases} ([M]\alpha\eta)(a) & \text{if } a := [N]\alpha\eta \neq \bot \\ \bot & \text{otherwise} \end{cases}$$
$$[b]\alpha\eta = b \quad (b \in \{0, \#\mathsf{t}, \#\mathsf{f}\})$$
$$[\mathsf{S}(M)]\alpha\eta = \begin{cases} n+1 & \text{if } n := [M]\alpha\eta \neq \bot \\ \bot & \text{otherwise} \end{cases}$$
$$[\text{if } P \text{ then } M \text{ else } N]\alpha\eta = \begin{cases} [M]\alpha\eta & \text{if } [P]\alpha\eta = \#\mathsf{t} \\ [N]\alpha\eta & \text{if } [P]\alpha\eta = \#\mathsf{f} \\ \bot & \text{otherwise} \end{cases}$$

We let D be the coalesced sum of the domains D_ρ, i.e. all bottoms of the D_ρ are identified. We define $\mathsf{val}_{\mathcal{C}}: D^{\mathcal{C}} \to D^{\mathsf{T}(\mathcal{C})}$ by $\mathsf{val}_{\mathcal{C}}(\alpha)(M) := [M]\alpha'\bot$ where $\bot(x^\rho) := \bot^\rho$ and $\alpha'(c^\rho) := \alpha(c^\rho)$ if $\alpha(c^\rho) \in D_\rho$, and $\alpha'(c^\rho) := \bot^\rho$ otherwise. It is easy to see that $\mathsf{val}_{\mathcal{C}}$ is natural in $\mathcal{C}$.

3   Strong Normalization by Approximation

Definition 3. Let $\mathsf{T}: \mathsf{Const} \to \mathsf{CountSet}$ be a term system and (D, val) a model of T. A *rewrite structure* for T and (D, val) is a triple $\mathcal{R} = (\mathcal{C}, \to, \alpha)$ where $\mathcal{C}$ is a constant system, $\rightarrow$ a binary relation on $\mathsf{T}(\mathcal{C})$ and $\alpha \in D^{\mathcal{C}}$. $\mathcal{R}$ is *monotone* if $M \rightarrow N$ implies $[M]\alpha \sqsubseteq [N]\alpha$ for all $M, N \in \mathsf{T}(\mathcal{C})$. $\mathcal{R}$ is *strongly normalizing* if every term $M \in \mathsf{T}(\mathcal{C})$ with $[M]\alpha \neq \perp$ is strongly normalizing w.r.t. $\rightarrow$, i.e. there is no infinite reduction sequence beginning with M.

Note that M is strongly normalizing w.r.t. $\rightarrow$ iff the restriction of $\rightarrow$ to the set $\{N \mid M \rightarrow^* N\}$ is wellfounded. Therefore it makes sense to speak of 'induction on the strong normalizability of M'.

Example 3. Continuing examples 1 and 2, we fix a set of typed constants $\mathcal{C} \in$ Const. Consider a set $\mathcal{E}$ of equations of the form $c\mathbf{P} = M$ where each P_i is either a variable, or one of #t, #f, 0, or of the form $\mathsf{S}(x)$, and such that all equations are left linear, the left-hand sides of the equations are mutually non-unifiable, and every free variable in a right-hand side also occurs in the corresponding left-hand side (see [2, 3]). Then $\mathcal{E}$ defines a constant interpretation $\alpha \colon \mathcal{C} \rightarrow D$ in a canonical way. Take, for example, the constant $<\colon \mathsf{nat} \rightarrow \mathsf{nat} \rightarrow \mathsf{boole}$ with the equations $x < 0 = \#\mathsf{f}$, $0 < \mathsf{S}(y) = \#\mathsf{t}$, $\mathsf{S}(x) < \mathsf{S}(x) = x < y$. Then $\alpha(<) \in D_{\mathsf{nat} \rightarrow \mathsf{nat} \rightarrow \mathsf{boole}}$ is recursively defined by

$$
k\ \alpha(<)\ m = \begin{cases} \#\mathsf{f} & \text{if } m = 0 \\ \#\mathsf{t} & \text{if } k = 0 \text{ and } m > 0 \\ (k-1)\ \alpha(<)\ (m-1) & \text{if } k > 0 \text{ and } m > 0 \\ \perp & \text{otherwise (i.e. } k = \perp \text{ or } m = \perp) \end{cases}
$$

On the other hand $\mathcal{E}$ also defines a reduction relation $\rightarrow$ between terms through the rules below, where $M[N/x]$ denotes (variable capture avoiding) substitution and $\mathsf{Inst}(\mathcal{E})$ is the set of all substitution instances of equations in $\mathcal{E}$.

$$
\mathcal{E}\ \frac{M = N \in \mathsf{Inst}(\mathcal{E})}{M \rightarrow N} \qquad \beta\ \frac{}{(\lambda x\, M)N \rightarrow M[N/x]}
$$

$$
\text{If}\ \frac{P \rightarrow P'}{\text{if } P \text{ then } M \text{ else } N \rightarrow \text{if } P' \text{ then } M \text{ else } N}
$$

$$
\text{If-}\#\mathsf{t}\ \frac{}{\text{if } \#\mathsf{t} \text{ then } M \text{ else } N \rightarrow M} \qquad \text{If-}\#\mathsf{f}\ \frac{}{\text{if } \#\mathsf{f} \text{ then } M \text{ else } N \rightarrow N}
$$

$$
\text{App-L}\ \frac{M \rightarrow M'}{MN \rightarrow M'N} \qquad \text{App-R}\ \frac{N \rightarrow N'}{MN \rightarrow MN'} \qquad \text{Abst}\ \frac{M \rightarrow M'}{\lambda x\, M \rightarrow \lambda x\, M'}
$$

Altogether we have defined a rewrite structure $\mathcal{R} = (\mathcal{C}, \rightarrow, \alpha)$ (restricting $\rightarrow$ to closed terms). It is easy to see that $\mathcal{R}$ is monotone: by induction on the definition of $\rightarrow$ one easily proves that $M \rightarrow N$ implies $[M]\alpha\eta \sqsubseteq [N]\alpha\eta$ for all variable environments. For the β-rule one needs that $[M[N/x]]\alpha\eta = [M]\alpha\eta_x^{[N]\alpha\eta}$, which can be proven by a straightforward induction on M. In Example 4 we will prove that that $\mathcal{R}$ is also strongly normalizing.

Definition 4. Let $\mathcal{R}_\omega = (\mathcal{C}_\omega, \rightarrow_\omega, \alpha_\omega)$ and $\mathcal{R} = (\mathcal{C}, \rightarrow, \alpha)$ be rewrite structures. We say that $\mathcal{R}_\omega$ *approximates* $\mathcal{R}$ if there exists a constant substitution $\theta \colon \mathcal{C}_\omega \rightarrow \mathcal{C}$ and a sequence of constant substitutions $\theta_n \colon \mathcal{C} \rightarrow \mathcal{C}_\omega$ such that the following *approximation conditions* are satisfied.

1. $\theta \circ \theta_n = \mathrm{id}_{\mathcal{C}}$ for all n.
2. For all n, $\alpha_\omega \circ \theta_n \sqsubseteq \alpha_\omega \circ \theta_{n+1}$ and $\bigsqcup_{n \in \mathbb{N}} \alpha_\omega \circ \theta_n = \alpha$.
3. For all $A \in \mathsf{T}(\mathcal{C}_\omega)$ and $N \in \mathsf{T}(\mathcal{C})$, if $A\theta \to N$ and $[A]\alpha_\omega \neq \bot$, then $A \to_\omega B$ for some $B \in \mathsf{T}(\mathcal{C}_\omega)$ with $B\theta = N$.

Theorem 1. *Let $\mathcal{R}_\omega = (\mathcal{C}_\omega, \to_\omega, \alpha_\omega)$, $\mathcal{R} = (\mathcal{C}, \to, \alpha)$, be rewrite structures for a term system T and a model (D, val) of T. If $\mathcal{R}_\omega$ is monotone, strongly normalizing and approximates $\mathcal{R}$, then $\mathcal{R}$ is strongly normalizing.*

Proof. Let $\mathcal{R}_\omega$ approximate $\mathcal{R}$ via $\theta : \mathcal{C}_\omega \to \mathcal{C}$ and $\theta_n : \mathcal{C} \to \mathcal{C}_\omega$, and assume that $\mathcal{R}_\omega$ is monotone and strongly normalizing. To show that $\mathcal{R}$ is strongly normalizing, assume $[M]\alpha \neq \bot$. We have

$$[M]\alpha = [M]\bigsqcup_{n \in \mathbb{N}} \alpha_\omega \circ \theta_n = \bigsqcup_{n \in \mathbb{N}} [M](\alpha_\omega \circ \theta_n) = \bigsqcup_{n \in \mathbb{N}} [M\theta_n]\alpha_\omega$$

by approximation condition 2, continuity and the substitution lemma. Hence $[M\theta_n]\alpha_\omega \neq \bot$ for some $n \in \mathbb{N}$. Since $M\theta_n\theta = M$, by approximation condition 1, it suffices to show the following claim: for any $A \in \mathsf{T}(\mathcal{C}_\omega)$, if $[A]\alpha_\omega \neq \bot$, then $A\theta$ is strongly normalizing. Since $\mathcal{R}_\omega$ is strongly normalizing we may use induction on the strong normalizability of A for proving the claim. Assume $[A]\alpha_\omega \neq \bot$ and $A\theta \to N$. We have to show that N is strongly normalizing. By approximation condition 3 we have $A \to_\omega B$ for some B with $B\theta = N$. Since $\mathcal{R}_\omega$ is monotone we have $[B]\alpha_\omega \neq \bot$. Hence N is strongly normalizing, by induction hypothesis.

Remark 1. Under the additional assumption that for every $d \in \mathcal{C}_\omega$ there exists n such that $\theta_n(\theta(d)) = d$ (which holds in the example below) one can prove that monotonicity of $\mathcal{R}_\omega$ implies monotonicity of $\mathcal{R}$.

Example 4. We define an approximation $\mathcal{R}_\omega = (\mathcal{C}_\omega, \to_\omega, \alpha_\omega)$ of the rewrite structure $\mathcal{R} = (\mathcal{C}, \to, \alpha)$ of Example 3 as follows. $\mathcal{R}_\omega$ is constructed from a constant set $\mathcal{C}_\omega$ and a set $\mathcal{E}_\omega$ in the same way as $\mathcal{R}$ was constructed from $\mathcal{C}$ and $\mathcal{E}$. Therefore it suffices to define $\mathcal{C}_\omega$ and $\mathcal{E}_\omega$. We set

$$\mathcal{C}_\omega := \{c_n \mid c \in \mathcal{C}, \ n \in \mathbb{N}\}$$

(c_n is just c with label n attached), and

$$\mathcal{E}_\omega := \{c_{n+1}\mathbf{P} = M\theta_n \mid c\mathbf{P} = M \in \mathcal{E}\}$$

where $\theta_n(c) := c_n$. Since there is no equation for c_0 we set $\alpha_\omega(c_0) := \bot$ (one could also argue that this follows from the general way α_ω is defined). Because the equations in $\mathcal{E}_\omega$ are free of recursive calls, it is easy to see that all terms in $\mathsf{T}(\mathcal{C}_\omega)$ are strongly normalizing with respect to $\to_\omega$: the usual strong normalization proof for the simply typed λ-calculus with β-reduction via computability predicates á la Tait can be easily extended to $\to_\omega$ (see [2] for details). In particular $\mathcal{R}_\omega$ is strongly normalizing (and monotone, according to Example 3).

Now we show that $\mathcal{R}_\omega$ approximates $\mathcal{R}$ via the constant substitutions θ_n defined above and $\theta : \mathcal{C}_\omega \to \mathcal{C}$, $\theta(c_n) := c$. Approximation condition 1 clearly holds. For approximation condition 2 we need to show that $\alpha_\omega(c_n) \sqsubseteq \alpha_\omega(c_{n+1})$ and $\alpha(c) = \bigsqcup_{n\in\mathbb{N}} \alpha_\omega(c_n)$. The constant interpretation α is defined as the least fixed point of a continuous functional $\Gamma: D^C \to D^C$, i.e. $\alpha = \bigsqcup_{n\in\mathbb{N}} \Gamma^n(\bot)$. An easy induction on n shows that $\alpha_\omega(c_n) = \Gamma^n(\bot)(c)$ (see [2] for details). Since $\Gamma^n(\bot) \sqsubseteq \Gamma^{n+1}(\bot)$ we are done. Finally we show that approximation condition 3 holds. Let $A \in \mathsf{T}(\mathcal{C}_\omega)$ such that $[A]\alpha_\omega \neq \bot$ and $A\theta \to N$. The constant c_0 cannot occur in A since otherwise the strictness of the semantics (easily proven by induction on terms) and the fact that $\alpha_\omega(c_0)$ would imply $[A]\alpha_\omega = \bot$. But then $A \to_\omega B$ for $B \in \mathsf{T}(\mathcal{C}_\omega)$ such that $B\theta = N$ as one easily proves by induction on the definition of $A\theta \to N$.

Since we have shown that $\mathcal{R}_\omega$ is monotone, strongly normalizing and approximates $\mathcal{R}$, it follows, by Theorem 1, that $\mathcal{R}$ is strongly normalizing.

4 Application: Termination of Higher-Order Rewrite Systems

The results of the previous section, in particular their application described in Example 4, give us a convenient method for proving strong normalization of higher-order rewrite systems in the format of Example 3: it suffices to prove that every term has a defined ($\neq \bot$) value in a strict domain-theoretic semantics. Now we apply this to prove strong normalization for a group of higher-order rewrite systems emerging from problems in proof theory. Since the proofs are similar in all cases we will carry this out in detail for one particular example only.

In [8] the axiom scheme of *open induction*

$$\text{OI} \quad \forall f\, (\forall g <_{\mathsf{lex}} f\, U(g) \to U(f)) \to \forall f\, U(f)$$

was used to interpret classical analysis in finite types in a corresponding intuitionistic system. In this axiom scheme U ranges over open predicates on $\mathbb{N} \to \rho$ (where 'open' refers to the $\mathbb{N}$-fold product of the discrete topology on ρ) and $g <_{\mathsf{lex}} f := \exists n\, (\forall k < n\, gk = fk \wedge gn <_\rho fn)$ with some wellfounded relation $<_\rho$ on ρ. Open induction was introduced in a slightly different form in [12] and analysed intuitionistically in [13, 14]. Classically, open induction can be proven using Nash-Williams' minimal-bad-sequence argument [15]. It was shown in [8] that intuitionistic arithmetic plus OI, can be (modified) realizability interpreted by *open recursion*

$$\text{OR} \quad \mathsf{R}^\circ Ff =_\mathbb{N} Ff(\lambda n, y, h.\text{if } y<_\rho fn \text{ then } \mathsf{R}^\circ F(f_n^y h) \text{ else } 0)$$

where

$$f_n^y h := \lambda k. \begin{cases} fk & \text{if } k < n \\ y & \text{if } k = n \\ hk & \text{if } k > n \end{cases}$$

To see the connection with open induction observe that $\{g \mid g <_{\mathsf{lex}} f\} = \{f_n^y h \mid n\!:\!\mathbb{N},\ y\!:\!\rho,\ h\!:\!\mathbb{N} \to \rho,\ y <_\rho fn\}$.

We let $\mathcal{C}$ consists of constants for Gödel primitive recursive functionals (for example $<:$ nat $\rightarrow$ nat $\rightarrow$ boole) and the constants $\mathsf{R}^{\circ}: \sigma \rightarrow \rho^{\omega} \rightarrow$ nat where $\rho^{\omega} :=$ nat $\rightarrow \rho$ and $\sigma := \rho^{\omega} \rightarrow$ (nat $\rightarrow \rho \rightarrow \rho^{\omega} \rightarrow$ nat) $\rightarrow$ nat.

$\mathcal{E}$ consists of the usual defining equations for the primitive recursive constants (see example 3) and the defining equation OR for the constants R° (setting, e.g. $f_n^y h := \lambda k.$if $k < n$ then fk else if $k < \mathsf{S}(n)$ then y else hk).

Theorem 2. *Gödel's system T extended by open recursion is strongly normalizing.*

Proof. Since, according to Example 4, $\mathcal{R}$ is strongly normalizing, it suffices to show that $[M]\alpha \neq \perp$ for every term $M \in \mathsf{T}(\mathcal{C})$. This can be done using the notion of totality. For every type ρ the total elements of D_ρ are defined in the obvious way by recursion on ρ. For example, a function $f \in D_{\rho\rightarrow\sigma}$ is total if it maps total arguments to total values. Obviously, $\perp$ is not total. Furthermore, by induction on terms it follows that $[M]\alpha$ is total provided all constants are, i.e. $\alpha(c^\rho)$ is total in D_ρ for all $c^\rho \in \mathcal{C}$. Our problem therefore reduces to showing that all constants are total. For the primitive recursive constants this follows by straightforward induction on the natural numbers. For R° one applies open induction, using the fact that for any continuous function $F: D_{\rho\rightarrow\mathsf{nat}} \rightarrow D_{\mathsf{nat}}$ the set $U := \{f \in D_{\rho\rightarrow\mathsf{nat}} \mid Ff \neq \perp\}$ is open.

In a similar way one can show that various forms of bar recursion [4–8, 16] lead to strongly normalizing extensions of Gödel's T [2]. In [3] it is furthermore shown that the domain-theoretic model can be modified so as to work for polymorphic second-order types (system F [17]) instead simple types.

The restriction of the rewrite relation defined in Example 3, which is due to the presence of the if-then-else construct, can be avoided by replacing if-then-else by pattern matching using an auxiliary function. For open recursion this would result in the following.

$$\mathsf{R}^{\circ}Ff = Ff(\lambda n, y, h.\tilde{\mathsf{R}}^{\circ}Ffnyh(y<_\rho fn))$$
$$\tilde{\mathsf{R}}^{\circ}Ffnyh\#\mathsf{t} = \mathsf{R}^{\circ}F(f_n^y h)$$
$$\tilde{\mathsf{R}}^{\circ}Ffnyh\#\mathsf{f} = 0$$

For this variant strong normalization, with respect to unrestricted reduction, can be proven with the same method (see [2] for the analogous case of bar recursion).

5   Conclusion

We introduced a general domain-theoretic method for proving termination for a wide class of term-rewriting systems. The main result, Theorem 1, is formulated more abstractly than the corresponding results in [2] and [3] making it clear that the essence of the method is independent of typing disciplines, the particular structure of terms, or particular rewrite strategies.

In summary, the method reduces the strong normalization proof to the following two tasks.

1. Construct a domain-theoretic model that interprets the constants according to the given rewrite rules (this will ensure monotonicity of the rewrite structure) and such that there is a strict dependency of the value of a term on the interpretation of a constant occurring at a position where the given strategy allows for a reduction (this will ensure approximation condition 3 – the approximation conditions 1 and 2 are automatic if one follows the construction given in Example 4).
2. Prove that all constants are total.

Note that the proof-theoretic strength necessary to prove termination for a particular rewrite system goes entirely into the proof of totality. Therefore one can say that our method reduces termination proofs to (technically much simpler) totality proofs. The gain in simplicity through this method becomes apparent if one compares it with weak normalization proofs for related systems given e.g. in [6, 16, 18]. In addition, the method not only simplifies existing proofs, but also leads to new results, such as the strong normalization proof for open recursion presented here.

Since our normalization results for higher-order rewrite systems depend on a given proof of strong normalization for the underlying typed λ-calculus our method does not compete with generic approaches to strong normalization for type theories [19, 20]. It is however conceivable that our method, because of its generality, can be extended to also prove strong normalization for the underlying typed λ-calculus.

References

1. Plotkin, G.: LCF considered as a programming language. Theoretical Computer Science **5** (1977) 223–255
2. Berger, U.: Continuous semantics for strong normalization. In Cooper, S., Löwe, B., Torenvliet, L., eds.: CiE 2005: New Computational Paradigms. Volume 3526 of Lecture Notes in Computer Science. Springer (2005) 23–34
3. Berger, U.: Strong normalization for applied lambda calculi. Submitted to: Logical Methods in Computer Science, January 2005 (2005)
4. Spector, C.: Provably recursive functionals of analysis: a consistency proof of analysis by an extension of principles in current intuitionistic mathematics. In Dekker, F.D.E., ed.: Recursive Function Theory: Proc. Symposia in Pure Mathematics. Volume 5., American Mathematical Society, Providence, Rhode Island (1962) 1–27
5. Howard, W.A.: Functional interpretation of bar induction by bar recursion. Composito Mathematicae **20** (1968) 107–124
6. Berardi, S., Bezem, M., Coquand, T.: On the computational content of the axiom of choice. Journal of Symbolic Logic **63** (1998) 600–622
7. Berger, U., Oliva, P.: Modified bar recursion and classical dependent choice. In: Logic Colloquium 2001. Springer (2005)
8. Berger, U.: A computational interpretation of open induction. In Titsworth, F., ed.: Proceedings of the Ninetenth Annual IEEE Symposium on Logic in Computer Science, IEEE Computer Society (2004) 326–334
9. Scott, D.S.: Outline of a mathematical theory of computation. In: 4th Annual Princeton Conference on Information Sciences and Systems. (1970) 169–176

10. Griffor, E., Lindström, I., Stoltenberg-Hansen, V.: Mathematical theory of domains. Cambridge University Press (1993)
11. Abramsky, S., Jung, A.: Domain theory. In Abramsky, S., Gabbay, D.M., Maibaum, T.S.E., eds.: Handbook of Logic in Computer Science. Volume 3. Clarendon Press (1994) 1–168
12. Raoult, J.C.: Proving open properties by induction. Information processing letters **29** (1988) 19–23
13. Coquand, T.: A note on the open induction principle (1997)
14. Mahboubi, A.: An induction principle over real numbers. Submitted to *Archive for Mathematical Logic* (2004)
15. Nash-Williams, C.: On well-quasi-ordering finite trees. Proc. Cambridge Phil. Soc. **59** (1963) 833–835
16. Tait, W.: Normal form theorem for barrecursive functions of finite type. In Fenstad, J., ed.: Proceedings of the Second Scandinavian Logic Symposium, North–Holland, Amsterdam (1971) 353–367
17. Girard, J.Y.: Interprétation functionelle et élimination des coupures de l'arithmétique d'ordre supérieur. PhD thesis, Université Paris VII (1972)
18. Bezem, M.: Strong normalization of barrecursive terms without using infinite terms. Archive for Mathematical Logic **25** (1985) 175–181
19. Ong, L., Ritter, E.: A generic normalisation argument: Application to the calculus of constructions. In Börger, E., Gurevich, Y., Meinke, K., eds.: Computer Science Logic (Proceedings of the Seventh CSL Conference). Number 832 in LNCS, Springer Verlag, Berlin, Heidelberg, New York (1993) 261–279
20. Hyland, J., Ong, C.H.: Modified realizability semantics and strong normalization proofs. In Bezem, M., Groote, J., eds.: Typed Lambda Calculi and Applications, Springer Lecture Notes in Computer Science Vol. 664 (1993) 179–194

On Bunched Polymorphism

Extended Abstract

Matthew Collinson[1], David Pym[1], and Edmund Robinson[2]

[1] University of Bath, BA2 7AY, UK
[2] Queen Mary, University of London, E1 4NS, UK

Abstract. We describe a polymorphic extension of the substructural lambda calculus $\alpha\lambda$ associated with the logic of bunched implications. This extension is particularly novel in that both variables and type variables are treated substructurally, being maintained through a system of zoned, bunched contexts. Polymorphic universal quantifiers are introduced in both additive and multiplicative forms, and then metatheoretic properties, including subject-reduction and normalization, are established. A sound interpretation in a class of indexed category models is defined and the construction of a generic model is outlined, yielding completeness. A concrete realization of the categorical models is given using pairs of partial equivalence relations on the natural numbers. Polymorphic existential quantifiers are presented, together with some metatheory. Finally, potential applications to closures and memory-management are discussed.

1 Introduction

In recent years, substructural logics and type systems have become firmly established as fundamental tools in the analysis of programming languages. The most prominent are linear logics and types [5], but there are more ad hoc systems, designed for low-level languages and memory management, for example [18].

The logic of bunched implications, **BI**, as exposed in [10], [11], [13] is a substructural logic of growing importance. **BI** provides a logic of resource, which treats the sharing of resource, rather than the number of uses treated by linear logic. The resource-sensitive aspect of **BI** has led to it being adopted as the basis of the assertion language of new program logics, notably separation logic [15], which allow for safe-reasoning about imperative languages with pointers.

BI has several well-understood classes of models, both truth-functional and categorical, and like linear logics, has an elegant proof-theory. In particular, there is an associated lambda calculus, $\alpha\lambda$, giving a propositions-as-types correspondence. The calculus is presented using derivations of typing judgements in which contexts of typed variables are certain trees, called bunches. The way to understand $\alpha\lambda$ is through a reading of the terms known as the sharing interpretation which emphasizes the use of some computational resource. As an example of this, $\alpha\lambda$ has both additive and multiplicative function types. A function of the additive kind may make use of the same computational resource as its argument,

L. Ong (Ed.): CSL 2005, LNCS 3634, pp. 36–50, 2005.

but this is not the case for the multiplicative. In [10], $\alpha\lambda$ was used to unify the Algol-like languages Syntactic Control of Interference (SCI) and Idealized Algol (IA), which had hitherto appeared to have irreconcilable features.

Whilst it has been demonstrated that **BI** has applications to program logic for imperative programming and to type systems for small, idealized languages, the full power of the type-system remains unexploited. The possibility exists to build a functional programming language along the lines of ML, but based on bunched rather than simple types. The typing of a program should then make guarantees about the use of resources (for example, memory, in the presence of references) as well as the compatibility of sub-expressions. This paper takes some of the first steps in that direction.

Polymorphism must be added to $\alpha\lambda$ in order to give a language with the expressivity of ML. We present a calculus which bears the same relationship to $\alpha\lambda$ as the Girard-Reynolds polymorphic lambda calculus $\lambda 2$ [3], [14] does to the simply-typed lambda calculus. Adding ordinary, impredicative polymorphism to $\alpha\lambda$ amounts to adding a further zone to typing contexts which manages the use of type variables. In this paper we take a further step, by considering a calculus in which the type variable zone consists of a bunch. This gives extra flexibility in the type system, for it allows us to consider both additive and multiplicative polymorphism. The additive polymorphism allows us to recover all standard uses of polymorphism, whilst the multiplicative polymorphism enforces non-sharing of resources associated with type variables. Multiplicative quantification closely resembles the freshness quantifier of Pitts and Gabbay [6]. Further steps and features are required before we have a genuinely ML-like type system, including predicative polymorphism, recursive types, references and typechecking.

In §2, we add polymorphic universal quantifiers to $\alpha\lambda$. We follow this with some of the more important metatheoretical results in §3. In §4, we describe an extension of the usual notion of categorical model. The additives are modelled in the usual way, and in a similar way, the multiplicatives are modelled by the right-adjoints to certain substitutions. In §5, we give an instance of such a model using the category *PER* of partial equivalence relations on the natural numbers.

In §6, we introduce polymorphic existential quantification. The desire to extend the sharing interpretation, together with metatheoretic concerns, governs the design of the multiplicative quantifier. The multiplicative existential is less semantically neat than the universal, but hints strongly at a number of applications, for it enables the hiding not just of a type, but also of the resources that accompany it. Thus there is an appealing intuition for multiplicative existentials as a kind of closure. We discuss connections to work on type systems for memory-management, specifically alias types [18] and regions [17], [19], where the use of location and region variables leads to forms of polymorphism. For alias types, this polymorphism appears to be multiplicative.

The work reported herein was carried out under the project 'Bunched ML', funded by the United Kingdom EPSRC. We acknowledge help and suggestions given by our collaborators, Josh Berdine and Peter O'Hearn of Queen Mary University London. We also thank the anonymous referees.

2   The Calculus

The calculus, which we shall call $\alpha 2\lambda 2$, has three levels of judgement. A first level judgement $X \vdash \tau$ gives a type τ over a bunch of type variables X. The second level, which has judgements of the form $X \vdash \Gamma$ generates the contexts Γ of ordinary variables over X. The third level comprises judgements $X \mid \Gamma \vdash M : \tau$ which show that a term M is well-typed with τ, given X and Γ.

Assume a countable collection of *type variables* $\alpha, \beta, \ldots$ to be given. The types used in the calculus are generated by

$$\tau := \top \mid I \mid \alpha \mid \tau \wedge \tau \mid \tau * \tau \mid \tau \rightarrow \tau \mid \tau \rightarrow\!\!* \tau \mid \forall \alpha.\tau \mid \forall_* \alpha.\tau \ ,$$

where α is any type variable. The connectives $\top$, $\wedge$, $\rightarrow$ and $\forall$ are the additive unit, product, function space and polymorphic universal quantifier, respectively. There are multiplicative unit I, product $*$, function space $\rightarrow\!\!*$ and universal $\forall_*$ connectives. We allow the letters σ, τ to range over types.

A *hub* is a bunch of type variables, generated as follows

$$X := \emptyset \mid \alpha \mid X, X \mid X; X \ ,$$

subject to the restriction that every type-variable may occur at most once in a bunch. Let X, Y, Z range over hubs.

Assume a countable collection of *variables* $x, y, z, \ldots$ to be given. A *(typing) context* is a bunch of typed variables, generated by

$$\Gamma := \emptyset \mid \emptyset_* \mid x : \tau \mid \Gamma, \Gamma \mid \Gamma; \Gamma \ ,$$

where x is a variable, τ is a type and any variable occurs at most once. The units $\emptyset$ and $\emptyset_*$ are distinct from the unit $\emptyset$ for hubs. The typing contexts are nothing more than the contexts of $\alpha\lambda$, but such that types may contain type variables.

Bunches are always subject to a pair of equivalence relations [13]. The first equivalence $\equiv$ on bunches is used to build structural rules that allow us to permute variables in hubs or contexts. It is given by commutative monoid rules for ";", for "," and by a congruence to ensure that the monoid rules can be applied at arbitrary depth in any bunch. The second relation $\cong$ is used to control contraction rules. The equivalence $\cong$ on hubs is simply renaming of type variables: $X \cong Y$ if Y can be obtained from X by renaming bijectively with type variables. The relation $\Gamma \cong \Delta$ between contexts holds just when Δ can be obtained by relabelling the variables of the leaves of Γ in a type preserving way: any leaf $x : \tau$ of Γ must correspond to a node $y : \tau$ of Δ.

There is an obvious notion of sub-bunch of a bunch. Let $B(B_1 \mid \ldots \mid B_n)$ be the notation for a bunch B with distinct, distinguished sub-bunches $B_1, \ldots, B_n$. Write $B[B_1'/B_1, \ldots B_n'/B_n]$ for the bunch formed by replacing each bunch B_i in B with B_i'.

The rules for generating *type formation judgements*, which specify types which are well-formed over hubs, are shown in Figure 1. A critical design decision is evident at this level. The formation rules for $\wedge$, $\rightarrow$, $*$ and $\rightarrow\!\!*$ are kept as simple as possible, in that formation takes place over a single, fixed hub.

(TAx) $\dfrac{}{\alpha \vdash \alpha}$ $(T\top)$ $\dfrac{}{\emptyset \vdash \top}$ $\dfrac{}{\emptyset \vdash I}$ (TI)

$(T\odot)$ $\dfrac{X \vdash \sigma \quad X \vdash \tau}{X \vdash \sigma \odot \tau}$ $(\odot$ is any of $\times, \rightarrow, *, -\!\!*)$

$(T\forall)$ $\dfrac{X; \alpha \vdash \tau}{X \vdash \forall \alpha.\tau}$ $\dfrac{X, \alpha \vdash \tau}{X \vdash \forall_* \alpha.\tau}$ $(T\forall_*)$

(TC) $\dfrac{X(Y; Y') \vdash \tau}{X(Y) \vdash \tau[Y/Y']}$ $(Y \cong Y')$ (TW) $\dfrac{Y \vdash \tau}{X(Y) \vdash \tau}$ $(Z \equiv Z')$ $\dfrac{Z \vdash \tau}{Z' \vdash \tau}$ (TE)

Fig. 1. Type formation rules

The construction of contexts which are valid over hubs is generated from the type-formation judgements. These are presented as judgements of the form $X \vdash \Gamma$ where X is a hub and Γ is a context and are characterised by: $X \vdash \Gamma$ holds if and only if $X \vdash \tau$ for each variable $x : \tau$ in Γ.

The terms of the language are given by the following grammar

$$
\begin{aligned}
M := \ & x \mid \top \mid I \mid \text{ let } I \text{ be } M \text{ in } M \\
& \mid \langle M, M \rangle \mid \pi_1 M \mid \pi_2 M \mid M * M \mid \text{ let } (x, y) \text{ be } M \text{ in } M \\
& \mid \lambda x : \tau.M \mid \text{app}(M, M) \mid \lambda_* x : \tau.M \mid \text{app}_*(M, M) \\
& \mid \Lambda \alpha.M \mid \text{App}(M, X, \tau) \mid \Lambda_* \alpha.M \mid \text{App}_*(M, X, \tau) \ ,
\end{aligned}
$$

where α is a type variable, τ is a type, X is a hub and x is a variable.

Let $FV(-)$ be the set of variables which are in a context $(-)$ or free (not bound by a lambda abstraction) in a term $(-)$. We use the notation $FTV(-)$ for the set of type variables which occur free in a bunch $(-)$, type $(-)$, the types of the variables in the context $(-)$ or the type of the term $(-)$, respectively. In a term $\text{App}(M, X, \tau)$ or $\text{App}_*(M, X, \tau)$, the type variables of X are free, so substitution must take account of this.

We introduce a syntactic measure μ which assigns to each term the set of type variables which are free and which occur in some application of the multiplicative universal quantifier. Formally, this is given by a recursive definition, where

$$\mu(\Lambda \alpha.M) = \mu(\Lambda_* \alpha.M) = \mu(M) \smallsetminus \{\alpha\} \qquad \mu(\text{App}_*(M, X, \tau)) = \mu(M) \cup FTV(X)$$

are the informative clauses.

The typing of terms uses the term and context formation judgements. The *term formation judgements* are derived according to a system of rules, a sample of which are shown in Figure 2. In addition to the rules shown, there are introduction and elimination rules for rules for additive $(\top)$ and multiplicative (I) units, additive $(\wedge)$ and multiplicative $(*)$ conjunction, additive lambda abstraction $(\rightarrow)$, contraction (C) and equivalence (E) for contexts. All of the rules other than the quantifier rules and the hub structurals use a fixed hub X. That is to say, they are essentially the familiar rules for $\alpha\lambda$, but parameterised by the hub. The elimination rules $(\wedge E)$, $(*E)$, $(\rightarrow E)$, $(-*E)$ are each subject to a side-condition

$$\mu(N) \cap FTV(M) = \emptyset \qquad (\dagger)$$

which requires the separation of certain of the free type variables present.

$$(Ax)\ \dfrac{X \vdash x : \tau}{X \mid x : \tau \vdash x : \tau} \qquad\qquad (W)\ \dfrac{X \mid \Gamma(\Delta) \vdash M : \tau \quad X \vdash \Delta'}{X \mid \Gamma(\Delta; \Delta') \vdash M : \tau}$$

$$(-*I)\ \dfrac{X \mid \Gamma, x : \sigma \vdash M : \tau}{X \mid \Gamma \vdash \lambda_* x : \sigma.M : \sigma -* \tau} \qquad (\dagger)\dfrac{X \mid \Gamma \vdash N : \sigma -* \tau \quad X \mid \Delta \vdash M : \sigma}{X \mid \Gamma, \Delta \vdash \mathrm{app}_*(N, M) : \tau}\ (-*E)$$

$$(\forall I)\ \dfrac{X; \alpha \mid \Gamma \vdash M : \tau}{X \mid \Gamma \vdash \Lambda\alpha.M : \forall\alpha.\tau} \quad (\alpha \notin FTV(\Gamma)) \qquad \dfrac{X, \alpha \mid \Gamma \vdash M : \tau}{X \mid \Gamma \vdash \Lambda_*\alpha.M : \forall_*\alpha.\tau}\ (\forall_* I)$$

$$(\forall E)\ \dfrac{X \mid \Gamma \vdash M : \forall\alpha.\tau \quad Y \vdash \sigma}{X; Y \mid \Gamma \vdash \mathrm{App}(M, Y, \sigma) : \tau[\sigma/\alpha]} \qquad \dfrac{X \mid \Gamma \vdash M : \forall_*\alpha.\tau \quad Y \vdash \sigma}{X, Y \mid \Gamma \vdash \mathrm{App}_*(M, Y, \sigma) : \tau[\sigma/\alpha]}\ (\forall_* E)$$

$$(FW)\ \dfrac{Y \mid \Gamma \vdash M : \tau}{X(Y) \mid \Gamma \vdash M : \tau} \qquad\qquad (X \equiv Z)\ \dfrac{X \mid \Gamma \vdash M : \tau}{Z \mid \Gamma \vdash M : \tau}\ (FE)$$

$$(FC)\ \dfrac{X(Y; Y') \mid \Gamma \vdash M : \tau}{X(Y) \mid \Gamma[Y/Y'] \vdash M[Y/Y'] : \tau[Y/Y']}\ (Y \cong Y')$$

Fig. 2. Sample of the term formation rules

The usual rules for $\beta\eta\zeta$-conversions for $\alpha\lambda$ are retained, see [13]. In addition, we have four conversions for quantifiers,

$$\begin{aligned}
\mathrm{App}(\Lambda\alpha.M, X, \alpha) \rightarrow_\beta M \qquad &\Lambda\alpha.\mathrm{App}(M, X, \alpha) : \tau \rightarrow_\eta M \\
\mathrm{App}_*(\Lambda_*\alpha.M, X, \alpha) \rightarrow_\beta M \qquad &\Lambda_*\alpha.\mathrm{App}_*(M, X, \alpha) : \tau \rightarrow_\eta M \ ,
\end{aligned}$$

where these terms are all typed over the same hub X and context Γ such that α is not free in Γ. Let $\twoheadrightarrow$ be the reduction relation generated by the single step conversions. As usual, these relations give rise to a system of $\beta\eta\zeta$-equalities.

3   Metatheory

Many of the standard properties of a lambda calculus hold for $\alpha 2\lambda 2$. In particular, the fact that hubs are affine (weakening is allowed around ',' and additive and multiplicative units are identified) yields admissible substitution rules.

Proposition 1. *(Substitution Laws)*

1. *If $X \mid \Gamma(x : \sigma) \vdash N : \tau$ and $X \mid \Delta \vdash M : \sigma$ are derivable and the condition $\mu(N) \cap FTV(M) = \emptyset$ holds then $X \mid \Gamma[\Delta/x] \vdash N[M/x] : \tau$.*
2. *If $Y \mid \Gamma \vdash M : \tau$ and $Z \vdash \sigma$ then $Y[Z/\alpha] \mid \Gamma[\sigma/\alpha] \vdash M[(Z, \sigma)/\alpha] : \tau[\sigma/\alpha]$.*

The side-condition on the first part is essential, because the derivation of N may have used $(\forall_* E)$. This makes the side-condition (†) on the elimination laws necessary for subject-reduction.

Proposition 2. *The four rules below are admissible.*

$$\frac{X \mid \Gamma \vdash \lambda x : \sigma.M : \sigma \to \tau}{X \mid \Gamma; x : \sigma \vdash M : \tau} \qquad \frac{X \mid \Gamma \vdash \lambda_* x : \sigma.M : \sigma \mathrel{-\!*} \tau}{X \mid \Gamma, x : \sigma \vdash M : \tau}$$

$$\frac{X \mid \Gamma \vdash \Lambda\alpha.M : \forall\alpha.\tau}{X; \alpha \mid \Gamma \vdash M : \tau} \qquad \frac{X \mid \Gamma \vdash \Lambda_*\alpha.M : \forall_*\alpha.\tau}{X, \alpha \mid \Gamma \vdash M : \tau}$$

The propositions above can be used to prove subject-reduction.

Theorem 1. *If $X \mid \Gamma \vdash M : \tau$ and $M \twoheadrightarrow N$ then $X \mid \Gamma \vdash N : \tau$ is derivable.*

All reductions of the calculus terminate, as is shown by translation into the polymorphic lambda calculus $\lambda 2$.

Theorem 2. *The calculus is strongly normalizing.*

The reduction relation can be extended to include ζ-reductions (commuting conversions) for $*$, following [13], and the subject-reduction and normalization theorems continue to hold. Similarly, the extension of $\alpha 2\lambda 2$ with the additive disjunction $\vee$ of $\alpha\lambda$ causes no difficulties.

4 Categorical Semantics

We now give a categorical semantics to $\alpha 2\lambda 2$. This is a hybrid of the indexed category semantics of $\lambda 2$ with the doubly closed category semantics of $\alpha\lambda$.

Before giving the modified version of hyperdoctrine, we introduce some terminology for a certain structure on a category. Consider a symmetric monoid $(\otimes, I, a, l, r, s)$ on a category $\mathbb{B}$. Let $1_{\mathbb{B}} : \mathbb{B} \longrightarrow \mathbb{B}$ be the identity functor. The monoid $\otimes$ is a *pseudoproduct* if for every object B in $\mathbb{B}$ there is a (first) *pseudoprojection*, that is, a natural transformation $\psi_B^1 : 1_{\mathbb{B}} \otimes B \Longrightarrow 1_{\mathbb{B}}$ satisfying the two coherence diagrams given below.

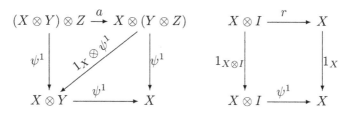

We write the component at an object A as $\psi^1_{A,B} : A \otimes B \longrightarrow A$. Using the symmetry isomorphisms s, it is easy to construct a second pseudoprojection $\psi^2_A : A \otimes 1_{\mathbb{B}} \Longrightarrow 1_{\mathbb{B}}$ with components $\psi^2_{A,B}$ where A, B are any objects of $\mathbb{B}$. We frequently omit both subscripts and superscripts on pseudoprojections.

All products are pseudoproducts, but not vice versa. The category $Set_\perp$ of pointed sets $X_\perp$ and functions which preserve the distinguished element $\perp$ has a pseudoproduct given by the coproduct. A pseudoprojection from $X_\perp + Y_\perp$ to $X_\perp$ may be taken to be $\perp, y \mapsto \perp, x \mapsto x$ for all $x \in X, y \in Y$.

A *cartesian doubly closed category* (CDCC) is a category with a pair of symmetric monoidal closed structures, one of which is cartesian. A functor between CDCC's is *strict* if it preserves both the cartesian closed and the monoidal closed structure on-the-nose. Let **CDCC** be the category of cartesian doubly closed categories and strict functors.

A *split indexed category* consists of a contravariant functor from a base category $\mathbb{B}$ to some category of categories, see [8] for a detailed account. A *hyperdoctrine* [16] is a categorical model of $\lambda 2$ consisting of a split indexed category with certain properties, including a system of adjunctions for modelling quantification. It also requires a distinguished base object Ω, called the *generic object*, which is characterized by the property that there is a natural bijection $\iota_J : (PJ)_0 \xrightarrow{\cong} \text{hom}(J, \Omega)$, where $\text{hom}(-, \Omega)$ is the contravariant hom functor for $\mathbb{B}$ and $(PJ)_0$ is the set of objects of the fibre PJ.

An *α2λ2-hyperdoctrine* is a split indexed category $P : \mathbb{B}^{op} \longrightarrow \mathbf{CDCC}$ with: generic object; finite products and binary pseudoproducts in the base; the unit of $\otimes$ is $\top$, the terminal object; for any projection π in the base, the functor $P(\pi)$ has a right-adjoint Π which satisfies the Beck-Chevalley condition; for any pseudoprojection ψ in the base, the functor $P(\psi)$ has a right-adjoint Ψ which satisfies an appropriate, weak form of the Beck-Chevalley condition.

Interpret hubs X as objects in the base $\mathbb{B}$, with

$$[\![\emptyset]\!] = \top \qquad [\![\alpha]\!] = \Omega \qquad [\![X; Y]\!] = [\![X]\!] \times [\![Y]\!] \qquad [\![X, Y]\!] = [\![X]\!] \otimes [\![Y]\!] .$$

Interpret type formations as objects $[\![X \vdash \tau]\!]$ of the fibre $P([\![X]\!])$. An instructive fragment of the interpretation is given by

$$[\![\emptyset \vdash T : T]\!] = \top \qquad [\![\emptyset \vdash I : I]\!] = I \qquad [\![\alpha \vdash \alpha]\!] = \iota_\Omega^{-1}(1_\Omega)$$

$$[\![X \vdash \tau * \tau']\!] = [\![X \vdash \tau]\!] \otimes [\![X \vdash \tau']\!] \qquad [\![X \vdash \tau -\!\!* \tau']\!] = [\![X \vdash \tau]\!] \multimap [\![X \vdash \tau']\!]$$

$$[\![X \vdash \forall \alpha.\tau]\!] = \Pi [\![X; \alpha \vdash \tau]\!] \qquad [\![X \vdash \forall_* \alpha.\tau]\!] = \Psi [\![X, \alpha \vdash \tau]\!] ,$$

where $\top$, I, $\otimes$ and $\multimap$ are from the doubly closed structure of fibres, Π is adjoint to $\pi : [\![X]\!] \times \Omega \longrightarrow [\![X]\!]$, and Ψ is adjoint to $\psi : [\![X]\!] \otimes \Omega \longrightarrow [\![X]\!]$. The interpretations of the omitted rules are quite standard. In particular, the interpretation of the rule (TW) makes use of projections and pseudoprojections.

Interpret contexts as objects $[\![X \vdash \Gamma]\!]$ of $P([\![X]\!])$ by extension of the interpretation of types, using the product and monoidal structure of the fibre.

Morphisms $[\![X \mid \Gamma \vdash M : \tau]\!] : [\![X \vdash \Gamma]\!] \longrightarrow [\![X \vdash \tau]\!]$ in $P([\![X]\!])$ are used to interpret term formations. A fragment of the interpretation is given below.

$$[\![X \mid x : \tau \vdash x : \tau]\!] = 1_{[\![X \vdash \tau]\!]} \qquad [\![X \mid \emptyset \vdash \top : \top]\!] = 1_\top \qquad [\![X \mid \emptyset_* \vdash I : I]\!] = 1_I$$

$$\frac{[\![X \mid \Gamma, x : \phi \vdash M : \psi]\!] = f : [\![X \vdash \Gamma]\!] \otimes [\![X \vdash \phi]\!] \longrightarrow [\![X \vdash \psi]\!]}{[\![X \mid \Gamma \vdash \lambda_* X : \phi.M : \phi \mathrel{-\!*} \psi]\!] = f^\wedge : [\![X \vdash \Gamma]\!] \longrightarrow ([\![X \vdash \phi]\!] \multimap [\![X \vdash \psi]\!])}$$

$$\frac{[\![X; \alpha \mid \Gamma \vdash M : \tau]\!] = g : [\![X; \alpha \vdash \Gamma]\!] \longrightarrow [\![X; \alpha \vdash \tau]\!]}{[\![X \mid \Gamma \vdash \Lambda\alpha.M : \forall\alpha.\tau]\!] = g^\Delta : [\![X \vdash \Gamma]\!] \longrightarrow [\![X \vdash \forall\alpha.\tau]\!]}$$

$$\frac{[\![X, \alpha \mid \Gamma \vdash M : \tau]\!] = h : [\![X, \alpha \vdash \Gamma]\!] \longrightarrow [\![X, \alpha \vdash \tau]\!]}{[\![X \mid \Gamma \vdash \Lambda_*\alpha.M : \forall_*\alpha.\tau]\!] = h^\triangleleft : [\![X \vdash \Gamma]\!] \longrightarrow [\![X \vdash \forall_*\alpha.\tau]\!]}$$

$$\frac{[\![X \mid \Gamma \vdash M : \forall\alpha.\tau]\!] = m : [\![X \vdash \Gamma]\!] \longrightarrow [\![X \vdash \forall\alpha.\tau]\!] \quad [\![Y \vdash \rho]\!] = B \in P([\![Y]\!])}{[\![X, Y \mid \Gamma \vdash \mathrm{App}(M, Y, \rho)]\!] = P(1_{[\![X]\!]} \times \iota(B))(m^\triangledown)}$$

$$\frac{[\![X \mid \Gamma \vdash M : \forall_*\alpha.\tau]\!] = m : [\![X \vdash \Gamma]\!] \longrightarrow [\![X \vdash \forall_*\alpha.\tau]\!] \quad [\![Y \vdash \rho]\!] = B \in P([\![Y]\!])}{[\![X, Y \mid \Gamma \vdash \mathrm{App}_*(M, Y, \rho)]\!] = P(1_{[\![X]\!]} \otimes \iota(B))(m^\triangleright)}$$

Here, $f^\wedge$ is the linear exponential mate of f, $(-)^\Delta$ and $(-)^\triangledown$ give the transposes of $\pi^* \dashv \Pi$, $(-)^\triangleleft$ and $(-)^\triangleright$ give the transposes of $\psi^* \dashv \Psi$, and the morphism $\iota(B) : [\![Y]\!] \longrightarrow \Omega$ in the base arises from the fact that Ω is generic.

4.1   Soundness and Completeness

In any $\alpha 2\lambda 2$-hyperdoctrine, every judgement can be interpreted.

Proposition 3. *(Weak Soundness) Every judgement $X \mid \Gamma \vdash M : \tau$ has an interpretation as a morphism $[\![X \vdash \Gamma]\!] \longrightarrow [\![X \vdash \tau]\!]$ in $P([\![X]\!])$.*

Substitution of a term for a variable takes place in a fixed hub, so its interpretation is modelled in the corresponding CDCC as in [13]. The interpretation of substitution for type variables uses reindexing functors and the generic object.

Proposition 4. *(Equational Soundness) If $X \mid \Gamma \vdash M = M' : \tau$ is derivable then $[\![X \mid \Gamma \vdash M : \tau]\!] = [\![X \mid \Gamma \vdash M' : \tau]\!]$ holds.*

The syntactic equalities are generated by the $\beta\eta\zeta$-conversions. All of these take place over a fixed hub, except for the reductions for the quantifiers. We know that the equalities over any hub are all validated in the corresponding CDCC. The β- and η-rules for the multiplicative quantifier are witnessed, respectively, by the equations

$$(P(1_{[\![X]\!]} \otimes 1_\Omega))((m^\triangleleft)^\triangleright) = m \qquad ((P(1_{[\![X]\!]} \otimes 1_\Omega))(n^\triangleright))^\triangleleft = n \ ,$$

given interpretations $[\![X, \alpha \mid \Gamma \vdash M : \tau]\!] = m : [\![X, \alpha \vdash \Gamma]\!] \longrightarrow [\![X, \alpha \vdash \tau]\!]$ and $[\![X \mid \Gamma \vdash N : \forall_*\alpha.\tau]\!] = n : [\![X \vdash \Gamma]\!] \longrightarrow [\![X \vdash \forall_*\alpha.\tau]\!]$. These equalities follow because the indexed category is split. The relevant equalities for additive quantification follow by the obvious modifications.

Completeness with respect to $\alpha 2\lambda 2$-hyperdoctrines is established by the usual method. That is, we build a generic model from the syntax such that if an equation holds between interpreted terms then it must also hold in the theory. The main novelty here is the construction of the base category, although this follows essentially the same pattern as the construction for $\lambda 2$ hyperdoctrines: objects are (bunches of) type variables and morphisms are substitutions derived from type formation judgements.

We construct the base $\mathbb{B}$ from the syntax of hubs and type formations. The objects of $\mathbb{B}$ are taken to be the equivalence classes of hubs under the congruence relation $\cong$, which handles α-conversion of type variables. Throughout this construction, we use hubs as representatives of equivalence classes. Let Ω be the equivalence class of α and $\top$ be the equivalence class of $\emptyset$.

The congruence $\cong$ on hubs extends to type formation judgements using substitution: $(X \vdash \tau) \cong (Y \vdash \tau') \iff (X \cong Y)$ & $(\tau' = \tau[Y/X])$ for all hubs X and Y. Again, we will tend to use representatives for equivalence classes in what follows. Define a mapping $\overline{(-)} : (X \vdash \tau) \mapsto \tau$ from type formations to types.

The morphisms of $\mathbb{B}$ from X to Y are certain trees with the same shape (internal node structure) as Y and with equivalence classes of type formations at the leaves. These morphisms are generated by an inductive definition.

There are a number of parts to the base case. These are identity, terminal, diagonal, projection, pseudoprojection, right unit, associativity, associativity inverse, symmetry. For brevity, we give only the diagonal and pseudoprojection clauses below. From these, the forms of the other cases may be easily inferred. Diagonal: for every X there is a morphism $\Delta_X : X \longrightarrow X; X'$, where X' is any hub which is disjoint from X and with $X' \cong X$. The morphism is given by $f_X; f_{X'}$ where f_X is formed by replacing every leaf α of X with $X \vdash \alpha$. Pseudoprojection: for all X and Y there is an arrow $\psi^1 : X, Y \longrightarrow X$ formed by replacing each leaf α of X with $X, Y \vdash \alpha$.

The inductive definition has three step cases: product, pseudoproduct and composite. Product: if there are arrows $f : X \longrightarrow Y$ and $g : X \longrightarrow Y'$ then there is a morphism $X; X' \xrightarrow{f;g} Y; Y'$. It is formed as the tree $f'; g'$ where f' is formed from f by replacing each leaf $X \vdash \tau$ with $X; X' \vdash \tau$, and similarly for g'. Pseudoproduct: If there is a morphism $f : X \longrightarrow Y$ and there is a morphism $g : X' \longrightarrow Y'$ then there is a morphism $X, X' \xrightarrow{f,g} Y, Y'$. It is formed as the tree f', g' where f' is formed from f by replacing each leaf $X \vdash \tau$ with $X, X' \vdash \tau$, and similarly for g'. Composite: the composite in $\mathbb{B}$ of a pair of arrows $f : X \longrightarrow Y$, $g : Y \longrightarrow Z$ is an arrow $g \circ f : X \longrightarrow Z$ constructed by replacing each leaf $Y \vdash \rho$ of g with the leaf $X \vdash \rho[\overline{f}/Y]$, where the mapping $\overline{(-)}$ is extended to trees in the obvious way.

Some comments and observations about the above definition are in order. In a number of the clauses above we have formed a morphism from X to Y using some words like "replace any variable α of Y with the judgement $X \vdash \tau$" and it is to be understood that any leaves of Y which are units $\emptyset$ should be replaced by the judgement $X \vdash \top$. Composition is a well-defined operation, independent of choices of representatives. The hom-sets of $\mathbb{B}$ are guaranteed to remain small.

It is a matter of lengthy calculation to verify that $\mathbb{B}$ is a category, has finite limits and has a symmetric monoid which is a pseudoproduct. These structures are suggested by the notation in the recursive definition.

Write $P(X)$ for the fibre over the equivalence class of X. The construction of each $P(X)$ follows the construction of a CDCC from $\alpha\lambda$, see [13]. *Objects* are equivalence classes of type formations $X \vdash \tau$, represented by pairs (X, τ). A *morphism* from (X, σ) to (X, τ) is an equivalence class of term formations $X \mid x : \sigma \vdash M : \tau$, where the equivalence is generated by α-equality for variables, the $\beta\eta\zeta$-rules (without the quantifier cases) and the congruence extended from the congruence $\cong$ on hubs.

Every arrow $u : X \longrightarrow Y$ of $\mathbb{B}$ yields a functor $P(u) : P(Y) \longrightarrow P(X)$ between fibres. The functor acts as $P(u)(Y, \tau) = (X, \tau[\overline{u}/Y])$ on any object (Y, τ) in $P(Y)$. The arrow assignment is given by $P(u)(Y, x, M) = (X, x, M[\overline{u}/Y])$ for any arrow (Y, x, M) in $P(Y)$. Both the object and arrow assignments can be verified to be well-defined and calculuations can be performed to show that $P(u)$ is indeed functorial.

Further calculations show that the functors $P(u)$ preserve the CDCC structure on-the-nose. Moreover, the functors induced by projections $\pi : X \times \Omega \longrightarrow X$ and pseudoprojections $\psi : X \otimes \Omega \longrightarrow X$ can be shown to have right-adjoints which satisfy the Beck-Chevalley conditions. The identity gives a natural bijection between the hom-sets $\mathbb{B}(X, \Omega)$ and fibres $P(X)$.

Theorem 3. *The functor $P : \mathbb{B}^{op} \longrightarrow \mathbf{CDCC}$ is an $\alpha 2\lambda 2$-hyperdoctrine.*

The completeness theorem follows as a corollary, since P is constructed from the syntax and each term is interpreted, essentially, by itself.

Corollary 1. *(Completeness) If $[\![X \mid \Gamma \vdash M : \tau]\!] = [\![X \mid \Gamma \vdash M' : \tau]\!]$ holds in every $\alpha 2\lambda 2$-hyperdoctrine then $X \mid \Gamma \vdash M = M' : \tau$ is derivable in the calculus.*

5   A PER Model

Partial equivalence relations on the natural numbers give rise to one of the simplest and most elegant models of the polymorphic lambda calculus [4]. We show how to produce a PER model for $\alpha 2\lambda 2$.

A *partial equivalence relation*, PER for short, consists of a symmetric, transitive, binary relation $R \subseteq \mathbb{N} \times \mathbb{N}$ on the natural numbers. Define the *domain* of R to be $dom(R) = \{n \in \mathbb{N} \mid nRn\}$. A map between PERs consists of an equivalence class of codes for recursive functions that track from the source PER to the target PER, that is, functions which preserve the relation. Let *PER* be the category of partial equivalence relations and PER_0 be its set of objects. The category is cartesian closed. It also has binary coproducts: embed isomorphically the two given PERs into PERs with disjoint domains, then take the union of the relations.

Since PER is cartesian closed and has a symmetric monoid (given by the coproduct) we might think that we can use these two structures to model $\alpha\lambda$. However, the monoid fails to be closed. This can be remedied by moving to a model based on pairs of PERs, motivated by a similar construction for sets. The category $Set \times Set$ of pairs of sets is a CDCC, see [11], [13]. Finite products and exponentials are given pointwise. Moreover, there is an additional symmetric monoidal closed structure with

$$(A^0, A^1) \otimes (B^0, B^1) = ((A^0 \times B^0) + (A^1 \times B^1), (A^0 \times B^1) + (A^1 \times B^0))$$
$$(A^0, A^1) \multimap (B^0, B^1) = ((A^0 \to B^0) \times (A^1 \to B^1), (A^0 \to B^1) \times (A^1 \to B^0)) \ ,$$

for all $A^0, A^1, B^0, B^1 \in Set$, where $A + B$ is the coproduct of A and B in Set. This can be viewed as an instance $Set^{\mathbf{2}}$ of Day's closure construction [1], [2], where $\mathbf{2} = \{0, 1\}$ is the discrete category with monoid given by addition modulo two. Now $PER \times PER$ can be viewed as $PER^{\mathbf{2}}$ and so is doubly closed by [1]. Its operations are defined in the same way as those of $Set \times Set$, remembering that the $+$ in the definition of $\otimes$ is now the coproduct in PER. For any pair (A^0, A^1) of PERs let $(A^0, A^1)^0 = A^0$ and $(A^0, A^1)^1 = A^1$. Extend the notion of domain to pairs of PERS with $dom(A, B) = dom(A) \times dom(B)$ for any PERs A and B. For any function $f : A \longrightarrow B$ and $C \subseteq A$ let $f\lceil_C$ be the restriction of f to C.

Let X be a bunch of type variables. Let $dom(\rho) = \bigcup_{\alpha \in FTV(X)} \rho(\alpha)$ for any function $\rho : FTV(X) \longrightarrow PER_0 \times PER_0$. An *environment* for X is a function $\rho : FTV(X) \longrightarrow PER_0 \times PER_0$ such that if any (Y, Z) is a sub-bunch of X then $dom(\rho\lceil_Y) \cap dom(\rho\lceil_Z) = \emptyset$ holds. Let $Env(X)$ be the set of environments for X.

A *semantic type* (over X) is a function $\tau : Env(X) \longrightarrow PER_0 \times PER_0$ from environments to pairs of PERs. These definitions give a natural generalization of the ordinary PER model of polymorphism, in which an environment consists of a tuple of PERs and a semantic type consists of a map from environments to PERs. A map from τ to τ' (over X) is an equivalence class $[e]$ of codes for pairs of codes, $([e^0], [e^1])$, where the recursive function corresponding to each e^i tracks from $(\tau\rho)^i$ to $(\tau'\rho)^i$ for all environments ρ. This gives a category $P(X)$ of semantic types over X.

Let $\alpha_1, \ldots, \alpha_n$ be the variables of X. A *substitution* $(-)[\tau_1/\alpha_1, \ldots, \tau_n/\alpha_n]$ for X consists of semantic types $\tau_1, \ldots, \tau_n$ over some bunch Y such that: if X has a sub-bunch (W, Z), where W has type variables with $\alpha_{i_1}, \ldots, \alpha_{i_p}$ and Z has type variables with $\alpha_{j_1}, \ldots, \alpha_{j_q}$ then $(-)[\tau_{i_1}/\alpha_{i_1}, \ldots, \tau_{i_p}/\alpha_{i_p}, \tau_{j_1}/\alpha_{j_1}, \ldots, \tau_{j_q}/\alpha_{j_q}]$ is a substitution for (W, Z) if $(-)[\tau_{i_1}/\alpha_{i_1}, \ldots, \tau_{i_p}/\alpha_{i_p}]$ is a substitution for W and $(-)[\tau_{j_1}/\alpha_{j_1}, \ldots, \tau_{j_q}/\alpha_{j_q}]$ is a substitution for Z and $dom(\tau_{i_l}(\rho)) \cap dom(\tau_{j_m}(\rho)) = \emptyset$ for all $\rho \in Env(Y)$ and $1 \leq l \leq p$ and $1 \leq m \leq q$. A map from Y to X is just such a substitution. This gives a category Bun of bunches of type variables.

If ρ is an environment for X and $A \in PER_0 \times PER_0$ then define a function $\rho^A : FTV(X) \cup \{\alpha\} \longrightarrow PER_0 \times PER_0$ by $\alpha \mapsto A$ and $\beta \mapsto \rho(\beta)$ for $\beta \neq \alpha$. Now ρ^A is an environment for $X; \alpha$. If A satisfies $A \cap dom(\rho) = \emptyset$ then ρ^A is also an environment for X, α.

Define semantic types τ_{α_i} over X, α by $\tau_{\alpha_i}(\rho) = \rho(\alpha_i)$, for each $1 \le i \le n$. Now $(-)[\tau_{\alpha_1}/\alpha_1, \ldots, \tau_{\alpha_n}/\alpha_n]$ defines a map from $X; \alpha$ to X, called π, and also a map from X, α to X, called ψ. Each of these induces a functor, with $P(\pi)(\tau)(\rho) = \tau(\rho\restriction_{FTV(X)})$ for $\rho \in Env(X; \alpha)$ and $P(\psi)(\tau)(\rho) = \tau(\rho\restriction_{FTV(X)})$ for $\rho \in Env(X, \alpha)$, respectively.

If τ' is a semantic type over $X; \alpha$ or, respectively, X, α then a semantic type over X is given, respectively, by

$$(\varPi\tau')\rho = \bigcap_{A \in PER_0 \times PER_0} \tau'(\rho^A) \qquad (\varPsi\tau')\rho = \bigcap_{\substack{A \in PER_0 \times PER_0 \\ A \cap dom(\rho) = \emptyset}} \tau'(\rho^A) \ ,$$

for each environment ρ for X. These assignments, which illustrate the distinction between additive and multiplicative quantification, extend to functors which are right-adjoints to π and ψ, respectively.

Let τ be a semantic type over X and τ' be a semantic type over $X; \alpha$ or X, α respectively. In the first case, a map from τ to $\varPi(\tau')$ is precisely the same thing as a map from $P(\pi)(\tau)$ to τ'. In the second case, a map from τ to $\varPsi(\tau')$ is precisely the same thing as a map from $P(\psi)(\tau)$ to τ'. We therefore have natural bijections between arrows

$$\frac{\tau \longrightarrow \varPi(\tau')}{P(\pi)(\tau) \longrightarrow \tau'} \qquad\qquad \frac{\tau \longrightarrow \varPsi(\tau')}{P(\psi)(\tau) \longrightarrow \tau'} \ ,$$

given by identity maps.

The above model is not quite a categorical model as described in the previous section. We produce an $\alpha 2\lambda 2$-hyperdoctrine by taking a suitable quotient on bunches to make the interpretation of all type variables identical.

6 Existential Quantifiers

Existential quantifiers may be defined in the polymorphic lambda calculus $\lambda 2$ and are closely connected to the concept of abstract data type [9]. In this section, we describe existential quantification in the bunched polymorphic setting, leading to both additive and multiplicative existentials.

First-order additive and multiplicative existential quantifiers have been studied in [11], [13]. Proof-theoretic considerations drive the design of the polymorphic existentials, just as they do in the first-order case.

Additive existential quantification, $\exists$, is quite straightforward to add to the system $\alpha 2\lambda 2$. However, the multiplicative quantifier, $\exists_*$, is very delicate. In particular, it requires a number of side-conditions which can interfere with the side-condition (†) used for $\alpha 2\lambda 2$. Rather than describing such a system in its full complexity, we first remove the universal quantifiers and instances of (†) before adding the existentials. However, in general, both universals and existentials can be considered together.

The grammars generating types and terms are extended with

$$\tau ::= \dots \mid \exists \alpha.\tau \mid \exists_* \alpha.\tau$$
$$M ::= \dots \mid \langle \phi, M \rangle \mid \textbf{unpack } M \textbf{ as } \langle \alpha, x \rangle \textbf{ in } M$$
$$\mid \langle Y, \phi, M \rangle_* \mid \textbf{unpack}_* \ M \textbf{ as } \langle \alpha, x \rangle \textbf{ in } M \ ,$$

where α and x are bound in **unpack** and **unpack$_*$** terms.

Just as with the multiplicative universal quantifier, we are forced to use an additional syntactic measure with the multiplicative existential. The set $WR(M)$ of *witnessing resources* of a term M is the set of type-variables which occur in in the left component Y of any sub-term $\langle Y, \phi, N \rangle_*$. This can be made precise with a recursive definition.

The rules for existentials, which follow the generalized forms for natural deduction introduced by Prawitz [12], are presented in Figure 3. Both of $(\exists E)$ and $(\exists_* E)$ are subject to the side-condition $\alpha \notin FTV(\Delta) \cup FTV(\sigma)$, which is standard for the elimination of existentials. In addition, both are subject to the side-condition $WR(M) \cap WR(N) = \emptyset$, because of the presence of the multiplicative. Furthermore, the condition $\alpha \notin WR(N)$ is required for $(\exists_* E)$.

$$(T\exists) \quad \frac{X; \alpha \vdash \tau}{X \vdash \exists \alpha.\tau} \qquad \frac{X \mid \Gamma \vdash M : \exists \alpha.\tau \quad X; \alpha \mid \Delta(x : \tau) \vdash N : \sigma}{X \mid \Delta(\Gamma) \vdash \textbf{unpack } M \textbf{ as } \langle \alpha, x \rangle \textbf{ in } N : \sigma} \quad (\exists E)$$

$$(\exists I) \quad \frac{X \mid \Gamma \vdash (M : \tau)[\phi/\alpha] \quad X \vdash \exists \alpha.\tau}{X \mid \Gamma \vdash \langle \phi, M \rangle : \exists \alpha.\tau}$$

$$(T\exists_*) \quad \frac{X, \alpha \vdash \tau}{X \vdash \exists_* \alpha.\tau} \qquad \frac{X \mid \Gamma \vdash M : \exists_* \alpha.\tau \quad X, \alpha \mid \Delta(x : \tau) \vdash N : \sigma}{X \mid \Delta(\Gamma) \vdash \textbf{unpack}_* \ M \textbf{ as } \langle \alpha, x \rangle \textbf{ in } N : \sigma} \quad (\exists_* E)$$

$$(\exists_* I) \quad \frac{X, Y(Z) \mid \Gamma \vdash (M : \tau)[\phi/\alpha] \quad Y(Z) \vdash \phi \quad X, Z \vdash \Gamma \quad X \vdash \exists_* \alpha.\tau}{X, Z \mid \Gamma \vdash \langle Y(Z), \phi, M \rangle_* : \exists_* \alpha.\tau}$$

Fig. 3. Existential rules

The additive quantifier behaves essentially as the standard polymorphic existential. The multiplicative is more unusual. This partially hides the resources (type variables) used in its formation. The work on first-order **BI** suggests a form in which Y is completely hidden. This rule is derivable from the one given. The more general version is adopted in order to give a corresponding η-rule.

The $\beta\eta$-conversions for existentials are

$$(X \mid \textbf{unpack } \langle \phi, M \rangle \textbf{ as } \langle \alpha, x \rangle \textbf{ in } N) \rightarrow_\beta (X \mid N[M/x][\phi/\alpha])$$
$$(X \mid \textbf{unpack } M \textbf{ as } \langle \alpha, x \rangle \textbf{ in } (N[\langle \alpha, x \rangle/z])) \rightarrow_\eta (X \mid N[M/z])$$
$$(X \mid \textbf{unpack}_* \ M \textbf{ as } \langle \alpha, x \rangle \textbf{ in } (N[\langle \alpha, \alpha, x \rangle_*/z])) \rightarrow_\eta (X \mid N[M/z])$$
$$(X, Z \mid \textbf{unpack}_* \ \langle Y(Z), \phi, M \rangle_* \textbf{ as } \langle \alpha, x \rangle \textbf{ in } N) \rightarrow_\beta (X, Y(Z) \mid N[M/x][\phi/\alpha])$$

and suitable ζ-conversions for existentials are also possible, provided no universal quantifiers are present. Notice how the hub changes in the β-conversion for the multiplicative. Let $\twoheadrightarrow$ be the reduction relation generated by $\rightarrow_\beta$ and $\rightarrow_\eta$.

Most of the metatheory goes through as it did for the system with universals rather than existentials. In particular, strong normalization can again be proved by the translation method. However, there are a few important changes, notably to substitution and subject-reduction.

Proposition 5. *If* $X \mid \Gamma(x : \tau) \vdash N : \sigma$ *and* $X \mid \Delta \vdash M : \tau$ *are both derivable and* $WR(M) \cap WR(N) = \emptyset$ *then* $X \mid \Gamma(\Delta) \vdash N[M/x] : \sigma$ *is derivable.*

The condition on the substitution law forces us to place the side-condition $WR(M) \cap WR(N) = \emptyset$ on the binary elimination rules $(\wedge E)$, $(\rightarrow E)$, $(*E)$, $(-\!*E)$ and is the reason why we need the same condition for the existentials.

Proposition 6. *If* $X \mid \Gamma \vdash M : \tau$ *is derivable and* $(X \mid M) \twoheadrightarrow (Y \mid N)$ *then* $Y \mid \Gamma \vdash N : \tau$ *is derivable.*

The existential does not have a simple $\alpha 2\lambda 2$-hyperdoctrine interpretation and, in particular, we cannot just use a left-adjoint to the pseudoprojection substitution. However, an interpretation can be given to each judgement by requiring the existence of certain assignments and arrows.

The introduction rule $(\exists_* I)$ for the multiplicative existential hides not only the representation type, but also the resources associated with the representation type. Once hidden, these resources are not visible to terms formed over the same hub (see the substitution rule) and are only revealed by a subsequent use of the elimination rule $(\exists_* E)$, leading to a hub-changing β-conversion, as above. In this respect the formation of multiplicative existentials is reminiscent of the formation of function closures. Furthermore, the elimination of $\exists_*$ is reminiscent of the application of function closures, though perhaps with some side-effects.

We conjecture that bunched polymorphism is an appropriate setting to develop type systems for memory-management. One approach to this is alias typing [18] which allows the programmer to issue instructions that safely allocate and deallocate chunks of memory, known as locations. Locations are used as parameters in types, for example $x : ptr(l)$, which asserts that a program variable is a pointer to the location l. A form of polymorphism is introduced through the use of location variables, which range over locations. Instructions are typed in contexts of *aliasing constraints*: these specifiy the types of entities contained in certain locations and location variables. It is difficult to formalize direct translations of such systems into the bunched setting because of the complexity of their type systems. However, it seems relatively clear that what the authors intend to enforce are non-sharing (anti-aliasing) constraints on chunks of memory. Consider, for example, the following statement, taken from [18]:

> The existential $\exists[\rho : Loc \mid \{\rho \mapsto \tau_1\}].\tau_2$ may be read "there exists some location ρ, different from all others in the program, such that ρ contains an object of type τ_1, and the value contained in this data structure has type τ_2.

What is intended to be different is surely not the location variable ρ itself, but rather the memory assigned to it by the environment. Under such a reading, it would seem more appropriate to use bunching rather than linearity as a foundation for the type system. Bunched alternatives to the linear approaches to type systems for *regions* [17], [19], should be equally interesting.

References

1. B.J. Day. On closed categories of functors. In *Lecture Notes in Mathematics* 137, pages 1–38. Springer-Verlag, Berlin-New York, 1970.
2. B.J. Day. An embedding theorem for closed categories. In *Lecture Notes in Mathematics* 420, pages 55–65. Springer-Verlag, Berlin, 1973.
3. J.-Y. Girard. Une extension de l'interprétation de Gödel à l'analyse et son application à l'élimination des coupures dans l'analyse et la théorie des types. In *Proc. 2nd Scandinavian Logic Symposium*, pages 63–92. North-Holland, 1971.
4. J.-Y. Girard. *Interprétation fonctionelle et élimination des coupures dans l'arithmétique d'ordre supérieur*. PhD thesis, Université Paris VII, 1972.
5. J.-Y. Girard. Linear logic. *Theoretical Computer Science* 50, pages 1–102, 1987.
6. M.J. Gabbay and A.M.Pitts A new approach to abstract syntax and variable binding. *Formal Aspects of Computing* 13, pages 341–363, 2002.
7. J.M.E. Hyland. A small complete category. *Annals of Pure and Applied Logic*, 40:135–165, 1988.
8. B. Jacobs. *Categorical Logic and Type Theory*, volume 141 of *Studies in Logic and the Foundations of Mathematics*. Elsevier, 1999.
9. J.C. Mitchell and G.D. Plotkin. Abstract types have existential type. *ACM Transactions on Programming Languages and Systems*, 10:470–502, 1988.
10. P. O'Hearn. On bunched typing. *J. Functional Programming*, 13:747–796, 2003.
11. P. O'Hearn and D. Pym. The logic of bunched implications. *Bulletin of Symbolic Logic*, 5(2):215–244, 1999.
12. D. Prawitz. Proofs and the meaning and completeness of logical constants. In *Essays on mathematical and philosophical logic*, pages 25–40. D. Reidel, 1978.
13. D.J. Pym. *The Semantics and Proof Theory of the Logic of Bunched Implications*, volume 26 of *Applied Logic Series*. Kluwer Academic Publishers, 2002. Errata at: `http://www.cs.bath.ac.uk/~pym/BI-monograph-errata.pdf`.
14. J.C. Reynolds. Towards a theory of type structure. In *Lecture Notes in Computer Science* 19, pages 408–425. Springer, 1974.
15. J.C. Reynolds. Separation logic: a logic for shared mutable data structure. In *Proc. LICS '02*, pages 55–74. IEEE Computer Science Press, 2002.
16. R.A.G. Seely. Categorical semantics for higher order polymorphic lambda calculus. *Journal of Symbolic Logic*, 52:969–989, 1987.
17. M. Tofte and J.-P. Talpin Region-based memory management. *Information and Computation*, 132(2):109–176, 1997.
18. D. Walker and J.G. Morrisett. Alias types for recursive data structures. In *Lecture Notes in Computer Science* 2071, pages 177–206. Springer-Verlag, 2001.
19. D. Walker and K. Watkins On regions and linear types In *Proc. International Conference on Functional Programming*, 181-192. 2001.

Distributed Control Flow
with Classical Modal Logic*

Tom Murphy VII, Karl Crary, and Robert Harper

Carnegie Mellon University
{tom7,crary,rwh}@cs.cmu.edu

Abstract. In previous work we presented a foundational calculus for spatially distributed computing based on intuitionistic modal logic. With the modalities $\Box$ and $\Diamond$ we were able to capture two key invariants: the mobility of portable code and the locality of fixed resources. This work investigates issues in distributed control flow through a similar propositions-as-types interpretation of *classical* modal logic. The resulting programming language is enhanced with the notion of a network-wide continuation, through which we can give computational interpretation of classical theorems (such as $\Box A \equiv \neg\Diamond\neg A$). Such continuations are also useful primitives for building higher-level constructs of distributed computing. The resulting system is elegant, logically faithful, and computationally reasonable.

1 Introduction

This paper is an exploration of distributed control flow using a propositions-as-types interpretation of classical modal logic. We build on our previous intuitionistic calculus, Lambda 5 [8], which is a simple programming language (and associated logic) for distributed computing. Lambda 5 focuses particularly on the spatial distribution of programs, and allows the programmer to express the *place* in which computation occurs using modal typing judgments. Through the modal operators $\Box$ and $\Diamond$ we are then able to express invariants about mobility and locality of resources. Our new calculus, C5, extends Lambda 5 with network-wide continuations, which arise naturally from the underlying classical logic. These continuations create a new relationship between the modalities $\Box$ and $\Diamond$, which we see with several examples, and serve as building blocks for other useful primitives. Before we introduce C5, we begin with a short reprise of Lambda 5.

Lambda 5. The Lambda 5 programming model is a network with many different *places*, or *nodes*. In order to be faithful to this model, we use a style of logic that has the ability to reason simultaneously from multiple perspectives, namely, modal logic. Compared to propositional logic, which is concerned with

---

* The ConCert Project is supported by the National Science Foundation under grant ITR/SY+SI 0121633: "Language Technology for Trustless Software Dissemination".

L. Ong (Ed.): CSL 2005, LNCS 3634, pp. 51–69, 2005.

truth, modal logic deals with truth from the perspective of different *worlds*. These worlds are related by an *accessibility relation*, which affects the strength of the modal connectives; different assumptions about accessibility give rise to different modal logics. For modeling a network where the worlds are nodes, we choose Intuitionistic S5 [14], whose relation is reflexive, symmetric, and transitive – every world is related to every other world. Therefore, except when comparing it to other systems, we essentially dispense with the accessibility relation altogether. This leads to a simpler explanation of the judgments and connectives.

A true @ω is the basic judgment, meaning that the proposition A is true at the world ω (we abbreviate this to A@ω). There are two new proposition forms for quantifying over worlds. $\Box A$ is the statement that A is true at every world. $\Diamond A$ means that A is true at *some* world. Because we think of these worlds as places in the network, operationally we interpret type $\Box A$ as representing mobile code or data of type A, and the type $\Diamond A$ as an address of a value of type A.

Propositions must be situated at a world in order to be judged true, so it is important to distinguish between the proposition $\Box A$ and the judgment $\Box A$@ω, the latter meaning that A is true in every world *from the perspective of ω*. In S5, every world has the same perspective with regard to statements about *all* or *some* world(s). But operationally this will be significant, as there is no true "global" code, only mobile code that currently exists at some world.

Though the logic distinguishes between $\Box A$@ω and $\Box A$@ω', both have precisely the same immediate consequences. The typical rule for eliminating $\Box$, for instance as given by Simpson [14] is

$$\frac{\Box A@\omega}{A@\omega'} \ \Box\text{E \scriptsize (Simpson)}$$

With this rule, it *never really matters* where $\Box A$ exists, since we can eliminate it instantly to any world. However, we do care operationally where mobile code resides, and so we adjust the natural deduction rules to reflect this bias. The logic features a novel *decomposition* into locally-acting introduction and elimination rules as well as *motion* rules for moving between worlds, i.e.

$$\frac{\Box A@\omega}{A@\omega} \ \Box\text{Elim} \qquad\qquad \frac{\Box A@\omega}{\Box A@\omega'} \ \Box\text{Move}$$

We argue [8] that this results in a more appropriate operational interpretation. Our classical system also features this decomposition, and like Lambda 5, we are able to retain a crisp connection to the underlying logic.

Although distributed computing problems are often thought of as being concurrent, both Lambda 5 and our new calculus are sequential. We consider concurrency an orthogonal issue, although we give remarks on how it can be accomplished in Section 5.

Classical Control Flow. The notion that control operators such as Scheme's call/cc or Felleisen's $\mathcal{C}$ can be given logical meaning via classical logic is well known. Essentially, if we interpret the type $\neg A$ as a continuation expecting a value of type A, then the types of these operators are classical tautologies. Griffin first proposed this in 1990 [4] with later refinements by (for example) Murthy [9]. Parigot's $\lambda\mu$-calculus [10] takes this idea and develops it into a full-fledged natural deduction system for classical logic[1]. It soon became clear that this was no accident – classical logic *is* the logic of control flow.

Therefore, a natural next step is to look at *classical* S5 to see what kind of programming language it gives us, which is the topic of this paper. We find that the notion of a network-wide continuation arises naturally, giving a computational explanation to (intuitionistically ridiculous) classical theorems such as $\Box A \equiv \neg \Diamond \neg A$. We also believe that such primitives are useful for building distributed computing mechanisms such as asynchronous message passing.

The paper proceeds as follows. We first present classical S5 judgmentally, giving a natural deduction system and intuition for its operational behavior. Next we give proof terms for some classical theorems, to elucidate the new connection between $\Box$ and $\Diamond$ made possible by network-wide continuations. In order to make these intuitions concrete, we then give an operational semantics based on an abstract network. We follow with some ideas about concurrency and how network-wide continuations can be used by distributed applications, and conclude with a discussion of related work. The appendix contains a proof that C5 really is classical S5 (along with establishing the existence of normal forms), by relating it to a sequent calculus that admits cut.

All of the proofs in this paper have been formalized in the Twelf system [11] and mechanically verified by its metatheorem checker [13][2]. Extended discussion of some of the proofs can be found in the accompanying technical report [7].

2   Classical S5

We wish to take a propositions-as-types interpretation of modal logic, so a judgmental proof theory for our logic is critical. In this section we give such a presentation of Classical S5.

Because modal logic is concerned with truth relativized to worlds, our judgments must reflect that. We have two main judgments in our proof theory.

$$A \text{ true } @ \omega \qquad\qquad A \text{ false } \star \omega$$

The first simply states that the proposition A is true at the world ω, as we had in Lambda 5. The second, which is new, says that the proposition A is *false* at the world ω. Although these two judgments are dual, the natural deduction system is deliberately biased towards deducing that propositions are true. We will only make assumptions about falsehood for the purpose of deriving a contradiction. As is standard, we reify the hypotheses about truth and falsehood into contexts (eliding true and false), and the central judgment of our proof theory becomes

$$\Gamma; \Delta \vdash A @ \omega$$

---

[1] Our calculus is quite similar to his (extended to the modal case!), although we prefer to present it with an emphasis on *truth* and *falsehood* judgments.

[2] They can be found at http://www.cs.cmu.edu/~concert/.

$$\frac{\Gamma,\omega';\Delta \vdash M : A@\omega'}{\Gamma;\Delta \vdash \mathtt{box}\ \omega'.M : \Box A@\omega}\ \Box I \qquad \frac{\Gamma;\Delta \vdash M : \Box A@\omega}{\Gamma;\Delta \vdash \mathtt{unbox}\ M : A@\omega}\ \Box E$$

$$\frac{}{\Gamma,x{:}A@\omega,\Gamma';\Delta \vdash x : A@\omega}\ \mathrm{hyp} \qquad \frac{\Gamma;\Delta \vdash M : \Box A@\omega' \quad \Gamma \vdash \omega'}{\Gamma;\Delta \vdash \mathtt{get}_\Box[\omega']M : \Box A@\omega}\ \Box M$$

$$\frac{\Gamma;\Delta \vdash M : A@\omega}{\Gamma;\Delta \vdash \mathtt{here}\ M : \Diamond A@\omega}\ \Diamond I \qquad \frac{\Gamma;\Delta \vdash M : \Diamond A@\omega' \quad \Gamma \vdash \omega'}{\Gamma;\Delta \vdash \mathtt{get}_\Diamond[\omega']M : \Diamond A@\omega}\ \Diamond M$$

$$\frac{\begin{array}{c}\Gamma;\Delta \vdash M : \Diamond A@\omega \\ \Gamma,\omega',x{:}A@\omega';\Delta \vdash N : B@\omega\end{array}}{\Gamma;\Delta \vdash \mathtt{letd}\,\omega'.x = M \mathrm{\ in\ } N : B@\omega}\ \Diamond E \qquad \frac{\begin{array}{c}\Gamma;\Delta \vdash N : A@\omega \\ \Gamma;\Delta \vdash M : A \supset B@\omega\end{array}}{\Gamma;\Delta \vdash MN : B@\omega}\ \supset E$$

$$\frac{\Gamma,x{:}A@\omega;\Delta \vdash M : B@\omega}{\Gamma;\Delta \vdash \lambda x.M : A \supset B@\omega}\ \supset I \qquad \frac{\Gamma;\Delta,u{:}A{\star}\omega \vdash M : A@\omega}{\Gamma;\Delta \vdash \mathtt{letcc}\,u \mathrm{\ in\ } M : A@\omega}\ \mathrm{bc}$$

$$\frac{\Gamma,\Delta,u{:}A{\star}\omega,\Delta' \vdash M : A@\omega}{\Gamma,\Delta,u{:}A{\star}\omega,\Delta' \vdash \mathtt{throw}\ M \mathrm{\ to}\ u : C@\omega'}\ \# \qquad \frac{\Gamma;\Delta \vdash M : \bot@\omega' \quad \Gamma \vdash \omega'}{\Gamma;\Delta \vdash \mathtt{go}[\omega']M : C@\omega}\ \bot E$$

$$\frac{\Gamma;\Delta \vdash M : A@\omega \quad \Gamma;\Delta \vdash N : B@\omega}{\Gamma;\Delta \vdash \langle M,N\rangle : A \land B@\omega}\ \land I \qquad \frac{\Gamma;\Delta \vdash M : A_1 \land A_2@\omega}{\Gamma;\Delta \vdash \pi_i M : A_i@\omega}\ \land E_i$$

Fig. 1. Classical S5 natural deduction ("C5")

where we deduce that A is true at world ω under truth assumptions of the form $B@\omega'$ appearing in Γ and falsehood assumptions of the form $C \star \omega''$ appearing in Δ. We also have hypotheses about the existence of worlds. It is cumbersome to write a separate context of world hypotheses, so these assumptions (written merely as ω) appear in Γ as well. We also take the common shortcut of only permitting mention of worlds that exist. Therefore, all judgments are hypothetical in at least *some* world (the world at which the conclusion is formed), until we introduce world constants in Section 4.

Operationally, we will think of a falsehood assumption $A{\star}\omega$ as a continuation, living at world ω, that expects something of type A.

Our natural deduction system appears in Fig. 1. These rules include proof terms, which we will explain shortly. Aside from the falsehood context, the rules for $\Box$, $\Diamond$ and $\supset$ are the same as in Lambda 5. The new connectives $\bot$ (discussed below) and $\land$ are treated as they would be in the intuitionistic case. The major additions are the structural rules *bc* (by contradiction) and # (contradict), which enable classical reasoning.

The *bc* rule is read as follows: In order to prove $A@\omega$, we can assume that A is false at ω. This corresponds directly to the classical axiom $(\neg A \supset A) \supset A$. Operationally, this names the current continuation – we use a distinct class of "falsehood" or "continuation" variables u for this. The # rule may be alarming at first glance, because it requires the assumption $:A \star \omega$ to appear in the conclusion. This is because the # rule is actually the *hypothesis* rule for falsehood assumptions, and will have a corresponding substitution principle[3]. The

---

[3] A theory of *hypothetical hypotheticals* would be able to express this in a less awkward – but perhaps no less alarming – way. Abel [1] for instance gives such a third-order encoding of the $\lambda\mu$-calculus.

rule simply states that if we have the assumption that A is false and are able to prove that A is true (at the same world), then we can deduce a contradiction and thus any proposition. The # rule is realized operationally as a `throw` of an expression (not a value, even though this is a call-by-value language) to a matching continuation. Note that continuations are *global* – we can throw from any world to a remote continuation $A\star\omega$, provided that we are able to construct a proof of $A@\omega$.

The rules for $\square$ and $\diamond$ are key to the system. $\square$ elimination is the easiest to understand: If we know that $\square A$ is true at some world, then we know A is true at the same world. To prove $\square A$, we must prove A at a hypothetical world about which nothing is known (rule $\square I$). Operationally, we realize $\square A$ as a piece of suspended code, with the hypothetical world ω' bound within it. Introduction of $\diamond$ is simple; if we know A then we know that A is true *somewhere* (namely here). Operationally this will record the value in a table and return an address that witnesses its existence. Elimination of $\diamond$ is as follows: if we know $\diamond A$, then we know there is some world where A is true (but we don't know anything else about it). Call this world ω' and assume $A@\omega'$ in order to continue reasoning. Finally, we provide *motion* rules (as per our decomposition) $\square M$ and $\diamond M$. Both simply allow knowledge of $\square A$ or $\diamond A$ at one world to be transported to another. Operationally these move the values between worlds.

Bottom has no introduction form, but we allow the *remote elimination* of it (rule $\perp E$). This is similar to the motion rules for $\square$ and $\diamond$, but is called **go** to indicate a transfer of control with no return[4].

Despite the fact that our proof theory is specially constructed to give rise to a good operational semantics, it really embodies classical S5. To see this, we observe that it is equivalent to a symmetric multiple-conclusion sequent calculus that is more straightforwardly classical S5. The sequent calculus has the subformula property and admits (a dual form of) cut, which also establishes the existence of normal forms for our proof terms. The argument is mostly similar to the one used for our previous calculus, and is not the focus of this paper. Interested readers can find this material in the Appendix; otherwise, we'll begin to motivate the operational semantics of our calculus with some examples.

3   Examples

In this section we give proof terms showing the new connection between $\square$ and $\diamond$ made possible by network-wide continuations. A full operational semantics is forthcoming in Section 4.1, but let us review our informal interpretation of the modal connectives now.

A value of type $\square A$ is a suspended expression that makes sense anywhere. We call such values *boxes*, and we can open them at any world using the **unbox** primitive, which begins evaluating the expression. A value of type $\diamond A$ is an address of a value that has been published in a table at some world. In order to

---

[4] We could have equivalently had a **get**$_\perp$ and a local **abort**, but there appears to be no practical use to this decomposition.

make addresses, we use the **here** construct to publish a value in the local table and generate a new address for it. We have the ability to travel and move certain data between worlds by using the **get** and **go** constructs.

Finally, because our examples involve negation ($\neg A$), we first briefly explain how we treat it.

Negation. Although we have not given the rules for the negation connective, it is easily added to the system. Here we take the standard shortcut of treating $\neg A$ as an abbreviation for $A \supset \perp$. We computationally read $\neg A @ \omega$ as a continuation expecting A, although this should be distinguished from a primitive continuation assumption $u{:}A \star \omega$: the former is introduced by lambda abstraction and eliminated by application, while the latter is formed with **letcc** and eliminated by a **throw** to it. The two are related in that we can reify a falsehood assumption $u{:}A \star \omega$ as a negated formula $\neg A @ \omega$ by forming a function that throws to it: $\lambda a.\,\texttt{throw}\,a\,\texttt{to}\,u$. Likewise, we can create a falsehood assumption from a term $M : \neg A @ \omega$, namely $M(\texttt{letcc}\,u\,\texttt{in}\ldots)$.

Classical Axioms. As examples, we give proof terms for several classical axioms. To implement one of these axioms, the programmer engages in a little theorem proving puzzle. Because we are dealing with classical logic, we have two sorts of resources in solving the puzzle: *values* of type A, as in intuitionistic logic, but also *contexts expecting* terms of type A. We can capture such contexts with **letcc**, so sometimes we go out of our way to create them; thus the the *need for* a value of some type can be as useful as the *presence of* one.

Our first example comes from the standard practice in classical modal logic of defining $\Box$ in terms of $\Diamond$ through the equivalence $\Box A \equiv \neg \Diamond \neg A$. From left to right the implication is intuitionistically valid, so we'll look at the proof of the implication right to left. In C5, the proof term tells an interesting story:

$$\lambda d.\,\texttt{box}\,\omega'. \qquad (d : (\Diamond \neg A) \supset \perp @ \omega;\,\text{need to return }A @ \omega')$$
$$\quad \texttt{letcc}\,u\,\texttt{in}\,\texttt{go}[\omega] \qquad (\text{applying } d \text{ will yield } \perp)$$
$$\quad d(\texttt{get}_\Diamond[\omega'](\texttt{here}(\lambda a.\,\texttt{throw}\,a\,\texttt{to}\,u)))$$

In each example, we'll assume that the whole term lives at the world ω. Operationally, the reading of $\neg \Diamond \neg A \supset \Box A$ is that given a continuation d (expecting the address of an A continuation), we will return a boxed A that is well-formed anywhere. It is easiest to understand this term from the perspective of the consumer of the resulting $\Box A$. When it is unboxed at some world ω', it grabs the current continuation u, which expects an A. It then publishes this continuation (reified as a function); the address is what we require as an argument for d. (What happens next depends on what d does with its argument!) The intervening **go** and $\texttt{get}_\Diamond$ accomplish the transfer of control between the two worlds.

Dually we can define $\Diamond$ in terms of $\Box$. Again, one direction is intuitionistically valid. The other, $\neg \Box \neg A \supset \Diamond A$, is asked to conjure up an address of an arbitrary A given a continuation (that expects a boxed A continuation). It is implemented by the following proof term:

$$\lambda b.\, \texttt{letcc}\, u \,\texttt{in} \qquad\qquad (b : (\Box\neg A) \supset \bot @ \omega; u : \Diamond A \star \omega)$$
$$\texttt{go}[\omega]\, b(\texttt{box}\; \omega'.\lambda a. \qquad (a : A @ \omega')$$
$$\texttt{throw}(\texttt{get}_\Diamond[\omega'](\texttt{here}\, a))\,\texttt{to}\, u)$$

Here, we immediately grab the $\Diamond A$ continuation with $\texttt{letcc}$. Since we will be calling b (proving $\bot$ and never returning), we "go" to the current world. We then form a $\texttt{box}$ to pass to the function b. It contains a function of type $A \supset \bot$, which takes the address of its argument and throws it to the saved continuation u. Thus the location of A that we ultimately return is any world that calls the $\neg A$ function that we've boxed up.

Excluded "Modal." The following example uses disjunction, which we've left out of our calculus so far. A description of some ways it can be added is given in Section 6, but for now we will be somewhat less formal and simply assume that we have constructors $\texttt{inl}$ and $\texttt{inr}$ for forming proofs of $A \lor B$.

Our example is a modal version of the excluded middle axiom: $\Box A \lor \Diamond\neg A$. We will again return a box that does something when opened.

$$\texttt{letcc}\, u_o \,\texttt{in} \qquad\qquad (u_o : \Box A \lor \Diamond\neg A \star \omega)$$
$$\texttt{inl}(\texttt{box}\; \omega'.\, \texttt{letcc}\, u \,\texttt{in} \qquad (u : A \star \omega')$$
$$\texttt{throw}(\texttt{inr}(\texttt{get}_\Diamond[\omega']\, \texttt{here}(\lambda a.\, \texttt{throw}\, a \,\texttt{to}\, u)))$$
$$\texttt{to}\, u_o)$$

First, we save the current continuation as u_o, since we will need to "change our minds" and return multiple different disjuncts. When asked for $\Box A \lor \Diamond\neg A$, the program initially says $\Box A$.If the box is opened, the program uses context expecting an A to produce a $\Diamond\neg A$, *time travels* back to when it was asked about the disjunction, and returns this different answer.If that $\neg A$ continuation is ever invoked, the program goes back and uses the A to fulfill the outstanding request for an A at the world where the box was opened.

In the style of sci-fi storytelling popular when describing such things, we conclude our examples with the following fable (with apologies to Wadler [15]):

A magician who purports to be from the future is making bold claims. Asking for a volunteer, he offers the following prize to anyone who comes on stage:

"I'm going to hand you a box that has *you* inside it! Either that, or I'll give you the address of a place with a magical time travelling portal."

Being questionably brave, you volunteer and walk onto the stage. The magician hands you your prize – a large cardboard box. Noting your skepticism, he adds, "You can open it anywhere, and you'll be inside."

You decide to take the box home. It's much too light to have anything in it, let alone yourself! You open the box and look inside, wondering what sort of gag he has planned. But suddenly you find that the box has vanished, and you're standing on stage waiting for him to tell you what you've won, again.

"The address of the time-travelling portal is," he begins, rattling off your home address. You are startled that he could have known your address, but when you later arrive home, you see an open cardboard box waiting. Is this

world vars	ω	world names	$\mathbf{w}$	labels	ℓ
value vars	x, y	cont labs	$\mathbf{k}$	cont vars	u

types $A, B ::= A \supset B \mid \Box A \mid \Diamond A \mid A \wedge B \mid \bot$

networks $\mathbb{N} ::= \mathbb{W}; R$ world exps $\mathbf{w} ::= \omega \mid \mathbf{w}$

configs $\mathbb{W} ::= \{\mathbf{w}_1 : \langle \chi_1, b_1 \rangle, \cdots \}$

cursors $R ::= \mathbf{w} : [k \prec v] \mid \mathbf{w} : [k \succ M]$

tables $b ::= \bullet \mid b, \ell = v$ cont tables $\chi ::= \bullet \mid \chi, \mathbf{k} = k$

config types $\Sigma ::= \{\mathbf{w}_1 : \langle X_1, \beta_1 \rangle, \cdots \}$

table types $\beta ::= \bullet \mid \beta, \ell : A$ ctable types $X ::= \bullet \mid X, \mathbf{k} : A$

cont exps $Z ::= \mathbf{w}.\mathbf{k} \mid u$

conts $k ::= \mathtt{return}\, Z \mid \mathtt{finish} \mid \mathtt{abort} \mid k \lhd f$

values $v ::= \lambda x.M \mid \mathtt{box}\, \omega.M \mid \mathbf{w}.\ell \mid \langle v, v' \rangle$

frames $f ::= \circ\, N \mid v \circ \mid \mathtt{here}\,\circ \mid \mathtt{unbox}\,\circ$
             $\mid \mathtt{letd}\, \omega.x = \circ\, \mathtt{in}\, N \mid \pi_n \circ \mid \langle \circ, N \rangle \mid \langle v, \circ \rangle$

exps $M, N ::= v \mid MN \mid x \mid \ell \mid \mathtt{get}_\Box[\mathbf{w}]M \mid \mathtt{here}\, M \mid \mathtt{get}_\Diamond[\mathbf{w}]M$
                 $\mid \mathtt{unbox}\, M \mid \mathtt{letd}\, \omega.x = M\, \mathtt{in}\, N \mid \mathtt{throw}\, M\, \mathtt{to}\, Z$
                 $\mid \mathtt{go}[\mathbf{w}]M \mid \mathtt{letcc}\, u\, \mathtt{in}\, M \mid \langle M, N \rangle \mid \pi_n M$

Fig. 2. Syntax of type system

supposed to be the portal? Knowing it to be harmless, but insisting on proving the magician to be a fraud, you step into it.

A hot flash of embarrassment passes over you as you realize that you are now standing in a cardboard box, in your house, as promised.

4 Type System and Operational Semantics

Our deductive proof theory corresponds to a natural programming language whose syntax is the proof terms from Fig. 1. In order to give this language an operational interpretation, we need to introduce a number of syntactic constructs, which are given in Fig. 2.

As in Lambda 5, the behavior of a program is specified in terms of an abstract network that steps from state to state. The network is built out of a fixed number of worlds, whose names we write as bold $\mathbf{w}$. Because we can now mention specific worlds in addition to hypothetical worlds ω, we introduce world expressions, which are written with a Roman w. A network state $\mathbb{N}$ has two parts. First is a world configuration $\mathbb{W}$ which identifies two tables with each world $\mathbf{w}_i$ present. The first table χ_i stores network-wide continuations by mapping continuation labels $\mathbf{k}$ to literal continuations k. The second table b_i maps value labels ℓ to values in order to store values whose address we have published. These tables have types X and β respectively (which map labels $\mathbf{k}$ and ℓ to types), and so we can likewise construct the type of an entire configuration, written Σ.

Aside from the current world configuration, a network state also contains a *cursor* denoting the current focus of computation. The cursor either takes the form $\mathbf{w} : [k \prec v]$ (returning the value v to the continuation k) or $\mathbf{w} : [k \succ M]$ (evaluating the expression M in continuation k). In either case it selects a world $\mathbf{w}$ where the computation is taking place.

$\Sigma; \Gamma; \Delta \vdash M : A@\text{w}$ The expression M has type A at world w
$\Sigma \vdash k : A\star\text{w}$ The continuation k expects a value of type A at world w
$\Sigma; \Delta \vdash Z : A\star\text{w}$ The continuation expression Z is well-formed with type A at w
$\Sigma \vdash b@\mathbf{w}$ The value table b is well-formed at the world named $\mathbf{w}$
$\Sigma \vdash \chi\star\mathbf{w}$ The continuation table χ is well-formed at the world named $\mathbf{w}$
$\Sigma \vdash R$ The cursor is well-formed
$\Sigma \vdash \mathbb{N}$ The network is well-formed

Fig. 3. Index of judgments. In each judgment Σ is a configuration typing, Γ is a context of truth hypotheses, and Δ is a context of falsehood hypotheses

Continuations themselves are stacks of frames (expressions with a "hole," written ○) with a bottommost **return**, **finish** or **abort**. The **finish** continuation represents the end of computation, so a network state whose cursor is returning a value to **finish** is called *terminal*. The **abort** continuation will be unreachable, and **return** will send the received value to a remote continuation.

Most of the expressions and values are straightforward. As in Lambda 5, the canonical value for □ abstracts over the hypothetical world and leaves its body unevaluated (**box** $\omega'.M$). The canonical form for ◇ is a pair of a world name and a label $\mathbf{w}.\ell$, which addresses a table entry at that world. Such an address is well-formed anywhere (assuming that $\mathbf{w}$'s table has a label ℓ containing a value of type A) and has type $\Diamond A@\text{w}'$. On the other hand we have another sort of label, written just ℓ, which is disembodied from its world. These labels arise from the **letd** construct, which deconstructs an address $\mathbf{w}.\ell$ into its components $\mathbf{w}$ and ℓ (see the $\Diamond E$ rule from Fig. 1). Disembodied labels only make sense at a single world – here ℓ would have type $A@\mathbf{w}$.

Although the external language only allows a **throw** to a continuation variable, intermediate states of evaluation require that these be replaced with the continuation value $\mathbf{w}.\mathbf{k}$, which pairs a continuation label with the world at which it lives. These continuation values are filled in by **letcc**.

The type system is given in Fig. 4 (we omit for space the rules that are the same as in Fig. 1 except for the configuration typing Σ). The index of judgments in Fig. 3 may be a useful reference in understanding them.

The rules *addr* and *lab* are used to type run-time artifacts of address publishing. In either case, we look up the type in the appropriate table typing β. As mentioned, *throw* allows a continuation expression Z, which is either a variable (typed with $hyp^\star$, as in the logic) or an address into a continuation table.

Typing of literal continuations k is fairly unsurprising. Note that the judgment $\Sigma \vdash k : A\star\text{w}$ means that the continuation k *expects* a value of type A at w. The **return** continuation arises only from a $\mathbf{get}_\Diamond$ or $\mathbf{get}_\square$, and so it allows only values of type $\Diamond A$ or $\square A$. We use the network continuation mechanism to name the the outstanding $\mathbf{get}_\Diamond$ or $\mathbf{get}_\square$ request on the remote machine.

For an entire network to be well-formed (rule *net*), all of the tables must have the type indicated by the configuration type Σ, which means that they must have exactly the same labels, and the values or continuations must be well-typed at the specified types (rules b and χ). Finally, the cursor must be

$$\frac{\Sigma(\mathbf{w}) = \langle X, \beta \rangle \quad \beta(\ell) = A}{\Sigma; \Gamma; \Delta \vdash \mathbf{w}.\ell : \diamond A @ \mathbf{w}'} \; \text{addr} \qquad \frac{\Sigma(\mathbf{w}) = \langle X, \beta \rangle \quad \beta(\ell) = A}{\Sigma; \Gamma; \Delta \vdash \ell : A @ \mathbf{w}} \; \text{lab}$$

$$\frac{\Sigma; \Gamma; \Delta \vdash M : A @ \mathbf{w} \quad \Sigma; \Delta \vdash Z : A \star \mathbf{w}}{\Sigma; \Gamma; \Delta \vdash \mathbf{throw} \; M \; \mathbf{to} \; Z : C @ \mathbf{w}'} \; \text{throw} \qquad \frac{\Sigma; \Gamma; \Delta, u : A \star \mathbf{w} \vdash M : A @ \mathbf{w}}{\Sigma; \Gamma; \Delta \vdash \mathbf{letcc} \; u \; \mathbf{in} \; M : A @ \mathbf{w}} \; \text{letcc}$$

$$\frac{\Sigma(\mathbf{w}) = \langle X, \beta \rangle \quad X(\mathbf{k}) = A}{\Sigma; \Delta \vdash \mathbf{w}.\mathbf{k} : A \star \mathbf{w}} \; \text{addr}^{\star} \qquad \frac{}{\Sigma; \Delta, u : A \star \mathbf{w} \vdash u : A \star \mathbf{w}} \; \text{hyp}^{\star}$$

$$\frac{\Sigma \vdash k : B \star \mathbf{w} \quad \Sigma; \cdot; \cdot \vdash N : A @ \mathbf{w}}{\Sigma \vdash k \lhd \circ N : A \supset B \star \mathbf{w}} \; \text{kapp}_1 \qquad \frac{}{\Sigma \vdash \mathbf{finish} : A \star \mathbf{w}} \; \text{kfinish}$$

$$\frac{\Sigma \vdash k : B \star \mathbf{w} \quad \Sigma; \cdot; \cdot \vdash v : A \supset B @ \mathbf{w}}{\Sigma \vdash k \lhd v \circ : A \star \mathbf{w}} \; \text{kapp}_2 \qquad \frac{}{\Sigma \vdash \mathbf{abort} : \bot \star \mathbf{w}} \; \text{kabort}$$

$$\frac{\Sigma \vdash k : C \star \mathbf{w} \quad \Sigma; \omega, x : A @ \omega; \cdot \vdash N : C @ \mathbf{w}}{\Sigma \vdash k \lhd \mathbf{letd} \, \omega.x = \circ \mathbf{in} \; N : \diamond A \star \mathbf{w}} \; \text{kletd} \qquad \frac{\Sigma \vdash k : \diamond A \star \mathbf{w}}{\Sigma \vdash k \lhd \mathbf{here} \, \circ : A \star \mathbf{w}} \; \text{khere}$$

$$\frac{\Sigma \vdash k : A \wedge B \star \mathbf{w} \quad \Sigma; \cdot; \cdot \vdash N : B @ \mathbf{w}}{\Sigma \vdash k \lhd \langle \circ, N \rangle : A \star \mathbf{w}} \; \text{k}\wedge_1 \qquad \frac{\Sigma \vdash k : A \star \mathbf{w}}{\Sigma \vdash k \lhd \mathbf{unbox} \, \circ : \Box A \star \mathbf{w}} \; \text{kunbox}$$

$$\frac{\Sigma \vdash k : A \wedge B \star \mathbf{w} \quad \Sigma; \cdot; \cdot \vdash v : A @ \mathbf{w}}{\Sigma \vdash k \lhd \langle v, \circ \rangle : B \star \mathbf{w}} \; \text{k}\wedge_2 \qquad \frac{A = \Box A' \; \text{or} \; \diamond A' \quad \Sigma; \cdot \vdash Z : A \star \mathbf{w}'}{\Sigma \vdash \mathbf{return} \; Z : A \star \mathbf{w}} \; \text{kret}$$

$$\frac{\beta = (\ell_1 : A_1, \ldots) \quad \Sigma; \cdot; \cdot \vdash v_1 : A_1 @ \mathbf{w} \quad \ldots}{\underbrace{\{\cdots, \mathbf{w} : \langle X, \beta \rangle, \cdots\}}_{\Sigma} \vdash \underbrace{\ell_1 = v_1, \ldots}_{b} @ \mathbf{w}} \; b \qquad \frac{\mathbf{w} \in \mathrm{dom}(\Sigma)}{\Sigma; \cdot; \cdot \vdash v : A @ \mathbf{w} \quad \Sigma \vdash k : A \star \mathbf{w}}{\Sigma \vdash \mathbf{w} : [k \prec v]} \; \text{ret}$$

$$\frac{X = (\mathbf{k}_1 : A_1, \ldots) \quad \Sigma \vdash k_1 : A_1 \star \mathbf{w} \quad \ldots}{\underbrace{\{\cdots, \mathbf{w} : \langle X, \beta \rangle, \cdots\}}_{\Sigma} \vdash \underbrace{\mathbf{k}_1 = k_1, \ldots}_{\chi} \star \mathbf{w}} \; \chi \qquad \frac{\mathbf{w} \in \mathrm{dom}(\Sigma)}{\Sigma; \cdot; \cdot \vdash M : A @ \mathbf{w} \quad \Sigma \vdash k : A \star \mathbf{w}}{\Sigma \vdash \mathbf{w} : [k \succ M]} \; \text{eval}$$

$$\frac{\Sigma \vdash R \quad \Sigma \vdash \chi_i @ \mathbf{w}_i \; \ldots \quad \Sigma \vdash b_i @ \mathbf{w}_i \; \ldots}{\Sigma \vdash \{\mathbf{w}_1 : \langle \chi_1, b_1 \rangle, \cdots, \mathbf{w}_m : \langle \chi_m, b_m \rangle\}; R} \; \text{net}$$

Fig. 4. Type system

well-formed: it must select a world that exists in the network, and there must exist a type A such that its continuation and value or expression both have type A and are closed.

Having set up the syntax and type system, we can now give the operational semantics and type safety theorem. After the following section we remark on how the semantics can be made concurrent, and give some thoughts on applications of distributed continuations.

4.1   Operational Semantics

The operational semantics of our language is given in Fig. 5, as a binary relation $\mapsto$ between network states. The semantics evaluates programs sequentially, though we give a concurrent semantics in Section 5.

$\supset_e$-p $\quad \mathbb{W}; \mathbf{w} : [k \succ MN]$ $\qquad\qquad\qquad\qquad \mapsto \mathbb{W}; \mathbf{w} : [k \lhd \circ N \succ M]$

$\supset_e$-s $\quad \mathbb{W}; \mathbf{w} : [k \lhd \circ N \prec v]$ $\qquad\qquad\qquad\quad \mapsto \mathbb{W}; \mathbf{w} : [k \lhd v \circ \succ N]$

$\supset_e$-r $\quad \mathbb{W}; \mathbf{w} : [k \lhd (\lambda x.M)\circ \prec v]$ $\qquad\qquad \mapsto \mathbb{W}; \mathbf{w} : [k \succ [v/x]M]$

value $\quad \mathbb{W}; \mathbf{w} : [k \succ v]$ $\qquad\qquad\qquad\qquad\qquad\; \mapsto \mathbb{W}; \mathbf{w} : [k \prec v]$

$\Diamond_i$-p $\quad \mathbb{W}; \mathbf{w} : [k \succ \mathbf{here}\, M]$ $\qquad\qquad\qquad\; \mapsto \mathbb{W}; \mathbf{w} : [k \lhd \mathbf{here}\, \circ \succ M]$

$\Diamond_i$-r $\quad \{\mathbf{w} : \langle \chi, b\rangle, \cdots\}; \mathbf{w} : [k \lhd \mathbf{here}\, \circ \prec v] \qquad \mapsto$
$\qquad\quad \{\mathbf{w} : \langle \chi, (b, \ell = v)\rangle, \cdots\}; \mathbf{w} : [k \prec \mathbf{w}.\ell] \qquad (\ell \text{ fresh})$

ℓ-r $\quad \{\mathbf{w} : \langle \chi, b\rangle, \cdots\}; \mathbf{w} : [k \succ \ell] \qquad\qquad\quad \mapsto$
$\qquad\quad \{\mathbf{w} : \langle \chi, b\rangle, \cdots\}; \mathbf{w} : [k \prec v] \qquad\qquad\;\; (b(\ell) = v)$

$\Diamond_e$-p $\quad \mathbb{W}; \mathbf{w} : [k \succ \mathbf{letd}\, \omega.x = M \text{ in } N] \quad \mapsto \mathbb{W}; \mathbf{w} : [k \lhd \mathbf{letd}\, \omega.x = \circ \text{ in } N \succ M]$

$\Diamond_e$-r $\quad \mathbb{W}; \mathbf{w} : [k \lhd \mathbf{letd}\, \omega.x = \circ \text{ in } N \prec \mathbf{w}'.\ell] \mapsto \mathbb{W}; \mathbf{w} : [k \succ [\ell/x][\mathbf{w}'/\omega]N]$

$\Box_e$-p $\quad \mathbb{W}; \mathbf{w} : [k \succ \mathbf{unbox}\, M]$ $\qquad\qquad\quad \mapsto \mathbb{W}; \mathbf{w} : [k \lhd \mathbf{unbox}\, \circ \succ M]$

$\Box_e$-r $\quad \mathbb{W}; \mathbf{w} : [k \lhd \mathbf{unbox}\, \circ \prec \mathbf{box}\, \omega.M] \mapsto \mathbb{W}; \mathbf{w} : [k \succ [\mathbf{w}/\omega]M]$

letcc $\quad \{\mathbf{w} : \langle \chi, b\rangle, \cdots\}; \mathbf{w} : [k \succ \mathbf{letcc}\, u \text{ in } M] \qquad \mapsto$
$\qquad\quad \{\mathbf{w} : \langle (\chi, \mathbf{k} = k), b\rangle, \cdots\}; \mathbf{w} : [k \succ [\mathbf{w}.\mathbf{k}/u]M] \quad (\mathbf{k} \text{ fresh})$

throw $\quad \{\mathbf{w}' : \langle \chi, b\rangle, \cdots\}; \mathbf{w} : [k \succ \mathbf{throw}\, M \text{ to } \mathbf{w}'.\mathbf{k}] \quad \mapsto$
$\qquad\quad \{\mathbf{w}' : \langle \chi, b\rangle, \cdots\}; \mathbf{w}' : [k' \succ M] \qquad\qquad (\chi(\mathbf{k}) = k')$

rpc $\quad \mathbb{W}; \mathbf{w} : [k \succ \mathbf{go}[\mathbf{w}']M] \qquad\qquad\qquad\qquad \mapsto$
$\qquad \mathbb{W}; \mathbf{w}' : [\mathbf{abort} \succ M] \qquad\qquad\qquad\qquad (\mathbf{w} \in \mathrm{dom}(\mathbb{W}))$

$\Box_m$ $\quad \{\mathbf{w} : \langle \chi, b\rangle, \cdots\}; \mathbf{w} : [k \succ \mathbf{get}_\Diamond[\mathbf{w}']M] \qquad \mapsto$
$\qquad \{\mathbf{w} : \langle (\chi, \mathbf{k} = k), b\rangle, \cdots\}; \mathbf{w}' : [\mathbf{return}\, \mathbf{w}.\mathbf{k} \succ M]$ (k fresh)

$\Diamond_m$ $\quad \{\mathbf{w} : \langle \chi, b\rangle, \cdots\}; \mathbf{w} : [k \succ \mathbf{get}_\Box[\mathbf{w}']M] \qquad \mapsto$
$\qquad \{\mathbf{w} : \langle (\chi, \mathbf{k} = k), b\rangle, \cdots\}; \mathbf{w}' : [\mathbf{return}\, \mathbf{w}.\mathbf{k} \succ M]$ (k fresh)

ret $\quad \{\mathbf{w} : \langle \chi, b\rangle, \cdots\}; \mathbf{w}' : [\mathbf{return}\, \mathbf{w}.\mathbf{k} \prec v] \qquad \mapsto$
$\qquad \{\mathbf{w} : \langle \chi, b\rangle, \cdots\}; \mathbf{w} : [k \prec v] \qquad\qquad\qquad (\chi(\mathbf{k}) = k)$

Fig. 5. Selected rules from the operational semantics

Not surprisingly, the semantics is continuation-based. At any step, the cursor is selecting a world and continuation, with a value to return to it or an expression to evaluate. The rules generally fall into a few categories, as exemplified by the (standard) rules for $\supset$: There are (p)ush rules, in which we begin evaluating a subexpression of some M, pushing the context into the continuation, (s)wap rules, where we have finished evaluating one sub-expression and move onto the next, and (r)eduction rules, where we finally have a value and eliminate it. Every well-typed machine state will be closed with respect to truth, falsehood, and world hypotheses, so we don't have rules for variables.

The first interesting rule is $\Diamond_i$-r. It publishes the value v by generating a new label ℓ, mapping that label to v within its value table, and returning the pair $\mathbf{w}.\ell$, where $\mathbf{w}$ is the current world. Whenever we try to evaluate a label (rule ℓ-r), we look it up in the current world's value table in order to find the value. A key consequence of type safety (Theorems 1, 2) is that labels are only evaluated in the correct world. To eliminate an address (rule $\Diamond_e$-r) we substitute the constituent world and label through the body of the letd. Note that this step is slightly non-standard, because we substitute the *expression* ℓ for a variable rather than some value. But because the variable is in general at a different world, we are not in a position to get its value yet. We instead wait until the expression ℓ is sent to its home world (perhaps as part of some larger expression) to be looked up. The rules for $\Box$ are much simpler: box $\omega.M$ is already a value (rule $\Box_i$-v), and to unbox we simply substitute the current world for the hypothetical one (rule $\Box_e$-r).

When encountering a letcc, we grab the current continuation k. Because the continuation may be referred to from elsewhere in the network, we publish it in a table and form a global address for it (of the form $\mathbf{w.k}$), just as we did for $\Diamond$ addresses. This value is substituted for the falsehood variable u.

Throwing to a continuation (rule *throw*) is handled straightforwardly. The continuation expression will be closed, and therefore of the form $\mathbf{w'.k}$. We look up the label $\mathbf{k}$ in $\mathbf{w'}$ – or rather, *cause* $\mathbf{w'}$ to look it up – and pass the expression M to it. Note that we do not evaluate the argument before throwing it to the remote continuation. In general we *can not* evaluate it, because it is only well-typed at the remote world, which may be different from the world we're in.

Finally, we have the rules that move between worlds. The rule for go is easiest; since the target world expression must be closed it will be a world constant in the domain of $\mathbb{W}$. We simply move the cursor to that world (destroying the current continuation, which can never be reached), and begin evaluating the expression M under the unreachable continuation abort. The rules for $\text{get}_\Diamond$ and $\text{get}_\Box$ work similarly, but they need to save the current continuation since they will be returned to! These steps push a return frame, which reduces like throw. In contrast, however, the argument (of type $\Box A$ or $\Diamond A$) will be eagerly evaluated, because such values are portable. (After all, the whole point is to create the box at one world and then move it to another.)

In order for our language to make sense it must be type safe; any well-typed program must have a well-defined meaning as a sequence of steps in the abstract network. Type safety is stated as usual in terms of progress and preservation:

Theorem 1 (Progress)
If $\Sigma \vdash \mathbb{N}$ then either $\mathbb{N}$ is terminal   or $\exists \mathbb{N'}.\mathbb{N} \mapsto \mathbb{N'}$.

Theorem 2 (Preservation)
If $\Sigma \vdash \mathbb{N}$ and $\mathbb{N} \mapsto \mathbb{N'}$ then $\exists \Sigma'. \Sigma' \supseteq \Sigma$ and $\Sigma' \vdash \mathbb{N'}$.

Progress says that any well-formed network state can take another step, or is done. (Recall a *terminal* network is one where the cursor is returning a value to a finish continuation.) Preservation says that any well-typed network state that takes a step results in another well-typed state (perhaps in an extended[5] configuration typing Σ'). By iterating alternate applications of these theorems we see that any well-typed program is able to step repeatedly and remain well-formed, or else eventually comes to rest in a terminal state.

---

[5] $\Sigma' \supseteq \Sigma$ iff Σ' and Σ each describe the same set of worlds, and for each world, if $X(\mathbf{k}) = A$ then $X'(\mathbf{k}) = A$, and likewise for β and β'.

5   Concurrency and Communication

Many distributed computing problems benefit from concurrency, with one or more processes running on each node in the network. This section gives some brief thoughts on concurrency in our classical calculus.

First-class continuations are often used in the implementation of coroutines. With primitives for recursion and state we could also implement coroutines in C5, however, such an implementation is silly because it would require the implementation of a global scheduler, and would anyway defeat the purpose of concurrency on multiple nodes – only one coroutine would be running at any given time!

Fortunately, our operational semantics admits ad hoc concurrency easily. If we simply replace the cursor R in our network state "$\mathbb{W}; R$" with a multiset of cursors $\mathfrak{R}$, then we can permit a step on any one of these cursors essentially according to the old rules:

$$\frac{\mathbb{W}; R \;\mapsto\; \mathbb{W}'; R'}{\mathbb{W}; \{R\} \uplus \mathfrak{R} \;\mapsto^c\; \mathbb{W}'; \{R'\} \uplus \mathfrak{R}}$$

We can then add primitives as desired to spawn new cursors. A very simple one evaluates M and N in parallel and returns each one to the same continuation.

$$\frac{\Gamma; \Delta \vdash M : A@\mathrm{w} \quad \Gamma; \Delta \vdash N : A@\mathrm{w}}{\Gamma; \Delta \vdash M|N : A@\mathrm{w}} \;\; \mathrm{par}$$

$$\mathbb{W}; \mathfrak{R} \uplus \{\mathbf{w}{:}[k \succ M|N]\} \;\mapsto^c\; \mathbb{W}; \mathfrak{R} \uplus \{\mathbf{w}{:}[k \succ M]\} \uplus \{\mathbf{w}{:}[k \succ N]\}$$

A suitable extension of type safety holds for $\mapsto^c$.

With concurrency in place we can implement asynchronous CML-style channels [12] with the help of continuations (and a few other features for developing mutable recursive structures). The type of a channel that allows sending and receiving of values of type A could be

$$A \text{ chan} \;\doteq\; \Diamond(A \text{ queue} \wedge (\neg A) \text{ queue})$$

Here a channel is represented as the address of a pair of queues. In order to send to this channel, the sender must be able to bring a value of type A to the world where the channel lives. Therefore it must be a box or diamond type itself (although the class of types that are mobile in this way can be easily extended; see the technical report for details [7]). The first queue holds the values that have been sent on the channel and not yet received; the second holds the continuations of outstanding `recieves`. To implement `recieve` (assuming no values are waiting in the first queue), we grab the current continuation, enqueue it, and abort.

This is a standard technique; our point is to emphasize the utility of continuations as primitives for implementing useful distributed computing features.

6   Disjunction

To add disjunction to C5, we need to use the following elimination form in order to preserve the correspondence with classical S5:

$$\frac{\Gamma;\Delta \vdash M : A \vee B @ \omega' \quad \begin{array}{l} \Gamma, x{:}A@\omega';\Delta \vdash N_1 : C@\omega \\ \Gamma, x{:}B@\omega';\Delta \vdash N_2 : C@\omega \end{array}}{\Gamma;\Delta \vdash \mathtt{case}\, M \,\mathtt{of}\, \mathtt{inl}\, x \Rightarrow N_1 \mid \mathtt{inr}\, x \Rightarrow N_2 : C@\omega} \; \vee E$$

This rule is completely unsurprising except that the case object M is at a *different world*, ω'. In our logic we've tried hard to avoid this sort of *action-at-a-distance*, instead preferring to have our introduction and elimination rules compute locally. However, a motion rule for disjunction is out of the question, because it is unsound: it is not the case that if $\Gamma;\Delta \vdash A \vee B@\omega$ then necessarily $\Gamma;\Delta \vdash A \vee B@\omega'$. In our previous paper we speculated that the remote case analysis could be implemented nonetheless by sending back merely a *bit* telling the case-analyzing world which branch it should enter, but this requires some suspicious operational machinery. The same is true in the classical case, which is why we have avoided treating disjunction so far.

As it turns out, support for disjunction and remote disjunction elimination is already present in C5, via one of de Morgan's laws. We define $A \vee B$ as $\neg(\neg A \wedge \neg B)$, and $A \vee B$ thus becomes a continuation that takes *two* continuations: one if the disjunct is A, and one if the disjunct is B. This technique is well-known for CPS conversion, and first-class continuations let us employ it without having to CPS-convert the entire program. Encoding the injections is easy:

$$\mathtt{inl}\, M \;\doteq\; \lambda x.(\pi_1 x)M \qquad\qquad \mathtt{inr}\, M \;\doteq\; \lambda x.(\pi_2 x)M$$

By grabbing the continuation at the point of case analysis, we can allow ourselves to move to a remote world (via **go**) to do the case analysis and rely on **throw** to get us back:

$$\begin{array}{l} \mathtt{case}\, M \,\mathtt{of}\, \mathtt{inl}\, x \Rightarrow N_1 \\ \quad\mid\; \mathtt{inr}\, x \Rightarrow N_2 \end{array} \;\doteq\; \begin{array}{l} \mathtt{letcc}\, u \,\mathtt{in}\, \mathtt{go}[\omega']M\langle \lambda x.\, \mathtt{throw}\, N_1 \,\mathtt{to}\, u, \\ \qquad\qquad\qquad\qquad\quad \lambda x.\, \mathtt{throw}\, N_2 \,\mathtt{to}\, u\rangle \end{array}$$

This has exactly the same typing conditions as the remote rule above; x is bound to the remote type $A@\omega'$, even though the expression N_1 is evaluated at ω.

Classical logic is ripe with possibilities for definition. It is interesting to consider their implications. Recall that in Section 3 we proved $\Diamond A$ equivalent to $\neg\Box\neg A$. This means that, like classical logicians, we could then just consider $\Diamond A$ a derived form. This would amount to a roundabout way of using the continuation table to publish values rather than the value table. Clearly, we could also take the even stranger route of defining $\Box A$ in terms of $\Diamond$, which gives us a mobile code "server" that sends code to our continuation whenever we like.

7   Conclusions

Related Work. Parigot's $\lambda\mu$-calculus has inspired many computational proof systems for classical logic, including Wadler's dual calculus [15]. The calculus is sequent-oriented and contains *cut* as a computational primitive, emphasizing the duality of computing with values and covalues (continuations). For programming in C5, we choose a natural deduction system which is deliberately *non*-dual. We bias the logic towards truth, which corresponds to computing mainly with values (as is typical) rather than covalues. Nevertheless, we expect that a dual version of classical S5 could be easily made to work, perhaps starting from the sequent calculus presented in the Appendix.

Because our calculus extends Lambda 5 [8], it is also related to the same mobile calculi, for example Moody's distributed S4 calculus [6], and Jia and Walker's S5-like hybrid logic [5]. Both calculi employ the $\Box$ and $\Diamond$ connectives with similar interpretations, though aspects of the underlying logics differ. Both give operational interpretations via concurrent process calculi with passive synchronization, and both systems use non-local introduction and elimination forms. In contrast, we achieve explicit active synchronization (in the form of $\mathbf{get}_\Diamond$, etc.) along with what we feel are more primitive operations for constructing and deconstructing objects of the modal types. With regard to the classical extensions, we know of no prior modal system that features distributed continuations.

Future Work. Our language now has a full arsenal of connectives and control operators, each connected to logic. Much work remains before C5 can be a practical programming language rather than exploratory calculus. Some are routine – adding extra-logical primitives like recursion and references – and some are difficult – compilation of mobile code fragments, distributed garbage collection, failure recovery, and certification.

Although we believe that C5 accommodates concurrency easily, it would be nice to have a logically-inspired account of it. Some other directions remain open to try. Proof search in linear logic sequent calculus [3] is known to admit an interpretation as concurrent computation [2]. Perhaps linear S5 in sequent style would be able to elegantly express both spatial properties and concurrency in logic?

We have presented a proof theory and corresponding programming language, C5, based on the classical modal logic S5. By exploiting the modalities we are able to give a logical account of mobility and locality, and thus an expressive programming language for distributed computing. From the logic's classical nature we derive the mechanism of distributed continuations, which creates a new connection between the $\Box$ and $\Diamond$ connectives, and forms a basis for the implementation of distributed computing primitives.

References

1. Andreas Abel. A third-order representation of the $\lambda\mu$-Calculus. In S.J. Ambler, R.L. Crole, and A. Momigliano, editors, *Electronic Notes in Theoretical Computer Science*, volume 58. Elsevier, 2001.

2. Jean-Yves Girard. Towards a geometry of interaction. *Contemporary Mathematics*, 92:69–108.
3. Jean-Yves Girard. Linear logic. *Theoretical Computer Science*, 50(1):1–102, January 1987.
4. Timothy G. Griffin. The formulae-as-types notion of control. In *Conf. Record 17th Annual ACM Symp. on Principles of Programming Languages, POPL'90, San Francisco, CA, USA, 17–19 Jan 1990*, pages 47–57. ACM Press, New York, 1990.
5. Limin Jia and David Walker. Modal proofs as distributed programs. *13th European Symposium on Programming*, pages 219–223, March 2004.
6. Jonathan Moody. Modal logic as a basis for distributed computation. Technical Report CMU-CS-03-194, Carnegie Mellon University, Oct 2003.
7. Tom Murphy, VII, Karl Crary, and Robert Harper. Distribed control flow with classical modal logic (technical report). Technical Report CMU-CS-04-177, Carnegie Mellon University, Dec 2004.
8. Tom Murphy, VII, Karl Crary, Robert Harper, and Frank Pfenning. A symmetric modal lambda calculus for distributed computing. In *Proceedings of the 19th IEEE Symposium on Logic in Computer Science (LICS 2004)*. IEEE Press, July 2004.
9. Chetan Murthy. Classical proofs as programs: How, what and why. Technical Report TR91-1215, Cornell University, 1991.
10. Michel Parigot. $\lambda\mu$-Calculus: An algorithmic interpretation of classical natural deduction. In Andrei Voronkov, editor, *Logic Programming and Automated Reasoning, International Conference LPAR'92, St. Petersburg, Russia, July 15-20, 1992, Proceedings*, volume 624 of *Lecture Notes in Computer Science*. Springer, 1992.
11. Frank Pfenning and Carsten Schürmann. System description: Twelf – a metalogical framework for deductive systems. In Harald Ganzinger, editor, *Proceedings of the 16th International Conference on Automated Deduction*, pages 202–206, Trento, Italy, July 1999. Springer-Verlag. LNAI 1632.
12. John H. Reppy. *Concurrent Programming in ML*. Cambridge University Press, Cambridge, England, 1999.
13. Carsten Schürmann and Frank Pfenning. A coverage checking algorithm for LF. In D. Basin and B. Wolff, editors, *Proceedings of the 16th International Conference on Theorem Proving in Higher Order Logics (TPHOLs 2003)*, pages 120–135, Rome, Italy, September 2003. Springer-Verlag LNCS 2758.
14. Alex Simpson. *The Proof Theory and Semantics of Intuitionistic Modal Logic*. PhD thesis, University of Edinburgh, 1994.
15. Philip Wadler. Call-by-value is dual to call-by-name. In *Proceedings of the 8th International Conference on Functional Programming (ICFP)*. ACM Press, August 2003.

Appendix

This appendix contains sketches of the proofs relating C5 to a classical S5 sequent calculus. This serves two purposes. First, because the sequent calculus is purely logical and does not feature our decomposition of the $\Box$ and $\Diamond$ rules, it is more obviously S5. Second, because the sequent calculus has the subformula property and admits cut, we get some standard results for our proof theory, such as the existence of normal forms and soundness. To begin, we need a few substitution theorems for our natural deduction system, one of which is interesting.

$$\frac{}{\Gamma, A@\omega \ \# \ A\star\omega, \Delta} \ \text{contra} \qquad\qquad \frac{}{\Gamma, \bot@\omega \ \# \ \Delta} \ \bot T$$

$$\frac{\Gamma, A \supset B@\omega, B@\omega \ \# \ \Delta \qquad \Gamma, A \supset B@\omega \ \# \ A\star\omega, \Delta}{\Gamma, A \supset B@\omega \ \# \ \Delta} \ \supset T \qquad \frac{\Gamma, A@\omega \ \# \ B\star\omega, A \supset B\star\omega, D}{\Gamma \ \# \ A \supset B\star\omega, D} \ \supset F$$

$$\frac{\Gamma, \Box A@\omega, A@\omega' \ \# \ \Delta}{\Gamma, \Box A@\omega \ \# \ \Delta} \ \Box T \qquad \frac{\Gamma, \omega' \ \# \ A\star\omega', \Box A\star\omega, \Delta}{\Gamma \ \# \ \Box A\star\omega, \Delta} \ \Box F$$

$$\frac{\Gamma, \omega', \Diamond A@\omega, A@\omega' \ \# \ \Delta}{\Gamma, \Diamond A@\omega \ \# \ \Delta} \ \Diamond T \qquad \frac{\Gamma \ \# \ A\star\omega', \Diamond A\star\omega, \Delta}{\Gamma \ \# \ \Diamond A\star\omega, \Delta} \ \Diamond F$$

$$\frac{\Gamma, A \land B@\omega, A@\omega, B@\omega \ \# \ \Delta}{\Gamma, A \land B@\omega \ \# \ \Delta} \ \land T \qquad \frac{\Gamma \ \# \ A\star\omega, A \land B\star\omega, \Delta \qquad \Gamma \ \# \ B\star\omega, A \land B\star\omega, \Delta}{\Gamma \ \# \ A \land B\star\omega, \Delta} \ \land F$$

Fig. 6. Classical S5 sequent calculus

Falsehood Substitution. For each sort of hypothesis we have a substitution theorem. Worlds can be substituted for hypothetical worlds, and substitution $[M/x]N$ for truth hypotheses is defined in the standard way. Substitution for falsehood hypotheses warrants special attention, however:

Theorem 3 (Falsehood Substitution)
If $\forall C, \omega''. \ \Gamma, x{:}A @\omega; \Delta \vdash M : C @\omega''$
and $\Gamma; \Delta, u{:}A\star\omega \vdash N : B @\omega'$
then $\Gamma; \Delta \vdash [\![x.M/u]\!]N : B @\omega'$.

This principle is dual to the # rule just as truth substitution is dual to the *hyp* rule. The # rule contradicts an $A\star\omega$ with an $A@\omega$, so when substituting for a falsehood assumption, we are able to assume $A@\omega$ and must produce another contradiction. We write falsehood substitution as $[\![x.M/u]\!]N$ where x is a binder (with scope through M) that stands for the value thrown to u. Just like truth substitution, it is defined pointwise on N except for the appropriate variable case (rule #):

$$[\![x.M/u]\!] \, \text{throw} \, N' \, \text{to} \, u \doteq [N'/x]M$$

Operationally, we see this as replacing the throw with some other handler for A. Since the new handler must have parametric type, typically it is a **throw** to some other continuation, perhaps after performing some computation on the proof of A.

Sequent Calculus. Our sequent calculus is motivated by simplicity and duality alone, because we will not give it a computational interpretation. One traditional way of doing classical theorem proving is to negate the target formula and prove a contradiction from it. Our sequent calculus (Fig. 6) is based on this view: the sequent $\Gamma \ \# \ \Delta$ means that the truth assumptions in Γ and the falsehood assumptions in Δ are mutually contradictory[6]. We treat contexts as unordered

---

[6] Our rules are also consistent with the more traditional multiple-conclusion reading, "if all of Γ are true, then one of Δ is true."

multisets, so the *action* can occur anywhere in either context. World hypotheses are placed in Γ, although to get a notationally dual system, we would place them in a third context "in the middle" of the sequent.

These rules should be read bottom-up, as if during proof search. The *contra* rule allows us to form a contradiction whenever a proposition is both true and false at the same world. The $\Box T$ rule says that if we know $\Box A @ \omega$, then we know $A @ \omega'$ for any ω' that exists. On the other hand, if we know that $\Box A$ is false, then we know A is false at some world ω'. However, we must treat this world as hypothetical and fresh since we don't know which one it is. The rules for $\Diamond$ are perfect mirror images of the rules for $\Box$. The treatment of implication is standard, and follows from the classical truth tables.

We then wish to prove that the natural deduction and sequent calculus are equivalent (Theorem 5). The translation from natural deduction to the sequent calculus requires a lemma. In an intuitionistic calculus this would be *cut*; for the symmetric classical calculus it turns out to be the familiar classical notion of *excluded middle*.

Theorem 4 (Excl. Middle) *If* $\Gamma, A @ \omega \# \Delta$ *and* $\Gamma \# A \star \omega, \Delta$ *then* $\Gamma \# \Delta$.

Proof of Theorem 4 is by lexicographic induction on the proposition A and then simultaneously on the two derivations.   □

Theorem 5 (Equivalence)

(a)   If $\Gamma; \Delta \vdash M : A @ \omega$ *then* $\Gamma \# A \star \omega, \Delta$.

(b)   If $\Gamma \# \Delta$ *then* $\exists M. \forall C, \omega. \Gamma; \Delta \vdash M : C @ \omega$.

It is easy to see why 5(b) is the right statement. Since we think of $\Gamma \# \Delta$ as a proof of contradiction, this corresponds to a natural deduction derivation that proves any proposition at any world. Theorem 5(a) is more subtle. We show that if A is true under assumptions Γ and Δ, then A being false at the same world is contradictory with those assumptions. Computationally, we can think of this as the "final continuation" to which the result computed in natural deduction is passed. Putting these two theorems together, we have that $\Gamma; \Delta \vdash M : A @ \omega$ gives $\Gamma \# A \star \omega, \Delta$, which then gives $\forall C, \omega'. \Gamma; \Delta, u{:}A \star \omega \vdash M' : C @ \omega'$. In particular, we choose $C = A$ and $\omega' = \omega$, and then by application of bc we have the original judgment (with a normalized proof term $\mathtt{letcc}\, u\, \mathtt{in}\, M'$). Thus $\vdash$ and $\#$ are really equivalent.

The proof of Theorem 5(a) is by straightforward induction on the derivation, using Theorem 4 where necessary. (The structural rules bc and $\#$ just become uses of contraction and weakening in the sequent calculus.)   □

Proof of 5(b) is interesting because of its manipulation of continuations through the use of falsehood substitution (Theorem 3). Uses of T rules are easy; they correspond directly to the elimination rules[7] in natural deduction. But since our natural deduction is biased towards manipulating truth rather than false-hood, the F rules are more difficult and make nontrivial use of the falsehood substitution theorem. For instance, in the $\wedge F$ case we have by induction:

---

[7] Except for implication, which is phrased differently in the sequent calculus.

$$\Gamma; \Delta, u_p{:}A \wedge B{\star}\omega, u_a{:}A{\star}\omega \vdash N_1 : C @ \omega' \; (\forall C, \omega')$$
$$\Gamma; \Delta, u_p{:}A \wedge B{\star}\omega, u_b{:}B{\star}\omega \vdash N_2 : C @ \omega' \; (\forall C, \omega')$$

By two applications of Theorem 3, we get that the following proof term has any type at any world:

$$\Big[\!\!\Big[x.\big[\!\big[y.\, \mathtt{throw}\, \langle x, y \rangle\, \mathtt{to}\, u_p / u_b \big]\!\big] N_2 \, / u_a \Big]\!\!\Big] N_1$$

We form an innermost $\mathtt{throw}$ of the pair $\langle x, y \rangle$ to our pair continuation u_p. This has free truth hypotheses $x : A$ and $y : B$. Therefore, we can use it to substitute away the u_b continuation in N_2 (any throw of M to u_b becomes a throw of $\langle x, M \rangle$ to u_p). Finally, we can use this new term to substitute away u_a in N_1, giving us a term that depends only on the pair continuation u_p. This pattern of *prepending* work onto continuations through substitution is characteristic of this proof.

The case for $\Box F$ is interesting because it uses $\mathtt{letcc}$ [8]. By induction we have:

$$\forall \omega'.\; \Gamma; \Delta, u{:}A{\star}\omega', u_b{:}\Box A{\star}\omega \vdash N : C @ \omega'' \;\; (\forall C, \omega'')$$

Then the proof term witnessing the theorem here is:

$$\mathtt{throw(box}\, \omega'.\, \mathtt{letcc}\, u\, \mathtt{in}\, N)\, \mathtt{to}\, u_b$$

It is not possible to use falsehood substitution on u in this case. To do so we would need to turn a term of type $A @ \omega'$ into a $\Box A @ \omega$ to throw to u_b. Although at a meta-level we know that we can choose any ω', it won't be possible to internalize this in order to create a $\Box A$. Instead we must introduce a new box, and choose ω' to be the new hypothetical world that the $\Box I$ rule introduces. At that point we use $\mathtt{letcc}$ to create a real $A{\star}\omega'$ assumption to discharge u. The remaining cases are similar or straightforward, and can be found in full detail in the Twelf code[9].

$\Box$

---

[8] In fact, this is the only place in the proof where a $\mathtt{letcc}$ is necessary. This gives a normal form for natural deduction terms where $\mathtt{letcc}$ appears only once at the outermost scope and immediately inside each $\mathtt{box}$.

[9] The most natural LF encoding of falsehood is 3rd-order [1]; we use a 2nd-order encoding in our proofs (proving the falsehood substitution theorem by hand) because third-order metatheorem checking is not yet available in the distribution.

A Logic of Coequations

Jiri Adámek*

Institute of Theoretical Computer Science
Technical University, Braunschweig, Germany
adamek@iti.cs.tu-bs.de

Abstract. By Rutten's dualization of the Birkhoff Variety Theorem, a collection of coalgebras is a covariety (i.e., is closed under coproducts, subcoalgebras, and quotients) iff it can be presented by a subset of a cofree coalgebra. We introduce inference rules for these subsets, and prove that they are sound and complete. For example, given a polynomial endofunctor of a signature Σ, the cofree coalgebra consists of colored Σ-trees, and we prove that a set T of colored trees is a logical consequence of a set S iff T contains every tree such that all recolorings of all its subtrees lie in S. Finally, we characterize covarieties whose presentation needs only n colors.

1 Introduction

In the theory of systems as coalgebras (in the category of sets) presented for example by Jan Rutten [14], cofree coalgebras $C(k)$ consist of "possible behavior patterns" of states of systems colored by k (observable) colors. Given a system A and a coloring, the corresponding homomorphism from A to $C(k)$ assigns to every state its behavior pattern. J. Rutten used subsets S of cofree coalgebras as a means of presentation of systems: a system A *satisfies* S iff every homomorphism $f\colon A \longrightarrow C(k)$ factorizes through $S \lhook\joinrel\longrightarrow C(k)$. And he proved the dual of the famous Birkhoff Variety Theorem: a collection of systems has a presentation via subsets of $C(k)$ iff it is a covariety, i.e., it is closed under coproducts, subsystems, and quotients. This holds for systems as coalgebras on an arbitrary k-accessible functor.

Several authors studied logical properties of subsets $S \subseteq C(k)$. Peter Gumm [10] observed that one can restrict to the coatomic subsets $C(k) - \{t\}$, for which we use the notation $\boxtimes t$ (read: avoid t). A system A satisfies $\boxtimes t$ iff under an arbitrary coloring all states avoid the behavior pattern t. Every subset S is logically equivalent to the conjunction of all $\boxtimes t$ where t ranges through the complement $C(k) - S$. Whereas subsets of cofree coalgebras dualize the concept

$$\text{collection of equations} = \text{quotient of a free algebra},$$

the coatomic subsets $\boxtimes t$ dualize the concept

* Grant MSM 6840770014 of the Ministry of Education of Czech Republic is acknowledged.

L. Ong (Ed.): CSL 2005, LNCS 3634, pp. 70–86, 2005.

equation = atomic quotient of a free algebra.

In fact, in the lattice of congruences of a free algebra every atom is a congruence generated by a single equation. We call the expressions $\boxtimes t$ *coequations*.

Peter Gumm and Tobias Schröder showed in [12] that for presentations of systems we can restrict ourselves to subcoalgebras of $C(k)$, and Steve Awodey and James Hughes [7] then proved that (dually to Birkhoff's characterization of equational theories in [9]) invariant subcoalgebras are precisely the coequational theories. They based their result on the theory of invariance of predicates developed by Bart Jacobs [13].

In the present paper we formulate two simple inference rules for coequations, and prove that they form a sound and complete system. We do this in three steps:

(1) Logic for polynomial functors H_Σ (of a k-ary signature Σ, where k is an infinite cardinal): Recall that a cofree H_Σ-coalgebra $C_\Sigma(k)$ is the algebra of all k-colored Σ-trees, see [5]. The two inference rules for coequations $\boxtimes t$ are: if a tree s is (a) a subtree of t, or (b) a recoloring of t, then $\boxtimes t$ is a logical consequence of $\boxtimes s$. We conclude that for every subset S of $C_\Sigma(k)$, the logical consequences of S are precisely those sets T of trees which contain every tree t such that all recolorings of all subtrees of t lie in S.

(2) Logic for k-accessible functors H: We express H as a quotient of H_Σ for a k-ary signature and conclude that the cofree coalgebra $C(k)$ is a canonical quotient of $C_\Sigma(k)$. Thus, elements of $C(k)$ are congruence classes of Σ-trees. We prove that $\boxtimes t$ is a consequence of $\boxtimes s$ iff every tree congruent to t is a recoloring of a subtree of a tree congruent to s. Unlike the previous case, we see no way how to formulate the inference rules for general subsets of $C(k)$. This is the reason why, in contrast to the authors cited above, we concentrate on coequations, rather than general subsets, in our paper (in spite of the fact that the negative way a coequation formulates properties of systems makes it less intuitive for applications).

(3) Logic for arbitrary endofunctors of **Set**: Here we use *generalized coequations*, i.e., transfinite chains of "approximations of coequations", introduced in [3], where we proved that every covariety has a presentation by generalized coequations. We now derive a logic of generalized coequations analogous to that for accessible functors. In the proof we use the extension of **Set** to the category **Class** of classes, and the fact that every endofunctor of **Set** has a unique extension to **Class**, as established in [4].

In the final section we characterize, for every cardinal n, those covarieties which can be presented by coequations using n colors. For $n = 1$ these are precisely the covarieties closed under bisimulation, as proved by Peter Gumm and Tobias Schröder [12]. Our characterization is analogous: we call two H-coalgebras equipped with a coloring by n colors n-*color-bisimilar* if they are bisimilar as coalgebras of $H(-) \times n$. An we prove that covarieties presentable by n-color-coequations are precisely those closed under n-color-bisimilarity.

2 Logic for Polynomial Functors

2.1. Recall from the expository paper of J. Rutten [14] that for every endofunctor H of **Set** a *coalgebra* $A = (Q, \alpha)$ is a system given by a set Q of states and a structure map $\alpha \colon Q \longrightarrow HQ$. A *homomorphism* from (Q, α) to a coalgebra (Q', α') is a function $f \colon Q \longrightarrow Q'$ with $\alpha' \cdot f = Hf \cdot \alpha$. For example, a deterministic system with a binary input and halting states is expressed as a coalgebra of $HQ = Q \times Q + 1$. Given a coalgebra $\alpha \colon Q \longrightarrow Q \times Q + 1$ and a state q in it, if $\alpha q = (q_0, q_1)$, then q_i is the next state of q for the input $i = 0, 1$; if $\alpha q \in 1$, then q is a halting state.

A *covariety* is a collection of coalgebras closed under coproducts, subcoalgebras, and quotient coalgebras. For k-accessible functors J. Rutten proved that covarieties are precisely the classes of coalgebras which have a presentation by a subset of a cofree coalgebra as follows:

Given a set k (of colors, finite or infinite), a *cofree coalgebra* on k is a coalgebra $C(k)$ with a structure map

$$\tau_k \colon C(k) \longrightarrow HC(k)$$

and a universal "coloring" map

$$\gamma_k \colon C(k) \longrightarrow k \ .$$

The universal property states that for every coalgebra $A = (Q, \alpha)$ and every coloring $f \colon Q \longrightarrow k$ there exists a unique homomorphism

$$f^{\#} \colon A \longrightarrow C(k) \qquad \text{with} \qquad f = \gamma_k f^{\#}.$$

2.2 Definition. (i) *Suppose that a subset $m \colon S \longhookrightarrow C(k)$ of the cofree coalgebra is given. We say that a coalgebra A **satisfies** S provided that for every coloring $f \colon Q \longrightarrow k$ the homomorphism $f^{\#}$ factorizes through m. For example, given $t \in C(k)$, then A satisfies $\boxtimes t$ iff for every coloring $f \colon A \longrightarrow k$ all states a fulfil $f^{\#}(a) \neq t$.*

(ii) *A class of coalgebras is* presented *by $m \colon S \longhookrightarrow C(k)$ if it contains precisely those coalgebras which satisfy S.*

(iii) *By a **logical consequence** of S is meant any subset $m' \colon S' \longhookrightarrow C(k)$ such that whenever a coalgebra satisfies S, then it also satisfies S'. Notation:*

$$S \models S'$$

2.3. Recall that a set functor H is called k-*accessible* if it preserves k-filtered colimits (for an infinite cardinal k); or, equivalently, if every element of HX lies in Hm $[HM]$ for some subset $m \colon M \longhookrightarrow X$ of cardinality smaller than k. If H is k-accessible then every covariety requests only one subset of $C(k)$ for its presentation:

Theorem. *For a k-accessible set functor H, a class of coalgebras is a covariety iff it can be presented by a subset of $C(k)$.*

This theorem was stated by J. Rutten [14] for functors bounded at k and weakly preserving pullbacks, but the latter assumption can be left out, see [12], and "bounded at k" (which is equivalent to being k^+-accessible, see [5]) can be weakened to k-accessible.

2.4 Example. Let Σ be a k-ary signature, i.e., all arities are cardinals smaller than k (k an infinite cardinal). Then the *polynomial functor* H_Σ defined on object X by

$$H_\Sigma X = \coprod_{\sigma \in \Sigma} X^n \qquad (n = \text{arity of } \sigma)$$

is k-accessible.

Recall (e.g. from [5]) that a terminal coalgebra (= cofree coalgebra on one color) can be described as the coalgebra of all Σ-*trees*, i.e., trees, finite or infinite, whose nodes are labeled by Σ in such a way that every node with an n-ary label has precisely n children. Trees are considered up to isomorphism throughout the paper. Analogously, a cofree coalgebra $C_\Sigma(k)$ is the coalgebra of all k-colored Σ-trees. That is, elements are Σ-trees whose nodes are additionally labelled by colors i (where $i < k$ is an ordinal). The coalgebraic structure τ_k^Σ assigns to every tree t whose root carries an n-ary operation label σ the n-tuple of its children t_i in the σ-summand of $H_\Sigma C_\Sigma(k)$, notation:

$$\tau_k^\Sigma(t) = \sigma(t_i)_{i<n}$$

And the universal coloring $\gamma_k^\Sigma : C_\Sigma(k) \longrightarrow k$ assigns to every tree the color of its root.

A Σ-coalgebra A is a set Q of states together with a structure map $\alpha \colon Q \longrightarrow H_\Sigma Q$ assigning to every $a \in Q$ an n-tuple $\sigma(a_i)_{i<n}$ in the σ-summand Q^n for some n-ary operation $\sigma \in \Sigma$. Given a coloring $f \colon Q \longrightarrow k$, the homomorphism $f^\# \colon A \longrightarrow C_\Sigma(k)$ takes every node a to the Σ-tree-unfolding of a using the colors given by f.

2.5 Example. The functor $HQ = Q \times Q + 1$ of 2.1 above is finitary, i.e., ω-accessible ($k = \omega$). The cofree coalgebra $C(\omega)$ can be described as the coalgebra of all binary trees, finite or infinite, whose nodes are colored by natural numbers. The structure map $\tau_\omega \colon C(\omega) \longrightarrow C(\omega) \times C(\omega) + 1$ has as halting states all singleton trees, and it assigns to every non-singleton tree the pair of its children, while $\gamma_\omega \colon C(\omega) \longrightarrow \omega$ is the color of the root.

The coequation $\boxtimes t_0$ where t_0 is a singleton tree presents all systems without halting states. The coequation $\boxtimes t_1$ where t_1 is the one-colored complete binary tree presents all systems which starting from every state can halt in finitely many steps. The coequation presents all systems such that for no state both successors are halting states. The coequation

presents all systems such that for every state whose both successors are halting states, these successors must be equal.

2.6 Example. Deterministic acceptors with the input set S are precisely the coalgebras of the polynomial functor

$$HX = X^S \times \textbf{bool} \ .$$

In fact, given a set Q of states with the next state function $\delta \colon Q \times S \longrightarrow Q$ (expressed in the curried from $\hat{\delta} \colon Q \longrightarrow Q^S$) and a predicate $accept \colon Q \longrightarrow \textbf{bool}$, we obtain a coalgebra structure map

$$\alpha = \langle \hat{\delta}, accept \rangle \colon Q \longrightarrow Q^S \times \textbf{bool} \ .$$

The functor H is ω-accessible if S is finite, and k-accessible for $k = (\textbf{card } S)^+$ (the cardinal successor of $\textbf{card } S$) otherwise.

The terminal coalgebra is the coalgebra $C(1) = \mathscr{P}(S^*)$ of all languages over S, see [6]: for every acceptor A the unique homomorphism into $C(1)$ takes a state q to the language $L_q(A)$ accepted by A with q as the initial state. Thus, every language $L \subseteq S^*$ presents a class $\boxtimes L$ of acceptors A: those with $L_q(A) \neq L$ for all states q. Examples:

(i) $\boxtimes S^*$: all acceptors having a non-accepting state;
(ii) $\boxtimes \{\varepsilon\}$: all acceptors such that from any accepting state we can reach an accepting state in $n > 0$ steps.

2.7 Definition. *Two nodes of a k-colored tree $t \in C_\Sigma(k)$ are called **equivalent** provided the two k-colored subtrees of t they represent are isomorphic.*

2.8 Lemma. *For two trees $t, s \in C_\Sigma(k)$ the following conditions are equivalent:*

(i) *$s = h(t)$ for some coalgebra homomorphism $h \colon C_\Sigma(k) \longrightarrow C_\Sigma(k)$ and*
(ii) *s and t have the same underlying Σ-tree, and any two equivalent nodes of t are also equivalent in s.*

Proof. (i) $\longrightarrow$ (ii). The coalgebra $T_\Sigma \simeq C_\Sigma(1)$ of all (uncolored) Σ-trees is terminal, and the unique homomorphism $u \colon C_\Sigma(k) \longrightarrow T_\Sigma$ assigns to every colored Σ-tree the underlying Σ-tree. The equality of the underlying Σ-trees of s and t follows from the fact that $u = u \cdot h \colon C_\Sigma(k) \longrightarrow T_\Sigma$. Furthermore, if the root of t is labelled by (σ, i) where σ is an n-ary operation, then since h is a homomorphism, the root of s is labelled by (σ, j), and h takes the m-th child of t to the m-th child of s. Consequently, for every node x of t, the corresponding subtree $t|x$ is taken to the subtree $s|x$ of s:

$$h(t|x) = s|x \qquad \text{for all nodes } x \text{ of } t.$$

Therefore, if two nodes are equivalent in t, they are equivalent in s.

(ii) $\longrightarrow$ (i). Choose a function $h_0 : C(k) \longrightarrow k$ by the following rule, where $t|x$ denotes the subtree of t at a node x:

$$h_0(r) = \begin{cases} \text{color of } x \text{ in } s & \text{if } r = t|x \text{ for some node } x \\ \text{arbitrary} & \text{else.} \end{cases}$$

By (ii) such a function exists. The unique homomorphism $h : C_\Sigma(k) \longrightarrow C_\Sigma(k)$ with $h_0 = \gamma_k \cdot h$ fulfils

$$h(t) = s \ .$$

In fact, since $u(t) = u(s)$, all we have to verify is that for every node x of t the colors of x in the trees $h(t)$ and s are equal. This is clear for the root: the root color of $h(t)$ is $h_0(t)$, which is the root color of s by definition of h_0. For nodes x of depth 1 this follows from h being a homomorphism: suppose $t = \sigma(t_i)_{i<n}$, then $s = \sigma(s_i)_{i<n}$ (because $u(t) = u(s)$) and $h(t) = \sigma(h(t_i))_{i<n}$ has the j-th child $h(t_j)$, whose root color $h_0(t_j)$ is the color of the j-th child of s by definition of h_0. A formal proof by induction on the depth of x is left to the reader. $\qquad\square$

2.9 Definition. *By a* **recoloring** *of a colored Σ-tree t is meant any tree s satisfying the equivalent conditions of 2.8. Given colored Σ-trees t, t' we write*

$$t' \sqsubseteq t$$

provided that t' is a recoloring of a subtree of t.

2.10 Remark. The relation $\sqsubseteq$ is a preorder: it is obviously reflexive, and transitivity follows from the fact that "recoloring of" is transitive due to 2.8(i), and "subtree of" is clearly transitive too. The relation $\sqsubseteq$ is, however, not antireflexive:

2.11 Example. The following trees

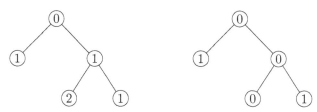

are recolorings of each other. And each of them has a recoloring having the same shape and all nodes of equal color.

2.12 Lemma. *For every k-colored Σ-tree t let A_t be the H_Σ-coalgebra of all equivalence classes $[x]$ of nodes x of t (see 2.7) with $\alpha : A_t \longrightarrow H_\Sigma A_t$ defined by*

$$\alpha[x] = \sigma([x_i])_{i<n}$$

if x has label (σ, u) and x_i are the children of x. Then

(i) $A_t \nvDash \boxtimes t$, and

(ii) $f^\#[x] \sqsubseteq t$ for every coloring $f \colon A_t \longrightarrow k$ and every state $[x] \in A_t$.

The above algebras A_t are analogous to the tree coalgebras of [5].

Proof. The above α is well-defined because given equivalent nodes x and y, the operations labelling x and y are equal, and the corresponding children are again equivalent. For every coloring $f \colon A_t \longrightarrow k$ the homomorphism $f^\# \colon A_t \longrightarrow C_\Sigma(k)$ is defined by the following rule: $f^\#[x]$ is the subtree $t|x$ of t at x recolored by f. (In fact, this function $f^\#$ is easily seen to be a homomorphism: suppose $\alpha[x] = \sigma([x_i])_{i<n}$, then

$$H_\Sigma f^\# \cdot \alpha[x] = \sigma\big(f^\#[x_i]\big)_{i<n} = \tau_k^\Sigma\big(f^\#[x]\big) \ ,$$

see 2.4. Moreover, $\gamma_k^\Sigma \cdot f^\# = f$.) Therefore, (ii) is clear. For (i), use the original coloring, more precisely, the function $g \colon A_t \longrightarrow k$ assigning to $[x]$ the color of x in t. By the above rule applied to $f = g$ we see that $g^\#$ assigns to the equivalence class of the root of t the value t. Therefore, $A \nvDash \boxtimes t$.   □

2.13 Theorem. *For a polynomial endofunctor a coequation $\boxtimes t$ is a logical consequence of coequations $\boxtimes s_i$ ($i \in I$) iff some s_j is a recoloring of a subtree of t. Shortly:*

$$\{\boxtimes s_i; i \in I\} \vDash \boxtimes t \quad iff \quad s_j \sqsubseteq t \quad for \ some \quad j \in I.$$

Proof. Denote by k^* the set of all words formed by ordinals smaller than k. Then a colored Σ-tree t can be formalized as a partial function from k^* to $\Sigma \times k$ such that the set $\mathrm{Def}\, t$ of all words where t is defined has the following properties:

(i) $\mathrm{Def}\, t$ contains the empty word ε,

(ii) $\mathrm{Def}\, t$ is prefix-closed, i.e., if t is defined in xy, then it is defined in x, and

(iii) if $t(x) = (\sigma, u)$ for an n-ary operation symbol σ, then $xi \in \mathrm{Def}\, t$ for all ordinals $i < n$ and $xi \notin \mathrm{Def}\, t$ for any $i \geq n$.

In particular, a subtree $t|x$ of t at the node $x \in \mathrm{Def}\, t$ is characterized as follows:

$$\mathrm{Def}(t|x) = \{y \in k^*; xy \in \mathrm{Def}\, t\}$$

and

$$t|x(y) = t(xy) \qquad \text{for all } y \in \mathrm{Def}(t|x).$$

(1) Sufficiency. We only need to prove that for two colored trees s and t we have $\boxtimes s \vDash \boxtimes t$ whenever s is either a recoloring of t or a subtree of t.

(1a) Let s be a subtree of t. Without loss of generality, we can assume that s is a child (i.e., a maximum proper subtree) of t – by repeating the same argument finitely many times we obtain the general case.

We have $\tau_k^\Sigma(t) = \sigma(t_r)_{r<n}$ and $s = t_{r_0}$ for some $\sigma \in \Sigma_n$ and some $r_0 < n$. Let A be a coalgebra satisfying $\boxtimes s$. If for some coloring $f \colon A \longrightarrow k$ we had $f^\#(a) = t$ then, since $f^\#$ is a homomorphism, $\alpha(a)$ would have the form $\sigma(a_r)_{r<n}$ where $f^\#(a_r) = t_r$ – this is impossible because then $f^\#(a_{r_0}) = s$.

(1b) Let $s = h(t)$ be a recoloring of t, where h is an endomorphism of $C_\Sigma(k)$. For every coalgebra A and every coloring $f: A \longrightarrow k$ we have

$$h \cdot f^\# = (\gamma_k \cdot h \cdot f^\#)^\# : A \longrightarrow C_\Sigma(k)$$

In fact, this follows from the universal property of γ_k: h is a homomorphism, thus, so is $h \cdot f^\#$, and both sides composed with γ_k yield $\gamma_k \cdot h \cdot f^\#$. If A satisfies $\boxtimes s$, then $h \cdot f^\#(a) \neq s = h(t)$ which implies $f^\#(a) \neq t$ for all $a \in A$. Therefore A satisfies $\boxtimes t$.

(2) Necessity. Assuming $\{\boxtimes s_i; i \in I\} \vDash \boxtimes t$, the coalgebra A_t of Lemma 2.12 does not satisfy all of $\boxtimes s_i$ for $i \in I$. Thus, there exists $j \in I$ with $f^\#[x] = s_j$ for some coloring f and some $[x] \in A_t$. Then Lemma 2.12 implies $s_j \sqsubseteq t$.   □

2.14 Corollary. *Given a polynomial functor H_Σ, the following deduction rules are sound and complete for the logical deduction of coequations:*
(1) *Child Rule*

$$\frac{\boxtimes t_j}{\boxtimes t} \qquad (\textit{if } t_j \textit{ is the } j\textit{-th child of } s)$$

and
(2) *Recoloring Rule*

$$\frac{\boxtimes s}{\boxtimes t} \qquad (\textit{if } s \textit{ is a recoloring of } t).$$

In fact, for each subtree s of t by applying the Child Rule finitely many times, we derive that $\boxtimes t$ is a logical consequence of $\boxtimes s$.

2.15 Corollary. *Given sets S and T of k-colored Σ-trees, then $S \vDash T$ iff T contains every tree t such that*

$$\textit{every recoloring of every subtree of } t \textit{ lies in } S. \tag{1}$$

In fact, recall that $S \subseteq C_\Sigma(k)$ is logically equivalent to the conjunction of $\boxtimes s_i$, $i \in I$, where $\{s_i; i \in I\}$ is the complement of S; analogously with T. Now it is easy to see that $S \vDash T$ holds iff for every tree t we have

$$T \vDash \boxtimes t \qquad \text{implies} \qquad S \vDash \boxtimes t . \tag{2}$$

From Theorem 2.13 we know $T \vDash \boxtimes t$ iff there exists a tree $s \sqsubseteq t$ with $s \in C_\Sigma(k) - T$; analogously for $S \vDash \boxtimes t$. Thus, (2) tells us that

$$(\exists s)(s \sqsubseteq t \wedge s \notin T) \implies (\exists s)(s \sqsubseteq t \wedge s \notin S) .$$

Or, equivalently,

$$(\forall s)(s \not\sqsubseteq t \vee s \in S) \implies (\forall s)(s \not\sqsubseteq t \vee s \in T) .$$

The premise of the last implication is only true for trees t with the above property (1) – and for such trees t the conslusion is that $t \in T$: in fact, since $t \sqsubseteq t$ we conclude $t \in T$ (and, moreover, $s \in T$ whenever $s \sqsubseteq t$ – but this can be left out due to the universal quantification of t).

3 Coequational Logic for Accessible Functors

3.1 Assumption. Throughout this section H denotes an accessible endofunctor of **Set**, i.e., one that preserves k-filtered colimits for some infinite regular cardinal k. As shown in [5], this is equivalent to the statement that H is a quotient of a k-ary polynomial endofunctor. That is, a k-ary signature Σ and a natural transformation

$$\varepsilon\colon H_\Sigma \longrightarrow H$$

with surjective components exist.

3.2 Example. (i) The *finite-power-set functor* $\mathscr{P}_f$, given on objects by $X \longmapsto \{A \subseteq X; A \text{ finite}\}$ is finitary ($k = \omega$). Let Σ be the signature with a single n-ary symbol σ_n for every natural number n. Then we have a "canonical" presentation $\mathscr{P}_f$ as a quotient of H_Σ: ε takes every n-tuple $\sigma_n(x_0, \ldots, x_{n-1})$ to the subset $\{x_0, \ldots, x_{n-1}\}$.

$\mathscr{P}_f$-coalgebras are precisely the finitely branching graphs.

(ii) For the *countable-power-set functor* $\mathscr{P}_c$, given on objects by $X \longmapsto \{A \subseteq X; A \text{ countable}\}$ we can use the signature with one nullary symbol and one ω-ary symbol. Here

$$\varepsilon_X\colon 1 + X^{\mathbb{N}} \longrightarrow \mathscr{P}_c X$$

takes the left-hand summand to $\emptyset$ and every sequence $f\colon \mathbb{N} \longrightarrow X$ to the image of f. $\mathscr{P}_c$-coalgebras are precisely the countably branching graphs.

3.3 Remark. The presentation of H as a quotient of H_Σ yields a presentation of the cofree coalgebra $\tau_k\colon C(k) \longrightarrow HC(k)$ of H as a quotient of the cofree coalgebra of H_Σ. In fact, the unique homomorphism $\hat{\varepsilon}$ of H-coalgebras with $\gamma_k\hat{\varepsilon} = \gamma_k^\Sigma$ is surjective:

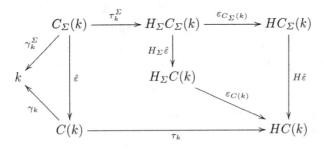

(Proof: choose $u\colon HC(k) \longrightarrow H_\Sigma C(k)$ with $\varepsilon_{C(k)}u = \text{id}$. The unique homomorphism $u^\#\colon C(k) \longrightarrow C_\Sigma(k)$ of H_Σ-coalgebras with $\gamma_k = \gamma_k^\Sigma u^\#$, where we use the structure map $u\tau_k$, splits $\hat{\varepsilon}$. In fact, $\hat{\varepsilon}u^\#$ is an endomorphism of $C(k)$ commuting with γ_k, thus, $\hat{\varepsilon}u^\# = \text{id}$.)

Consequently, the elements of $C(k)$ are the congruence classes $[t]$ of Σ-trees t modulo the kernel congruence of $\hat{\varepsilon}$. In case H is finitary, this congruence has a concise description: one applies ε-equations fintely or infinitely many times, where ε-equations are equations between terms in $H_\Sigma X$ merged by ε_X. See [2].

3.4 Example. For the functor $\mathscr{P}_f$ the terminal coalgebra $T = C(1)$ has been described by M. Barr [8] by presenting the corresponding congruence on Σ-trees. J. Worell presents a direct description in [15]: T is the coalgebra of all finitely branching, strongly extensional trees. Here a (rooted, nonordered) tree is called *strongly extensional* provided that for every pair of distinct children of any node the two subtrees are not bisimilar. The coalgebra structure $T \longrightarrow \mathscr{P}_f T$ takes every tree to the set of its children. Analogously, for a set k of colors we have a coalgebra $C(k)$ of all finitely branching k-colored nonordered trees which are strongly extensional. The latter means that for every pair of distinct children of any node the two k-colored subtrees are not bisimilar (as colored trees).

Examples of coequations:

(i) The class of all finitely branching graphs without leaves (i.e., nodes having no neighbour) is presented by $\boxtimes t$ where t is the single-node tree.

(ii) The coequation $\boxtimes t$ where t is a one-colored infinite path presents all graphs such that from every node a path leads into a leaf.

(iii) The coequation $\boxtimes t$ where t is an infinite path colored one-toone with $\mathbb{N}$ presents all graphs such that every node b from which no leaf is reachable has two paths of unequal lengths from b to a common target.

3.5 Theorem. *A coequation $\boxtimes t$ is a logical consequence of coequations $\boxtimes s_i$ ($i \in I$) iff for every Σ-tree t' with $t = [t']$ there exists a Σ-tree $r \sqsubseteq t'$ with $s_j = [r]$ for some $j \in I$.*

Proof. (1) We prove a preliminary result first. Let $A \xrightarrow{\alpha} HA$ be an H-coalgebra. Choose $u \colon HA \longrightarrow H_\Sigma A$ with

$$\varepsilon_A u = \mathrm{id}$$

and consider the corresponding H_Σ-coalgebra $u\alpha \colon A \longrightarrow H_\Sigma A$. For every coloring $f \colon A \longrightarrow k$ we have the unique homomorphisms $f^\# \colon A \longrightarrow C(k)$ and $f_\Sigma^\# \colon A \longrightarrow C_\Sigma(k)$, and they are related by

$$f^\# = \hat{\varepsilon} f_\Sigma^\#.$$

In fact, this follows from the universal property of γ_k since the right-hand side fulfils

$$\gamma_k(\hat{\varepsilon} f_\Sigma^\#) = \gamma_k^\Sigma f_\Sigma^\# = f$$

and is a homomorphism of H-coalgebras: using Remark 3.3 we see that the diagram

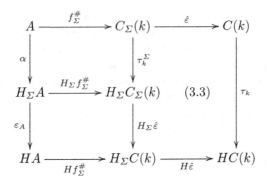

commutes.

(2) Sufficiency: suppose that for every tree t' with $t = [t']$ the above condition holds. Let A be an H-coalgebra satisfying every $\boxtimes s_i$ ($i \in I$). If A does not satisfy $\boxtimes t$, we derive a contradiction as follows. Let

$$t = f^{\#}(a) \qquad \text{for some } f\colon A \longrightarrow k \text{ and } a \in A.$$

The tree $t' = f_\Sigma^{\#}(a)$ fulfils $t = [t']$, since $f^{\#} = \hat{\varepsilon} f_\Sigma^{\#}$, therefore there exists $r \sqsubseteq t'$ with $s_j = [r]$ for some $j \in I$. Since A as a Σ-coalgebra does not satisfy $\boxtimes t'$, it follows from Theorem 2.13 that it does not satisfy $\boxtimes r$. Thus, there exists a coloring g and a state $b \in A$ with $g_\Sigma^{\#}(b) = r$. Then $g^{\#}(b) = \hat{\varepsilon} g_\Sigma^{\#}(b) = [r] = s_j$, a contradiction.

(3) Necessity. Suppose $\{\boxtimes s_i; i \in I\} \models \boxtimes t$. For every tree t' with $t = [t']$ form the H_Σ-coalgebra

$$A = A_{t'}$$

of Lemma 2.12. The corresponding H-coalgebra

$$A \xrightarrow{\;\alpha\;} H_\Sigma A \xrightarrow{\;\varepsilon_A\;} HA$$

fulfils $f^{\#} = \hat{\varepsilon} f_\Sigma^{\#}$ for every coloring $f\colon A \longrightarrow k$. In fact, this follows easily from the universal property of γ_k: by Remark 3.3 $\hat{\varepsilon} f_\Sigma^{\#}$ is a homomorphism of H-algebras with $f = \gamma_k \cdot \hat{\varepsilon} f_\Sigma^{\#}$. Since A does not satisfy $\boxtimes t'$, see Lemma 2.12, we conclude that the H-coalgebra A does not satisfy $\boxtimes t$. Therefore, there exists $j \in I$ and a coloring $g\colon A \longrightarrow k$ with

$$g^{\#}(a) = s_j \qquad \text{for some } a \in A.$$

The tree $r = g_\Sigma^{\#}(a)$ fulfils $s_j = [r]$, and by Lemma 2.12 we have $r \sqsubseteq t$. $\qquad \square$

3.6 Remark. We do not know how to formulate a corollary analogous to 2.15 here: the trouble is that in Theorem 3.5 different representatives t' can lead to different choices of $j \in I$.

4   Arbitrary Set Functors

4.1 Cofree-Coalgebra Chain. For an arbitrary endofunctor H of **Set** we cannot work with $C(k)$ because cofree coalgebras need not exist. Instead, we work with a transfinite chain $W(k)\colon \mathbf{Ord^{op}} \longrightarrow \mathbf{Set}$ "approximating" $C(k)$, dual to the free-algebra chain introduced in [1]. It is the essentially unique chain such that for its objects W_p (p an ordinal) and connecting morphisms $w_{pq}\colon W_p \longrightarrow W_q$ ($p \geq q$) the following transfinite induction holds:

$$W_0 = 1,$$
$$W_{p+1} = HW_p \times k \quad \text{and} \quad w_{p+1,q+1} = Hw_{p,q} \times id_k;$$

and for every limit ordinal q

$$W_q = \lim_{p<q} W_p \quad \text{with the limit cone } w_{qp}\colon W_q \longrightarrow W_p \ (p < q).$$

We call $W(k)$ the *cofree-coalgebra chain* of H. Given a collection $t_p \in W_p$ ($p \in$ **Ord**) of elements, we call it *compatible* if $w_{pq}(t_p) = t_q$ for all ordinals $p \geq q$.

4.2 Notation. Given a coalgebra $HA \overset{\alpha}{\longrightarrow} A$ and a coloring $f\colon A \longrightarrow k$ define the cone $(f_p^{\#})_{p\in\mathbf{Ord}}$ of the chain $W(k)$ ("approximating" the homomorphism $f^{\#}$) to be the unique cone $f_p^{\#}\colon A \longrightarrow W_p$, $p \in$ **Ord**, for which

$$f_{p+1}^{\#} = \langle Hf_p^{\#} \cdot \alpha, f \rangle\colon A \longrightarrow W_p \times k \qquad \text{(for all } p \in \mathbf{Ord}).$$

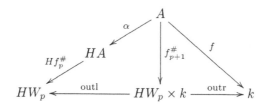

4.3 Definition. *A* ***generalized coequation*** *in k colors is an expression $\boxtimes t$ where*

$$t = (t_i)_{i\in\mathbf{Ord}} \qquad \text{with } t_i \in W_i$$

is a compatible collection of elements of the cofree-coalgbra chain.

    A coalgebra A is said to ***satisfy*** *the generalized coequation $\boxtimes t$ if for every coloring $f\colon A \longrightarrow k$ and every state a the compatible collections $f_i^{\#}(a)$ is non-equal to t. That is: for every state a there exists $p \in$ **Ord** with $f_p^{\#}(a) \neq t_p$.*

4.4 Theorem (see [3]). *For every endofunctor H of **Set** a collection of coalgebras can be presented by generalized coequations iff it is a covariety.*

4.5 Example. Let us consider the power-set functor $\mathscr{P}$. Its coalgebras are graphs. We have a compatible collection t with $t_{p+1} = \emptyset$ ($\in \mathscr{P}W_p$) for all ordinals p. The generalized coequation $\boxtimes t$ describes all graphs without leaves. Another compatible collection is s with $s_{p+1} = W_p$ for all p. Here, obviously, every graph A satisfies $\boxtimes s$: the cardinalities of W_p strictly grow with growing p, thus, there exists p with $\mathbf{card}\, W_p > \mathbf{card}\, A$. It follows that $f^{\#}_{p+1}(a) \neq s_{p+1}$ for all $a \in A$.

4.6 Remark. We follow [4] and assume that we work in ZFC (Zermelo-Fraenkel set theory with Axiom of Choice); we denote by **Class** the category of classes and functions. We proved in [4] that

(i) every endofunctor H of **Set** has an extension $\widehat{H}\colon$ **Class** $\longrightarrow$ **Class** unique up to natural isomorphism,

and

(ii) every endofunctor of **Class** has cofree coalgebras.

Thus, given a set functor H and a (small) set k of colors we have the cofree coalgebra $C(k)$ of $\widehat{H}$ in **Class**. The universal coloring $\gamma_k\colon C(k) \longrightarrow k$ yields a cone[1]

$$g_p = \left(\gamma_k\right)^{\#}_p\colon C(k) \longrightarrow W_p \qquad (p \in \mathbf{Ord})$$

of the cofree-coalgebra chain $W(k)$ of H.

4.7 Example. Let Σ be a *large signature*, i.e., a class $\Sigma = (\Sigma_n)_{n \in \mathbf{Card}}$ of operation symbols each having a prescribed arity n, which is a small cardinal. We obtain the *polynomial endofunctor* $H_\Sigma X = \coprod_{\sigma \in \Sigma} X^n$ of **Class**. For every small cardinal k the cofree coalgebra $C_\Sigma(k)$ is, precisely as in Example 2.4, the coalgebra of all k-colored Σ-trees. Observe that each such tree is an object within **Set**. But the collection $C_\Sigma(k)$ of all of them is a proper class.

4.8 Observation. For every endofunctor H of **Set** the functor $\widehat{H}$ is a quotient of a polynomial functor, $\varepsilon\colon H_\Sigma \longrightarrow \widehat{H}$. A cofree $\widehat{H}$-coalgebra $C(k)$ is a quotient of the Σ-tree coalgebra $C_\Sigma(k)$ modulo $\ker \hat{\varepsilon}$ for $\hat{\varepsilon}$ defined in Remark 3.3.

In fact, define Σ by $\Sigma_n = H(n)$ for all small cardinals n, then the Yoneda Lemma yields a natural transformation $\varepsilon\colon H_\Sigma \longrightarrow \widehat{H}$ with surjective components. And then apply Remark 3.3.

4.9 Notation. Given a presentation $\varepsilon\colon H_\Sigma \longrightarrow H$ as above, for every k-colored Σ-tree $s \in C_\Sigma(k)$ we denote by $[s] \in W(k)$ the compatible collection whose p-th component is the image of s under $C_\Sigma(k) \xrightarrow{\hat{\varepsilon}} C(k) \xrightarrow{g_p} W_p(k)$, see 4.6:

$$[s]_p = g_p\big(\hat{\varepsilon}(s)\big) \qquad (p \in \mathbf{Ord})$$

4.10 Theorem. *A generalized coequation $\boxtimes t$ is a logical consequence of generalized coequations $\boxtimes s_i$ $(i \in I)$ iff for every Σ-tree t' with $t = [t']$ there exists a Σ-tree $r \sqsubseteq t'$ with $s_j = [r]$ for some $j \in I$.*

---

[1] **Card** and **Ord** denote the classes of all small cardinals and small ordinals, respectively.

The proof of this theorem is completely analogous to that of Theorem 3.5 except that we do not claim that t has the form $[t']$ for some Σ-tree t'. However, if t does not have that form, then $\boxtimes t$ is *trivial*, i.e., satisfied by every coalgebra – and there is nothing to prove then. In fact, whenever $\boxtimes t$ is not trivial, we choose a coalgebra A and a coloring $f\colon A \longrightarrow k$ with $f^{\#}(a) = t$ for some $a \in A$. Then arguing as in 3.5 we get $f^{\#} = \hat{e} \cdot f_{\Sigma}^{\#}\colon A \longrightarrow C(k)$ in **Class**. Moreover, $f_p^{\#} = g_p \cdot f^{\#}$ (easy induction on p) which implies that

$$t_p = g_p \cdot f^{\#}(a) = [t']_p \qquad (p \in \mathbf{Ord})$$

for the Σ-tree $t' = f_{\Sigma}^{\#}(a)$. Consequently, $t = [t']$.

5 How Colorful Are Covarieties?

Throughout this section H denotes a k-accessible endofunctor of **Set**. All examples of covarieties above used one or two colors for the coequational presentation. However, there are simple covarieties requiring infinitely many colors:

5.1 Example. A finitary covariety which does not fulfil any coequation $\boxtimes t$ such that t lies in $C(n)$ for n finite. We consider again the functor $HQ = Q \times Q + 1$ from 2.1. A state q in a coalgebra is called *1-based* provided that the repeated input 1 leads from q $(= q_0)$ to non-deadlock states $q_1, q_2, q_3, \ldots$, but the input 0 leads from any q_k to a deadlock $\bar{q}_k$ for $k = 0, 1, 2, \ldots$. Let $\mathscr{A}$ be the covariety of all coalgebras in which for every 1-based state q there exist $k \neq \ell$ with $\bar{q}_k = \bar{q}_\ell$. It is easy to see that $\mathscr{A}$ is presented by the single coequation

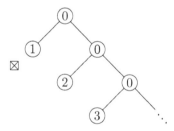

(In fact, if A satisfies the above coequation $\boxtimes s$ and q is a 1-based state in A, then the deadlock states $\bar{q}_k$ are not pairwise distinct – otherwise by coloring $\bar{q}_k$ with $k+1$ and all other states by 0 we obtain $f\colon A \longrightarrow \omega$ with $f^{\#}(q) = s$, a contradiction. Thus, $A \in \mathscr{A}$. The converse is obvious.)

In this covariety $\mathscr{A}$ for every finite-colored tree t there exists a coalgebra $B_t \in \mathscr{A}$ not satisfying $\boxtimes t$. In fact, suppose first that t has a 1-based node, q_0. Since t is finitely colored, some of the colors i has the property that infinitely many of the deadlock states $\bar{q}_k$ have color i; let $\sim$ be the equivalence on the nodes of t whose one class is formed by all the i-colored leaves of t, and all other classes are singleton sets. Then we obtain a coalgebra B_t whose states are the equivalence classes of $\sim$ with the obvious coalgebra structure and an obvious coloring $f: B_t \longrightarrow \omega$ obtained from the coloring of t. It is easy to see that $f^{\#}(q_0) = t$ and $B_t \in \mathscr{A}$: every 1-based state of B_t has the form q_k and we have two different deadlock states of t of depth bigger than k. Conversely, if no node of t is 1-based, then the coalgebra B_t obtained from the nodes of t (ignoring the coloring) lies in $\mathscr{A}$, and the coloring of t yields $f: B_t \longrightarrow \omega$ with $f^{\#}(\varepsilon) = t$.

Consequently, the covariety $\mathscr{A}$ cannot be presented by finite-colored coequations.

5.2 Remark. One-colored covarieties have a beautiful characterization: they are precisely the covarieties closed under bisimilarity, as proved by Peter Gumm and Tobias Schröder in [12]. Recall that two coalgebras A and A' are called *bisimilar* if there exists a bisimulation between them such that every state of A is bisimilar to a state of A', and vice versa. In particular, given an epimorphic homomorphism $e: A \longrightarrow A'$, then A and A' are bisimilar.

For n-colored covarieties, where $n \leq k$ is any cardinal, the appropriate concept is the following:

5.3 Definition. (1) *An H-coalgebra $A \overset{\alpha}{\longrightarrow} HA$ equipped with a coloring $f: A \longrightarrow n$ is considered as a coalgebra of $H(-) \times n$ via $\langle \alpha, f \rangle: A \longrightarrow HA \times n$. Given another H-coalgebra A' with a coloring in n, we call A and A' n-color bisimilar provided that they are bisimilar as $H(-) \times n$-coalgebras.*

(2) *A covariety $\mathscr{A}$ is said to be* closed under n-color bisimilarity *provided that it contains every H-coalgebra A with the following property: for every coloring $f: A \longrightarrow n$ there exists a coalgebra $A' \in \mathscr{A}$ and a coloring $g: A' \longrightarrow n$ such that A and A' are n-color bisimilar.*

5.4 Proposition. *A covariety can be presented by n-color coequations, i. e., by $\boxtimes t$ for $t \in C(n)$, iff it is closed under n-color bisimilarity.*

Proof. Without loss of generality we can assume that H preserves monomorphisms: if we change the value of H at $\emptyset$ to be $\emptyset$, the new functor preserves monomorphisms and has the category of coalgebras isomorphic to **Coalg** H.

(1) Let $\mathscr{A}$ be a covariety closed under n-color bisimilarity. Denote by $M \subseteq C(n)$ the union of all images of homomorphisms $f^{\#}: A \longrightarrow C(n)$, where A is a coalgebra in $\mathscr{A}$ and f is a coloring of A. We prove that $\mathscr{A}$ is presented by the subobject $m: M \longrightarrow C(n)$. Since every coalgebra in $\mathscr{A}$ clearly satisfies M, we only have to verify the converse. We first observe that M is a coalgebra in $\mathscr{A}$. More precisely, let us choose, for every element x of M, a homomorphism $f_x^{\#}: A_x \longrightarrow C(n)$ with $A_x \in \mathscr{A}$ and x lying in $f_x^{\#}[A_x]$, then the induced homomorphism $h: \coprod_{x \in M} A_x \longrightarrow C(n)$ has image M, i.e., $h = m \cdot k$ for some epimorphism k. Here $B = \coprod A_x$ carries the coalgebra structure $\beta: B \longrightarrow HB$ of a coproduct of coalgebras. Since H preserves monomorphisms, the image of a homomorphism is a subcoalgebra of the codomain: there is a unique $\mu: M \longrightarrow HM$ making k and m coalgebra homomorphisms. And $M \in \mathscr{A}$ because it is a quotient of $\coprod_{x \in X} A_x \in \mathscr{A}$.

Let A be a coalgebra satisfying M, then for every coloring $f: A \longrightarrow n$ we are to find a coalgebra $A' \in \mathscr{A}$ and a coloring $g: A' \longrightarrow n$ such that A and A'

are n-color bisimilar (thus proving that A lies in $\mathscr{A}$). We factorize the homomorphism $f^{\#}$ into an epimorphism $e\colon A \longrightarrow A'$ followed by a monomorphism $i\colon A' \longrightarrow C(n)$, then A' carries the unique structure $\alpha'\colon A' \longrightarrow HA$ of a coalgebra such that e and i are homomorphisms:

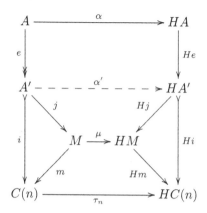

Since A satisfies M, the subobject i is contained in m, say, $i = m \cdot j$, and j is a homomorphism because i is one and Hm is a monomorphism. Therefore A' is a subcoalgebra of M, which proves $A' \in \mathscr{A}$. The coloring $g = \gamma_n \cdot i\colon A' \longrightarrow n$ defines an n-color coalgebra which is n-color bisimilar to (A, f). In fact, the epimorphic homomorphism $e\colon A \longrightarrow A'$ fulfils $f = g \cdot e$, since $i \cdot e = f^{\#}$. Therefore, e is a homomorphism of the corresponding $H(-) \times n$-coalgebras. Being epimorphic, it defines the the required bisimulation. This proves $A \in \mathscr{A}$.

(2) Let $\mathscr{A}$ be a covariety presented by n-color coequations $\boxtimes t_i$, $i \in I$. Given a coalgebra A with the property of the above definition, we prove $A \in \mathscr{A}$. In fact, given a coloring $f\colon A \longrightarrow n$, there is $A' \in \mathscr{A}$ and a coloring $g\colon A' \longrightarrow n$ such that (A, f) and (A', g) are bisimilar $H(-) \times n$-coalgebras. Let $R \subseteq A \times A'$ be a bisimulation such that every $a \in A$ is related to some $a' \in A'$. The corresponding homomorphism $h\colon R \longrightarrow C(n)$ has the property that $f^{\#} = h \cdot \pi$ and $g^{\#} = h \cdot \pi'$ for the projections π, π'. Therefore, given $a \in A$, we have $f^{\#}(a) = g^{\#}(a')$. And $g^{\#}(a) \neq t_i$, since A' satisfies $\boxtimes t_i$. This proves that A satisfies $\boxtimes t_i$. Consequently, $A \in \mathscr{A}$. $\qquad\square$

Acknowledgement

The author wishes to express his gratitude to the referees: their comments improved the presentation of the paper considerably.

References

1. Adámek, J.: Free algebras and automata realizations in the language of categories. Commentationes Mathematicae Universitatis Carolinae **15** (1974) 589–602

2. Adámek, J.: On a description of terminal coalgebras and iterative theories. Electronic Notes in Theoretical Computer Science **82.1** (2003), full version in Information and Computation (to appear)
3. Adámek, J.: Birkhoff's covariety theorem without limitation. Commentationes Mathematicae Universitatis Carolinae **46** (2005) 197–215
4. Adámek, J., Milius, S., Velebil, J.: On coalgebra based on classes. Theoretical Computer Science **316** (2004) 3–23
5. Adámek, J., Porst, H.-E.: On tree coalgebras and coalgebra presentations. Theoretical Computer Science **311** (2004) 257–283
6. Arbib, M. A., Manes, E. G.: Parametrized data types do not need highly constrained parameters. Information and Control **52** (1982) 130–158
7. Awodey, S., Hughes, J.: Modal operators and the formal dual of Birkoff's completeness theorem. Mathematical Structures in Computer Science **13** (2003) 233–258
8. Barr, M.: Terminal coalgebras in well-founded set theory. Theoretical Computer Science **124** (1984) 182–192
9. Birkhoff, G.: On the structure of abstract algebras. Proceedings of the Cambridge Philosophical Society **31** (1935) 433–454
10. Gumm, H. P.: Elements of the general theory of coalgebras, preprint 1999
11. Gumm, H. P.: Birkoff's variety theorem for coalgebras. Contributions to General Algebra **13** (2000) 159–173
12. Gumm, H. P., Schröder, T.: Covarieties and complete covarieties. Electronic Notes in Theoretical Computer Science **11** (1998)
13. Jacobs, B.: The temporal logic of coalgebras via Galois algebras. Mathematical Structures in Computer Science **12** (2002) 875–903
14. Rutten, J.: Universal coalgebra: a theory of systems. Theoretical Computer Science **249** (2000) 3–80
15. Worrell J.: On the final sequence of a finitary set functor, Theoretical Computer Science **338** (2005) 184–199

A Semantic Formulation of ⊤⊤-Lifting
and Logical Predicates for Computational Metalanguage

Shin-ya Katsumata

Research Institute for Mathematical Sciences, Kyoto University
sinya@kurims.kyoto-u.ac.jp

Abstract. A semantic formulation of Lindley and Stark's ⊤⊤-lifting is given. We first illustrate our semantic formulation of the ⊤⊤-lifting in **Set** with several examples, and apply it to the logical predicates for Moggi's computational metalanguage. We then abstract the semantic ⊤⊤-lifting as the lifting of strong monads across bifibrations with lifted symmetric monoidal closed structures.

1 Introduction

Logical predicates are a method for extracting submodels of the pure simply typed lambda calculus ($\lambda^{\Rightarrow}$ for short) by induction on type. Logical predicates are widely applied to the reasoning of the properties of $\lambda^{\Rightarrow}$ [9, 16, 23, 24].

We are interested in extending logical predicates to *Moggi's computational metalanguage* (λ_{ml} for short) [18], which has additional types $T\tau$ called *monadic type*. To do so, we need to consider a scheme to calculate a predicate at type $T\tau$ from a predicate at type τ. Recently, Lindley and Stark develop the *leapfrog method* and show the strong normalisation of λ_{ml} in the style of Tait-Girard reducibility [11, 12]. The novelty of the leapfrog method is the operation called ⊤⊤-*lifting*, which calculates a reducibility predicate at type $T\tau$ from a reducibility predicate at type τ.

However, Lindley and Stark's ⊤⊤-lifting is defined with respect to the syntactic structure of λ_{ml}, and is designed for the proof of the strong normalisation. This paper attempts to provide a semantic aspect of their ⊤⊤-lifting. The main contribution of this paper is twofolds:

1. We provide a semantic formulation of Lindley and Stark's ⊤⊤-lifting in set theory (section 3). This formulation is carried out by finding a semantic counterpart for each of the building block in ⊤⊤-lifting. We instanciate ⊤⊤-liftings with well-known strong monads over **Set**, and show that the logical predicates using the semantic ⊤⊤-lifting implies the *basic lemma* of logical predicates.
2. We re-formulate the above semantic ⊤⊤-lifting as a construction of *liftings* of strong monads, and give a categorical account of this construction within the framework of fibred category theory (section 4). We then show that the above semantic ⊤⊤-lifting and Abadi's ⊤⊤-closure operation are instances of ⊤⊤-lifting.

2 Preliminaries

Moggi's Computational Metalanguage
We begin with the syntax of λ_{ml}. We define the set of types $\mathbf{Typ}_{ml}$ by the following BNF (we consider a single base type b for simplicity):

L. Ong (Ed.): CSL 2005, LNCS 3634, pp. 87–102, 2005.

$$\mathbf{Typ}_{ml} \ni \tau ::= b \mid \tau \Rightarrow \tau \mid T\tau.$$

Monadic types $T\tau$ are for the programs yielding values of type τ with some computational effect. A *typing context* (ranged over by Γ) is simply a finite sequence of variable-type pairs without any duplication of variables.

The calculus λ_{ml} has two new term constructions related to monadic types: $[-]$ and "let x^τ be M in N". Their typing rules are the following:

$$\frac{\Gamma \vdash M : \tau}{\Gamma \vdash [M] : T\tau} \qquad \frac{\Gamma \vdash M : T\tau \quad \Gamma, x : \tau \vdash N : T\tau'}{\Gamma \vdash \text{let } x^\tau \text{ be } M \text{ in } N : T\tau'}$$

The term $[M]$ expresses the value of M involving the trivial computational effect. The term "let x^τ be M in N" expresses a sequential computation of M and N; the term M is first computed, its value is then bound to x^τ and next the term N is computed.

Equational theory of λ_{ml} extends $\beta\eta$ axioms of $\lambda^\Rightarrow$ with the following axioms:

$$\text{let } x^\tau \text{ be } [M] \text{ in } N = N[M/x] \quad (T.\beta)$$
$$\text{let } x^\tau \text{ be } M \text{ in } [x^\tau] = M \quad (T.\eta)$$
$$\text{let } x^\tau \text{ be } (\text{let } y^{\tau'} \text{ be } L \text{ in } M) \text{ in } N = \text{let } y^{\tau'} \text{ be } L \text{ in let } x^\tau \text{ be } M \text{ in } N \quad (T.assoc)$$

Categorical Semantics of λ_{ml}

A categorical semantics of λ_{ml} is given in a Cartesian closed category $\mathbb{C}$ equipped with a strong monad $\mathcal{T} = (T, \eta, \mu, \theta)$. We omit the formal definition of strong monads; see e.g. [18]. For a morphism $f : A \to TB$ in $\mathbb{C}$, we write $f^\#$ for the morphism $\mu_B \circ Tf : TA \to TB$.

Let B be an object in $\mathbb{C}$. We first assign to each type τ an object $[\![\tau]\!]$ in $\mathbb{C}$ by induction on type:

$$[\![b]\!] = B, \quad [\![\tau \Rightarrow \tau']\!] = [\![\tau]\!] \Rightarrow [\![\tau']\!], \quad [\![T\tau]\!] = T[\![\tau]\!].$$

We extend this assignment to typing contexts by

$$[\![x_1 : \tau_1, \cdots, x_n : \tau_n]\!] = [\![\tau_1]\!] \times \cdots \times [\![\tau_n]\!].$$

The semantics of λ_{ml} in $\mathbb{C}$ is an extension of the standard categorical semantics of $\lambda^\Rightarrow$ with the following rules:

- For a well-formed term $\Gamma \vdash [M] : T\tau$, we define

$$[\![[M]]\!] = \eta_{[\![\tau]\!]} \circ [\![M]\!].$$

- For a well-formed term $\Gamma \vdash \text{let } x^\tau \text{ be } M \text{ in } N : T\tau'$, we define

$$[\![\text{let } x^\tau \text{ be } M \text{ in } N]\!] = [\![N]\!]^\# \circ \theta_{[\![\Gamma]\!],[\![\tau]\!]} \circ \langle \text{id}_{[\![\Gamma]\!]}, [\![M]\!] \rangle$$

3   A Semantic Formulation of ⊤⊤-Lifting

In [12], Lindley and Stark prove the strong normalisation of λ_{ml} by extending the reducibility predicate technique. The novelty of their method is the operation called ⊤⊤-lifting, which calculates a reducibility predicate at a monadic type from that at an ordinary type.

Definition 3.1 ([12], section 3.1).

1. *We define the set of* raw continuations *by the following BNF:*

$$K ::= \mathrm{Id} \mid K \circ (x^\tau.N)$$

   where the notation $(x^\tau.N)$ indicates that N is a term with a distinguished free variable x^τ.
   A judgement for a raw continuation is a triple $T\tau \vdash_C K : T\tau'$. Raw continuations are typed by the following rules:

$$\frac{}{T\tau \vdash_C \mathrm{Id} : T\tau} \qquad \frac{x : \tau \vdash N : T\tau' \quad T\tau' \vdash_C K : T\tau''}{T\tau \vdash_C K \circ (x^\tau.N) : T\tau''}$$

   We write $T\tau \vdash_C K$ to mean that there exists a (unique) type $T\tau'$ such that $T\tau \vdash_C K : T\tau'$ is derived from the above rules.
2. *We define an* application $K@M$ *of a term M to a continuation K by*

$$\mathrm{Id}@M = M, \quad (K \circ (x^\tau.N))@M = K@(\mathsf{let}\ x^\tau\ \mathsf{be}\ M\ \mathsf{in}\ N).$$

3. *Given a set P of terms of type τ, we define a set $P^{\top\top}$ of terms of type $T\tau$ by*

$$P^\top = \{T\tau \vdash_C K \mid \forall M \in P\ .\ K@[M] \in SN\}$$
$$P^{\top\top} = \{M : T\tau \mid \forall K \in P^\top\ .\ K@M \in SN\}$$

   where SN is the set of strongly normalising terms.

From this point, we let $\mathcal{T} = (T, \eta, \mu, \theta)$ be a strong monad over **Set**, and fix a categorical semantics of λ_{ml} with respect to the strong monad $\mathcal{T}$ and the evident CCC structure in **Set**. We give a semantic formulation of the syntactic ⊤⊤-lifting by finding semantic counterparts of continuations, applications and the set SN. This formulation is carried out with respect to the strong monad $\mathcal{T}$. We introduce the following notation: for subsets $X \subseteq I$ and $Y \subseteq J$, by $X \Rightarrow Y$ we mean the subset $\{f \mid \forall x \in X\ .\ f(x) \in Y\}$ of $I \Rightarrow J$.

To simplify the situation, we assume that all continuations in definition 3.1 have the same type $T\rho$ (this restriction will be relaxed in section 5). We let $R = [\![\rho]\!]$.

Continuation   We formulate a continuation as a function

$$f \in [\![\tau]\!] \Rightarrow TR.$$

We explain the idea of this formulation below. We notice that a continuation $T\tau \vdash_C$ Id $\circ (x^\tau.M) : T\rho$ is equivalent to a context let x^τ be $-$ in M, and an application of a term to the continuation is equivalent to plugging the term in the hole of the context. The essential information of the context is the body M, and it has the following typing:

$$x : \tau \vdash M : T\rho.$$

Our formulation represents this information as a function $f \in [\![\tau]\!] \Rightarrow TR$.

Application We define an application of an element $x \in [\![T\tau]\!]$ to a continuation $f \in [\![\tau]\!] \Rightarrow TR$ to be $f^\# x$.

The Set SN The set SN is hard-coded in the definition of $P^\top$ and $P^{\top\top}$ since the syntactic $\top\top$-lifting is designed for the proof of the strong normalisation of λ_{ml}. We replace SN with some subset $S \subseteq TR$, and call S a *result predicate*.

We also relax the condition that the set R is given by $[\![\rho]\!]$ with some type ρ; we simply allow R to be any set and call R a *result type*.

Once continuations, applications and the set SN are semantically formulated, it is straightforward to define $P^\top$ and $P^{\top\top}$. We summarise the above discussion as follows:

Definition 3.2. *Let R be a set (called* result type*) and $S \subseteq TR$ be a subset (called* result predicate*).*

1. *A continuation is a function $f \in [\![\tau]\!] \Rightarrow TR$.*
2. *We define an application of $x \in [\![T\tau]\!]$ to a continuation $f \in [\![\tau]\!] \Rightarrow TR$ to be $f^\# x$.*
3. *Let $P \subseteq [\![\tau]\!]$ be a subset. We define a subset $P^{\top\top} \subseteq [\![T\tau]\!]$ by*

$$P^\top = \{f \in [\![\tau]\!] \Rightarrow TR \mid \forall x \in P . f(x) \in S\} = P \Rightarrow S$$
$$P^{\top\top} = \{x \in [\![T\tau]\!] \mid \forall f \in P^\top . f^\#(x) \in S\},$$

which is equivalent to

$$P^{\top\top} = \{x \in [\![T\tau]\!] \mid \forall f \in P \Rightarrow S . f^\#(x) \in S\}.$$

We call the operation $(-)^{\top\top}$ the $\top\top$-lifting of T with R and $S \subseteq TR$.

We can also consider the semantic $\top\top$-lifting for binary relations (*binary $\top\top$-lifting* for short) over the semantics of λ_{ml}. Let R be a set and $S \subseteq (TR)^2$ be a subset. A continuation is a pair (f, g) of functions from $[\![\tau]\!]$ to TR. An application of $(x, y) \in [\![T\tau]\!]^2$ to a continuation (f, g) is defined to be $(f^\# x, g^\# y)$. For a binary relation $P \subseteq [\![\tau]\!]^2$, we define $P^{\top\top}$ as follows:

$$P^\top = \{(f, g) \in ([\![\tau]\!] \Rightarrow TR)^2 \mid \forall(x, y) \in P . (fx, gy) \in S\}$$
$$P^{\top\top} = \{(x, y) \in [\![T\tau]\!]^2 \mid \forall(f, g) \in P^\top . (f^\# x, g^\# y) \in S\}.$$

Examples of Semantic $\top\top$-Liftings

An interesting point is that we can obtain $\top\top$-liftings for various strong monads and result type/predicate pairs. We see some concrete examples of the semantic $\top\top$-lifting below.

Example 3.3. We consider the *lifting monad* $T_\perp$, which simply adjoins an extra element $\perp$ to a given set. We calculate a $\top\top$-lifting of $T_\perp$ with the following data:

- The result type R is $\{*\}$ (thus $T_\perp R = \{*, \perp\}$).
- The result predicate S is $\{*\}$.

For a subset $P \subseteq [\![\tau]\!]$, we have $P^{\top\top} = P$.

Example 3.4. We consider the *state monad* T_s whose functor part is given by $T_s I = M \Rightarrow I \times M$ for some set M. We let $M_0 \subseteq M$ be a subset and calculate a $\top\top$-lifting of T_s with the following data:

- The result type R is some set.
- The result predicate S is $M_0 \Rightarrow R \times M_0$, the set of functions $f \in T_s R$ such that $\forall x \in M_0 \,.\, f(x) \in M_0 \times R$.

For a subset $P \subseteq [\![\tau]\!]$, we expand the definition of $P^{\top\top}$ and obtain

$$P^{\top\top} = \{f \in T_s[\![\tau]\!] \mid \forall g \in P \times M_0 \Rightarrow R \times M_0 \,.\, g \circ f \in M_0 \Rightarrow R \times M_0\}.$$

In fact, $P^{\top\top}$ can be characterised as follows:

$$P^{\top\top} = \begin{cases} M_0 \Rightarrow P \times M_0 & (\emptyset \subsetneq R \times M_0 \subsetneq R \times M) \\ T_s[\![\tau]\!] & \text{(otherwise)} \end{cases}$$

Below we prove the first case of this characterisation; the second case is trivial. We first prove

$$P \times M_0 = \{i \in [\![\tau]\!] \times M \mid \forall g \in P \times M_0 \Rightarrow R \times M_0 \,.\, g(i) \in R \times M_0\}.$$

($\subseteq$) Easy. ($\supseteq$) Let $x \notin P \times M_0$. From the assumption on $R \times M_0$, we can take two elements $s \in R \times M_0$ and $s' \in (R \times M)\backslash(R \times M_0)$. We then define the following function $g \in [\![\tau]\!] \times M \Rightarrow R \times M$:

$$g(x) = \begin{cases} s & (x \in P \times M_0) \\ s' & (x \notin P \times M_0) \end{cases}$$

which is clearly included in $P \times M_0 \Rightarrow R \times M_0$. However $g(x) \notin R \times M_0$, so we conclude that $x \notin (r.h.s.)$. Therefore

$$f \in M_0 \Rightarrow P \times M_0$$
$$\Longleftrightarrow \forall x \in M_0 \,.\, \forall g \in P \times M_0 \Rightarrow R \times M_0 \,.\, g(f(x)) \in R \times M_0$$
$$\Longleftrightarrow f \in P^{\top\top}.$$

Example 3.5. We calculate a binary $\top\top$-lifting of the lifting monad $T_\perp$ with the following data:

- The result type R is a one-point set $\{*\}$. We have $T_\perp R = \{\perp, *\}$.
- The result predicate $S \subseteq (T_\perp R)^2$ is $\{(x, y) \in (T_\perp R)^2 \mid (x = * \implies y = *)\}$.

For a subset $P \subseteq [\![\tau]\!]$, we obtain $P^{\top\top} = P \cup \{(\bot, \bot)\}$.

Example 3.6. We consider the *finite powerset monad* $\mathcal{T}_p$, whose functor part is given by $T_p(X) = \{x \subseteq X \mid x \text{ is a finite set}\}$. We calculate a binary $\top\top$-lifting wf $\mathcal{T}_p$ with the following data:

- The result type R is a one-point set $\{*\}$. We have $T_p R = \{\emptyset, R\}$.
- The result predicate $S \subseteq (T_p R)^2$ is $\{(x, y) \in (T_p R)^2 \mid x = R \implies y = R\}$.

We identify a function $f \in [\![\tau]\!] \Rightarrow T_p R$ and a subset (written with the capital letter of the function) $F = \{x \in [\![\tau]\!] \mid f(x) = R\} \subseteq [\![\tau]\!]$. Under this identification, for each $x \in T_p[\![\tau]\!]$, we have

$$f^\# x = R \iff \bigcup_{e \in x} fe = R \iff \exists e \in x . e \in F.$$

For a subset $P \subseteq [\![\tau]\!]$, we expand the definition of $P^{\top\top}$ and obtain

$$P^{\top\top} = \{(p, q) \in (T_p[\![\tau]\!])^2 \mid \forall F, G \subseteq [\![\tau]\!] . (\forall (x, y) \in P . x \in F \implies y \in G) \implies \forall e \in p . e \in F \implies \exists e' \in q . e' \in G\}.$$

This is not intuitive, but interestingly we have the following characterisation of $P^{\top\top}$:

$$P^{\top\top} = \{(p, q) \mid \forall a \in p . \exists b \in q . (a, b) \in P\}. \tag{1}$$

This appears in the pattern of defining *pre-bisimulation relation* in concurrency.

The rest of this example is the proof of equation 1. ($\subseteq$) Let $(p, q) \in P^{\top\top}$ and $a \in p$. We show $\exists b \in q . (a, b) \in P$. We supply $\{a\}$ and $\{b \mid (a, b) \in P\}$ to F and G in the definition of $(p, q) \in P^{\top\top}$. We obtain

$$(\forall (x, y) \in P . x = a \implies (a, y) \in P\})$$
$$\implies (\forall e \in p . e = a \implies \exists e' \in q . (a, e') \in P\})$$

whose premise part is trivially true. By letting e be a in the conclusion part of the above formula, we obtain $\exists e' \in q . (a, e') \in P$. ($\supseteq$) We take $p, q \in T_p[\![\tau]\!]$ such that $\forall a \in p . \exists b \in q . (a, b) \in P$. Let $F, G \subseteq [\![\tau]\!]$, $e \in p$ and assume $\forall (x, y) \in P . x \in F \implies y \in G$ (we call this assumption (*)) and $e \in F$. We show $\exists e' \in q . e' \in G$. Since $e \in p$, there exists $e' \in q$ such that $(e, e') \in P$. From (*), we have $e \in F \implies e' \in G$. Thus e' gives a witness of $\exists e' \in q . e' \in G$.

Logical Predicates for λ_{ml} Using $\top\top$-Lifting

The semantic $\top\top$-lifting constructs a subset of $[\![T\tau]\!]$ from a subset of $[\![\tau]\!]$. This construction is suitable for extending the concept of *logical predicates* to λ_{ml}. We show that a logical predicate using the semantic $\top\top$-lifting extract a submodel of λ_{ml}. We fix a result type R and a result predicate $S \subseteq TR$, and consider the $\top\top$-lifting determined by R and S.

Definition 3.7. *A* $\top\top$-*logical predicate is a type-indexed family* $\{P^\tau \subseteq [\![\tau]\!]\}_{\tau \in \mathbf{Typ}_{ml}}$ *of subsets satisfying*

$$P^{T\tau} = (P^\tau)^{\top\top}, \qquad P^{\tau \Rightarrow \tau'} = P^\tau \Rightarrow P^{\tau'}.$$

For a typing context $\Gamma = x_1 : \tau_1, \cdots, x_n : \tau_n$, by P^Γ we mean the product $P_1^\tau \times \cdots \times P_n^\tau$, which is a subset of $[\![\Gamma]\!]$.

Theorem 3.8 (Basic Lemma). *Let P be a* $\top\top$-*logical predicate. For any well-formed term $\Gamma \vdash M : \tau$, we have* $[\![M]\!] \in P^\Gamma \Rightarrow P^\tau$.

Proof. We show the following properties on the $\top\top$-lifting. Let $X \subseteq I$ and $Y \subseteq J$ be subsets.

1. $\eta_I \in X \Rightarrow X^{\top\top}$. Let $x \in X$. Then for any $f \in X \Rightarrow S$, we have $f^\#(\eta_I(x)) = f(x) \in S$. Therefore $\eta_I(x) \in X^{\top\top}$.
2. $\mu_I \in (X^{\top\top})^{\top\top} \Rightarrow X^{\top\top}$. Let $x \in (X^{\top\top})^{\top\top}$ and $f \in X \Rightarrow S$. We show $f^\#(\mu_I(x)) \in S$. It is easy to show that $f \in X \Rightarrow S$ implies $f^\# \in X^{\top\top} \Rightarrow S$, hence $(f^\#)^\# \in (X^{\top\top})^{\top\top} \Rightarrow S$. Notice that $f^\#(\mu_I(x)) = (f^\#)^\#(x)$. Therefore $f^\#(\mu_I(x)) \in S$.
3. $\theta_{I,J} \in X \times Y^{\top\top} \Rightarrow (X \times Y)^{\top\top}$. Let $a \in X, b \in Y^{\top\top}$ and $f \in X \times Y \Rightarrow S$. We show $f^\# \circ \theta_{I,J}(a,b) \in S$. We note that the strength $\theta_{I,J}$ is given by $\theta_{I,J}(a,b) = T(\lambda b \in B . (a,b))(b)$ as **Set** is a well-pointed category (see e.g. [18]). Thus $f^\# \circ \theta_{I,J}(a,b) = (\lambda b \in B . f(a,b))^\#(b)$. Since $\lambda b \in B . f(a,b) \in Y \Rightarrow S$, for each $b \in Y^{\top\top}$ we have $(\lambda b \in B . f(a,b))^\#(b) \in S$. Therefore $f^\# \circ \theta_{I,J}(a,b) \in S$
4. $f \in X \Rightarrow Y$ implies $Tf \in X^{\top\top} \Rightarrow Y^{\top\top}$. Let $x \in X^{\top\top}$ and $g \in Y \Rightarrow S$. We show $g^\#(Tf(x)) = (g \circ f)^\#(x) \in S$. This holds from $g \circ f \in X \Rightarrow S$ and the definition of $x \in X^{\top\top}$.
5. From 2 and 4, $f \in X \Rightarrow Y^{\top\top}$ implies $f^\# \in X^{\top\top} \Rightarrow Y^{\top\top}$.

We prove the theorem by induction on derivation of a well-formed term $\Gamma \vdash M : \tau$. We omit the cases for the syntax constructions inherited from $\lambda^\Rightarrow$; see e.g. [2]. The cases new to λ_{ml} is the following.

- Case $\Gamma \vdash [M] : T\tau$. From IH, we have $[\![M]\!] : P^\Gamma \Rightarrow P^\tau$. From 1, we have $[\![[M]]\!] = \eta_{[\![\tau]\!]} \circ [\![M]\!] : P^\Gamma \Rightarrow P^{T\tau}$.
- Case $\Gamma \vdash$ let x^τ be M in $N : T\tau'$ with well-formed terms $\Gamma \vdash M : T\tau$ and $\Gamma, x : \tau \vdash N : T\tau'$. From IH, $[\![M]\!] : P^\Gamma \Rightarrow P^{T\tau}$ and $[\![N]\!] : P^\Gamma \times P^\tau \Rightarrow P^{T\tau'}$. From 3 and 5, we have $[\![N]\!]^\# \circ \theta_{[\![\Gamma]\!],[\![\tau]\!]} : P^\Gamma \times P^{T\tau} \Rightarrow P^{T\tau'}$. Therefore $[\![$let x^τ be M in $N]\!] = [\![N]\!]^\# \circ \theta_{[\![\Gamma]\!],[\![\tau]\!]} \circ \langle \mathrm{id}_{[\![\Gamma]\!]}, [\![M]\!] \rangle : P^\Gamma \Rightarrow P^{T\tau'}$.

$\square$

4   A Categorical Generalisation of $\top\top$-Lifting

In the proof of theorem 3.8, we notice that the operation $(-)^{\top\top}$ resembles an endofunctor (claim 4) equipped with morphisms constituting a strong monad (claim 1,2,3). It is indeed a strong monad over the category $\mathbf{Sub(Set)}$ of subsets and functions respecting subsets (example 4.3). Furthermore, the strong monad $(-)^{\top\top}$ makes the following diagram commute:

$$
\begin{array}{ccc}
\mathbf{Sub(Set)} & \xrightarrow{\ (-)^{\top\top}\ } & \mathbf{Sub(Set)} \\
{\scriptstyle \pi}\big\downarrow & & \big\downarrow{\scriptstyle \pi} \\
\mathbf{Set} & \xrightarrow[\ T\ ]{} & \mathbf{Set}
\end{array}
$$

where $\pi : \mathbf{Sub(Set)} \to \mathbf{Set}$ is the evident forgetful functor. This suggests that we can understand the semantic $\top\top$-lifting as a *construction* of such a strong monad from T.

We give a categorical generalisation of this construction using fibrations and symmetric monoidal closed structures. We replace π with a bifibration $p : \mathbb{E} \to \mathbb{B}$ equipped with a lifted symmetric monoidal closed structure (definition 4.2). We then capture the semantic $\top\top$-lifting as a construction of a strong monad over $\mathbb{E}$ from that over $\mathbb{B}$.

We borrow some notations from 2-category theory. We use $\bullet$ and $*$ for the vertical and horizontal compositions of natural transformations, respectively. We overload $\circ$ with the notation for the composition of functors, as well as for the composition of a functor and a natural transformation.

4.1 Preliminaries

Symmetric Monoidal Close Category. We assume that the reader is familiar with *symmetric monoidal closed categories*. We reserve symbols $\mathbf{I}, \otimes, -\!\circ$ for unit objects, tensor products and exponentials. A *symmetric monoidal functor* is a functor $F : \mathbb{C} \to \mathbb{D}$ between symmetric monoidal categories $\mathbb{C}, \mathbb{D}$ together with morphisms $m_{\mathbf{I}} : \mathbf{I}_{\mathbb{D}} \to F\mathbf{I}_{\mathbb{C}}$ and $m_{X,Y} : FX \otimes_{\mathbb{D}} FY \to F(X \otimes_{\mathbb{C}} Y)$ satisfying certain coherence laws (see e.g. [14]).

Example 4.1. 1. The category $\mathbf{Set}$ has a symmetric monoidal closed structure given by a chosen CCC structure.
  2. The category $\omega\mathbf{CPPO}$ of pointed ω-CPOs and strict ω-continuous functions has a symmetric monoidal closed structure given by Sierpinski space $\mathbf{O} = \{\bot \sqsubseteq \top\}$, smash products and strict ω-continuous function spaces.
  3. The functor $\times : (\omega\mathbf{CPPO})^2 \to \mathbf{Set}$ sending a pair (X, Y) of pointed ω-CPOs to the binary product $X \times Y$ of carrier sets is a symmetric monoidal functor.

Strong Monad. A *strong monad* T over a symmetric monoidal category $\mathbb{B}$ is a tuple (T, η, μ, θ) such that (T, η, μ) is an ordinary monad over $\mathbb{B}$ and $\theta_{X,Y} : X \otimes TY \to T(X \otimes Y)$ is a natural transformation called *tensorial strength* satisfying certain coherence laws (see e.g. [10]). A *strong monad morphism* from $T = (T, \eta, \mu, \theta)$ to $T' = (T', \eta', \mu', \theta')$ is a natural transformation $\sigma : T \to T'$ satisfying

$$
\mu' \bullet (\sigma * \sigma) = \sigma \bullet \mu, \qquad \eta' = \sigma \bullet \eta, \qquad \theta'_{X,Y} \circ (X \otimes \sigma_Y) = \sigma_{X \otimes Y} \circ \theta_{X,Y}.
$$

Fibration. We assume that the reader is familiar with preliminaries on fibration. A good reference is [7].

Definition 4.2. *A functor* $p : \mathbb{E} \to \mathbb{B}$ *is a bifibration with a lifted symmetric monoidal closed structure if p is a preordered bifibration, $\mathbb{E}$ and $\mathbb{B}$ are symmetric monoidal closed categories and p strictly preserves the symmetric monoidal closed structure in $\mathbb{E}$. We use dot notation $\dot{\mathbf{I}}$, $\dot{\otimes}$, $\dot{\multimap}$ to denote the symmetric monoidal closed structure in $\mathbb{E}$ which are sent to the symmetric monoidal closed structure $\mathbf{I}, \otimes, \multimap$ in $\mathbb{B}$ by p.*

Example 4.3. We define a category $\mathbf{Sub}(\mathbf{Set})$ by the following data: an object is a pair (X, I) where X is a subset of I, and a morphisms from (X, I) to (Y, J) is a function in $X \Rightarrow Y$. The category $\mathbf{Sub}(\mathbf{Set})$ has the following CCC structure:

$$\dot{\mathbf{1}} = (\{*\}, \{*\})$$
$$(X, I) \mathbin{\dot{\times}} (Y, J) = (\{(i, j) \mid i \in X \wedge j \in Y\}, I \times J)$$
$$(X, I) \Rightarrow (Y, J) = (X \Rightarrow Y, I \Rightarrow J).$$

(here the reader should not worry about the confusion caused by a clash of the notation $\Rightarrow$). This structure is strictly preserved by the evident forgetful functor π : $\mathbf{Sub}(\mathbf{Set}) \to \mathbf{Set}$, which is actually a partial-order bifibration. Therefore π is a bifibration with a lifted symmetric monoidal closed structure.

One good property of the class of bifibrations with lifted symmetric monoidal closed structures is the closure under change-of-base along symmetric monoidal functors.

Proposition 4.4 (e.g. [5]). *Let p : $\mathbb{E} \to \mathbb{B}$ be a bifibration with a lifted symmetric monoidal closed structure and $F : \mathbb{C} \to \mathbb{B}$ be a symmetric monoidal functor. Then the change-of-base of p along F is again a bifibration with a lifted symmetric monoidal closed structure.*

Example 4.5. We consider the following change-of-base of π along $\times$:

$$
\begin{array}{ccc}
\mathbf{Rel}(\omega\mathbf{CPPO}) & \longrightarrow & \mathbf{Sub}(\mathbf{Set}) \\
{\scriptstyle \pi_2} \downarrow & \lrcorner & \downarrow {\scriptstyle \pi} \\
(\omega\mathbf{CPPO})^2 & \xrightarrow{\times} & \mathbf{Set}
\end{array}
$$

From proposition 4.4, π_2 is again a bifibration with a lifted symmetric monoidal closed structure. An object in $\mathbf{Rel}(\omega\mathbf{CPPO})$ is a triple (X, I, J) where I, J are pointed ω-CPOs and X is an arbitrary subset of $I \times J$, that is, a binary relation between I and J. A morphism in $\mathbf{Rel}(\omega\mathbf{CPPO})$ from (X, I, J) to (X', I', J') is a pair $(f : I \to I', g : J \to J')$ of strict ω-continuous functions such that $f \times g \in X \Rightarrow X'$. We can similarly derive the category of n-ary relations between ω-CPOs by change-of-base.

4.2   ⊤⊤-Lifting as a Construction of Liftings of Strong Monads

We fix a bifibration p : $\mathbb{E} \to \mathbb{B}$ with a lifted symmetric monoidal closed structure. We define a fibration of *lifted strong monads* which is suitable for characterising the ⊤⊤-lifting.

Definition 4.6. *1. We say that a strong monad $\dot{T} = (\dot{T}, \dot{\eta}, \dot{\mu}, \dot{\theta})$ over $\mathbb{E}$ is a lifting of a strong monad $T = (T, \eta, \mu, \theta)$ over $\mathbb{B}$ if the following holds:*

$$p \circ \dot{T} = T \circ p, \quad p \circ \dot{\eta} = \eta \circ p, \quad p \circ \dot{\mu} = \mu \circ p, \quad p(\dot{\theta}_{X,Y}) = \theta_{pX,pY}.$$

2. *We write $\mathbf{Mon}(\mathbb{B})$ for the category of strong monads over $\mathbb{B}$ and strong monad morphisms between them.*
3. *We define a category $\mathbf{Mon}_l(\mathbb{E})$ using the following data:*
   - *An object in $\mathbf{Mon}_l(\mathbb{E})$ is a pair of a strong monad $\dot{T}$ over $\mathbb{E}$ and a strong monad T over $\mathbb{B}$ such that $\dot{T}$ is a lifting of T. We sometimes represent an object in $\mathbf{Mon}_l(\mathbb{E})$ simply by a strong monad over $\mathbb{E}$ when its underlying strong monad over $\mathbb{B}$ is clear from the context.*
   - *A morphism in $\mathbf{Mon}_l(\mathbb{E})$ is a pair of strong monad morphisms $\dot{\alpha} : \dot{T} \to \dot{T}'$ and $\alpha : T \to T'$ such that $p \circ \dot{\alpha} = \alpha \circ p$.*
4. *We write $\mathbf{Mon}(p) : \mathbf{Mon}_l(\mathbb{E}) \to \mathbf{Mon}(\mathbb{B})$ for the following forgetful functor:*

$$\mathbf{Mon}(p)(\dot{T}, T) = T, \quad \mathbf{Mon}(p)(\dot{\alpha}, \alpha) = \alpha.$$

Theorem 4.7. $\mathbf{Mon}(p)$ *is a fibration.*

Proof. See appendix A.1 □

We are ready to give a categorical account of the semantic $\top\!\top$-lifting. We capture the $\top\!\top$-lifting as a construction of a lifting of a strong monad over $\mathbb{E}$ from that over $\mathbb{B}$. For this construction, *continuation monads* play a crucial role. We observe the following facts.

- For each object I in $\mathbb{B}$, an endofunctor $(- \multimap I) \multimap I$ over $\mathbb{B}$ is a strong monad (called *continuation monad*). Particularly, for a strong monad T over $\mathbb{B}$ and an object R in $\mathbb{B}$, we have a continuation monad $(- \multimap TR) \multimap TR$ and a strong monad morphism

$$\sigma : T \longrightarrow (- \multimap TR) \multimap TR$$

whose component at an object I in $\mathbb{B}$ is given by the following transposition (object annotations are omitted):

$$\frac{TI \otimes (I \multimap TR) \xrightarrow{\ s\ } (I \multimap TR) \otimes TI \xrightarrow{\ \theta\ } T((I \multimap TR) \otimes I) \xrightarrow{\ @^{\#}\ } TR}{\sigma_I = \lambda(@^{\#} \circ \theta \circ s) : \ TI \longrightarrow (I \multimap TR) \multimap TR}$$

where s and $@$ are a symmetry and an evaluation morphisms in $\mathbb{B}$, respectively.
- Let S be an object in $\mathbb{E}$ above TR and consider a continuation monad $(- \dot{\multimap} S) \dot{\multimap} S$ over $\mathbb{E}$. It is a lifting of $(- \multimap TR) \multimap TR$ since p strictly preserves the symmetric monoidal closed structure in $\mathbb{E}$.

The following diagram summarises these facts in $\mathbf{Mon}(p)$:

$$\begin{array}{cc}
(- \dot{\multimap} S) \dot{\multimap} S & \mathbf{Mon}_l(\mathbb{E}) \\
 & \downarrow{\scriptstyle \mathbf{Mon}(p)} \\
T \xrightarrow{\ \sigma\ } (- \multimap TR) \multimap TR & \mathbf{Mon}(\mathbb{B})
\end{array}$$

We now consider a Cartesian lifting of σ.

$$
\begin{array}{ccc}
\sigma^*((-\stackrel{.}{\multimap} S)\stackrel{.}{\multimap} S) \xdashrightarrow{\ \overline{\sigma}\ } (-\stackrel{.}{\multimap} S)\stackrel{.}{\multimap} S & & \mathbf{Mon}_l(\mathbb{E}) \\[2mm]
& & \Big\downarrow \mathbf{Mon}(p) \\[2mm]
\mathcal{T} \xrightarrow{\quad \sigma \quad} (-\multimap TR)\multimap TR & & \mathbf{Mon}(\mathbb{B})
\end{array}
$$

We claim that the vertex $\sigma^*((-\stackrel{.}{\multimap} S)\stackrel{.}{\multimap} S)$, which is by definition a lifting of $\mathcal{T}$, gives the ⊤⊤-lifting of $\mathcal{T}$. There are two sets of evidence supporting our claim.

- The set-theoretic ⊤⊤-lifting in section 3 is an instance of this generalised ⊤⊤-lifting. We work in the fibration $\pi : \mathbf{Sub(Set)} \to \mathbf{Set}$ from example 4.3. Subsequently, for any strong monad $\mathcal{T}$ and subsets $X \subseteq I$ and $S \subseteq TR$, we have:

$$
\begin{aligned}
\sigma^*((X \Rightarrow S) \Rightarrow S) &= \{x \in TI \mid \sigma^*(x) \in ((X \Rightarrow S) \Rightarrow S)\} \\
&= \{x \in TI \mid \forall f \in X \Rightarrow S . \sigma^*(x)(f) \in S\} \\
&= \{x \in TI \mid \forall f \in X \Rightarrow S . f^\# x \in S\} \\
&= X^{\top\top}.
\end{aligned}
$$

- Let D, E be pointed ω-CPOs and R be an arbitrary subset of $D \times E$. In [1], Abadi considered the following closure operation $(-)^{\top\top}$ as a semantic abstraction of Pitts' syntactic ⊤⊤-closure operation [21]:

$$
\begin{aligned}
R^\top &= \{(f,g) \in [D \to_\perp \mathbf{O}] \times [E \to_\perp \mathbf{O}] \mid \forall (x,y) \in R . fx = gy\} \\
R^{\top\top} &= \{(x,y) \in D \times E \mid \forall (f,g) \in R^\top . fx = gy\}
\end{aligned}
$$

where $[- \to_\perp -]$ denotes strict ω-continuous function spaces.

The above closure operation is an instance of our semantic ⊤⊤-lifting. We work in the fibration $\pi_2 : \mathbf{Rel}(\omega\mathbf{CPPO}) \to (\omega\mathbf{CPPO})^2$ from example 4.5. The ⊤⊤-lifting of the *identity monad* over $(\omega\mathbf{CPPO})^2$ with the following data coincides with Abadi's ⊤⊤-closure operation.
  - The result type R is $(\mathbf{O}, \mathbf{O})$.
  - The result predicate S is $(\{(\perp, \perp), (\top, \top)\}, (\mathbf{O}, \mathbf{O}))$.

We write $\mathcal{T}^{\top\top}$ for $\sigma^*((-\stackrel{.}{\multimap} S)\stackrel{.}{\multimap} S)$.

5   Multiple Result Types

We relax the restriction we imposed on the result type in section 3. Let $p : \mathbb{E} \to \mathbb{B}$ be a bifibration with a lifted symmetric monoidal closed structure and $\mathcal{T}$ be a strong monad over $\mathbb{B}$.

Theorem 5.1. *If p has fibred (finite/small) products, then so does* $\mathbf{Mon}(p)$.

Proof. See appendix A.2.   □

Let $\{(S_k, R_k)\}_{k \in K}$ be a set of pairs of objects in $\mathbb{E}$ and $\mathbb{B}$ such that $pS_k = TR_k$ for all $k \in K$. For each $k \in K$, the pair (S_k, R_k) determines a $\top\top$-lifting $T^{\top\top_k}$. They are all liftings of T, so we consider the following fibred product in $\mathbf{Mon}_l(\mathbb{E})_T$:

$$\bigwedge_{k \in K} T^{\top\top_k}$$

which is again a lifting of T.

Example 5.2. We flip the relation S in example 3.6 and obtain the following $\top\top$-lifting:

$$P^{\top\top'} = \{(p, q) \mid \forall b \in q \ . \ \exists a \in p \ . \ (a, b) \in P\}.$$

The intersection

$$P^{\top\top} \wedge P^{\top\top'} = \{(p, q) \mid (\forall b \in q \ . \ \exists a \in p \ . \ (a, b) \in P) \wedge (\forall a \in p \ . \ \exists b \in q \ . \ (a, b) \in P)\}$$

coincides with the pattern of bisimulation.

6   Related Work

This work has been inspired by Lindley and Stark's paper [12] and Lindley's thesis [11]. Lindley and Stark introduce the syntactic $\top\top$-lifting for λ_{ml} and prove the strong normalisation of λ_{ml}. In the latter part of [12], they also discuss an extension of the syntactic $\top\top$-lifting to other types such as sum types. However, this extension has not been covered here.

Operations which are similar to Lindley and Stark's $\top\top$-lifting have previously appeared in several other studies. Some examples of these studies are: the reducibility technique for linear logic by Girard [4], Parigot's work on the second order classical natural deduction [20], Pitts' $\top\top$-closure operation [21] and Melliès and Vouillon's *biorthogonality* [15]. In addition, Abadi gives a semantic formulation of Pitts' $\top\top$-closure operation and discusses the relationship between $\top\top$-closed relations (those which satisfy $R = R^{\top\top}$) and admissibility [1]. The $\top\top$-closed relations are applied to the verification of the correctness of program transformations [8, 19], and to the characterisation of the observational equivalence for a language with local states [22].

Categorical study of logical predicates established in [13, 17] is generalised by Hermida using fibrational category theory [6]. The key observation of his generalisation is that logical predicates with respect to a fibration $p : \mathbb{E} \to \mathbb{B}$ employ a CCC structure in $\mathbb{E}$ which is strictly preserved by p. This observation leads us to consider liftings of strong monads and bifibrations with lifted symmetric monoidal closed structures.

In general, there are many liftings of a strong monad. In [3], Larrecq, Lasota and Nowak propose a construction method of liftings of strong monads using factorisation systems. Their method appears to be fundamentally different from our semantic $\top\top$-lifting. However, some of their examples of liftings of strong monads over **Set** can also be calculated with our method. It will be interesting to establish a formal relationship between their lifting of strong monads and the semantic $\top\top$-lifting developed by us.

7   Conclusion

We semantically formulated Lindley and Stark's $\top\top$-lifting and showed that it provides a satisfactory construction method of logical predicates for λ_{ml}. We also examined several examples of the semantic $\top\top$-lifting of strong monads over **Set**.

We then categorically re-formulated the $\top\top$-lifting as a lifting of a monad along a bifibration with a symmetric monoidal closed structure using continuation monads. This generalisation subsumes the set-theoretic $\top\top$-lifting in section 3 and Abadi's $\top\top$-lifting.

Acknowledgement

I am grateful to Don Sannella, Samuel Lindley, Masahito Hasegawa, Miki Tanaka and anonymous referees for their valuable advice. Most of this work was carried out in Edinburgh university under an LFCS studentship.

References

1. M. Abadi. $\top\top$-closed relations and admissibility. *MSCS*, 10(3):313–320, 2000.
2. R. Amadio and P.-L. Curien. *Domains and Lambda-Calculi*, volume 46 of *Cambridge Tracts in Theoretical Computer Science*. Cambridge University Press, 1998.
3. J. G.-Larrecq, S. Lasota, and D. Nowak. Logical relations for monadic types. In *Proc. CSL*, volume 2471 of *LNCS*, pages 553–568. Springer, 2002.
4. J. Y. Girard. Linear logic. *Theor. Comp. Sci.*, 50:1–102, 1987.
5. M. Hasegawa. Categorical glueing and logical predicates for models of linear logic. Technical Report RIMS-1223, Research Institute for Mathematical Sciences, Kyoto University, 1999.
6. C. Hermida. *Fibrations, Logical Predicates and Indeterminants*. PhD thesis, University of Edinburgh, 1993.
7. B. Jacobs. *Categorical Logic and Type Theory*. Elsevier, 1999.
8. P. Johann. Short cut fusion is correct. *J. Funct. Program.*, 13(4):797–814, 2003.
9. A. Jung and J. Tiuryn. A new characterization of lambda definability. In *Proc. TLCA*, volume 664 of *LNCS*, pages 245–257. Springer, 1993.
10. A. Kock. Strong functors and monoidal monads. *Archiv der Mathematik*, 23:113–120, 1970.
11. S. Lindley. *Normalisation by Evaluation in the Compilation of Typed Functional Programming Languages*. PhD thesis, University of Edinburgh, 2004.
12. S. Lindley and I. Stark. Reducibility and $\top\top$-lifting for computation types. In *TLCA*, pages 262–277, 2005.
13. Q. Ma and J. Reynolds. Types, abstractions, and parametric polymorphism, part 2. In *Proc. MFPS 1991*, volume 598 of *LNCS*, pages 1–40. Springer, 1992.
14. S. MacLane. *Categories for the Working Mathematician (Second Edition)*, volume 5 of *Graduate Texts in Mathematics*. Springer, 1998.
15. P.-A. Melliès and J. Vouillon. Recursive polymorphic types and parametricity in an operational framework. In *Proc. LICS 2005*. To appear.
16. J. Mitchell. Representation independence and data abstraction. In *Proc. POPL*, pages 263–276, 1986.

17. J. Mitchell and A. Scedrov. Notes on sconing and relators. In *Proc. CSL 1992*, volume 702 of *LNCS*, pages 352–378. Springer, 1993.

18. E. Moggi. Notions of computation and monads. *Information and Computation*, 93(1):55–92, 1991.

19. S. Nishimura. Correctness of a higher-order removal transformation through a relational reasoning. In *APLAS*, volume 2895 of *LNCS*, pages 358–375. Springer, 2003.

20. M. Parigot. Proofs of strong normalisation for second order classical natural deduction. *Journal of Symbolic Logic*, 62(4):1461–1479, 1997.

21. A. Pitts. Parametric polymorphism and operational equivalence. *Mathematical Structures in Computer Science*, 10(3):321–359, 2000.

22. A. Pitts and I. Stark. Operational reasoning for functions with local state. In A. D. Gordon and A. M. Pitts, editors, *Higher Order Operational Techniques in Semantics*, Publications of the Newton Institute, pages 227–273. Cambridge University Press, 1998.

23. G. Plotkin. Lambda-definability in the full type hierarchy. In *"To H.B. Curry: Essays on Combinatory Logic, Lambda Calculus and Formalism"*, pages 367–373. Academic Press, San Diego, 1980.

24. W. Tait. Intensional interpretation of functionals of finite type I. *Journal of Symbolic Logic*, 32, 1967.

A   Proof

A.1   Proof of Theorem 4.7

When $p : \mathbb{E} \to \mathbb{B}$ is a fibration, $p \circ - : [\mathbb{E}, \mathbb{E}] \to [\mathbb{E}, \mathbb{B}]$ is also a fibration. Then an endofunctor F over $\mathbb{E}$ is a lifting of an endofunctor G over $\mathbb{B}$ if and only if F is above $G \circ p$ in the fibration $p \circ -$.

Let T, T' be strong monads over $\mathbb{B}$, $\alpha : T \to T'$ be a strong monad morphism and $\dot{T}'$ be a strong monad over $\mathbb{E}$ which is a lifting of T'. We construct a monad $\dot{T} = (\dot{T}, \dot{\eta}, \dot{\mu}, \dot{\theta})$ together with a strong monad morphism $\dot{\alpha} : \dot{T} \to \dot{T}'$ which is Cartesian above α.

- We define the endofunctor $\dot{T} : \mathbb{E} \to \mathbb{E}$ to be the vertex $(\alpha \circ p)^* \dot{T}'$ of the following Cartesian lifting of $\alpha \circ p$ in the fibration $p \circ -$:

$$(\alpha \circ p)^* \dot{T}' \xdashrightarrow{\overline{(\alpha \circ p)}(\dot{T}')} \dot{T}'$$

$$T \circ p \xrightarrow{\;\;\alpha \circ p\;\;} T' \circ p$$

We define $\dot{\alpha} = \overline{(\alpha \circ p)}(\dot{T}')$.

- We define the unit $\dot{\eta}$ and the multiplication $\dot{\mu}$ by the morphisms obtained from the universal property of the Cartesian morphism $\dot{\alpha}$ in the fibration $p \circ -$:

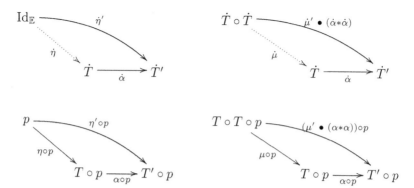

- For objects X, Y in $\mathbb{E}$ above objects I, J in $\mathbb{B}$ respectively, we define the strength $\dot{\theta}_{X,Y}$ as follows:

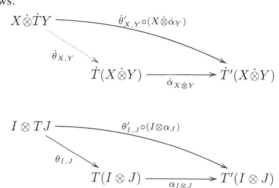

We can easily verify that $\dot{\eta}, \dot{\mu}, \dot{\theta}$ satisfy the law of strong monad using the fact that p is faithful (since p is a preordered fibration). For example, to show $\dot{\mu}_X \circ \dot{T}(\dot{\eta}_X) = \mathrm{id}_X$ for each object X in $\mathbb{E}$, we calculate:

$$p(\dot{\mu}_X \circ \dot{T}(\dot{\eta}_X)) = \mu_{pX} \circ T(\eta_{pX}) = \mathrm{id}_{pX} = p(\mathrm{id}_X).$$

Since p is faithful, we conclude that $\dot{\mu}_X \circ \dot{T}(\dot{\eta}_X) = \mathrm{id}_X$.

The morphism $\dot{\alpha}$ is clearly a monad morphism from the construction of $\dot{\eta}, \dot{\mu}, \dot{\theta}$.

To see that $\dot{\alpha}$ is a Cartesian morphism, we consider a situation in $\mathbf{Mon}(p)$ described in the left diagram:

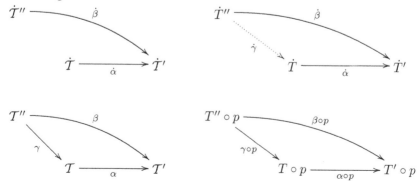

This situation induces the right diagram in $p \circ -$. From the universal property of $\dot{\alpha}$, we obtain a unique morphism $\dot{\gamma} : \dot{T}'' \to \dot{T}$ above $\gamma \circ p$ satisfying $\dot{\alpha} \bullet \dot{\gamma} = \dot{\beta}$. To verify that $\dot{\gamma}$ is a strong monad morphism, we use the universal property of $\dot{\alpha}$. We show $\dot{\gamma} \bullet \dot{\eta}'' = \dot{\eta}$ as an example. First, $\dot{\gamma} \bullet \dot{\eta}''$ and $\dot{\eta}$ are above $\eta \circ p$ in the fibration $p \circ -$. Next, we have

$$\dot{\alpha} \bullet \dot{\gamma} \bullet \dot{\eta}'' = \dot{\beta} \bullet \dot{\eta}'' = \dot{\eta}' = \dot{\alpha} \bullet \dot{\eta}$$

From the universal property of $\dot{\alpha}$, we have $\dot{\gamma} \bullet \dot{\eta}'' = \dot{\eta}$. We can similarly verify the other equations of the law of strong monad morphism. □

A.2   Proof of Theorem 5.1

(Sketch) Let $T = (T, \eta, \mu, \theta)$ be a strong monad over $\mathbb{B}$, K be a (finite) set and suppose that we have a lifting $\dot{T}_k = (\dot{T}_k, \dot{\eta}_k, \dot{\mu}_k, \dot{\theta}_k)$ of T for each $k \in K$.

The fibred product $\hat{T} = (\hat{T}, \hat{\eta}, \hat{\mu}, \hat{\theta})$ of $\dot{T}_k$ is given as follows.

- The functor part is defined by $\hat{T}X = \bigwedge_{k \in K} \dot{T}_k X$. We write $\pi_X^k : \hat{T}X \to \dot{T}_k X$ for the k-th projection.
- We observe that for objects X, Y in $\mathbb{E}$ and a morphism $f : pX \to pY$ in $\mathbb{B}$, we have the following natural isomorphism:

$$\mathbb{E}_f(X, \hat{T}Y) \cong \mathbb{E}_{pX}(X, f^*(\hat{T}Y)) \cong \mathbb{E}_{pX}\left(X, \bigwedge_{k \in K} f^* \dot{T}_k\right) \cong \prod_{k \in K} \mathbb{E}_f(X, \dot{T}_k Y).$$

We write ϕ for the right-to-left part of the above isomorphism. The unit, multiplication and strength is then defined by:

$$\hat{\eta}_X = \phi\langle(\dot{\eta}_k)_X\rangle_{k \in K}$$
$$\hat{\mu}_X = \phi\langle(\dot{\mu}_k)_X \circ \dot{T}_k(\pi_X^k) \circ \pi_X^k\rangle_{k \in K}$$
$$\hat{\theta}_{X,Y} = \phi\langle(\dot{\theta}_k)_{X,Y} \circ (X \otimes \pi_Y^k)\rangle_{k \in K}$$

The reader can verify that $\hat{T}$ is indeed a strong monad, and is a fibred product of $\{\dot{T}_k\}_{k \in K}$.

Order Structures on Böhm-Like Models

Paula Severi and Fer-Jan de Vries

Department of Computer Science, University of Leicester, UK

Abstract. We are interested in the question whether the models induced by the infinitary lambda calculus are orderable, that is whether they have a partial order with a least element making the context operators monotone. The first natural candidate is the prefix relation: a prefix of a term is obtained by replacing some subterms by $\bot$. We prove that six models induced by the infinitary lambda calculus (which includes Böhm and Lévy-Longo trees) are orderable by the prefix relation. The following two orders we consider are the compositions of the prefix relation with either transfinite η-reduction or transfinite η-expansion. We prove that these orders make the context operators of the η-Böhm trees and the $\infty\eta$-Böhm trees monotone. The model of Berarducci trees is not monotone with respect to the prefix relation. However, somewhat unexpectedly, we found that the Berarducci trees are orderable by a new order related to the prefix relation in which subterms are not replaced by $\bot$ but by a lambda term O called *the ogre* which devours all its inputs. The proof of this uses simulation and coinduction. Finally, we show that there are 2^c unorderable models induced by the infinitary lambda calculus where c is the cardinality of the continuum.

1 Introduction

In this paper we give order structure to some models induced by the infinitary lambda calculi. Our starting point are lambda calculi that extend finite lambda calculus with infinite terms and transfinite reduction. The β and η reduction rules apply to infinite terms in much the same way as they apply to finite terms. However, characteristic for these calculi is that they contain a $\bot$-rule that maps a certain set $\mathcal{U}$ of meaningless terms to $\bot$. Without such an addition the extension of finite lambda calculus with infinite terms and reductions immediately would result in loss of confluence [8]. All infinite calculi that we consider have the same set of finite and infinite terms $\Lambda_\bot^\infty$. The variation comes from the choice of the set $\mathcal{U}$ and the strength of extensionality.

Figure 1 summarises the infinitary lambda calculi studied so far [3, 7–9, 13, 15]. An interesting aspect of infinitary lambda calculus is the possibility of capturing the notion of tree (such as Böhm and Lévy-Longo trees) as a normal form. These trees were originally defined for finite lambda terms only, but in the infinitary lambda calculus we can also consider normal forms of infinite terms. The three infinitary lambda calculi mentioned in the first three rows of Figure 1 capture the well-known cases of Böhm, Lévy-Longo and Berarducci trees [3, 8, 9]. In the fourth row, there is an uncountable class of infinitary lambda calculi with

L. Ong (Ed.): CSL 2005, LNCS 3634, pp. 103–118, 2005.

REDUCTION RULES	NORMAL FORMS	NF
Beta and $\perp$ for terms without tnf	Berarducci trees	$\mathsf{BerT} = \mathsf{P}_{\overline{\mathcal{TN}}}$
Beta and $\perp$ for terms without whnf	Lévy-Longo trees	$\mathsf{LLT} = \mathsf{P}_{\overline{\mathcal{WN}}}$
Beta and $\perp$ for terms without hnf	Böhm trees	$\mathsf{BT} = \mathsf{P}_{\overline{\mathcal{HN}}}$
Beta, $\perp$ parametric on $\mathcal{U}$	Parametric trees	$\mathsf{P}_{\mathcal{U}}$
Beta, $\perp$ for terms w.o. hnf and Eta	η-Böhm trees	$\eta\mathsf{BT}$
Beta, $\perp$ for terms w.o. hnf and EtaBang	$\infty\eta$-Böhm trees	$\infty\eta\mathsf{BT}$

Fig. 1. Infinitary Lambda Calculi

a $\perp$-rule parametrised by a set $\mathcal{U}$ of meaningless terms [7, 10]. By changing the parameter set $\mathcal{U}$ of the $\perp$-rule, we obtain different infinitary lambda calculi. If $\mathcal{U}$ is the set of terms without head normal form, we capture the notion of Böhm tree. If $\mathcal{U}$ is the set of terms without weak head normal form we obtain the Lévy-Longo trees. And if $\mathcal{U}$ is the set of terms without top head normal form to $\perp$, we recover the Berarducci trees. The infinitary lambda calculus sketched in the one but last row incorporates the η-rule [13]. This calculus captures the notion of η-Böhm tree. The last row in Figure 1 mentions the infinitary lambda calculus incorporating the η!-rule, a strengthened form of the η-rule [15]. The normal forms in this calculus capture the notion of $\infty\eta$-Böhm trees. In this paper we give some new examples of parametric trees.

When the infinite extensions are confluent and normalising (normal forms can now be infinite too!) they induce a function $\mathsf{NF} : \Lambda^\infty_\perp \to \Lambda^\infty_\perp$ mapping a term to its unique normal form. The normal form functions NF induce λ-models (models of the finite lambda calculus): just interpret a term M by its normal form $\mathsf{NF}(M)$ and application $M \cdot N$ of two terms M and N by $\mathsf{NF}(MN)$.

Figure 2 summarizes the results proved in this paper. The first order we consider is the prefix relation $\preceq$. This is a natural order on terms. If terms are represented as trees, prefixes of a tree are obtained by pruning some of its subtrees and replacing them by $\perp$. Whereas application in the model of Böhm trees is well-known to be continuous with respect to the Scott topology induced by the prefix relation, it is perhaps less well-known that in case of the model of Berarducci trees, the normal form function $\mathsf{BerT} : \Lambda^\infty_\perp \to \Lambda^\infty_\perp$ and the application operator are not even monotone [6] and it is not clear how to define a domain-theoretic model whose local structure is represented by Berarducci trees, though some attempts have been made via types and filter models [4]. We prove that $\mathsf{P}_{\mathcal{U}} : \Lambda^\infty_\perp \to \Lambda^\infty_\perp$ preserves $\preceq$ provided $\mathcal{U}$ is quasi-regular and $\perp P$ is equal to $\perp$. This generalizes the proof of monotonicity of BT and LLT given in [14]. We, then, conclude that the prefix relation makes the context operators of six models monotone including the models of Böhm and Lévy-Longo trees.

We also define two orders for the extensional models and prove that they make the context operators monotone. The partial order $\preceq_\eta$ on the set of η-Böhm trees is the composition of the prefix relation with transfinite η-reduction and it corresponds to the order on D^*_∞ [5]. The partial order $\preceq_{\eta!}$ on the set of $\infty\eta$-Böhm trees is the composition of the prefix relation with transfinite η!-reduction and it corresponds to the order on Scott's model D_∞.

The next step is to find an order for Berarducci trees. We prove that the least element of an arbitrary orderable model induced by NF should be either $\bot$ or a term O called *the ogre* which eats all its inputs. In case the least element is $\bot$ then $\bot P$ should reduce to $\bot$ for all $P \in \Lambda_\bot^\infty$. Hence, $\bot$ cannot be the least element of an order on Berarducci trees and the only possible candidate is O. The term O is the solution to the recursive equation $O = \lambda x.O$ and it can be obtained by applying any fixed point operator to the combinator $K = \lambda xy.x$. In the lambda model induced by Böhm trees, the ogre is interpreted as bottom. But there are many other lambda models such as the ones induced by Lévy-Longo and Berarducci trees that give a different interpretation to ogre. In these models, O is identified with the infinite sequence of abstractions $\lambda x_1.\lambda x_2.\lambda x_3 \ldots$. We consider an order called $\trianglelefteq$ on terms related to the prefix relation in which subterms are not replaced by $\bot$ but by the term O. We prove that the parametric trees $\mathsf{P}_\mathcal{U} : \Lambda_\bot^\infty \rightarrow \Lambda_\bot^\infty$ preserve $\trianglelefteq$ provided $\mathcal{U}$ is quasi-regular and $O \in \mathsf{P}_\mathcal{U}(\Lambda_\bot^\infty)$ using simulations and coinduction. We, then, conclude that $\trianglelefteq$ makes the context operators monotone of five models including the model of Berarducci trees. We can see in Figure 2 that the relations $\preceq$ and $\trianglelefteq$ make the context operators of some models simultaneously monotone.

Finally, we show that there are 2^c unorderable models induced by the infinitary lambda calculus where c is the cardinality of the continuum. In [12] Salibra proves that there is a continuum of unorderable λ-models by considering the equation $\Omega M M = \Omega$. This idea does not work for infinitary lambda calculus because this equation interpreted as a reduction rule is not left linear and adding it to the infinitary lambda calculus of Berarducci trees would destroy confluence, as can be seen with help of a variant of Klop's counterexample in [11]. In our case, the trick consists in equating $\bot P$ sometimes to $\bot$ and sometimes not. We consider the set $\mathcal{B}^0$ of closed Böhm trees without $\bot$ which has cardinality c and construct infinitary lambda calculi whose normal form functions U_X are indexed on $X \subseteq \mathcal{B}^0$ by stating that $\bot P$ reduces to $\bot$ if $P \in X$.

2   Infinite Lambda Calculi

We will now briefly recall some notions and facts of infinite lambda calculus from our earlier work [7–9, 13, 15]. We assume familiarity with basic notions and notations from [1]. Let Λ be the set of λ-terms and $\Lambda_\bot$ be the set of finite λ-terms with $\bot$ given by the inductive grammar:

$$M ::= \bot \mid x \mid (\lambda x M) \mid (MM)$$

where x is a variable from some fixed set of variables $\mathcal{V}$. We follow the usual conventions on syntax. Terms and variables will respectively be written with (super- and subscripted) letters M, N and x, y, z. Terms of the form $(M_1 M_2)$ and $(\lambda x M)$ will respectively be called applications and abstractions. A context $C[\ ]$ is a term with a hole in it, and $C[M]$ denotes the result of filling the hole by the term M, possibly by capturing some free variables of M. If $\sigma : \mathcal{V} \rightarrow \Lambda^\infty$ then M^σ is the simultaneous substitution of the variables in M by σ.

Normal forms NF	Prefix $\preceq$	Ogre order $\trianglelefteq$	Prefix up to η $\preceq_\eta$	Prefix up to $\eta!$ $\preceq_{\eta!}$	Orderable models
$\infty\eta\mathsf{BT}$	−	−	−	+	+
$\eta\mathsf{BT}$	−	−	+	−	+
$\mathsf{BT} = P_{\overline{\mathcal{HN}}}$	+	−	−	−	+
$P_{\overline{\mathcal{HN}}-\mathcal{O}}$	+	+	−	−	+
$P_{\mathcal{HAUO}}$	+	−	−	−	+
$P_{\mathcal{HA}}$	+	+	−	−	+
$\mathsf{LLT} = P_{\overline{\mathcal{WN}}}$	+	+	−	−	+
$P_{\mathcal{SA}}$	+	+	−	−	+
U_X	−	−	−	−	−
$\mathsf{BerT} = P_{\overline{\mathcal{TN}}}$	−	+	−	−	+

Fig. 2. Orderability of the models induced by NF

The set $\Lambda_\perp^\infty$ of finite and infinite λ-terms is defined by coinduction using the same grammar as for $\Lambda_\perp$. This set contains the three sets of Böhm, Lévy-Longo and Berarducci trees. In [7, 9, 10], an alternative definition of the set $\Lambda_\perp^\infty$ is given using a metric. The coinductive and metric definitions are equivalent [2]. In this paper we consider only one set of λ-terms, namely $\Lambda_\perp^\infty$, in contrast to the formulations in [9, 10] where several sets (which are all subsets of $\Lambda_\perp^\infty$) are considered. The paper [7] shows that the infinitary lambda calculi can be formulated using a common set $\Lambda_\perp^\infty$, confluence and normalisation still hold since the extra terms added by the superset $\Lambda_\perp^\infty$ are meaningless and equated to $\perp$.

We define several rules used to define different infinite lambda calculi. The β, η and η^{-1}-rules are extensions of the rules for finite lambda calculus to infinite terms. The $\eta!$-rule does not appear in the finite lambda calculus. The $\perp$-rule is parametric on a set $\mathcal{U} \subset \Lambda^\infty$ of meaningless terms [7, 10] where Λ^∞ is the set of terms in $\Lambda_\perp^\infty$ that do not contain $\perp$ (see Section 4).

Definition 1. We define the following rewrite rules on $\Lambda_\perp^\infty$:

$$(\lambda x.M)N \to M[x := N] \ (\beta) \qquad \frac{M[\perp := \Omega] \in \mathcal{U} \quad M \neq \perp}{M \to \perp} \ (\perp)$$

$$\frac{x \notin FV(M)}{\lambda x.Mx \to M} \ (\eta) \qquad \frac{x \notin FV(M)}{M \to \lambda x.Mx} \ (\eta^{-1}) \qquad \frac{x \twoheadrightarrow_{\eta^{-1}} N \quad x \notin FV(M)}{\lambda x.MN \to M} \ (\eta!)$$

In this paper we need various rewrite relations constructed from these rules on the set $\Lambda_\perp^\infty$. These are defined in the standard way, eg. $\to_{\beta\perp\eta!}$ is the smallest binary relation containing the β, $\perp$ and $\eta!$-rules which is closed under contexts. Reduction sequences can be of any transfinite ordinal length α: $M_0 \to M_1 \to M_2 \to \ldots M_\omega \to M_{\omega+1} \to \ldots M_{\omega+\omega} \to M_{\omega+\omega+1} \to \ldots M_\alpha$. This makes sense if the limit terms $M_\omega, M_{\omega+\omega}, \ldots$ in such sequence are all equal to the corresponding Cauchy limits, $\lim_{\beta\to\lambda} M_\beta$, in the underlying metric space for any limit ordinal $\lambda \leq \alpha$. If this is the case, the reduction is called *Cauchy converging*. We need the stronger concept of a *strongly converging* reduction that in addition satisfies that the depth of the contracted redexes goes to infinity at each limit term: $\lim_{\beta \to \lambda} d_\beta = \infty$ for each limit ordinal $\lambda \le \alpha$, where d_β is the depth in M_β of the contracted redex in $M_\beta \to M_{\beta+1}$. Any finite reduction is, then, strongly converging. We use the following notation:

1. $M \to N$ denotes a one step reduction from M to N;
2. $M \twoheadrightarrow N$ denotes a finite reduction from M to N;
3. $M \twoheadrightarrow\!\!\!\!\twoheadrightarrow N$ denotes a strongly converging reduction from M to N.

Variations on the reduction rules give rise to different calculi (see Figure 1). The resulting infinite lambda calculus $(\Lambda_\perp^\infty, \to_\rho)$ we will denote by λ_ρ^∞ for any $\rho \in \{\beta\perp, \beta\perp\eta, \beta\perp\eta!\}$. Since the $\perp$-rule is parametric, each set $\mathcal{U}$ of meaningless terms gives a different infinitary lambda calculus $\lambda_{\beta\perp}^\infty$.

Definition 2. 1. We say that a term M in λ_ρ^∞ is in ρ-*normal form* if there is no N in λ_ρ^∞ such that $M \to_\rho N$.
2. We say that λ_ρ^∞ is *confluent (Church-Rosser)* if $(\Lambda_\perp^\infty, \twoheadrightarrow_\rho)$ satisfies the *diamond property*, i.e. $_\rho\!\!\twoheadleftarrow \circ \twoheadrightarrow_\rho \subseteq \twoheadrightarrow_\rho \circ {}_\rho\!\!\twoheadleftarrow$.
3. We say that λ_ρ^∞ is *normalising* if for all $M \in \Lambda_\perp^\infty$ there exists an N in ρ-normal form such that $M \twoheadrightarrow_\rho N$.

Theorem 3. *[7, 9, 10] Let $\mathcal{U}$ be a set of meaningless terms. The calculi $\lambda_{\beta\perp}^\infty$ with a parametric $\perp$-rule on the set $\mathcal{U}$ are confluent, normalising and satisfy postponement of $\perp$ over β.*

In [7] confluence of the parametric calculi is proved for Cauchy converging reduction as well as for strongly converging reduction.

Theorem 4. *[13, 15] The infinite lambda calculi of $\infty\eta$-Böhm and η-Böhm trees are confluent and normalising.*

Assumption. In the rest of the paper whenever we refer to $\mathsf{NF} : \Lambda_\perp^\infty \to \Lambda_\perp^\infty$, we are assuming that the infinitary lambda calculus in question is confluent and normalising and that NF is the function that maps a term to its unique normal form. We denote by $M =_{\mathsf{NF}} N$ if $\mathsf{NF}(M) = \mathsf{NF}(N)$.

3   Basic Forms

In this section we introduce new forms of terms analogous to the notions of head, weak head and top normal forms and define certain specific subsets of Λ^∞ (terms of $\Lambda_\perp^\infty$ without $\perp$) containing the respective forms.

Definition 5. Let $M \in \Lambda_\perp^\infty$. We define that

1. M is a *head normal form* (hnf) if $M = \lambda x_1 \ldots x_n.y P_1 \ldots P_k$.
2. M is a *weak head normal form* (whnf) if M is a hnf or $M = \lambda x.N$.
3. A term M is a *top normal form* (tnf) if it is either a whnf or an application (NP) if there is no Q such that $N \twoheadrightarrow_\beta \lambda x.Q$.

4. M is a *rootactive form* (with respect to β) if for all $M \twoheadrightarrow_\beta N$ there exists a redex $(\lambda x.P)Q$ such that $N \twoheadrightarrow_\beta (\lambda x.P)Q$.
5. M is a *head bottom form* (hbf) if $M = \lambda x_1 \ldots x_n.\bot P_1 \ldots P_k$.
6. M is a *head active form* (haf) if $M = \lambda x_1 \ldots x_n.RP_1 \ldots P_k$ and R is rootactive.
7. M is a *strong active form* (saf) if $M = RP_1 \ldots P_k$ and R is rootactive.
8. M is a *strong active form relative* to X (X-saf) if $M = RP_1 \ldots P_k$ and R is rootactive and $P_1, \ldots, P_k \in X$.
9. M is an *infinite left spine form* (ilsf) if $M = \lambda x_1 \ldots x_n.((\ldots P_2)P_1$.
10. M is a *strong infinite left spine form* (silsf) if $M = ((\ldots P_2)P_1$.
11. M is a *basic form* if it is either a head normal form, a head bottom form, a head active form, an infinite left spine or the ogre.

We now define some subsets of Λ^∞ for the previous defined forms.

Definition 6. We define the following subsets of Λ^∞:

$$\mathcal{HN} = \{M \in \Lambda^\infty \mid M \twoheadrightarrow_\beta N \text{ and } N \text{ in head normal form}\}$$
$$\mathcal{WN} = \{M \in \Lambda^\infty \mid M \twoheadrightarrow_\beta N \text{ and } N \text{ in weak head normal form}\}$$
$$\mathcal{TN} = \{M \in \Lambda^\infty \mid M \twoheadrightarrow_\beta N \text{ and } N \text{ in top normal form}\}$$

By $\overline{\mathcal{HN}}$, $\overline{\mathcal{WN}}$ and $\overline{\mathcal{TN}}$ we denote their respective complements.

Definition 7.   1. The *basic sets* are the following subsets of Λ^∞:

$$\mathcal{HA} = \{M \in \Lambda^\infty \mid M \twoheadrightarrow_\beta N \text{ and } N \text{ is head active}\}$$
$$\mathcal{IL} = \{M \in \Lambda^\infty \mid M \twoheadrightarrow_\beta N \text{ and } N \text{ is an infinite left spine form}\}$$
$$\mathcal{O} = \{M \in \Lambda^\infty \mid M \twoheadrightarrow_\beta O\}$$

2. The *strongly basic sets* are the following subsets of Λ^∞:

$$\mathcal{R} = \{M \in \Lambda^\infty \mid M \text{ is rootactive}\} = \overline{\mathcal{TN}}$$
$$\mathcal{SA} = \{M \in \Lambda^\infty \mid M \twoheadrightarrow_\beta N \text{ and } N \text{ is strong active }\}$$
$$\mathcal{SIL} = \{M \in \Lambda^\infty \mid M \twoheadrightarrow_\beta N \text{ and } N \text{ is a strong infinite left spine form }\}$$

3. Finally we define a family of subsets of Λ^∞ depending on some $X \subseteq \Lambda^\infty$:

$$\mathcal{SA}_X = \{M \in \Lambda^\infty \mid M \twoheadrightarrow_\beta N \text{ and } N \text{ is a strong active form relative to } X\}$$

Note that $R[\bot := \Omega] \in \mathcal{R}$ iff R is $\bot$ or R is rootactive with respect to β.

Definition 8. The *skeleton* of a term $M \in \Lambda^\infty_\bot$ is defined by coinduction:

$\mathsf{skel}(M) = y$	if $M \twoheadrightarrow_\beta y$
$\mathsf{skel}(M) = \bot$	if $M \twoheadrightarrow_\beta \bot$
$\mathsf{skel}(M) = \lambda x.\mathsf{skel}(N)$	if $M \twoheadrightarrow_\beta \lambda x.N$
$\mathsf{skel}(M) = \mathsf{skel}(N)\,\mathsf{skel}(P)$	if $M \twoheadrightarrow_\beta NP$ and $N \not\twoheadrightarrow_\beta \lambda x.Q$ for any Q
$\mathsf{skel}(M) = M$	if M does not have a top normal form

The skeleton of a term is essentially the Berarducci tree of a term but instead of replacing rootactive terms by $\bot$, we leave rootactive terms untouched.

Lemma 9. *Let* $M \in \Lambda^\infty_\bot$. *Then* $M \twoheadrightarrow_\beta \mathsf{skel}(M)$ *and* $\mathsf{skel}(M)$ *is a basic form.*

4 Axioms of Meaningless Terms

In this section we recall the axioms of meaningless terms [7, 10] and give new examples of parametric infinite lambda calculi. Let $\mathcal{U} \subseteq \Lambda^\infty$ be an arbitrary set. The axioms of meaningless terms on the set $\mathcal{U}$ are:

1. Closure under β-reduction. If $M \in \mathcal{U}$ and $M \twoheadrightarrow_\beta N$ then $N \in \mathcal{U}$.
2. Overlap. If $\lambda x.M \in \mathcal{U}$ then $(\lambda x.M)N \in \mathcal{U}$.
3. Closure under substitution. If $M \in \mathcal{U}$ then $M^\sigma \in \mathcal{U}$.
4. Rootactiveness. $\mathcal{R} \subseteq \mathcal{U}$.
5. Indiscernibility. Let $M \overset{u}{\leftrightarrow} N$ denote that if N is obtained from M by replacing some (possibly infinitely many) subterms in $\mathcal{U}$ by other terms in $\mathcal{U}$. Then, $M \in \mathcal{U}$ iff $N \in \mathcal{U}$.

Definition 10. A set $\mathcal{U} \subseteq \Lambda^\infty$ of meaningless terms is a set that satisfies the five axioms of meaningless terms.

Hence, the parametric infinitary lambda calculi are the calculi $\lambda^\infty_{\beta\perp}$ with a parametric $\perp$-rule on a set $\mathcal{U}$ satisfying the axioms of meaningless terms given above. The normal form of these calculi is denoted by $\mathsf{P}_\mathcal{U}$. If $\mathcal{U} = \Lambda^\infty$ then $M =_{\mathsf{P}_\mathcal{U}} \perp$ for all $M \in \Lambda^\infty_\perp$ and $\mathsf{P}_\mathcal{U}$ induces the trivial theory. Since indiscernibility is not easy to prove, we will reduce it to some property which is easier to prove. For this, we need the following properties on a set $\mathcal{U} \subseteq \Lambda^\infty$:

1. Closure under β-expansion. If $N \in \mathcal{U}$ and $M \twoheadrightarrow_\beta N$ then $M \in \mathcal{U}$.
2. Indiscernibility on skeletons. Let P be a skeleton such that $P \preceq_\mathcal{U} M$ and $P \preceq_\mathcal{U} N$. Then, $M \in \mathcal{U}$ iff $N \in \mathcal{U}$.

Definition 11. A set $\mathcal{U}$ of strongly meaningless terms is a set that satisfies: closure under β-reduction, overlap, closure under substitution, rootactiveness, closure under β-expansion and indiscernibility on skeletons.

Theorem 12. $[7, 10]$ $\overline{\mathcal{HN}}$, $\overline{\mathcal{WN}}$ and $\overline{\mathcal{TN}} = \mathcal{R}$ are sets of meaningless terms.

Definition 13. Let $\mathcal{U} \subseteq \Lambda^\infty$, $M, N \in \Lambda^\infty_\perp$. Then, $M \preceq_\mathcal{U} N$ if M is obtained from N by replacing some subterms of N which belong to $\mathcal{U}$ by $\perp$.

Lemma 14. Let $\mathcal{U}$ be closed under substitution. If $M \preceq_\mathcal{U} N$ and $M \twoheadrightarrow_\beta M'$ then $N \twoheadrightarrow_\beta N'$ and $M' \preceq_\mathcal{U} N'$ for some N'.

Proof. This is proved by induction on the length of the reduction sequence. $\square$

The following lemma may not hold for terms that are not rootactive. For instance, take $(\lambda x.\Omega) \in \mathcal{U}$, $M = \perp P$ and $N = (\lambda x.\Omega)P$. Then $M \preceq_\mathcal{U} N$ and $N \to_\beta N' = \Omega$ but there is no M' such that $M \twoheadrightarrow_\beta M' \preceq_\mathcal{U} N'$.

Lemma 15. Let $\mathcal{U}$ be closed under substitution and M rootactive. If $M \preceq_\mathcal{U} N$ and $N \to_\beta N'$ then there exists M' such that $M \twoheadrightarrow_\beta M'$ and $M' \preceq_\mathcal{U} N'$.

Proof. We do only one step of β-reduction. Since M is rootactive, we then have that $M = (\lambda x.M_0)M_1 \ldots M_k$. But then $N = (\lambda x.N_0)N_1 \ldots N_k$ and $M_i \preceq_{\mathcal{U}} N_i$. We contract the β-redex in the head position in N and in M. Since $\mathcal{U}$ is closed under substitution, $M_0[x := M_1]M_2 \ldots M_k \preceq_{\mathcal{U}} N_0[x := N_1]N_2 \ldots N_k$.   □

Lemma 16. *Let $\mathcal{U}$ be closed under substitution. If $M \preceq_{\mathcal{U}} N$ and M rootactive then N is rootactive.*

Proof. Suppose now that N is not rootactive, then there exists a top normal form N' such that $N \twoheadrightarrow_\beta N'$ by contracting only head redexes. Then, by Lemma 15 there exists M' such that $M \twoheadrightarrow_\beta M'$ and $M' \preceq_{\mathcal{U}} N'$. If N' is a top normal form then so is M'.   □

Theorem 17. *If $\mathcal{U} \subset \Lambda^\infty$ is a set of strongly meaningless terms then it is also a set of meaningless terms.*

Proof. Both definitions have the first four axioms in common. We prove indiscernibility. Let $M \overset{\mathcal{U}}{\leftrightarrow} N$. Then there exists P such that $P \preceq_{\mathcal{U}} M$ and $P \preceq_{\mathcal{U}} N$. By Lemma 9 and Lemma 14, we have that $\mathsf{skel}(P) \preceq_{\mathcal{U}} M'$ and $\mathsf{skel}(P) \preceq_{\mathcal{U}} N'$ for some M', N' such that $M \twoheadrightarrow_\beta M'$ and $N \twoheadrightarrow_\beta N'$. By indiscernibility on skeletons $M' \in \mathcal{U}$ iff $N' \in \mathcal{U}$. Since $\mathcal{U}$ is closed under β-reduction and β-expansion, we have that $M \in \mathcal{U}$ iff $N \in \mathcal{U}$.   □

Theorem 18. *The following sets are sets of strongly meaningless terms:*

1. *$\mathcal{HA}$, $\mathcal{SA}$, $\mathcal{HA} \cup \mathcal{IL}$ and $\mathcal{HA} \cup \mathcal{O}$*
2. *$\mathcal{SA}_X$ if X is a subset of closed terms in $\mathsf{BerT}(\Lambda_\perp^\infty)$ without $\perp$.*

Proof. The first five axioms are not difficult to prove. We prove indiscernibility on skeletons for $\mathcal{SA}_X$. Suppose P is a skeleton and $P \preceq_{\mathcal{SA}_X} M, N$.

1. If P is either a head normal form, the ogre or an infinite left spine so are M and N. Hence, $M, N \notin \mathcal{SA}_X$.
2. If $P = \lambda x_1 \ldots x_n.RP_1 \ldots P_k$ is a head active form. By Lemma 16, M and N are also head active forms. Then $M = \lambda x_1 \ldots x_n.R'M_1 \ldots M_k$, $N = \lambda x_1 \ldots x_n.R''N_1 \ldots N_k$. and $P_i \preceq_{\mathcal{SA}_X} M_i, N_i$ for $1 \le i \le k$. If $M \in \mathcal{SA}_X$ then $n = 0$ and $M_i = \mathsf{BerT}(M_i) \in X \subseteq \Lambda^\infty$. Since $M_i = \mathsf{BerT}(M_i)$, we have that M_i does not contain subterms in $\mathcal{SA}_X$ and hence $P_i = M_i$. Then, P_i does not contain $\perp$ and also $P_i = N_i$. Clearly, $N_i \in X$ and $N \in \mathcal{SA}_X$.
3. Suppose $P = \lambda x_1 \ldots x_n.\perp P_1 \ldots P_k$ is a head bottom form. The bottom in the head of P has to be replaced by some term in $\mathcal{SA}_X$ to get M and N. Then, we proceed as in the previous part to prove that $P_i = M_i = N_i \in X$.   □

5   Regular and Quasi-regular Sets

In this section we define and give examples of regular and quasi-regular sets of meaningless terms. Figure 3 summarizes and shows all these sets. ordered by inclusion. We use the notation $\mathcal{U} \to \mathcal{U}$ if $\mathcal{U} \supset \mathcal{U}'$.

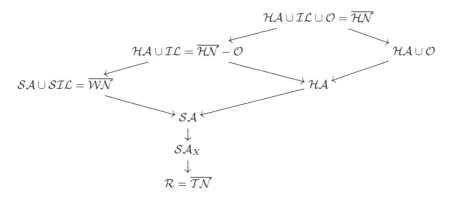

Fig. 3. Sets of meaningless terms ordered by inclusion

Definition 19. Let $\mathcal{U} \subseteq \Lambda^\infty$ be a set of meaningless terms.

1. $\mathcal{U}$ is regular if for all basic sets X, if $X \cap \mathcal{U} \neq \emptyset$ then $X \subseteq \mathcal{U}$.
2. $\mathcal{U}$ is quasi-regular if for all strongly basic sets X, if $X \cap \mathcal{U} \neq \emptyset$ then $X \subseteq \mathcal{U}$.

If a set is regular then it is quasi-regular. The sets $\mathcal{SA}_X$ are neither regular nor quasi-regular provided $X \neq \emptyset$ and $X \neq \Lambda^\infty$.

Theorem 20. *Let $\mathcal{U}$ be a set of meaningless terms.*

1. *If $\lambda x.M \in \mathcal{U}$ then $M \in \mathcal{U}$.*
2. *If $\lambda x.M \in \mathcal{U}$ then $\mathcal{HA} \subseteq \mathcal{U}$. In particular, if $\mathcal{O} \in \mathcal{U}$ then $\mathcal{HA} \subset \mathcal{U}$.*
3. *If $\mathcal{SIL} \subseteq \mathcal{U}$ then $\mathcal{SA} \subseteq \mathcal{U}$.*
4. *If $\mathcal{IL} \subseteq \mathcal{U}$ then $\mathcal{HA} \subseteq \mathcal{U}$.*
5. *If a head normal form is in $\mathcal{U}$ then $\mathcal{U} = \Lambda^\infty$.*

Proof. We only prove the first three parts. The rest are similar.

1. By the overlap and closure under β-reduction axioms, $(\lambda x.M)x \rightarrow_\beta M \in \mathcal{U}$.
2. By the overlap axiom, $(\lambda x.M)Q \in \mathcal{U}$ for all $Q \in \Lambda^\infty$. By indiscernibility we have that $RQ \in \mathcal{U}$ for $R \in \mathcal{R}$ and also $RQ_1 \ldots Q_k \in \mathcal{U}$ for all $Q_i \in \Lambda^\infty$. By the previous part and indiscernibility, $\lambda x.R \in \mathcal{U}$ and hence $\lambda x_1 \ldots x_n.RQ_1 \ldots Q_k \in \mathcal{U}$.
3. Let $({}^\omega Q) = ((\ldots)Q)Q)$. We have that $({}^\omega Q) = ({}^\omega Q)Q \in \mathcal{U}$ By indiscernibility, $RQ \in \mathcal{U}$ for any $R \in \mathcal{R}$ and also $RQ_1 \ldots Q_k \in \mathcal{U}$ for all $Q_i \in \Lambda^\infty$. □

Corollary 21. *The regular sets are: $\mathcal{HA} \cup \mathcal{IL} \cup \mathcal{O} = \overline{\mathcal{HN}}$, $\mathcal{HA} \cup \mathcal{IL} = \overline{\mathcal{HN}} - \mathcal{O}$, $\mathcal{HA} \cup \mathcal{O}$ and $\mathcal{HA}$. The quasi-regular sets are the regular ones and the sets $\mathcal{SA} \cup \mathcal{SIL} = \overline{\mathcal{WN}}$, $\mathcal{SA}$ and $\mathcal{R} = \overline{\mathcal{TN}}$.*

6   Explicit Definition of the Normal Forms

Figure 4 shows the difference between the normal forms of the different parametric infinitary lambda calculi considered in this paper.

In the figure we make the abbreviations: $\lambda \boldsymbol{x}.M = \lambda x_1 \ldots x_n.M$ and $M\boldsymbol{P} = MP_1 \ldots P_k$. For simplicity we assume that $P_i \in \mathsf{P}_\mathcal{U}(\Lambda_\perp^\infty)$ for all i. The case of head bottom forms is not shown in the table but it is as the case of head active forms where $\perp$ plays the role of the rootactive term R. The cases $\mathcal{U} = \overline{\mathcal{HN}}, \overline{\mathcal{WN}}$ and $\overline{\mathcal{TN}}$ correspond to the cases of Böhm, Lévy-Longo and Berarducci trees respectively.

SET $\mathcal{U}$	HEAD NORMAL FORM $\mathsf{P}_\mathcal{U}(\lambda \boldsymbol{x}.y\boldsymbol{P})$	OGRE $\mathsf{P}_\mathcal{U}(\mathsf{O})$	HEAD ACTIVE FORM $\mathsf{P}_\mathcal{U}(\lambda \boldsymbol{x}.R\boldsymbol{P})$	INF LEFT SPINE FORM $\mathsf{P}_\mathcal{U}(\lambda \boldsymbol{x}.((\ldots P_2)P_1))$
$\overline{\mathcal{HN}}$	$\lambda \boldsymbol{x}.y\boldsymbol{P}$	$\perp$	$\perp$	$\perp$
$\overline{\mathcal{HN}} - \mathcal{O}$	$\lambda \boldsymbol{x}.y\boldsymbol{P}$	O	$\perp$	$\perp$
$\mathcal{HA} \cup \mathcal{O}$	$\lambda \boldsymbol{x}.y\boldsymbol{P}$	$\perp$	$\perp$	$\lambda \boldsymbol{x}.((\ldots P_2)P_1)$
$\mathcal{HA}$	$\lambda \boldsymbol{x}.y\boldsymbol{P}$	O	$\perp$	$\lambda \boldsymbol{x}.((\ldots P_2)P_1)$
$\overline{\mathcal{WN}}$	$\lambda \boldsymbol{x}.y\boldsymbol{P}$	O	$\lambda \boldsymbol{x}.\perp$	$\lambda \boldsymbol{x}.\perp$
$\mathcal{SA}$	$\lambda \boldsymbol{x}.y\boldsymbol{P}$	O	$\lambda \boldsymbol{x}.\perp$	$\lambda \boldsymbol{x}.((\ldots P_2)P_1)$
$\mathcal{SA}_X$	$\lambda \boldsymbol{x}.y\boldsymbol{P}$	O	$\begin{cases} \lambda \boldsymbol{x}.\perp & \text{if } \boldsymbol{P} \in X \\ \lambda \boldsymbol{x}.\perp\boldsymbol{P} & \text{otherwise} \end{cases}$	$\lambda \boldsymbol{x}.((\ldots P_2)P_1)$
$\overline{\mathcal{TN}}$	$\lambda \boldsymbol{x}.y\boldsymbol{P}$	O	$\lambda \boldsymbol{x}.\perp\boldsymbol{P}$	$\lambda \boldsymbol{x}.((\ldots P_2)P_1)$

Fig. 4. Definition of $\mathsf{P}_\mathcal{U}(M)$ when M is a skeleton

7   Models Induced by **NF**

There are many ways of making models of lambda calculus, i.e. λ-models. In this paper we will emphasise yet another method where the lambda calculus itself does the job. The idea is simple: any confluent and normalising extension of lambda calculus gives rise to a model: namely the set of normal forms. Taking the normal form of the application of two normal forms then is the application for this semantics.

Definition 22. The model induced by NF, denoted by $\mathcal{M}(\mathsf{NF})$, is the applicative structure $(\mathsf{NF}(\Lambda_\perp^\infty), \, . \, , [\![\,]\!])$ defined as follows:

1. $M.N = \mathsf{NF}(MN)$ for all $M, N \in \mathsf{NF}(\Lambda_\perp^\infty)$,
2. $[\![M]\!]_\sigma = \mathsf{NF}(M^\sigma)$ for all $M \in \Lambda$.

It is easy to prove that $\mathcal{M}(\mathsf{NF})$ is a λ-model using confluence and normalization (see Definition 5.2.7, Definition 5.3.1 and Theorem 5.3.6 in [1]).

Definition 23. A partial order $\sqsubseteq$ on a set A is a relation on A that reflexive, transitive and antisymmetric. If the partial order $\sqsubseteq$ on A has a least element we say that $\sqsubseteq$ is a pointed poset on A.

We consider partial orders on the set $\Lambda_\perp^\infty$ or $\mathsf{NF}(\Lambda_\perp^\infty)$. If M is the least element of a pointed poset $\sqsubseteq$ on $\mathsf{NF}(\Lambda_\perp^\infty)$ then, obviously, M is in normal form. Domain Theory usually follows the convention of denoting the least element by $\perp$. In our case, $\perp$ is a special constant in the syntax which equates the undefined or meaningless terms but we will see that it is not necessarily the least element. In some cases, the least element could be the ogre O (if $\mathsf{O} \in \mathsf{NF}(\Lambda_\perp^\infty)$).

Definition 24. Let $C[\,]$ be a context in $\Lambda_\perp^\infty$. The context operator $C[\,]$ restricted to NF is the function $\lambda\!\!\!\lambda M{\in}\mathsf{NF}(\Lambda_\perp^\infty).\mathsf{NF}(C[M]) : \mathsf{NF}(\Lambda_\perp^\infty) \to \mathsf{NF}(\Lambda_\perp^\infty)$.

For the models induced by NF, it makes sense to define a notion of monotonicity that considers all context operators and not only the application.

Definition 25. The partial order $\sqsubseteq$ makes the context operators of $\mathcal{M}(\mathsf{NF})$ monotone if the following hold:

1. $(\mathsf{NF}(\Lambda_\perp^\infty), \sqsubseteq)$ is a pointed poset and
2. the context operators $C[\,]$ restricted to NF are monotone in $(\mathsf{NF}(\Lambda_\perp^\infty), \sqsubseteq)$ for all context $C[\,] \in \Lambda_\perp^\infty$.

Definition 26. We say that $\mathcal{M}(\mathsf{NF})$ is orderable (by $\sqsubseteq$) if there exists a partial order $\sqsubseteq$ on $\mathsf{NF}(\Lambda_\perp^\infty)$ that makes the context operators monotone. We say that $\mathcal{M}(\mathsf{NF})$ is unorderable if it is not orderable.

8   The Prefix Relation

Definition 27. Let $M, N \in \Lambda_\perp^\infty$. We say that M is a prefix of N (we write $M \preceq N$) if M is obtained from N by replacing some subterms of N by $\perp$.

The prefix relation $\preceq$ is a pointed poset on $\mathsf{NF}(\Lambda_\perp^\infty)$ with $\perp$ as least element.

Lemma 28. If $M \preceq N$ then there exists N' such that $\mathsf{P}_\mathcal{U}(M) \preceq N'$ and $N \twoheadrightarrow_\beta N'$.

Proof. Using Lemma 9 and by taking $\mathcal{U} = \Lambda^\infty$ in Lemma 14, we have that $\mathsf{skel}(M) \preceq N'$ for some N' such that $N \twoheadrightarrow_\beta N'$. Hence $\mathsf{P}_\mathcal{U}(M) \preceq \mathsf{skel}(M) \preceq N'$. $\qquad\square$

The following theorem is a generalization of the proof of monotonicity of BT and LLT given in [14]. It is possible to give an alternative proof of this theorem using a simulation similar to Theorem 42.

Theorem 29. *Let $\mathcal{U}$ be quasi-regular and $\mathcal{SA} \subseteq \mathcal{U}$. Then, $\mathsf{P}_{\mathcal{U}} : \Lambda_{\perp}^{\infty} \to \Lambda_{\perp}^{\infty}$ is monotone in $(\Lambda_{\perp}^{\infty}, \preceq)$.*

Proof. Let $M, N \in \Lambda_{\perp}^{\infty}$ such that $M \preceq N$. We prove that $P = \mathsf{P}_{\mathcal{U}}(M) \preceq \mathsf{P}_{\mathcal{U}}(N)$. By Lemma 28 we have that $P \preceq Q$ and $N \twoheadrightarrow_{\beta} Q$ for some Q. It is enough to prove that $P^n \preceq \mathsf{P}_{\mathcal{U}}(Q)$ (where P^n denotes the truncation of P at depth n). Then, $P = \bigcup_{n \in \omega} P^n \preceq \mathsf{P}_{\mathcal{U}}(Q)$. We prove $P^n \preceq \mathsf{P}_{\mathcal{U}}(Q)$ fo all n by induction.

1. $P = \lambda x_1 \ldots x_n.y P_1 \ldots P_m$. Then $Q = \lambda x_1 \ldots x_n.y Q_1 \ldots Q_m$ and for all i, $P_i \preceq Q_i$. Hence, $\mathsf{P}_{\mathcal{U}}(Q) = \lambda x_1 \ldots x_n.y \mathsf{P}_{\mathcal{U}}(Q_1) \ldots \mathsf{P}_{\mathcal{U}}(Q_m)$. By induction hypothesis, $(P_i)^h \preceq \mathsf{P}_{\mathcal{U}}(Q_i)$ for all $h < n$. It is easy to see that $P^n \preceq \mathsf{P}_{\mathcal{U}}(Q)$.
2. $P = \mathsf{O}$. Then $P = Q = \mathsf{O}$.
3. $P = \lambda x_1 \ldots x_n.\perp P_1 \ldots P_m$. Then, $Q = \lambda x_1 \ldots x_n.Q_0$. Since $\mathcal{SA} \subseteq \mathcal{U}$, we have that $m = 0$. If $n > 0$ then by Theorem 20 no abstraction belongs to $\mathcal{U}$ and hence $\mathsf{P}_{\mathcal{U}}(Q) = \lambda x_1 \ldots x_n.\mathsf{P}_{\mathcal{U}}(Q_0)$.
4. $P = \lambda x_1 \ldots x_n.((\ldots)P_2)P_1$. Then, $Q = \lambda x_1 \ldots x_n.((\ldots)Q_2)Q_1$. Suppose towards a contradiction that $Q \in \mathcal{U}$. Then $((\ldots)Q_2)Q_1 \in \mathcal{U}$ by Theorem 20. Since $\mathcal{U}$ is quasi-regular, all infinite left spine should belong to $\mathcal{U}$ and contradicts the fact that P is an infinite left spine in $\beta\perp$-normal form. Hence, $\mathsf{P}_{\mathcal{U}}(Q) = \lambda x_1 \ldots x_n.((\ldots)\mathsf{P}_{\mathcal{U}}(Q_2))\mathsf{P}_{\mathcal{U}}(Q_1)$. By induction hypothesis, $(\mathsf{P}_{\mathcal{U}}(P_i))^h \preceq \mathsf{P}_{\mathcal{U}}(Q_i)$ for all $h < n$. It is easy to see that $(P)^n \preceq \mathsf{P}_{\mathcal{U}}(Q)$. $\square$

The next corollary is deduced from Corollary 21 and the previous theorem.

Corollary 30. *The functions $\mathsf{NF} \in \{\mathsf{BT}, \mathsf{P}_{\overline{\mathcal{HN}}-\mathcal{O}}, \mathsf{P}_{\mathcal{HAUO}}, \mathsf{P}_{\mathcal{HA}}, \mathsf{LLT}, \mathsf{P}_{\mathcal{SA}}\}$ are monotone in $(\Lambda_{\perp}^{\infty}, \preceq)$.*

Theorem 31. *If $\mathsf{NF} : \Lambda_{\perp}^{\infty} \to \Lambda_{\perp}^{\infty}$ is monotone in $(\Lambda_{\perp}^{\infty}, \preceq)$ then the prefix relation $\preceq$ makes the context operators of $\mathcal{M}(\mathsf{NF})$ monotone.*

Proof. If $M \preceq N$ then $C[M] \preceq C[N]$. Since $\mathsf{P}_{\mathcal{U}}$ is monotone, we have that $\mathsf{P}_{\mathcal{U}}(C[M]) \preceq \mathsf{P}_{\mathcal{U}}(C[N])$. $\square$

Corollary 32. *The prefix relation $\preceq$ makes the context operators of $\mathcal{M}(\mathsf{NF})$ monotone for $\mathsf{NF} \in \{\mathsf{BT}, \mathsf{P}_{\overline{\mathcal{HN}}-\mathcal{O}}, \mathsf{P}_{\mathcal{HAUO}}, \mathsf{P}_{\mathcal{HA}}, \mathsf{LLT}, \mathsf{P}_{\mathcal{SA}}\}$.*

Corollary 33. *The models induced by $\mathsf{BT}, \mathsf{P}_{\overline{\mathcal{HN}}-\mathcal{O}}, \mathsf{P}_{\mathcal{HAUO}}, \mathsf{P}_{\mathcal{HA}}, \mathsf{LLT}$ and $\mathsf{P}_{\mathcal{SA}}$ are all orderable.*

We show some examples in which the prefix relation does not make all the context operators monotone:

1. The prefix relation $\preceq$ does not make the application monotone of $\mathcal{M}(\mathsf{BerT})$, though it makes the abstraction monotone. Take $M = \perp$, $N = \lambda x.\perp$ and $P = y$. Then $M \preceq N$ but $M \cdot P \not\preceq N \cdot P$.
2. The prefix relation $\preceq$ does not make either the abstraction or the application of $\mathcal{M}(\eta\mathsf{BT})$ and $\mathcal{M}(\infty\eta\mathsf{BT})$ monotone.
   (a) Take $M = y\perp$ and $N = yx$. Then $M \preceq N$ but $\lambda x.M \not\preceq \lambda x.N$.
   (b) Take $M = \lambda xy.z(x\perp y)y$, $N = \lambda xy.z(xyy)y$ and $P = (\lambda x.x)$. Then $M \preceq N$ but $M \cdot P \not\preceq N \cdot P$.

9   Orders for Extensionality

We define two partial orders for which the context operators of the extensional models will be monotone.

Definition 34. 1. Let $M, N \in \eta\mathsf{BT}(\Lambda^\infty_\perp)$. Then, $M \preceq_\eta N$ if $M \; _\eta\!\twoheadleftarrow P \preceq Q \twoheadrightarrow_\eta N$ for some $P, Q \in \mathsf{BT}(\Lambda^\infty_\perp)$.
2. Let $M, N \in \infty\eta\mathsf{BT}(\Lambda^\infty_\perp)$. Then, $M \preceq_{\eta!} N$ if $M \; _{\eta!}\!\twoheadleftarrow P \preceq Q \twoheadrightarrow_{\eta!} N$ for some $P, Q \in \mathsf{BT}(\Lambda^\infty_\perp)$.

Lemma 35. *[13, 15] Let $M, N \in \Lambda^\infty_\perp$. If $M \twoheadrightarrow_\eta N$, then $\mathsf{BT}(M) \twoheadrightarrow_\eta \mathsf{BT}(N)$. And if $M \twoheadrightarrow_{\eta!} N$, then $\mathsf{BT}(M) \twoheadrightarrow_{\eta!} \mathsf{BT}(N)$.*

Theorem 36. 1. $\preceq_\eta$ *makes the context operators of $\mathcal{M}(\eta\mathsf{BT})$ monotone.*
2. $\preceq_{\eta!}$ *makes the context operators of $\mathcal{M}(\infty\eta\mathsf{BT})$ monotone.*

Proof. We only prove (1). The proof of (2) is similar. Suppose that $M \preceq_\eta N$. Then $\mathsf{BT}(M) \twoheadrightarrow_\eta P \preceq Q \; _\eta\!\twoheadleftarrow \mathsf{BT}(N)$. By Lemma 35 and monotonicity of BT (Corollary 30), $\mathsf{BT}(C[M]) \twoheadrightarrow_\eta \mathsf{BT}(C[P]) \preceq \mathsf{BT}(C[Q]) \; _\eta\!\twoheadleftarrow \mathsf{BT}(C[N])$.   □

Corollary 37. *The models induced by $\eta\mathsf{BT}$ and $\infty\eta\mathsf{BT}$ are orderable.*

10   Ogre as Least Element

In order to make the application of Berarducci trees monotone, the ogre should be the least element and not $\perp$. This is a consequence of the following theorem:

Theorem 38. *If $\sqsubseteq$ makes the application of $\mathcal{M}(\mathsf{NF})$ monotone then we have that:*

1. *either $\perp$ is the least element of $\sqsubseteq$ and $\perp P \to_\perp \perp$ for all $P \in \Lambda^\infty_\perp$ or*
2. *O is the least element of $\sqsubseteq$.*

Proof. Suppose that $M \in \mathsf{NF}(\Lambda^\infty_\perp)$ is the least element. Then $M \sqsubseteq \lambda x.M$ and we choose $x \notin fv(M)$. If application is monotone then $M \cdot P \sqsubseteq (\lambda x.M) \cdot P =_{\mathsf{NF}} M$ and hence $MP =_{\mathsf{NF}} M$ for all P for all $P \in \mathsf{NF}(\Lambda^\infty_\perp)$. Now either $M = \perp$ in which case $\perp P \to_\perp \perp$ for all $P \in \Lambda^\infty_\perp$. Or $M \neq \perp$ and then $Mx = M$ for all x. Hence M is the solution of the recursive equation $M = \lambda x.M$ and so $M = \mathsf{O}$.   □

We define a partial order making the model of Berarducci trees monotone:

Definition 39. Let $\mathsf{O} \in \mathsf{NF}(\Lambda^\infty_\perp)$. We define $\trianglelefteq$ on $\mathsf{NF}(\Lambda^\infty_\perp)$ as follows: $M \trianglelefteq N$ if M is obtained from N by replacing some subterms of N by O.

It is easy to see that $\trianglelefteq$ is partial order and that O is the least element.

Definition 40. An ogre simulation is a relation $\mathcal{S}$ on $\Lambda^\infty_\perp$ such that $M\mathcal{S}N$ implies:

1. If $M = \lambda x_1 \ldots x_n.y$ then $N = \lambda x_1 \ldots x_n.y$.
2. If $M = \lambda x_1 \ldots x_n.\bot$ then $N = \lambda x_1 \ldots x_n.\bot$.
3. If $M = \lambda x_1 \ldots x_n.PQ$ then $N = \lambda x_1 \ldots x_n.P'Q'$, $P \mathcal{S} P'$ and $Q \mathcal{S} Q'$.

The relation $\trianglelefteq$ is the maximal ogre simulation.

Lemma 41. *Let* $M \trianglelefteq N$.

1. *If* $M \twoheadrightarrow_\beta M'$ *then there exists* N' *such that* $M' \trianglelefteq N'$ *and* $N \twoheadrightarrow_\beta N'$.
2. *If* $N \twoheadrightarrow_\beta N'$ *then there exists* M' *such that* $M' \trianglelefteq N'$ *and* $M \twoheadrightarrow_\beta M'$.
3. *If* M *is rootactive then* N *is rootactive.*

Proof. The first two parts are proved by induction on the length of the reduction sequence. The last part uses the second one. □

Theorem 42. *Let* $O \in P_{\mathcal{U}}(\Lambda_\bot^\infty)$. *If* $\mathcal{U}$ *is quasi-regular then* $P_{\mathcal{U}} : \Lambda_\bot^\infty \to \Lambda_\bot^\infty$ *is monotone in* $(\Lambda_\bot^\infty, \trianglelefteq)$.

Proof. Let $M, N \in \Lambda_\bot^\infty$ such that $M \trianglelefteq N$. We prove that $P_{\mathcal{U}}(M) \trianglelefteq P_{\mathcal{U}}(N)$. Let $U = \mathsf{skel}(M)$. By Lemma 41 we have that $U \trianglelefteq V$ and $N \twoheadrightarrow_\beta V$ for some V. We define $\mathcal{S}$ as the set of pairs $(P_{\mathcal{U}}(P), P_{\mathcal{U}}(Q))$ such that P and Q are subterms of respectively U and V at the same position p and they are not subterms of rootactive terms. Note that if $U \trianglelefteq V$ then $P \trianglelefteq Q$. We prove that $\mathcal{S}$ is an ogre simulation. Suppose $(P, Q) \in \mathcal{S}$. Then,

1. $P = \lambda x_1 \ldots x_n.yP_1 \ldots P_m$. Then $Q = \lambda x_1 \ldots x_n.yQ_1 \ldots Q_m$ and for all i, $P_i \trianglelefteq Q_i$. Hence, $P_{\mathcal{U}}(P) = \lambda x_1 \ldots x_n.yP_{\mathcal{U}}(P_1) \ldots P_{\mathcal{U}}(P_m)$ and $P_{\mathcal{U}}(Q) = \lambda x_1 \ldots x_n.yP_{\mathcal{U}}(Q_1) \ldots P_{\mathcal{U}}(Q_m)$. By definition of $\mathcal{S}$, $(P_{\mathcal{U}}(P_i), P_{\mathcal{U}}(Q_i)) \in \mathcal{S}$.
2. $P = O$. Then $P_{\mathcal{U}}(P) = O$.
3. $P = \lambda x_1 \ldots x_n.RP_1 \ldots P_m$. Then, $Q = \lambda x_1 \ldots x_n.Q_0Q_1 \ldots Q_m$, also $R \trianglelefteq Q_0$ and $P_i \trianglelefteq Q_i$ for $1 \leq i \leq m$. By Lemma 41, if R is rootactive so is Q_0. Hence, $P_{\mathcal{U}}(P) = \lambda x_1 \ldots x_n.\bot P_{\mathcal{U}}(P_1) \ldots P_{\mathcal{U}}(P_m)$ and $P_{\mathcal{U}}(Q) = \lambda x_1 \ldots x_n.\bot P_{\mathcal{U}}(Q_1) \ldots P_{\mathcal{U}}(Q_m)$. By definition of $\mathcal{S}$, we have that $(P_{\mathcal{U}}(P_i), P_{\mathcal{U}}(Q_i)) \in \mathcal{S}$.
4. $P = \lambda x_1 \ldots x_n.\bot P_1 \ldots P_m$. Similar to the previous case.
5. $P = \lambda x_1 \ldots x_n.((\ldots)P_2)P_1$. Then $Q = \lambda x_1 \ldots x_n.((\ldots)Q_2)Q_1$. We have two cases:
   (a) If $P_{\mathcal{U}}(P) = \lambda x_1 \ldots x_n.\bot$ then $P_{\mathcal{U}}(Q) = \lambda x_1 \ldots x_n.\bot$ by Theorem 20 and the fact that $\mathcal{U}$ is quasi-regular.
   (b) $P_{\mathcal{U}}(P) = \lambda x_1 \ldots x_n.((\ldots)P_{\mathcal{U}}(P_2))P_{\mathcal{U}}(P_1)$. By Theorem 20 and since $\mathcal{U}$ is quasi-regular, we have that $P_{\mathcal{U}}(Q) = \lambda x_1 \ldots x_n.((\ldots)P_{\mathcal{U}}(Q_2))P_{\mathcal{U}}(Q_1)$. By definition of $\mathcal{S}$, $(P_{\mathcal{U}}(P_i), P_{\mathcal{U}}(Q_i)) \in \mathcal{S}$. □

The next corollary is deduced from Corollary 21 and the previous theorem.

Corollary 43. BerT, $P_{\mathcal{SA}}$, LLT, $P_{\mathcal{HA}}$ *and* $P_{\overline{\mathcal{HN}}-\mathcal{O}}$ *are monotone in* $(\Lambda_\bot^\infty, \trianglelefteq)$.

Theorem 44. *If* $\mathsf{NF} : \Lambda_\bot^\infty \to \Lambda_\bot^\infty$ *is monotone in* $(\Lambda_\bot^\infty, \trianglelefteq)$ *then* $\trianglelefteq$ *makes the context operators of* $\mathcal{M}(\mathsf{NF})$ *monotone.*

Proof. If $M \trianglelefteq N$ then $C[M] \trianglelefteq C[N]$. Since $\mathsf{P}_{\mathcal{U}}$ is monotone, we have that $\mathsf{P}_{\mathcal{U}}(C[M]) \trianglelefteq \mathsf{P}_{\mathcal{U}}(C[N])$. $\qquad\square$

Corollary 45. $\trianglelefteq$ *makes the context operators monotone of the models induced by* BerT, $\mathsf{P}_{\mathcal{SA}}$, LLT, $\mathsf{P}_{\mathcal{HA}}$ *and* $\mathsf{P}_{\overline{\mathcal{HN}}-\mathcal{O}}$.

Corollary 46. *The model induced by* BerT *is orderable.*

The order $\trianglelefteq$ does not make the context operators of the models induced by $\mathsf{P}_{\mathcal{SA}_X}$ monotone if $X \neq \emptyset$ and $X \neq \Lambda^\infty$. For instance, if $X = \{I\}$ then $\bot\mathsf{O} \ntrianglelefteq \bot =_{\mathsf{P}_{\mathcal{SA}_X}} \bot I$.

11   Unorderable Models

In this section we construct 2^c unorderable models induced by the infinitary lambda calculus where c is the cardinality of the continuum. We consider the set $\mathcal{B}^0$ of closed terms in $\mathsf{BT}(\Lambda_\bot^\infty)$ without $\bot$ which has the cardinality c of the continuum. For each subset X of $\mathcal{B}^0$, we construct an infinitary lambda calculus as follows. By Theorem 18, $\mathcal{SA}_{(X \cup \mathcal{O})}$ is a set of meaningless terms and $\mathsf{P}_{\mathcal{SA}_{(X \cup \mathcal{O})}}$ is a parametric tree which we abbreviate as U_X.

Theorem 47. *Let $X \subseteq \mathcal{B}^0$ be non-empty. The models induced by the parametric trees U_X are unorderable.*

Proof. Suppose there exists a partial order $\sqsubseteq$ that makes the context operators of $\mathcal{M}(\mathsf{U}_X)$ monotone. By Theorem 38, we have that O is the least element of $\sqsubseteq$. Since X is non-empty, there exists $M \in X$ and $M = \lambda x_1 \ldots x_n.x_i M_1 \ldots M_k$. Take $N = \lambda x_1 \ldots x_n.x_i \mathsf{O} \ldots \mathsf{O}$. On one hand, both head bottom forms $\bot\mathsf{O}$ and $\bot M$ reduce to $\bot$. On the other hand, the head bottom form $\bot N$ does not reduce to $\bot$. We have that $N \notin X \cup \mathcal{O}$ because the terms in $X \subseteq \mathcal{B}^0$ are Böhm trees that have a head normal form at any depth. Hence, $\bot\mathsf{O} =_{\mathsf{U}_X} \bot \sqsubseteq \bot N \sqsubseteq \bot =_{\mathsf{U}_X} \bot M$. $\qquad\square$

Corollary 48. *There are 2^c unorderable models induced by the infinitary lambda calculus where c is the cardinality of the continuum.*

Acknowledgements

We thank Mariangiola Dezani-Ciancaglini and Alexander Kurz for helpful comments and discussions.

References

1. H. P. Barendregt. *The Lambda Calculus: Its Syntax and Semantics.* North-Holland, Amsterdam, Revised edition, 1984.

2. M. Barr. Terminal coalgebras for endofunctors on sets. *Theoretical Computer Science*, 114(2):299–315, 1999.

3. A. Berarducci. Infinite λ-calculus and non-sensible models. In *Logic and algebra (Pontignano, 1994)*, pages 339–377. Dekker, New York, 1996.

4. A. Berarducci and M. Dezani-Ciancaglini. Infinite λ-calculus and types. *Theoretical Computer Science*, 212(1-2):29–75, 1999. Gentzen (Rome, 1996).

5. M. Coppo, M. Dezani-Ciancaglini, and M. Zacchi. Type theories, normal forms, and D_∞-lambda-models. *Information and Computation*, 72(2):85–116, 1987.

6. M. Dezani-Ciancaglini, P. Severi, and F. J. de Vries. Infinitary lambda calculus and discrimination of Berarducci trees. *Theoretical Computer Science*, 298(2)(275–302), 2003.

7. J. Kennaway and F. J. de Vries. Infinitary rewriting. In Terese, editor, *Term Rewriting Systems*, volume 55 of *Cambridge Tracts in Theoretical Computer Science*, pages 668–711. Cambridge University Press, 2003.

8. J. R. Kennaway, J. W. Klop, M. R. Sleep, and F. J. de Vries. Infinite lambda calculus and Böhm models. In *Rewriting Techniques and Applications*, volume 914 of *LNCS*, pages 257–270. Springer-Verlag, 1995.

9. J. R. Kennaway, J. W. Klop, M. R. Sleep, and F. J. de Vries. Infinitary lambda calculus. *Theoretical Computer Science*, 175(1):93–125, 1997.

10. J. R. Kennaway, V. van Oostrom, and F. J. de Vries. Meaningless terms in rewriting. *J. Funct. Logic Programming*, Article 1:35 pp, 1999.

11. J. W. Klop. *Combinatory Reduction Systems*, volume 127 of *Mathematical centre tracts*. Mathematisch Centrum, 1980.

12. A. Salibra. Topological incompleteness and order incompleteness of the lambda calculus. *ACM Transactions on Computational Logic*, 4(3):379–401, 2003. (Special Issue LICS 2001).

13. P. Severi and F. J. de Vries. An extensional Böhm model. In *Rewriting Techniques and Applications*, volume 2378 of *LNCS*, pages 159–173. Springer-Verlag, 2002.

14. P. Severi and F. J. de Vries. Continuity and discontinuity in lambda calculus. In *Typed Lambda Calculus and Applications*, volume 3461 of *LNCS*. Springer-Verlag, 2005.

15. P. Severi and F. J. d. Vries. A Lambda Calculus for D_∞. Technical report, University of Leicester, 2002.

Higher-Order Matching and Games

Colin Stirling

School of Informatics, University of Edinburgh
cps@inf.ed.ac.uk

Abstract. We provide a game-theoretic characterisation of higher-order matching. The idea is suggested by model checking games. We then show that some known decidable instances of matching can be uniformly proved decidable via the game-theoretic characterisation.

Keywords: games, higher-order matching, typed lambda calculus.

1 The Matching Problem

Assume simply typed lambda calculus with base type $\mathbf{0}$ and the definitions of α-equivalence, β and η-reduction. A type is $\mathbf{0}$ or $A \to B$ where A and B are types. A type A always has the form $(A_1 \to (\ldots A_n \to \mathbf{0})\ldots)$ which is usually written $A_1 \to \ldots \to A_n \to \mathbf{0}$. We also assume a standard definition of *order*: the order of $\mathbf{0}$ is 1 and the order of $A_1 \to \ldots \to A_n \to \mathbf{0}$ is $k+1$ where k is the maximum of the orders of the A_is.

Terms are built from a countable set of variables $x, y, \ldots$ and constants, $a, f, \ldots$: each variable and constant is assumed to have a unique type. The set of simply typed terms is the smallest set T such that if x (f) has type A then $x : A \in T$ $(f : A \in T)$, if $t : B \in T$ and $x : A \in T$, then $\lambda x.t : A \to B \in T$, and if $t : A \to B \in T$ and $u : A \in T$ then $tu : B \in T$. The *order* of a typed term is the order of its type. A typed term is *closed* if it does not contain free variables.

A *matching problem* has the form $v = u$ where $v, u : A$ for some type A, and u is closed. The *order* of the problem is the maximum of the orders of the free variables $x_1, \ldots, x_n$ in v. A *solution* of a matching problem is a sequence of terms $t_1, \ldots, t_n$ such that $v\{t_1/x_1, \ldots, t_n/x_n\} =_{\beta\eta} u$. The decision question is: given a matching problem, does it have a solution? The problem is conjectured to be decidable in [3]. However, if it is decidable then its complexity is non-elementary [9, 11]. Decidability has been proven for the general problem up to order 4 and for various special cases [5, 6, 8]. Loader proved that the matching problem is undecidable for the variant definition of solution that uses just β-equality [4]. An excellent source of information about the problem is [2].

Throughout, we slightly change the syntax of terms and types. The type $A_1 \to \ldots \to A_n \to \mathbf{0}$ is rewritten $(A_1, \ldots, A_n) \to \mathbf{0}$ and we assume that all terms in normal form are in η-*long form*. That is, if $t : \mathbf{0}$ then it either has the form $u : \mathbf{0}$ where u is a constant or a variable, or has the form $u(t_1, \ldots, t_k)$ where $u : (B_1, \ldots, B_k) \to \mathbf{0}$ is either a constant or a variable and each $t_i : B_i$ is in η-long form. And if $t : (A_1, \ldots, A_n) \to \mathbf{0}$ then t has the form $\lambda y_1 \ldots y_n.t_0$

L. Ong (Ed.): CSL 2005, LNCS 3634, pp. 119–134, 2005.

where $t_0 : \mathbf{0}$ is a term in η-long form. A term is *well-named* if each occurrence of a variable y within a λ abstraction is unique.

An interpolation equation has the form $x(v_1, \ldots, v_n) = u$ where each v_i is a closed term in normal form and $u : \mathbf{0}$ is also in normal form. The *type* of the equation is the type of the free variable x, which has the form $(A_1, \ldots, A_n) \to \mathbf{0}$ where $v_i : A_i$. An *interpolation problem* P is a finite family of interpolation equations $x(v_1^i, \ldots, v_n^i) = u_i$, $i : 1 \le i \le m$, all with the same free variable x. The *type* of P is the type A of the variable x and the *order* of P is the order of A. A *solution* of P of type A is a closed term $t : A$ such that $t(v_1^i, \ldots, v_n^i) =_\beta u_i$ for each i. We write $t \models P$ if the closed term t solves the problem P.

An interpolation problem reduces to matching: there is the equivalent problem $f(x(v_1^1, \ldots, v_n^1), \ldots, x(v_1^m, \ldots, v_n^m)) = f(u_1, \ldots, u_m)$, when $f : \mathbf{0}^m \to \mathbf{0}$. Schubert shows the converse, that a matching problem of order n is reducible to an interpolation problem of order at most $n + 2$ [7]. A *dual* interpolation problem includes inequations $x(v_1^i, \ldots, v_n^i) \ne u_i$. Padovani proved that a matching problem of order n is reducible to dual interpolation of the same order [6]. In the following we concentrate on the interpolation problem for orders greater than 1. If P has order 1 then it has the form $x = u_i$, $1 \le i \le m$. Consequently, P only has a solution if $u_i = u_j$ for each i and j.

In the following we develop a game-theoretic characterisation of $t \models P$. The idea is inspired by model-checking games (such as in [10]) where a structure, a transition graph, is navigated relative to a property and players make choices at appropriate positions. In section 2 we define some preliminary notions and in section 3 we present the term checking game and prove its correctness. Unlike transition graphs, terms t involve binding which results in moves that jump around t. The main virtue of using games is that they allow one to understand little "pieces" of a solution term t in terms of subplays and how they thereby contribute to solving P. In section 4 we identify regions of a term t that we call "tiles" and define their subplays. In section 5 we introduce four transformations on tiles that preserve a solution term: these transformations are justified by analysing subplays. In section 6 we then show that the transformations provide simple proofs of decidability for known instances of the interpolation problem via the small model property: if $t \models P$ then $t' \models P$ for some small term t'.

2   Preliminaries

A right term u of an interpolation equation may contain bound variables: an example is $f(a, \lambda x_1 \ldots x_4.x_1(x_1(x_2)))$. Let $X = \{x_1, \ldots, x_k\}$ be the set of bound variables in u. Assume a fresh set of constants $C = \{c_1, \ldots, c_k\}$ such that each c_i has the same type as x_i.

Definition 1 The *ground closure* of a closed term w, whose bound variables belong to X, with respect to C, written $\mathrm{Cl}(w, X, C)$, is defined inductively:

1. if $w = a : \mathbf{0}$, then $\mathrm{Cl}(w, X, C) = \{a\}$
2. if $w = f(w_1, \ldots, w_n)$, then $\mathrm{Cl}(w, X, C) = \{w\} \cup \bigcup \mathrm{Cl}(w_i, X, C)$
3. if $w = \lambda x_{j_1} \ldots x_{j_n}.u$, then $\mathrm{Cl}(w, X, C) = \mathrm{Cl}(u\{c_{j_1}/x_{j_1}, \ldots, c_{j_n}/x_{j_n}\}, X, C)$

The ground closure of $u = f(a, \lambda x_1 \ldots x_4.x_1(x_1(x_2)))$ with respect to $\{c_1, \ldots, c_4\}$ is the set of ground terms $\{u, a, c_1(c_1(c_2)), c_1(c_2), c_2\}$.

Next, we wish to identify subterms of the left-hand terms v_j of an interpolation equation relative to a finite set of constants C.

Definition 2 The *subterms* of w relative to C, written $\mathrm{Sub}(w, C)$, is defined inductively using an auxiliary set $\mathrm{Sub}'(w, C)$:

1. if w is a variable or a constant, then $\mathrm{Sub}(w, C) = \mathrm{Sub}'(w, C) = \{w\}$
2. if w is $x(w_1, \ldots, w_n)$ then $\mathrm{Sub}(w, C) = \mathrm{Sub}'(w, C) = \{w\} \cup \bigcup \mathrm{Sub}(w_i, C)$
3. if w is $f(w_1, \ldots, w_n)$, then $\mathrm{Sub}(w, C) = \mathrm{Sub}'(w, C) = \{w\} \cup \bigcup \mathrm{Sub}'(w_i, C)$
4. if w is $\lambda y_1 \ldots y_n.v$, then $\mathrm{Sub}(w, C) = \{w\} \cup \mathrm{Sub}(v, C)$
5. if w is $\lambda y_1 \ldots y_n.v$, then $\mathrm{Sub}'(w, C) = \bigcup \mathrm{Sub}(v\{c_{i_1}/y_1, \ldots, c_{i_n}/y_n\}, C)$ where each $c_{i_j} \in C$ has the same type as y_j

For the remainder of the paper we assume a fixed interpolation problem P of type A whose order is greater than 1. P has the form $x(v_1^i, \ldots, v_n^i) = u_i$, $1 \leq i \leq m$, where each v_j^i and u_i are in long normal form. We also assume that terms v_j^i and u_i are well-named and that no pair share bound variables. For each i, let X_i be the (possibly empty) set of bound variables in u_i and let C_i be a corresponding set of new constants (that do not occur in P), the *forbidden constants*. We are interested in when $t \models P$ and t does not contain forbidden constants.

Definition 3 Assume $P : A$ is the fixed interpolation problem:

1. T is the set of subtypes of A and the subtypes of subterms of u_i
2. for each i, the right subterms are $\mathsf{R}_i = \mathrm{Cl}(u_i, X_i, C_i)$
3. for each i, the left subterms are $\mathsf{L}_i = \bigcup \mathrm{Sub}(v_j^i, C_i) \cup C_i$

3   Tree-Checking Games

Using ideas suggested by model-checking we present a characterisation of interpolation. This is not the first time that such techniques have been applied to higher-order matching. Comon and Jurski define (bottom-up) tree automata for the 4th-order case that characterise all solutions to a problem [1]. The states of the automata essentially depend on Padovani's representation of the observational equivalence classes of terms up to 4th-order [6]. The existence of such an automaton not only guarantees decidability, but also shows that the set of all solutions is regular.

We now introduce a game-theoretic characterisation of interpolation for *all* orders. The idea is inspired by model-checking games where a model (a transition graph) is traversed relative to a property and players make choices at appropriate positions. Similarly, in the following game the model is a putative solution term t that is traversed relative to the interpolation problem. However, because of binding play may jump here and there in t. Consequently, our games lack the simple control structure of Comon and Jurski's automata where flow starts at the leaves of t and proceeds to its root. Moreover, the existence of the game does not assure decidability. Its purpose is to provide a mechanism for understanding how small pieces of a solution term contribute to solving the problem.

A. $t_m = \lambda y_1 \ldots y_j$ and $t_m \downarrow_1 t'$ and $q_m = q[(l_1, \ldots, l_j), r]$. So, $t_{m+1} = t'$ and $\theta_{m+1} = \theta_m\{l_1\eta_m/y_1, \ldots, l_j\eta_m/y_j\}$ and q_{m+1} and η_{m+1} are by cases on t_{m+1}.
   1. $a : \mathbf{0}$. So, $\eta_{m+1} = \eta_m$. If $r = a$ then $q_{m+1} = q[\exists]$ else $q_{m+1} = q[\forall]$.
   2. $f : (B_1, \ldots, B_k) \to \mathbf{0}$. So, $\eta_{m+1} = \eta_m$. If $r = f(s_1, \ldots, s_k)$ then $q_{m+1} = q_m$ else $q_{m+1} = q[\forall]$.
   3. $y : B$. If $\theta_{m+1}(y) = l\eta_i$, then $q_{m+1} = q[l, r]$ and $\eta_{m+1} = \eta_i$.
B. $t_m = f : (B_1, \ldots, B_k) \to \mathbf{0}$ and $q_m = q[(l_1, \ldots, l_j), f(s_1, \ldots, s_k)]$. So, $\theta_{m+1} = \theta_m$ and $\eta_{m+1} = \eta_m$ and q_{m+1} and t_{m+1} are decided as follows.
   1. $\forall$ chooses a direction $i' : 1 \leq i' \leq k$ and $t_m \downarrow_{i'} t'$. So, $t_{m+1} = t'$. If $s_{i'} : \mathbf{0}$, then $q_{m+1} = q[(\ ), s_{i'}]$. If $s_{i'}$ is $\lambda x_{i_1} \ldots x_{i_n}.s$ then $q_{m+1} = q[(c_{i_1}, \ldots, c_{i_n}), s\{c_{i_1}/x_{i_1}, \ldots, c_{i_n}/x_{i_n}\}]$.
C. $t_m = y$ and $q_m = q[l, r]$. If $l = \lambda z_1 \ldots z_j.w$ and $t_m \downarrow_i t'_i$, $1 \leq i \leq j$, then $\eta_{m+1} = \eta_m\{t'_1\theta_m/z_1, \ldots, t'_j\theta_m/z_j\}$ else $\eta_{m+1} = \eta_m$. The remaining components t_{m+1}, q_{m+1} and η_{m+1} are by cases on l.
   1. $a : \mathbf{0}$ or $\lambda\overline{z}.a$. So, $t_{m+1} = t_m$ and $\theta_{m+1} = \theta_m$. If $r = a$ then $q_{m+1} = q[\exists]$ else $q_{m+1} = q[\forall]$.
   2. $c : (B_1, \ldots, B_k) \to \mathbf{0}$. So, $\theta_{m+1} = \theta_m$. If $r \neq c(s_1, \ldots, s_k)$ then $t_{m+1} = t_m$ and $q_{m+1} = q[\forall]$. If $r = c(s_1, \ldots, s_k)$ then $\forall$ chooses a direction $i' : 1 \leq i' \leq k$ and $t_m \downarrow_{i'} t'$. So, $t_{m+1} = t'$. If $s_{i'} : \mathbf{0}$, then $q_{m+1} = q[(\ ), s_{i'}]$. If $s_{i'}$ is $\lambda x_{i_1} \ldots x_{i_n}.s$ then $q_{m+1} = q[(c_{i_1}, \ldots, c_{i_n}), s\{c_{i_1}/x_{i_1}, \ldots, c_{i_n}/x_{i_n}\}]$.
   3. $f(w_1, \ldots, w_k)$ or $\lambda\overline{z}.f(w_1, \ldots, w_k)$. So, $t_{m+1} = t_m$ and $\theta_{m+1} = \theta_m$. If $r \neq f(s_1, \ldots, s_k)$, then $q_{m+1} = q[\forall]$. If $r = f(s_1, \ldots, s_k)$ then $\forall$ chooses a direction $i' : 1 \leq i' \leq k$. If $s_{i'} : \mathbf{0}$ then $q_{m+1} = q[w_{i'}, s_{i'}]$. If $w_{i'} = \lambda z'_1 \ldots z'_n.w$ and $s_{i'} = \lambda x_{i_1} \ldots x_{i_n}.s$, then $q_{m+1} = q[w\{c_{i_1}/z'_1, \ldots, c_{i_n}/z'_n\}, s\{c_{i_1}/x_{i_1}, \ldots, c_{i_n}/x_{i_n}\}]$.
   4. $z'(l_1, \ldots, l_k)$ or $\lambda\overline{z}.z'(l_1, \ldots, l_k)$. If $\eta_{m+1}(z') = t'\theta_i$ then $\theta_{m+1} = \theta_i$ and $t_{m+1} = t'$ and $q_{m+1} = q[(l_1, \ldots, l_k), r]$.

Fig. 1. Game moves

We assume that a potential solution term t for P has the right type, is in long normal form, is well-named (with variables that are disjoint from variables in P) and does not contain forbidden constants. The term t is represented as a tree, tree(t). If t is $y : \mathbf{0}$ or $a : \mathbf{0}$ then tree(t) is the single node labelled with t. In the case of $u(v_1, \ldots, v_k)$ when u is a variable or a constant, we assume that a dummy λ with the empty sequence of variables is placed before any subterm $v_i : \mathbf{0}$ in the tree representation. With this understanding, if t is $u(v_1, \ldots, v_n)$, then tree(t) consists of the root node labelled u and n-successor nodes labelled with tree(v_i). We use the notation $u \downarrow_i t'$ to represent that tree t' is the ith successor of the node u. If t is $\lambda\overline{y}.v$, where $\overline{y}$ is a possibly empty sequence of variables $y_1 \ldots y_n$, then tree(t) consists of the root node labelled $\lambda\overline{y}$ and a single successor node tree(v): in this case we assume $\lambda\overline{y} \downarrow_1$ tree(v). We also assume that each node labelled with an occurrence of a variable y_j has a backward arrow $\uparrow^j$ to the $\lambda \overline{y}$ that binds it: the index j tells us which element is y_j in $\overline{y}$.

The tree representation of $\lambda y_1 y_2 . f(f(y_2, y_2), y_1(y_2))$ is tantamount to the syntax tree of $\lambda y_1 y_2 . f(\lambda . f(\lambda . y_2, \lambda . y_2), \lambda . y_1(\lambda . y_2))$. In the following we use t to be the λ-term t, or its λ-tree or the label (a constant, variable or $\lambda \overline{y}$) at its root node.

The tree-checking game $\mathsf{G}(t, P)$ is played by one participant, player $\forall$, the refuter who attempts to show that t is not a solution of P. The game appeals to a finite set of states involving elements of L_i and R_i. There are three kinds of states: argument, value and final states. Argument states have the form $q[(l_1, \ldots, l_k), r]$ where each $l_j \in \mathsf{L}_i$ (and k can be 0) and $r \in \mathsf{R}_i$. Value states have the form $q[l, r]$ where $l \in \mathsf{L}_i$ and $r \in \mathsf{R}_i$. A final state is either $q[\forall]$, the winning state for $\forall$, or $q[\exists]$, the losing state for $\forall$.

The game appeals to a sequence of supplementary look-up tables θ_j and η_j, $j \geq 1$: θ_j is a partial map from variables in t to elements $w\eta_k$ where $w \in \mathsf{L}_i$ and $k < j$, and η_j is a partial map from variables in L_i to elements $t'\theta_k$ where t' is a node of the tree t and $k < j$. The initial elements θ_1 and η_1 are both the empty table.

A *play* of $\mathsf{G}(t, P)$ is a sequence of positions $t_1 q_1 \theta_1 \eta_1, \ldots, t_n q_n \theta_n \eta_n$ where each t_i is a node of t and $t_1 = \lambda \overline{y}$ is the root of t, and each q_i is a state, and q_n is a final state. A node t' of the tree t may repeatedly occur in a play. The initial state is decided as follows: $\forall$ chooses an equation $x(v_1^i, \ldots, v_n^i) = u_i$ from P and $q_1 = q[(v_1^i, \ldots, v_n^i), u_i]$. If the current position is $t_m q_m \theta_m \eta_m$ and q_m is not a final state, then the next position $t_{m+1} q_{m+1} \theta_{m+1} \eta_{m+1}$ is determined by a move of Figure 1.

Moves are divided into three groups that depend on t_m. Group A covers the case when $t_m = \lambda \overline{y}$, group B when $t_m = f$ and group C when $t_m = y$. We assume standard updating notation for θ_{m+1} and η_{m+1}: $\beta\{\alpha_1/y_1, \ldots, \alpha_m/y_m\}$ is the function similar to β except that $\beta(y_i) = \alpha_i$. Moreover, in the case of rules B1, C2 and C3 we assume that the constants c_{i_j} belong to the forbidden sets C_i. The look-up tables are used in rules A3 and C4. If $t_m = \lambda \overline{y}$ and $t_m \downarrow_1 t_{m+1} = y$, then η_{m+1} and q_{m+1} are determined by the entry for y in θ_{m+1}: if the entry is $l\eta_i$, then l is the left element of q_{m+1} and $\eta_{m+1} = \eta_i$. In the case of C4, if $t_m = y$ and $q_m = q[l, r]$ and $l = z'(l_1, \ldots, l_k)$ or $\lambda \overline{z}.z'(l_1, \ldots, l_k)$, then θ_{m+1} and t_{m+1} are determined by the entry for z' in the table η_{m+1}: if the entry is $t'\theta_i$ then $t_{m+1} = t'$ and $\theta_{m+1} = \theta_i$. It is this rule that allows the next move to be a jump around the term tree (to a node labelled with a λ). The moves A1-A3, B1 and C2 traverse down the term tree while C1 and C3 remain at the current node.

Example 1 Let P be the problem $x(v) = u$ where $v = \lambda z.z$ and $u = f(\lambda x.x)$. Let $X = \{x\}$ and $C = \{c\}$ and let t be the term $\lambda y.y(y(f(\lambda y_1.y_1)))$ and so, $\mathrm{tree}(t)$ is

$$(t_1)\lambda y \downarrow_1 (t_2)y \downarrow_1 (t_3)\lambda \downarrow_1 (t_4)y \downarrow_1 (t_5)\lambda \downarrow_1 (t_6)f \downarrow (t_7)\lambda y_1 \downarrow_1 (t_8)y_1$$

There is just one play of $\mathsf{G}(t, P)$, as follows.

$$t_1\, q[(\lambda z.z), f(\lambda x.x)]\, \theta_1\, \eta_1$$
$$t_2\, q[\lambda z.z, f(\lambda x.x)]\, \theta_2 \eta_2 \quad \theta_2 = \theta_1\{(\lambda z.z)\eta_1/y\} \quad \eta_2 = \eta_1 \qquad\qquad A3$$
$$t_3\, q[(\ ), f(\lambda x.x)]\, \theta_3 \eta_3 \quad \theta_3 = \theta_2 \qquad\qquad\qquad \eta_3 = \eta_2\{t_3\theta_2/z\} \quad C4$$
$$t_4\, q[\lambda z.z, f(\lambda x.x)]\, \theta_4 \eta_4 \quad \theta_4 = \theta_3 \qquad\qquad\qquad \eta_4 = \eta_1 \qquad\qquad A3$$
$$t_5\, q[(\ ), f(\lambda z.z)]\, \theta_5\, \eta_5 \quad \theta_5 = \theta_4 \qquad\qquad\qquad \eta_5 = \eta_4\{t_5\theta_4/z\} \quad C4$$
$$t_6\, q[(\ ), f(\lambda z.z)]\, \theta_6\, \eta_6 \quad \theta_6 = \theta_5 \qquad\qquad\qquad \eta_6 = \eta_5 \qquad\qquad A2$$
$$t_7\, q[(c), c]\, \theta_7\, \eta_7 \quad\quad \theta_7 = \theta_6 \qquad\qquad\qquad \eta_7 = \eta_6 \qquad\qquad B1$$
$$t_8\, q[c, c]\, \theta_8\, \eta_8 \quad\quad\; \theta_8 = \theta_7\{c\eta_7/y_1\} \qquad\quad \eta_8 = \eta_7 \qquad\qquad A3$$
$$t_8\, q[\exists]\, \theta_9\, \eta_9 \quad\quad\;\; \theta_9 = \theta_8 \qquad\qquad\qquad \eta_9 = \eta_8 \qquad\qquad C1$$

The game rule applied to produce a move is also given. □

A partial play of $\mathsf{G}(t, P)$ finishes when a final state, $q[\forall]$ or $q[\exists]$, occurs. Player $\forall$ *loses* a play if the final state is $q[\exists]$ and $\forall$ loses the game $\mathsf{G}(t, P)$ if she loses every play. The following result provides a characterisation of $t \models P$.

Theorem 1 $\forall$ *loses* $\mathsf{G}(t, P)$ *if, and only if,* $t \models P$.

Proof. For any position $t_i q_i \theta_i \eta_i$ of a play of $\mathsf{G}(t, P)$ we say that it *m-holds* (*m-fails*) if $q = q[\exists]$ ($q = q[\forall]$) and when q_i is not final, by cases on t_i and q_i (and look-up tables become delayed substitutions)

- if $t_i = \lambda\overline{y}$ and $q_i = q[(l_1, \ldots, l_k), r]$ and t' is $(t_i\theta_i)(l_1\eta_i, \ldots, l_k\eta_i)$ then $t' = r$ ($t' \neq r$) and t' normalises with m β-reductions
- if $t_i = f$ and $q_i = q[(l_1, \ldots, l_k), r]$ and t' is $t_i\theta_i$ then $t' = r$ ($t' \neq r$) and t' normalises with m β-reductions
- if $t_i = z$ and $q_i = q[l, r]$ and $t_i \downarrow_j t'_j$ and t' is $l\eta_i(t'_1\theta_i, \ldots, t'_k\theta_i)$ then $t' = r$ ($t' \neq r$) and t' normalises with m β-reductions.

The following are easy to show by case analysis.

1. if $t_i q_i \theta_i \eta_i$ *m*-holds then $q_i = q[\exists]$ or for any next position $t_{i+1} q_{i+1}\theta_{i+1}\eta_{i+1}$ it m'-holds, $m' < m$, or it m'-holds, $m' \leq m + 1$, and the right-term in q_{i+1} is smaller than in q_i
2. if $t_i q_i \theta_i \eta_i$ *m*-fails then $q_i = q[\forall]$ or there is a next position $t_{i+1} q_{i+1}\theta_{i+1}\eta_{i+1}$ and it m'-fails, $m' < m$, or it m'-fails, $m' \leq m + 1$, and the right-term in q_{i+1} is smaller than in q_i

For instance, assume $t_i q_i \theta_i \eta_i$ *m*-holds and $t_i = \lambda y_1 \ldots y_k$ and $t_i \downarrow_1 t_{i+1} = y$ and $t_{i+1} \downarrow_j t'_j$ and $q_i = q[(l_1, \ldots, l_k), r]$. So, $\theta_{i+1} = \theta_i\{\overline{l_j\eta_i}/\overline{y_j}\}$ and $q_{i+1} = q[l, r]$ if $\theta_{i+1}(y) = l\eta_m$ and $\eta_{i+1} = \eta_n$. So, $t_i = \lambda y_1 \ldots y_k.y(t'_1, \ldots, t'_m)$ and by assumption $(t_i\theta_i)(l_1\eta_i, \ldots, l_k\eta_i) = r$. With a β-reduction we get $\theta_{i+1}(y)(t'_1\theta_{i+1}, \ldots, t'_m\theta_{i+1})$ which is $(l\eta_{i+1})(t'_1\theta_{i+1}, \ldots, t'_m\theta_{i+1})$ and so position $t_{i+1} q_{i+1}\theta_{i+1}\eta_{i+1}$ $(m-1)$-holds. Next, assume $t_i q_i \theta_i \eta_i$ *m*-holds, $t_i = f$, $q_i = q[(l_1, \ldots, l_j), f(s_1, \ldots, s_k)]$ and $t_i \downarrow_j t'_j$. By assumption, $f(t'_1, \ldots, t'_k)\theta_i = f(s_1, \ldots, s_k)$. So, $t'_j\theta_i = s_j$. Consider any choice of next position. If $s_j : \mathbf{0}$ then $q_{i+1} = q[(\ ), s_j]$, $t_{i+1} = t'_j$ and $\theta_{i+1} = \theta_i$. Therefore, $t'_j\theta_{i+1} = s_j$ and so this next position either m'-holds, $m' < m$ or m-holds and s_j is smaller than $f(s_1, \ldots, s_k)$. Alternatively, $s_j = \lambda\overline{x}.s$. Therefore, $t'_j = \lambda\overline{z}.t'$ and $t'\theta_i\{\overline{c_i}/\overline{z_i}\} = s\{\overline{c_i}/\overline{x_i}\}$ where the c_is are new, m'-holds for $m' \leq m$. And so $t'_j\theta_i(c_1, \ldots, c_n) = s\{\overline{c_i}/\overline{x_i}\}$ $(m' + 1)$-holds, as required. Assume $t_i q_i \theta_i \eta_i$

m-holds and $t_i = y$, $q_i = q[l, r]$, $l = \lambda z_1 \ldots z_k.w$, $w = z(l_1, \ldots, l_m)$, $t_i \downarrow_j t'_j$ and $t_{i+1}\theta_{i+1} = \eta_{i+1}(z)$. By assumption, $(\lambda z_1 \ldots z_k.w)\eta_i(t'_1\theta_i, \ldots, t'_k\theta_i) = r$. With one β-reduction $\eta_{i+1}(z)(l_1\eta_{i+1}, \ldots, l_m\eta_{i+1}) = r$, that is $t'_{i+1}\theta_{i+1}((l_1\eta_{i+1}, \ldots, l_m\eta_{i+1})$ $= r$ and so the next position $(m-1)$-holds. All other cases of 1 are similar to one of these three, and the proof of 2 is also very similar.

The result follows from 1 and 2: if $t \models P$ then for each initial position there is an m such that it m-holds and if $t \not\models P$ then there is an initial position that m-fails. □

The tree checking game can be easily extended to characterise dual interpolation by including a second player $\exists$ who is responsible for choices involving inequations.

Assume that $t_0 \models P$, so $\forall$ loses the game $\mathsf{G}(t_0, P)$. The number of plays is the number of branches in the right terms of P. We can index each play with $i\alpha$ when α is a branch of the right-term of the ith equation of P containing forbidden constants: $\pi^{i\alpha}$ is the play where all $\forall$ choices are dictated by α. This means that two plays $\pi^{i\alpha}$, $\pi^{i\beta}$ have a common prefix and differ after a position involving a $\forall$ choice, when the branches α and β diverge.

We also allow π to range over *subplays* which are consecutive subsequences of positions of any play of $\mathsf{G}(t_0, P)$. The length of π, $|\pi|$, is the number of positions in π. We let $\pi(i)$ be the ith position of π, $\pi(i, j)$ be the interval $\pi(i), \ldots, \pi(j)$ and π_i be its ith suffix, the interval $\pi(i, |\pi|)$. For ease of notation, we write $t \in \pi(i)$, $q \in \pi(i)$, $\theta \in \pi(i)$ and $\eta \in \pi(i)$ if $\pi(i) = tq\theta\eta$ and $t \notin \pi(i)$ means that $\pi(i) = t'q\theta\eta$ and $t \neq t'$. If $q = q[(l_1, \ldots, l_k), r]$ or $q[l, r]$ then its *right-term* is r.

Definition 1 A subplay π is *ri, right-term invariant*, if $q \in \pi(1)$ and $q' \in \pi(|\pi|)$ share the same right-term r.

Definition 2 Table θ' *extends* θ if for all $y \in \mathrm{dom}(\theta)$, $\theta'(y) = \theta(y)$. Similarly, η' extends η if for all $z \in \mathrm{dom}(\eta)$, $\eta'(z) = \eta(z)$.

We widen the usage of "extends" to positions: $\pi(j)$ θ-extends $\pi(i)$ if $\theta' \in \pi(j)$ extends $\theta \in \pi(i)$, $\pi(j)$ η-extends $\pi(i)$ if $\eta' \in \pi(j)$ extends $\eta \in \pi(i)$ and $\pi(j)$ extends $\pi(i)$ if $\pi(j)$ θ-extends and η-extends $\pi(i)$.

If $\pi(i)$'s look-up table is called when move A3 or C4 produces $\pi(j)$ then $\pi(j)$ is a child of $\pi(i)$.

Definition 3 Assume $\pi \in \mathsf{G}(t_0, P)$. If $\pi(i) = t\,q[(l_1, \ldots, l_k), r]\,\theta\,\eta$, $\pi(j) = t'q[l_m, r']\theta'\eta$, $\theta'(t') = l_m\eta$ and $t' \uparrow^m t$, then $\pi(j)$ is a *child* of $\pi(i)$. If $\pi(i) = y\,q[\lambda z_1 \ldots \lambda z_k.w, r]\,\theta\,\eta$, $\pi(j-1) = y'\,q[l, r']\,\theta'\,\eta'$, $l = \lambda\bar{x}.z_m(\bar{l})$ or $\lambda\bar{x}.z_m$ or $z_m(\bar{l})$ or z_m and $\eta'(z_m) = t'\eta$ and $y \downarrow_m t'$, then $\pi(j)$ is a *child* of $\pi(i)$.

Fact 1 If $\pi(j)$ is a child of $\pi(i)$ then $\pi(j)$ *extends* $\pi(i)$.

4   Tiles and Subplays

Assume that $t_0 \models P$. We would like to identify regions of the tree t_0. For this purpose, we define *tiles* that are partial trees.

Definition 1 Assume $B = (B_1, \ldots, B_k) \to \mathbf{0} \in \mathsf{T}$.

1. λ is an *atomic leaf* of type $\mathbf{0}$
2. if $x_j : B_j$ then $\lambda x_1 \ldots x_k$ is an *atomic leaf* of type B
3. $a : \mathbf{0}$ is a *constant* tile
4. if $f : B$ and $t_j : B_j$ are atomic leaves then $f(t_1, \ldots, t_k)$ is a *constant* tile
5. $y : \mathbf{0}$ is a *simple* tile
6. if $y : B$ and $t_j : B_j$ are atomic leaves then $y(t_1, \ldots, t_k)$ is a *simple* tile

A region of t_0 can be identified with a constant or simple tile. A leaf $u : \mathbf{0}$ of t_0 is the tile u. If $B \neq \mathbf{0}$ then an occurrence of $u : B$ in t_0, $u = f$ or y, with its immediate children $\lambda \overline{x}_1, \ldots, \lambda \overline{x}_k$, where $\overline{x}_i$ may be empty, is the tile $u(\lambda \overline{x}_1, \ldots, \lambda \overline{x}_k)$ in t_0.

Tiles in t_0 induce subplays of $\mathsf{G}(t_0, P)$. A play on $t = f(\lambda \overline{x}_1, \ldots, \lambda \overline{x}_k)$ is a pair of positions $\pi(i, i+1)$ with $t \in \pi(i)$: $q[(l_1, \ldots, l_m), r] \in \pi(i)$, $r = f(s_1, \ldots, s_k)$, $\lambda \overline{x}_j \in \pi(i+1)$ is a leaf of t and $q[(\,), s_j]$ or $q[(c_1, \ldots, c_n), s_j\{\overline{c}_{i'}/\overline{z}_{i'}\}]$ is the state in $\pi(i+1)$, depending on the type of s_j.

Definition 2 A subplay π is a *play* on $y(\lambda \overline{x}_1, \ldots, \lambda \overline{x}_k)$ in t_0 if $y \in \pi(1)$ and $\pi(|\pi|)$ is a child of $\pi(1)$. It is a *j-play* if $\lambda \overline{x}_j \in \pi(|\pi|)$.

A play π on $y(\lambda \overline{x}_1, \ldots, \lambda \overline{x}_k)$ in t_0 can have arbitrary length. It starts at y and finishes at a leaf $\lambda \overline{x}_i$. In between, flow of control can be almost anywhere in t_0 (including y). Crucially, $\pi(|\pi|)$ extends $\pi(1)$: the free variables in the sub-tree of t_0 rooted at y preserve their values, and the free variables in w when $q[\lambda z_1 \ldots z_k.w, r] \in \pi(1)$ also preserve their values. If $\pi \in \mathsf{G}(t_0, P)$ and $y \in \pi(i)$ then there can be numerous plays $\pi(i, j)$ on $y(\lambda \overline{x}_1, \ldots, \lambda \overline{x}_k)$ in t_0, including no plays at all. We now examine some pertinent properties of plays

Proposition 1 *Assume* $\pi \in \mathsf{G}(t_0, P)$, $\pi(i, m)$ *and* $\pi(i, n)$, $n > m$, *are plays on* $y(\lambda \overline{x}_1, \ldots, \lambda \overline{x}_k)$ *and* $\lambda \overline{x}_j \in \pi(m)$.

1. *There is a position* $\pi(m')$, $m' < n$, *that is a child of* $\pi(m)$.
2. *If* $\pi(m')$ *is the first position that is a child of* $\pi(m)$, $t' \in \pi(m')$, y_1 *occurs on the branch between* $\lambda \overline{x}_j$ *and* t', t' *is an* i'-*descendent of* y_1 *and* $y_1 \downarrow_{i'} \lambda \overline{z}_{i'}$, *then there is an* i'-*play* $\pi(m_1, n_1)$ *on* $y_1(\lambda \overline{z}_1, \ldots, \lambda \overline{z}_{k'})$ *such that* $m < m_1$ *and* $n_1 < m'$.
3. *If* $\pi(m + m')$ *is the first position that is a child of* $\pi(m)$, $\pi(m, m + m')$ *is ri and* $\pi(i, n)$ *is a j-play then* $\pi(n + m')$ *is the first position that is a child of* $\pi(n)$, $\pi(n, n + m')$ *is ri and for all* $n' \leq m'$, $t \in \pi(m + n')$ *iff* $t \in \pi(n + n')$.
4. *If* $\pi(m + m')$ *is the first position that is a child of* $\pi(m)$, $\pi(m, m + m')$ *is not ri and* $\pi(i, n)$ *is a j-play then there is a* $\pi' \in \mathsf{G}(t_0, P)$ *with* $\pi'(n) = \pi(n)$, $\pi'(n + m')$ *is the first position that is a child of* $\pi'(n)$, $\pi'(n, n + m')$ *is not ri and for all* $n' \leq m'$, $t \in \pi(m + n')$ *iff* $t \in \pi'(n + n')$.

Proof. 1. Assume $\pi(i) = y \, q[\lambda z_1 \ldots z_k.w, r] \, \theta \, \eta_i$ and $\pi(i, m)$, $\pi(i, n)$ are plays on $y(\lambda \overline{x}_1, \ldots, \lambda \overline{x}_k)$ with $\lambda \overline{x}_j \in \pi(m)$. The table $\eta = \eta_i\{\lambda \overline{x}_1\theta/z_1, \ldots, \lambda \overline{x}_k\theta/z_k\}$ belongs to $\pi(i+1)$ and positions $\pi(m-1)$, $\pi(n-1)$ both η-extend $\pi(i+1)$.

because $\pi(m)$, $\pi(n)$ are children of $\pi(i)$. No look-up table $\eta_l \in \pi(l)$, $l < i+1$, has these entries $\eta(z_{i'}) = \lambda\overline{x}_{i'}\theta$. Consider the first position $\pi(m_1)$ after $\pi(m)$ that is at a variable $y_1 \in \pi(m_1)$. Clearly, y_1 is a descendent of $\lambda\overline{x}_j$ in t_0. If y_1 is bound by $\lambda\overline{x}_j$ then $\pi(m_1)$ is a child of $\pi(m)$ and the result is proved. Otherwise, there are two cases $\pi(m_1)$ is a child of $\pi(l)$, $l < i$, and, so, by move A3 its look-up table η' cannot extend η. Play may jump anywhere in t_0 by move C4. If there is not a play $\pi(m_1, n_1)$ on the simple tile headed with y_1 then for all later positions $\pi(m_2)$, $m_2 > m_1$, $\pi(m_2)$ cannot η-extend $\pi(i + 1)$ which is a contradiction. Therefore, play must continue with a position $\pi(n_1)$ that is a child of $\pi(m_1)$. Secondly, y_1 is bound by a $\lambda\overline{y}$ that is below $\lambda\overline{x}_j$. But then y_1 is bound to a leaf of a constant tile that occurs between $\lambda\overline{x}_j$ and y_1 and so move C3 must apply and play proceeds to a child of y_1. This argument is now repeated for the next position after $\pi(n_1)$ that is at a variable $y_2 \in \pi(m_2)$: y_2 must be a descendent of $\lambda\overline{x}_j$. The argument proceeds as above, except there is the new case that $\pi(m_2)$ is a child of $\pi(n_1)$. However, by move A3, $\pi(m_2)$ cannot η-extend $\pi(i + 1)$. Therefore, eventually play must reach a child of $\pi(m)$.

2. This follows from the proof of 1.

3. Assume $\pi(m+m')$ is the first position that is a child of $\pi(m)$, $\pi(m, m+m')$ is ri and $\pi(i, n)$ is a j-play. Consequently, $\pi(m) = \lambda\overline{x}_j\, q\,\theta\,\eta$ and $\pi(n) = \lambda\overline{x}_j\, q'\,\theta\,\eta'$ and both η-extend $\pi(i + 1)$ because they are both children of $\pi(i)$. Consider positions $\pi(m + 1)$, $\pi(n + 1)$. If $m' = 1$ the result follows. Otherwise, by move A3, $\pi(m + 1) = y_1\, q[l, r]\,\theta_1\,\eta_1$ and $\pi(n + 1) = y_1\, q[l, r']\,\theta_1'\,\eta_1$. These positions have the same look-up table η_1, the same left-terms in their state, and θ_1, θ_1' only differ in their values for the variables that are bound by $\lambda\overline{x}_j$. Therefore, play must continue from both positions in the same way until a child of $\pi(m)$ and $\pi(n)$ is reached.

4. Assume $\pi = \pi^{i\alpha}$. The argument is similar to 3 except that the same $\forall$ choices in the non ri play $\pi(m, m + m')$ need to be made. Therefore, there must be a $\pi' = \pi^{i\beta}$ such that $\pi'(n) = \pi(n)$ and the same $\forall$ choices are made in $\pi'(n, n + m')$.   □

Tiles can be composed to form composite tiles. A (possibly composite) tile is a partial tree which can be extended at any atomic leaf. If $t(\lambda\overline{x})$ is a tile with leaf $\lambda\overline{x}$ and t' is a constant or simple tile, then $t(\lambda\overline{x}.t')$ is the composite tile that is the result of placing t' directly beneath $\lambda\overline{x}$ in t. Throughout, we assume that tiles are well-named. We now define a salient kind of simple or composite tile.

Definition 3 A tile is *basic* if it contains one occurrence of a free variable and does not contain any constants. A tile is an (extended) constant tile if it contains one occurrence of a constant and no occurrences of a free variable.

The single occurrence of a free variable in a basic tile must be its head variable and the single occurrence of a constant in a constant tile must be its head occurrence.

A contiguous region of t_0 can be identified with a basic or constant tile: a node y with its children and some, or all, of their children and so on (as long as children of a variable $y' : B \neq 0$ are included) is a larger region that is a basic tile if y is its only free variable and it contains no constants. We write $t(\lambda \overline{x}_1, \ldots, \lambda \overline{x}_k)$ if t is a basic tile with atomic leaves $\lambda \overline{x}_1, \ldots, \lambda \overline{x}_k$. A basic or constant tile in t_0 induces subplays of $\mathsf{G}(t_0, P)$ that are compositions of plays of its component tiles.

Definition 4 A subplay π is a *play* on $t(\lambda \overline{x}_1, \ldots, \lambda \overline{x}_k)$ in t_0 if $t \in \pi(1)$, for some i, $\lambda \overline{x}_i \in \pi(|\pi|)$, there is the branch $t = y_1 \downarrow_{j_1} \lambda \overline{x}_{j_1}^1 \downarrow_1 y_2 \ldots y_n \downarrow_{j_n} \lambda \overline{x}_{j_n}^n = \lambda \overline{x}_i$ and π can be split into plays $\pi(i_m, j_m)$ on $y_m(\lambda \overline{x}_1^m, \ldots \lambda \overline{x}_{k_m}^m)$ where $i_1 = 1$, $i_{m+1} = j_m + 1$ and $j_n = |\pi|$. It is a *j-play* if $\lambda \overline{x}_j \in \pi(|\pi|)$.

The definition for constant tiles is similar. Properties of plays of simple tiles lift to plays of basic tiles.

Corollary 1 *Assume* $\pi \in \mathsf{G}(t_0, P)$, $\pi(i, m')$ *and* $\pi(i, n')$, $n' > m'$, *are plays on* $t(\lambda \overline{x}_1, \ldots, \lambda \overline{x}_k)$ *and* $\lambda \overline{x}_j \in \pi(m')$, $t = y_1 \downarrow_{j_1} \lambda \overline{x}_{j_1}^1 \downarrow_1 y_2 \ldots y_n \downarrow_{j_n} \lambda \overline{x}_{j_n}^n = \lambda \overline{x}_j$ *and* $\pi(i, m')$ *is split into plays* $\pi(i_m, j_m)$ *on* $y_m(\lambda \overline{x}_1^m, \ldots \lambda \overline{x}_{k_m}^m)$ *where* $i_1 = i$, $i_{m+1} = j_m + 1$ *and* $j_n = m'$.

1. $\pi(m')$ *extends* $\pi(i)$.
2. *There is a position* $\pi(m_1)$, $m' < m_1 < n'$, *that is a child of* $\pi(j_i)$ *for some* i.
3. *If* $\pi(m_1)$ *is the first position that is a child of* $\pi(j_i)$ *for some* i, $t' \in \pi(m_1)$, y' *occurs on the branch between* $\lambda \overline{x}_j$ *and* t', t' *is an* i'-*descendent of* y' *and* $y' \downarrow_{i'} \lambda \overline{z}_{i'}$, *then there is an* i'-*play* $\pi(m_2, n_2)$ *on* $y'(\lambda \overline{z}_1, \ldots, \lambda \overline{z}_{k'})$ *such that* $m' < m_2$ *and* $n_2 < m_1$.
4. *If* $\pi(m' + m_1)$ *is the first position that is a child of* $\pi(j_i)$, *for some* i, $\pi(m', m'+m_1)$ *is ri and* $\pi(i, n')$ *is a* j-*play then* $\pi(n'+m_1)$ *is the first position that is a child of any position* $\pi(n'')$ *such that* $\lambda \overline{x}_{j_i}^i \in \pi(n'')$, $\pi(n', n + m_1)$ *is ri and for all* $n_1 \leq m_1$, $t \in \pi(m' + n_1)$ *iff* $t \in \pi(n' + n_1)$.
5. *If* $\pi(m' + m_1)$ *is the first position that is a child of* $\pi(j_i)$, *for some* i, $\pi(m', m' + m_1)$ *is not ri and* $\pi(i, n')$ *is a* j-*play then then there is a* $\pi' \in \mathsf{G}(t_0, P)$ *with* $\pi'(n') = \pi(n')$ *and* $\pi'(n' + m_1)$ *is the first position that is a child of any position* $\pi'(n'')$ *such that* $\lambda \overline{x}_{j_i}^i \in \pi'(n'')$, $\pi'(n', n + m_1)$ *is not ri and for all* $n_1 \leq m_1$, $t \in \pi(m' + n_1)$ *iff* $t \in \pi'(n' + n_1)$.

Definition 5 Assume π is a j-play (play) on t. It is a *shortest* j-play (play) if no proper prefix of π is a j-play (play) and it is an *ri* j-play (play) if π is also ri. It is a *canonical* j-play (play) if each $t' \in \pi(i)$ is a node of t. Two plays π and π' on t are *independent* if one is not contained in the other: that is, $\pi \neq \pi_1 \pi' \pi_2$ and $\pi' \neq \pi_1 \pi \pi_2$.

Definition 6 Two basic tiles t and t' in t_0 are *equivalent*, written $t \equiv t'$ if they are the same basic tiles with the same free variable y (bound to the same $\lambda \overline{y}$). A tile t' is a *j-descendent* of $t(\lambda \overline{x}_1, \ldots, \lambda \overline{x}_k)$ in t_0 if there is a branch in t_0 from $\lambda \overline{x}_j$ to t'.

Definition 7 The tile $t(\lambda \overline{x}_1, \ldots, \lambda \overline{x}_k)$ is *j-end* in t_0, if every free variable below $\lambda \overline{x}_j$ in t_0 is bound above t. It is an *end* tile if it is j-end for all j. The tile $t(\lambda \overline{x}_1, \ldots, \lambda \overline{x}_k)$ is a *top* tile in t_0 if its free variable y is bound by the initial lambda $\lambda \overline{y}$ of t_0.

A shortest play on a top tile is canonical. The following is a simple consequence of Corollary 1.

Fact 1 *If $\pi \in G(t_0, P)$ and t is a j-end tile and $t \in \pi(i)$, then there is at most one j-play $\pi(i, m)$ on t.*

We also want to classify tiles according to their plays.

Definition 8 The tile $t(\lambda\overline{x}_1, \ldots, \lambda\overline{x}_k)$ is sri if every shortest play on t is ri. It is j-ri if every shortest j-play on it is ri.

Definition 9 Assume $t(\lambda\overline{x}_1, \ldots, \lambda\overline{x}_k)$ is a basic tile in t_0 and π is a subplay. We inductively define when t is j-*directed in* π

1. if $t \notin \pi(i)$ for all i, then t is j-directed in π
2. if $\pi(i)$ is the first position with $t \in \pi(i)$ and there is a shortest j-play $\pi(i, m)$ on t and $\pi(i, m)$ is ri and t is j-directed in π_{m+1}, then t is j-directed in π.

Definition 10 Tile t is j-*directed* in t_0 if it is j-directed in every $\pi \in G(t_0, P)$.

If t is j-directed in t_0 then $\pi \in G(t_0, P)$ is partitioned uniquely into a sequence of ri inner regions $\pi(i_k, m_k)$ which are shortest j-plays on t.

$$\pi(1) \ldots \underset{t}{\pi(i_1)} \ldots \underset{\lambda\overline{x}_j}{\pi(m_1)} \ldots \underset{t}{\pi(i_n)} \ldots \underset{\lambda\overline{x}_j}{\pi(m_n)} \ldots \pi(|\pi|)$$

By definition, t cannot occur outside these regions. If $\pi = \pi^{i\alpha}$ then any play $\pi^{i\beta}$ will have the same intervals $\pi^{i\beta}(i_k, m_k)$ until the point that $\pi^{i\alpha}, \pi^{i\beta}$ diverge (which is outside a region). A tile can be j-directed in t_0 for multiple j.

We now pick out an interesting feature about embedded end tiles.

Proposition 2 *If $t_1 \equiv t_2$ are end tiles in t_0 and t_2 is a j-descendent of t_1, then either t_2 is j-directed in t_0 or there are $\pi, \pi' \in G(t_0, P)$ and j-plays $\pi(m_1, n_1)$ on t_1, $\pi'(m_2, n_2)$ on t_2 that are not ri and $m_2 > n_1$.*

Proof. Assume $t_1 \equiv t_2$ are end tiles and t_2 is a j-descendent of t_1. Both t_1 and t_2 have the same head variable bound to the same $\lambda\overline{y}$ above t_1 in t_0. Let $\pi \in G(t_0, P)$. Consider the first position $t_2 \in \pi(m)$. There must be an earlier position $t_1 \in \pi(i)$ such that $\pi(m)$ extends $\pi(i)$ and a j-play $\pi(i, i + k)$ on t_1. If this play is ri then because $t_1 \equiv t_2$ are end tiles there is the same j-play on t_2, $\pi(m, m + k)$. This argument is repeated for subsequent plays or until the j-play on t_1 is not ri. If the play on t_1 is not ri then for some play π' with $\pi'(m) = \pi(m)$ there is the same j-play $\pi'(m, m + k)$ on t_2. $\qquad \square$

5   Transformations

In this section we define four transformations. A transformation $\mathbf{T}$ changes a tree s into a tree t, written $s\,\mathbf{T}\,t$. Each transformation preserves the crucial property: if $s\,\mathbf{T}\,t$ and $s \models P$ then $t \models P$ which is proved using the game-theoretic characterisation. The first transformation is easy. Let t' be a subtree of t_0 whose root node is a variable y or a constant $f : B \neq \mathbf{0}$. $\mathsf{G}(t_0, P)$ *avoids* t' if $t' \notin \pi(i)$ for all positions and plays $\pi \in \mathsf{G}(t_0, P)$. Let $t_0[a/t']$ be the result of replacing t' in t_0 with the constant $a : \mathbf{0}$.

T1 If $\mathsf{G}(t_0, P)$ avoids t' then transform t_0 to $t_0[a/t']$

Assume that $t_0 \models P$. The other transformations involve basic tiles. If a j-end tile is j-directed then it is redundant and can be removed from t_0.

T2 Assume $t(\lambda \overline{x}_1, \ldots, \lambda \overline{x}_k)$ is a j-directed, j-end tile in t_0 and t' is the subtree of t_0 rooted at t. If t_j is the subtree directly beneath $\lambda \overline{x}_j$ then transform t_0 to $t_0[t_j/t']$.

The next transformation separates plays.

Definition 1 Assume $t = t(\lambda \overline{x}_1, \ldots, \lambda \overline{x}_k)$ is a basic sri tile in t_0 that is not an end tile. Tile t is a *separator* if there are two independent shortest plays that end at different leaves of t.

T3 If $t(\lambda \overline{x}_1, \ldots, \lambda \overline{x}_k)$ is a separator in t_0 and t' is the subtree of t_0 rooted at t then transform t_0 to $t_0[t(\lambda \overline{x}_1.t', \ldots, \lambda \overline{x}_k.t'/t')]$.

Here, we have added an extra copy of t directly below each $\lambda \overline{x}_j$: we assume that the head variable of this copy of t is bound by the $\lambda \overline{y}$ that binds the head variable of the original t and we assume that all variables below $\lambda \overline{x}_j$ that are bound in t in t_0 are now bound in the copy of t: this means that the original t becomes an end tile.

The next transformation, in effect, allows tiles to be "lowered" in t_0.

Definition 2 Assume $t(\lambda \overline{x}_1, \ldots, \lambda \overline{x}_k)$ is j-ri and not j-end in t_0 and directly below $\lambda \overline{x}_j$ is the constant or basic tile $u(\lambda \overline{z}_1, \ldots, \lambda \overline{z}_m)$ whose head variable, if there is one, is not bound in t. Tile t is j-*permutable with* u in t_0 if whenever $\pi(i, m)$ is a shortest j-play on t then either (1) there are no other j-plays $\pi(i, m')$ on t or (2) $\pi(m+1, n)$ is a shortest play on u and it is ri and u is an end tile.

T4 Assume $t(\lambda \overline{x}_1, \ldots, \lambda \overline{x}_k)$ is j-permutable with $u(\lambda \overline{z}_1, \ldots, \lambda \overline{z}_m)$ in t_0 and t' is the subtree rooted at u in t_0. If t_i and t'_i are the subtrees of t_0 directly below $\lambda \overline{x}_i$ and $\lambda \overline{z}_i$ then transform t_0 to $t_0[u(\lambda \overline{z}_1.w_1, \ldots, \lambda \overline{z}_m.w_m)/t']$ where $w_i = t(\lambda \overline{x}_1.t_1, \ldots, \lambda \overline{x}_{j-1}.t_{j-1}, \lambda \overline{x}_j.t'_i, \lambda \overline{x}_{j+1}.t_{j+1}, \ldots, \lambda \overline{x}_k.t_k)$.

The tile t is copied below u: however, in the copy of t below $\lambda \overline{z}_i$ of u t'_i (and not t_i) occurs below $\lambda \overline{x}_j$ of t. We assume that the free variables of t_i and t'_i retain their binders in the transformed term and that the copies of t below u bind the free x_j.

Consider the case when the j-ri tile t is not j-permutable with the constant tile $f(\lambda \overline{z}_1, \ldots, \lambda \overline{z}_m)$. There is a shortest j-play $\pi(i, m)$ on t and another j-play $\pi(i, n)$ on t.

$$\begin{array}{ccccccc} \pi(i) & \ldots & \pi(m) & \pi(m+1) & \ldots & \pi(n) & \pi(n+1) \\ t & & \lambda \overline{x}_j & f & & \lambda \overline{x}_j & f \end{array}$$

Consequently, permuting t with f is not permitted: the transformed term would exclude the extra play on f.

In an application of **T4**, if t is a top j-ri tile and every shortest j-play is canonical then after its application t will be j-end and j-directed, and therefore can be removed by **T2**. In this case, the tile t does percolate down the term tree t_0.

We now show that the four transformations preserve interpolation.

Proposition 1 *For $1 \leq i \leq 4$, if $s\,\mathbf{T}i\,t$ and $s \models P$ then $t \models P$.*

Proof. This is clear when $i = 1$. Consider $i = 2$. Assume $t(\lambda\overline{x}_1, \ldots, \lambda\overline{x}_k)$ is a j-directed, j-end tile in t_0, t' is the subtree of t_0 rooted at t and t_j is the subtree directly beneath $\lambda\overline{x}_j$, $t'_0 = t_0[t_j/t']$ and $t_0 \models P$. We shall convert $\pi = \pi^{i\alpha} \in \mathsf{G}(t_0, P)$ into the play $\sigma = \sigma^{i\alpha} \in \mathsf{G}(t'_0, P)$ that $\forall$ loses. The play π is split uniquely into regions.

$$\pi(1) \ldots \underset{t}{\pi(i_1)} \ldots \underset{\lambda\overline{x}_j}{\pi(m_1)} \ldots \underset{t}{\pi(i_2)} \ldots \underset{\lambda\overline{x}_j}{\pi(m_2)} \ldots \underset{t}{\pi(i_n)} \ldots \underset{\lambda\overline{x}_j}{\pi(m_n)} \ldots \pi(|\pi|)$$

The play σ is just the outer subplays (modulo minor changes to the look-up tables) because each $\pi(m_k)$ extends $\pi(i_k)$.

$$\pi(1) \ldots \pi(i_1 - 1)\pi(m_1 + 1) \ldots \pi(i_n - 1)\pi(m_n + 1) \ldots \pi(|\pi|)$$

We show, using a similar argument as is used in Proposition 1.1 of Section 4, that if s is a node in t or is a descendent of a leaf $\lambda\overline{x}_m$, $m \neq j$, of t then s cannot occur in any outer subplay of π. If s were to appear in an outer subplay then move C4 must have applied: there is then a variable y and a position in an outer subplay $y \in \pi(n)$ and $\theta \in \pi(n)$ and $\theta(y) = l\eta$ and there is a free variable z in l such that $\eta(z) = s\theta'$. However, this is impossible. Consider $\theta_1 \in \pi(i_1)$: clearly, there is no free variable in the subtree rooted at t with this property. When play reaches $\pi(m_1)$ because t is a j-end tile and because $\pi(m_1)$ extends $\pi(i_1)$ there cannot be a free variable in the subtree t_j with this property either. This argument is now repeated for subsequent positions $\pi(i_k)$ and $\pi(m_k)$.

Let $i = 3$. Assume $t(\lambda\overline{x}_1, \ldots, \lambda\overline{x}_k)$ is a separator in t_0, t' is the subtree of t_0 rooted at t and $t'_0 = t_0[t(\lambda\overline{x}_1.t', \ldots, \lambda\overline{x}_k.t')/t']$. We shall convert $\pi = \pi^{i\alpha} \in \mathsf{G}(t_0, P)$ into $\sigma = \sigma^{i\alpha} \in \mathsf{G}(t'_0, P)$ that $\forall$ loses. Consider any shortest play on t in $\pi^{i\alpha}$, $\pi(m, k)$ and assume it is a j-play. By definition this play is ri. Therefore, this interval is transformed into the following interval for t'_0.

$$\underset{t}{\pi(m)} \ldots \underset{\lambda\overline{x}_j}{\pi(k)} \underset{t}{\pi(m)} \ldots \underset{\lambda\overline{x}_j}{\pi(k)}$$

where the second t is the copy of t directly beneath $\lambda\overline{x}_j$ in t'_0.

Finally, $i = 4$. Assume $t(\lambda\overline{x}_1, \ldots, \lambda\overline{x}_k)$ is j-permutable with $u(\lambda\overline{z}_1, \ldots, \lambda\overline{z}_m)$ in t_0, t' is the subtree rooted at u in t_0, t_i and t'_i are the subtrees of t_0 directly below $\lambda\overline{x}_i$ and $\lambda\overline{z}_i$ and $t'_0 = t_0[u(\lambda\overline{z}_1.w_1, \ldots, \lambda\overline{z}_m.w_m)/t']$ where w_i is as in **T4**. We shall convert $\pi = \pi^{i\alpha} \in \mathsf{G}(t_0, P)$ into $\sigma = \sigma^{i\alpha} \in \mathsf{G}(t'_0, P)$ that $\forall$ loses. The play π can be divided into non-overlapping regions $\pi(i_k, m_k)$.

$$\pi(1) \ldots \underset{t}{\pi(i_1)} \ldots \underset{\lambda\overline{x}_j}{\pi(m_1)} \underset{u}{\pi(m_1 + 1)} \ldots \underset{t}{\pi(i_n)} \ldots \underset{\lambda\overline{x}_j}{\pi(m_n)} \underset{u}{\pi(m_n + 1)} \ldots \pi(|\pi|)$$

where $\pi(i_k, m_k)$ are shortest j-plays: such a region may also contain other shortest j-plays on t:

$$\ldots \underset{t}{\pi(i_k)} \ldots \underset{t}{\pi(i')} \ldots \underset{\lambda\overline{x}_j}{\pi(m')} \ldots \underset{\lambda\overline{x}_j}{\pi(m_k)} \ldots$$

If $u = f(\lambda\overline{z}_1, \ldots, \lambda\overline{z}_n)$ is a constant tile then (1) of Definition 2 applies: so each $\pi(i_k, m_k)$ only contains a single occurrence of $\lambda\overline{x}_j$ because the play is ri. Moreover, there are no further j-plays $\pi(i_k, m')$ on t. Therefore, σ includes the following change to π for each interval $\pi(i_k, m_k)$ where we ignore the minor changes to look-up tables

$$\underset{t}{\pi(i_k)} \ldots \underset{\lambda\overline{x}_j}{\pi(m_k)} \underset{f}{\pi(m_k+1)} \underset{\lambda\overline{z}_{k_i}}{\pi(m_k+2)} \underset{t}{\pi(i_k)} \ldots \underset{\lambda\overline{x}_j}{\pi(m_k)} \underset{t'_{k_i}}{\pi(m_k+3)} \ldots$$

where $t'_{k_i} \in \pi(i_k)$ is the copy of t directly beneath $\lambda\overline{z}_{k_i}$ in t'_0.

Next, let u be a basic tile. To obtain σ we iteratively do additions and deletions to π starting with $\pi(i_1, m_1)$ and then recursively transforming inner j-plays on t within this region. Let π be the result of the changes to the initial π for the intervals $\pi(i_j, m_j)$, $j < k$. Consider the interval $\pi(i_k, m_k)$. Consider case (1) of Definition 2. Let $\pi(m_k + 1, n_k^i)$ be all plays on $u \in \pi(m_k + 1)$. If there are no plays then π is initially unchanged. Otherwise, π has the following structure:

$$\ldots \underset{t}{\pi(i_k)} \ldots \underset{\lambda\overline{x}_j}{\pi(m_k)} \underset{u}{\pi(m_k+1)} \ldots \underset{\lambda\overline{z}_{k_i}}{\pi(n_k^i)} \underset{t'_{k_i}}{\pi(n_k^i+1)} \ldots$$

To obtain the new π, we do the following addition for each i

$$\underset{t}{\pi(i_k)} \ldots \underset{\lambda\overline{x}_j}{\pi(m_k)} \underset{u}{\pi(m_k+1)} \ldots \underset{\lambda\overline{z}_{k_1}}{\pi(n_k^1)} \ldots \underset{\lambda\overline{z}_{k_i}}{\pi(n_k^i)} \underset{t}{\pi(i_k)} \ldots \underset{\lambda\overline{x}_j}{\pi(m_k)} \underset{t'_{k_i}}{\pi(n_k^i+1)} \ldots$$

where t immediately after $\pi(n_k^i)$ is its copy in t'_0 directly beneath $\lambda\overline{z}_{k_i}$.

Finally, we consider the case that u is an end tile. Let $\pi(m_k + 1, m_k + n)$ be the unique play on u with $\lambda\overline{z}_i \in \pi(m_k + n)$. Consider all j-plays $\pi(i_k, m_k^i)$ on $t \in \pi(i_k)$ where $m_k^1 = m_k$:

$$\underset{t}{\pi(i_k)} \ldots \underset{\lambda\overline{x}_j}{\pi(m_k^1)} \ldots \underset{\lambda\overline{x}_j}{\pi(m_k^i)} \underset{u}{\pi(m_k^i+1)} \ldots \underset{\lambda\overline{z}_i}{\pi(m_k^i+n)} \underset{t'_i}{\pi(m_k^i+n+1)} \ldots$$

There must be the same play on u at each $\pi(m_k^i + 1)$ because the value of the head variable of u is always the same and u is an end tile. So initially we do the following addition

$$\underset{t}{\pi(i_k)} \ldots \underset{\lambda\overline{x}_j}{\pi(m_k^1)} \underset{u}{\pi(m_k^1+1)} \ldots \underset{\lambda\overline{z}_i}{\pi(m_k^1+n)} \underset{t}{\pi(i_k)} \ldots \underset{\lambda\overline{x}_j}{\pi(m_k^1)} \underset{t'_i}{\pi(m_k^1+n+1)}$$

where the second $t \in \pi(i_k)$ is the copy of t directly below $\lambda\overline{z}_i$ in t'_0, and for subsequent $i > 1$ we delete the ri region $\pi(m_k^i + 1, m_k^i + n)$. To complete the argument, we recursively apply this technique to shortest j-plays on t within $\pi(i_k, m_k^1)$: note that j-plays on t below $\lambda\overline{z}_i$ within $\pi(i_k, m_k^1)$ will include additional ri plays on on t and on u. $\qquad\square$

6   Decidable Instances

We now briefly sketch how the the game-theoretic characterisation of matching provides uniform decidability proofs for two instances of interpolation that are known to be decidable, the 4th-order problem and the atoms case where in each equation $x(v_1, \ldots, v_n) = u$ the term u is a constant $a : \mathbf{0}$ [5, 6]. In both cases the proof establishes the *small model property* (if $t_0 \models P$ then there is a small $t \models P$) via the transformations of the previous section. In neither case do we need to appeal to observational equivalence.

Figure 2 presents the algorithm for both cases. The procedure is initiated by marking all leaves of t_0 and recursively proceeds towards its root. At each stage, a lowest marked node u is examined for transformations: the algorithm has, therefore, already ascended all branches below u.

Assume $t_0 \models P$

1. mark all leaves $u : \mathbf{0}$ of t_0
2. choose a marked node u such that no descendent of u is marked
3. if $t_0 \mathbf{T1} t'$ at u then $t_0 = t'$ and unmark all nodes and return to 1
4. identify basic or constant tile $t = t(\lambda \overline{x}_1, \ldots, \lambda \overline{x}_k)$ rooted at u
5. if $t_0 \mathbf{Ti} t'$ at t for $i \in \{2, 3\}$ then $t_0 = t'$ and unmark all nodes and return to 1
6. identify successor basic or constant tiles t_i below $\lambda \overline{x}_i$
7. if $t_0 \mathbf{T4} t'$ at t and a successor then $t_0 = t'$ and unmark all nodes and return to 1.
8. if $u' \downarrow_{i_1} \lambda \overline{y} \downarrow_1 u$ then unmark u and mark u' and return to 2
9. finish

Fig. 2. The algorithm

Clearly, the procedure must terminate with $t_0 \models P$ and where no transformation applies anywhere in t_0. Assume t_0 is such a term.

Proposition 1 *If t' is a subterm of t_0 such that t' only contains sri tiles, leaves $y : \mathbf{0}$ and $a : \mathbf{0}$ then t' consists of sri end tiles and leaves $a : \mathbf{0}$.*

Proof. By a simple induction. A leaf u may be a constant or a variable. Consider u' such that $u' \downarrow_{i_1} \lambda \overline{y} \downarrow_1 u$. By repeating the argument for other directions i_j from u', the tile rooted at u' will be an end tile. Consider the first time that a tile isnt an end tile. Either $\mathbf{T3}$ or $\mathbf{T4}$ must apply, which is a contradiction.

Hence for the atoms case, as all tiles are sri, every end tile is also a top tile. There can be at most m separators where m is the number of equations. Finally, Proposition 2 of Section 4 provides a simple upper bound both on the size of an end tile in t_0 and the number of embedded end tiles. The details are straightforward.

Next we consider the 4th-order case. The term t_0 consists of top tiles, leaves and constant tiles. Shortest plays on a top tile are canonical. The number of top tiles that are not sri is bounded (by the sum of the sizes of the sets R_i of section 2). Again there can be at most m separators. Now, the crucial property is that given a sequence of sri top tiles $t_i(\lambda \overline{x}_1^i, \ldots, \lambda \overline{x}_{k_i}^i)$ such that for each i, t_{i+1} is directly below $\lambda \overline{x}_{j_i}^i$ then most of the tiles t_i are n_i-end and n_i-directed for some n_i which follows easily from Proposition 1 of Section 4. (If a shortest ri j-play on t_i, $\pi(k, m)$, is such that there is a child $\pi(m')$ of $\pi(m)$, so $y : \mathbf{0} \in \pi(m')$, then every j-play $\pi(k, n)$ of t_i is such that there is a child $\pi(n')$ of $\pi(n)$ and $y \in \pi(n')$ or $\pi(k, m')$ is not ri and for some n', $\pi(k, n')$ is also not ri.)

References

1. Comon, H. and Jurski, Y. Higher-order matching and tree automata. *Lecture Notes in Computer Science*, **1414**, 157-176, (1997).
2. Dowek, G. Higher-order unification and matching. In A. Robinson and A. Voronkov ed. *Handbook of Automated Reasoning*, Vol 2, 1009-1062, North-Holland, 2001.
3. Huet, G. *Rèsolution d'èquations dans les langages d'ordre 1, 2, ... ω*. Thèse de doctorat d'ètat, Universitè Paris VII, (1976).
4. Loader, R. Higher-order β-matching is undecidable, *Logic Journal of the IGPL*, **11(1)**, 51-68, (2003).
5. Padovani, V. Decidability of all minimal models. *Lecture Notes in Computer Science*, **1158**, 201-215, (1996).
6. Padovani, V. Decidability of fourth-order matching. *Mathematical Structures in Computer Science*, **10(3)**, 361-372, (2001).
7. Schubert, A. Linear interpolation for the higher-order matching problem. *Lecture Notes in Computer Science*, **1214**, 441-452.
8. Schmidt-Schauβ, M. Decidability of arity-bounded higher-order matching. *Lecture Notes in Artificial Intelligence*, **2741**, 488-502, (2003).
9. Statman, R. The typed λ-calculus is not elementary recursive. *Theoretical Computer Science*, **9**, 73-81, (1979).
10. Stirling, C. *Modal and Temporal Properties of Processes*, Texts in Computer Science, Springer, 2001.
11. Wierzbicki, T. Complexity of higher-order matching. *Lecture Notes in Computer Science*, **1632**, 82-96, (1999).

Decidability of Type-Checking in the Calculus of Algebraic Constructions with Size Annotations

Frédéric Blanqui

Laboratoire Lorrain de Recherche en Informatique et Automatique (LORIA)
Institut National de Recherche en Informatique et Automatique (INRIA)
615 rue du Jardin Botanique, BP 101, 54602 Villers-lès-Nancy, France
blanqui@loria.fr

Abstract. Since Val Tannen's pioneering work on the combination of simply-typed λ-calculus and first-order rewriting [11], many authors have contributed to this subject by extending it to richer typed λ-calculi and rewriting paradigms, culminating in the Calculus of Algebraic Constructions. These works provide theoretical foundations for type-theoretic proof assistants where functions and predicates are defined by oriented higher-order equations. This kind of definitions subsumes usual inductive definitions, is easier to write and provides more automation.

On the other hand, checking that such user-defined rewrite rules, when combined with β-reduction, are strongly normalizing and confluent, and preserve the decidability of type-checking, is more difficult. Most termination criteria rely on the term structure. In a previous work, we extended to dependent types and higher-order rewriting, the notion of "sized types" studied by several authors in the simpler framework of ML-like languages, and proved that it preserves strong normalization.

The main contribution of the present paper is twofold. First, we prove that, in the Calculus of Algebraic Constructions with size annotations, the problems of type inference and type-checking are decidable, provided that the sets of constraints generated by size annotations are satisfiable and admit most general solutions. Second, we prove the latter properties for a size algebra rich enough for capturing usual induction-based definitions and much more.

1 Introduction

The notion of "sized type" was first introduced in [20] and further studied by several authors [1, 3, 19, 29] as a tool for proving the termination of ML-like function definitions. It is based on the semantics of inductive types as fixpoints of monotone operators, reachable by transfinite iteration. For instance, natural numbers are the limit of $(S_i)_{i<\omega}$, where S_i is the set of natural numbers smaller than i (inductive types with constructors having functional arguments require ordinals bigger than ω). The idea is then to reflect this in the syntax by adding size annotations on types indicating in which subset S_i a term is. For instance, subtraction on natural numbers can be assigned the type $- : nat^\alpha \Rightarrow nat^\beta \Rightarrow nat^\alpha$, where α and β are implicitly universally quantified, meaning that the size

L. Ong (Ed.): CSL 2005, LNCS 3634, pp. 135–150, 2005.

of its output is not bigger than the size of its first argument. Then, one can ensure termination by restricting recursive calls to arguments whose size – by typing – is smaller. For instance, the following ML-like definition of $\lceil \frac{x}{y+1} \rceil$:

```
letrec div x y = match x with
  | 0 -> 0
  | S x' -> S (div (x' - y) y)
```

is terminating since, if x is of size at most α and y is of size at most β, then x' is of size at most $\alpha - 1$ and $(x' - y)$ is of size at most $\alpha - 1 < \alpha$.

The Calculus of Constructions (CC) [16] is a powerful type system with polymorphic and dependent types, allowing to encode higher-order logic. The Calculus of Algebraic Constructions (CAC) [8] is an extension of CC where functions are defined by higher-order rewrite rules. As shown in [10], it subsumes the Calculus of Inductive Constructions (CIC) [17] implemented in the Coq proof assistant [14], where functions are defined by induction. Using rule-based definitions has numerous advantages over induction-based definitions: definitions are easier (*e.g.* Ackermann's function), more propositions can be proved equivalent automatically, one can add simplification rules like associativity or using rewriting modulo AC [5], etc. For proving that user-defined rules terminate when combined with β-reduction, [8] essentially checks that recursive calls are made on structurally smaller arguments.

In [6], we extended the notion of sized type to CAC, giving the Calculus of Algebraic Constructions with Size Annotations (CACSA). We proved that, when combined with β-reduction, user-defined rules terminate essentially if recursive calls are made on arguments whose size – by typing – is strictly smaller, by possibly using lexicographic and multiset comparisons. Hence, the following rule-based definition of $\lceil \frac{x}{y+1} \rceil$:

$$
\begin{aligned}
0 \,/\, y &\;\rightarrow\; 0 \\
(s\,x) \,/\, y &\;\rightarrow\; s\,((x - y) \,/\, y)
\end{aligned}
$$

is terminating since, in the last rule, if x is of size at most α and y is of size at most β, then $(s\,x)$ is of size at most $\alpha + 1$ and $(x - y)$ is of size at most $\alpha < \alpha + 1$. Note that this rewrite system cannot be proved terminating by criteria only based on the term structure, like RPO or its extensions to higher-order terms [21, 27]. Note also that, if a term t is structurally smaller than a term u, then the size of t is smaller than the size of u. Therefore, CACSA proves the termination of any induction-based definition like CIC/Coq, but also definitions like the previous one. To our knowledge, this is the most powerful termination criterion for functions with polymorphic and dependent types like in Coq. The reader can find other convincing examples in [6].

However, [6] left an important question open. For the termination criterion to work, we need to make sure that size annotations assigned to function symbols are valid. For instance, if subtraction is assigned the type $- : nat^\alpha \Rightarrow nat^\beta \Rightarrow nat^\alpha$, then we must make sure that the definition of $-$ indeed outputs a term whose size is not greater than the size of its first argument. This amounts to check that, for every rule in the definition of $-$, the size of the right hand-side is not greater than the size of the left hand-side. This can be easily verified by hand if, for instance, the definition of $-$ is as follows:

$$
\begin{aligned}
0 - x &\rightarrow 0 \\
x - 0 &\rightarrow x \\
(s\ x) - (s\ y) &\rightarrow x - y
\end{aligned}
$$

The purpose of the present work is to prove that this can be done automatically, by inferring the size of both the left and right hand-sides, and checking that the former is smaller than the latter.

$$
\begin{aligned}
nil &: (A : \star)list^\alpha A\ 0 \\
cons &: (A : \star)A \Rightarrow (n : nat)list^\alpha A\ n \Rightarrow list^{s\alpha} A\ (sn) \\
if\_in\_then\_else &: bool \Rightarrow (A : \star)A \Rightarrow A \Rightarrow A \\
insert &: (A : \star)(\leq : A \Rightarrow A \Rightarrow bool)A \Rightarrow (n : nat)list^\alpha A\ n \Rightarrow list^{s\alpha} A\ (sn) \\
sort &: (A : \star)(\leq : A \Rightarrow A \Rightarrow bool)(n : nat)list^\alpha A\ n \Rightarrow list^\alpha A\ n
\end{aligned}
$$

$$
\begin{aligned}
if\ true\ in\ A\ then\ u\ else\ v &\rightarrow u \\
if\ false\ in\ A\ then\ u\ else\ v &\rightarrow v \\
insert\ A \leq x\_(nil\_) &\rightarrow cons\ A\ x\ 0\ (nil\ A) \\
insert\ A \leq x\_(cons\_y\ n\ l) &\rightarrow if\ x \leq y\ in\ list\ A\ (s\ (s\ n)) \\
& then\ cons\ A\ x\ (s\ n)\ (cons\ A\ y\ n\ l) \\
& else\ cons\ A\ y\ (s\ n)\ (insert\ A \leq x\ n\ l) \\
sort\ A \leq \_(nil\_) &\rightarrow nil\ A \\
sort\ A \leq \_(cons\_x\ n\ l) &\rightarrow insert\ A \leq x\ n\ (sort\ A \leq n\ l)
\end{aligned}
$$

Fig. 1. Insertion sort on polymorphic and dependent lists

We now give an example with dependent and polymorphic types. Let $\star$ be the sort of types and $list : \star \Rightarrow nat \Rightarrow \star$ be the type of polymorphic lists of fixed length whose constructors are nil and $cons$. Without ambiguity, s is used for the successor function both on terms and on size expressions. The functions $insert$ and $sort$ defined in Figure 1 have size annotations satisfying our termination criterion. The point is that $sort$ preserves the size of its list argument and thus can be safely used in recursive calls. Checking this automatically is the goal of this work.

An important point is that the ordering naturally associated with size annotations implies some subtyping relation on types. The combination of subtyping and dependent types (without rewriting) is a difficult subject which has been studied by Chen [12]. We reused many ideas and techniques of his work for designing CACSA and proving important properties like β-subject reduction (preservation of typing under β-reduction) [7].

Another important point is related to the meaning of type inference. In ML, type inference means computing a type of a term in which the types of free and bound variables, and function symbols (letrec's in ML), are unknown. In other words, it consists in finding a simple type for a pure λ-term. Here, type inference means computing a CACSA type, hence dependent and polymorphic (CACSA contains Girard's system F), of a term in which the types and size annotations of free and bound variables, and function symbols, are known. In dependent type theories, this kind of type inference is necessary for type-checking [15]. In other words, we do not try to infer relations between the sizes of the arguments of a function and the size of its output like in [4, 13]. We try to check that, with the annotated types declared by the user for its function symbols, rules satisfy the termination criterion described in [6].

Moreover, in ML, type inference amounts to solve equality constraints in the type algebra. Here, type inference amounts to solve equality and ordering constraints in the size algebra. The point is that the ordering on size expressions is not anti-symmetric: it is a quasi-ordering. Thus, we have a combination of unification and symbolic quasi-ordering constraint solving.

Finally, because of the combination of subtyping and dependent typing, the decidability of type-checking requires the existence of minimal types [12]. Thus, we must also prove that a satisfiable set of equality and ordering constraints has a smallest solution, which is not the case in general. This is in contrast with non-dependently typed frameworks.

Outline. In Section 2, we define terms and types, and study some properties of the size ordering. In Section 3, we give a general type inference algorithm and prove its correctness and completeness under general assumptions on constraint solving. Finally, in Section 4, we prove that these assumptions are fulfilled for the size algebra introduced in [3] which, although simple, is rich enough for capturing usual inductive definitions and much more, as shown by the first example above. Missing proofs are given in [9].

2   Terms and Types

Size Algebra. Inductive types are annotated by *size expressions* from the following algebra $\mathcal{A}$:

$$a ::= \alpha \mid sa \mid \infty$$

where $\alpha \in \mathcal{Z}$ is a *size variable*. The set $\mathcal{A}$ is equipped with the quasi-ordering $\leq_{\mathcal{A}}$ defined in Figure 2. Let $\simeq_{\mathcal{A}} = \leq_{\mathcal{A}} \cap \geq_{\mathcal{A}}$ be its associated equivalence.

Let $\varphi, \psi, \rho, \ldots$ denote size substitutions, *i.e.* functions from $\mathcal{Z}$ to $\mathcal{A}$. One can easily check that $\leq_{\mathcal{A}}$ is stable by substitution: if $a \leq_{\mathcal{A}} b$ then $a\varphi \leq_{\mathcal{A}} b\varphi$. We extend $\leq_{\mathcal{A}}$ to substitutions: $\varphi \leq_{\mathcal{A}} \psi$ iff, for all $\alpha \in \mathcal{Z}$, $\alpha\varphi \leq_{\mathcal{A}} \alpha\psi$.

We also extend the notion of "more general substitution" from unification theory as follows: φ is *more general than* ψ, written $\varphi \sqsubseteq \psi$, iff there is φ' such that $\varphi\varphi' \leq_{\mathcal{A}} \psi$.

Terms. We assume the reader familiar with typed λ-calculi [2] and rewriting [18]. Details on CAC(SA) can be found in [6, 8]. We assume given a set $\mathcal{S} = \{\star, \square\}$ of *sorts* ($\star$ is the sort of types and propositions; $\square$ is the sort of predicate types),

$$\text{(refl)} \ \ a \leq_{\mathcal{A}} a \qquad \text{(trans)} \ \ \frac{a \leq_{\mathcal{A}} b \quad b \leq_{\mathcal{A}} c}{a \leq_{\mathcal{A}} c}$$

$$\text{(mon)} \ \ \frac{a \leq_{\mathcal{A}} b}{sa \leq_{\mathcal{A}} sb} \qquad \text{(succ)} \ \ \frac{a \leq_{\mathcal{A}} b}{a \leq_{\mathcal{A}} sb} \qquad \text{(infty)} \ \ a \leq_{\mathcal{A}} \infty$$

Fig. 2. Ordering on size expressions

a set $\mathcal{F}$ of function or predicate *symbols*, a set $\mathcal{CF}^{\square} \subseteq \mathcal{F}$ of *constant predicate symbols*, and an infinite set $\mathcal{X}$ of *term variables*. The set $\mathcal{T}$ of terms is:

$$t ::= \mathbf{s} \mid x \mid C^a \mid f \mid [x : t]t \mid (x : t)t \mid tt$$

where $\mathbf{s} \in \mathcal{S}$, $x \in \mathcal{X}$, $C \in \mathcal{CF}^{\square}$, $a \in \mathcal{A}$ and $f \in \mathcal{F} \setminus \mathcal{CF}^{\square}$. A term $[x : t]u$ is an *abstraction*. A term $(x : T)U$ is a *dependent product*, simply written $T \Rightarrow U$ when x does not occur in U. Let $\boldsymbol{t}$ denote a sequence of terms $t_1, \ldots, t_n$ of length $|\boldsymbol{t}| = n$.

Every term variable x is equipped with a sort $\mathbf{s}_x$ and, as usual, terms equivalent modulo sort-preserving renaming of bound variables are identified. Let $\mathcal{V}(t)$ be the set of size variables in t, and $\mathrm{FV}(t)$ be the set of term variables free in t. Let $\theta, \sigma, \ldots$ denote term substitutions, *i.e.* functions from $\mathcal{X}$ to $\mathcal{T}$. For our previous examples, we have $\mathcal{CF}^{\square} = \{nat, list, bool\}$ and $\mathcal{F} = \mathcal{CF}^{\square} \cup \{0, s, /, nil, cons, insert, sort\}$.

Rewriting. Terms only built from variables and symbol applications $f\boldsymbol{t}$ are said to be *algebraic*. We assume given a set $\mathcal{R}$ of *rewrite rules* $l \rightarrow r$ such that l is algebraic, $l = f\boldsymbol{l}$ with $f \notin \mathcal{CF}^{\square}$ and $\mathrm{FV}(r) \subseteq \mathrm{FV}(l)$. Note that, while left hand-sides are algebraic and thus require syntactic matching only, right hand-sides may have abstractions and products. β-reduction and rewriting are defined as usual: $C[[x : T]u \ v] \rightarrow_\beta C[u\{x \mapsto v\}]$ and $C[l\sigma] \rightarrow_\mathcal{R} C[r\sigma]$ if $l \rightarrow r \in \mathcal{R}$. Let $\rightarrow \ = \ \rightarrow_\beta \cup \rightarrow_\mathcal{R}$ and $\rightarrow^*$ be its reflexive and transitive closure. Let $t \downarrow u$ iff there exists v such that $t \rightarrow^* v \ ^* \!\!\leftarrow u$.

Typing. We assume that every symbol f is equipped with a sort $\mathbf{s}_f$ and a type $\tau_f = (\boldsymbol{x} : \boldsymbol{T})U$ such that, for all rules $f\boldsymbol{l} \rightarrow r \in \mathcal{R}$, $|\boldsymbol{l}| \leq |\boldsymbol{T}|$ (f is not applied to more arguments than the number of arguments given by τ_f). Let $\mathcal{F}^{\mathbf{s}}$ (resp. $\mathcal{X}^{\mathbf{s}}$) be the set of symbols (resp. variables) of sort $\mathbf{s}$. As usual, we distinguish the following classes of terms where t is any term:

- objects: $o ::= x \in \mathcal{X}^* \mid f \in \mathcal{F}^* \mid [x : t]o \mid ot$
- predicates: $p ::= x \in \mathcal{X}^{\square} \mid C^a \in \mathcal{CF}^{\square} \mid f \in \mathcal{F}^{\square} \setminus \mathcal{CF}^{\square} \mid [x : t]p \mid (x : t)p \mid pt$
- kinds: $K ::= \star \mid (x : t)K$

Examples of objects are the constructors of inductive types $0, s, nil, cons, \ldots$ and the function symbols $-, /, insert, sort, \ldots$. Their types are predicates: inductive types $bool, nat, list, \ldots$, logical connectors $\wedge, \vee, \ldots$, universal quantifications $(x : T)U, \ldots$. The types of predicates are kinds: $\star$ for types like $bool$ or nat, $\star \Rightarrow nat \Rightarrow \star$ for $list$, $\ldots$

An *environment* Γ is a sequence of variable-term pairs. An environment is *valid* if a term is typable in it. The typing rules of CACSA are given in Figure 4 and its subtyping rules in Figure 3. In (symb), φ is an arbitrary size substitution. This reflects the fact that, in type declarations, size variables are implicitly universally quantified, like in ML. In contrast with [12], subtyping uses no sorting judgment. This simplification is justified in [7].

In comparison with [7], we added the side condition $\mathcal{V}(t) = \emptyset$ in (size). It does not affect the properties proved in [7] and ensures that the size ordering is compatible with subtyping (Lemma 2). By the way, one could think of taking the more general rule $C^a t \leq C^b u$ with $t \simeq_\mathcal{A} u$. This would eliminate the need for equality constraints and thus simplify a little bit the constraint solving procedure. More generally, one could think in taking into account the monotony of type constructors by having, for instance, *list* $nat^a \leq$ *list* nat^b whenever $a \leq_\mathcal{A} b$. This requires extensions to Chen's work [12] and proofs of many non trivial properties of [7] again, like Theorem 1 below or subject reduction for β.

$$\text{(refl)} \quad T \leq T \qquad \text{(size)} \quad C^a t \leq C^b t \quad (C \in \mathcal{CF}^\square, \ a \leq_\mathcal{A} b, \ \mathcal{V}(t) = \emptyset)$$

$$\text{(prod)} \quad \frac{U' \leq U \quad V \leq V'}{(x:U)V \leq (x:U')V'} \qquad \text{(conv)} \quad \frac{T' \leq U'}{T \leq U} \quad (T \downarrow T', \ U' \downarrow U)$$

$$\text{(trans)} \quad \frac{T \leq U \quad U \leq V}{T \leq V}$$

Fig. 3. Subtyping rules

$$\text{(ax)} \quad \vdash \star : \square \qquad \text{(prod)} \quad \frac{\Gamma \vdash U : \mathbf{s} \quad \Gamma, x:U \vdash V : \mathbf{s}'}{\Gamma \vdash (x:U)V : \mathbf{s}'}$$

$$\text{(size)} \quad \frac{\vdash \tau_C : \square}{\vdash C^a : \tau_C} \quad (C \in \mathcal{CF}^\square, \ a \in \mathcal{A}) \qquad \text{(symb)} \quad \frac{\vdash \tau_f : \mathbf{s}_f}{\vdash f : \tau_f \varphi} \quad (f \notin \mathcal{CF}^\square)$$

$$\text{(var)} \quad \frac{\Gamma \vdash T : \mathbf{s}_x}{\Gamma, x:T \vdash x : T} \ (x \notin \mathrm{dom}(\Gamma)) \qquad \text{(weak)} \quad \frac{\Gamma \vdash t : T \quad \Gamma \vdash U : \mathbf{s}_x}{\Gamma, x:U \vdash t : T} \ (x \notin \mathrm{dom}(\Gamma))$$

$$\text{(abs)} \quad \frac{\Gamma, x:U \vdash v : V \quad \Gamma \vdash (x:U)V : \mathbf{s}}{\Gamma \vdash [x:U]v : (x:U)V} \qquad \text{(app)} \quad \frac{\Gamma \vdash t : (x:U)V \quad \Gamma \vdash u : U}{\Gamma \vdash tu : V\{x \mapsto u\}}$$

$$\text{(sub)} \quad \frac{\Gamma \vdash t : T \quad \Gamma \vdash T' : \mathbf{s}}{\Gamma \vdash t : T'} \quad (T \leq T')$$

Fig. 4. Typing rules

∞-**Terms.** An ∞-*term* is a term whose only size annotations are ∞. In particular, it has no size variable. An ∞-*environment* is an environment made of ∞-terms. This class of terms is isomorphic to the class of (unannotated) CAC terms. Our goal is to be able to infer annotated types for these terms, by using the size annotations given in the type declarations of constructors and function symbols $0, s, /, nil, cons, insert, sort, \ldots$

Since size variables are intended to occur in object type declarations only, and since we do not want matching to depend on size annotations, we assume that rules and type declarations of predicate symbols $nat, bool, list, \ldots$ are made of ∞-terms. As a consequence, we have:

Lemma 1. – If $t \to_{\mathcal{R}} t'$ then, for all φ, $t\varphi \to_{\mathcal{R}} t'\varphi$.
– If $\Gamma \vdash t : T$ then, for all φ, $\Gamma\varphi \vdash t\varphi : T\varphi$.

We make three important assumptions:

(1) $\mathcal{R}$ preserves typing: for all $l \to r \in \mathcal{R}$, Γ, T and σ, if $\Gamma \vdash l\sigma : T$ then $\Gamma \vdash r\sigma : T$. It is generally not too difficult to check this by hand. However, as already mentioned in [6], finding sufficient conditions for this to hold in general does not seem trivial.

(2) $\beta \cup \mathcal{R}$ is confluent. This is for instance the case if $\mathcal{R}$ is confluent and left-linear [23], or if $\beta \cup \mathcal{R}$ is terminating and $\mathcal{R}$ is locally confluent.

(3) $\beta \cup \mathcal{R}$ is terminating. In [6], it is proved that $\beta \cup \mathcal{R}$ is terminating essentially if, in every rule $fl \to r \in \mathcal{R}$, recursive calls in r are made on terms whose size – by typing – are smaller than l, by using lexicographic and multiset comparisons. Note that, with type-level rewriting, confluence is necessary for proving termination [8].

Important Remark. One may think that there is some vicious circle here: we assume the termination for proving the decidability of type-checking, while type-checking is used for proving termination! The point is that termination checks are done incrementally. At the beginning, we can check that some set of rewrite rules $\mathcal{R}_1$ is terminating in the system with β only. Indeed, we do not need to use $\mathcal{R}_1$ in the type conversion rule (conv) for typing the terms of $\mathcal{R}_1$. Then, we can check in $\beta \cup \mathcal{R}_1$ that some new set of rules $\mathcal{R}_2$ is terminating, and so on. . .

Various properties of CACSA have already been studied in [7]. We refer the reader to this paper if necessary. For the moment, we just mention two important and non trivial properties based on Chen's work on subtyping with dependent types [12]: subject reduction for β and transitivity elimination:

Theorem 1 ([7]). $T \leq U$ iff $T{\downarrow} \leq_s U{\downarrow}$, where $\leq_s$ is the restriction of $\leq$ to (refl), (size) and (prod).

We now give some properties of the size and substitution orderings. Let $\to_{\mathcal{A}}$ be the confluent and terminating relation on $\mathcal{A}$ generated by the rule $s\infty \to \infty$.

Lemma 2. Let $a{\downarrow}$ be the normal form of a w.r.t. $\to_{\mathcal{A}}$.
– $a \simeq_{\mathcal{A}} b$ iff $a{\downarrow} = b{\downarrow}$.
– If $\infty \leq_{\mathcal{A}} a$ or $s^{k+1}a \leq_{\mathcal{A}} a$ then $a{\downarrow} = \infty$.
– If $a \leq_{\mathcal{A}} b$ and $\varphi \leq_{\mathcal{A}} \psi$ then $a\varphi \leq_{\mathcal{A}} b\psi$.
– If $\varphi \leq_{\mathcal{A}} \psi$ and $U \leq V$ then $U\varphi \leq V\psi$.

Note that ∞-terms are in $\mathcal{A}$-normal form. The last property (compatibility of size ordering wrt subtyping) follows from the restriction $\mathcal{V}(t) = \emptyset$ in (size).

3   Decidability of Typing

In this section, we prove the decidability of type inference and type-checking for ∞-terms under general assumptions that will be proved in Section 4. We begin with some informal explanations.

How to do type inference? The critical cases are (symb) and (app). In (symb), a symbol f can be typed by any instance of τ_f, and two different instances may be necessary for typing a single term (*e.g.* $s(sx)$). For type inference, it is therefore necessary to type f by its most general type, namely a renaming of τ_f with fresh variables, and to instantiate it later when necessary.

Assume now that we want to infer the type of an application tu. We naturally try to infer a type for t and a type for u using distinct fresh variables. Assume that we get T and U' respectively. Then, tu is typable if there is a size substitution φ and a product type $(x : P)Q$ such that $T\varphi \leq (x : P)Q$ and $U'\varphi \leq P$.

After Theorem 1, checking whether $A \leq B$ amounts to check whether $A{\downarrow} \leq_s B{\downarrow}$, and checking whether $A \leq_s B$ amounts to apply the (prod) rule as much as possible and then to check that (refl) or (size) holds. Hence, $T\varphi \leq (x : P)Q$ only if $T{\downarrow}$ is a product. Thus, the application tu is typable if $T{\downarrow} = (x : U)V$ and there exists φ such that $U'{\downarrow}\varphi \leq_s U\varphi$. Finding φ such that $A\varphi \leq_s B\varphi$ amounts to apply the (prod) rule on $A \leq_s B$ as much as possible and then to find φ such that (refl) or (size) holds. So, a subtyping problem can be transformed into a constraint problem on size variables.

We make this precise by first defining the constraints that can be generated.

Definition 1 (Constraints). Constraint problems *are defined as follows:*

$$\mathcal{C} ::= \bot \mid \top \mid \mathcal{C} \wedge \mathcal{C} \mid a = b \mid a \leq b$$

where $a, b \in \mathcal{A}$, $=$ is commutative, $\wedge$ is associative and commutative, $\mathcal{C} \wedge \mathcal{C} = \mathcal{C} \wedge \top = \mathcal{C}$ and $\mathcal{C} \wedge \bot = \bot$. A finite conjunction $\mathcal{C}_1 \wedge \ldots \wedge \mathcal{C}_n$ is identified with $\top$ if $n = 0$. A constraint problem is in canonical form if it is neither of the form $\mathcal{C} \wedge \top$, nor of the form $\mathcal{C} \wedge \bot$, nor of the form $\mathcal{C} \wedge \mathcal{C} \wedge \mathcal{D}$. In the following, we always assume that constraint problems are in canonical form. An equality (resp. inequality) problem is a problem having only equalities (resp. inequalities). An inequality $\infty \leq \alpha$ is called an ∞-inequality. An inequality $s^p\alpha \leq s^q\beta$ is called a linear inequality. Solutions to constraint problems are defined as follows:

- $S(\bot) = \emptyset$,
- $S(\top)$ *is the set of all size substitutions,*
- $S(\mathcal{C} \wedge \mathcal{D}) = S(\mathcal{C}) \cap S(\mathcal{D})$,
- $S(a = b) = \{\varphi \mid a\varphi = b\varphi\}$,
- $S(a \leq b) = \{\varphi \mid a\varphi \leq_{\mathcal{A}} b\varphi\}$.

Let $S^\ell(\mathcal{C}) = \{\varphi \mid \forall \alpha, \alpha\varphi{\downarrow} \neq \infty\}$ be the set of linear solutions.

We now prove that a subtyping problem can be transformed into constraints.

Lemma 3. *Let $S(U, V)$ be the set of substitutions φ such that $U\varphi \leq_s V\varphi$. We have $S(U, V) = S(\mathcal{C}(U, V))$ where $\mathcal{C}(U, V)$ is defined as follows:*

- $\mathcal{C}((x:U)V,(x:U')V') = \mathcal{C}(U',U) \wedge \mathcal{C}(V,V')$,
- $\mathcal{C}(C^a\boldsymbol{u}, C^b\boldsymbol{v}) = a \leq b \wedge \mathcal{E}^0(u_1,v_1) \wedge \ldots \wedge \mathcal{E}^0(u_n,v_n)$ $\textit{if } |\boldsymbol{u}| = |\boldsymbol{v}| = n$,
- $\mathcal{C}(U,V) = \mathcal{E}^1(U,V)$ $\textit{in the other cases}$,

$\textit{and } \mathcal{E}^i(U,V) \textit{ is defined as follows:}$

- $\mathcal{E}^i((x{:}U)V,(x{:}U')V') = \mathcal{E}^i([x{:}U]V,[x{:}U']V') = \mathcal{E}^i(UV,U'V')$
  $= \mathcal{E}^i(U,U') \wedge \mathcal{E}^i(V,V')$,
- $\mathcal{E}^1(C^a,C^b) = a = b$,
- $\mathcal{E}^0(C^a,C^b) = a = b \wedge \infty \leq a$,
- $\mathcal{E}^i(c,c) = \top \textit{ if } c \in \mathcal{S} \cup \mathcal{X} \cup \mathcal{F} \setminus \mathcal{CF}^\square$,
- $\mathcal{E}^i(U,V) = \bot \textit{ in the other cases.}$

$\textit{Proof.}$ First, we clearly have $\varphi \in S(\mathcal{E}^1(U,V))$ iff $U\varphi = V\varphi$, and $\varphi \in S(\mathcal{E}^0(U,V))$ iff $U\varphi = V\varphi$ and $\mathcal{V}(U\varphi) = \emptyset$. Thus, $S(U,V) = S(\mathcal{C}(U,V))$. $\qquad\square$

$$(\text{ax}) \quad \Gamma \vdash^{\mathcal{y}}_{\text{a}} \star : \square \qquad (\text{prod}) \quad \frac{\Gamma \vdash^{\mathcal{y}}_{\text{a}} U : \mathbf{s}_x \quad \Gamma, x : U \vdash^{\mathcal{y}}_{\text{a}} V : \mathbf{s}'}{\Gamma \vdash^{\mathcal{y}}_{\text{a}} (x:U)V : \mathbf{s}'}$$

$$(\text{size}) \quad \Gamma \vdash^{\mathcal{y}}_{\text{a}} C^\infty : \tau_C \quad (C \in \mathcal{CF}^\square) \qquad (\text{symb}) \quad \Gamma \vdash^{\mathcal{y}}_{\text{a}} f : \tau_f \rho_{\mathcal{y}} \quad (f \notin \mathcal{CF}^\square)$$

$$(\text{var}) \quad \Gamma \vdash^{\mathcal{y}}_{\text{a}} x : x\Gamma \quad (x \in \text{dom}(\Gamma)) \qquad (\text{abs}) \quad \frac{\Gamma \vdash^{\mathcal{y}}_{\text{a}} U : \mathbf{s}_x \quad \Gamma, x : U \vdash^{\mathcal{y}}_{\text{a}} v : V}{\Gamma \vdash^{\mathcal{y}}_{\text{a}} [x:U]v : (x:U)V} \ (V \neq \square)$$

$$(\text{app}) \quad \frac{\Gamma \vdash^{\mathcal{y}}_{\text{a}} t : T \quad \Gamma \vdash^{\mathcal{y}\cup\mathcal{V}(T)}_{\text{a}} u : U'}{\Gamma \vdash^{\mathcal{y}}_{\text{a}} tu : V\varphi\rho_{\mathcal{y}}\{x \mapsto u\}} \quad \begin{array}{l} (T{\downarrow} = (x:U)V, \mathcal{C} = \mathcal{C}(U'{\downarrow},U), \\ S(\mathcal{C}) \neq \emptyset, \varphi = mgs(\mathcal{C})) \end{array}$$

Fig. 5. Type inference rules

For renaming symbol types with variables outside some finite set of already used variables, we assume given a function ρ which, to every finite set $\mathcal{Y} \subseteq \mathcal{Z}$, associates an injection $\rho_{\mathcal{y}}$ from $\mathcal{Y}$ to $\mathcal{Z}\setminus\mathcal{Y}$. In Figure 5, we define a type inference algorithm $\vdash^{\mathcal{y}}_{\text{a}}$ parametrized by a finite set $\mathcal{Y}$ of (already used) variables under the following assumptions:

(1) It is decidable whether $S(\mathcal{C})$ is empty or not.
(2) If $S(\mathcal{C}) \neq \emptyset$ then $\mathcal{C}$ has a most general solution $mgs(\mathcal{C})$.
(3) If $S(\mathcal{C}) \neq \emptyset$ then $mgs(\mathcal{C})$ is computable.

It would be interesting to try to give a modular presentation of type inference by clearly separating constraint generation from constraint solving, as it is done for ML in [24] for instance. However, for dealing with dependent types, one at least needs higher-order pattern unification. Indeed, assume that we have a constraint generation algorithm which, for a term t and a type (meta-)variable X, computes a set $\mathcal{C}$ of constraints on X whose solutions provide valid instances of X, $\textit{i.e.}$ valid types for t. Then, in (app), if the constraint generation gives $\mathcal{C}_1$ for $t : Y$ and $\mathcal{C}_2$ for $u : Z$, then it should give something like $\mathcal{C}_1 \wedge \mathcal{C}_2 \wedge (\exists U.\exists V. Y =_{\beta\eta} (x:U)Vx \wedge Z \leq U \wedge X =_{\beta\eta} Vu)$ for $tu : X$.

We now prove the correctness, completeness and minimality of $\vdash_{\mathsf{a}}^{y}$, assuming that symbol types are well sorted ($\vdash \tau_f : \mathsf{s}_f$ for all f).

Theorem 2 (Correctness). *If Γ is a valid ∞-environment and $\Gamma \vdash_{\mathsf{a}}^{y} t : T$, then $\Gamma \vdash t : T$, t is an ∞-term and $\mathcal{V}(T) \cap y = \emptyset$.*

Proof. By induction on $\vdash_{\mathsf{a}}^{y}$. We only detail the (app) case.

(app) By induction hypothesis, $\Gamma \vdash t : T$, $\Gamma \vdash u : U'$ and t and u are ∞-terms. Thus, tu is an ∞-term. By Lemma 1, $\Gamma \vdash t : T\varphi$ and $\Gamma \vdash u : U'\varphi$. Since $T\varphi\downarrow = (x : U\varphi)V\varphi$, we have $T\varphi \neq \square$ and $\Gamma \vdash T\varphi : \mathsf{s}$. By subject reduction, $\Gamma \vdash (x : U\varphi)V\varphi : \mathsf{s}$. Hence, by (sub), $\Gamma \vdash t : (x : U\varphi)V\varphi$. By Lemma 3, $S(\mathcal{C}) = S(U'\downarrow, U)$ and $U'\downarrow \varphi \leq_s U\varphi$. Since $\Gamma \vdash U\varphi : \mathsf{s}'$, by (sub), $\Gamma \vdash u : U\varphi$. Therefore, by (app), $\Gamma \vdash tu : V\varphi\{x \mapsto u\}$ and $\Gamma \vdash tu : V\varphi\rho_{y}\{x \mapsto u\}$ since $\mathcal{V}(u) = \emptyset$. $\square$

Theorem 3 (Completeness and minimality). *If Γ is an ∞-environment, t is an ∞-term and $\Gamma \vdash t : T$, then there are T' and ψ such that $\Gamma \vdash_{\mathsf{a}}^{y} t : T'$ and $T'\psi \leq T$.*

Proof. By induction on $\vdash$. We only detail some cases.

(symb) Take $T' = \tau_f \rho_{y}$ and $\psi = \rho_{y}^{-1}\varphi$.

(app) By induction hypothesis, there exist T, ψ_1, U' and ψ_2 such that $\Gamma \vdash_{\mathsf{a}}^{y} t : T$, $T\psi_1 \leq (x : U)V$, $\Gamma \vdash_{\mathsf{a}}^{y \cup \mathcal{V}(T)} u : U'$ and $U'\psi_2 \leq U$. By Lemma 2, $\mathcal{V}(U') \cap \mathcal{V}(T) = \emptyset$. Thus, $\mathrm{dom}(\psi_1) \cap \mathrm{dom}(\psi_2) = \emptyset$. So, let $\psi = \psi_1 \uplus \psi_2$. By Lemma 1, $T\downarrow \psi \leq_s (x : U\downarrow)V\downarrow$. Thus, $T\downarrow = (x : U_1)V_1$, $U\downarrow \leq U_1\psi$ and $V_1\psi \leq V\downarrow$. Since $U'\psi \leq U$ and $U\downarrow \leq U_1\psi$, we have $U'\downarrow \psi \leq U_1\psi$ and, by Lemma 1, $U'\downarrow \psi \leq_s U_1\psi$. Thus, $\psi \in S(U'\downarrow, U_1)$. By Lemma 3, $S(U'\downarrow, U_1) = S(\mathcal{C})$ with $\mathcal{C} = \mathcal{C}(U'\downarrow, U_1)$. Thus, $S(\mathcal{C}) \neq \emptyset$ and there exists $\varphi = mgs(\mathcal{C})$. Hence, $\Gamma \vdash_{\mathsf{a}}^{y} tu : V_1\varphi\rho_{y}\theta$ where $\theta = \{x \mapsto u\}$. We are left to prove that there exists φ' such that $V_1\varphi\rho_{y}\theta\varphi' \leq V\theta$. Since $\varphi = mgs(\mathcal{C})$, there exists ψ' such that $\varphi\psi' \leq_{\mathcal{A}} \psi$. So, let $\varphi' = \rho_{y}^{-1}\psi'$. Since $\mathcal{V}(u) = \emptyset$, θ commutes with size substitutions. Since $V_1\psi \leq V\downarrow \leq V$, by Lemma 2, $V_1\varphi\rho_{y}\theta\varphi' = V_1\varphi\psi'\theta \leq V_1\psi\theta \leq V\theta$. $\square$

Theorem 4 (Decidability of type-checking). *Let Γ be an ∞-environment, t be an ∞-term and T be a type such that $\Gamma \vdash T : \mathsf{s}$. Then, the problem of knowing whether there is ψ such that $\Gamma \vdash t : T\psi$ is decidable.*

Proof. The decision procedure consists in (1) trying to compute the type T' such that $\Gamma \vdash_{\mathsf{a}}^{y} t : T'$ by taking $y = \mathcal{V}(T)$, and (2) trying to compute $\psi = mgs(\mathcal{C}(T', T))$. Every step is decidable.

We prove its correctness. Assume that $\Gamma \vdash_{\mathsf{a}}^{y} t : T'$, $y = \mathcal{V}(T)$ and $\psi = mgs(\mathcal{C}(T', T))$. Then, $T'\psi \leq T\psi$ and, by Theorem 2, $\Gamma \vdash t : T'$. By Lemma 1, $\Gamma \vdash t : T'\psi$. Thus, by (sub), $\Gamma \vdash t : T\psi$.

We now prove its completeness. Assume that there is ψ such that $\Gamma \vdash t : T\psi$. Let $y = \mathcal{V}(T)$. Since Γ is valid and $\mathcal{V}(\Gamma) = \emptyset$, by Theorem 3, there are T' and φ such that $\Gamma \vdash_{\mathsf{a}}^{y} t : T'$ and $T'\varphi \leq T\psi$. This means that the decision procedure cannot fail ($\psi \uplus \varphi \in S(T', T)$). $\square$

4   Solving Constraints

In this section, we prove that the satisfiability of constraint problems is decidable, and that a satisfiable problem has a smallest solution. The proof is organized as follows. First, we introduce simplification rules for equalities similar to usual unification procedures (Lemma 4). Second, we introduce simplification rules for inequalities (Lemma 5). From that, we can deduce some general result on the form of solutions (Lemma 7). We then prove that a conjunction of inequalities has always a linear solution (Lemma 8). Then, by using linear algebra techniques, we prove that a satisfiable inequality problem has always a smallest solution (Lemma 11). Finally, all these results are combined in Theorem 5 for proving the assumptions of Section 3.

Let a *state* $\mathbb{S}$ be $\bot$ or a triplet $\mathcal{E}|\mathcal{E}'|\mathcal{C}$ where $\mathcal{E}$ and $\mathcal{E}'$ are conjunctions of equalities and $\mathcal{C}$ a conjunction of inequalities. Let $S(\bot) = \emptyset$ and $S(\mathcal{E}|\mathcal{E}'|\mathcal{C}) = S(\mathcal{E} \wedge \mathcal{E}' \wedge \mathcal{C})$ be the solutions of a state. A conjunction of equalities $\mathcal{E}$ is in *solved form* if it is of the form $\alpha_1 = a_1 \wedge \ldots \wedge \alpha_n = a_n$ $(n \geq 0)$ with the variables α_i distinct from one another and $\mathcal{V}(a) \cap \{\alpha\} = \emptyset$. It is identified with the substitution $\{\alpha \mapsto a\}$.

$$
\begin{array}{ll}
(1) & \mathcal{E} \wedge sa = sb \mid \mathcal{E}' \mid \mathcal{C} \;\rightsquigarrow\; \mathcal{E} \wedge a = b \mid \mathcal{E}' \mid \mathcal{C} \\
(2) & \mathcal{E} \wedge a = a \mid \mathcal{E}' \mid \mathcal{C} \;\rightsquigarrow\; \mathcal{E} \mid \mathcal{E}' \mid \mathcal{C} \\
(3) & \mathcal{E} \wedge a = s^{k+1}a \mid \mathcal{E}' \mid \mathcal{C} \;\rightsquigarrow\; \bot \\
(4) & \mathcal{E} \wedge \infty = s^{k+1}a \mid \mathcal{E}' \mid \mathcal{C} \;\rightsquigarrow\; \bot \\
(5) & \mathcal{E} \wedge \alpha = a \mid \mathcal{E}' \mid \mathcal{C} \;\rightsquigarrow\; \mathcal{E}\{\alpha \mapsto a\} \mid \mathcal{E}'\{\alpha \mapsto a\} \wedge \alpha = a \mid \mathcal{C}\{\alpha \mapsto a\} \text{ if } \alpha \notin \mathcal{V}(a)
\end{array}
$$

Fig. 6. Simplification rules for equalities

The simplification rules on equalities given in Figure 6 correspond to the usual simplification rules for first-order unification [18], except that substitutions are propagated into the inequalities.

Lemma 4. *The relation of Figure 6 terminates and preserves solutions: if $\mathbb{S}_1 \rightsquigarrow \mathbb{S}_2$ then $S(\mathbb{S}_1) = S(\mathbb{S}_2)$. Moreover, any normal form of $\mathcal{E}|\top|\mathcal{C}$ is either $\bot$ or of the form $\top|\mathcal{E}'|\mathcal{C}'$ with $\mathcal{E}'$ in solved form and $\mathcal{V}(\mathcal{C}') \cap \mathrm{dom}(\mathcal{E}') = \emptyset$.*

We now introduce a notion of graphs due to Pratt [25] that allows us to detect the variables that are equivalent to ∞. In the following, we use other standard techniques from graph combinatorics and linear algebra. Note however that we apply them on symbolic constraints, while they are generally used on numerical constraints. What we are looking for is substitutions, not numerical solutions. In particular, we do not have the constant 0 in size expressions (although it could be added without having to change many things). Yet, for proving that satisfiable problems have most general solutions, we will use some isomorphism between symbolic solutions and numerical ones (see Lemma 10).

Definition 2 (Dependency graph). *To a conjunction of linear inequalities $\mathcal{C}$, we associate a graph $G_{\mathcal{C}}$ on $\mathcal{V}(\mathcal{C})$ as follows. To every constraint $s^p\alpha \leq s^q\beta$,* *we associate the labeled edge* $\alpha \xrightarrow{p-q} \beta$. *The* cost *of a path* $\alpha_1 \xrightarrow{p_1} \ldots \xrightarrow{p_k} \alpha_{k+1}$ *is* $\Sigma_{i=1}^{k} p_i$. *A cyclic path (i.e. when* $\alpha_{k+1} = \alpha_1$) *is* increasing *if its cost is* > 0.

$$\begin{aligned}
&(1) \quad \mathcal{C} \wedge a \leq s^k \infty \rightsquigarrow \mathcal{C} \\
&(2) \qquad\qquad \mathcal{C} \wedge \mathcal{D} \rightsquigarrow \mathcal{C} \wedge \{\infty \leq \alpha \mid \alpha \in \mathcal{V}(\mathcal{D})\} \quad \text{if } G_{\mathcal{D}} \text{ is increasing} \\
&(3) \quad \mathcal{C} \wedge s^k \infty \leq s^l \alpha \rightsquigarrow \mathcal{C}\{\alpha \mapsto \infty\} \wedge \infty \leq \alpha \qquad \text{if } \alpha \in \mathcal{V}(\mathcal{C})
\end{aligned}$$

Fig. 7. Simplification rules for inequalities

A conjunction of inequalities $\mathcal{C}$ is in *reduced form* if it is of the form $\mathcal{C}_\infty \wedge \mathcal{C}_\ell$ with $\mathcal{C}_\infty$ a conjunction of ∞-inequalities, $\mathcal{C}_\ell$ a conjunction of linear inequalities with no increasing cycle, and $\mathcal{V}(\mathcal{C}_\infty) \cap \mathcal{V}(\mathcal{C}_\ell) = \emptyset$.

Lemma 5. *The relation of Figure 7 on inequality problems terminates and preserves solutions. Moreover, any normal form is in reduced form.*

Lemma 6. *If $\mathcal{C}$ is a conjunction of inequalities then $S(\mathcal{C}) \neq \emptyset$. Moreover, if $\mathcal{C}$ is a conjunction of ∞-inequalities then $S(\mathcal{C}) = \{\varphi \mid \forall \alpha \in \mathcal{V}(\mathcal{C}), \alpha\varphi{\downarrow} = \infty\}$.*

Lemma 7. *Assume that $\mathcal{E}|\top|\mathcal{C}$ has normal form $\top|\mathcal{E}'|\mathcal{C}'$ by the rules of Figure 6, and $\mathcal{C}'$ has normal form $\mathcal{D}$ by the rules of Figure 7. Then, $S(\mathcal{E} \wedge \mathcal{C}) \neq \emptyset$, $\mathcal{E}' = mgs(\mathcal{E})$ and every $\varphi \in S(\mathcal{E} \wedge \mathcal{C})$ is of the form $\mathcal{E}'(v \uplus \psi)$ with $v \in S(\mathcal{D}_\infty)$ and $\psi \in S(\mathcal{D}_\ell)$.*

Proof. The fact that, in this case, $S(\mathcal{E}) \neq \emptyset$ and $\mathcal{E}' = mgs(\mathcal{E})$ is a well known result on unification [18]. Since $S(\mathcal{E} \wedge \mathcal{C}) = S(\mathcal{E}' \wedge \mathcal{D})$, $\mathcal{V}(\mathcal{E}') \cap \mathcal{V}(\mathcal{D}) = \emptyset$ and $S(\mathcal{D}) \neq \emptyset$, we have $S(\mathcal{E} \wedge \mathcal{C}) \neq \emptyset$. Furthermore, every $\varphi \in S(\mathcal{E} \wedge \mathcal{C})$ is of the form $\mathcal{E}'\varphi'$ since $S(\mathcal{E}' \wedge \mathcal{D}) \subseteq S(\mathcal{E}')$. Now, since $\mathcal{V}(\mathcal{D}_\infty) \cap \mathcal{V}(\mathcal{D}_\ell) = \emptyset$, $\varphi' = v \uplus \psi$ with $v \in S(\mathcal{D}_\infty)$ and $\psi \in S(\mathcal{D}_\ell)$. $\qquad\square$

Hence, the solutions of a constraint problem can be obtained from the solutions of the equalities, which is a simple first-order unification problem, and from the solutions of the linear inequalities resulting of the previous simplifications.

In the following, let $\mathcal{C}$ be a conjunction of K linear inequalities with no increasing cycle, and L be the biggest label in absolute value in $G_{\mathcal{C}}$. We first prove that $\mathcal{C}$ has always a linear solution by using Bellman-Ford's algorithm.

Lemma 8. $S^\ell(\mathcal{C}) \neq \emptyset$.

Proof. Let $succ(\alpha) = \{\beta \mid \alpha \xrightarrow{p} \beta \in G_{\mathcal{C}}\}$ and $succ^*$ be the reflexive and transitive closure of $succ$. Choose $\gamma \in \mathcal{Z} \setminus \mathcal{V}(\mathcal{C})$, a set R of vertices in $G_{\mathcal{C}}$ such that $succ^*(R)$ covers $G_{\mathcal{C}}$, and a minimal cost $q_\beta \geq KL$ for every $\beta \in R$. Let the cost of a vertex α_{k+1} along a path $\alpha_1 \xrightarrow{p_1} \alpha_2 \xrightarrow{p_2} \ldots \alpha_{k+1}$ with $\alpha_1 \in R$ be $q_{\alpha_1} + \Sigma_{i=1}^{k} p_i$. Now, let ω_β be the maximal cost for β along all the possible paths from a vertex in R. We have $\omega_\beta \geq 0$ since there is no increasing cycle. Hence, for all edge $\alpha \xrightarrow{p} \beta \in G_{\mathcal{C}}$, we have $\omega_\alpha + p \leq \omega_\beta$. Thus, the substitution $\varphi = \{\alpha \mapsto s^{\omega_\alpha} \gamma \mid \alpha \in \mathcal{V}(\mathcal{C})\} \in S^\ell(\mathcal{C})$. $\qquad\square$

We now prove that any solution has a more general linear solution. This implies that inequality problems are always satisfiable and that the satisfiability of a constraint problem only depends on its equalities.

Lemma 9. *If $\varphi \in S(\mathcal{C})$ then there exists $\psi \in S^\ell(\mathcal{C})$ such that $\psi \leq_A \varphi$.*

We now prove that $S^\ell(\mathcal{C})$ has a smallest element. To this end, assume that inequalities are ordered and that $\mathcal{V}(\mathcal{C}) = \{\alpha_1, \ldots, \alpha_n\}$. We associate to $\mathcal{C}$ an adjacency-like matrix $M = (m_{i,j})$ with K lines and n columns, and a vector $v = (v_i)$ of length K as follows. Assume that the i-th inequality of $\mathcal{C}$ is of the form $s^p \alpha_j \leq s^q \alpha_k$. Then, $m_{i,j} = 1$, $m_{i,k} = -1$, $m_{i,l} = 0$ if $l \notin \{j,k\}$, and $v_i = q - p$. Let $P = \{z \in \mathbb{Q}^n \mid Mz \leq v, z \geq 0\}$ and $P' = P \cap \mathbb{Z}^n$.

To a substitution $\varphi \in S^\ell(\mathcal{C})$, we associate the vector z^φ such that z_i^φ is the natural number p such that $\alpha_i \varphi = s^p \beta$.

To a vector $z \in P'$, we associate a substitution φ_z as follows. Let $\{G_1, \ldots, G_s\}$ be the connected components of $G_{\mathcal{C}}$. For all i, let c_i be the component number to which α_i belongs. Let $\beta_1, \ldots, \beta_s$ be variables distinct from one another and not in $\mathcal{V}(\mathcal{C})$. We define $\alpha_i \varphi_z = s^{z_i} \beta_{c_i}$.

We then study the relations between symbolic and numerical solutions.

Lemma 10.
- *If $\varphi \in S^\ell(\mathcal{C})$ then $z^\varphi \in P'$. Furthermore, if $\varphi \leq_A \varphi'$ then $z^\varphi \leq z^{\varphi'}$.*
- *If $z \in P'$ then $\varphi_z \in S^\ell(\mathcal{C})$. Furthermore, if $z \leq z'$ then $\varphi_z \leq_A \varphi_{z'}$.*
- *$z^{\varphi_z} = z$ and $\varphi_{z^\varphi} \sqsubseteq \varphi$.*

Finally, we are left to prove that P' has a smallest element. The proof uses techniques from linear algebra.

Lemma 11. *There is a unique $z^* \in P'$ such that, for all $z \in P'$, $z^* \leq z$.*

An efficient algorithm for computing the smallest solution of a set of linear inequalities with at most two variables per inequality can be found in [22]. A more efficient algorithm can perhaps be obtained by taking into account the specificities of our problems.

Gathering all the previous results, we get the decidability.

Theorem 5 (Decidability). *Let $\mathcal{C}$ be a constraint problem. Whether $S(\mathcal{C})$ is empty or not can be decided in polynomial time w.r.t. the size of equalities in $\mathcal{C}$. Furthermore, if $S(\mathcal{C}) \neq \emptyset$ then $S(\mathcal{C})$ has a smallest solution that is computable in polynomial time w.r.t. the size of inequalities.*

5   Conclusion and Related Works

In Section 3, we give a general algorithm for type inference with size annotations based on constraint solving, that does not depend on the size algebra. For having completeness, we require satisfiable sets of constraints to have a computable most general solution. In Section 4, we prove that this is the case if the size algebra is built from the symbols s and ∞ which, although simple, captures usual inductive definitions (since then the size corresponds to the number of constructors) and much more (see the introduction and [6]).

A natural extension would be to add the symbol $+$ in the size algebra, for typing list concatenation in a more precise way for instance. We think that the techniques used in the present work can cope with this extension. However, without restrictions on symbol types, one may get constraints like $1 \leq \alpha + \beta$ and loose the unicity of the smallest solution. We think that simple and general restrictions can be found to avoid such constraints to appear. Now, if symbols like $\times$ are added to the size algebra, then we lose linearity and need more sophisticated mathematical tools.

The point is that, because we consider dependent types and subtyping, we are not only interested in satisfiability but also in minimality and unicity, in order to have completeness of type inference [12]. There exist many works on type inference and constraint solving. We only mention some that we found more or less close to ours: Zenger's indexed types [30], Xi's Dependent[1] ML [28], Odersky *et al*'s ML with constrained types [24], Abel's sized types [1], and Barthe *et al*'s staged types [4]. We note the following differences:

Terms. Except [4], the previously cited works consider λ-terms *à la* Curry, *i.e.* without types in λ-abstractions. Instead, we consider λ-terms *à la* Church, *i.e.* with types in λ-abstractions. Note that type inference with λ-terms *à la* Curry and polymorphic or dependent types is not decidable. Furthermore, they all consider functions defined by fixpoint and matching on constructors. Instead, we consider functions defined by rewrite rules with matching both on constructor and defined symbols (*e.g.* associativity and distributivity rules).

Types. If we disregard constraints attached to types, they consider simple or polymorphic types, and we consider fully polymorphic and dependent types. Now, our data type constructors carry no constraints: constraints only come up from type inference. On the other hand, the constructors of Zenger's indexed data types must satisfy polynomial equations, and Xi's index variables can be assigned boolean propositions that must be satisfiable in some given model (*e.g.* Presburger arithmetic). Explicit constraints allow a more precise typing and more function definitions to be accepted. For instance (see [6]), in order for *quicksort* to have type $list^\alpha \Rightarrow list^\alpha$, we need the auxiliary *pivot* function to have type $nat^\infty \Rightarrow list^\alpha \Rightarrow list^\beta \times list^\gamma$ with the constraint $\alpha = \beta + \gamma$. And, if *quicksort* has type $list^\infty \Rightarrow list^\infty$ then a rule like $f\ (cons\ x\ l) \to g\ x\ (f\ (quicksort\ l))$ is rejected since $(quicksort\ l)$ cannot be proved to be smaller than $(cons\ x\ l)$. The same holds in [1, 4].

Constraints. In contrast with Xi and Odersky *et al* who consider the constraint system as a parameter, giving DML(C) and HM(X) respectively, we consider a fixed constraint system, namely the one introduced in [3]. It is close to the one considered by Abel whose size algebra does not have ∞ but whose types have explicit bounded quantifications. Inductive types are indeed interpreted

---

[1] By "dependent", Xi means constrained types, not full dependent types.

in the same way. We already mentioned also that Zenger considers polynomial equations. However, his equivalence on types is defined in such a way that, for instance, $list^\alpha$ is equivalent to $list^{2\alpha}$, which is not very natural. So, the next step in our work would be to consider explicit constraints from an abstract constraint system. By doing so, Odersky $et\ al$ get general results on the completeness of inference. Sulzmann [26] gets more general results by switching to a fully constrained-based approach. In this approach, completeness is achieved if every constraint can be represented by a type. With term-based inference and dependent types, which is our case, completeness requires minimality which is not always possible [12].

Constraint Solving. In [4], Barthe $et\ al$ consider system F with ML-like definitions and the same size annotations. Since they have no dependent type, they only have inequality constraints. They also use dependancy graphs for eliminating ∞, and give a specific algorithm for finding the most general solution. But they do not study the relations between linear constraints and linear programming. So, their algorithm is less efficient than [22], and cannot be extended to size annotations like $a + b$, for typing addition or concatenation.

Inference of Size Annotations. As already mentioned in the introduction, we do not infer size annotations for function symbols like [4, 13]. We just check that function definitions are valid wrt size annotations, and that they preserve termination. However, finding annotations that satisfy these conditions can easily be expressed as a constraint problem. Thus, the techniques used in this paper can certainly be extended for inferring size annotations too. For instance, if we take $-\ :\ nat^\alpha \Rightarrow nat^\beta \Rightarrow nat^X$, the rules of $-$ given in the introduction are valid whenever $0 \leq X$, $\alpha \leq X$ and $X \leq X$, and the most general solution of this constraint problem is $X = \alpha$.

Acknowledgments

I would like to thank very much Miki Hermann, Hongwei Xi, Christophe Ringeissen and Andreas Abel for their comments on a previous version of this paper.

References

1. A. Abel. Termination checking with types. *Theoretical Informatics and Applications*, 38(4):277–319, 2004.
2. H. Barendregt. Lambda calculi with types. In S. Abramsky, D. Gabbay, and T. Maibaum, editors, *Handbook of logic in computer science*, volume 2. Oxford University Press, 1992.
3. G. Barthe, M. J. Frade, E. Giménez, L. Pinto, and T. Uustalu. Type-based termination of recursive definitions. *Mathematical Structures in Computer Science*, 14(1):97–141, 2004.
4. G. Barthe, B. Grégoire, and F. Pastawski. Practical inference for type-based termination in a polymorphic setting. In *Proc. of TLCA'05*, LNCS 3461.

5. F. Blanqui. Rewriting modulo in Deduction modulo. In *Proc. of RTA'03*, LNCS 2706.
6. F. Blanqui. A type-based termination criterion for dependently-typed higher-order rewrite systems. In *Proc. of RTA'04*, LNCS 3091.
7. F. Blanqui. Full version of [6]. See http://www.loria.fr/~blanqui/.
8. F. Blanqui. Definitions by rewriting in the Calculus of Constructions. *Mathematical Structures in Computer Science*, 15(1):37–92, 2005.
9. F. Blanqui. Full version. See http://www.loria.fr/~blanqui/.
10. F. Blanqui. Inductive types in the Calculus of Algebraic Constructions. *Fundamenta Informaticae*, 65(1-2):61–86, 2005.
11. V. Breazu-Tannen. Combining algebra and higher-order types. In *Proc. of LICS'88*.
12. G. Chen. *Subtyping, Type Conversion and Transitivity Elimination*. PhD thesis, Université Paris VII, France, 1998.
13. W. N. Chin and S. C. Khoo. Calculating sized types. *Journal of Higher-Order and Symbolic Computation*, 14(2-3):261–300, 2001.
14. Coq-Development-Team. *The Coq Proof Assistant Reference Manual - Version 8.0*. INRIA Rocquencourt, France, 2004. http://coq.inria.fr/.
15. T. Coquand. An algorithm for testing conversion in type theory. In G. Huet and G. Plotkin, ed., *Logical Frameworks*, p 255–279. Cambridge University Press, 1991.
16. T. Coquand and G. Huet. The Calculus of Constructions. *Information and Computation*, 76(2-3):95–120, 1988.
17. T. Coquand and C. Paulin-Mohring. Inductively defined types. In *Proc. of COLOG'88*, LNCS 417.
18. N. Dershowitz and J.-P. Jouannaud. Rewrite systems. In J. van Leeuwen, editor, *Handbook of Theoretical Computer Science*, vol B, ch 6. North-Holland, 1990.
19. E. Giménez. Structural recursive definitions in type theory. In *Proc. of ICALP'98*, LNCS 1443.
20. J. Hughes, L. Pareto, and A. Sabry. Proving the correctness of reactive systems using sized types. In *Proc. of POPL'96*.
21. J.-P. Jouannaud and A. Rubio. The Higher-Order Recursive Path Ordering. In *Proc. of LICS'99*.
22. G. Lueker, N. Megiddo, and V. Ramachandran. Linear programming with two variables per inequality in poly-log time. *SIAM Journal on Computing*, 19(6):1000–1010, 1990.
23. F. Müller. Confluence of the lambda calculus with left-linear algebraic rewriting. *Information Processing Letters*, 41(6):293–299, 1992.
24. M. Odersky, M. Sulzmann, and M. Wehr. Type inference with constrained types. *Theory and Practice of Object Systems*, 5(1):35–55, 1999.
25. V. Pratt. Two easy theories whose combination is hard. Technical report, MIT, United States, 1977.
26. M. Sulzmann. A general type inference framework for Hindley/Milner style systems. In *Proc. of FLOPS'01*, LNCS 2024.
27. D. Walukiewicz-Chrząszcz. Termination of rewriting in the Calculus of Constructions. *Journal of Functional Programming*, 13(2):339–414, 2003.
28. H. Xi. *Dependent types in practical programming*. PhD thesis, Carnegie-Mellon, Pittsburgh, United States, 1998.
29. H. Xi. Dependent types for program termination verification. *Journal of Higher-Order and Symbolic Computation*, 15(1):91–131, 2002.
30. C. Zenger. Indexed types. *Theoretical Computer Science*, 187(1-2):147–165, 1997.

On the Role of Type Decorations in the Calculus of Inductive Constructions

Bruno Barras[1] and Benjamin Grégoire[2]

[1] INRIA Futurs, France
Bruno.Barras@inria.fr
[2] INRIA Sophia-Antipolis, France
Benjamin.Gregoire@sophia.inria.fr

Abstract. In proof systems like Coq [16], proof-checking involves comparing types modulo β-conversion, which is potentially a time-consuming task. Significant speed-ups are achieved by compiling proof terms, see [9]. Since compilation erases some type information, we have to show that convertibility is preserved by type erasure. This article shows the equivalence of the Calculus of Inductive Constructions (formalism of Coq) and its domain-free version where parameters of inductive types are also erased. It generalizes and strengthens significantly a similar result by Barthe and Sørensen [5] on the class of functional Domain-free Pure Type Systems.

1 Introduction

In proof systems based on the Curry-Howard isomorphism, proof-checking boils down to type-checking in a system with dependent types. Such systems usually include a conversion rule of the form:

$$\frac{\Gamma \vdash t : \tau \quad \tau \simeq_\beta \tau'}{\Gamma \vdash t : \tau'} [\text{conv}]$$

where $\simeq_\beta$ stands for β-convertibility. This rule can be used to make complex computation. Examples of that usage include reflection tactics [6] in Coq and the proof of the four-colors theorem. This conversion rule is generally implemented in a purely interpretative way[1], because it is a hard task to perform strong β-reduction (reduction occurs also under binders) in a compiled setting. In [9] Grégoire and Leroy show how to strongly normalize and how to decide β-equivalence on terms, by compiling proof-terms towards an abstract machine (a slightly modified version of OCaml's ZAM) and analyzing computed values with a readback procedure.

This scheme raises a problem: compilation has the effect of erasing type annotations (used to ensure the decidability of type checking and so the impossibility

---

[1] By interpretative, we mean algorithms that perform the conversion test by explicitly manipulating proof terms represented as trees.

L. Ong (Ed.): CSL 2005, LNCS 3634, pp. 151–166, 2005.

of runtime error). So, while conversion is defined over Church-style terms (abstractions carry a type annotation $\lambda x : \tau . t$), the abstract machine based version of conversion works on Curry-style terms ($\lambda x . t$, also called *domain-free* terms). The correctness of such compilation scheme with respect to the original formalism relies on a general issue of proving the equivalence of a given type system with its domain-free counterpart. This problem has already been studied in the case of Pure Type Systems [2] (PTS, Church-style) and their domain-free version, the Domain-Free Pure Type System [5] (DFPTS) by Barthe and Sørensen. The authors prove an equivalence theorem under the assumption that the system is normalizing (so we can reason on normal terms) and functional. The latter condition is used to ensure type uniqueness. Earlier, Streicher [15] proved this result for Calculus of Constructions, still based on normalization and type uniqueness.

Our paper enhances previous work in two ways. Firstly, we extend the results of [5] to a richer class of systems that feature cumulativity[2] and inductive types. A notable point of our notion of type erasure is that we erase parameters of constructors[3], since they do not participate in the computation. Our results apply to the Calculus of Inductive Constructions (CIC), and yield an efficient sound and complete convertibility test for Coq.

Secondly, our results do not rely on type uniqueness, which does not hold any more due to subtyping (even without subtyping, type uniqueness does not hold for any PTS). Instead, we introduce an equivalence on types to recover a loose notion of type uniqueness.

This equivalence theorem also has consequences on implementation. Many proof systems prefer to use Church-style λ-terms, in particular because type inference is decidable under simple conditions and it is often easier to build a set theoretic model of those formalisms. On the other hand, Curry-style terms reflect the computational behavior of λ-terms better. This is related to the fact that pure λ-calculus is the execution model of the core of many functional languages. But type inference is generally not decidable, and type checking fails on non-normal terms. The equivalence theorem shows that we can have a system with good properties such as type decidability, and compare terms as Curry-style terms, allowing compilation techniques. Regarding inductive types, parameters can be erased in constructors, which leads to the same representation as in a compiled language like OCaml [10].

We prove the equivalence between a type annotated system where conversion compares type decorations (we call it β) and a second type system (we call it ϵ) where annotations are in the syntax but conversion ignores them. Then it is trivial to define a third system (the domain free version) where type decorations are not in the syntax and then prove the equivalence with the system ϵ. This way, we separate the problems of changing the term representation and that of actually changing the conversion.

---

[2] Cumulativity is a simple notion of subtyping that reduce need to duplicate definition across the various universes of the PTS.

[3] For instance, the first argument of the ternary constructor of lists cons : $\forall \alpha, \alpha \rightarrow$ list $\alpha \rightarrow$ list α can be erased.

For explanatory purposes, we will distinguish the strengthening of Barthe's and Sørensen's result (removing the functionality hypothesis) and the extension to a broader class of systems. Section 2 introduces Cumulative Type Systems (CTS), which are PTS with cumulativity and an abstract notion of conversion. Then, we briefly give metatheoretical properties of CTS. Section 3 shows how both ω and ϵ systems can be represented by instantiating this abstract conversion in two ways. It ends by proving Preservation of Equational Theory (PET), Preservation of Subtyping (PS) and Preservation of Typing (PT) between both systems. These properties simply state that conversion, subtyping and typing are equivalent notions. Then, we will extend these results to inductive types (Sect. 4) and conclude.

2 A Generic Version of Cumulative Type Systems (CTS)

Pure Type Systems (PTS, [2]) are a generalization of several type systems such as simply typed λ-calculus, system F, Calculus of Constructions, etc. Since some systems have dependent types (type parameterized by expressions or programs), they use the same syntax for terms and types and types are also subject to a type discipline.

2.1 Syntax of Terms

As for PTS, Cumulative Type Systems [3] are generated from specifications. To the three parameters of the PTS, we add two extra parameters. The first one $\prec$ allows subtyping over sorts: if $s_1 \prec s_2$, then any type of s_1 is also a type of s_2, without any explicit coercion. This is called cumulativity. The second extra parameter $\simeq$, called conversion, is a relation between types indicating which types are identified. In the rest of this paper we will instantiate this parameter with different relations. This follows the same idea as in [12, 14].

Let us make this more precise by simultaneously defining the syntax of terms ($\mathscr{T}$) and specifications of CTS (**S**). Let $\mathscr{V}$ be an infinite set of variables.

Definition 1 (term and specification).
A specification is a tuple $\mathbf{S} = (\mathscr{S}, \mathscr{A}, \mathscr{R}, \prec, \simeq)$ *where*
- $\mathscr{S}$ *is a set of sorts.*
- $\mathscr{A} \subseteq \mathscr{S} \times \mathscr{S}$ *is a set of axioms.*
- $\mathscr{R} \subseteq \mathscr{S} \times \mathscr{S} \times \mathscr{S}$ *is a set of rules*
- $\prec \subseteq \mathscr{S} \times \mathscr{S}$ *is an inclusion relation between sorts.*
- $\simeq \subseteq \mathscr{T} \times \mathscr{T}$ *is an equivalence relation between terms. It should be a congruence:* $\forall x, M, N, N'. \ N \simeq N' \ \Rightarrow \ M\{x \leftarrow N\} \simeq M\{x \leftarrow N'\}$

The set $\mathscr{T}$ *of expressions (over* **S***) is given by the abstract syntax :*
$$\mathscr{T} ::= \mathscr{V} \mid \mathscr{S} \mid \Pi \mathscr{V} : \mathscr{T}. \mathscr{T} \mid \lambda \mathscr{V} : \mathscr{T}. \mathscr{T} \mid \mathscr{T} \ \mathscr{T}$$

We use t, A, B, M, N, T, U, V, etc. to denote elements of $\mathscr{T}$; x, y, z, etc. to denote elements of $\mathscr{V}$; s, s', etc. to denote elements of $\mathscr{S}$. The substitution of variable x for a term N in M will be written $M\{x \leftarrow N\}$. As usual, we consider β-reduction on terms, written $\rightarrow$. We write $\xrightarrow{*}$ its reflexive and transitive closure, and $\simeq_\beta$ the smallest equivalence relation including $\rightarrow$ (β-conversion). PTS are a special case of CTS where $\prec$ is $\emptyset$ and $\simeq$ is β-conversion.

2.2 Cumulativity

As already stated, cumulativity introduces some kind of subtyping. Let us now define the subtyping relation induced by our CTS parameters:

Definition 2 (cumulativity). *The one step subtyping relation $\preceq$ over an equivalence relation $\simeq$ and an inclusion relation between sorts $\prec$ is given by the rules below.*

$$\frac{T \simeq U}{T \preceq U} \qquad \frac{s_1 \prec^* s_2}{s_1 \preceq s_2} \qquad \frac{T \simeq T' \qquad U \preceq U'}{\Pi x{:}T.U \preceq \Pi x{:}T'.U'}$$

This relation is also named cumulativity. We write $\preceq_S$ to refer to the equivalence relation and the inclusion relation between sorts of S. When $\prec$ is fixed, we use the notation $\preceq_\simeq$ or just $\preceq$ if $\simeq$ is clear from the context.

Note that following Luo's Extended Calculus of Constructions [11], the subtyping relation is not contravariant w.r.t. the domain of functions (the domains of a function type and its subtype are convertible). Contravariance is rejected because it would invalidate our proof as we shall in section 3.3.

At that point, we define several properties of relations related to abstract rewriting systems.

Definition 3 (commutation, reducibility). *Let R_1, R_2 be two binary relations.*

- R_1, R_2 *commute, written $(R_1, R_2) \in \mathcal{C}$, iff*

$$\forall x, x_1, x_2.\ x\ R_1\ x_1 \wedge x\ R_2\ x_2 \Rightarrow \exists y.\ x_2\ R_1\ y \wedge x_1\ R_2\ y$$

- R_1 *is reducible to R_2 modulo β-reduction, written $R_1 \in \mathcal{R}_{R_2}$, iff*

$$\forall t, u.\ t\ R_1\ u \Rightarrow \exists t', u'.\ t \overset{*}{\to} t' \wedge u \overset{*}{\to} u' \wedge t'\ R_2\ u'$$

Lemma 1. *For any equivalence relation R, cumulativity preserves commutation with β-reduction*

$$(R, \overset{*}{\to}) \in \mathcal{C} \Rightarrow (\preceq_R, \overset{*}{\to}) \wedge (\preceq_R^{-1}, \overset{*}{\to}) \in \mathcal{C}$$

Lemma 2. *Cumulativity preserves reducibility to any equivalence relation commuting with β-reduction:*

$$(R_2, \overset{*}{\to}) \in \mathcal{C} \wedge R_1 \in \mathcal{R}_{R_2} \Rightarrow \preceq_{R_1}^* \in \mathcal{R}_{\preceq_{R_2}^*}$$

Proof: See appendix A.

2.3 Typing

Definition 4 (Typing judgment). *Let $\mathbf{S}$ be the specification $(\mathscr{S}, \mathscr{A}, \mathscr{R}, \prec, \simeq)$*

- *A context is a list:* $\Gamma ::= [\,] \mid \Gamma; (x{:}T)$
  $(x{:}T)$ *denotes a local declaration of a variable x of type T.*

$$\frac{}{\mathcal{WF}([])}[\text{WE}] \qquad \frac{\Gamma \vdash T : s \qquad s \in \mathscr{S}}{\mathcal{WF}(\Gamma; (x{:}T))}[\text{WS} - \text{LOCAL}]$$

$$\frac{\mathcal{WF}(\Gamma) \qquad (s_1, s_2) \in \mathscr{A}}{\Gamma \vdash s_1 : s_2}[\text{SORT}] \qquad \frac{\mathcal{WF}(\Gamma) \qquad (x{:}T) \in \Gamma}{\Gamma \vdash x : T}[\text{VAR}]$$

$$\frac{\Gamma \vdash T : s1 \qquad \Gamma; (x{:}T) \vdash U : s2 \qquad (s1, s2, s3) \in \mathscr{R}}{\Gamma \vdash \Pi x{:}T.U : s3}[\text{PROD}]$$

$$\frac{\Gamma \vdash \Pi x{:}T.U : s \quad \Gamma; (x{:}T) \vdash M : U}{\Gamma \vdash \lambda x{:}T.M : \Pi x{:}T.U}[\text{LAM}] \qquad \frac{\Gamma \vdash M : \Pi x{:}T.U \quad \Gamma \vdash N : T}{\Gamma \vdash M\ N : U\{x \leftarrow N\}}[\text{APP}]$$

$$\frac{\Gamma \vdash M : T \quad \Gamma \vdash U : s \quad T \preceq^* U}{\Gamma \vdash M : U}[\text{CONV}] \qquad \frac{\Gamma \vdash M : T \quad T \preceq^* s}{\Gamma \vdash M : s}[\text{CONV}_s]$$

Fig. 1. Typing rules for CTS

– *The typing relation $\vdash$ is given by the rules in Fig. 1. There is also a judgment $\mathcal{WF}()$ to mean that a context is well formed. Both two judgments are simultaneously defined by mutual induction. We occasionally write $\vdash_S$ to explicit the dependency with the specification $\mathbf{S}$.*

The rules are the same as for PTS except that in our CTS, CONV should rather be seen as a subsumption rule, and CONV_s is necessary when cumulativity is used towards a non-typable sort.

Figure 2 lists the fundamental meta-theoretical properties of CTS. They are easy generalizations of PTS's properties. First equation expresses that type derivations of CTS are preserved by substitution. The second one shows that typing is preserved by well-typed context narrowing. Equation (3), that a type is a sort or typable by a sort. Then we have the well-known subject reduction property. The last one is the inversion lemma. We will not give their proofs since they have been formally checked using Coq in [3][4].

Later on, we will study the relation between different CTS which differ on the inclusion between sorts and conversion. We say that $\mathbf{S}_1$ is included in $\mathbf{S}_2$ if they are included component-wise. In that case $\preceq_{\mathbf{S}_1} \subseteq \preceq_{\mathbf{S}_2}$ and $\vdash_{\mathbf{S}_1} \subseteq \vdash_{\mathbf{S}_2}$. Put it in another way, subtyping and typing are monotonic w.r.t. the specification.

3 β-Conversion and Conversion Modulo Type Annotations

Now we have this general framework of CTS, we can instantiate it with the parameters corresponding to the considered logical formalisms. For the rest of this paper we suppose that $\mathscr{S}, \mathscr{A}, \mathscr{R}$ and $\prec$ are fixed. In the case of typeful systems, terms are identified modulo β:

---

[4] The complete source of that formalization is available online at http://logical. inria.fr/~barras/pts_proofs/PTS/main.html. All subsequent URLs will be relative to http://logical.inria.fr/~barras/pts_proofs/PTS/.

Substitution $\Gamma \vdash N : T \wedge \Gamma; (x{:}T) \vdash M : U \Rightarrow \Gamma \vdash M\{x \leftarrow N\} : U\{x \leftarrow N\}$ (1)

Metatheory.html#substitution

Context conversion $\Gamma \vdash M : T \wedge \Delta \preceq^* \Gamma \Rightarrow \Delta \vdash M : T$ (2)

Metatheory.html#subtype_in_env

Correctness of types $\Gamma \vdash A : B \Rightarrow B \in \mathscr{S} \vee \exists s \in \mathscr{S}.\ \Gamma \vdash B : s$ (3)

Metatheory.html#type_correctness

Subject reduction $\Gamma \vdash t : T \wedge t \xrightarrow{*} t' \Rightarrow \Gamma \vdash t' : T$ (4)

LambdaSound.html#beta_sound

Inversion lemmas

$$\Gamma \vdash s_1 : T \Rightarrow \exists s_2.\ (s_1, s_2) \in \mathscr{A} \wedge s_2 \preceq^* T$$
$$\Gamma \vdash x : T \Rightarrow \exists T'.\ (x{:}T') \in \Gamma \wedge T' \preceq^* T$$
$$\Gamma \vdash \lambda x{:}A.\,M : T \Rightarrow \exists B, s.\ \Gamma; (x{:}A) \vdash M : B \wedge \Gamma \vdash \Pi x{:}A.\,B : s \wedge \Pi x{:}A.\,B \preceq^* T$$
$$\Gamma \vdash M\ N : T \Rightarrow \exists A, B.\ \Gamma \vdash M : \Pi x{:}A.\,B \wedge \Gamma \vdash N : A \wedge B\{x \leftarrow N\} \preceq^* T$$
$$\Gamma \vdash \Pi x{:}A.\,B : T \Rightarrow \exists (s_1, s_2, s_3) \in \mathscr{R}.\ \Gamma \vdash A : s_1 \wedge \Gamma; (x{:}A) \vdash B : s_2 \wedge s_3 \preceq^* T$$

Metatheory.html#inversion_lemma

Fig. 2. Meta-theoretical properties of CTS

Definition 5 (specification β). *Since β-conversion is a congruence, we can build a CTS upon it. Let β be the specification $(\mathscr{S}, \mathscr{A}, \mathscr{R}, \prec, \simeq_\beta)$.*

Lemma 3. $\preceq_\beta^*$ *is reducible to* $\preceq_=$.
Proof: Since $\preceq_=$ is transitive, we only have to show $\preceq_\beta^* \in \mathcal{R}_{\preceq_=^*}$, which is a consequence of Lemma 2 and the well known Church-Rosser property of β-conversion.

3.1 ϵ-Conversion

To define the notion of convertibility that do not take type annotations into account we first define equality modulo type annotation. It captures the essence of domain-free conversion but within a type-carrying syntax.

Definition 6 (ϵ-equality, ϵ-convertibility). *Two terms are ϵ-equal if they are equal modulo type annotations. We write $=_\epsilon$ this equality. ϵ-convertibility is the smallest equivalence relation including ϵ-equality and β-reduction. We write $\simeq_\epsilon$ this relation.*

$$\frac{}{x =_\epsilon x} \qquad \frac{}{s =_\epsilon s} \qquad \frac{T =_\epsilon T' \quad U =_\epsilon U'}{T\ U =_\epsilon T'\ U'} \qquad \frac{T =_\epsilon T'}{\lambda x{:}A.\,T =_\epsilon \lambda x{:}A'.\,T'}$$

$$\frac{T =_\epsilon T' \quad U =_\epsilon U'}{\Pi x{:}T.\,U =_\epsilon \Pi x{:}T'.\,U'} \qquad \frac{T =_\epsilon U}{T \simeq_\epsilon U} \qquad \frac{T \simeq_\epsilon U \quad U \to V}{T \simeq_\epsilon V} \qquad \frac{T \simeq_\epsilon U \quad V \to U}{T \simeq_\epsilon V}$$

Definition 7. *It is trivial to see that ϵ-equality and $\simeq_\epsilon$ are equivalence relations and congruences. Let ϵ be the specification $(\mathscr{S}, \mathscr{A}, \mathscr{R}, \prec, \simeq_\epsilon)$.*

This conversion enjoys reducibility results similar to β:

Lemma 4. $\simeq_\epsilon$ *is reducible to* $=_\epsilon$ *and* $\preceq^*_\epsilon$ *is reducible to* $\preceq_{=_\epsilon}$.

Proof: First prove $(=_\epsilon, \overset{*}{\rightarrow}) \in \mathcal{C}$ and using the confluence of β-reduction, we extend the result to $\simeq_\epsilon$. For the second statement, we prove $\preceq^*_\epsilon \in \mathcal{R}_{\preceq^*_{=_\epsilon}}$ using Lemma 2, and remark that $\preceq_{=_\epsilon}$ is reflexive and transitive.

3.2   Uniqueness of Types

It is well known that any functional PTS (a PTS where $\mathcal{A}$ and $\mathcal{R}$ are functional relations) enjoys the type uniqueness property:

$$\Gamma \vdash M : T \;\wedge\; \Gamma \vdash M : T' \;\Rightarrow\; T \simeq_\beta T'$$

This has been already formally proved in Lego by Pollack [14]. Unfortunately, non trivial subtyping (including cumulativity) breaks this property. Let alone CTS, the property does not hold for non functional PTS, which include the well-known (and useful) Calculus of Constructions with universes.

However, we can remark that we only need type uniqueness regarding the *domain types* of functions. This relaxed uniqueness notion holds for CTS because subtyping can occur only on sorts in the codomain (as Luo already noticed for the Extended Calculus of Constructions [11]). This uniqueness of domain types is formalized by a relation $\approx_\beta$ which ensures convertibility of domain types, but allows any change of sort in the codomain. It reuses Definition 2.

Definition 8. *We write* $\approx_\beta$ *the reflexive and transitive closure of the cumulativity relation derived from* $\simeq_\beta$ *and* $\mathscr{S} \times \mathscr{S}$, *and* $\approx_=$ *the cumulativity relation derived from* $\mathscr{S} \times \mathscr{S}$ *and* $=$. *We say that* t_1 *is close to* t_2 *if* $t_1 \approx_\beta t_2$.

The important facts are $\approx_\beta$ is an equivalence relation, $\approx_=$ is transitive and $\approx_\beta$ is reducible to $\approx_=$ (Lemma 2).

Lemma 5 (type uniqueness modulo $\approx_\beta$). *Specification* β *has type uniqueness modulo* $\approx_\beta$:

$$\Gamma \vdash_\beta t : T \;\wedge\; \Gamma \vdash_\beta t : T' \;\Rightarrow\; T \approx_\beta T'$$

Proof: by induction on $\Gamma \vdash_\beta t : T$, then inversion on $\Gamma \vdash_\beta t : T'$. In all cases, we use the fact that $\preceq^*_\beta \subset \approx_\beta$ (by monotonicity) and that $\approx_\beta$ is a symmetric and transitive relation. Application case uses the fact that $\Pi x{:}T.\, U \approx_\beta \Pi x{:}T'.\, U' \Rightarrow T \simeq_\beta T' \;\wedge\; U \approx_\beta U'$.

3.3   Equivalence of ϵ and β

Proving $\Gamma \vdash_\beta t : T \Rightarrow \Gamma \vdash_\epsilon t : T$ is trivial by monotonicity of typing. The converse is more difficult to derive. The first idea is to proceed by induction on the typing judgment. The only difficulty is with the conversion rules. It is of course false that $\preceq^*_\epsilon$ is included in $\preceq^*_\beta$, even for well typed terms: take $\lambda x{:}A.\, x :$ $A \rightarrow A$ and $\lambda x{:}B.\, x : B \rightarrow B$. So we have to do some induction loading. We can remark that if we compare only objects of same type, then we would necessarily have $A \simeq_\beta B$ (note that it would not be the case if cumulativity was contravariant w.r.t. the domain of functions). In order to have the weakest invariant we only assume that their types are close. But this invariant has to propagate to subterms. Consider $c : C$ and

$$(\lambda f{:}A \to A.\, c)\, (\lambda x{:}A.\, x) \qquad (\lambda f{:}B \to B.\, c)\, (\lambda x{:}B.\, x).$$

Both terms have type C, but not their subterms: arguments respectively have type $A \to A$ and $B \to B$. This example shows that β-redexes break the proposed invariant. If we assume our terms are in normal form, λ-abstractions are found only as argument of a variable. Since two ϵ-equal variables are equal, domain type uniqueness can establish the invariant that ϵ-convertible abstraction are compared only when we know their types are convertible, hence close. So, we can prove:

Lemma 6.

$$\left.\begin{array}{ll} \Gamma \vdash_\beta t : T & \Gamma \vdash_\beta t' : T' \\ t =_\epsilon t' & t, t' \in \mathcal{NF} \\ t \notin \lambda \ \vee \ T \approx_\beta T' \end{array}\right\} \Rightarrow t = t'$$

where λ is the set of lambda abstraction terms.

Proof: by induction on t, inversion of hypotheses $t =_\epsilon t'$, $\Gamma \vdash_\beta t : T$ and $\Gamma \vdash_\beta t' : T'$. We do only the interesting cases.

- Cases for sorts, variables and products are trivial.
- Case $t = \lambda x{:}A.\, M$ we have $t' = \lambda x{:}A'.\, M'$; $A, M, A', M' \in \mathcal{NF}$; $M =_\epsilon M'$. Inversion of type judgments yields:
  $\Gamma \vdash_\beta \Pi x{:}A.\, B : s$; $\quad \Gamma; (x{:}A) \vdash_\beta M : B$;
  $\Gamma \vdash_\beta \Pi x{:}A'.\, B' : s'$; $\quad \Gamma; (x{:}A') \vdash_\beta M' : B'$;
  $\Pi x{:}A.\, B \preceq^*_\beta T \quad \Pi x{:}A'.\, B' \preceq^*_\beta T'$
  Thanks to last premise, we get $\Pi x{:}A.\, B \approx_\beta \Pi x{:}A'.\, B'$, so $A \simeq_\beta A'$ and $B \approx_\beta B'$. Since A, A' are in normal form, they are equal (Lemma 3). Equality of bodies holds using the induction hypothesis.
- Case $t = M\, N$ and $t' = M'\, N'$; inversion of type judgments yields: $\Gamma \vdash_\beta M : \Pi x{:}A.\, B$; $\Gamma \vdash_\beta N : A$; $\Gamma \vdash_\beta M' : \Pi x{:}A'.\, B'$; $\Gamma \vdash_\beta N' : A'$
  Since $t \in \mathcal{NF}$, M is not an abstraction, so by induction hypothesis $M = M'$. By uniqueness of type (Lemma 5) $\Pi x{:}A.\, B \approx_\beta \Pi x{:}A'.\, B'$ so $A \simeq_\beta A'$. Equality of arguments holds using the induction hypothesis.

The invariant is weaker than in Barthe and Sørensen [5], and leads to a simpler proof.

Theorem 1 (PET ϵ w.r.t. β). *If specification β is normalizing, $\Gamma \vdash_\beta M : T$ and $\Gamma \vdash_\beta M' : T$,*

$$M \simeq_\beta M' \Leftrightarrow M \simeq_\epsilon M'$$

Proof: $M \simeq_\beta M' \Rightarrow M \simeq_\epsilon M'$ holds by monotonicity. Now assume $M \simeq_\epsilon M'$. Since specification β is normalizing, M (resp. M') has a normal form N (resp. N') which has type T by subject reduction. We have $N \simeq_\epsilon N'$ and also $N =_\epsilon N'$ by reducibility. (Lemma 4). Lemma 6 proves $N = N'$, so $M \simeq_\beta M'$.

In fact, we can replace the two premises of this theorem by $\Gamma \vdash_\beta M : T$ and $\Gamma \vdash_\beta M' : T'$ and $T \approx_\beta T'$. We only need $T \approx_\beta T'$ to apply Lemma 6.

We extend Lemma 6 to the cumulative subtyping relation:

Lemma 7.

$$\left. \begin{array}{cc} \Gamma \vdash_\beta t : T & \Gamma \vdash_\beta t' : T' \\ t \preceq_{=_\epsilon} t' & t, t' \in \mathcal{NF} \\ t \notin \lambda \vee T \approx_\beta T' & \end{array} \right\} \Rightarrow t \preceq_= t'$$

Proof: By induction over $t \preceq_{=_\epsilon} t'$. Obviously we use Lemma 6.

Corollary 1 (PS ϵ w.r.t. β). *If specification β is normalizing, $\Gamma \vdash_\beta T : s$ and $\Gamma \vdash_\beta U : s'$, then*

$$T \preceq^*_\epsilon U \Leftrightarrow T \preceq^*_\beta U$$

Note that the normalization hypothesis is required for the same reason as for Theorem 1.

Lemma 8. *If specification β is normalizing then*
$$\Gamma \vdash_\epsilon t : T \Rightarrow \Gamma \vdash_\beta t : T \text{ and } \mathcal{WF}_\epsilon(\Gamma) \Rightarrow \mathcal{WF}_\beta(\Gamma)$$

Proof: by mutual induction over $\Gamma \vdash_\epsilon t : T$, all cases are trivial but (CONV) and (CONV$_s$).

- Case of (CONV$_s$): by induction hypothesis $\Gamma \vdash_\beta t : T$. Reducibility property (Lemma 4) yields $T \stackrel{*}{\to} T' \preceq_{=_\epsilon} s$, so by inversion $T \stackrel{*}{\to} T' = s' \prec s$, we conclude $T \preceq^*_\beta s$
- Case of (CONV): by induction hypothesis we get $\Gamma \vdash_\beta t : T$ and $\Gamma \vdash_\beta U : s$. By correctness of type, either T is a sort s_1 and by an argument similar to (CONV$_s$) we prove $s_1 \prec s' \stackrel{*}{\leftarrow} U$ and conclude, or there exists a sort s' such that $\Gamma \vdash_\beta T : s'$. Preservation of subtyping entails $T \preceq^*_\beta U$ and we can conclude.

Theorem 2 (PT ϵ w.r.t. β). *If specification β is normalizing then*

$$\Gamma \vdash_\epsilon t : T \Leftrightarrow \Gamma \vdash_\beta t : T$$

4 Extension to Calculus of Inductive Constructions

The goal of this section is to extend preservation of typing results about CTS to the Calculus of Inductive Constructions (CIC). The extra features are inductive types, which are a generalisation of ML's datatypes. To be precise, CIC is not parameterized by a sort hierarchy, but since the latter has very few impact on the syntactic metatheory, we do not define it, but rather use it abstractly. See [16] for a precise definition.

The proof follows exactly the same steps, so we will only mention places where there are additional cases. Since we still consider conversion as a parameter we will be able to share many properties between the usual CIC and its ϵ counterpart. Let us first define the syntax, reduction rules and typing rules of this common core.

4.1   Syntax of CIC

Inductive types allow to build (well founded) data structures with variants using *constructors*. It is also possible to analyze variants and access constructors arguments by shallow *pattern-matching*. Finally, there is a facility to build recursive functions (*fixpoints*). Some care is needed not to break the logical consistency of the formalism.

Definition 9 (Terms and specifications of CIC). *Let $\mathscr{I}$ be a set of names. We extend expressions with inductive constructions:*

$$Terms : \mathscr{T} ::= \ldots \mid \mathscr{I} \mid C^i_{\mathscr{I}}(\mathscr{T}, \mathscr{T}) \mid \langle \mathscr{T} \rangle \mathsf{case}\ \mathscr{T}\ \mathsf{of}\ \overrightarrow{\mathscr{V} \Rightarrow \mathscr{T}} \mid \mathsf{fix}_n(\mathscr{V} : \mathscr{T} := \mathscr{T})$$

We use I to denote elements of $\mathscr{I}$. Notation $\boldsymbol{X}$ denotes a vector of X (#($\boldsymbol{v}$) is the length of $\boldsymbol{v}$).

   Specifications have a sixth field ELIM $\subseteq \mathscr{I} \times \mathscr{S}$ *that controls the range of pattern-matching for each inductive type.*

Set $\mathscr{I}$ is the set of names of inductive types (e.g. list, prod, etc.). Constructors are not identified by name, but by a couple formed of the inductive type it belongs to, and a number identifying which variant it builds: $C^i_I(\boldsymbol{p}, \boldsymbol{a})$ is the i-th constructor of inductive type I. Since they represent datastructures, we enforce that they are always fully applied to arguments ($\boldsymbol{a}$). Vector $\boldsymbol{p}$ is the value of the parameters, they can be thought of as the explicit instantiation of polymorphic parameters of ML datatypes. They are syntactically separated from "real arguments" for convenience. A built-in **case** construct allows shallow pattern-matching on terms of inductive types, in the construction $\langle P \rangle \mathsf{case}\ M\ \mathsf{of}\ \overrightarrow{\boldsymbol{x} \Rightarrow b}$, M is the term to destruct, $\boldsymbol{x_i}$ are bounded variables for each branch b_i, and denote the arguments of the i-th constructor. P is called a *predicate* and is here only to ensure decidability of type-checking in the case of dependent elimination. This will be explained later on.

   The reduction rule for **case** construct allows to select the branch corresponding to the constructor of an object. If the latter is a constructor then a reduction can occur:

$$\langle P \rangle \mathsf{case}\ C^i_I(\boldsymbol{p}, \boldsymbol{a})\ \mathsf{of}\ \overrightarrow{\boldsymbol{x} \Rightarrow t}\ \rightarrow\ t_i\{\boldsymbol{x_i} \leftarrow \boldsymbol{a}\} \qquad \text{if } \#(\boldsymbol{x_i}) = \#(\boldsymbol{a})$$

where $t\{\boldsymbol{x} \leftarrow \boldsymbol{a}\}$ is the parallel substitution of terms $\boldsymbol{a}$ for variables $\boldsymbol{x}$ in t.

   Note that P, I and $\boldsymbol{p}$ do not participate in the reduction, so they would be erased at compile-time. We will show that I and $\boldsymbol{p}$ can be erased, but not P.

   Finally, the calculus supports recursive functions via guarded fixpoints

$$\mathsf{fix}_n(f : T := M)$$

T represents the type of the fixpoint, M its body; f is the name of the variable used in M to make recursive calls, and n is the position of the recursive argument. The usual reduction rule for fixpoints is $F \rightarrow M\{f \leftarrow F\}$ for $F = \mathsf{fix}_n(f : T := M)$, but such definition instantly breaks strong normalization.

To avoid infinite unrolling of the fixpoint, reduction is allowed only when the n-th argument is in constructor form. This *guard* associated to a typing condition that ensures that M makes a structural recursion over its n-th argument will preserve normalization. The guarded reduction is

$$F\ t\ \rightarrow\ (M\{f \leftarrow F\})\ t \quad \text{if } \#(t) = n \wedge t_n = \mathsf{C}_I^i(p, a) \wedge F = \mathsf{fix}_n(f : T := M)$$

Here we can also remark that T does not participate in the reduction, but as for the case predicate, type of fixpoints will not be erasable.

4.2   Typing

Before defining the typing rules of the inductive constructions, we introduce a new judgment $\Gamma \vdash T @ u \vartriangleright A$ to type-check n-ary applications . It should read: in context Γ, an expression of type T can be applied to arguments u, and this application has return type A.

In traditional presentation of CIC, typing rules are configured by a signature Σ which contains declarations of inductive types, that is a family name I with his type and a type for each constructor of I.

Definition 10 (signature).

$$\Sigma ::= []\ |\ \Sigma;\ \mathit{Ind}(I[\Delta_p] : \Pi \Delta_a . s := \overrightarrow{\Pi \Delta_i . I\ \mathit{Dom}(\Delta_p)\ t_i})$$

Context Δ_p declares the parameters of the inductive definition. They are global to the definition and constructor can refer to them. Context Δ_a is the type of "real" arguments of I. s is the sort where the inductive objects lie. Then for each constructor, Δ_i gives the type of arguments of the i-th constructor. The inductive name I may appear in Δ_i. Finally, t_i defines which instance of the "real" arguments of I the constructor builds.

The same way contexts are subject to a typing judgment $\mathcal{WF}()$, there is a judgment to check that inductive declarations are well-formed. It includes type-checking of the various components of the declaration and a syntactic criterion called *positivity* to ensure strong normalization and consistency, but its exact definition does not matter here. See [13] for details.

Definition 11 (typing of CIC). *Typing rules for CTS are extended with the new rule defined in Fig. 3.*

Rule (IND) is like that of variables. Rule (CONSTR) is a combination of a variable rule (i-th constructor has type $\Pi \Delta_p . T_i$) and n-ary application (we do as if it was applied to pa). The side condition ensures that parameters and arguments are splitted correctly. Rule (FIX) is as usual except there is a side condition (GUARDED) that ensures that the fixpoint proceeds by structural induction over its n-th argument. It is a syntactic criterion we will not detail here.

Rule (CASE) is the most complicated. Because of dependent types, branches may have different types. The type of the i-th branch is equal to P instantiated with the particular instance of the i-th constructor. And the type of the

$$\frac{\mathcal{WF}(\Gamma)}{\Gamma \vdash A @ [] \triangleright A}[\text{VNIL}] \qquad \frac{\Gamma \vdash t : T \quad \Gamma \vdash U\{x \leftarrow t\} @ u \triangleright A}{\Gamma \vdash \Pi x{:}T.U @ tu \triangleright A}[\text{VCONS}]$$

$$\frac{\begin{array}{c}\mathcal{WF}(\Gamma)\\ \text{Ind}(I[\Delta_p] : A := \overrightarrow{T}) \in \Sigma\end{array}}{\Gamma \vdash I : \Pi\Delta_p.A}[\text{IND}] \qquad \frac{\begin{array}{c}\text{Ind}(I[\Delta_p] : A := \overrightarrow{T}) \in \Sigma\\ \#(\Delta_p) = \#(\boldsymbol{p})\\ \Gamma \vdash \Pi\Delta_p.T_i @ \boldsymbol{pa} \triangleright I\,\boldsymbol{p}\,\boldsymbol{u}\end{array}}{\Gamma \vdash \mathsf{C}_I^i(\boldsymbol{p}, \boldsymbol{a}) : I\,\boldsymbol{p}\,\boldsymbol{u}}[\text{CONSTR}]$$

$$\frac{\begin{array}{c}\text{Ind}(I[\Delta_p] : \Pi\Delta_a.\,s := \overrightarrow{\Pi\Delta_i.\,I\,\text{Dom}(\Delta_p)\,\boldsymbol{t}_i}) \in \Sigma \quad \Gamma \vdash M : I\,\boldsymbol{p}\,\boldsymbol{a}\\ \text{E}\textsc{lim}(I, s') \qquad \forall i.\;\boldsymbol{x}_i = \text{Dom}(\Delta_i)\\ \Gamma \vdash P : \Pi\Delta_a\{\text{Dom}(\Delta_p) \leftarrow \boldsymbol{p}\}.\,\Pi x{:}I\,\boldsymbol{p}\,\text{Dom}(\Delta_a).\,s'\\ \forall i.\;\Gamma\Delta_i\{\text{Dom}(\Delta_p) \leftarrow \boldsymbol{p}\} \vdash b_i : P\,\boldsymbol{t}_i\{\text{Dom}(\Delta_p) \leftarrow \boldsymbol{p}\}\,\mathsf{C}_I^i(\boldsymbol{p}, \boldsymbol{x}_i)\end{array}}{\Gamma \vdash \langle P \rangle \text{case}\;M\;\text{of}\;\overrightarrow{\boldsymbol{x} \Rightarrow \boldsymbol{b}} : P\,\boldsymbol{a}\,M}[\text{CASE}]$$

$$\frac{\Gamma;(f{:}T) \vdash M : T \qquad \text{G}\textsc{uarded}(\text{fix}_n(f{:}T := M))}{\Gamma \vdash \text{fix}_n(f{:}T := M) : T}[\text{FIX}]$$

Fig. 3. New Typing rules for CIC

expression is P instantiated with the instance of the matched term (M). Side condition $\text{E}\textsc{lim}(I, s')$ is used to restrict the class of objects that can be built by pattern-matching. It may be necessary to be restricitve to avoid paradoxes. However, its definition is not relevent to our purposes.

The metatheory of CIC has been studied by various authors. It was first introduced by Paulin [13]. Substitution lemma, type correctness, subject reduction and type uniqueness also hold. Inversion lemma has to be extended to the case of the inductive constructions. We do not define it here but it always follow the same scheme. In his Ph.D., Barras [3] formalized the syntactic metatheory (strong normalization excluded) of an alternative presentation of CIC in Coq[5]. In particular, CIC enjoys the type uniqueness property modulo $\approx_\beta$ (the proof is the same as 5).

Hypothesis 1 (strong normalization) *CIC is normalizing.*

The above hypothesis can be seen either as a claim that CIC is normalizing (Werner [17] showed the strong normalization of CIC but with a subset of the sort hierarchy[6]) or as an assumption on the sort hierarchy for the subsequent lemmas to hold, if we see CIC as a general framework parameterized like CTSs.

4.3 ϵ-Equality

For CIC, apart from erasing domain types of functions, ϵ-equality also ignores parameters and inductive names of constructors; the only relevant information

---

[5] Of course, Gödel's second incompleteness theorem shows that if CIC is consistent, it is not possible to show this consistency within CIC.

[6] Yet the trickiest part: it includes non degenerated impredicativity and strong elimination...

for constructors are their constructor number and *real arguments*. As before we can ensure the equality of the erased part of constructors from the equality of their types. For instance, 0 and `false` are convertible and can have the same representation (their constructor number 0). Moreover, lists (cons nat 0 (nil nat)) and (cons bool false (nil bool)) are also convertible because their parameters (here the polymorphic arguments `nat` and `bool`) are ignored. This is what we call *Calculus of Inductive Constructions with Implicit Parameters*.

In this calculus, the conversion algorithm can safely implement constructors by a pair formed with a constructor number and a list of *real arguments*. It worths mentioning that it corresponds pretty well to how datatypes are compiled in languages of the ML family.

Definition 12 (ϵ-equality). *We extend ϵ-equality and ϵ-convertibility (Def. 6) with the following rules:*

$$\frac{}{I =_\epsilon I} \qquad \frac{P =_\epsilon P' \qquad M =_\epsilon M' \qquad \forall i,\ x_i = x'_i \qquad t_i =_\epsilon t'_i}{\langle P \rangle \mathsf{case}\ M\ \mathsf{of}\ \overrightarrow{x \Rightarrow t} =_\epsilon \langle P' \rangle \mathsf{case}\ M'\ \mathsf{of}\ \overrightarrow{x' \Rightarrow t'}}$$

$$\frac{i = i' \qquad a =_\epsilon a'}{C^i_I(p, a) =_\epsilon C^{i'}_{I'}(p', a')} \qquad \frac{T =_\epsilon T' \qquad M =_\epsilon M'}{\mathsf{fix}_n(f : T := M) =_\epsilon \mathsf{fix}_n(f : T' := M')}$$

Remark that we do not erase type information of fixpoints and cases. This is because it breaks Preservation of Equational Theory. For example terms

$$\mathsf{fix}_2(f : (B \to B) \to A \to A := \lambda g : B \to B.\, \lambda x : A.\, x)\ \lambda x : B.\, x$$
$$\mathsf{fix}_2(f : (C \to C) \to A \to A := \lambda g : C \to C.\, \lambda x : A.\, x)\ \lambda x : C.\, x$$

have the same type, are in normal form (they have no second argument), and are not convertible, but would become equal if we ignore information of fixpoints. The key point is that guarded fixpoints can behave as a non-reducible β-redex (when partially applied or when recursive argument is not a constructor). We can find some similar counter-examples where a non-reducible case blocks a β-redex, so we cannot ignore case's predicate.

4.4   Equivalence of CICϵ w.r.t. CIC

In Sect. 3 the proof relies on the ability to first infer the type of a head term in normal form and second to verify the convertibility of abstractions with close types. As a preliminary, we can extend the result of uniqueness of typing. And as for product, we have a kind of inversion for close inductive types :

$$I\ a \approx_\beta I'\ a' \Rightarrow I = I' \wedge a \simeq_\beta a'$$

The premise regarding the type constraint is changed since we must know that the types must be $\approx_\beta$ also in the case of constructors. Firstly regarding ϵ-equal normal forms:

Lemma 9.

$$\left.\begin{array}{cc} \Gamma \vdash_\beta t : T & \Gamma \vdash_\beta t' : T' \\ t =_\epsilon t' & t, t' \in \mathcal{NF} \\ t \notin \{\lambda, C\} \vee T \approx_\beta T' \end{array}\right\} \Rightarrow t = t'$$

where C is the set of constructor terms.

Proof: The interesting cases are those for constructors and pattern-matching. Convertibility of constructors do not imply convertibility of their parameters, but last premise entails that their types are close (as for abstractions), so they belong to the same inductive type with the same parameters.

For pattern-matching, we first prove the equality of arguments by induction hypothesis, which can be neither a constructor (t is in normal form) nor an abstraction (t is well-typed). By uniqueness of types, their types are close, which implies that both t and t' are pattern-matching over the same inductive with same parameters. So, both predicates have close types. By induction hypothesis, they are equal. So branches have close types pairwise. Finally we can prove the equality of branches.

Again, the rest of the proof goes exactly as in Section 3.3, and we can conclude to the equivalence of both systems:

Theorem 3 (PET,PS,PT for CIC). *If specification β is normalizing then*

$$\begin{array}{ll} (PET) & \Gamma \vdash_\beta M : T \wedge \Gamma \vdash_\beta M' : T \Rightarrow M \simeq_\beta M' \Leftrightarrow M \simeq_\epsilon M' \\ (PS) & \Gamma \vdash_\beta T : s \wedge \Gamma \vdash_\beta U : s' \Rightarrow T \preceq_\epsilon^* U \Leftrightarrow T \preceq_\beta^* U \\ (PT) & \Gamma \vdash_\epsilon t : T \Leftrightarrow \Gamma \vdash_\beta t : T \end{array}$$

It is easy to show that CIC_ϵ is equivalent to is "the Calculus of Inductive Constructions with Implicit Parameters" (defined has CIC where type decorations and inductive parameters are remove from the syntax).

5   Conclusion and Future Work

We have introduced an (almost) type-free version of the Calculus of Inductive Constructions. In this new formalism, conversion test is more efficient, and moreover is compatible with compilation of proof terms as in [9]. We have shown that it is equivalent to CIC (provided that the latter is normalizing), by generalizing Barthe's and Sørensen's proof [5]. We can not get rid the normalization hypothesis since, as shown by Barthe and Coquand [4], system U^- (a non normalizing PTS) is not equivalent to its domain-free version.

Our equivalence proof can be turned into an algorithm that reannotates a type-free term in normal form given its (unannotated) type. This is useful for toplevels to display the result of a normalization step. This algorithm has been integrated to the current development version of Coq.

A first direction to investigate is what happens if we remove the predicate of pattern-matching expressions and the type of fixpoints. We have shown that preservation of typing does not hold in the way we stated it. Nonetheless, it would be interesting to see how the formalism is affected regarding for instance expressivity and consistency.

Another direction is to study the case of contravariant subtyping. The problem here is that contravariance breaks our type uniqueness property and so our main lemma 6. So, again, equational theory is not exactly preserved, but we conjecture the equiconsistency of both systems. However, contravariant subtyping may radically change the way the proof works, so let us mention that adding subtyping to depend types has already been studied by Aspinall and Compagnoni [1] and by Castagna and Chen [7], and by Chen [8] for the Calculus of Constructions. We might need some of the proof techniques developed there.

References

1. D. Aspinall and A. Compagnoni. Subtyping dependent types. *Theor. Comput. Sci.*, 266(1-2):273–309, 2001.
2. H. Barendregt. Lambda calculi with types. In *Handbook of Logic in Computer Science*, volume 2. Abramsky & Gabbay & Maibaum (Eds.), Clarendon, 1992.
3. B. Barras. *Auto-validation d'un système de preuves avec familles inductives*. PhD thesis, Université Paris 7, 1999.
4. G. Barthe and T. Coquand. On the equational theory of non-normalizing pure type systems. *Journal of Functional Programming*, 14(2):191–209, Mar. 2004.
5. G. Barthe and M. Sørensen. Domain-free pure type systems. *In Journal of Functional Programming*, 10(5):412–452, September 2000.
6. S. Boutin. Using reflection to build efficient and certified decision procedures. In *TACS*, pages 515–529, 1997.
7. G. Castagna and G. Chen. Dependent types with subtyping and late-bound overloading. *INFCTRL: Information and Computation (formerly Information and Control)*, 168, 2001.
8. G. Chen. Subtyping calculus of construction. In *22nd International Symposium on Mathematical Foundations of Computer Science (MFCS)*, 1997.
9. B. Grégoire and X. Leroy. A compiled implementation of strong reduction. In *International Conference on Functional Programming 2002*, pages 235–246. ACM Press, 2002.
10. X. Leroy and J. V. D. Doligez. *The Objective Caml System*. Institut National de Recherche en Informatique et en Automatique, August 2004. Software and documentation available on the Web, http://caml.inria.fr/.
11. Z. Luo. *An Extended Calculus of Constructions*. PhD thesis, University of Edinburgh, 1990.
12. J. McKinna and R. Pollack. Some lambda calculus and type theory formalized. *Journal of Automated Reasoning*, 23(3–4), Nov. 1999.
13. C. Paulin-Mohring. *Extraction de programmes dans le Calcul des Constructions*. Ph.d. thesis, Paris 7, January 1989.
14. R. Pollack. *The Theory of LEGO: A Proof Checker for the Extended Calculus of Constructions*. PhD thesis, Univ. of Edinburgh, 1994.
15. T. Streicher. *Semantics of Type Theory: Correctness, Completeness, and Independence Results*. Birkhauser, 1991.
16. The Coq development team. The coq proof assistant reference manual v7.2. Technical Report 255, INRIA, France, march 2002. http://coq.inria.fr/doc8/main.html.
17. B. Werner. *Une Théorie de Constructions Inductives*. Ph.d. thesis, Université Paris 7, May 1994.

A   Proof of Lemma 2

First we prove by induction on $\preceq_{R_1}$

$$R_1 \in \mathcal{R}_{R_2} \Rightarrow \preceq_{R_1} \in \mathcal{R}_{\preceq_{R_2}} \qquad (*)$$

Then by induction on the number of steps in $T \preceq^*_{R_1} U$. The base cases are trivial. The inductive case is explained in the following diagram:

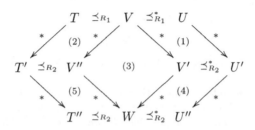

(1) existence of V', U' by induction hypothesis
(2) existence of T', V'' by (*)
(3) existence of W by confluence of β-reduction
(4) existence of U'' by $(R_2, \xrightarrow{*}) \in \mathcal{C}$ and Lemma 1
(5) existence of T'' by $(R_2, \xrightarrow{*}) \in \mathcal{C}$, Lemma 1 and symmetry of R_2

L-Nets, Strategies and Proof-Nets

Pierre-Louis Curien[1] and Claudia Faggian[2,*]

[1] CNRS - Université Paris 7
[2] Universitá di Padova

Abstract. We consider the setting of L-nets, recently introduced by Faggian and Maurel as a game model of concurrent interaction and based on Girard's Ludics. We show how L-nets satisfying an additional condition, which we call logical L-nets, can be sequentialized into traditional tree-like strategies, and vice-versa.

1 Introduction

In the context of Game Semantics several proposals are emerging – with different motivations – towards strategies where sequentiality is relaxed to capture a more parallel form of interaction, or where the order between moves in a play is not totally specified. Such strategies appear as graphs, rather then more traditional tree-like strategies. We are aware of work by Hyland, Schalk, Melliès, McCusker and Wall. Here we will consider the setting of L-nets, recently introduced by Faggian and Maurel [8] as a game model of concurrent interaction, based on Girard's Ludics.

The idea underlying L-nets (as well as other approaches) is to not completely specify the order in which the actions should be performed, while still being able to express constraints. Certain tasks may have to be performed before other tasks. Other actions can be performed in parallel, or scheduled in any order.

More traditional strategies, and in particular Hyland-Ong innocent strategies [13], are trees. In this paper we are interested in relating some representatives of these two kinds of strategies. We show how strategies represented by graphs, with little ordering information, can be sequentialized into tree-like strategies; conversely, sequential (tree) strategies can be relaxed into more asynchronous ones.

Two Flavours of Views. It is known that tree strategies (innocent strategies) can be presented as sets of views with certain properties. A view is a linearly ordered sequence of moves (again with certain properties), and the set of views forms a tree. Any interaction (play) results into a totally ordered set of moves.

A graph strategy (an L-net) is a set of partially ordered views (p.o. views), where a p.o. view is a partially ordered set of moves, which expresses an enabling relation, or a scheduling among moves. The set of such p.o. views forms a directed acyclic graph. Any interaction (play) results into a partially ordered set of moves.

In our setting a tree strategy is, in particular, a graph strategy. Hence we have an homogeneous space, inside which we can move, applying our procedures Seq and $Deseq$, which respectively add or relax dependency (sequentiality).

---

* Research partially supported by Cooperation project CNR-CNRS Italy-France 2004-2005 (Interaction et complexité, project No 16251).

L. Ong (Ed.): CSL 2005, LNCS 3634, pp. 167–183, 2005.

From Graph Strategies to Tree Strategies and Vice-Versa. The graph strategies we will consider are (a class of) L-nets. The tree-like strategies we will consider are Girard's designs [11] (syntactically, designs are particular sorts of Curien's abstract Böhm trees [4, 5]). As a computational object a design is a Hyland-Ong innocent strategy on a universal arena, as discussed in [7]. An L-net is a graph strategy *on the same arena.*

We will show how to associate a design to certain L-net, in such a way that all constraints expressed by the L-net are preserved. This is not possible for an arbitrary L-net; it is easy to build a counter-example taking inspiration from Gustave function (a well-known example of a non-sequential function, see e.g. [1]). For this reason, we first introduce the notion of *logical L-nets*, which are L-nets satisfying a condition called cycles condition[1]. We then make the following constructions: in section 4, we show how to obtain a set of designs $seq(\mathfrak{D})$ from a logical L-net $\mathfrak{D}$, while in section 5, we show how to obtain a logical L-net $deseq(\mathfrak{D})$ from a design $\mathfrak{D}$, in such a way that for all designs $\mathfrak{D}$ we have $\mathfrak{D} \in seq(deseq(\mathfrak{D}))$.

The Proof-Net Experience. Tree strategies can be seen as abstract (and untyped) sequent calculus derivations. By contrast, L-nets are graphs which can be seen as abstract multiplicative-additive proof-nets. Indeed, there are two standard ways to handle proofs in Linear Logic: either as sequent calculus derivations, or as proof-nets. Sequent calculus derivations can be mapped onto proof-nets, by forgetting some of the order between the rules, and conversely proof-nets can be sequentialized into proofs. In this paper we use similar techniques in the framework of game semantics. It is a contribution of the paper to transfer the use of proof-net technologies to the semantic setting of Game strategies. This appears to be a natural consequence of a general direction bringing together syntax and semantics.

2   Tree Strategies (Designs) and Sequent Calculus Derivations

Designs, introduced in [11], have a twofold nature: they are at the same time semantic structures (an innocent strategy, presented as a set of views) and syntactic structures, which can be understood as abstract sequent calculus derivations (in a focusing calculus, which we will introduce next).

While we do not recall the standard definitions of view and innocent strategy, in the following we review in which sense a tree strategy is a sequent calculus derivation, and viceversa.

2.1   Focalization and Synthetic Connectives

Multiplicative and additive connectives of Linear Logic separate into two families: positives $(\otimes, \oplus, 1, 0)$ and negatives $(\mathfrak{N}, \&, \perp, \top)$. A formula is positive (negative) if its outermost connective is positive (negative).

A cluster of connectives with the same polarity can be seen as a single connective (called a *synthetic connective*), and a "cascade" of decompositions with the same

---

[1] This condition is a simplified version Hughes and Van Glabbeek's toggling condition [12].

polarity as a single step (rule). This corresponds to a property known as focalization, discovered by Andreoli (see [2]), and which provides a strategy in proof-search: (i) negative connectives, if any, are decomposed immediately, (ii) we choose a positive focus, and persistently decompose it up to its negative sub-formulas.

Shift. To these standard connectives, it is convenient to add two new (dual) connectives, called Shift[2]: $\downarrow$ (positive) and $\uparrow$ (negative). The role of the Shift operators is to change the polarity of a formula: if N is negative, $\downarrow N$ is positive. When decomposing a positive connective into its negative subformulas (or viceversa), the shift marks the polarity change. The shift is the connective which *captures "time"* (or sequentiality): it marks a step in computation.

Focusing Calculus. Focalization is captured by the following sequent calculus, originally introduced by Girard in [10], and closely related to the focusing calculus by Andreoli (see [2]). We refer to those papers for more details.

Axioms: $\vdash x^{\perp}, x$
We assume, by convention, that all atoms x are positive (hence $x^{\perp}$ is negative).

Any positive (resp. negative) cluster of connectives can be written as a $\oplus$ of $\otimes$ (resp. a & of $\invamp$), modulo distributivity and associativity. The rules for synthetic connectives are as follows. Notice that each rule has labels; rather than more usual labels such as $\otimes L$, $\otimes R$, etc., we label the rules with the active formulas, in the way we describe below.

Positive Connectives: Let $P(N_1, \ldots, N_n) = \bigoplus_{I \in \mathcal{N}} (\bigotimes_{i \in I} (\downarrow N_i))$, where I and $\mathcal{N}$ are index sets. Each $\bigotimes_{i \in I} (\downarrow N_i)$ is called an additive component. In the calculus, there is an introduction rule for each additive component. Let us write N_I for $\bigotimes_{i \in I} (\downarrow N_i)$.

$$\frac{\ldots \quad \vdash N_i, \Delta_i \quad \vdash N_{i'}, \Delta_{i'} \ldots}{\vdash P, \Delta} \ (P, N_I)$$

A positive rule is labelled with a pair: (i) the focus and (ii) the $\otimes$ of subformulas which appear in the premises (that is, the additive component we are using).

Negative Connectives: Let $N(P_1, \ldots, P_n) = \&_{I \in \mathcal{N}} (\invamp_{i \in I} (\uparrow P_i))$. We have a premise for each additive component. Let us write P_I for $\invamp_{i \in I} (\uparrow P_i)$.

$$\frac{\ldots \quad \vdash P_I, \Delta \quad \vdash P_J, \Delta \ldots}{\vdash N, \Delta} \ \ldots, (N, P_I), (N, P_J), \ldots$$

A negative rule is labelled by a set of pairs: a pair of the form (focus, $\invamp$ of subformulas) for each premise.

---

[2] The Shift operators have been introduced by Girard as part of the decomposition of the exponentials.

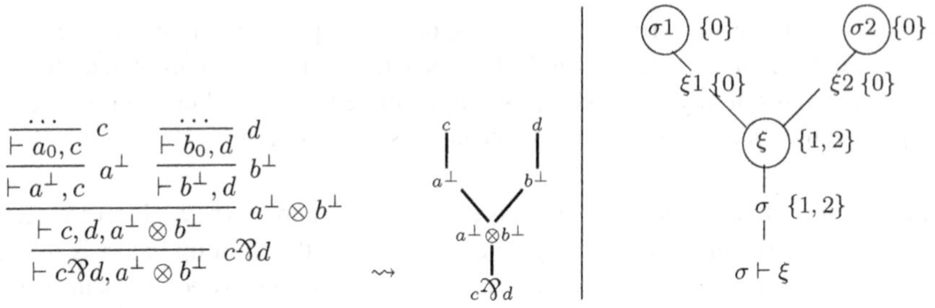

Fig. 1.

We call each of the pairs we used in the labels an *action*. (If a proof does not use &, to each rule corresponds an action. Otherwise, there is an action for each additive component.)

It is important to notice the *duality* between positive and negative rules: to each negative premise corresponds a positive rule. For each action in a negative rule, there is a corresponding positive action, which characterizes a positive rule.

2.2 Designs as (Untyped) Focusing Proofs

Given a focusing proof, we can associate to it a design (forgetting the types). Conversely, given a tree of actions which is a design, we have the "skeleton" of a sequent calculus derivation. This skeleton becomes a concrete (typed) derivation as soon as we are able to decorate it with types. Let us sketch this using an example.

First Example. Consider the (purely multiplicative) derivation on the l.h.s. of Figure 1. Each rule is labelled by the active formula. $a^\perp, b^\perp$ denote negative formulas which respectively decompose into a_0, b_0. Notice that we deal with Shift implicitly, writing $a^\perp \otimes b^\perp$ for $\downarrow a^\perp \otimes \downarrow b^\perp$, and so on.

Now we forget everything in the sequent derivation, but the labels. We obtain the tree of labels (actions) depicted in Figure 1.

This formalism is more concise than the original sequent proof, but still carries all relevant information. To retrieve the sequent calculus counterpart is immediate. Rules and active formulas are explicitly given. Moreover we can *retrieve the context dynamically*. For example, when we apply the Tensor rule, we know that the context of $a^\perp \otimes b^\perp$ is c, d, because they are used afterwards (above). After the decomposition of $a^\perp \otimes b^\perp$, we know that c (resp. d) is in the context of $a^\perp$ because it is used after $a^\perp$ (resp. $b^\perp$).

Addresses (Loci). One of the essential features of Ludics is that proofs do not manipulate formulas, but *addresses*. An address is a sequence of natural numbers, which could be thought of as a name, a channel, or as the address in the memory where an *occurrence of a formula* is stored. If we give address ξ to an occurrence of a formula, its (immediate) subformulas will receive addresses $\xi i, \xi j$, etc. Let $a = ((p_1 \otimes\!\!\!\!\!\!{\scriptstyle 2}\, p_2) \oplus q^\perp) \otimes r^\perp$. If we locate a at the address ξ, we can locate $p_1 \otimes\!\!\!\!\!\!{\scriptstyle 2}\, p_2, q, r$ respectively in $\xi 1, \xi 2, \xi 3$ (the choice of addresses is arbitrary, as long as each occurrence receives a distinct address).

Let us consider an *action* (P, N_I), where $N_I = \bigotimes_{i \in I} (\downarrow N_i)$ is (ξ, K). Its translation is (ξ, K), where ξ is the address of P, and K is the set of natural numbers corresponding to the relative addresses of the subformulas N_i.

First Example, Continuation. Coming back to our example (Figure 1), let us abstract from the type annotation (the formulas), and work with addresses. We locate $a^\perp \otimes b^\perp$ at the address ξ; for its subformulas a and b we choose the subaddresses $\xi 1$ and $\xi 2$. In the same way, we locate $c \,\aftergroup 9 d$ at the address σ and so on for its subformulas.

To indicate the polarity, in the pictures we circle positive actions (to remind that they are clusters of $\otimes$ and $\oplus$). Our example leads to the tree of actions on the r.h.s. of Figure 1, which is an actual design.

2.3   Understanding the Additives (Slices)

The treatment of the additive structure is based on the notion of slice.

A &-rule must be thought of as the superposition of two unary rules, $\&_L, \&_R$. We write the two components of the rule which introduces $a\&b$ as $(a\&b, a)$ and $(a\&b, b)$. Given a sequent calculus derivation in Multiplicative Additive Linear Logic (MALL), if for each &-rule we select one of the premises, we obtain a derivation where all &-rules are unary. This is called a *slice* [9]. For example, the derivation on the l.h.s. below, can be decomposed into the slices on the r.h.s..

$$
\dfrac{\dfrac{\vdash a, c \quad \vdash b, c}{\vdash a\&b, c}}{\vdash (a\&b) \oplus d, c}
\quad \leadsto \quad
\dfrac{\dfrac{\vdash a, c}{\vdash a\&b, c}\,(a\&b, a)}{\vdash (a\&b) \oplus d, c}
\quad \text{and} \quad
\dfrac{\dfrac{\vdash b, c}{\vdash a\&b, c}\,(a\&b, b)}{\vdash (a\&b) \oplus d, c}
$$

An &-rule is a set (the superposition) of unary rules on the *same formula*. For this reason, we will write $a\&b$ also as $\{(a\&b, a), (a\&b, b)\}$.

A More Structured Example. Let $a = (m \otimes n) \oplus c$,
$m = (p_1^\perp \aftergroup 9 p_2^\perp)\&(q_1^\perp \aftergroup 9 q_2^\perp)\&r^\perp$, $n = b_1^\perp \aftergroup 9 b_2^\perp \aftergroup 9 b_3^\perp$, with p_i, q_i, b_i positive formulas. Consider the following derivation, where the set of labels R_1 is
$\{(m, p_1^\perp \aftergroup 9 p_2^\perp), (m, q_1^\perp \aftergroup 9 q_2^\perp), (m, r^\perp)\}$ and R_2 is $\{(n, b_1^\perp \aftergroup 9 b_2^\perp \aftergroup 9 b_3^\perp)\}$.

$$
\dfrac{\dfrac{\dfrac{\dfrac{\cdots}{\vdash p_1, p_2}\,p_1 \quad \dfrac{\cdots}{\vdash q_1, q_2}\,q_2 \quad \dfrac{\cdots}{\vdash r}\,r}{\vdash m}\,R_1 \quad \dfrac{\dfrac{\cdots}{\vdash b_1, b_2, b_3}}{\vdash n}\,R_2}{\vdash (m \otimes n) \oplus c}}{\quad} a, m \otimes n
$$

It is immediate to obtain the corresponding typed design:

Let us now give addresses to the subformulas of A. The counterpart of the previous tree is the following one, which is actually a design.

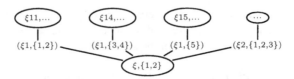

Bipoles (Reading a Design). It is very natural to read a design (or an L-net) as built out of *bipoles*, which are the groups formed by a positive action (say, on address ξ) and all the negative actions which follow it (all being at immediate subaddresses ξi of ξ). *Each address corresponds to a formula occurrence.* The positive action corresponds to a positive connective. The negative actions are partitioned according to the addresses: each address corresponds to a formula occurrence, each action on that address corresponds to an additive component.

Towards Proof-Nets. Let us consider a multiplicative design (a slice). We are given *two partial orders,* which correspond to two kinds of information on each action $\kappa = (\sigma, I)$: (i) a time relation (*sequential order*); (ii) a space relation (*prefix order*), corresponding to the relation of being subaddress (the arena dependency in Game Semantics).

Let us look again at our first example of design. We make explicit the relation of being a subaddress with a dashed arrow, as follows:

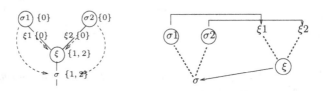

If we emphasize the prefix order rather than the sequential order, we recognize something similar to a proof-net (see [6]), with some additional information on sequentialization. Taking forward this idea of proof-nets leads us to L-nets.

3   Logical L-Nets

In this section, we recall the notion of L-net of Faggian and Maurel [8], but we replace the acyclicity condition by the stronger cycles condition.

Actions (Arena and Moves). An *action* is either the special symbol † (called daimon) or (cf. above) a pair $k = (\xi, I)$ given by an address ξ and a finite set I of indices. When not ambiguous, we write just ξ for the action (ξ, I). In the following, the letters k, a, b, c, d vary on actions.

We say that σ is a *subaddress* of ξ if ξ is a prefix of σ (written $\xi \sqsubseteq \sigma$). We say that an action (ξ, I) *generates* the addresses ξi, for all $i \in I$, and write $a \sqsubseteq_1 b$ if the action a generates the address of the action b (a is the parent of b). We will write $a \sqsubseteq b$ for the transitive closure of this relation. Actions together with the relation $\sqsubseteq_1$ define what could be called a *universal arena*.

A *polarized action* is given by an action k together with a *polarity*, positive (k^+) or negative (k^-). The action † is defined to be positive. When clear from the context, or not relevant, we omit the explicit indication of the polarity.

L-Nets (Graph Strategies). L-nets have an internal structure, described by a directed acyclic graph (d.a.g.) on polarized actions, and an interface, providing the names on which the L-net can communicate with the rest of the world.

An *interface* is a pair of disjoint sets Ξ, Λ of addresses (names), which we write as a sequent $\Xi \vdash \Lambda$. We call Λ the positive (or outer) names, and Ξ the negative (or inner) names. Ξ is either empty or a singleton. We think of the inner names as passive, or receiving, and of the outer names as active or sending.

Directed graphs and notations. We consider any directed acyclic graph G up to its transitive closure, and in fact we only draw the skeleton (the minimal graph whose transitive closure is the same as that of G). We write $a \leftarrow b$ if there is an edge from b to a. In all our pictures, the edges are oriented downwards. We use $\overset{*}{\leftarrow}$ for $\leftarrow \leftarrow \dots \leftarrow$.

A node n of G is called *minimal* (resp. *maximal*) if there is no node a such that $a \leftarrow n$ (resp. $n \leftarrow a$). Given a node n, we denote by $\ulcorner n \urcorner_G$ (the *view* of n) the sub-graph induced by restriction of G on $\{n\} \cup \{n', n' \overset{*}{\leftarrow} n\}$ (we omit to indicate G whenever possible).

It is standard to represent a strict partial order as a d.a.g., where we have an edge from b to a whenever $a < b$. Conversely, (the transitive closure of) a d.a.g. is a strict partial order.

Definition 1 (pre L-nets). *A pre L-net is given by:*

- *An interface $\Xi \vdash \Lambda$.*
- *A set A of nodes which are labelled by polarized actions[3].*
- *A structure on A of* directed acyclic bipartite graph *(if $k \leftarrow k'$, the two actions have opposite polarity) such that:*

    i. Parents. *For any action $a = (\sigma, J)$, either σ belongs to the interface (and then its polarity is as indicated by the base), or it has been generated by a preceding action $c \overset{*}{\leftarrow} a$ of opposite polarity. Moreover, if a is negative, then $c \leftarrow a$.*

    ii. Views. *For each action k, in $\ulcorner k \urcorner$ each address only appears once, i.e. all a's such that $a \overset{*}{\leftarrow} k$ are on distinct addresses.*

    iii. Sibling. *Negative actions with the same predecessor are all distinct.*

    iv. Positivity. *If a is maximal w.r.t. $\overset{*}{\leftarrow}$, then it is positive.*

To complete the definition of logical L-nets, we still need (i) a notion allowing us to deal with multiple copies of the same action induced by the additive structure and (ii) a correctness criterion on graphs. We first give a few definitions.

---

[3] Hence nodes are *occurrences* of actions, but we freely speak of actions for brevity.

Bipoles and Rules. The positive actions induce a partition of the d.a.g. $\mathfrak{D}$ just described. A *bipole* (cf. previous section) is the tree we obtain when restricting $\mathfrak{D}$ either (i) to a positive action and the actions which immediately follow it, or (ii) to the negative actions which are initial (degenerated case).

Let us partition each bipole according to the addresses. A *rule* is a maximal set $\{(\xi, K_j)\}$ of actions which have the same address, and belong to the same bipole. A rule is positive or negative according to the polarity of its actions. When a rule is not a singleton, we call it an *additive rule* (think of each action as an additive component). An *additive pair* is a pair $(\xi, J)^-, (\xi, J')^-$ belonging to an additive rule. Observe that if a rule is not a singleton, it must be negative. If we look at the bipole in the following picture, we have two rules: $R_1 = \{(\sigma 1, J)\}$ and $R_2 = \{(\sigma 2, J'), (\sigma 2, J'')\}$.

Paths. An edge is an *entering edge* of the action a if it has a as target. If R is a negative rule and e an entering edge of an action $a \in R$, we call e a *switching edge* of R. A *path* is a sequence of nodes $k_1, ...k_n$ belonging to distinct rules, and such that for each i either $k_i \rightarrow k_{i+1}$ (the path is going down) or $k_i \leftarrow k_{i+1}$ (the path is going up). A *switching path* on a pre L-net is a path which uses at most one switching edge for each negative rule. A *switching cycle* is a cycle (on a sequence of nodes $k_1, ...k_n$ belonging to distinct rules) which contains at most one switching edge for any negative rule.

Definition 2 (logical L-net). *A logical L-net is a pre L-net such that*

- **Additives.** *Given two positive actions $k_1 = (\xi, K_1), k_2 = (\xi, K_2)$ on the same address, there is an additive pair w_1, w_2 such that $k_1 \overset{*}{\leftarrow} w_1$, and $k_2 \overset{*}{\leftarrow} w_2$.*
- **Cycles.** *Given a non-empty union C of switching cycles, there is an additive rule W not intersecting C, and a pair $w_1, w_2 \in W$ such that for some nodes $c_1, c_2 \in C$, $w_1 \overset{*}{\leftarrow} c_1$, and $w_2 \overset{*}{\leftarrow} c_2$.*

L-Nets as Sets of Views / Chronicles. We call *chronicle* (view) a set $\mathfrak{c}$ of actions equipped with a partial order, such that: $\mathfrak{c}$ has a unique maximal element (the apex), and satisfies the (analog of the) parent condition.

Any node k in a pre L-net $\mathfrak{D}$ defines a chronicle, which is $\overline{\ulcorner k \urcorner}$, where the overlining operation is defined on directed acyclic graphs G whose nodes are injectively labelled, as follows: replace all nodes of G by their labels, yielding a graph G' isomorphic to G, Then $\overline{G}$ is the transitive closure of G', i.e., G' viewed as a strict partial order (cf. above). We can associate to each L-net $\mathfrak{D}$ a set of chronicles $\phi(\mathfrak{D})$, as follows:

$$\phi(\mathfrak{D}) = \{\overline{\ulcorner n \urcorner} \mid n \text{ is a node of } \mathfrak{D}\}$$

The set $\phi(\mathfrak{D})$ is closed downwards, in the following sense: if $\mathfrak{c} \in \phi(\mathfrak{D})$, if k is the maximal action of $\mathfrak{c}$, and if $k' \in \mathfrak{c}$ is such that k covers k', i.e., $k' < k$ and there exists no $k'' \in \mathfrak{c}$ such that $k' < k'' < k$, then $\ulcorner k' \urcorner$ (taken with respect to $\mathfrak{c}$) belongs to $\phi(\mathfrak{D})$.

Conversely, given a set Δ of chronicles which is closed downwards, we define a directed graph $\psi(\Delta)$ as follows: the nodes are the elements of Δ and the edges are all the pairs of the form (c', c) such that, if k, k' are the maximal actions of c, c', respectively, then $k' \in c'$ and k covers k' (in c). It is easy to see that for any downwards closed set of chronicles Δ we have $\phi(\psi(\Delta)) = \Delta$. Conversely, given an L-net $\mathfrak{D}$, we have that $\psi(\phi(\mathfrak{D}))$ is isomorphic as a graph to $\mathfrak{D}$.

The functions ϕ and ψ are inverse bijections (up to graph-isomorphisms of L-nets) between the collection of L-nets and the set of downward closed sets of chronicles Δ such that $\psi(\Delta)$ is an L-net.

In this paper, we will largely rely on the presentation of L-nets as sets of chronicles (views). This in particular allows us to treat easily the superposition of two L-nets as the union of the two sets of chronicles (see section 5.2). We shall write write $c \in \mathcal{S}$ and $\mathcal{S} \subseteq \mathfrak{D}$ for $c, \mathcal{S}, \mathfrak{D}$ respectively a chronicle, a set of chronicles and an L-net.

Slices. A *slice* is an L-net in which there is no additive pair (or, equivalently, no repetition of addresses). A slice $\mathfrak{S}$ of an L-net $\mathfrak{D}$ is a maximal subgraph of $\mathfrak{D}$ which is closed under view ($\ulcorner k \urcorner_{\mathfrak{S}} = \ulcorner k \urcorner_{\mathfrak{D}}$) and it is a slice.

L-Nets and Logical L-Nets. Our definition of logical L-net differs from the defininition of L-nets in [8] in the cycles condition, which replaces the acyclicity condition of L-nets, which asserts that there are no switching cycles in a slice. It is immediate that our *cycles condition* implies the acyclicity condition. Hence, a logical L-net is, in particular, an L-net. Notice that while acyclicity is a property of a slice, the new condition speaks of cycles which traverse slices.

Designs. The designs of [11], can be regarded as a special case of L-nets: they are those L-nets such that each positive node is the source of at most one negative node, and each negative node has a single entering edge. Equivalently, the L-nets corresponding to designs are those which are trees that branch only on positive nodes.

4   Sequentializing a Graph Strategy

A node in an L-net should be thought of as a cluster of operations which can be performed at the same time. An edge states a dependency, an enabling relation, or a precedence among actions. Let us consider a very simple example: a chronicle c, i.e. a partially ordered view (p.o. view). A sequentialization of c is a linear extension of the partial order. That is, we add sequentiality (edges) to obtain a total order. A total order which extends c will define a complete scheduling of the tasks, in such a way that each action is performed only after all of its constraints are satisfied.

Dependency between the actions of a slice, and of sets of slices (L-nets) is more subtle, as there are also global constraints.

The aim of this section is to provide a procedure, which takes an L-net and adds sequentiality in such a way that the constraints specified by the L-net are respected. In particular, all actions in a p.o. view of $\mathfrak{D}$ will be contained in a (totally ordered) view of the tree $Seq(\mathfrak{D})$. The process of sequentialization is non-deterministic, as one can expect, i.e. there are different ways to produce a design from a logical L-net.

As we have both multiplicative and additive structure, when sequentializing we will perform two tasks: 1. add sequentiality (sequential links) until the order in each chronicle is completely determined, 2. separate slices which are shared through additive superposition.

The key point in sequentialization is to select a rule which does not depend on others. This is the role of the Splitting lemma.

Lemma 1 (Splitting lemma). *Given an L-net $\mathfrak{D}$ which satisfies the cycles condition, if $\mathfrak{D}$ has a negative rule, then it has a splitting negative rule. A negative rule $W = \{\ldots, w_i, \ldots\}$ is splitting if either it is conclusion of the L-net (each w_i is a root), or if deleting all the edges $w_i \rightarrow w$ there is no more connection (i.e., no path) between any of the w_i and w.*

The proof is an adaptation to our setting of the proof of the similar lemma in [12]. Moreover, the proof implies that

Proposition 1. *The splitting negative rule W can always be chosen of minimal height: either it is conclusion of the L-net, or it is above a positive action, which is conclusion.*

Remark 1. A consequence of the previous proposition is that, when applying the splitting lemma, we are always able to work "bottom up".

4.1 Sequentialization

An L-net does not need to be connected. This is a natural and desirable feature if we want both parallelism and partial proofs, that is proofs which can be completed into a proper proof. Actually, non-connectedness is an ingredient of Andreoli's concurrent proof construction. On the logical side, non-connectedness corresponds to the mix rule.

There is no special problem for sequentializing non-connected L-nets, except that we need to admit the mix-rule. But as the (controversial) mix rule is refused by designs, we distinguish logical L-nets which are connected.

Given an L-net $\mathfrak{D}$ and a slice $\mathfrak{S} \subseteq \mathfrak{D}$, a *switching graph* of $\mathfrak{S}$ is a subgraph obtained from $\mathfrak{S}$ by choosing a single edge for each negative node, and deleting all the other ones. A slice is *S-connected* if all its possible switching graphs are connected. Finally, we call an L-net S-connected if all its maximal slices are.

Proposition 2. *A logical L-net $\mathfrak{D}$ which is S-connected can be sequentialized into a design, or (equivalently) into its sequent calculus presentation.*

Remark 2. If we admit mix, it is easy to adapt the procedure below to sequentialize any logical L-net.

Proof. The proof is by induction on the number N of negative nodes of the L-net $\mathfrak{D}$.

Case 1: $N = 0$. $\mathfrak{D}$ consists of a single positive action k, which does not need further sequentialization.

Case 2: $N > 0$ and There Are Negative Initial Nodes. By definition of L-net, all negative nodes which are initial belong to the same rule $W = \{\ldots, w_i, \ldots\}$.

Let $\mathfrak{D}_i$ be the union of all slices $\mathfrak{S} \subseteq \mathfrak{D}$ such that $w_i \in \mathfrak{S}$. That is, $\mathfrak{D}_i$ is the maximal L-net obtained as set of all chronicles $\mathfrak{c}$ such that $w_j \notin \mathfrak{c}$, for any $w_j \neq w_i$. It is immediate that, operationally, $\mathfrak{D}_i$ is the graph obtained from $\mathfrak{D}$ following these two steps: (i) delete all nodes c such that $w_j \leftarrow^* c$, for $j \neq i$; (ii) delete any negative node which has become a leaf.

$\mathfrak{D}_i$ is S-connected. Let $\mathfrak{D}'_i$ be the tree obtained from $\mathfrak{D}_i$ by removing w_i and by

sequentializing the resulting L-net. $\mathfrak{C}_i = \begin{matrix} \mathfrak{D}'_i \\ | \\ w_i \end{matrix}$ is a design. The forest given by the union of all $\mathfrak{C}_i$ is a design:

$$\frac{\overset{\cdots}{\vdash \xi_I, \Delta} \quad \overset{\cdots}{\vdash \xi_J, \Delta}}{\xi \vdash \Delta} \ W$$

Case 3: $N > 0$ and There Are No Negative Initial Nodes. We select a splitting negative rule $X = \{x_1 = (\xi i, J_1), \ldots, x_n = (\xi i, J_n)\}$. This rule is part of a bipole, with root $k = (\xi, I)$ and possibly other negative rules Y_j. We delete the edges from $x \in X$ to k, disconnecting $\mathfrak{D}$.

Let us call G_X the part of the graph containing X, and G_k the other part. Let us check that the cycles condition is preserved for both G_X and G_k (preservation of all other properties is immediate). In the case of G_k it is obvious, in the case of G_X it comes from the fact that k determines a "bottle-neck" in the graph, as any path going down from G_X to G_k must traverse k. Let us assume that there are switching cycles in G_X, hence a fortiori in $\mathfrak{D}$. The cycles condition for $\mathfrak{D}$ implies that there is an additive pair w_1, w_2 such that each w_i is below a node c_i in one of the cycles. If w_1, w_2 were in G_k, any path going down from c_i to w_i should traverse k. This would mean that there is a path down from k to w_i for each w_i, and hence that both w_i belong to $\ulcorner k \urcorner$, which is against the definition of L-net.

We conclude by applying induction. G_k will sequentialize into a design containing the node k. G_X will sequentialize into a set of trees of roots respectively $x_1, \ldots, x_n$. We obtain a design by having each of these trees pointing to k.

$$\frac{\overline{\xi i \vdash \Delta_i} \ X \quad \cdots \quad \overline{\xi j, \Delta} \ Y_j}{\vdash \xi, \Delta} (\xi, I) \quad \cdots$$

4.2   Examples of Sequentialization

Let us consider the following L-net $\mathfrak{R}$, where we have two negative rules, both splitting:

$X = \{(\xi 0, I), (\xi 0, J)\}$ and $A = \{(\alpha 0, \{0\})\}$.

If we choose X, we obtain the two trees on the left-hand side of Figure 2, and then the design $\mathfrak{X}$. Instead, choosing A we obtain the design $\mathfrak{A}$ (on the r.h.s.).

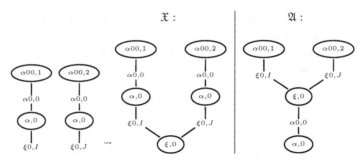

Fig. 2.

5   Desequentializing a Tree Strategy

Beyond the fact that an action can be seen as a cluster of operations that can be performed together thanks to focalization, in a design (actually, in any tree strategy) remains a lot of artificial sequentiality, just as in sequent calculus proofs for Linear Logic. In the case of proofs, the solution has been to develop proof-nets, a theory which has revealed itself extremely fruitful.

We want to apply similar techniques to designs. Our aim in this section is to remove some artificial sequentialization, while preserving essential sequentialization, namely that allowing to recover axioms and to deal with additives.

All dependency (sequentialization) which is taken away by desequentialization can be (non-deterministically) restored through sequentialization (Theorem 1).

5.1   Desequentialization

It is rather immediate to move from designs to an explicit sequent calculus style representation. We already sketched this with an example, and refer to [11] for the details (notice that, because of weakening, there are several sequent calculus representations of a design). To each node k in a design we can associate a sequent of addresses, corresponding to the sequent on which the action is performed. We choose an algorithm which performs weakening as high as possible in the derivation, pushing it to the leaves.

Leaves. For each leaf k in a design, we can recover the sequent of addresses corresponding to the sequent on which that action is performed.

Given a leaf k in the design, its translation k^* is the same node k, to which we explicitly associate a set of addresses, which we call $link(k)$, in the following way: if k is either the action of address ξ on the sequent $\overline{\vdash \xi, \Gamma}$ $k = (\xi, I)$ or the special action † on $\overline{\vdash \Gamma}$ $k = \dagger$, we have $link(k) = \Gamma$.

Positive Conclusion. Let us consider a design whose root is a positive action (ξ, I), and call Π_i the forest of subtrees whose conclusions have address ξi. The design translates into the L-net

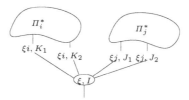

in the following way. Associate the L-net Π_i^* to each Π_i. Take the union of all Π_i^*. Add $(\xi, I)^+$ to the nodes, and extend the set of edges with a relation $(\xi, I) \leftarrow k$ for each action k of address ξi.

Negative Conclusion. Let us consider a design having as conclusion the negative rule $X = \{x_i = (\xi, I)^-, x_2 = (\xi, J)^-, ...\}$. Let us call Π_I the subtree above (ξ, I). A design of negative conclusion translates into an L-net in the following way.

1. For each subtree (premiss) Π_I do the following.
   - Associate the L-net Π_I^* to Π_I.
   - Add $(\xi, I)^-$ to the nodes of Π_I^*.
   - Extend the set of edges with a relation $(\xi, I)^- \leftarrow k$ for each action k such that:
     - k has address ξi ($i \in I$), or
     - k is a leaf such that $\xi i \in link(k)$.

   Let us call $\mathfrak{D}_I$ the resulting graph (which is an L-net).
2. Consider $\mathfrak{D}_I, \mathfrak{D}_J, \dots$. Obtain $\mathfrak{D}_I', \mathfrak{D}_J', \dots$ by extending the set of edges of each $\mathfrak{D}_I$ with a relation $(\xi, I)^- \leftarrow k$ for each positive node k such that $\ulcorner k \urcorner \in \mathfrak{D}_I, \ulcorner k \urcorner \notin \mathfrak{D}_J$, for some $J \neq I$.
3. Superpose $\mathfrak{D}_I', \mathfrak{D}_J', \dots$. Superposition is obtained by taking the union of the chronicles (see [8] and the examples below).

   Superposition is the only step which can introduce cycles. However, if a new cycle C is introduced we find a node $c > x_i$ and a node $c' > x_j$, for $x_i, x_j \in X$.

We have the following result, relating desequentialization and sequentialization.

Theorem 1. *Given (a sequent calculus representation of) a design $\mathfrak{D}$, let us desequentialize it into the L-net $\mathfrak{R}$. There exists a strategy of sequentialization (section 4.1) which allows us to sequentialize $\mathfrak{R}$ into $\mathfrak{D}$.*

The proof comes from the fact that for each step in the desequentialization there is a step of sequentialization which reverses it.

5.2   Examples of Superposition

The superposition of two L-nets is their union as sets of chronicles. Let us see an example. Consider the two L-nets $\mathfrak{D}_1, \mathfrak{D}_2$ in Figure 3. The superposition of $\mathfrak{D}_1$ and $\mathfrak{D}_2$ produces the L-net $\mathfrak{D} = \mathfrak{D}_1 \bigcup \mathfrak{D}_2$.

In fact, the set of chronicles of $\mathfrak{D}_1$ is the set of chronicles $\ulcorner \kappa \urcorner$ defined by each of its actions κ, that is:

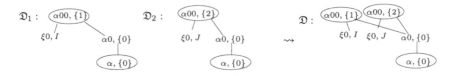

Fig. 3.

$$\{\overset{\alpha 0,0}{\underset{\alpha,0}{\bullet}}\ \overset{}{\underset{\alpha,0}{\bullet}},(\xi 0,I),\ulcorner(\alpha 00,\{1\})\urcorner=\mathfrak{D}_1\}.$$ The set of chronicles of $\mathfrak{D}_2$ is:

$$\{\overset{\alpha 0,0}{\underset{\alpha,0}{\bullet}}\ \overset{}{\underset{\alpha,0}{\bullet}},(\xi 0,J),\ulcorner(\alpha 00,\{2\})\urcorner=\mathfrak{D}_2\}.$$ The resulting union is:

$$\{\overset{\alpha 0,0}{\underset{\alpha,0}{\bullet}}\ \overset{}{\underset{\alpha,0}{\bullet}},(\xi 0,I),(\xi 0,J),\mathfrak{D}_1,\mathfrak{D}_2\},$$ which corresponds to $\mathfrak{D}$.

5.3 Examples of Desequentialization

Example 1. Desequentializing either of the designs $\mathfrak{A}$ or $\mathfrak{X}$ in our previous example of sequentialization yields the original L-net $\mathfrak{R}$ (cf. section 4.2).

Example 2. Let us consider the design in Figure 4, where we just omit an obvious negative action at the place of . . . .

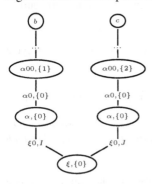

Following the procedure for desequentializing given above, a few easy steps produce the two L-nets $\mathfrak{D}_1, \mathfrak{D}_2$, represented in Figure 5. Observe that we have a chronicle for each node; $\mathfrak{D}_1 \cap \mathfrak{D}_2$ is equal to $\{\ulcorner(\alpha,\{0\})\urcorner,\ulcorner(\alpha 0,\{0\})\urcorner\}$. We obtain $\mathfrak{D}_1'$ by adding the relation $(\xi 0,I) \leftarrow (\alpha 00,\{1\})$, and $\mathfrak{D}_2'$ in a similar way. Remember that we consider each chronicle in the graph modulo its underlying partial order, that is why it is not necessary to explicitly write the edge $(\xi 0, b)$. The union $\mathfrak{D}_1' \bigcup \mathfrak{D}_2'$ produces the L-net on the right-hand side of Figure 5.

Fig. 4.

5.4 A Typed Example: Additives

The following (typical) example with additives illustrates what it means to have more parallelism. Assume we have derivations $\Pi_1, \Pi_2, \Pi_3, \Pi_4$ of (respectively) $\vdash A, C, \vdash A, D, \vdash B, C, \vdash B, D$. In the sequent calculus (and in proof-nets with boxes) there are two distinct ways to derive $\vdash A\&B, C\&D$, and the two derivations differ only by commutations of the rules.

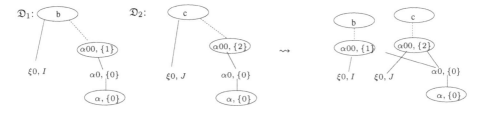

Fig. 5.

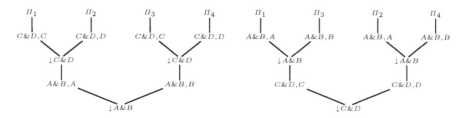

The same phenomenon can be reproduced in the setting of designs, or in the setting of polarized linear logic. Very similar to the above derivations are the two following (typed) designs, where we introduced some $\downarrow$ to have distinct binary connectives. We write formulas instead of addresses, to make the example easier to grasp.

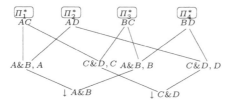

The desequentialization of either of the trees above is the following L-net $\mathfrak{R}$:

Conversely, when sequentializing $\mathfrak{R}$, we get back either one or the other, depending on whether we choose to start from $A\&B$ or from $C\&D$. Notice that both $A\&B$ and $C\&D$ are splitting.

6   Discussion and Further Work

We can isolate two classes of L-nets, those of maximal sequentiality (the tree strategies), which are idempotent with respect to *Seq* and those of minimal sequentiality. Notice that while *Seq* applies to arbitrary L-nets, here we have defined *Deseq* only on trees. This is still enough to characterize also the class of L-nets of minimal sequentiality, as those for which we have $Deseq(Seq(\mathfrak{D})) = \mathfrak{D}$, for any choice in $Seq(\mathfrak{D})$.

We expect to be able to define the desequentialization of arbitrary L-nets, by using the splitting Lemma. Moreover, we believe that sequentialization and desequentialization can be extended to infinite L-nets, by working bottom-up lazily, or stream-like.

In the setting we presented, if we have just enough sequentiality to recover axioms and dependencies from the additives, we obtain (an abstract counter-part of) MALL proof-nets. At the other extreme, all sequentiality can be made explicit, and we have designs "à la locus solum" [11] (or abstract polarized MALL $\downarrow\uparrow$ proof nets as in [14]). L-nets allow us to vary between these extremes, and hence provide us with a framework in which we can graduate sequentiality.

Here we are strongly inspired by a proposal by Girard, to move from proof-nets to their sequentialization (sequent calculus derivation) in a continuum, by using jumps. It must be noticed that edges inducing sequentiality in L-nets actually correspond to Girard's jumps.

We need to understand better this gradient of sequentiality. (i) In this paper we saturate L-nets to maximal sequentiality. We intend to study ways to perform sequentialization gradually, adding sequential edges progressively. (ii) We would like to have a more precise understanding of what it means to have maximal or minimal sequentiality, and to investigate the extent of our desequentialization.

In future work, we wish to investigate a typed setting. The immediate typed counter-part of logical L-nets should be focusing proof-nets [3]. While previous work on focusing proof-nets was limited to multiplicative linear logic, our framework extends to additive connectives.

Acknowledgments

We would like to thank Olivier Laurent for crucial discussions on MALL proof nets, and also Dominic Hughes and Rob van Glabbeek for fruitful exchanges on the technique of domination.

References

1. R. Amadio and P.-L. Curien. *Domains and Lambda-calculi*. Cambridge University Press, 1998.
2. J.-M. Andreoli. Focussing and proof construction. *Annals of Pure and Applied Logic*, 2001.
3. J.-M. Andreoli. Focussing proof-net construction as a middleware paradigm. In *Proceedings of Conference on Automated Deduction (CADE)*, 2002.
4. P.-L. Curien. Abstract bohm trees. *MSCS*, 8(6), 1998.
5. P.-L. Curien. Introduction to linear logic and ludics, part ii. to appear in Advances of Mathematics, China, available at www.pps.jussieu.fr/curien, 2004.
6. C. Faggian. Travelling on designs: ludics dynamics. In *CSL'02*, volume 2471 of *LNCS*. Springer Verlag, 2002.
7. C. Faggian and M. Hyland. Designs, disputes and strategies. In *CSL'02*, volume 2471 of *LNCS*. Springer Verlag, 2002.
8. C. Faggian and F. Maurel. Ludics nets, a game model of concurrent interaction. In *Proc. of LICS (Logic in Computer Science)*. IEEE Computer Society Press, 2005.
9. J.-Y. Girard. Linear logic. *Theoretical Computer Science*, (50):1–102, 1987.

10. J.-Y. Girard. On the meaning of logical rules i: syntax vs. semantics. In Berger and Schwicht-enberg, editors, *Computational logic*, NATO series F 165, pages 215–272. Springer, 1999.
11. J.-Y. Girard. Locus solum. *MSCS*, 11:301–506, 2001.
12. D. Hughes and R. van Glabbeek. Proof nets for unit-free multiplicative-additive linear logic. *ACM Transactions on Computational Logic*, 2005.
13. M. Hyland and L. Ong. On full abstraction for PCF. *Information and Computation*, 2000.
14. O. Laurent. *Etude de la polarisation en logique*. PhD thesis, 2002.

Permutative Logic[*]

Jean-Marc Andreoli[1,3], Gabriele Pulcini[2], and Paul Ruet[3]

[1] Xerox Research Centre Europe, 38240 Meylan, France
Jean-Marc.Andreoli@xrce.xerox.com
[2] Facoltà di Lettere e Filosofia, Università Roma Tre, 00146 Roma, Italy
pulcini@iml.univ-mrs.fr
[3] CNRS - Institut de Mathématiques de Luminy
13288 Marseille Cedex 9, France
ruet@iml.univ-mrs.fr

Abstract. Recent work establishes a direct link between the complexity of a linear logic proof in terms of the exchange rule and the topological complexity of its corresponding proof net, expressed as the minimal rank of the surfaces on which the proof net can be drawn without crossing edges. That surface is essentially computed by sequentialising the proof net into a sequent calculus which is derived from that of linear logic by attaching an appropriate structure to the sequents. We show here that this topological calculus can be given a better-behaved logical status, when viewed in the variety-presentation framework introduced by the first author. This change of viewpoint gives rise to permutative logic, which enjoys cut elimination and focussing properties and comes equipped with new modalities for the management of the exchange rule. Moreover, both cyclic and linear logic are shown to be embedded into permutative logic. It provides the natural logical framework in which to study and constrain the topological complexity of proofs, and hence the use of the exchange rule.

1 Introduction

In order to study proofs as topological objects, notably proofs of linear logic [7], one is naturally led to view proof nets as surfaces on which the usual proofs are drawn without crossing edges [5, 13, 14]. Recent work by Métayer [14] establishes a direct link between the complexity of a linear logic proof in terms of the exchange rule and the topological complexity of its corresponding proof net, expressed as the minimal rank of the compact oriented surfaces with boundary on which the proof net can be drawn without crossing edges and with the conclusions of the proof on the boundary. For instance, cyclic linear logic proofs [19] are drawn on disks since they are purely non-commutative, and the standard proof of $\vdash (A \otimes B) \multimap (B \otimes A)$ is drawn on a torus with a single hole. In general, exchange rules introduce handles or disconnect the boundary.

Gaubert [6] shows that that surface can be computed by sequentialising the proof net into a sequent calculus, proposed by the third author, which is derived from that of linear logic by incorporating an appropriate structure to the

---

[*] Research partly supported by Italy-France CNR-CNRS cooperation project 16251.

L. Ong (Ed.): CSL 2005, LNCS 3634, pp. 184–199, 2005.
© Springer-Verlag Berlin Heidelberg 2005

sequents. Indeed, the above surfaces turn out to be oriented, and it is standard that any oriented compact surface is homeomorphic to a connected sum of tori (see, e.g., [12]). On the other hand, the conclusions of the proofs are drawn on disjoint oriented circles, hence the appropriate structure in [6] is that of a permutation (product of disjoint cycles) together with a natural number (number of tori), actually a complete topological invariant of the surface.

Interestingly, these structures and the operations which are performed on them constitute an instance of the variety-presentation framework introduced in [3]: the varieties we consider in the present paper are the structures used in [6], our presentations are simply varieties with a distinguished point, and both are related by simple axioms, which sort of generalise the properties of partial orders and order varieties in non-commutative logic [2].

We show that the calculus in [6] can be given a better-behaved logical status, when viewed in this framework. This change of viewpoint gives rise to permutative logic, PL for short, where connectives are presentations together with a polarity (positive or negative): the usual pair $\otimes, \otimes$ of linear logic is naturally extended with new modalities $\#, \flat$ for (dis)connecting cycles and new constants $h, \hbar$ for the management of handles. The sequents of PL are varieties and the sequent calculus comes with structural rules, also considered in Melliès' planar logic [13]. The sequent calculus of PL enjoys cut elimination and the focussing [4, 15] property; these properties do not hold in [6] because the two par rules are not reversible. Moreover, both cyclic and linear logic are shown to be embedded into PL.

Unlike [6, 14] which enable to quantify the exchange and topological complexities of a proof, PL provides control mechanisms and is the natural logical framework in which to study and constrain these complexities. We believe in particular that PL should be of interest to concurrent programming and computational linguistics, two fields in which these issues matter [1, 8, 10, 11, 18].

2   Surfaces and Permutations

2.1   Q-Permutations

Surfaces (with or without boundary) are connected 2-dimensional topological manifolds, and it is standard that any compact surface is homeomorphic to a connected sum of tori and projective planes. In the case of orientable compact surfaces, the above homeomorphism is simply with a connected sum of tori. For instance, the sphere corresponds to a sum of 0 torus, etc. For a classical textbook on algebraic topology, we refer the reader to, e.g., [12].

We consider here oriented compact surfaces *with decomposed boundary*, i.e., triples (S, X, ι) where S is a compact surface with boundary and a given orientation, X is a finite set and $\iota : X \to \partial S$ is an injective map from X into to the boundary ∂S of S, such that any hole (i.e., connected component of the boundary) contains at least one distinguished point (i.e., a point in the image of ι). Since holes are circles topologically, the last condition says exactly that X induces a cell decomposition of ∂S (into one or several edges).

These surfaces with decomposed boundary are the objects of a category, which is simply a subcategory of the category of pairs of CW-complexes considered for instance in relative homology: a morphism (resp. isomorphism) from (S, X, ι) to (S', X, ι') is an orientation-preserving continuous map (resp. homeomorphism) $f : S \to S'$ such that $f(\iota(x)) = \iota'(x)$ for each $x \in X$.

Now, an oriented compact surface with decomposed boundary (S, X, ι) induces a cyclic order on each subset of X which is the inverse image by ι of a hole of S. By taking the product of these disjoint cycles, we obtain a permutation $\sigma \in \mathfrak{S}(X)$. On the other hand, S comes with a natural number d called the genre of S, the number of tori (handles) in the connected sum forming S. This leads to the following definition of a q-permutation (where q is meant to remind that a quantity, here a natural number, is attached to the permutation).

Definition 1 (q-permutation). *A* q-permutation *is a triple* (X, σ, d) *where* X *is a finite set,* σ *is a permutation on* X *and* d *is a natural number.*

Hence, to each oriented compact surface with decomposed boundary (S, X, ι) having d handles is associated the q-permutation (X, σ, d) with σ defined as above. For instance, the surface with decomposed boundary illustrated in the above figure induces the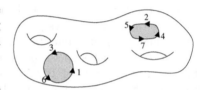
q-permutation $(X, \{(1, 3, 6), (2, 5, 7, 4)\}, 3)$ on $X = \{1, \dots, 7\}$. It is clear that (X, σ, d) is invariant under isomorphism: the number of handles is a topological invariant, and so is the cyclic order on each hole because orientation is preserved. We actually have a complete invariant: (S, X, ι) is isomorphic to (S', X, ι') if, and only if, the associated q-permutations are equal. In the sequel, all the operations we define can be interpreted either in terms of q-permutations or in terms of oriented compact surfaces with decomposed boundary up to isomorphism.

2.2   The Variety-Presentation Framework of q-Permutations

Q-permutations form a variety-presentation framework as defined in [3]. We give here the ingredients of the variety-presentation framework of q-permutations, i.e., the support set operator, the promotion, composition and decomposition operators and the relaxation relation[1].

We assume given an arbitrary countably infinite set $\mathcal{P}$, the elements of which are called places, and a distinguished element $\mathbf{0} \notin \mathcal{P}$. Now, a variety (resp. a presentation) is simply a q-permutation on a finite subset of $\mathcal{P} \cup \{\mathbf{0}\}$ which does not contain (resp. contains) $\mathbf{0}$. This is consistent with the usual view of presentations as varieties with a distinguished place, which is generic to all variety-presentation frameworks.

Definition 2 (support set, promotion, void presentation). *For any q-permutation* $\mu = (X, \sigma, d)$, *its* support set *is defined by* $|\mu| = X \cap \mathcal{P}$. *Any place*

---

[1] In fact, we adopt a slight variant in the presentation w.r.t. [3] as to the status of places and of the support set operator.

$x \in \mathcal{P}$ can be associated with a presentation, called its promotion, which is the q-permutation $(\{\mathbf{0}, x\}, \chi_{\mathbf{0},x}, 0)$, where $\chi_{a,b}$ denotes the transposition exchanging a and b. By abuse of notation, it will be denoted by x so that $|x| = \{x\}$. Finally, the void presentation $\bigcirc$ is the q-permutation $(\{\mathbf{0}\}, \emptyset, 0)$; obviously $|\bigcirc| = \emptyset$.

The topological interpretation of the promotion of x (resp. of the void presentation) is a disk the border of which is labelled by $\mathbf{0}$ and x (resp. $\mathbf{0}$ alone).

Definition 3 (composition). Let $\omega = (X, \sigma, d)$ and $\tau = (Y, \theta, e)$ be presentations such that $|\omega| \cap |\tau| = \emptyset$ (i.e., $X \cap Y = \{\mathbf{0}\}$). Then $\omega * \tau$ is the variety (Z, ξ, f) where

- $Z = (X \cup Y) \setminus \{\mathbf{0}\}$,
- if $\sigma_1, \ldots, \sigma_p, (\mathbf{0}, \gamma)$ are the disjoint cycles of σ and $\theta_1, \ldots, \theta_q, (\mathbf{0}, \delta)$ are the disjoint cycles of θ (here, γ and δ are ordered lists of places), then the disjoint cycles of ξ are
  - either $\sigma_1, \ldots, \sigma_p, \theta_1, \ldots, \theta_q, (\gamma, \delta)$ when γ or δ is non-empty
  - or $\sigma_1, \ldots, \sigma_p, \theta_1, \ldots, \theta_q$ when both γ and δ are empty, i.e. when $\sigma(\mathbf{0}) = \theta(\mathbf{0}) = \mathbf{0}$,
- $f = d + e$.

The permutation ξ above is obtained by gluing at $\mathbf{0}$ the orbits of $\mathbf{0}$ in σ and θ. In terms of surfaces, the composition operator is the amalgamated sum of the two surfaces over a small interval around $\mathbf{0}$ on the boundary. Standard topology of surfaces ensures that the result is indeed an oriented surface. This can be visu-

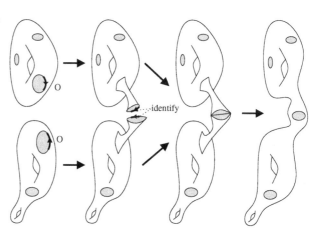

alised in the above figure. The number of holes in the output surface is the sum of the numbers of holes in the input ones, decreased by one (the two holes containing $\mathbf{0}$ have been merged into one) or two (if the two holes containing $\mathbf{0}$ contain no other distinguished point).

Definition 4 (decomposition). Let $\alpha = (X, \sigma, d)$ be a variety and $x \in |\alpha| = X$. The presentation $(\alpha)_x$ is defined as the triple $(X \setminus \{x\} \cup \{\mathbf{0}\}, \sigma', d)$ where σ' is obtained from σ by replacing x by $\mathbf{0}$.

The topological interpretation of this operation is quite straightforward: it does not change the surface, simply relabels x as $\mathbf{0}$.

Definition 5 (relaxation). The relaxation relation is the smallest reflexive transitive relation $\preccurlyeq$ on q-permutations such that:

- divide: $(X, \sigma, d) \preccurlyeq (X, \theta, d)$, where σ is obtained from θ by dividing one cycle (γ, δ) into two: (γ) and (δ),
- merge: $(X, \sigma, d+1) \preccurlyeq (X, \theta, d)$, where σ is obtained from θ by merging two cycles (γ) and (δ) into one (γ, δ),
- degenerate merge: $(X, \sigma, d+1) \preccurlyeq (X, \sigma, d)$.

The degenerate merge rule is in fact obtained by taking (γ) or (δ) empty in the merge rule, but we make it a separate case since the cycles of a permutation are, by definition, non-empty. Since both divide and merge increment the rank $2d + p - 1$ of a q-permutation (X, σ, d) where σ is a permutation with p cycles, we have:

Proposition 1. *Relaxation is a partial order on q-permutations.*

The topological interpretation of relaxation is simply an amalgamated sum: given an oriented compact surface with decomposed boundary (S, X, ι), take two intervals u' and u'' on ∂S and not containing any distinguished point of S. Orient u' in the direction induced by S and u'' in the opposite direction and identify the oriented edges thus obtained. When u' and u'' are on the same connected component (hole) of ∂S, this is a divide; otherwise, this is a merge, and results in a new handle, as is illustrated below:

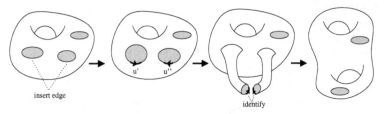

Theorem 1 (variety-presentation framework). *Q-permutations, together with the above operators, satisfy the axioms of variety-presentation frameworks.*

Proof. These axioms, recalled in Appendix A, are almost trivial, and result from direct application of the definitions. The Composition axiom for example essentially expresses that edge identifications in a surface can be performed in any order. □

Following [3], q-permutations, as any variety-presentation framework, define a coloured logic in which connectives are presentations together with a polarity, and sequents are varieties. We explicit that logic, called Permutative Logic (PL), in Section 3.

A Note on the Categorical Interpretation of q-Permutations. It is worth observing that q-permutations on initial segments of $\mathbb{N}^*$ are also the morphisms of a traced symmetric tensor category [9] where objects are natural numbers, tensor is the sum, and the trace is determined by the feedback $\mathrm{tr}_{n,1}(\sigma, d) : n \to n$ on a single wire (for a permutation $\sigma : n+1 \to n+1$), which is defined as follows: $\mathrm{tr}_{n,1}(\sigma, d) = (\sigma\!\restriction_{1,\dots,n}, d+1)$ if $\sigma(n+1) = n+1$; otherwise $\mathrm{tr}_{n,1}(\sigma, d) = (\sigma', d)$, with $\sigma'(i) = \sigma(i)$ when $\sigma(i) \leq n$, and $\sigma'(i) = \sigma(n+1)$ when $\sigma(i) = n+1$. This category is essentially obtained from the category of tangles with same number of inputs and outputs by forgetting over- and under-crossings. Hence, the trace determines a kind of "restriction operator" on q-permutations, but not the one we are interested in here, which is motivated by the topological interpretation of q-permutations and can be computed as in any variety-presentation framework using the composition, decomposition operations and the void presentation by:

$$\alpha\restriction_D = (((\alpha)_{x_1} * \bigcirc)_{x_2} * \bigcirc \ldots)_{x_n} * \bigcirc \text{ where } |\alpha| \setminus D = \{x_1, \ldots, x_n\}$$

The restriction of a variety (X, σ, d) to a set $Y \subseteq X$ is clearly the variety (Y, τ, d) where the cycles of τ are those of σ from which the elements outside Y are removed. The topological interpretation of restriction is simply the composition of $\iota : X \to \partial S$ with the inclusion map $Y \subseteq X$. For instance, the restriction of the above-mentioned q-permutation $(X, \{(1, 3, 6), (2, 5, 7, 4)\}, 3)$ to $Y = \{1, \ldots, 5\}$ is $(Y, \{(1, 3), (2, 5, 4)\}, 3)$. If id_X denotes the identity on X, $X = \{1, 2, 3\}$ and $Y = \{1, 2\}$, then $(\mathrm{id}_X, 0)\restriction_Y = (\mathrm{id}_Y, 0)$ whereas $\mathrm{tr}_{3,1}(\mathrm{id}_X, 0) = (\mathrm{id}_Y, 1)$.

2.3 Computing Relaxation

A permutation $\sigma \in \mathfrak{S}_k = \mathfrak{S}(\{1, \ldots, k\})$ can be written as a product of transpositions, and the following result is standard:

Lemma 1. *If $\sigma \in \mathfrak{S}_k$, then the smallest number n of transpositions $\tau_1, \ldots, \tau_n$ such that $\sigma = \tau_1 \cdots \tau_n$ is given by $n = k - \sigma^\bullet$, where $\sigma^\bullet$ denotes the number of cycles of σ.*

Observe that the effect of both divide and merge on the permutation σ of a given variety (σ, d) is a composition by a transposition. Indeed, divide amounts to taking a cycle (a, Γ, b, Δ) of σ and split it into the two cycles (a, Γ) and (b, Δ), leading to a permutation θ; conversely, merge amounts to taking two cycles (a, Γ) and (b, Δ) and merge them into a single cycle (a, Γ, b, Δ), leading to a permutation θ: in both cases, $\theta = \chi_{a,b} \circ \sigma$, where $\chi_{a,b}$ denotes the transposition exchanging a and b.

Theorem 2. *Given varieties (σ, d) and (θ, e) with $\sigma, \theta \in \mathfrak{S}_k$, $(\theta, e) \preccurlyeq (\sigma, d)$ if, and only if, $m(\sigma, \theta) \leq e - d$, where:*

$$m(\sigma, \theta) = \frac{k - (\theta\sigma^{-1})^\bullet - \theta^\bullet + \sigma^\bullet}{2}.$$

Proof. By Lemma 1, a sequence of divides and merges, say i divides and j merges, from (σ, d) to (θ, e) gives rise to a decomposition of $\theta\sigma^{-1}$ as a product of at least $k - (\theta\sigma^{-1})^\bullet$ transpositions. On the other hand, each occurrence of divide increments the number of cycles and each occurrence of merge decrements it, so $i - j = \theta^\bullet - \sigma^\bullet$. From $i + j \geq k - (\theta\sigma^{-1})^\bullet$, we deduce $j \geq m(\sigma, \theta)$.

Now, $e - d$ is the maximum number of merges in a sequence from (σ, d) to (θ, e). Therefore, if $(\theta, e) \preccurlyeq (\sigma, d)$ and $m(\sigma, \theta) > e - d$, we have $j > e - d$, which is impossible. Conversely, if $m(\sigma, \theta) \leq e - d$, consider a decomposition of $\theta\sigma^{-1}$ as a product of exactly $k - (\theta\sigma^{-1})^\bullet$ transpositions: the number of merges then equals $m(\sigma, \theta)$ and $(\theta, e) \preccurlyeq (\sigma, d)$. $\square$

Table 1. The sequent calculus of permutative logic

Identities

$$\text{axiom} \ \frac{}{\vdash_0 (A, A^\perp)} \qquad \text{cut} \ \frac{\vdash_d \Sigma, (\Gamma, A) \qquad \vdash_e \Theta, (\Delta, A^\perp)}{\vdash_{d+e} \Sigma, \Theta, (\Gamma, \Delta)}$$

Structural rules

$$\text{divide} \ \frac{\vdash_d \Sigma, (\Gamma, \Delta)}{\vdash_d \Sigma, (\Gamma), (\Delta)} \qquad \text{merge} \ \frac{\vdash_d \Sigma, (\Gamma), (\Delta)}{\vdash_{d+1} \Sigma, (\Gamma, \Delta)}$$

Logical rules

$$\text{par} \ \frac{\vdash_d \Sigma, (\Gamma, A, B)}{\vdash_d \Sigma, (\Gamma, A \,\mathbin{\rotatebox[origin=c]{180}{\&}}\, B)} \qquad \text{tensor} \ \frac{\vdash_d \Sigma, (\Gamma, A) \qquad \vdash_e \Theta, (\Delta, B)}{\vdash_{d+e} \Sigma, \Theta, (\Delta, \Gamma, A \otimes B)}$$

$$\text{flat} \ \frac{\vdash_d \Sigma, (\Gamma), (A)}{\vdash_d \Sigma, (\Gamma, \flat A)} \qquad \text{sharp} \ \frac{\vdash_d \Sigma, (\Gamma, A)}{\vdash_d \Sigma, (\Gamma), (\# A)}$$

$$\text{hbar} \ \frac{\vdash_{d+1} \Sigma, (\Gamma)}{\vdash_d \Sigma, (\Gamma, \hbar)} \qquad \text{h} \ \frac{}{\vdash_1 (h)}$$

$$\text{bottom} \ \frac{\vdash_d \Sigma, (\Gamma)}{\vdash_d \Sigma, (\Gamma, \perp)} \qquad \text{one} \ \frac{}{\vdash_0 (1)}$$

3 Formulas, Sequents, and Inference Rules

Definition 6 (formula). Formulas *of PL are obtained from a fixed countable set of* negative atoms $p, q, \ldots$ *and their positive duals* $p^\perp, q^\perp, \ldots$, *by means of the binary connectives* $\mathbin{\rotatebox[origin=c]{180}{\&}}, \otimes$, *the unary connectives* $\flat, \#$, *and the constants* $\hbar, h, \perp, 1$.

The involutive duality is given by De Morgan rules:

$$\begin{array}{llll}
(A \mathbin{\rotatebox[origin=c]{180}{\&}} B)^\perp = B^\perp \otimes A^\perp & (\flat A)^\perp = \# A^\perp & \hbar^\perp = h & \perp^\perp = 1 \\
(A \otimes B)^\perp = B^\perp \mathbin{\rotatebox[origin=c]{180}{\&}} A^\perp & (\# A)^\perp = \flat A^\perp & h^\perp = \hbar & 1^\perp = \perp
\end{array}$$

In general, in a variety-presentation framework, there are two n-ary connectives τ^+ and τ^- of opposite polarities for each presentation τ with a normalised support set $\{1, \ldots, n\}$, hence an infinite set of connectives. However, most of the time, this set is redundant, and it is sufficient to restrict to the connectives derived from a finite set of presentations from which all the others can be reconstructed using the operations of the framework (composition, decomposition). More precisely, given Ω a set of presentations, define the set Ω^* of presentations *generated* by Ω to be the smallest set of presentations X containing places, Ω and satisfying: $(\omega * \tau)_x \in X$ for any $\omega, \tau \in X$ and $x \in |\omega| \cup |\tau|$. Furthermore, Ω is said to be *spanning* when Ω^* is the set of all presentations and to be a *basis* when it is spanning and none of its strict subsets is. It is not difficult to check the following for q-permutations:

Proposition 2. *A basis for q-permutations is the set of 4 presentations below.*

Presentation τ	Neg. conn. τ^-	Pos. conn. τ^+
$\{(\mathbf{0},1,2)\},0$	$\oslash$	$\otimes$
$\{(\mathbf{0}),(1)\},0$	$\flat$	$\#$
$\{(\mathbf{0})\},1$	$\hbar$	h
$\{(\mathbf{0})\},0$	$\bot$	1

Definition 7 (sequent). *A sequent is a variety together with a mapping from its support set into the set of formulas, modulo renaming of the support set consistent with the mapping to formulas.*

It is convenient to represent a sequent as a list of lists of formulas, indexed by a natural number, denoted $\vdash_d (\Gamma_1), \dots, (\Gamma_q)$ where d is the number and $\Gamma_1, \dots, \Gamma_q$ are the lists of formulas. It corresponds to the presentation (σ, d) where σ is the permutation whose cycles are precisely $(\Gamma_1), \dots, (\Gamma_q)$, each inner list being taken modulo cyclic exchange and the outer list being taken modulo unrestricted exchange (it is a multiset). Note that if a list Γ_i is empty, it is simply ignored (it does not correspond to a cycle in σ). In other words, we have the implicit equalities:

$$\vdash_d \Sigma, \Sigma_1, \Sigma_2, \Sigma' = \vdash_d \Sigma, \Sigma_2, \Sigma_1, \Sigma'$$
$$\vdash_d \Sigma, (\Gamma, \Delta) = \vdash_d \Sigma, (\Delta, \Gamma)$$
$$\vdash_d \Sigma, () = \vdash_d \Sigma$$

Modulo these identities, there is a one-to-one correspondence between sequents and their representations as indexed lists of lists. Using this representation, the sequent calculus of PL is given in Table 1. The usual exchange rule of LL is decomposed in PL as follows:

$$\text{divide} \ \frac{\vdash_d \Sigma, (\Gamma, A, B)}{\vdash_d \Sigma, (\Gamma, A), (B)}$$

$$\text{merge} \ \frac{\vdash_d \Sigma, (A, \Gamma), (B)}{\vdash_{d+1} \Sigma, (A, \Gamma, B)}$$
$$\overline{\vdash_{d+1} \Sigma, (\Gamma, B, A)}$$

Dotted lines here indicate application of the identities on sequents. It is interesting to note that exchange is not involutive, even at the level of sequents, since a handle has been added.

The inference figure for an n-ary connective attached to a presentation τ (normalised so that $|\tau| = \{1, \dots, n\}$) is directly obtained from the generic pattern of variety-presentation frameworks:

$$\tau^- \ \frac{\omega * \tau(A_1, \dots, A_n)}{\omega * \tau^-(A_1, \dots, A_n)} \qquad \tau^+ \ \frac{\omega_1 * A_1 \qquad \cdots \qquad \omega_n * A_n}{\tau(\omega_n, \dots, \omega_1) * \tau^+(A_1, \dots, A_n)}$$

Let us detail, for example, how the inference figure for connective $\otimes$ is obtained. In that case, we have $\tau = (\{(\mathbf{0}, 1, 2)\}, 0)$ and the connective is positive. The conclusion of the corresponding inference is therefore $\tau(\omega_B, \omega_A) * A \otimes B$ and its premisses are $\omega_A * A$ and $\omega_B * B$. Now $\tau(\omega_B, \omega_A)$ is defined as presentation τ in which place 1 is substituted by ω_B and place 2 by ω_A. The substitution operation on a presentation in any variety-presentation framework is defined in general by

$$\tau((\omega_z)_{z \in |\tau|}) = (((\tau * x)_{x_1} * \omega_{x_1} \cdots)_{x_n} * \omega_{x_n})_x \qquad (1)$$

where x is an arbitrary place outside $|\tau|$ and $x_1, \ldots, x_n$ is an arbitrary enumeration of $|\tau|$ (the axioms of variety-presentations ensure that the result is independent of any choice for x and the enumeration of $|\tau|$). By distinguishing in ω_A and ω_B the cycle containing $\mathbf{0}$, we have that ω_A is of the form $\vdash_d \Sigma, (\Gamma, \mathbf{0})$ and ω_B of the form $\vdash_e \Theta, (\Delta, \mathbf{0})$. By applying (1), we get that $\tau(\omega_B, \omega_A)$ is the sequent $\vdash_{d+e} \Sigma, \Theta, (\Delta, \Gamma, \mathbf{0})$. Hence the inference figure for $\otimes$. Similarly, the inference figure for $\#$ is the positive inference associated with $\tau = (\{(\mathbf{0}), (\mathbf{1})\}, 0)$. Its conclusion is $\tau(\omega) * \#A$ and its premiss $\omega * A$. Now representing ω as $\vdash_d \Sigma, (\Gamma, \mathbf{0})$, we get that $\tau(\omega)$ is $\vdash_d \Sigma, (\Gamma), (\mathbf{0})$, and, composing ω with A and $\tau(\omega)$ with $\#A$ we obtain the result. The other inference figures are obtained in the same way.

3.1 Basic Properties

Let $A_1, \ldots, A_n \vdash B$ denote the sequent $\vdash_0 (A_1^\perp, \ldots, A_n^\perp, B)$, and $A \dashv\vdash B$ denote the two sequents $A \vdash B$ and $B \vdash A$.

Proposition 3. *The following sequents are provable in permutative logic:*

$A \vdash \flat A$ $\flat A \vdash A \otimes \hbar$

$(A \otimes B) \otimes C \dashv\vdash A \otimes (B \otimes C)$ $\flat\flat A \dashv\vdash \flat A$ $A \otimes \flat B \dashv\vdash \flat B \otimes A$

$A \otimes \perp \dashv\vdash A$ $\flat \perp \dashv\vdash \perp$ $\flat(A \otimes \flat B) \dashv\vdash \flat A \otimes \flat B$

$\perp \otimes A \dashv\vdash A$ $\flat \hbar \dashv\vdash \hbar$ $\flat(A \otimes B) \dashv\vdash \flat(B \otimes A)$.

Proof.

$A \vdash \flat A$		$\flat A \vdash A \otimes \hbar$	$A \otimes \flat B \dashv\vdash \flat B \otimes A$
		$\dfrac{\vdash_0 (A^\perp, A)}{}$	$\dfrac{\vdash_0 (B^\perp, B)}{}$
divide	$\dfrac{\vdash_0 (A^\perp, A)}{\vdash_0 (A^\perp), (A)}$	merge $\dfrac{\vdash_0 (\#A^\perp), (A)}{\vdash_1 (\#A^\perp, A)}$	$\dfrac{\vdash_0 (\#B^\perp), (B) \qquad \vdash_0 (A^\perp, A)}{\vdash_0 (\#B^\perp \otimes A^\perp, A), (B)}$
	$\dfrac{}{\vdash_0 (A^\perp, \flat A)}$	$\dfrac{\vdash_0 (\#A^\perp, A, \hbar)}{\vdash_0 (\#A^\perp, A \otimes \hbar)}$	$\dfrac{\vdash_0 (\#B^\perp \otimes A^\perp, \flat B, A)}{\vdash_0 (\#B^\perp \otimes A^\perp, \flat B \otimes A)}$

$\flat\flat A \vdash \flat A$	$\flat \perp \vdash \perp$	$\flat \hbar \vdash \hbar$	$\flat(A \otimes B) \dashv\vdash \flat(B \otimes A)$
$\dfrac{\vdash_0 (A^\perp, A)}{\vdash_0 (\#A^\perp), (A)}$	$\dfrac{}{\vdash_0 (1)}$	$\dfrac{}{\vdash_1 (h)}$	$\dfrac{\vdash_0 (B, B^\perp) \qquad \vdash_0 (A, A^\perp)}{\vdash_0 (B^\perp \otimes A^\perp, A, B)}$
$\dfrac{}{\vdash_0 (\#\#A^\perp), (), (A)}$	$\dfrac{\vdash_0 (\#1), ()}{\cdots\cdots\cdots\cdots}$	$\dfrac{\vdash_1 (\#h), ()}{\cdots\cdots\cdots\cdots}$	$\dfrac{}{\vdash_0 (\#(B^\perp \otimes A^\perp)), (A, B)}$
$\dfrac{\cdots\cdots\cdots\cdots\cdots}{\vdash_0 (\#\#A^\perp), (A)}$	$\dfrac{}{\vdash_0 (\#1)}$	$\dfrac{}{\vdash_1 (\#h)}$	$\dfrac{}{\vdash_0 (\#(B^\perp \otimes A^\perp)), (B \otimes A)}$
$\dfrac{}{\vdash_0 (\#\#A^\perp, \flat A)}$	$\dfrac{}{\vdash_0 (\#1, \perp)}$	$\dfrac{}{\vdash_0 (\#h, \hbar)}$	$\dfrac{}{\vdash_0 (\#(B^\perp \otimes A^\perp), \flat(B \otimes A))}$

$\flat(A ⅋ \flat B) \vdash \flat A ⅋ \flat B$	$\flat A ⅋ \flat B \vdash \flat(A ⅋ \flat B)$

$$\frac{\vdash_0 (B, B^\perp)}{\vdash_0 (B),(\#B^\perp)} \qquad \vdash_0 (A, A^\perp)$$
$$\frac{}{\vdash_0 (\#B^\perp \otimes A^\perp, A),(B)}$$
$$\frac{}{\vdash_0 (\#(\#B^\perp \otimes A^\perp)),(A),(B)}$$
$$\frac{}{\vdash_0 (\#(\#B^\perp \otimes A^\perp), \flat A),(B)}$$
$$\frac{}{\vdash_0 (\#(\#B^\perp \otimes A^\perp), \flat A, \flat B)}$$
$$\vdash_0 (\#(\#B^\perp \otimes A^\perp), \flat A ⅋ \flat B)$$

$$\frac{\vdash_0 (B, B^\perp)}{\vdash_0 (B),(\#B^\perp)} \qquad \frac{\vdash_0 (A, A^\perp)}{\vdash_0 (\#A^\perp),(A)}$$
$$\frac{}{\vdash_0 (\#B^\perp \otimes \#A^\perp),(A),(B)}$$
$$\frac{}{\vdash_0 (\#B^\perp \otimes \#A^\perp),(A, \flat B)}$$
$$\frac{}{\vdash_0 (\#B^\perp \otimes \#A^\perp),(A ⅋ \flat B)}$$
$$\vdash_0 (\#B^\perp \otimes \#A^\perp, \flat(A ⅋ \flat B))$$

$\qquad\qquad\square$

Corollary 1. *The following sequents are provable in permutative logic:*

$$\perp \vdash \hbar \qquad A ⅋ \hbar \dashv\vdash \hbar ⅋ A \qquad \flat(\hbar ⅋ A) \dashv\vdash \hbar ⅋ \flat A$$

Proof. Easy from the previous proposition. $\qquad\qquad\square$

As a consequence, we have the following corollary.

Corollary 2. *Any negative formula is equivalent to a formula of the form:*

$$N = \flat(P^1_⅋) ⅋ \cdots ⅋ \flat(P^k_⅋) ⅋ Q_1 ⅋ \cdots ⅋ Q_\ell ⅋ \hbar ⅋ \cdots ⅋ \hbar$$

with d occurrences of $\hbar$ and $k, \ell, d \geq 0$. For each $i = 1, \ldots, k$ the formula $P^i_⅋$ is of the form $P^i_1 ⅋ \cdots ⅋ P^i_{n_i}$ for some $n_i \geq 1$, and the P^i_j and Q_i are positive formulas. Explicit parentheses for associativity have been omitted.

As a special case, $\perp$ corresponds to the case where $k, \ell, d = 0$. Note that each of P^i_j and Q_i being positive, they can in turn be decomposed as above (by duality).

The following defined connectives $\dot⅋$ and $\dot\otimes$ are useful for the embedding of LL into PL (Theorem 3): $A \mathbin{\dot⅋} B = A ⅋ \flat B$. Its dual is: $A \mathbin{\dot\otimes} B = \#A \otimes B$. The following properties are straightforward.

$$A \mathbin{\dot⅋} (B \mathbin{\dot⅋} C) \dashv\vdash A \mathbin{\dot⅋} (C \mathbin{\dot⅋} B) \qquad\qquad A ⅋ B \vdash A \mathbin{\dot⅋} B$$
$$A ⅋ (B \mathbin{\dot⅋} C) \dashv\vdash (A ⅋ B) \mathbin{\dot⅋} C \dashv\vdash (A \mathbin{\dot⅋} C) ⅋ B \qquad\qquad A \mathbin{\dot⅋} \perp \dashv\vdash A$$

3.2   Subsystems

The surface associated by Métayer [14] to a proof net in multiplicative linear logic can be explicitly computed by the sequent calculus introduced by Gaubert [6]. In this calculus, there are no structural rules, and the logical rules only deal with the connectives $\otimes$ and $⅋$. The cycles correspond to the conclusions on the same border and the number attached to the sequent is the number of handles of the surface. Our $\otimes$ rule is the same, and the two $⅋$ rules in [6] are recovered as follows:

$$\text{divide} \quad \cfrac{\cfrac{\vdash_d \Sigma, (\Gamma, A, \Delta, B)}{\vdash_d \Sigma, (B, \Gamma, A, \Delta)}}{\cfrac{\vdash_d \Sigma, (B, \Gamma, A), (\Delta)}{\vdots}}$$

$$\text{par} \quad \cfrac{\vdash_d \Sigma, (\Gamma, A, B), (\Delta)}{\vdash_d \Sigma, (\Gamma, A \mathbin{\text{⅋}} B), (\Delta)}$$

$$\text{merge} \quad \cfrac{\vdash_d \Sigma, (\Gamma, A), (B, \Delta)}{\vdash_{d+1} \Sigma, (\Gamma, A, B, \Delta)}$$

$$\text{par} \quad \cfrac{\vdash_{d+1} \Sigma, (\Gamma, A, B, \Delta)}{\vdash_{d+1} \Sigma, (\Gamma, A \mathbin{\text{⅋}} B, \Delta)}$$

Observe that the two ⅋ rules in [6] are not reversible, hence calculus in [6] cannot have the focussing property. Melliès' planar logic [13] exactly corresponds to the $(\otimes, \text{⅋})$ fragment of PL restricted to proofs with 0 handle. The following theorem shows that PL is a conservative extension of cyclic linear logic [19] and linear logic [7].

Theorem 3. *Any formula A of CyLL (resp. LL) is turned into a formula A^{cy} (resp. A^{li}) of PL by $a^{\mathrm{cy}} = a^{\mathrm{li}} = a$ for an atom a and by:*

$$\begin{array}{ll}
(A \mathbin{\text{⅋}} B)^{\mathrm{cy}} = A^{\mathrm{cy}} \mathbin{\text{⅋}} B^{\mathrm{cy}} & (A \mathbin{\text{⅋}} B)^{\mathrm{li}} = A^{\mathrm{li}} \mathbin{\dot{\text{⅋}}} B^{\mathrm{li}} \\
(A \otimes B)^{\mathrm{cy}} = A^{\mathrm{cy}} \otimes B^{\mathrm{cy}} & (A \otimes B)^{\mathrm{li}} = A^{\mathrm{li}} \mathbin{\dot{\otimes}} B^{\mathrm{li}}
\end{array}$$

A formula A of CyLL (resp. LL) is provable in CyLL (resp. LL) if, and only if, A^{cy} (resp. A^{li}) is provable in PL.

Proof. The case of CyLL is an obvious induction: essentially, CyLL is the $(\otimes, \text{⅋})$ fragment of permutative logic with 1 cycle and 0 handle (i.e., rank 0).

For LL, we extend the translation to sequents, and we first show that if $\vdash A_1, \ldots, A_n$ is provable in LL then $\vdash_0 (A_1^{\mathrm{li}}), \ldots, (A_n^{\mathrm{li}})$ is provable in PL, by induction of a proof of $\vdash A_1, \ldots, A_n$ in LL. Since clearly $A^{\perp \mathrm{li}} = A^{\mathrm{li} \perp}$, axiom and the ⅋ and ⊗ rules of LL are translated as follows in PL:

$$\text{divide} \quad \cfrac{\cfrac{}{\vdash_0 (A^{\mathrm{li}}, A^{\perp \mathrm{li}})}}{\vdash_0 (A^{\mathrm{li}}), (A^{\perp \mathrm{li}})} \qquad \cfrac{\cfrac{\vdash_0 \Sigma^{\mathrm{li}}, (A^{\mathrm{li}}), (B^{\mathrm{li}})}{\vdash_0 \Sigma^{\mathrm{li}}, (A^{\mathrm{li}}, \flat B^{\mathrm{li}})}}{\vdash_0 \Sigma^{\mathrm{li}}, (A^{\mathrm{li}} \mathbin{\text{⅋}} \flat B^{\mathrm{li}})} \qquad \cfrac{\vdash_0 \Sigma^{\mathrm{li}}, (A^{\mathrm{li}}) \qquad \vdash_0 (B^{\mathrm{li}}), \Theta^{\mathrm{li}}}{\cfrac{\vdash_0 \Sigma^{\mathrm{li}}, (\# A^{\mathrm{li}})}{\vdash_0 \Sigma^{\mathrm{li}}, (\# A^{\mathrm{li}} \otimes B^{\mathrm{li}}), \Theta^{\mathrm{li}}}}$$

To show the converse, we associate to any formula, sequent, proof of PL its *linear skeleton* in the evident way, by forgetting the information specific to PL, i.e., by forgetting $\flat, \#$ and by mapping $\hbar, h$ to $\perp, 1$, by mapping the variety underlying a sequent to its support set, and by forgetting the $\flat, \#$ rules and the structural rules. It is straightforward to check that a proof in PL is thus mapped to a proof in LL, and this is enough to conclude. □

It is not obvious however that pomset calculus [17], non-commutative logic [2] or ordered calculus [16] are subsystems of PL.

By using the linear skeleton just defined, it is possible to show that PL behaves as a topological decoration of (essentially the multiplicative fragment of) LL. This observation should be a basis for a theory of proof nets for PL.

Theorem 4. *If $\vdash_d \Sigma$ is a sequent of PL and π is a proof of its skeleton in LL, then for some $e \geq 0$, there is a proof of $\vdash_{d+e} \Sigma$ in PL whose skeleton is π. In particular, if the linear skeleton of a sequent $\vdash_d \Sigma$ of PL is provable in LL, then for some $e \geq 0$, $\vdash_{d+e} \Sigma$ is provable in PL.*

Proof. We only need to show the first assertion. It is obtained by induction on π, together with the observation that a connective of PL in Σ which is forgotten by the skeleton operation can always be decomposed in PL, possibly at a certain cost (in terms of structural rules, hence in terms of handles). We omit the details here. □

4   Cut Elimination and Focussing

4.1   Cut Elimination

Theorem 5. *Any proof in PL can be transformed into a proof without cut.*

Proof. The proof follows the usual pattern [3], where cuts are eliminated by repetitive application of reduction rules to the proofs. There are three kinds of reductions: axiom case (when one of the premisses of the cut is an axiom), commutative conversion (when the principal formula in one of the premisses of the cut is not the cut-formula) and symmetric reductions (when the principal formula in both premisses of the cut is the cut formula). Some cases are detailed below, the other configurations being treated similarly. □

Symmetric reductions:

$$\text{cut}\ \dfrac{\dfrac{\vdash_d \Sigma, (\Gamma), (A)}{\vdash_d \Sigma, (\Gamma, \flat A)} \quad \dfrac{\vdash_e \Theta, (\Delta, A^\perp)}{\vdash_e \Theta, (\Delta), (\# A^\perp)}}{\vdash_{d+e} \Sigma, \Theta, (\Gamma), (\Delta)} \quad \leadsto \quad \text{cut}\ \dfrac{\vdash_d \Sigma, (\Gamma), (A) \quad \vdash_e \Theta, (\Delta, A^\perp)}{\vdash_{d+e} \Sigma, \Theta, (\Gamma), (\Delta)}$$

$$\text{cut}\ \dfrac{\dfrac{\vdash_d \Sigma, (\Gamma, A, B)}{\vdash_d \Sigma, (\Gamma, A \otimes B)} \quad \dfrac{\vdash_f \Xi, (\Lambda, B^\perp) \quad \vdash_e \Theta, (\Delta, A^\perp)}{\vdash_{e+f} \Theta, \Xi, (\Lambda, B^\perp \otimes A^\perp, \Delta)}}{\vdash_{d+e+f} \Sigma, \Theta, \Xi, (\Gamma, \Delta, \Lambda)} \quad \leadsto$$

$$\text{cut}\ \dfrac{\text{cut}\ \dfrac{\vdash_d \Sigma, (\Gamma, A, B) \quad \vdash_f \Xi, (\Lambda, B^\perp)}{\vdash_{d+f} \Sigma, \Xi, (\Gamma, A, \Lambda)} \quad \vdash_e \Theta, (\Delta, A^\perp)}{\vdash_{d+e+f} \Sigma, \Theta, \Xi, (\Gamma, \Delta, \Lambda)}$$

Commutative conversions:

$$\text{cut}\ \dfrac{\dfrac{\vdash_d \Sigma, (\Gamma, C), (A)}{\vdash_d \Sigma, (\Gamma, C, \flat A)} \quad \vdash_e \Theta, (\Delta, C^\perp)}{\vdash_{d+e} \Sigma, \Theta, (\Gamma, \Delta, \flat A)} \quad \leadsto \quad \text{cut}\ \dfrac{\dfrac{\vdash_d \Sigma, (\Gamma, C), (A) \quad \vdash_e \Theta, (\Delta, C^\perp)}{\vdash_{d+e} \Sigma, \Theta, (\Gamma, \Delta), (A)}}{\vdash_{d+e} \Sigma, \Theta, (\Gamma, \Delta, \flat A)}$$

$$\text{cut}\ \dfrac{\dfrac{\vdash_d \Sigma, (\Gamma), (\Delta, C)}{\vdash_{d+1} \Sigma, (\Gamma, \Delta, C)} \quad \vdash_e \Theta, (\Lambda, C^\perp)}{\vdash_{d+e+1} \Sigma, \Theta, (\Gamma, \Delta, \Lambda)} \quad \leadsto \quad \text{merge}\ \dfrac{\text{cut}\ \dfrac{\vdash_d \Sigma, (\Gamma), (\Delta, C) \quad \vdash_e \Theta, (\Lambda, C^\perp)}{\vdash_{d+e} \Sigma, \Theta, (\Gamma), (\Delta, \Lambda)}}{\vdash_{d+e+1} \Sigma, \Theta, (\Gamma, \Delta, \Lambda)}$$

4.2   Focussing

As with any coloured logic derived from a variety-presentation framework, the sequent calculus has a remarkable property called focussing, which eliminates irrelevant non-determinism in proof construction. It reflects general permutability properties of inferences: any positive (resp. negative) inference can be permuted

Table 2. The focussed sequent calculus of permutative logic

Identity

$$\text{Ax} \ \frac{}{\vdash_0 p \blacktriangleright p^{\perp}} \quad \text{if } p \text{ is a negative atom}$$

Structural rules

$$\text{unfocus} \ \frac{\vdash_d \Sigma, (\Gamma, A)}{\vdash_d \Sigma \blacktriangleright \Gamma, A} \qquad \text{focus1} \ \frac{\vdash_d \Sigma \blacktriangleright}{\vdash_d \Sigma} \qquad \text{focus2} \ \frac{\vdash_d \Sigma \blacktriangleright \Gamma, A}{\vdash_d \Sigma, (\Gamma, A) \blacktriangleright}$$
$$\text{if } A \text{ is negative} \qquad \text{if } \Sigma \text{ is reduced} \qquad \text{if } A \text{ is positive}$$

$$\text{divide} \ \frac{\vdash_d \Sigma, (\Gamma, \Delta) \blacktriangleright}{\vdash_d \Sigma, (\Gamma), (\Delta) \blacktriangleright} \qquad \text{merge} \ \frac{\vdash_d \Sigma, (\Gamma), (\Delta) \blacktriangleright}{\vdash_{d+1} \Sigma, (\Gamma, \Delta) \blacktriangleright}$$

Logical rules

$$\text{par} \ \frac{\vdash_d \Sigma, (\Gamma, A, B)}{\vdash_d \Sigma, (\Gamma, A \,\invamp\, B)} \qquad \text{tensor} \ \frac{\vdash_d \Sigma \blacktriangleright \Gamma, A \qquad \vdash_e \Theta \blacktriangleright \Delta, B}{\vdash_{d+e} \Sigma, \Theta \blacktriangleright \Delta, \Gamma, A \otimes B}$$

$$\text{flat} \ \frac{\vdash_d \Sigma, (\Gamma), (A)}{\vdash_d \Sigma, (\Gamma, \flat A)} \qquad \text{sharp} \ \frac{\vdash_d \Sigma \blacktriangleright \Gamma, A}{\vdash_d \Sigma, (\Gamma) \blacktriangleright \# A}$$

$$\text{hbar} \ \frac{\vdash_{d+1} \Sigma, (\Gamma)}{\vdash_d \Sigma, (\Gamma, \hbar)} \qquad \text{h} \ \frac{}{\vdash_1 \blacktriangleright h}$$

$$\text{bottom} \ \frac{\vdash_d \Sigma, (\Gamma)}{\vdash_d \Sigma, (\Gamma, \perp)} \qquad \text{one} \ \frac{}{\vdash_0 \blacktriangleright 1}$$

upward (resp. downward) if the active formulas of the lower inference are not principal in the upper inference. Thus, inferences of the same polarity can be grouped together. This can be captured in a variant of the sequent calculus called the focussing sequent calculus.

Definition 8. *The sequents of the focussing sequent calculus are of two types:*

- *standard sequents of the form $\vdash_d \Sigma$;*
- *focussed sequents of the form $\vdash_d \Sigma \blacktriangleright \Gamma$ where the list of formulas Γ has been singled out. Note that, here, Γ is not taken modulo cyclic exchange.*

A structure on formulas (eg. sequent) is said to be reduced *if it does not contain any negative compound formula.*

The focussing sequent calculus is given in Table 2. Its negative logical inferences are identical to those of the standard sequent calculus. Its positive logical inferences are also those of the standard calculus, except that the principal formula is syntactically distinguished as *focus*, and, when read bottom-up, the focus is passed to its active formulas as long as they remain positive. In the generic variety-presentation framework, the positive rule for an *n*-ary connective is:

$$\frac{\omega_1 \blacktriangleright A_1 \qquad \cdots \qquad \omega_n \blacktriangleright A_n}{\tau(\omega_n, \ldots, \omega_1) \blacktriangleright \tau^+(A_1, \ldots, A_n)}$$

The unfocus rule models the loss of focus due to a change of polarity of the focus (from positive to negative). The rules divide, merge, focus1 and focus2 are in fact a single rule in the generic variety-presentation framework:

$$\text{focus} \ \frac{\omega \blacktriangleright A}{\alpha} \qquad\qquad\qquad\qquad \text{focus} \ \frac{\vdash_e \Theta \blacktriangleright \Gamma, A}{\vdash_d \Sigma}$$

if α is reduced, A is positive $\qquad \overset{\text{in PL}}{\longrightarrow} \qquad$ if Σ is reduced, A is positive
and $\alpha \preccurlyeq \omega * A$ $\qquad\qquad\qquad\qquad\qquad$ and $\vdash_d \Sigma \ \preccurlyeq \vdash_e \Theta, (\Gamma, A)$

It is easy to show that the calculus with the focus rule is equivalent to that with the divide, merge, focus1 and focus2 rules. The latter has been adopted only to make explicit the use of the divide and merge rules, which are implicit in the side relaxation condition of the focus rule.

Theorem 6. *A standard sequent is provable in PL if and only if it is provable in the focussing sequent calculus of PL.*

Proof. It is straightforward to map any inference of the focussing calculus into an inference of the standard calculus (or a dummy inference): just drop the $\blacktriangleright$ sign when it appears. Hence, this ensures the soundness of the focussing calculus. Its completeness is much more involved. It relies exclusively on the axioms of variety-presentation frameworks. It is shown in three steps. First, the negative rules are shown to be invertible in the focussing calculus. Second, the "focus" rule is shown to hold even when Σ is not reduced (using the previous result). And third, the positive rules of the standard calculus are shown to hold in the focussing calculus (using the previous results). In fact, all these properties result from generic permutability properties between inferences, depending on their polarities. The interested reader is referred to [3] for details. $\qquad \square$

5   Future Work

Permutative logic opens new perspectives in the design of non-commutative logical systems. It not only quantifies the use of the structural rule of exchange but also allows to put constraints on that use. Many aspects of the logic have not been studied in this paper. For example, proof-nets in PL deserve a study of their own. It can be expected that a correctness criterion for PL should be found which extends that for CyLL: a cut free proof structure is CyLL-correct if it is LL-correct and planar. In PL, the planarity condition should be replaced by some condition involving more complex surfaces. Another interesting aspect is proof construction. It is very easy to show that, unlike the entropy of non-commutative logic [2], relaxation in PL cannot be optimised so that, during proof construction, the positive inferences perform only the "minimal" amount of relaxation that is strictly needed. However, it is conjectured that if proof-construction is viewed as a constraint propagation problem, this optimality can be recovered. Finally, we have not explored semantics issues. A trivial but uninformative phase semantics can be derived, as in any coloured logic (i.e., based on a variety-presentation framework). More work is needed to achieve interesting semantics interpretation of formulas and proofs.

Acknowledgements

We are grateful to Anne Pichon for helpful remarks on topological questions.

References

1. V. M. Abrusci. Non-commutative logic and categorial grammar: ideas and questions. In V. M. Abrusci and C. Casadio, editors, *Dynamic Perspectives in Logic and Linguistics*. Cooperativa Libraria Universitaria Editrice Bologna, 1999.
2. V. M. Abrusci and P. Ruet. Non-commutative logic I: the multiplicative fragment. *Annals of Pure and Applied Logic*, 101(1):29–64, 2000.
3. J.-M. Andreoli. An axiomatic approach to structural rules for locative linear logic. In *Linear logic in computer science*, volume 316 of *London Mathematical Society Lecture Notes Series*. Cambridge University Press, 2004.
4. J.-M. Andreoli and R. Pareschi. Linear objects: logical processes with built-in inheritance. *New Generation Computing*, 9, 1991.
5. G. Bellin and A. Fleury. Planar and braided proof-nets for multiplicative linear logic with mix. *Archive for Mathematical Logic*, 37(5-6):309–325, 1998.
6. C. Gaubert. Two-dimensional proof-structures and the exchange rule. *Mathematical Structures in Computer Science*, 14(1):73–96, 2004.
7. J.-Y. Girard. Linear logic. *Theoretical Computer Science*, 50(1):1–102, 1987.
8. L. Habert, J.-M. Notin, and D. Galmiche. Link: a proof environment based on proof nets. In Springer, editor, *Analytic Tableaux and Related Methods*, volume 2381 of *Lecture Notes in Computer Science*, pages 330–334. Springer, 2002.
9. A. Joyal, R. Street, and D. Verity. Traced monoidal categories. *Mathematical Proceedings of the Cambridge Philosophical Society*, 119:447–468, 1996.
10. J. Lambek. The mathematics of sentence structure. *American Mathematical Monthly*, 65(3):154–170, 1958.
11. A. Lecomte and Ch. Retoré. Pomset logic as an alternative categorial grammar. In Morrill Oehrle, editor, *Formal Grammar, Barcelona*, 1995.
12. W. S. Massey. *A basic course in algebraic topology*. Springer, 1991.
13. P.-A. Melliès. A topological correctness criterion for multiplicative non-commutative logic. In *Linear logic in computer science*, volume 316 of *London Mathematical Society Lecture Notes Series*. Cambridge University Press, 2004.
14. F. Métayer. Implicit exchange in multiplicative proofnets. *Mathematical Structures in Computer Science*, 11(2):261–272, 2001.
15. D. Miller, G. Nadathur, F. Pfenning, and A. Scedrov. Uniform proofs as a foundation for logic programming. *Annals Pure Appl. Logic*, 51:125–157, 1991.
16. J. Polakow and F. Pfenning. Natural deduction for intuitionistic non-commutative linear logic. In *Typed Lambda-Calculi and Applications*, volume 1581 of *Lecture Notes in Computer Science*. Springer, 1999.
17. C. Retoré. Pomset logic - A non-commutative extension of commutative linear logic. In *International Conference on Typed Lambda-Calculi and Applications*, volume 1210 of *Lecture Notes in Computer Science*. Springer, 1997.

18. P. Ruet and F. Fages. Concurrent constraint programming and non-commutative logic. In *Computer Science Logic 1997*, volume 1414 of *Lecture Notes in Computer Science*, pages 406–423. Springer, 1998.
19. D.N. Yetter. Quantales and (non-commutative) linear logic. *Journal of Symbolic Logic*, 55(1), 1990.

A   Axioms of Variety-Presentation Frameworks

The axioms of variety-presentation frameworks are the following (see [3]):

- **Composition**:
  For any presentations ω_1, ω_2,

$$|\omega_1| \cap |\omega_2| = \emptyset \;\Rightarrow\; \begin{cases} |\omega_1 * \omega_2| = |\omega_1| \cup |\omega_2| \\ \omega_1 * \omega_2 = \omega_2 * \omega_1 \end{cases}$$

- **Decomposition**:
  For any variety α, place x and presentation ω:

$$x \in |\alpha| \;\Rightarrow\; x \notin |(\alpha)_x| \;\wedge\; x * (\alpha)_x = \alpha$$
$$x \notin |\omega| \;\wedge\; \omega * x = \alpha \;\Rightarrow\; \omega = (\alpha)_x$$

  This implies, by composition, that if $x \in |\alpha|$ then $|(\alpha)_x| = |\alpha| \setminus \{x\}$. Hence, for a given x, the mappings $\alpha \mapsto (\alpha)_x$ (for any variety α having occurrence x) and $\omega \mapsto \omega * x$ (for any presentation ω not having occurrence x) are inverse of each other.

- **Commutation**:
  For any variety α, presentations ω_1, ω_2 and places x_1, x_2,

$$\left. \begin{array}{l} |\alpha| \cap (\{x_1\} \cup |\omega_1|) = \{x_1\} \\ |\alpha| \cap (\{x_2\} \cup |\omega_2|) = \{x_2\} \\ (\{x_1\} \cup |\omega_1|) \cap (\{x_2\} \cup |\omega_2|) = \emptyset \end{array} \right\} \;\Rightarrow\; ((\alpha)_{x_1} * \omega_1)_{x_2} * \omega_2 = ((\alpha)_{x_2} * \omega_2)_{x_1} * \omega_1$$

  From the previous axioms, it is easy to show that, under the stated condition, the two sides of the equality have the same occurrence set. This axiom asserts that they are equal.

- **Relaxation**:
  For any varieties α_1, α_2, presentation ω and place x,

$$\alpha_1 \preccurlyeq \alpha_2 \;\Rightarrow\; \begin{cases} |\alpha_1| = |\alpha_2| = D \\ (|\omega| \cup \{x\}) \cap D = \{x\} \;\Rightarrow\; (\alpha_1)_x * \omega \preccurlyeq (\alpha_2)_x * \omega \end{cases}$$

  Hence, relaxation applies only to varieties with the same occurrence set and is compatible with decomposition/composition.

Focusing the Inverse Method for Linear Logic

Kaustuv Chaudhuri and Frank Pfenning*

Department of Computer Science
Carnegie Mellon University
{kaustuv,fp}@cs.cmu.edu

Abstract. Focusing is traditionally seen as a means of reducing inessential non-determinism in backward-reasoning strategies such as uniform proof-search or tableaux systems. In this paper we construct a form of focused derivations for propositional linear logic that is appropriate for forward reasoning in the inverse method. We show that the focused inverse method conservatively generalizes the classical hyperresolution strategy for Horn-theories, and demonstrate through a practical implementation that the focused inverse method is considerably faster than the non-focused version.

1 Introduction

Strategies for automated deduction can be broadly classified as backward reasoning or forward reasoning. Among the backward reasoning strategies we find tableaux and matrix methods; forward reasoning strategies include resolution and the inverse method. The approaches seem fundamentally difficult to reconcile because the state of a backward reasoner is global, while a forward reasoner maintains locally self-contained state.

Both backward and forward approaches are amenable to reasoning in non-classical logics. This is because they can be derived from an inference system that defines a logic. The derivation process is systematic to some extent, but in order to obtain an effective calculus and an efficient implementation, we need to analyze and exploit deep proof-theoretic or semantic properties of each logic under consideration.

Some themes stretch across both backwards and forwards systems and even different logics. Cut-elimination and its associated subformula property, for example, are absolutely fundamental for both types of systems, regardless of the underlying logic. In this paper we advance the thesis that *focusing* is similarly universal. Focusing was originally designed by Andreoli [1, 2] to remove inessential non-determinism from backward proof search in classical linear logic. It has already been demonstrated [3] that focusing applies to other logics; here we show that focusing is an important concept for theorem proving in the forward direction.

As the subject of our study we pick propositional intuitionistic linear logic [4–6]. This choice is motivated by two considerations. First, it includes the propositional core of the Concurrent Logical Framework (CLF), so our theorem prover, and its first-order extension, can reason with specifications written in CLF; many such specifications, including Petri nets, the π-calculus and Concurrent ML, are described in [7]. For many

---

* This work has been supported by the Office of Naval Research (ONR) under grant MURI N00014-04-1-0724 and by the National Science Foundation (NSF) under grant CCR-0306313.

L. Ong (Ed.): CSL 2005, LNCS 3634, pp. 200–215, 2005.

of these applications, the intuitionistic nature of the framework is essential. Second, it is almost a worst-case scenario, combining the difficulties of modal logic, intuitionistic logic, and linear logic, where even the propositional fragment is undecidable. A treatment, for example, of classical linear logic without the lax modality can be given very much along the same lines, but would be simpler in several respects.

Our contributions are as follows. First, we show how to construct a non-focusing inverse method for intuitionistic linear logic. This follows a fairly standard recipe [8], although the resource management problem germane to linear logic has to be considered carefully. Second, we define focused derivations for intuitionistic linear logic. The focusing properties of the connectives turn out to be consistent with their classical interpretation, but completeness does not come for free because of the additional restrictions placed by intuitionistic (and modal) reasoning. The completeness proof is also somewhat different from ones we have found in the literature. Third, we show how to adapt focusing so it can be used in the inverse method. The idea is quite general and, we believe, can be adapted to other non-classical logics. Fourth, we demonstrate via experimental results that the focused inverse method is substantially faster than the non-focused one. Fifth, we show that refining the inverse method with focusing agrees exactly with classical hyperresolution on Horn formulas, a property which fails for non-focusing versions of the inverse method. This is practically significant, because even in the linear setting many problems or sub-problems may be non-linear and Horn, and need to be treated with reasonable efficiency.

In a related paper [9] we generalize our central results to first-order intuitionistic linear logic, provide more detail on the implementation choices, and give a more thorough experimental evaluation. Lifting the inverse method here to include quantification is far from straightforward, principally because of the rich interactions between linearity, weakening, and contraction in the presence of free variables. However, these considerations are orthogonal to the basic design of forward focusing which remains unchanged from the present paper.

Perhaps most closely related to our work is Tammet's inverse method prover for classical linear logic [10] which is a refinement of Mints' resolution system [11]. Some of Tammet's optimizations are similar in nature to focusing, but are motivated primarily by operational rather than by logical considerations. As a result, they are not nearly as far-reaching, as evidenced by the substantial speedups we obtain with respect to Tammet's implementation. Our examples were chosen so that the difference between intuitionistic and classical linear reasoning was inessential.

2 Backward Linear Sequent Calculus

We use a backward cut-free sequent calculus for propositions constructed out of the propositional linear connectives $\{\otimes, \mathbf{1}, \multimap, \&, \top, !\}$; the extension to first-order connectives using the recipe outlined in [9] is straightforward. To simplify the presentation we leave out $\oplus$ and $\mathbf{0}$, though the implementation supports them and some of the experiments in Sec. 5.2 use them. Propositions are written using uppercase letters A, B, C, with p standing for atomic propositions. The sequent calculus is a standard fragment of JILL [6], containing dyadic two-sided sequents of the form $\Gamma; \Delta \Longrightarrow C$: the zone Γ contains the unrestricted hypotheses and Δ contains the linear hypotheses. Both contexts are unordered. For the rules of this calculus we refer the reader to [6, page 14]. Also in [6] are the standard weakening and contraction properties for the unrestricted hypotheses, which means we can treat Γ as a set, and admissibility of cut by means of a simple lexicographic induction.

Definition 1 (subformulas). *A* decorated formula *is a tuple* $\langle A, s, w \rangle$ *where A is a proposition, s is a sign (+ or −) and w is a* weight *(h for heavy or l for light). The* subformula relation $\leq$ *is the smallest reflexive and transitive relation between decorated subformulas satisfying the following inequalities:*

$$\langle A, s, h \rangle \leq \langle !A, s, * \rangle \quad \langle A, \bar{s}, l \rangle \leq \langle A \multimap B, s, * \rangle \quad \langle B, s, l \rangle \leq \langle A \multimap B, s, * \rangle$$
$$\langle A_i, s, l \rangle \leq \langle A_1 \otimes A_2, s, * \rangle \quad \langle A_i, s, l \rangle \leq \langle A_1 \,\&\, A_2, s, * \rangle \qquad\qquad i \in \{1, 2\}$$

where $\bar{s}$ *is the opposite of s, and* $*$ *can stand for either h or l, as necessary. Decorations and the subformula relation are lifted to (multi)sets in the obvious way.*

Property 2 (subformula property). *In any sequent* $\Gamma'; \Delta' \Longrightarrow C'$ *used in a proof of* $\Gamma; \Delta \Longrightarrow C$: $\langle \Gamma', -, h \rangle \cup \langle \Delta', -, * \rangle \cup \{\langle C', +, * \rangle\} \leq \langle \Gamma, -, h \rangle \cup \langle \Delta, -, l \rangle \cup \{\langle C, +, l \rangle\}$.

$\square$

For the remainder of the paper, all rules are restricted to decorated subformulas of a given goal sequent. A right (resp. left) rule is applicable if the principal formula in the conclusion is a positive (resp. negative) subformula of the goal sequent. Of the judgmental rules (reviewed in the next section), init is restricted to atomic subformulas that are *both* positive and negative decorated subformulas, and the copy rule is restricted to negative heavy subformulas.

3   Forward Linear Sequent Calculus

In addition to the usual non-determinism in rule and sub-goal selection, the single-use semantics of linear hypotheses gives rise to *resource non-determinism* during backward search. Its simplest form is *multiplicative*, caused by binary multiplicative rules ($\otimes R$ and $\multimap L$), where the linear zone of the conclusion has to be distributed into the premises. In order to avoid an exponential explosion, backward search strategies postpone this split either by an input/output interpretation, where proving a sub-goal consumes some of the resources from the input and passes the remaining resources on as outputs [12], or via Boolean constraints on the occurrences of linear hypotheses [13]. Interestingly, multiplicative non-determinism is entirely absent in a forward reading of multiplicative rules: the linear context in the conclusion is formed simply by adjoining those of the premises. On the multiplicative-exponential fragment, for example, forward search has no resource management issues at all. Resource management problems remain absent even in the presence of binary additives (& and $\oplus$).

The only form of resource non-determinism for the forward direction arises in the presence of additive constants ($\top$ and $\mathbf{0}$). For example, the backward $\top R$ rule has an arbitrary linear context which we cannot guess in the forward direction. We therefore leave it empty (no linear assumptions are needed), but we have to remember that we can add linear assumptions if necessary. We therefore differentiate sequents whose linear context can be weakened and those whose can not.

judgmental	

$$\frac{}{\cdot\,;p \longrightarrow^0 p}\ \text{init} \qquad \frac{\Gamma;\Delta,A \longrightarrow^w C}{\Gamma\cup\{A\};\Delta \longrightarrow^w C}\ \text{copy}$$

multiplicative

$$\frac{\Gamma;\Delta \longrightarrow^w A \quad \Gamma';\Delta' \longrightarrow^{w'} B}{\Gamma\cup\Gamma';\Delta,\Delta' \longrightarrow^{w\vee w'} A\otimes B}\ \otimes R$$

$$\frac{\Gamma;\Delta,A,B \longrightarrow^w C}{\Gamma;\Delta,A\otimes B \longrightarrow^w C}\ \otimes L$$

$$\frac{\Gamma;\Delta,A_i \longrightarrow^1 C \quad (A_j\notin\Delta)}{\Gamma;\Delta,A_1\otimes A_2 \longrightarrow^1 C}\ \otimes L_i$$
$$(i,j)\in\{(1,2),(2,1)\}$$

$$\frac{}{\cdot\,;\cdot \longrightarrow^0 1}\ 1R \qquad \frac{\Gamma;\Delta \longrightarrow^0 C}{\Gamma;\Delta,1 \longrightarrow^0 C}\ 1L$$

$$\frac{\Gamma;\Delta,A \longrightarrow^w B}{\Gamma;\Delta \longrightarrow^w A\multimap B}\ \multimap R$$

$$\frac{\Gamma;\Delta \longrightarrow^1 B \quad (A\notin\Delta)}{\Gamma;\Delta \longrightarrow^1 A\multimap B}\ \multimap R'$$

$$\frac{\Gamma;\Delta,B \longrightarrow^w C \qquad \Gamma';\Delta' \longrightarrow^{w'} A \quad (w=0\vee B\notin\Delta')}{\Gamma\cup\Gamma';\Delta,\Delta',A\multimap B \longrightarrow^{w\vee w'} C}\ \multimap L$$

additive

$$\frac{\Gamma;\Delta \longrightarrow^w A \qquad \Gamma';\Delta' \longrightarrow^{w'} B \quad (\Delta/w\approx\Delta'/w')}{\Gamma\cup\Gamma';\Delta\cup\Delta' \longrightarrow^{w\wedge w'} A\&B}\ \&R$$

$$\frac{}{\cdot\,;\cdot \longrightarrow^1 \top}\ \top R \qquad \frac{\Gamma;\Delta,A_i \longrightarrow^w C}{\Gamma;\Delta,A_1\&A_2 \longrightarrow^w C}\ \&L_i$$
$$i\in\{1,2\}$$

exponential

$$\frac{\Gamma;\cdot \longrightarrow^w A}{\Gamma;\cdot \longrightarrow^0 !A}\ !R \qquad \frac{\Gamma,A;\Delta \longrightarrow^w C}{\Gamma;\Delta,!A \longrightarrow^w C}\ !L$$

$$\frac{\Gamma;\Delta \longrightarrow^0 C \quad (A\notin\Gamma)}{\Gamma;\Delta,!A \longrightarrow^0 C}\ !L'$$

Fig. 1. Forward linear sequent calculus

To distinguish forward from backward sequents, we shall use a single arrow ($\longrightarrow$), possibly decorated, but keep the names of the rules the same.

Definition 3 (forward sequents).

1. *A* forward sequent *is of the form* $\Gamma;\Delta \longrightarrow^w C$. Γ *and* Δ *hold the unrestricted and linear resources respectively, and* w *is a Boolean (0 or 1) called the* weak-flag. *Sequents with* $w=1$ *are called* weakly linear *or simply* weak, *and those with* $w=0$ *are* strongly linear *or* strong.
2. *The* correspondence relation $\prec$ *between forward and backward sequents is defined as follows:* $\left(\Gamma;\Delta \longrightarrow^w C\right) \prec \left(\Gamma';\Delta' \Longrightarrow C\right)$ *iff* $\Gamma\subseteq\Gamma'$, *and* $\Delta=\Delta'$ *or* $\Delta\subseteq\Delta'$ *depending on whether* $w=0$ *or* $w=1$, *respectively. The forward sequent* s *is* sound *if for every backward sequent* s' *such that* $s\prec s'$, s' *is derivable in the backward calculus.*
3. *The* subsumption relation $\leq$ *between forward sequents is the smallest relation to satisfy:*

$$\left.\begin{array}{c}\left(\Gamma;\Delta \longrightarrow^0 C\right) \leq \left(\Gamma';\Delta \longrightarrow^0 C\right)\\[4pt]\left(\Gamma;\Delta \longrightarrow^1 C\right) \leq \left(\Gamma';\Delta' \longrightarrow^w C\right)\end{array}\right\}\ \textit{where } \Gamma\subseteq\Gamma' \textit{ and } \Delta\subseteq\Delta'.$$

Note that strong sequents never subsume weak sequents.

Obviously, if $s_1\leq s_2$ and $s_2\prec s$, then $s_1\prec s$. It is easy to see that weak sequents model affine logic: this is familiar from embeddings into linear logic that translate affine implications $A \rightarrow B$ as $A \multimap (B \otimes \top)$. The collection of inference rules for the forward calculus is in fig. 1. The rules must be read while keeping the subformula restriction in mind; precisely, a rule applies only when the principal formula is a subformula of the goal sequent.

The trickiest aspect of these rules are the side conditions (given in parentheses) and the weakness annotations. In order to understand these, it may be useful to think in term of the following property, which we maintain for all rules in order to avoid redundant inferences.

Definition 4. *A rule with conclusion s and premisses $s_1, \ldots, s_n$ is said to satisfy the irredundancy property if for no $i \in \{1, \ldots, n\}$, $s_i \leq s$.*

In other words, a rule is irredundant if none of its premisses subsumes the conclusion. Note that this is a local property; we do not discuss here more global redundancy criteria.

The first immediate observation is that binary rules simply take the union of the unrestricted zone from the premisses. The action of the rules on the linear zone is also prescribed by linearity when the sequents are strong ($w = 0$).

The binary additive rule ($\&R$) is applicable in the forward direction when both premisses are weak ($w = 1$), regardless of their linear zone. This is because in this case the linear zones can always be weakened to make them equal. We therefore compute the upper bound ($\sqcup$) of the two multi-sets: if A occurs n times in Δ and m times in Δ', then it occurs $\max(n, m)$ times in $\Delta \sqcup \Delta'$.

If only one premiss of the binary additive rule is weak, the linear zone of the weak premiss must be included in the linear zone of the other strong premiss. If both premisses are strong, their linear zones must be equal. We abstract the four possibilities in the form of an additive compatibility test.

Definition 5 (additive compatibility). *Given two forward sequents $\Gamma; \Delta \longrightarrow^w C$ and $\Gamma'; \Delta' \longrightarrow^{w'} C$, their additive zones Δ and Δ' are additively compatible given their respective weak-flags, which we write as $\Delta/w \approx \Delta'/w'$, if the following hold:*

$$\Delta/0 \approx \Delta'/0 \quad \text{if } \Delta = \Delta' \qquad \Delta/0 \approx \Delta'/1 \quad \text{if } \Delta' \subseteq \Delta$$
$$\Delta/1 \approx \Delta'/1 \quad \text{always} \qquad \Delta/1 \approx \Delta'/0 \quad \text{if } \Delta \subseteq \Delta'$$

For binary multiplicative rules like $\otimes R$, the conclusion is weak if either of the premisses is weak; thus, the weak-flag of the conclusion is a Boolean-or of those of the premisses. Dually, for binary additive rules, the conclusion is weak if both premisses are weak, so we use a Boolean-and to conjoin the weak flags. Most unary rules are oblivious to the weakening decoration, which simply survives from the premiss to the conclusion. The exception is $!R$, for which it is unsound to have a weak conclusion; there is no derivation of $\cdot; \top \Longrightarrow !\top$, for example.

Left rules with weak premisses require some attention. It is tempting to write the "weak" $\otimes L$ rules as:

$$\frac{\Gamma; \Delta, A \longrightarrow^1 C}{\Gamma; \Delta, A \otimes B \longrightarrow^1 C} \otimes L_1 \qquad \frac{\Gamma; \Delta, B \longrightarrow^1 C}{\Gamma; \Delta, A \otimes B \longrightarrow^1 C} \otimes L_2.$$

(Note that the irredundancy property requires that at least one of the operands of $\otimes$ be present in the premiss.) This pair of rules, however, would allow redundant inferences such as:

$$\frac{\Gamma;\Delta,A,B \longrightarrow^1 C}{\Gamma;\Delta,A,A \otimes B \longrightarrow^1 C} \otimes L_2.$$

We might as well have consumed both A and B to form the conclusion, and obtained a stronger result. The sensible strategy is: when A and B are both present, they must *both* be consumed. Otherwise, only apply the rule when one operand is present in a weak sequent. A similar observation can be made about all such rules: there is one weakness-agnostic form, and some possible refined forms to account for weak sequents.

Property 6 (irredundancy). *All forward rules satisfy the irredundancy property.* □

The soundness and completeness theorems are both proven by structural induction; we omit the easy proofs. Note that the completeness theorem shows that the forward calculus infers a possibly stronger form of the goal sequent.

Theorem 7 (soundness). *If $\Gamma;\Delta \longrightarrow^w C$ is derivable, then it is sound.*

Theorem 8 (completeness). *If $\Gamma;\Delta \Longrightarrow C$ is derivable, then there exists a derivable forward sequent $\Gamma';\Delta' \longrightarrow^w C$ such that $\left(\Gamma';\Delta' \longrightarrow^w C\right) \prec \left(\Gamma;\Delta \Longrightarrow C\right)$.*

4   Focused Derivations

Search using the backward calculus can always apply invertible rules eagerly in any order as there always exists a proof that goes through the premises of the invertible rule. Andreoli pointed out [1] that a similar and dual feature exists for non-invertible rules also: it is enough for completeness to apply a sequence of non-invertible rules eagerly in one atomic operation, as long as the corresponding connectives are of the same *synchronous* nature. Andreoli's observation for classical linear logic was extended to intuitionistic linear logic by Howe [3], but his inference rules have some overlap and an imprecise treatment of the atomic propositions that gives rise to unnecessary cycles in derivations. Our focusing calculus can be seen as a refinement of Howe's calculus, by preventing this imprecision and overlap in the inference rules. As a result, our completeness proof is considerably simpler, being a direct consequence of cut-elimination.

Before we sketch our formulation, a brief note about the classification of connectives: in classical linear logic the synchronous or asynchronous nature of a given connective is identical to its polarity; the negative connectives ($\&$, $\top$, γ, $\bot$, $\forall$) are asynchronous, and the positive connectives ($\otimes$, $\mathbf{1}$, $\oplus$, $\mathbf{0}$, $\exists$) are synchronous. The nature of intuitionistic connectives, though, must be derived without an appeal to polarity, which is not only difficult to motivate given the asymmetry of intuitionistic logic, but is also alien to the judgmental philosophy underlying the intuitionistic reconstruction of linear logic [6]. We derive the nature of connectives by examining the rules and phases of search: an asynchronous connective is one for which decomposition is complete in the active phase; a synchronous connective is one for decomposition is complete in the focused phase. This derivation happens to coincide with polarities for classical linear logic, but does differ for other logics. For intuitionistic (non-linear) logic, for instance, the conjunction $\wedge$, a connective of negative polarity, is seen as *both* synchronous and asynchronous in our derivation. (See section 6.1.)

As our backward linear sequent calculus has two sides, we have left- and right-synchronous and asynchronous connectives. For non-atomic propositions a left-synchronous connective is right-asynchronous, and a left-asynchronous connective right-synchronous;

symbol	meaning
P	left-synchronous (&, $\top$, $-\circ$, p)
Q	right-synchronous ($\otimes$, $\mathbf{1}$, !, p)
L	left-asynchronous ($\otimes$, $\mathbf{1}$, !)
R	right-asynchronous (&, $\top$, $-\circ$)

this appears to be universal in well-behaved logics. We define the notations in the adjoining table. The backward focusing calculus consists of three kinds of sequents; *right-focal sequents* of the form $\Gamma ; \Delta \gg A$ (A under focus), *left-focal sequents* of the form $\Gamma ; \Delta ; A \ll Q$, and *active sequents* of the form $\Gamma ; \Delta ; \Omega \Longrightarrow C$. Γ indicates the unrestricted zone as usual, Δ contains *only* left-synchronous propositions, and Ω is an ordered sequence of propositions (of arbitrary nature).

The active phase is entirely deterministic: it operates on the right side of the active sequent until it becomes right-synchronous, i.e., of the form $\Gamma ; \Delta ; \Omega \Longrightarrow Q$. Then the propositions in Ω are decomposed in order from right to left. The order of Ω is used solely to avoid spurious non-deterministic choices. Eventually the sequent is reduced to the form $\Gamma ; \Delta ; \cdot \Longrightarrow Q$, which we call *neutral sequents*.

A focusing phase is launched from a neutral sequent by selecting a proposition from Γ, Δ or the right hand side. This focused formula is decomposed until the top-level connective becomes asynchronous. Then we enter an active phase for the previously focused proposition.

Atomic propositions and modal operators need a special mention. Andreoli observed in [1] that it is sufficient to assign arbitrarily a synchronous or asynchronous nature to the atoms as long as duality is preserved; here, the asymmetric nature of the intuitionistic sequents suggests that they should be synchronous. If the left-focal formula is an atom, then the sequent is initial iff the linear zone Δ is empty *and* the right hand side matches the focused formula; this gives the focused version of the "init" rule. If an atom has right-focus, however, it is not enough to simply check that the left matches the right, as there might be some pending decompositions; consider eg. $\cdot ; p \& q \gg q$. Focus is therefore blurred in this case, and we correspondingly disallow a right atom in a neutral sequent from gaining focus.

The other subtlety is with the $!R$ rule: although ! is right synchronous, the $!R$ rule cannot maintain focus on the operand. If this were forced, there could be no focused proof of $!(A \otimes B) -\circ !(B \otimes A)$, for example. This is because there is a hidden transition from the truth of $!A$ to the validity of A which in turn reduces to the truth of A (see [6]). The first is synchronous, the second asynchronous, so the exponential has aspects of both. Girard has made a similar observation that exponentials are composed of one micro-connective to change polarity, and another to model a given behavior [14, Page 114]; this observation extends to other modal operators, such as why-not (?) of JILL [6] or the lax modality of CLF [7].

The full set of rules is in fig. 2. Soundness of this calculus is rather an obvious property – forget the distinction between Δ and Ω, elide the focus and blur rules, and

right-focal

$$\dfrac{\Gamma;\Delta_1 \gg A \quad \Gamma;\Delta_2 \gg B}{\Gamma;\Delta_1,\Delta_2 \gg A \otimes B} \; \otimes R$$

$$\dfrac{}{\Gamma;\cdot \gg 1} \; 1R \qquad \dfrac{\Gamma;\cdot;\cdot \Longrightarrow A}{\Gamma;\cdot \gg !A} \; !R$$

left-focal

$$\dfrac{}{\Gamma;\cdot;p \ll p} \; \text{init} \qquad \dfrac{\Gamma;\Delta;A_i \ll Q}{\Gamma;\Delta;A_1 \& A_2 \ll Q} \; \& L_i$$

$$\dfrac{\Gamma;\Delta_1;B \ll Q \quad \Gamma;\Delta_2 \gg A}{\Gamma;\Delta_1,\Delta_2;A \multimap B \ll Q} \; \multimap R$$

focus

$$\dfrac{\Gamma;\Delta;P \ll Q}{\Gamma;\Delta,P;\cdot \Longrightarrow Q} \; \text{lf}$$

$$\dfrac{\Gamma,A;\Delta;A \ll Q}{\Gamma,A;\Delta;\cdot \Longrightarrow Q} \; \text{copy}$$

$$\dfrac{\Gamma;\Delta \gg Q \quad Q \text{ non-atomic}}{\Gamma;\Delta;\cdot \Longrightarrow Q} \; \text{rf}$$

right-active

$$\dfrac{\Gamma;\Delta;\Omega \Longrightarrow A \quad \Gamma;\Delta;\Omega \Longrightarrow B}{\Gamma;\Delta;\Omega \Longrightarrow A \& B} \; \& R$$

$$\dfrac{}{\Gamma;\Delta;\Omega \Longrightarrow \top} \; \top R \qquad \dfrac{\Gamma;\Delta;\Omega \cdot A \Longrightarrow B}{\Gamma;\Delta;\Omega \Longrightarrow A \multimap B} \; \multimap R$$

left-active

$$\dfrac{\Gamma;\Delta;\Omega \cdot A \cdot B \Longrightarrow Q}{\Gamma;\Delta;\Omega \cdot A \otimes B \Longrightarrow Q} \; \otimes L$$

$$\dfrac{\Gamma;\Delta;\Omega \Longrightarrow Q}{\Gamma;\Delta;\Omega \cdot 1 \Longrightarrow Q} \; 1L \qquad \dfrac{\Gamma,A;\Delta;\Omega \Longrightarrow Q}{\Gamma;\Delta;\Omega \cdot !A \Longrightarrow Q} \; !L$$

$$\dfrac{\Gamma;\Delta,P;\Omega \Longrightarrow Q}{\Gamma;\Delta;\Omega \cdot P \Longrightarrow Q} \; \text{act}$$

blur

$$\dfrac{\Gamma;\Delta;L \Longrightarrow Q}{\Gamma;\Delta;L \ll Q} \; \text{lb}$$

$$\dfrac{\Gamma;\Delta;\cdot \Longrightarrow R}{\Gamma;\Delta \gg R} \; \text{rb} \qquad \dfrac{\Gamma;\Delta;\cdot \Longrightarrow p}{\Gamma;\Delta \gg p} \; \text{rb}^*$$

Fig. 2. Backward linear focusing calculus

the original backward calculus appears. For completeness of the focusing calculus, we proceed by interpreting every backward sequent as an active sequent in the focusing calculus, then showing that the backward rules are admissible in the focusing calculus. This proof relies on admissibility of cut in the focusing calculus. Because a non-atomic left-synchronous proposition is right-asynchronous, a left-focal sequent needs to match only an active sequent in a cut; similarly for right-synchronous propositions. Active sequents should match other active sequents, however. Cuts destroy focus, as they generally require commutations spanning phase boundaries; the products of a cut are therefore active.

The proof needs two key lemmas: the first notes that permuting the ordered context doesn't affect provability, as the ordered context does not mirror any deep non-commutativity in the logic. This lemma thus allows cutting formulas from anywhere inside the ordered context, and also to re-order the context when needed.

Lemma 9. *If* $\Gamma;\Delta;\Omega \Longrightarrow C$, *then* $\Gamma;\Delta;\Omega' \Longrightarrow C$ *for any permutation* Ω' *of* Ω. $\quad\square$

The other lemma shows that left-active rules can be applied even if the right-hand side is not synchronous. This lemma is vital for commutative cuts.

Lemma 10. *The following variants of the left-active rules are admissible*

$$\dfrac{\Gamma;\Delta,P;\Omega \Longrightarrow C}{\Gamma;\Delta;\Omega \cdot P \Longrightarrow C} \qquad \dfrac{\Gamma;\Delta;\Omega \cdot A \cdot B \Longrightarrow C}{\Gamma;\Delta;\Omega \cdot A \otimes B \Longrightarrow C} \qquad \dfrac{\Gamma;\Delta;\Omega \Longrightarrow C}{\Gamma;\Delta;\Omega \cdot 1 \Longrightarrow C} \qquad \dfrac{\Gamma,A;\Delta;\Omega \Longrightarrow C}{\Gamma;\Delta;\Omega \cdot !A \Longrightarrow C}$$

Theorem 11 (cut). *If*

1. $\Gamma;\Delta \gg A$ *and:*
   (a) $\Gamma;\Delta';\Omega \cdot A \Longrightarrow C$ *then* $\Gamma;\Delta,\Delta';\Omega \Longrightarrow C$.
   (b) $\Gamma;\Delta',A;\Omega \Longrightarrow C$ *then* $\Gamma;\Delta,\Delta';\Omega \Longrightarrow C$.
2. $\Gamma;\cdot \gg A$ *and* $\Gamma,A;\Delta;\Omega \Longrightarrow C$ *then* $\Gamma;\Delta;\Omega \Longrightarrow C$.
3. $\Gamma;\Delta;\Omega \Longrightarrow A$ *and:*
   (a) $\Gamma;\Delta';A \ll Q$ *then* $\Gamma;\Delta,\Delta';\Omega \Longrightarrow Q$.
   (b) $\Gamma;\Delta';\Omega' \cdot A \Longrightarrow C$ *then* $\Gamma;\Delta,\Delta';\Omega \cdot \Omega' \Longrightarrow C$.
   (c) $\Gamma;\Delta',A;\Omega' \Longrightarrow C$ *then* $\Gamma;\Delta,\Delta';\Omega \cdot \Omega' \Longrightarrow C$.
4. $\Gamma;\cdot;\cdot \Longrightarrow A$ *and:*
   (a) $\Gamma,A;\Delta;\Omega \Longrightarrow C$ *then* $\Gamma;\Delta;\Omega \Longrightarrow C$.
   (b) $\Gamma,A;\Delta \gg B$ *then* $\Gamma;\Delta \gg B$.
5. $\Gamma;\Delta;B \ll A$ *and:*
   (a) $\Gamma;\Delta';A \Longrightarrow Q$ *then* $\Gamma;\Delta,\Delta';B \ll Q$.
   (b) $\Gamma;\Delta',A;\cdot \Longrightarrow Q$ *then* $\Gamma;\Delta,\Delta';B \ll Q$.
6. $\Gamma;\Delta;\cdot \Longrightarrow A$ *and* $\Gamma;\Delta',A \gg B$ *then* $\Gamma;\Delta,\Delta' \gg B$.

Proof (sketch). By lexicographic induction on the given derivations. The argument is lengthy rather than complex, and is an adaptation of similar structural cut-admissibility proofs in eg. [6]. □

Theorem 12 (completeness).
If $\Gamma;\Delta \Longrightarrow C$ *and* Ω *is any serialization of* Δ, *then* $\Gamma;\cdot;\Omega \Longrightarrow C$.

Proof (sketch). First show that all ordinary rules are admissible in the focusing system using cut. Proceed by induction on the derivation of $\mathcal{D} :: \Gamma;\Delta \Longrightarrow C$, splitting cases on the last applied rule, using cut and lem. 9 as required. A representative case (for $\otimes R$):

$$\mathcal{D} = \frac{\mathcal{D}_1 :: \Gamma;\Delta \Longrightarrow A \quad \mathcal{D}_2 :: \Gamma;\Delta' \Longrightarrow B}{\Gamma;\Delta,\Delta' \Longrightarrow A \otimes B} \otimes R$$

Let Ω and Ω' be serializations of Δ and Δ' respectively; by the induction hypothesis on $\mathcal{D}_1$ and $\mathcal{D}_2$, we have $\Gamma;\cdot;\Omega \Longrightarrow A$ and $\Gamma;\cdot;\Omega' \Longrightarrow B$. Now, it is easy to show that $\Gamma;\cdot;A \cdot B \Longrightarrow A \otimes B$. The result follows by the use of cut twice, for A and B in the active context respectively, to get $\Gamma;\cdot;\Omega \cdot \Omega' \Longrightarrow A \otimes B$, and then noting that any serialization of Δ,Δ' is a permutation of $\Omega \cdot \Omega'$. □

5   Forward Focusing

We now construct the forward version of the focusing calculus. Both the active and focal phases in the backward direction are *eager* in the sense that intermediate sequents are not important; instead, just the neutral sequents (i.e., of the form $\Gamma;\Delta;\cdot \Longrightarrow Q$) at the phase boundaries are important. One therefore thinks of the backward focusing calculus as one of neutral sequents. Analogously, the forward focusing system discards the intermediate focal and active sequents by means of calculating the corresponding derived rules for forward neutral sequents.

For any given synchronous subformula, the derived inferences for that subformula correspond to a single pair of focal and active phases. This observation is not new; Andreoli called them *bipoles* [2]. However, there are important differences between backward reasoning bipoles and their forward analogue: as shown in thm. 8, the forward calculus generates stronger forms of sequents than in the corresponding backward proof. Therefore, not every branch of the backward bipole will be available in the forward direction. The forward derived rules therefore need some additional mechanism in the internal nodes to handle these cases.

We still adapt the essential idea of bipoles of viewing every proposition as a relation between the conclusion of the bipole and its possible premises at the leaves of the bipole. This relational interpretation gives us the derived rules corresponding to the proposition; the premises and conclusions of these derived rules are neutral sequents, which we indicate by means of a double-headed sequent arrow ($\longrightarrow\!\!\!\!\rightarrow$).

Each relation R takes as input the premises of the bipole, $s_1 \cdot s_2 \cdots s_n$ (written Σ), and constructs the relevant portion of a conclusion sequent s; we write this as $R[\Sigma] \hookrightarrow s$. There are three classes of these relations:

1. Right focal relations for the focus formula A, written $\mathsf{foc}_{\Downarrow}^+(A)$.
2. Left focal relations for the focus formula A, written $\mathsf{foc}_{\Downarrow}^-(A)$.
3. Active relations, written $\mathsf{act}_{\Downarrow}(\Gamma;\Delta;\Omega \Longrightarrow \gamma)$, where γ is either $\cdot$ or C.

The focal relations are understood as defining derived rules corresponding to a given proposition. The conclusion of these derived rules are themselves neutral sequents. For a right focal relation $\mathsf{foc}_{\Downarrow}^+(Q)$, the corresponding derived rule is:

$$\frac{\Sigma \quad \left(\mathsf{foc}_{\Downarrow}^+(Q)[\Sigma] \hookrightarrow \Gamma;\Delta \longrightarrow\!\!\!\!\rightarrow^w \cdot\right)}{\Gamma;\Delta \longrightarrow\!\!\!\!\rightarrow^w Q} \; \mathsf{foc}_{\Downarrow}^+$$

Similarly, for negative propositions, we have two rules, depending on whether the focused proposition is a heavy subformula of the goal sequent or not.

$$\frac{\Sigma \quad \left(\mathsf{foc}_{\Downarrow}^-(P)[\Sigma] \hookrightarrow \Gamma;\Delta \longrightarrow\!\!\!\!\rightarrow^w Q\right)}{\Gamma;\Delta,P \longrightarrow\!\!\!\!\rightarrow^w Q} \; \mathsf{foc}_{\Downarrow}^- \qquad \frac{\Sigma \quad \left(\mathsf{foc}_{\Downarrow}^-(A)[\Sigma] \hookrightarrow \Gamma;\Delta \longrightarrow\!\!\!\!\rightarrow^w Q\right)}{\Gamma \cup \{A\};\Delta \longrightarrow\!\!\!\!\rightarrow^w Q} \; !\mathsf{foc}_{\Downarrow}^-$$

As before, these derived rules are understood to contain only signed subformulas of the goal sequent. The active relations essentially replay the active rules of the backward focusing calculus, except they also account for weak sequents as needed.

For lack of space we leave out the details of the definition of these relations; they can be found in the accompanying technical report [15]. Instead, we shall give an example. Consider the negative principal subformula $P = p \,\&\, q \multimap r \,\&\, (s \otimes t)$ and the three input sequents $\Gamma_1;\Delta_1 \longrightarrow\!\!\!\!\rightarrow^1 p$, $\Gamma_2;\Delta_2 \longrightarrow\!\!\!\!\rightarrow^0 q$, and $\Gamma_3;\Delta_3,s \longrightarrow\!\!\!\!\rightarrow^1 Q$, named s_1, s_2, and s_3 respectively. By the definition of $\mathsf{foc}_{\Downarrow}^-$:

$$\mathsf{foc}_{\Downarrow}^-(P)[s_3 \cdot s_1 \cdot s_2] \hookrightarrow \Gamma_3 \cup \Gamma_1 \cup \Gamma_2;\Delta_3,\Delta_2 \longrightarrow\!\!\!\!\rightarrow^1 Q \qquad \text{if } t \notin \Delta_3 \text{ and } \Delta_1 \subseteq \Delta_2$$

In other words, the instance of the full derived rule for P matched against the given sequents stands for the following derived rule of inference specialized to this scenario:

$$\frac{\Gamma_1;\Delta_1 \longrightarrow\!\!\!\!\rightarrow^1 p \quad \Gamma_2;\Delta_2 \longrightarrow\!\!\!\!\rightarrow^0 q \quad \Gamma_3;\Delta_3,s \longrightarrow\!\!\!\!\rightarrow^1 Q \quad (t \notin \Delta_3) \quad (\Delta_1 \subseteq \Delta_2)}{\Gamma_3 \cup \Gamma_1 \cup \Gamma_2;\Delta_3,\Delta_2,P \longrightarrow\!\!\!\!\rightarrow^1 Q}$$

The proofs of soundness and completeness of the forward focusing calculus with respect to the backward focusing calculus are are in [15]. Soundness is shown by simple structural induction on the $\text{foc}_{\Downarrow}^{+}$, $\text{foc}_{\Uparrow}^{-}$ and $\text{act}_{\Downarrow}$ derivations. Completeness is a rather more complex result because the forward and backward focused proofs are not in bijection. The essential idea of the proof is to define a complete calculus of backward derived rules, and prove the calculus of forward derived rules complete with respect to this intermediate calculus.

Theorem 13 (soundness). *If* $\Gamma ; \Delta \longrightarrow\!\!\!\twoheadrightarrow^{w} Q$ *is derivable, then it is sound.* □

Theorem 14 (completeness). *If* $\Gamma ; \Delta ; \cdot \Longrightarrow Q$ *is derivable, then there exists a derivable focused sequent* $\Gamma' ; \Delta' \longrightarrow\!\!\!\twoheadrightarrow^{w} Q$ *such that* $\left(\Gamma' ; \Delta' \longrightarrow\!\!\!\twoheadrightarrow^{w} Q \right) \prec \left(\Gamma ; \Delta \Longrightarrow Q \right)$. □

5.1 The Focused Inverse Method

What remains is to implement the inverse method search strategy that uses the forward focusing calculus. Before describing the focused inverse method, we briefly sketch the usual single-step inverse method here, eliding the implementation issues that are out of the scope of this paper[1]. The inverse method consists of three essential components: the *database* of computed sequents, the library of *rules* that can be applied to sequents to compute new sequents, and the main search loop or *engine*. Rules are constructed by naming all subformulas of the goal sequent with fresh propositional labels, and specializing the inference rules of the full logic to principal uses of the subformula labels; the general rules are then discarded. This procedure is key to giving the inverse method a goal direction, as the search space is constrained to subformulas of the goal. Traditionally the library of rules is considered static during a given search, but as we describe in [9], it is beneficial, especially in the first-order extension, to allow the library of rules to be extended during search with partial applications– a form of memoization. The inputs for these rules are drawn from the database of computed sequents. At the start of search, this database contains just the initial sequents, which are determined by considering all *atomic* subformulas that are both positively and negatively occurring in the goal sequent. The engine repeatedly selects sequents from the database, and applies rules from the library to generate new sequents; if these new sequents are not subsumed by any sequent derived earlier, they are inserted in to the database. Completeness of the search strategy is guaranteed by using a fair selection (i.e., equivalent to breadth-first search) of sequents from the database in order to generate new sequents.

The primary issue in the presence of focusing is what propositions to generate rules for. As the calculus of derived rules has only neutral sequents as premises and conclusions, we need only generate rules for propositions that occur in neutral sequents; we call them *frontier propositions*. To find the frontier propositions in a goal sequent, we simply abstractly replay the focusing and active phases to identify the phase transitions. Each transition from an active to a focal phase produces a frontier proposition. Formally, we define two generating functions, f (focal) and a (active), from signed propositions to multisets of frontier propositions. None of the logical constants are in

---

[1] See a related paper for notes on implementation [9].

$$f(p)^- = \emptyset \quad f(p)^+ = a(p)^\pm = \{p\} \quad f(\mathbf{1})^\pm = a(\mathbf{1})^\pm = \emptyset \quad f(\top)^\pm = a(\top)^\pm = \emptyset$$

$$f(A \otimes B)^- = a(A \otimes B)^- \qquad\qquad f(A \otimes B)^+ = f(A)^+, f(B)^+$$
$$a(A \otimes B)^- = a(A)^-, a(B)^- \qquad\qquad a(A \otimes B)^+ = f(A \otimes B)^+, A \otimes B$$

$$f(A \,\&\, B)^- = f(A)^-, f(B)^- \qquad\qquad f(A \,\&\, B)^+ = a(A \,\&\, B)^+$$
$$a(A \,\&\, B)^- = f(A \,\&\, B)^-, A \,\&\, B \qquad a(A \,\&\, B)^+ = a(A)^+, a(B)^+$$

$$f(A \multimap B)^- = f(A)^+, f(B)^- \qquad\qquad f(A \multimap B)^+ = a(A \multimap B)^+$$
$$a(A \multimap B)^- = f(A \multimap B)^-, A \multimap B \quad a(A \multimap B)^+ = a(A)^-, a(B)^+$$

$$f(!A)^- = a(!A)^- \quad f(!A)^+ = a(A)^+ \quad a(!A)^- = a(A)^- \quad a(!A)^+ = f(A)^+, !A$$

Fig. 3. Calculating frontier propositions

the frontier, for the conclusions of rules such as $\top R$ and $\mathbf{1}R$ are easy to predict, and can be generated as needed. Similarly we do not count a negative focused atomic proposition in the frontier as we know that the conclusion of the init rule needs to have the form $\Gamma; \cdot; p \ll p$; this restricts the collection of spurious initial sequents that are not possible in a focused proof. The steps in the calculation are shown in figure 3; as a simple example, $f(p \,\&\, q \multimap r \,\&\, (s \otimes t))^- = p, q, s, t$.

Definition 15 (frontier). *Given a goal* $\Gamma; \Delta; \cdot \Longrightarrow Q$ *(which is neutral), its frontier contains:*

i.   all (top-level) propositions in Γ, Δ, Q;
ii. for any $A \in \Gamma, \Delta$, *the collection* $f(A)^-$; *and*
iii. the collection $f(Q)^+$.

Property 16 (neutral subformula property). *In any backward focused proof, all neutral sequents consist only of frontier propositions of the goal sequent.* □

In the preparatory phase for the inverse method, we calculate the frontier propositions of the goal sequent. There is no need to generate initial sequents separately, as the executions of negative atoms in the frontier directly give us the necessary initial sequents. The general design of the main loop of the prover and the argument for its completeness are fairly standard [8, 10]; we use a lazy refinement of this basic design [9] that is ideal for multi-premiss rules.

5.2   Some Experimental Results

We have implemented an expanded version of the forward focusing calculus as a certifying[2] inverse method prover for intuitionistic linear logic, including the missing connectives $\oplus$, $\mathbf{0}$, and the lax modality[3]. Table 1 contains a running-time comparison of the focusing prover (**F**) against a non-focusing version (**NF**) of the prover (directly implementing the calculus of sec. 3), and Tammet's Gandalf "nonclassical" distribution

---

[2] By *certifying*, we mean that it produces independently verifiable proof objects.
[3] Available from the first author's web page at http://www.cs.cmu.edu/~kaustuv/

Table 1. Some experimental results

Test	NF	F	Gt	Gr
blocks-world	0.02 s	≤ 0.01 s	13.51 s	0.03 s
change	3.20 s	≤ 0.01 s	—	0.63 s
affine1	0.01 s	≤ 0.01 s	0.03 s	≤ 0.01 s
affine2	≈ 12 m	1.21 s	—	—
qbf1	0.03 s	≤ 0.01 s	—	2.40 s
qbf2	0.04 s	≤ 0.01 s	—	42.34 s
qbf3	≈ 35 m	0.53 s	—	—

All measurements are wall-clock times on an unloaded computer with a 2.80GHz
Pentium 4 processor, 512KB L1 cache and 1GB of main memory; "—" denotes
unsuccessful proof within ≈ ten hours.

that includes a pair of (non-certifying) provers for classical linear logic, one (**Gr**) using a refinement of Mints' resolution system for classical linear logic [10, 11], and the
other (**Gt**) using a backward Tableaux-based strategy. Neither of these provers incorporates focusing. The test problems ranged from simple stateful encodings such as blocks-
world or change-machines, to more complex problems such as encoding of affine logic
problems, and translations of various quantified Boolean formulas using the algorithm
in [16]. Focusing was faster in every case, with an average speedup of about three orders
of magnitude over the non-focusing version.

6   Embedding Non-linear Logics

6.1   Intuitionistic Logic

When we move from intuitionistic to intuitionistic linear logic, we gain a lot of expressive power. Nonetheless, many problems, even if posed in linear logic, have significant
non-linear components or sub-problems. Standard translations into linear logic, however, have the problem that any focusing properties enjoyed by the source are lost in
the translation. In a focusing system for intuitionistic logic, as hinted to by Howe [3]
and briefly considered below, a quite deterministic proof with, say, one phase of focusing, will be decomposed into many small phases, leading to a large loss in efficiency. Fortunately, it is possible to translate intuitionistic logic in a way that preserves focusing. To illustrate, consider a minimal intuitionistic propositional logic with
connectives $\{\wedge, \mathbf{t}, \supset\}$. The focusing system for this logic has three kinds of sequents,
$\Gamma \gg_I A$ (right-focal), $\Gamma; A \ll_I Q$ (left-focal), and $\Gamma; \Omega \Longrightarrow_I C$ (active), with $\supset$ treated as
right-synchronous, and $\wedge$ as *both* (right-) synchronous and asynchronous. The meta-
variables P, Q, L and R are used in the spirit of section 4; that is, P for left-synchronous
$\{\wedge, \mathbf{t}, \supset, p\}$, Q for right-synchronous $\{\wedge, \mathbf{t}, p\}$, L for left-asynchronous $\{\wedge, \mathbf{t}\}$, and R
for right-asynchronous $\{\wedge, \mathbf{t}, \supset\}$. Q^* means that Q is not atomic, i.e., just containing
$\{\wedge, \mathbf{t}\}$.

$$F(p)^- = p \quad F(p)^+ = p \quad A(p)^- = \,!p \quad A(p)^+ = p$$

$$F(A \wedge B)^- = F(A)^- \,\&\, F(B)^- \quad F(A \wedge B)^+ = F(A)^+ \otimes F(B)^+$$

$$A(A \wedge B)^- = A(A)^- \otimes A(B)^- \quad A(A \wedge B)^+ = A(A)^+ \,\&\, A(B)^+$$

$$F(\mathbf{t})^- = \top \quad F(\mathbf{t})^+ = 1 \quad A(\mathbf{t})^- = 1 \quad A(\mathbf{t})^+ = \top$$

$$F(A \supset B)^- = F(A)^+ \multimap F(B)^- \quad F(A \supset B)^+ = A(A \supset B)^+$$

$$A(A \supset B)^- = \,!F(A \supset B)^- \quad A(A \supset B)^+ = A(A)^- \multimap A(B)^+$$

Fig. 4. Embedding intuitionistic logic

$$
\frac{}{\Gamma; p \ll_I p}
\qquad
\frac{\Gamma; A_i \ll_I Q}{\Gamma; A_1 \wedge A_2 \ll_I Q}
\qquad
\frac{\Gamma; B \ll_I Q \quad \Gamma \gg_I A}{\Gamma; A \supset B \ll_I Q}
\qquad
\frac{\Gamma \gg_I A \quad \Gamma \gg_I B}{\Gamma \gg_I A \wedge B}
\qquad
\frac{}{\Gamma \gg_I \mathbf{t}}
$$

$$
\frac{\Gamma; \Omega \cdot A \cdot B \Longrightarrow_I Q}{\Gamma; \Omega \cdot A \wedge B \Longrightarrow_I Q}
\qquad
\frac{\Gamma; \Omega \Longrightarrow_I A \quad \Gamma; \Omega \Longrightarrow_I B}{\Gamma; \Omega \Longrightarrow_I A \wedge B}
\qquad
\frac{}{\Gamma; \Omega \Longrightarrow_I \mathbf{t}}
\qquad
\frac{\Gamma; \Omega \cdot A \Longrightarrow_I B}{\Gamma; \Omega \Longrightarrow_I A \supset B}
$$

$$
\frac{\Gamma, P; \Omega \Longrightarrow_I Q}{\Gamma; \Omega \cdot P \Longrightarrow_I Q} \; act
\qquad
\frac{\Gamma \gg_I Q^*}{\Gamma; \cdot \Longrightarrow_I Q^*}
\qquad
\frac{\Gamma; P \ll_I Q}{\Gamma, P; \cdot \Longrightarrow_I Q}
\qquad
\frac{\Gamma; \cdot \Longrightarrow_I R}{\Gamma \gg_I R}
\qquad
\frac{\Gamma; L \Longrightarrow_I Q}{\Gamma; L \ll_I Q}
$$

The translation is modal with two phases: A (active) and F (focal). A positive focal $\wedge$ is translated as $\otimes$, and the duals as $\&$. For every use of the act rule, the corresponding translation phase affixes an exponential; the phase-transitions in the image of the translation exactly mirror those in the source. The details of the translation are in figure 4.

It is easily shown that these translations preserve the focusing structure of proofs.

Property 17 (preservation of the structure of proofs).

1. *If* $\Gamma \gg_I A$, *then* $F(\Gamma)^-; \cdot \gg F(A)^+$.
2. *If* $\Gamma; A \ll_I Q$, *then* $F(\Gamma)^-; \cdot; F(A)^- \ll F(Q)^+$.
3. *If* $\Gamma; \Omega \Longrightarrow_I Q$, *then* $F(\Gamma)^-; \cdot; A(\Omega)^- \Longrightarrow F(Q)^+$.
4. *If* $\Gamma; \Omega \Longrightarrow_I R$, *then* $F(\Gamma)^-; \cdot; A(\Omega)^- \Longrightarrow A(R)^+$.   □

The reverse translation, written $-^o$, is trivial: simply erase all !s, rewrite $\&$ and $\otimes$ as $\wedge$, $\top$ and $\mathbf{1}$ as $\mathbf{t}$, and $\multimap$ as $\Rightarrow$.

Property 18 (soundness).

1. *If* $\Gamma; \cdot \gg A$, *then* $\Gamma^o \gg_I A^o$.
2. *If* $\Gamma; \cdot; A \ll Q$, *then* $\Gamma^o; A^o \ll_I Q^o$.
3. *If* $\Gamma; \cdot; \Omega \Longrightarrow C$, *then* $\Gamma^o; \Omega^o \Longrightarrow_I C^o$.   □

An important feature of this translation is that only (certain) negative atoms and implications are !-affixed; this is related to a similar observation by Dyckhoff that the ordinary propositional intuitionistic logic has a contraction-free sequent calculus that duplicates only negative atoms and implications [17]. It is also important to note that this translation extends easily to handle the disjunctions $\vee$ and $\perp$ (in the source) and $\oplus$ and $\mathbf{0}$ in the target logic; this naturality is not as obvious for Howe's synchronicity-aware translation [3].

6.2 Classical Horn Formulas

A related issue arises with respect to (non-linear) Horn logic. In complex specifications that employ linearity, there are often significant sub-specifications that lie in the Horn fragment. Unfortunately, the straightforward inverse method is quite inefficient on Horn formulas, something already noticed by Tammet [10]. So his prover switches between hyperresolution for Horn and near-Horn formulas and the inverse method for other propositions.

With focusing, this becomes entirely unnecessary. Our focused inverse method for intuitionistic linear logic, when applied to a classical, non-linear Horn formula, will exactly behave as classical hyperresolution. This remarkable property gives further evidence to the power of focusing as a technique for forward theorem proving.

A propositional Horn clause has the form $p_1 \supset \cdots \supset p_n \supset p$ where all p_i and p are atomic. A Horn theory Ψ is just a set of Horn clauses. This can easily be generalized to include conjunction and truth. The results in this section extend also to the first-order case, where Horn formulas allow outermost universal quantification.

The hyperresolution strategy on this framework is essentially just forward reasoning with rule set "hyper" for any $p_1 \supset \cdots \supset p_n \supset p \in \Psi$. Note that these

$$\frac{p_1 \quad p_2 \quad \cdots \quad p_n}{p} \; \text{hyper}$$

will be unit clauses if $n = 0$. If we translate every clause $p_1 \supset \cdots \supset p_n \supset p$ as $!(p_1 \multimap \cdots \multimap p_n \multimap p)$, it is easy to see that the derived rules associated with the results of the translation are exactly the hyperresolution rules.

7   Conclusion

We have presented the design of an inverse method theorem prover for propositional intuitionistic linear logic and have demonstrated through experimental results that focusing represents a highly significant improvement. Though elided here, the results persist in the presence of a lax modality [7], and extend to the first-order case as shown by the authors in a related paper [9], which also contains many more details on the implementation and a more thorough empirical evaluation.

Our methods derived from focusing can be applied directly and more easily to classical linear logic and (non-linear) intuitionistic logic, also yielding focused inverse method provers. While we do not have an empirical evaluation of such provers, the reduction in the complexity of the search space is significant. We therefore believe that focusing is a nearly universal improvement to the inverse method and should be applied as a matter of course, possibly excepting only (non-linear) classical logic.

In future work we plan to add higher-order and linear terms in order to obtain a theorem prover for all of CLF [7]. The main obstacles will be to develop feasible algorithms for unification and to integrate higher-order equational constraints. We are also interested in exploring if model-checking techniques could help to characterize the shape of the linear zone that could arise in a backward proof in order to further restrict forward inferences.

Finally, we plan a more detailed analysis of connections with a bottom-up logic programming interpreter for the LO fragment of classical linear logic [18]. This fragment, which is in fact affine, has the property that the unrestricted context remains constant throughout a derivation, and incorporates focusing at least partially via a back-chaining rule. It seems plausible that our prover might simulate their interpreter when LO specifications are appropriately translated into intuitionistic linear logic, similar to the translation of classical Horn clauses.

Acknowledgments

We thank Kevin Watkins for illuminating discussions on the topic of focused derivations, and the anonymous referees of this and an earlier version of this paper for numerous helpful suggestions.

References

1. Andreoli, J.M.: Logic programming with focusing proofs in linear logic. Journal of Logic and Computation **2** (1992) 297–347
2. Andreoli, J.M.: Focussing and proof construction. Annals of Pure and Applied Logic **107** (2001) 131–163
3. Howe, J.M.: Proof Search Issues in Some Non-Classical Logics. PhD thesis, University of St. Andrews (1998)
4. Girard, J.Y.: Linear logic. Theoretical Computer Science **50** (1987) 1–102
5. Barber, A.: Dual Intuitionistic Linear Logic. Technical Report ECS-LFCS-96-347, University of Edinburgh (1996)
6. Chang, B.Y.E., Chaudhuri, K., Pfenning, F.: A judgmental analysis of linear logic. Technical Report CMU-CS-03-131R, Carnegie Mellon University (2003)
7. Cervesato, I., Pfenning, F., Walker, D., Watkins, K.: A concurrent logical framework I & II. Technical Report CMU-CS-02-101 and 102, Department of Computer Science, Carnegie Mellon University (2002) Revised May 2003.
8. Degtyarev, A., Voronkov, A.: The Inverse Method. In: Handbook of Automated Reasoning. MIT Press (2001) 179–272
9. Chaudhuri, K., Pfenning, F.: A focusing inverse method theorem prover for first-order linear logic. In: Proceedings of the 20th International Conference on Automated Deduction (CADE-20). (2005) To appear.
10. Tammet, T.: Resolution, inverse method and the sequent calculus. In: Proceedings of KGC'97, Springer-Verlag LNCS 1289 (1997) 65–83
11. Mints, G.: Resolution calculus for the first order linear logic. Journal of Logic, Language and Information **2** (1993) 59–83
12. Cervesato, I., Hodas, J.S., Pfenning, F.: Efficient resource management for linear logic proof search. Theoretical Computer Science **232** (2000) 133–163 Special issue on Proof Search in Type-Theoretic Languages, D. Galmiche and D. Pym, editors.
13. Harland, J., Pym, D.J.: Resource-distribution via boolean constraints. In McCune, W., ed.: Proceedings of CADE-14, Springer-Verlag LNAI 1249 (1997) 222–236
14. Girard, J.Y.: Locus solum: from the rules of logic to the logic of rules. Mathematical Structures in Computer Science **11** (2001) 301–506
15. Chaudhuri, K., Pfenning, F.: Focusing the inverse method for linear logic. Technical Report CMU-CS-05-106, Carnegie Mellon University (2005)
16. Lincoln, P., Mitchell, J.C., Scedrov, A., Shankar, N.: Decision problems for propositional linear logic. Annals of Pure and Applied Logic **56** (1992) 239–311
17. Dyckhoff, R.: Contraction-free sequent calculi for intuitionistic logic. Journal of Symbolic Logic (1992) 795–807
18. Bozzano, M., Delzanno, G., Martelli, M.: Model checking linear logic specifications. TPLP **4** (2004) 573–619

Towards a Typed Geometry of Interaction

Esfandiar Haghverdi[1] and Philip J. Scott[2],[*]

[1] School of Informatics, Indiana University
Bloomington, IN 47406, USA
ehaghver@indiana.edu
http://xavier.informatics.indiana.edu/~ehaghver
[2] Dept. of Mathematics & Statistics
University of Ottawa, Ottawa, Ontario
Canada K1N 6N5
phil@mathstat.uottawa.ca
http://www.uottawa.ca/site/~phil

Abstract. Girard's Geometry of Interaction (GoI) develops a mathematical framework for modelling the dynamics of cut-elimination. We introduce a typed version of GoI, called Multiobject GoI (MGoI) for multiplicative linear logic without units in categories which include previous (untyped) GoI models, as well as models not possible in the original untyped version. The development of MGoI depends on a new theory of partial traces and trace classes, as well as an abstract notion of orthogonality (related to work of Hyland and Schalk) We develop Girard's original theory of types, data and algorithms in our setting, and show his execution formula to be an invariant of Cut Elimination. We prove Soundness and Completeness Theorems for the MGoI interpretation in partially traced categories with an orthogonality.

1 Introduction

Geometry of Interaction (GoI) is a novel interpretation of linear logic, introduced by Girard in a fundamental series of papers beginning in the late 80's [11–13] and continued recently in [14]. One striking feature of this work is that it provides a mathematical framework for modelling cut-elimination (normalization) as a dynamical process of information flow, independent of logical syntax. To these ends, Girard introduces methods from functional analysis and operator algebras to model proofs and their dynamical behaviour. At the same time, these methods allow GoI to provide new foundational insights into the theory of algorithms.

Girard's original framework, based on C*-algebras, was studied in detail in several works of Danos and Regnier (for example in [8]) and by Malacaria and Regnier [26]. The GoI program itself has been applied to the analysis of optimal reduction by Gonthier, Abadi, and Lévy [9], to complexity theory [6], to game semantics and token machines [5, 24], etc.

Let us briefly recall some aspects of Girard's original GoI. Traditional denotational semantics models normalization of proofs (or lambda terms) by static

---

[*] P. J. Scott's research is supported in part by an NSERC Discovery grant.

L. Ong (Ed.): CSL 2005, LNCS 3634, pp. 216–231, 2005.
© Springer-Verlag Berlin Heidelberg 2005

equalities: if Π, Π' are proofs and if Π reduces to Π' by cut-elimination, then in any appropriate model, $[\![\Pi]\!] = [\![\Pi']\!]$. Instead, in his GoI program, Girard considers proofs (or algorithms) as operators, pictured as I/O boxes: a proof of a sequent $\vdash \Gamma$ is interpreted as a box with input and output wires labelled by Γ. The formulas or types in Γ form the I/O-*interface* of the proof. Girard works in an *untyped* setting, so in fact the labels of the wires range over a space U satisfying various domain equations (see below). Now consider a proof Π of a sequent $\vdash [\Delta], \Gamma$, where Δ is a list of all the cut-formulas used. Girard associates to such a proof a pair of operators (u, σ), where u is a hermitian of norm at most 1, and σ is a partial symmetry representing the cuts Δ. The dynamics of cut-elimination may now be captured in a solution of a system of *feedback* equations, summarized in an operator $EX(u, \sigma)$ (the *Execution Formula*). We remark that our general categorical framework (based on partial traces) permits a structured approach to solving such feedback equations and deriving properties of the Execution formula. Finally, it can be shown ([12, 17]) that for denotations of proofs $(u = [\![\Pi]\!])$ of appropriate types in System F, $EX([\![\Pi]\!], \sigma)$ is an invariant of cut-elimination.

Categorical foundations of GoI were initiated in the 90's in lectures by M. Hyland and by S. Abramsky. An early categorical framework was given in [4]. Recent work has stressed the role of Joyal-Street-Verity's *traced monoidal categories* [23] (with additional structure). For example, Abramsky's *GoI situations* [1, 3, 15] provide a basic algebraic foundation for GoI for multiplicative, exponential linear logic (MELL). Recently, we used a special kind of GoI situation (with traced unique decomposition categories) to axiomatize the details of Girard's original GoI 1 paper [17].

In our previous papers, we emphasized several important aspects of Girard's seminal work (at least in GoI 1 and 2).

1. The original Girard framework is essentially *untyped*: there is a reflexive object U in the underlying model (with various retractions and/or domain isomorphisms, e.g. $U \otimes U \lhd U$).
2. Cut-elimination is interpreted by *feedback*, naturally represented in traced monoidal categories. The execution formula, defined via trace, provides an invariant for cut-elimination.
3. Girard introduced an *orthogonality* operation $\perp$ on endomaps of U together with the notion of *types* (as sets of endomaps equal to their biorthogonal).
4. There are notions of *data* and *algorithm* encoded into this dynamical setting, with fundamental theorems connecting types, algorithms, and the convergence of the execution formula.

Points (1) and (2) above were already emphasized in the Abramsky program, as well as in the work of Danos and Regnier [1, 3, 8, 17]. Orthogonalities have been studied abstractly by Hyland and Schalk [21]. The points (1)–(4) are critical to our view of GoI in [17, 18] and to the technical developments in this paper.

Alas, Girard's original GoI is not without its own share of syntactical bureaucracy: there are domain isomorphisms (of the reflexive object U) and an associated ∗-algebra of codings and uncodings. On the one hand, this means the original GoI interpretation of proofs is essentially untyped (i.e. categorically, proofs are interpreted in the monoid $Hom(U, U)$, using the above-mentioned algebra) (see [3, 17, 18]). On the other hand, this led Danos and Regnier ([8]) to study this algebra in detail in certain concrete models, leading to their extensive analysis of reduction paths in untyped lambda calculus.

Our aim in this paper is to move away from "uni-object GoI" to a typed version. This permits us to both generalize GoI and axiomatize its essential features. For example, by removing reflexive objects U, we also unlock the possibilities of generalizing Girard-style GoI to more general tensor categories including cases where the tensor is "product-like" in addition to "sum-like". We shall illustrate both of these styles in the examples below.

The contributions of this paper can be summarized as follows:

- We introduce an axiomatization for partially traced symmetric monoidal categories and provide examples based on **Vec$_{fd}$**, finite dimensional vector spaces, and **CMet**, complete metric spaces. Our axiomatization is different from that in [2], although related in spirit.
- We introduce an abstract orthogonality [21], appropriate for GoI, on our models.
- We introduce a multiobject version of Girard's GoI semantics (MGoI) in partially traced models with orthogonality. This includes Girard's notions of *types, datum, algorithm* and the *execution formula*. We give an MGoI interpretation for the multiplicative fragment of linear logic without units (MLL) and show that the execution formula is an invariant of cut-elimination (see Section 5 below). Recall that Girard's original GoI (as presented in [3]) requires a reflexive object $U \neq \{\mathbf{0}\}$, with a retraction $U \oplus U \lhd U$, which is impossible in **Vec$_{fd}$**.
- We prove a soundness and completeness theorem for our MGoI interpretation of MLL in arbitrary partially traced categories with an orthogonality relation. As an application, we can also prove a completeness result for untyped GoI semantics of MLL (see our [17]) in a traced UDC based GoI Situation; the latter result will appear in the journal version of this paper.

It is worth remarking that GoI does not work well with units. They are not part of the original interpretation ([12]), and fail to satisfy the properties demanded by the main theorems. In [18] we show that the "natural" category of types and associated morphisms in certain GoI-situations fails to have tensor and par units act correctly. We suspect the same is true for the MGoI case introduced here.

The rest of the paper is organized as follows. In Section 2 we introduce partially traced symmetric monoidal categories and discuss some examples. In Section 3 we introduce the abstract orthogonality relation in a partially traced symmetric monoidal category and discuss how it relates to the work in [21]. In Section 4 we introduce our new semantics, MGoI, and give an interpretation for MLL. Section 5 discusses the execution formula and the soundness theorem, while in Section 6 we prove a completeness theorem for the MGoI interpretation of MLL in a partially traced category with an orthogonality relation. Finally,

Section 7 contains some thoughts about possible future directions, projects and links to related work in the literature.

Note: The full proofs of the results here will appear in the journal version of this paper, available on our websites.

2   Trace Class

The notion of categorical trace was introduced by Joyal, Street and Verity in an influential paper [23]. The motivation for their work arose in algebraic topology and knot theory, although the authors were aware that such traces also have many applications in Computer Science, where they include such notions as feedback, fixedpoints, iteration theories, etc. For references and history, see [1, 3, 17].

In this paper we go one step further and look at partial traces. The idea of generalizing the abstract trace of [23] to the partial setting is not new. For example, partial traces were already studied in work of Abramsky, Blute, and Panangaden [2], in unpublished lecture notes of Gordon Plotkin [27], work of A. Jeffrey [22] (discussed below) and others. The guiding example in [2] is the relationship between trace class operators on a Hilbert space and Hilbert-Schmidt operators. This allows the authors to establish a close correspondence between trace and nuclear ideals in a tensor $*$-category. Plotkin's work develops a theory of Conway ideals on biproduct categories, and an associated categorical trace theory. Unfortunately none of these extant theories is appropriate for Girard's GoI. So we present an axiomatization for partial traces suitable for our purposes.

Recall, following Joyal, Street, and Verity [23], a (parametric) trace in a symmetric monoidal category $(\mathbb{C}, \otimes, I, s)$ is a family of maps $Tr_{X,Y}^U : \mathbb{C}(X \otimes U, Y \otimes U) \to \mathbb{C}(X, Y)$, satisfying various well-known naturality equations. A *partial* (parametric) trace requires instead that each $Tr_{X,Y}^U$ be a partial map (with domain denoted $\mathbb{T}_{X,Y}^U$) and satisfy various closure conditions.

Definition 1 (Trace Class). Let $(\mathbb{C}, \otimes, I, s)$ be a symmetric monoidal category. A *(parametric) trace class* in $\mathbb{C}$ is a choice of a family of subsets, for each object U of $\mathbb{C}$, of the form

$$\mathbb{T}_{X,Y}^U \subseteq \mathbb{C}(X \otimes U, Y \otimes U) \text{ for all objects } X, Y \text{ of } \mathbb{C}$$

together with a family of functions, called a *(parametric) partial trace*, of the form

$$Tr_{X,Y}^U : \mathbb{T}_{X,Y}^U \to \mathbb{C}(X, Y)$$

subject to the following axioms. Here the parameters are X and Y and a morphism $f \in \mathbb{T}_{X,Y}^U$, by abuse of terminology, is said to be *trace class*.

- **Naturality** in X and Y: For any $f \in \mathbb{T}_{X,Y}^U$ and $g : X' \to X$ and $h : Y \to Y'$,

$$(h \otimes 1_U)f(g \otimes 1_U) \in \mathbb{T}_{X',Y'}^U$$

and $\qquad Tr_{X',Y'}^U((h \otimes 1_U)f(g \otimes 1_U)) = h\,Tr_{X,Y}^U(f)\,g$

– **Dinaturality** in U: For any $f : X \otimes U \to Y \otimes U'$, $g : U' \to U$,

$$(1_Y \otimes g)f \in \mathbb{T}^U_{X,Y} \text{ iff } f(1_X \otimes g) \in \mathbb{T}^{U'}_{X,Y},$$

and $$Tr^U_{X,Y}((1_Y \otimes g)f) = Tr^{U'}_{X,Y}(f(1_X \otimes g)).$$

– **Vanishing I**: $\mathbb{T}^I_{X,Y} = \mathbb{C}(X \otimes I, Y \otimes I)$ and for $f \in \mathbb{T}^I_{X,Y}$

$$Tr^I_{X,Y}(f) = \rho_Y f \rho_X^{-1}.$$

Here $\rho_A : A \times I \to A$ is the right unit isomorphism of the monoidal category.

– **Vanishing II**: For any $g : X \otimes U \otimes V \to Y \otimes U \otimes V$, if $g \in \mathbb{T}^V_{X \otimes U, Y \otimes U}$, then

$$g \in \mathbb{T}^{U \otimes V}_{X,Y} \text{ iff } Tr^V_{X \otimes U, Y \otimes U}(g) \in \mathbb{T}^U_{X,Y},$$

and $$Tr^{U \otimes V}_{X,Y}(g) = Tr^U_{X,Y}(Tr^V_{X \otimes U, Y \otimes U}(g)).$$

– **Superposing**: For any $f \in \mathbb{T}^U_{X,Y}$ and $g : W \to Z$,

$$g \otimes f \in \mathbb{T}^U_{W \otimes X, Z \otimes Y},$$

and $$Tr^U_{W \otimes X, Z \otimes Y}(g \otimes f) = g \otimes Tr^U_{X,Y}(f).$$

– **Yanking**: $s_{UU} \in \mathbb{T}^U_{U,U}$,   and $Tr^U_{U,U}(s_{U,U}) = 1_U$.

A symmetric monoidal category $(\mathbb{C}, \otimes, I, s)$ with such a trace class is called a *partially traced category*, or a *category with a trace class*. If we let X and Y be I (the unit of the tensor), we get a family of operations $Tr^U_{I,I} : \mathbb{T}^U_{I,I} \to \mathbb{C}(I, I)$ defining what we call a *non-parametric trace*.

Remark 1. An early definition of a partial parametric trace is due to Abramsky, Blute and Panangaden in [2]. Our definition is different but related to theirs. First, we have used the Yanking axiom in Joyal, Street and Verity [23], whereas in [2] they use a conditional version of the so-called "generalized yanking"; that is, for $f : X \to U$ and $g : U \to Y$, $Tr^U_{X,Y}(s_{U,Y}(f \otimes g)) = gf$ *whenever* $s_{U,Y}(f \otimes g)$ is of trace class. It was shown in [15] that for traced monoidal categories the two axioms of yanking and generalized yanking are equivalent in the presence of all the other axioms. This equivalence remains valid for the partially traced categories introduced here. In our theory s_{UU} is traceable for all U; on the other hand, many examples in [2] do not have this property. Our Vanishing II axiom differs from and is weaker than the one proposed in [2]: it is a "conditional" equivalence. More importantly, we do not require one of the ideal axioms in [2]. Namely, we do **not** ask that for $f \in \mathbb{T}^U_{X,Y}$ and any $h : U \to U$, $(1_Y \otimes h)f$ and $f(1_X \otimes h)$ be in $\mathbb{T}^U_{X,Y}$. Indeed in the next section we prove that the categories (**Vec**$_{\mathbf{fd}}$, $\oplus$) of finite dimensional vector spaces, and (**CMet**, $\times$) of complete metric spaces are partially traced. It can be shown that in both categories the above ideal axiom and Vanishing II of [2] fail and hence they are not traced in the sense of ABP. In defense of not enforcing this ideal axiom, we observe that it is not required for any of the trace axioms. Any partially traced category in the sense of ABP for which the yanking axiom holds will be partially traced according to our definition. Finally, we observe that the nonparametric version of our partial trace is also different from the one in [2].

Other notions of categorical partial trace have been examined by Alan Jeffrey [22] and also by various category theorists. For example, Jeffrey cuts down the domain of the trace operator to admissible (traceable) objects U which form a full subcategory of the original category. This is not possible for us: our trace classes do not form subcategories. For example, in keeping with functional analysis on infinite dimensional spaces, the ABP theory of traced ideals [2], and with Girard's papers on GoI, we do not wish the identity map to be traced; nor are our trace classes necessarily closed under all possible compositions.

One is obliged to say that there are many different approaches to partial categorical traces and ideals; ours is geared to Girard's GoI. We should also note that our examples will not be partially traced categories according to Jeffrey's definition. It is not possible to capture our traceability conditions on morphisms using his approach, as they cannot be characterized as object properties.

2.1   Examples of Partial Traces

(a) Finite Dimensional Vector Spaces

The category $\mathbf{Vec_{fd}}$ of finite dimensional vector spaces and linear transformations is a symmetric monoidal, indeed an additive, category (see [25]), with monoidal product taken to be $\oplus$, the direct sum (biproduct). Hence, given $f : \oplus_I X_i \to \oplus_J Y_j$ with $|I| = n$ and $|J| = m$, we can write f as an $m \times n$ matrix $f = [f_{ij}]$ of its components, where $f_{ij} : X_j \to Y_i$ (notice the switch in the indices i and j).

We give a trace class structure on the category $(\mathbf{Vec_{fd}}, \oplus, 0)$ as follows. We shall say an $f : X \oplus U \to Y \oplus U$ is trace class iff $(I - f_{22})$ is invertible, where I is the identity matrix, and I and f_{22} have size $dim(U)$. In that case, we write

$$Tr^U_{X,Y}(f) = f_{11} + f_{12}(I - f_{22})^{-1} f_{21} \qquad (1)$$

This definition is motivated by a generalization of the fact that for a matrix A, $(I - A)^{-1} = \sum_i A^i$, whenever the infinite sum converges. Clearly this sum converges when the matrix norm of A is strictly less than 1, or when A is nilpotent, but in both cases the general idea is the desire to have $(I - A)$ invertible. If the infinite sum for $(I - f_{22})^{-1}$ exists, the above formula for $Tr^U_{X,Y}(f)$ becomes the usual "particle-style" trace in [1, 3, 17]. One advantage of formula (1) is that it does not *a priori* assume the convergence of the sum, nor even that $(I - f_{22})^{-1}$ be computable by iterative methods.

Proposition 1. $(\mathbf{Vec_{fd}}, \oplus, 0)$ *is partially traced, with trace class as above.*

The proof of Proposition 1 uses the following standard facts from linear algebra:

Lemma 1. *Let $M = \begin{bmatrix} A & B \\ C & D \end{bmatrix}$ be a partitioned matrix with blocks A $(m \times m)$,
B $(m \times n), C$ $(n \times m)$ and D $(n \times n)$. If D is invertible, then M is invertible iff
$A - BD^{-1}C$ (the Schur Complement of D) is invertible.*

Lemma 2. *Given A $(m \times n)$ and B $(n \times m)$, $(I_m - AB)$ is invertible iff $(I_n - BA)$
is invertible. Moreover $(I_m - AB)^{-1}A = A(I_n - BA)^{-1}$.*

(b) Other Finite Dimensional Examples

Proposition 1 remains valid for the category (**Hilb$_{\text{fd}}$** , $\oplus$) of finite dimensional
Hilbert spaces and bounded linear maps. As discussed in Remark 1, the category
(**Vec$_{\text{fd}}$**, $\oplus$) is *not* partially traced in the sense of ABP; nor is it traced in the
sense of A. Jeffrey, since (for example) the identity is not trace class.

(c) Metric Spaces

Consider the category **CMet** of complete metric spaces with non-expansive
maps. Define $f : (M, d_M) \to (N, d_N)$ to be *non-expansive* iff there is a fixed
$0 \le \alpha \le 1$ such that $d_N(f(x), f(y)) \le \alpha d_M(x, y)$, for all $x, y \in M$. Note that the
tempting collection of complete metric spaces and contractions ($\alpha < 1$) is not a
category: there are no identity morphisms! **CMet** has products, namely given
(M, d_M) and (N, d_N) we define $(M \times N, d_{M \times N})$ with $d_{M \times N}((m, n), (m', n')) =
max\{d_M(m, m'), d_N(n, n')\}$.

We define the trace class structure on **CMet** (where $\otimes = \times$) as follows. We
say that a morphism $f : X \times U \to Y \times U$ is in $\mathbb{T}^U_{X,Y}$ iff for every $x \in X$ the
induced map $\pi_2 \lambda u.f(x, u) : U \to U$ has a unique fixed point; in other words, iff
for every $x \in X$, there is a unique u, and a y, such that $f(x, u) = (y, u)$. Note
that in this case y is necessarily unique. Also, note that contractions have unique
fixed points, by the Banach fixed point theorem.

Suppose $f \in \mathbb{T}^U_{X,Y}$. We define $Tr^U_{X,Y}(f) : X \to Y$ by $Tr^U_{X,Y}(f)(x) = y$, where
$f(x, u) = (y, u)$ for the unique u. Equivalently, $Tr^U_{X,Y}(f)(x) = \pi_1 f(x, u)$ where
u is the unique fixed point of $\pi_2 \lambda t.f(x, t)$.

Proposition 2. (**CMet**, $\times$, $\{*\}$) *is a partially traced category with trace class
as above.*

Lemma 3. *Let A and B be complete metric spaces, $f : A \to B$ and $g : B \to A$.
Then, gf has a unique fixed point if and only if fg does. Moreover, let $a \in A$
be the unique fixed point of $gf : A \to A$ and $b \in B$ be the unique fixed point of
$fg : B \to B$. Then $f(a) = b$ and $g(b) = a$.*

Proposition 2 remains valid for the category (**Sets**, $\times$) of sets and mappings.
The latter then becomes a partially traced category with the same definition for
trace class morphisms as in **CMet**. However, this fails for the category (**Rel**, $\times$),
of sets and relations: consider the sets $A = \{a\}$, $B = \{b, b'\}$, and let $f =
\{(a, b), (a, b')\}$ and $g = \{(b, a), (b', a)\}$.

(d) Total Traces

Of course, all (totally-defined) traces in the usual definition of a traced monoidal category yield a trace class, namely the entire homset is the domain of Tr. In particular, all the examples in our previous work on uni-object GoI [17, 18], for example based on unique decomposition categories, still apply here.

Remark 2. [A Non-Example]
Consider the structure $(\mathbf{CMet}, \times)$. Defining the trace class morphisms as f such that $\pi_2 \lambda u. f(x, u) : U \to U$ is a contraction for every $x \in X$, does not yield a partially traced category: all axioms are true except for dinaturality and Vanishing II.

3   Orthogonality Relations

Girard originally introduced orthogonality relations into linear logic to model formulas (or types) as sets equal to their biorthogonal (e.g. in the phase semantics of the original paper [10] and in GoI 1 [11]). Recently M. Hyland and A. Schalk gave an abstract approach to orthogonality relations in symmetric monoidal closed categories [21]. They also point out that an orthogonality on a traced symmetric monoidal category $\mathbb{C}$ can be obtained by first considering their axioms applied to $Int(\mathbb{C})$, the compact closure of $\mathbb{C}$, and then translating them down to $\mathbb{C}$. Below we give this translation (not explicitly calculated in [21]), using the so-called "GoI construction" $\mathcal{G}(\mathbb{C})$ [1, 15] instead of $Int(\mathbb{C})$. The categories $\mathcal{G}(\mathbb{C})$ and $Int(\mathbb{C})$ are both compact closures of $\mathbb{C}$, and are shown to be isomorphic in [15]. Alas, we do not have the space to give the details of these constructions; however the reader can safely ignore the remarks above and use the definition below independently of its motivation. To understand the detailed constructions behind the definition, the interested reader is referred to the above references.

    As we are dealing with partial traces we need to take extra care in stating the axioms below; namely, an axiom involving a trace should be read with the proviso: "whenever all traces exist".

Definition 2. Let $\mathbb{C}$ be a traced symmetric monoidal category. An *orthogonality relation* on $\mathbb{C}$ is a family of relations $\perp_{UV}$ between maps $u : V \to U$ and $x : U \to V$

$$V \xrightarrow{u} U \perp_{UV} U \xrightarrow{x} V$$

subject to the following axioms:

(i) *Isomorphism:* Let $f : U \otimes V' \to V \otimes U'$ and $\hat{f} : U' \otimes V \to V' \otimes U$ be such that $Tr^{V'}(Tr^{U'}((1 \otimes 1 \otimes s_{U',V'})\alpha^{-1}(f \otimes \hat{f})\alpha))) = s_{U,V}$ and $Tr^V(Tr^U((1 \otimes 1 \otimes s_{U,V})\alpha^{-1}(\hat{f} \otimes f)\alpha))) = s_{U',V'}$. Here $\alpha = (1 \otimes 1 \otimes s)(1 \otimes s \otimes 1)$ with s at appropriate types. Note that this simply means that $f : (U, V) \to (U', V')$ and $\hat{f} : (U', V') \to (U, V)$ are inverses of each other in $\mathcal{G}(\mathbb{C})$. Then for all $u : V \to U$ and $x : U \to V$,

$$u \perp_{UV} x \quad \text{iff} \quad Tr^U_{V',U'}(s_{U,U'}(u \otimes 1_{U'})fs_{V',U}) \perp_{U'V'} Tr^V_{U',V'}((1_{V'} \otimes x)\hat{f})$$

(ii) *Tensor:* For all $u : V \rightarrow U$, $v : V' \rightarrow U'$ and $h : U \otimes U' \rightarrow V \otimes V'$,

$$u \perp_{UV} Tr_{U,V}^{U'}((1_V \otimes v)h) \text{ and } v \perp_{U'V'} Tr_{U',V'}^{U}(s_{U,V}(u \otimes 1_{V'})hs_{U',U})$$

$$\text{implies} \qquad (u \otimes v) \perp_{U \otimes U', V \otimes V'} h$$

(iii) *Implication:* For all $u : V \rightarrow U$, $y : U' \rightarrow V'$ and $f : U \otimes V' \rightarrow V \otimes U'$

$$u \perp_{UV} Tr_{U,V}^{V'}((1_V \otimes y)f) \text{ and } Tr_{V',U'}^{V}(s_{V,U'}f(u \otimes 1_{V'})s_{V',V}) \perp_{U'V'} y$$

$$\text{implies} \qquad f \perp_{V \otimes U', U \otimes V'} (u \otimes y)$$

(iv) *Identity:* For all $u : V \rightarrow U$ and $x : U \rightarrow V$

$$u \perp_{UV} x \text{ implies } 1_I \perp_{II} Tr_{I,I}^{V}(xu)$$

(v) *Symmetry:* For all $u : V \rightarrow U$ and $x : U \rightarrow V$

$$u \perp_{UV} x \text{ iff } x \perp_{VU} u$$

Remark 3. (i) It should be noted that for a (partially) traced symmetric monoidal category, the axioms for Tensor and Implication are equivalent in the presence of the other axioms: by dinaturality of trace we have $Tr_{V',U'}^{V}(s_{V,U'}f(u \otimes 1_{V'})s_{V',V}) = Tr_{V',U'}^{U}(s_{U,U'}(u \otimes 1_{U'})fs_{V',U}))$, then use the Symmetry axiom. Thus we shall drop the Implication axiom.

(ii) Our work on GoI reveals that one needs another axiom which we observe as the converse of the Tensor axiom and relaxation of one of the premises. This is related to abstract computation and the notion of datum in GoI. Hence, we shall replace the Tensor axiom by the following Strong Tensor axiom. Our Strong Tensor axiom is similar to, but *not* the same as the Precise Tensor axiom of [21]. The latter requires an additional property on the biconditional.

Strong Tensor: For all $u : V \rightarrow U$, $v : V' \rightarrow U'$ and $h : U \otimes U' \rightarrow V \otimes V'$,

$$v \perp_{U'V'} Tr_{U',V'}^{U}(s_{U,V}(u \otimes 1_{V'})hs_{U',U}) \quad \text{iff} \quad (u \otimes v) \perp_{U \otimes U', V \otimes V'} h,$$

whenever the trace exists. It can be shown that in the presence of the Strong Tensor, Isomorphism, and Symmetry axioms, $v \perp_{U'V'} Tr_{U',V'}^{U}(s_{U,V}(u \otimes 1_{V'})hs_{U',U})$ implies $u \perp_{UV} Tr_{U,V}^{U'}((1_V \otimes v)h)$, whenever all traces exist.

Definition 3. Let $\mathbb{C}$ be a traced symmetric monoidal category. A *strong orthogonality* relation is defined as in Definition 2 but with the Tensor axiom replaced by the Strong Tensor axiom above, and the Implication axiom dropped.

In the context of GoI, we will be working with strong orthogonality relations on endomorphism sets of objects in the underlying categories. Biorthogonally closed (i.e. $X = X^{\perp\perp}$) subsets of certain endomorphism sets are important as they define *types* (GoI interpretation of formulae.) We have observed that all the orthogonality relations that we work with in this paper can be characterized using trace classes. This suggests the following, which seems to cover many known examples.

Example 1 (Orthogonality as trace class) Let $(\mathbb{C}, \otimes, I, Tr)$ be a partially traced category where $\otimes$ is the monoidal product with unit I, and Tr is the partial trace operator as in Section 2. Let A and B be objects of $\mathbb{C}$. For $f : A \to B$ and $g : B \to A$, we can define an orthogonality relation by declaring $f \perp_{BA} g$ iff $gf \in \mathbb{T}^A_{I,I}$. It turns out[1] that this is a variation of the notion of *Focussed orthogonality* of Hyland and Schalk [21].

Hence, from our previous discussion on traces, we obtain the following examples:

- **Vec$_{fd}$** . For $A \in \mathbf{Vec_{fd}}$, $f, g \in End(A)$, define $f \perp g$ iff $I - gf$ is invertible.
- **CMet**. Let $M \in \mathbf{CMet}$. For $f, g \in End(M)$, define $f \perp g$ iff gf has a unique fixed point.

4   Multi-object GoI Interpretation

In this section we introduce the multiobject Geometry of Interaction semantics for MLL in a partially traced symmetric monoidal category $(\mathbb{C}, \otimes, I, Tr, \perp)$ equipped with an orthogonality relation $\perp$ as in the previous section. Here $\otimes$ is the monoidal product with unit I and Tr is the partial trace operator as in Section 2. We do not require that the category $\mathbb{C}$ have a reflexive object, so uni-object GoI semantics (as in [12, 17]) may not be possible to carry out in $\mathbb{C}$.

Interpreting formulae:

Let A be an object of $\mathbb{C}$ and let $f, g \in End(A)$. We say that f is *orthogonal to* g, denoted $f \perp g$, if $(f, g) \in \perp$. Also given $X \subseteq End(A)$ we define

$$X^\perp = \{f \in End(A) \mid \forall g \in X, f \perp g\}.$$

We now define an operator on the objects of $\mathbb{C}$ as follows: Given an object A, $\mathcal{T}(A) = \{X \subseteq End(A) \mid X^{\perp\perp} = X\}$. We shall also need the notion of a denotational interpretation of formulas. We define an interpretation map $[\![-]\!]$ on the formulas of MLL as follows. Given the value of $[\![-]\!]$ on the atomic propositions as objects of $\mathbb{C}$, we extend it to all formulas by:

- $[\![A^\perp]\!] = [\![A]\!]$
- $[\![A \,\mathbin{⅋}\, B]\!] = [\![A \otimes B]\!] = [\![A]\!] \otimes [\![B]\!]$.

We then define the MGoI-interpretation for formulas as follows. We use the notation $\theta(A)$ for this interpretation.

- $\theta(\alpha) \in \mathcal{T}([\![\alpha]\!])$, where α is an atomic formula.
- $\theta(\alpha^\perp) = \theta(\alpha)^\perp$
- $\theta(A \otimes B) = \{a \otimes b \mid a \in \theta(A), b \in \theta(B)\}^{\perp\perp}$
- $\theta(A \,\mathbin{⅋}\, B) = \{a \otimes b \mid a \in \theta(A)^\perp, b \in \theta(B)^\perp\}^\perp$

---

[1] We thank the anonymous referee for pointing out this connection.

Two easy consequences of the definition are: (i) for any formula A, $(\theta A)^\perp = \theta A^\perp$, and (ii) $\theta(A) \subseteq End(\llbracket A \rrbracket)$.

Interpretation of Proofs:

We define the MGoI interpretation for proofs of MLL without units, similarly to [17]. Every MLL sequent will be of the form $\vdash [\Delta], \Gamma$ where Γ is a sequence of formulas and Δ is a sequence of cut formulas that have already been made in the proof of $\vdash \Gamma$ (see [12, 17]). This device is used to keep track of the cuts in a proof of $\vdash \Gamma$. A proof Π of $\vdash [\Delta], \Gamma$ is represented by a morphism $\llbracket \Pi \rrbracket \in End(\otimes \llbracket \Gamma \rrbracket \otimes \llbracket \Delta \rrbracket)$. With $\Gamma = A_1, \cdots, A_n$, $\otimes \llbracket \Gamma \rrbracket$ stands for $\llbracket A_1 \rrbracket \otimes \cdots \otimes \llbracket A_n \rrbracket$, similarly for Δ. We drop the double brackets wherever there is no danger of confusion. We also define $\sigma = s \otimes \cdots \otimes s$ (m-copies) where s is the symmetry map at different types (omitted for convenience), and $|\Delta| = 2m$. The morphism σ represents the cuts in the proof of $\vdash \Gamma$, i.e. it models Δ. In the case where Δ is empty (that is for a cut-free proof), we define $\sigma : I \to I$ to be 1_I where I is the unit of the monoidal product in $\mathbb{C}$.

Let Π be a proof of $\vdash [\Delta], \Gamma$. We define the MGoI interpretation of Π, denoted by $\llbracket \Pi \rrbracket$, by induction on the length of the proof as follows.

1. Π is an *axiom* $\vdash A, A^\perp$, $\llbracket \Pi \rrbracket := s_{V,V}$ where $\llbracket A \rrbracket = \llbracket A^\perp \rrbracket = V$.

2. Π is obtained using the *cut* rule on Π' and Π'' that is

$$
\begin{array}{cc}
\Pi' & \Pi'' \\
\vdots & \vdots
\end{array}
$$

$$
\cfrac{\vdash [\Delta'], \Gamma', A \quad \vdash [\Delta''], A^\perp, \Gamma''}{\vdash [\Delta', \Delta'', A, A^\perp], \Gamma', \Gamma''} \; cut
$$

Define $\llbracket \Pi \rrbracket = \tau^{-1}(\llbracket \Pi' \rrbracket \otimes \llbracket \Pi'' \rrbracket)\tau$, where τ is the permutation
$$\Gamma' \otimes \Gamma'' \otimes \Delta' \otimes \Delta'' \otimes A \otimes A^\perp \xrightarrow{\tau} \Gamma' \otimes A \otimes \Delta' \otimes A^\perp \otimes \Gamma'' \otimes \Delta'',$$

3. Π is obtained using the *exchange* rule on the formulas A_i and A_{i+1} in Γ'. That is Π is of the form

$$
\Pi'
$$
$$
\vdots
$$

$$
\cfrac{\vdash [\Delta], \Gamma'}{\vdash [\Delta], \Gamma} \; exchange
$$

where $\Gamma' = \Gamma'_1, A_i, A_{i+1}, \Gamma'_2$ and $\Gamma = \Gamma'_1, A_{i+1}, A_i, \Gamma'_2$. Then, $\llbracket \Pi \rrbracket = \tau^{-1}\llbracket \Pi' \rrbracket \tau$, where $\tau = 1_{\Gamma'_1} \otimes s \otimes 1_{\Gamma'_2 \otimes \Delta}$.

4. Π is obtained using an application of the *par* rule, that is Π is of the form:

$$
\Pi'
$$
$$
\vdots
$$

$$
\cfrac{\vdash [\Delta], \Gamma', A, B}{\vdash [\Delta], \Gamma', A \,\bindnasrepma\, B} \; \bindnasrepma \qquad . \text{ Then } \llbracket \Pi \rrbracket = \llbracket \Pi' \rrbracket
$$

5. Π is obtained using an application of the *times* rule, that is Π is of the form:

$$
\begin{array}{cc}
\Pi' & \Pi'' \\
\vdots & \vdots
\end{array}
$$

$$\frac{\vdash [\Delta'], \Gamma', A \quad \vdash [\Delta''], \Gamma'', B}{\vdash [\Delta', \Delta''], \Gamma', \Gamma'', A \otimes B} \, \otimes$$

Then $[\![\Pi]\!] = \tau^{-1}([\![\Pi']\!] \otimes [\![\Pi'']\!])\tau$, where τ is the permutation $\Gamma' \otimes \Gamma'' \otimes A \otimes B \otimes \Delta' \otimes \Delta'' \xrightarrow{\tau} \Gamma' \otimes A \otimes \Delta' \otimes \Gamma'' \otimes B \otimes \Delta''$. This corresponds exactly to the definition of tensor product in Abramsky's $\mathcal{G}(\mathbb{C})$ (see [1, 15].)

Example 1. (a) Let Π be the following proof:

$$\frac{\vdash A, A^\perp \quad \vdash A, A^\perp}{\vdash [A^\perp, A], A, A^\perp} \, cut$$

Then the MGoI semantics of this proof is given by

$$[\![\Pi]\!] = \tau^{-1}(s \otimes s)\tau = s_{V \otimes V, V \otimes V}$$

where $\tau = (1 \otimes 1 \otimes s)(1 \otimes s \otimes 1)$ and $[\![A]\!] = [\![A^\perp]\!] = V$.
(b) Now consider the following proof

$$\frac{\dfrac{\dfrac{\dfrac{\dfrac{\vdash B, B^\perp \quad \vdash C, C^\perp}{\vdash B, C, B^\perp \otimes C^\perp}}{\vdash B, B^\perp \otimes C^\perp, C}}{\vdash B^\perp \otimes C^\perp, B, C}}{\vdash B^\perp \otimes C^\perp, B \,\mathbin{\bindnasrepma}\, C}}{}$$

Its denotation is $s_{V \otimes W, V \otimes W}$, where $[\![B]\!] = [\![B^\perp]\!] = V$ and $[\![C]\!] = [\![C^\perp]\!] = W$.

Proposition 3. *Let Π be an MLL proof of $\vdash [\Delta], \Gamma$ where $|\Delta| = 2m$ and $|\Gamma| = n$ (counting occurrences of propositional variables). Then $[\![\Pi]\!]$ is a fixed-point free involutive permutation on $n + 2m$ objects of $\mathbb{C}$. That is $[\![\Pi]\!] : V_1 \otimes \cdots \otimes V_{n+2m} \to V_1 \otimes \cdots \otimes V_{n+2m}$ induces a permutation π on $\{1, 2 \cdots, n + 2m\}$ and*

- $\pi^2 = 1$
- *For all $i \in \{1, 2, \cdots, n + 2m\}$, $\pi(i) \neq i$.*
- *For all $i \in \{1, 2, \cdots, n + 2m\}$, $V_i = V_{\pi(i)}$.*

4.1 Dynamics

Dynamics is at the heart of the GoI interpretation as compared to denotational semantics and it is hidden in the cut-elimination process. The mathematical model of cut-elimination is given by the so called *execution formula* defined as follows:

$$EX([\![\Pi]\!], \sigma) = Tr^{\otimes \Delta}_{\otimes \Gamma, \otimes \Gamma}((1 \otimes \sigma)[\![\Pi]\!]) \tag{2}$$

where Π is a proof of the sequent $\vdash [\Delta], \Gamma$, and $\sigma = s \otimes \cdots \otimes s$ (m times) models Δ. Note that $EX(\llbracket \Pi \rrbracket, \sigma)$ is a morphism from $\otimes \Gamma \to \otimes \Gamma$, when it exists. We shall prove below (see Theorem 2) that the execution formula always exists for any MLL proof Π.

Example 2. Consider the proof Π in Example 1 above. Recall also that $\sigma = s$ in this case ($m = 1$). Then $\quad EX(\llbracket \Pi \rrbracket, \sigma) = Tr((1 \otimes s_{V,V}) s_{V \otimes V, V \otimes V}) = s_{V,V}$.

Note that in this case we have obtained the MGoI interpretation of the cut-free proof of $\vdash A, A^{\perp}$, obtained by applying Gentzen's Hauptsatz to the proof Π.

5   Soundness of the Interpretation

In this section we state one of the main results of this paper: the soundness of the MGoI interpretation. We show that if a proof Π is reduced (via cut-elimination) to another proof Π', then $EX(\llbracket \Pi \rrbracket, \sigma) = EX(\llbracket \Pi' \rrbracket, \tau)$; that is, $EX(\llbracket \Pi \rrbracket, \sigma)$ is an invariant of reduction. In particular, if Π' is cut-free (i.e. a normal form) we have $EX(\llbracket \Pi \rrbracket, \sigma) = \llbracket \Pi' \rrbracket$. Intuitively this says that if one thinks of cut-elimination as computation then $\llbracket \Pi \rrbracket$ can be thought of as an algorithm. The computation takes place as follows: if $EX(\llbracket \Pi \rrbracket, \sigma)$ exists then it yields a datum (cf. cut-free proof). This intuition will be made precise below (Theorems 2 & 3).

The next fundamental lemma follows directly from our trace axioms:

Lemma 4 (Associativity of cut). *Let Π be a proof of $\vdash [\Gamma, \Delta], \Lambda$ and σ and τ be the morphisms representing the cut-formulas in Γ and Δ respectively. Then*

$$EX(\llbracket \Pi \rrbracket, \sigma \otimes \tau) = EX(EX(\llbracket \Pi \rrbracket, \tau), \sigma) = EX(EX((1 \otimes s)\llbracket \Pi \rrbracket (1 \otimes s), \sigma), \tau),$$

whenever all traces exist. (This is essentially the Church-Rosser Property).

Definition 4. Let $\Gamma = A_1, \cdots, A_n$ and $V_i = \llbracket A_i \rrbracket$.

- A *datum of type $\theta\Gamma$* is a morphism $M : \otimes_i V_i \to \otimes_i V_i$ such that for any $\beta_i \in \theta(A_i^{\perp})$, $\otimes_i \beta_i \perp M$ and $M \cdot \beta_1 := Tr^{V_1}(s_{\otimes_i V_i, V_1}^{-1}(\beta_1 \otimes 1_{V_2} \otimes \cdots \otimes 1_{V_n}) M s_{\otimes_i V_i, V_1})$ exists. (In Girard's notation [12], $M \cdot \beta_1$ corresponds to $ex(CUT(\beta_1, M))$.)
- An *algorithm of type $\theta\Gamma$* is a morphism $M : \otimes_i V_i \otimes \llbracket \Delta \rrbracket \to \otimes_i V_i \otimes \llbracket \Delta \rrbracket$ for some $\Delta = B_1, B_2, \cdots, B_{2m}$ with m a nonnegative integer and $B_{i+1} = B_i^{\perp}$ for $i = 1 \cdots, 2m - 1$, such that if $\sigma : \otimes_{j=1}^{2m} \llbracket B_j \rrbracket \to \otimes_{j=1}^{2m} \llbracket B_j \rrbracket$ is $\otimes_{j=1}^{2m-1} s_{\llbracket B_j \rrbracket, \llbracket B_{j+1} \rrbracket}$, $EX(M, \sigma)$ exists and is a datum of type $\theta\Gamma$. (Here σ is defined to be 1_I for $m = 0$.)

Lemma 5. *Let $\Gamma = A_2, \cdots, A_n$, $V_i = \llbracket A_i \rrbracket$, and $M : \otimes_i V_i \to \otimes_i V_i$, for $i = 1, \cdots, n$. Then, M is a datum of type $\theta(A_1, \Gamma)$ iff for every $a_1 \in \theta(A_1^{\perp})$, $M \cdot a_1$ (defined as above) exists and is in $\theta(\Gamma)$.*

Theorem 2 (Proofs as algorithms). *Let Π be an MLL proof of a sequent $\vdash [\Delta], \Gamma$. Then $\llbracket \Pi \rrbracket$ is an algorithm of type $\theta\Gamma$.*

Corollary 1 (Existence of Dynamics). *Let Π be an MLL proof of a sequent $\vdash [\Delta], \Gamma$. Then $Ex(\llbracket \Pi \rrbracket, \sigma)$ exists.*

Theorem 3 (EX is an invariant). *Let Π be an MLL proof of a sequent $\vdash [\Delta], \Gamma$. Then,*

- *If Π reduces to Π' by any sequence of cut-eliminations, then $EX(\llbracket \Pi \rrbracket, \sigma) = EX(\llbracket \Pi' \rrbracket, \tau)$. So $EX(\llbracket \Pi \rrbracket, \sigma)$ is an invariant of reduction.*
- *In particular, if Π' is any cut-free proof obtained from Π by cut-elimination, then $EX(\llbracket \Pi \rrbracket, \sigma) = \llbracket \Pi' \rrbracket$.*

6   Completeness

In this section we give a completeness theorem for MLL in a partially traced category equipped with an orthogonality relation, under MGoI semantics. Recall from Proposition 3 that the denotation of a proof $\llbracket \Pi \rrbracket$ induces a fixed-point free involutive permutation. We now seek a converse.

Theorem 4 (Completeness). *Let M be a fixed-point free involutive permutation from $V_1 \otimes \cdots \otimes V_n \to V_1 \otimes \cdots \otimes V_n$ (induced by a permutation μ on $\{1, 2, \cdots, n\}$) where $n > 0$ is an even integer, $V_i = \llbracket A_i \rrbracket$, and $V_i = V_{\mu(i)}$ for all $i = 1, \cdots, n$. Then there is a provable MLL formula φ built from the A_i, with a proof Π such that $\llbracket \Pi \rrbracket = M$.*

Motivated by this result, we can also prove a completeness theorem for MLL in any traced Unique Decomposition Category with a reflexive object, under (uni-object) GoI semantics [17]. This will appear in the full journal article.

7   Conclusion and Future Work

In this work we introduce a new semantics called multiobject Geometry of Interaction (MGoI). This semantics, while inspired by GoI, differs from it in significant points. Namely, we deal with many objects in the underlying category, we make use of a denotational semantics to define the interpretation of logical formulas and we develop the execution formula based on a new theory of partial traces and trace classes. Moreover, there is an orthogonality relation linked to the notion of trace class, which allows us to develop Girard's theory of types, data and algorithms in our setting. This permits a structured approach to Girard's concept of solving feedback equations [14], and an axiomatization of the critical features needed for showing that the execution formula is an invariant of cut-elimination. Computationally, GoI provides a kind of algorithm for normalization based on the execution formula. In future work, we hope to explore the algorithmic and convergence properties of the execution formula in various models, independently of the syntax.

An advantage of the approach taken here is that we are able to carry out our MGoI interpretation in categories of finite dimensional vector spaces and the other examples mentioned above. This is not possible for the earlier theory of uni-object GoI (for example, $\mathbf{Vec_{fd}}$ does not have non-trivial reflexive objects). Our examples illustrate that both "sum-style" and "product-style" GoI (as discussed in [3]) are compatible with our multiobject approach.

An obvious direction for future research is to extend our MGoI interpretation to the exponentials and additives of linear logic: this is under active development. As well, the thorny problem of how to handle the units (as mentioned in the Introduction) is being explored. New directions in GoI semantics now arise with the introduction of partial traces and abstract orthogonalities. For example, we are pursuing the correspondence of trace class/nuclear morphisms as achieved in [2] for their examples. We are also currently exploring MGoI interpretations in Banach spaces and related categories, to find appropriate trace class structures.

It is natural to seek examples of traces that are induced by more general notions of orthogonalities, especially those arising in functional analysis. We hope this may lead to new classes of MGoI models, perhaps connected to current work in operator algebras and general solutions to feedback equations, as in [14].

References

1. Abramsky, S. (1996), Retracing Some Paths in Process Algebra. In *CONCUR 96, Springer LNCS* **1119**, 1-17.
2. Abramsky, S., Blute, R. and Panangaden, P. (1999), Nuclear and trace ideals in tensored *-categories, *J. Pure and Applied Algebra* vol. 143, 3–47.
3. Abramsky, S., Haghverdi, E. and Scott, P.J. (2002), Geometry of Interaction and Linear Combinatory Algebras. *MSCS*, vol. 12(5), 2002, 625-665, CUP.
4. Abramsky, S. and Jagadeesan, R. (1994), New Foundations for the Geometry of Interaction. *Information and Computation* **111** (1), 53-119.
5. Baillot, P. (1995), Abramsky-Jagadeesan-Malacaria strategies and the geometry of interaction, mémoire de DEA, Universite Paris 7, 1995.
6. Baillot, P. and Pedicini, M. (2000), Elementary complexity and geometry of interaction, *Fundamenta Informaticae,* vol. 45, no. 1-2, 2001
7. Danos, V. (1990), *La logique linéaire appliquée à l'étude de divers processus de normalisation et principalement du λ-calcul.* PhD thesis, Université Paris VII.
8. Danos, V. and Regnier, L. (1995), Proof-nets and the Hilbert Space. In: *Advances in Linear Logic*, London Math. Soc. Notes, **222**, CUP, 307–328.
9. Gonthier, G., Abadi, M. and Lévy, J.-J. (1992), The geometry of optimal lambda reduction. In Proceedings of Logic in Computer Science, vol. 9 pp. 15-26.
10. Girard, J.-Y. (1987), Linear Logic. *Theoretical Computer Science* **50** (1), pp. 1-102.
11. Girard, J.-Y. (1988), Geometry of Interaction II: Deadlock-free Algorithms. In *Proc. of COLOG'88, LNCS* **417**, Springer, 76–93.
12. Girard, J.-Y. (1989a) Geometry of Interaction I: Interpretation of System F. In *Proc. Logic Colloquium 88,* North Holland, 221–260.
13. Girard, J.-Y. (1995), Geometry of Interaction III: Accommodating the Additives. In: *Advances in Linear Logic*, LNS **222**,CUP, 329–389,
14. Girard, J.-Y. (2004). Cours de Logique, Rome 2004. Forthcoming.
15. Haghverdi, E. *A Categorical Approach to Linear Logic, Geometry of Proofs and Full Completeness*, PhD Thesis, University of Ottawa, Canada 2000.

16. Haghverdi, E. Unique Decomposition Categories, Geometry of Interaction and combinatory logic, *Math. Struct. in Comp. Science*, vol. **10**, 2000, 205-231.
17. Haghverdi, E. and P.J.Scott. A categorical model for the Geometry of Interaction, to appear in *Theoretical Computer Science* (cf. ICALP 2004, Springer LNCS 3142).
18. Haghverdi, E. and P.J.Scott. From Geometry of Interaction to Denotational Semantics. Proceedings of CTCS2004. In ENTCS, vol. 122, pp. 67-87. Elsevier.
19. Hasegawa, M. (1997), Recursion from Cyclic Sharing : Traced Monoidal Categories and Models of Cyclic Lambda Calculus, *Springer LNCS* **1210**, 196-213.
20. Hines, P. (2003), A categorical framework for finite state machines, *Math. Struct. in Comp. Science*, vol. **13**, 451-480.
21. Hyland, M and Schalk, A. (2003), Glueing and Orthogonality for Models of Linear Logic. *Theoretical Computer Science* vol. 294, pp. 183–231.
22. Jeffrey, A.S.A. (1998), Premonoidal categories and a graphical view of programs. (see the webpage: http://klee.cs.depaul.edu/premon/). Also: *Electr. Notes Theor. Comput. Sci.* 10: (1997)
23. Joyal, A., Street, R. and Verity, D. (1996), Traced Monoidal Categories. *Math. Proc. Camb. Phil. Soc.* **119**, 447-468.
24. Laurent, O., (2001), A Token Machine for Full Geometry of Interaction. In TLCA '01, SLNCS 2044, pp. 283-297.
25. Mac Lane, S. (1998), *Categories for the Working Mathematician*, 2nd Ed. Springer.
26. Malacaria, P. and Regnier. L. (1991), Some Results on the Interpretation of λ-calculus in Operator Algebras. *Proc. LICS* pp. 63-72, IEEE Press.
27. Plotkin, G. Trace Ideals, MFPS 2003 invited lecture, Montreal.
28. Regnier, L. (1992), *Lambda-calcul et Réseaux*, PhD Thesis, Université Paris VII.

From Pebble Games to Tractability:
An Ambidextrous Consistency Algorithm
for Quantified Constraint Satisfaction

Hubie Chen and Víctor Dalmau

Departament de Tecnologia
Universitat Pompeu Fabra
Barcelona, Spain
{hubie.chen,victor.dalmau}@upf.edu

Abstract. The constraint satisfaction problem (CSP) and quantified constraint satisfaction problem (QCSP) are common frameworks for the modelling of computational problems. Although they are intractable in general, a rich line of research has identified restricted cases of these problems that are tractable in polynomial time. Remarkably, many tractable cases of the CSP that have been identified are solvable by a single algorithm, which we call here the consistency algorithm. In this paper, we give a natural extension of the consistency algorithm to the QCSP setting, by making use of connections between the consistency algorithm and certain two-person pebble games. Surprisingly, we demonstrate a variety of tractability results using the algorithm, revealing unified structure among apparently different cases of the QCSP.

1 Introduction

The *constraint satisfaction problem (CSP)* is widely acknowledged as a convenient framework for modelling search problems. An instance of the CSP is a closed primitive positive formula over a relational signature, that is, a formula of the form

$$\exists v_1 \ldots \exists v_n (R(v_{i_1}, \ldots, v_{i_k}) \wedge \ldots)$$

along with a relational structure over the same signature. The question is to decide whether or not the relational structure models the formula. In the constraint satisfaction literature, the atomic formulas $R(v_{i_1}, \ldots, v_{i_k})$ in a CSP instance are called constraints, and the CSP is usually phrased as the problem of deciding whether or not there is a variable assignment satisfying all constraints in a given set. The CSP can also be defined as the problem of deciding, given an ordered pair $(\mathbf{A}, \mathbf{B})$ of relational structures, whether or not there is a relational homomorphism from $\mathbf{A}$ to $\mathbf{B}$. As the many equivalent formulations of the CSP suggest, CSPs arise naturally in a wide variety of domains, including logic, algebra, database theory, artificial intelligence, and graph coloring.

The initially given definition of the CSP leads naturally to the generalization of the CSP called the *quantified constraint satisfaction problem (QCSP)*, where

L. Ong (Ed.): CSL 2005, LNCS 3634, pp. 232–247, 2005.

both universal and existential quantification are allowed. The greater generality of the QCSP permits the modelling of a variety of computational problems that cannot be expressed using the CSP, for instance, problems from the areas of verification, non-monotonic reasoning, planning, and game playing. The higher expressiveness of the QCSP, however, comes at the price of higher complexity: the QCSP is in general complete for the complexity class PSPACE, in contrast to the CSP, which is in general complete for the complexity class NP.

The general intractability of the CSP and the QCSP has motivated a large and rich body of research aimed at identifying and understanding restricted cases of these problems that are tractable, by which we mean decidable in polynomial time. Remarkably, many tractable cases of the CSP that have been identified are solvable by parameterizations of a single algorithm which we call the *consistency algorithm*. Intuitively speaking, the consistency algorithm performs inference by continually considering bounded-size subsets of the entire set of variables, of a CSP instance.

In this paper, we give an algorithm for the QCSP that naturally extends the consistency algorithm for the CSP. Surprisingly, we are able to obtain a variety of QCSP tractability results using our algorithm, revealing unified structure among apparently different cases of the QCSP. Because of its qualitative similarity to CSP consistency as well as its ability to solve a number of QCSP generalizations of CSP instances solvable by CSP consistency methods, we believe that our algorithm can reasonably be viewed as a consistency algorithm for the QCSP, and refer to it as such.

In the rest of this section, we briefly review related work on the CSP, give a general description of our algorithm, and then describe the tractable cases of the QCSP that we demonstrate to be solvable by our algorithm.

Restrictions on the CSP. To discuss the relevant restricted cases of the CSP that have been previously studied, we view the CSP as the "relational homomorphism" problem of deciding, given an ordered pair $(\mathbf{A}, \mathbf{B})$ of relational structures, whether or not there is a homomorphism from $\mathbf{A}$ to $\mathbf{B}$. By and large, the restrictions that have been identified and studied can be placed into one of two categories, which have become known as *left-hand side* restrictions and *right-hand side* restrictions. Left-hand side restrictions, also known as *structural restrictions*, arise by considering a prespecified class $\mathcal{A}$ of relational structures from which the left-hand side structure $\mathbf{A}$ must come, whereas right-hand side restrictions, arise by considering a prespecified class $\mathcal{B}$ of relational structures from which the right-hand side structure $\mathbf{B}$ must come. When $\mathcal{B} = \{\mathbf{B}\}$ is of size one, $\mathbf{B}$ is often called the *constraint language*.

In studies of the left-hand side, the restriction of *bounded treewidth* has played a major role. Treewidth is, intuitively, a graph-theoretic measure of how "tree-like" a graph is, and the treewidth of a relational structure $\mathbf{A}$ is simply the treewidth of the *Gaifman graph* of $\mathbf{A}$, or the graph having an edge between any two elements occurring together in a tuple. The tractability of collections of relational structures $\mathcal{A}$ having *bounded treewidth* was shown by Dechter and Pearl [11] and Freuder [13]. An alternative proof was later given by Kolaitis and Vardi

[19]. Dalmau et al. [9] building on ideas of Kolaitis and Vardi [19, 20] showed that consistency is an algorithm for the CSP under bounded treewidth. They also gave a wider condition that naturally expanded the condition of bounded treewidth, and showed that the wider condition is tractable, again, via the consistency algorithm. The optimality of this latter result of Dalmau et al. was demonstrated by Grohe [16], who proved that in the case that $\mathcal{A}$ has bounded arity, if $\mathcal{A}$ gives rise to a tractable case of the CSP, then it must fall into the natural expansion identified by Dalmau et al. [9]. There has also been work on the case of unbounded arity [15].

Study of right-hand side restrictions has its origins in the paper of Schaefer [21], who classified all constraint languages $\mathbf{B}$, with a two-element universe, giving rise to a tractable case of the CSP. Much research has been directed towards classifying all constraint languages with a finite universe giving rise to a tractable CSP, and recent years have seen some major classification results [2, 3] as well as other tractability results; in many of the tractability results, such as [1, 4, 10, 17], consistency is used to demonstrate tractability.

Quantified Pebble Games and Our New Consistency Algorithm. It has been demonstrated by Kolaitis and Vardi [20] that there are intimate connections between the consistency algorithm and certain combinatorial pebble games. It is from the vantage point of these pebble games that we develop our quantified analog of the consistency algorithm, and this paper overall takes strong methodological and conceptual inspiration from the paper [20] as well as from [9].

To explain some details, the consistency algorithm is a general method that adds more constraints to a CSP instance, and in some cases can detect an inconsistency (that is, that the CSP instance is not satisfiable). We have said that the consistency algorithm solves certain restricted cases of the CSP; by this, we mean that an instance is unsatisfiable if *and only if* an inconsistency is detected. In [20], consistency was linked to *existential pebble games* which are played by two players, the *Spoiler* and the *Duplicator*, on a pair of relational structures $(\mathbf{A}, \mathbf{B})$; these games were introduced by Kolaitis and Vardi [18] for analyzing the expressive power of the logic *Datalog*. The link established is that the Duplicator wins the existential pebble game if and only if no inconsistency is detected by the consistency algorithm: that is, deciding if an instance is *consistent* is equivalent to deciding if the Duplicator has a *winning strategy* for the existential pebble game.

We define a version of this pebble game which we call the *quantified pebble game* that is suited for the QCSP setting. Recall that an instance of the CSP can be viewed as a pair of relational structures $(\mathbf{A}, \mathbf{B})$; it turns out that an instance of the QCSP can be viewed as a pair of relational structures $(\mathbf{A}, \mathbf{B})$ along with a quantifier prefix $p = Q_1 a_1 \ldots Q_n a_n$ where the Q_i are quantifiers and $\{a_1, \ldots, a_n\}$ are the elements of $\mathbf{A}$. In the existential pebble game, the Spoiler places pebbles on elements of $\mathbf{A}$ and the Duplicator must respond by placing pebbles on elements of $\mathbf{B}$. A primary difference between the existential pebble game and our new pebble game is that whereas the Spoiler may, at any time, place a pebble on any element of **A** in the existential pebble game, the Spoiler must respect the quantifier prefix in placing down new pebbles in the quantified pebble game. In a sense that we will make precise, any new pebbles placed by the Spoiler must come "after" all existing pebbles in the quantifier prefix p.

After introducing our quantified pebble game and identifying some of its basic properties, we show how a consistency algorithm for the QCSP can be derived from our pebble game, analogously to the way in which the CSP consistency algorithm can be derived from the existential pebble game.

Fascinatingly and surprisingly, our QCSP consistency algorithm is ambidextrous in that it can be employed to obtain *both* left-hand and right-hand side tractability results, and hence reveals unified structure among apparently different cases of the QCSP. We now turn to describe these tractability results.

The Left-Hand Side: Bounded Treewidth. As we have discussed, bounded treewidth not only gives rise to a tractable CSP, but – under the assumption of bounded arity – lies behind every left-hand side restricted CSP. However, it has recently been revealed that bounded treewidth, at least applied straightforwardly, behaves differently in the QCSP. Chen [7] recently showed that bounded treewidth also guarantees tractability in the QCSP, but under the assumption that both the universe size of **B** and the number of quantifier alternations are constant. The question of whether or not any part of this assumption could be lifted while preserving tractability was recently addressed by Gottlob et al. [14], who proved that if there is no bound on the universe size of **B**, bounded treewidth instances are generally intractable. This seems to suggest a disconnect between bounded treewidth in the CSP and QCSP setting – since the tractability of bounded treewidth in the CSP setting is independent of any properties of the right-hand side structure **B** – and prompts the question of whether or not there is any "pure" left-hand side restriction, generalizing bounded treewidth in the CSP, that guarantees tractability in the QCSP.

In this paper, we answer this question in the affirmative by introducing a natural generalization of bounded treewidth (in the CSP) that guarantees QCSP tractability. This generalization is easy to describe from a high-level perspective: the treewidth of a relational structure **A** can be described as the minimization of a certain quantity over *all* possible orderings of the elements of **A**. Our new notion of treewidth can be described in the same way, but where the minimization is over all orderings of the **A**-elements that respect the quantifier prefix (in a natural way that we make precise in the paper).

We show that instances of the QCSP having bounded treewidth in our new sense are tractable via consistency, and (in the last section of the paper) also expand this tractability result in a way that is analogous to and generalizes the expansion carried out in the CSP setting [9]. It is worth emphasizing that, as in [9], we show that our new type of bounded treewidth (and its expansion) is tractable by consistency, a method that does *not* require the computation of any form of tree decomposition.

In [14], another tractable left-hand side restricted QCSP was identified; see also [12] for closely related work.

The Right-Hand Side. On the right-hand side, we give two tractability results via consistency. The first is the tractability of constraint languages **B** having a *near-unanimity polymorphism*; this class has been studied and shown to be tractable in the CSP [17], and is also known to be tractable in the QCSP [6]. The second concerns constraint languages **B** having a *set function polymorphism*. While such constraint languages are all tractable in the CSP setting [10], they give rise to two modes of behavior in the QCSP [8]. We show that any such constraint languages that are QCSP tractable are tractable by consistency. Both of these classes are fundamental to consistency in the CSP setting, as they both give exact characterizations of particular parameterizations of consistency that have been of interest; we refer the reader to the cited papers for more information.

2 Quantified Constraint Satisfaction

In this section, we define the quantified constraint satisfaction problem and associated notions to be used throughout the paper.

A *relational signature* is a finite set of relation symbols, each of which has an associated arity. A *relational structure* **A** (over signature σ) consists of a universe A and a relation $R^{\mathbf{A}}$ over A for each relation symbol R (of σ), such that the arity of $R^{\mathbf{A}}$ matches the arity associated to R. We refer to the elements of the universe of a relational structure **A** as **A**-elements. Throughout this paper, we assume that all relational structures under discussion are finite.

A *quantified constraint formula* over signature σ is a closed first-order formula having conjunction as its only propositional connective in prenex normal form. That is, a quantified constraint formula (over σ) is a formula of the form $Q_1 v_1 \ldots Q_n v_n \psi(v_1, \ldots, v_n)$ where each Q_i is a quantifier (either $\exists$ or $\forall$) and ψ is the conjunction of expressions of the form $R(v_{i_1}, \ldots, v_{i_k})$, where R is a relation symbol from σ and k is the arity of R.

An instance of the *quantified constraint satisfaction problem (QCSP)* consists of a quantified constraint formula ϕ and a relational structure **B** over the same signature; the question is to decide whether or not $\mathbf{B} \models \phi$.

It is known that any quantified constraint formula with only existential quantifiers, known as *conjunctive queries*, can be naturally associated with a relational structure [5]. It will be conceptually and terminologically useful for us to associate an object called a *quantified relational structure* to each quantified constraint formula. A *quantified relational structure* is a pair $(p, \mathbf{A})$ where **A** is a relational structure and p is an expression of the form $Q_1 v_1 \ldots Q_n v_n$ where each Q_i is a quantifier (either $\exists$ or $\forall$) and $v_1, \ldots, v_n$ are exactly the elements of the universe of **A**. The quantified relational structure $(p, \mathbf{A})$ associated to a quantified constraint formula ϕ is obtained by letting p be the quantifier prefix of ϕ and leeting $R^{\mathbf{A}}$ contain all tuples $(a_1, \ldots, a_k)$ such that $R(a_1, \ldots, a_k)$ appears as a constraint in ϕ. As an example, suppose that σ contains only a binary relation symbol E, and let ϕ be the quantified constraint formula

$$\forall v_1 \exists v_2 \forall v_3 \exists v_4 (E(v_1, v_2) \wedge E(v_2, v_3) \wedge E(v_3, v_4) \wedge E(v_4, v_1)).$$

The quantified relational structure associated to ϕ in this case is

$$(\forall v_1 \exists v_2 \forall v_3 \exists v_4, \mathbf{A})$$

where $\mathbf{A}$ has universe

$$\{v_1, v_2, v_3, v_4\}$$

and

$$E^{\mathbf{A}} = \{(v_1, v_2), (v_2, v_3), (v_3, v_4), (v_4, v_1)\}.$$

It is clearly also possible to inversely map a quantified relational structure to a quantified constraint formula, and we will speak of the quantified relational structure associated to a quantified constraint formula. Indeed, we will freely interchange between corresponding quantified constraint formulas and quantified relational structures.

We say that there is a *homomorphism* from a quantified relational structure $(p, \mathbf{A})$ to a relational structure $\mathbf{B}$ if $\mathbf{B} \models \phi$, where ϕ is the quantified constraint formula associated to $(p, \mathbf{A})$. This notion of homomorphism generalizes the usual notion of relational homomorphism. Let $\mathbf{A}$ and $\mathbf{B}$ be relational structures over the same signature; it is straightforward to verify that there is a relational homomorphism from $\mathbf{A}$ to $\mathbf{B}$ if and only if there is a homomorphism from $(p, \mathbf{A})$ to $\mathbf{B}$, where p existentially quantifies all elements of the universe of $\mathbf{A}$ (in any order). Recall that a mapping h from the universe of $\mathbf{A}$ to the universe of $\mathbf{B}$ is a relational homomorphism from $\mathbf{A}$ to $\mathbf{B}$ if for any relation symbol R and any tuple $(a_1, \ldots, a_k) \in R^{\mathbf{A}}$ of $\mathbf{A}$, it holds that $(h(a_1), \ldots, h(a_k)) \in R^{\mathbf{B}}$.

A quantifier prefix $p = Q_1 v_1 \ldots Q_n v_n$ can be viewed as the concatenation of *quantifier blocks* where quantifiers in each block are the same, and consecutive quantifier blocks have different quantifiers. For example, the quantifier prefix $\forall v_1 \forall v_2 \exists v_3 \forall v_4 \forall v_5 \exists v_6 \exists v_7 \exists v_8$, consists of four quantifier blocks: $\forall v_1 \forall v_2$, $\exists v_3$, $\forall v_4 \forall v_5$, and $\exists v_6 \exists v_7 \exists v_8$. We say that a variable v_j *comes after* a variable v_i in p if they are in the same quantifier block, or v_j is in a quantifier block following the quantifier block of v_i. Equivalently, the variable v_j comes after the variable v_i in p if one of the following conditions holds: (1) $j \geq i$, or (2) $j < i$ and all of the quantifiers $Q_j, \ldots, Q_i$ are of the same type. Notice that in a quantified constraint formula, variables in the same quantifier block can be interchanged without disrupting the semantics of the formula.

3   Pebble Games and Consistency

In previous work [9, 20] certain combinatorial two-player games called *existential pebble games* were shown to be strongly linked to and shed insight on consistency algorithms for constraint satisfaction, and were also used to understand and identify tractable cases of the CSP. Here, we introduce *quantified pebble games* that naturally generalize existential pebble games, and show that they can similarly be used to identify QCSP tractability results.

Our quantified pebble game is defined as follows. The game is played between two players, the *Spoiler* and the *Duplicator*, on a quantified relational structure $(p, \mathbf{A})$ and a relational structure $\mathbf{B}$, where $\mathbf{A}$ and $\mathbf{B}$ are over the same signature. Game play proceeds in rounds, and in each round one of the following occurs:

1. The Spoiler places a pebble on an existentially quantified $\mathbf{A}$-element a coming after all $\mathbf{A}$-elements that already have a pebble. In this case, the Duplicator must respond by placing a corresponding pebble, denoted by $h(a)$, on a $\mathbf{B}$-element.
2. The Spoiler places a pebble on a universally quantified $\mathbf{A}$-element a coming after all $\mathbf{A}$-elements that already have a pebble. The Spoiler then places a corresponding pebble, denoted by $h(a)$, on a $\mathbf{B}$-element.
3. The Spoiler removes a pebble from an $\mathbf{A}$-element a. In this case, the corresponding pebble $h(a)$ on $\mathbf{B}$ is removed.

When game play begins, there are no pebbles on any $\mathbf{A}$-elements, nor on any $\mathbf{B}$-elements, and so the first round is of one of the first two types. We assume that the Spoiler never places two pebbles on the same $\mathbf{A}$-element, so that h is a partial function (as opposed to a relation). The Duplicator wins the quantified pebble game if he can always ensure that h is a *projective homomorphism* from $\mathbf{A}$ to $\mathbf{B}$; otherwise, the Spoiler wins. A *projective homomorphism* (from $\mathbf{A}$ to $\mathbf{B}$) is a partial function h from the universe of $\mathbf{A}$ to the universe of $\mathbf{B}$ such that for any relation symbol R and any tuple $(a_1, \ldots, a_k) \in R^{\mathbf{A}}$ of $\mathbf{A}$, there exists a tuple $(b_1, \ldots, b^k) \in R^{\mathbf{B}}$ where $h(a_i) = b_i$ for all a_i on which h is defined.

We will be most interested in versions of the quantified pebble game where the number of pebbles that can be placed on $\mathbf{A}$-elements is bounded by a constant k, which we call the *quantified k-pebble game*. To emphasize the distinction between this bounded version and the general game, we will refer to the quantified pebble game with no restriction on the number of pebbles as the *general quantified pebble game*. Moreover, it will be useful to consider a version of the quantified pebble game called the *truth quantified pebble game*. In the truth quantified pebble game on $(Q_1 a_1 \ldots Q_n a_n, \mathbf{A})$ and $\mathbf{B}$, there are exactly n rounds; in the rth round, the Spoiler places down a pebble on the $\mathbf{A}$-element a_r, and then – depending on whether or not a_r is universally or existentially quantified – either the Spoiler or the Duplicator places down a corresponding pebble on a $\mathbf{B}$-element b_r.

Proposition 1. *The following four statements are equivalent:*

1. *There is a homomorphism from $(p, \mathbf{A})$ to $\mathbf{B}$.*
2. *The Duplicator wins the truth quantified pebble game on $(p, \mathbf{A})$ and $\mathbf{B}$.*
3. *The Duplicator wins the general quantified pebble game on $(p, \mathbf{A})$ and $\mathbf{B}$.*
4. *For all $k \geq 2$, the Duplicator wins the quantified k-pebble game on $(p, \mathbf{A})$ and $\mathbf{B}$.*

We now formalize what it means for the Duplicator to win the *quantified k-pebble game*, where at most k pebbles may be placed by the Spoiler on $\mathbf{A}$ at a time. When f is a partial function, we use $\mathsf{dom}(f)$ to denote the domain of f. A *k-projective homomorphism* is a projective homomorphism h with $|\mathsf{dom}(h)| \leq k$.

Define a *winning strategy* for the Duplicator for the quantified k-pebble game on $(p, \mathbf{A})$ and $\mathbf{B}$ to be a non-empty set H of k-projective homomorphisms (from $\mathbf{A}$ to $\mathbf{B}$) having the following properties, corresponding to the three types of rounds given above:

1. For every $h \in H$ with $|\mathsf{dom}(h)| < k$ and every existentially quantified $\mathbf{A}$-element $a \notin \mathsf{dom}(h)$ coming after all elements of $\mathsf{dom}(h)$, there exists a projective homomorphism $h' \in H$ extending h with $\mathsf{dom}(h') = \mathsf{dom}(h) \cup \{a\}$.
2. For every $h \in H$ with $|\mathsf{dom}(h)| < k$, every $\mathbf{B}$-element b, and every universally quantified $\mathbf{A}$-element $a \notin \mathsf{dom}(h)$ coming after all elements of $\mathsf{dom}(h)$, there exists a projective homomorphism $h' \in H$ extending h with $\mathsf{dom}(h') = \mathsf{dom}(h) \cup \{a\}$ and $h'(a) = b$.
3. H is closed under subfunctions, that is, if $h \in H$ and h extends h', then $h' \in H$.

By taking the just-given definition of a winning strategy for the quantified k-pebble-game and excluding the requirements "$|\mathsf{dom}(h)| < k$" in items (1) and (2), we obtain the definition of a winning strategy for the general quantified pebble game.

The following fact is immediate from the just-given definition.

Proposition 2. *Let H and H' be winning strategies for the quantified k-pebble game on a quantified relational structure $(p, \mathbf{A})$ and a relational structure $\mathbf{B}$. Then, $H \cup H'$ is also a winning strategy for the quantified k-pebble game on the pair of structures.*

Proposition 2 implies that in a quantified k-pebble game, there is a *maximal winning strategy*, equal to the union of all winning strategies.

We now show how to efficiently compute winning strategies for the quantified k-pebble game.

Proposition 3. *For every fixed k, it is possible to decide in polynomial time, given a quantified relational structure $(p, \mathbf{A})$ and relational structure $\mathbf{B}$, whether or not there is a winning strategy for the Duplicator in the quantified k-pebble game.*

Proof. A straightforward way to perform the decision is as follows. Compute the set of all k-projective homomorphisms from $\mathbf{A}$ to $\mathbf{B}$, and let H denote this set. Continually eliminate from H any $h \in H$ with $|\mathsf{dom}(h)| < k$ that cannot be extended as described in parts (1) and (2) of the definition of winning strategy; and also continually eliminate any $h \in H$ whose subfunctions are not all in H. Once this procedure stabilizes, the result is either an empty set or a winning strategy. Observe that all projective homomorphisms $h_1, h_2, \ldots$ that are eliminated by this procedure cannot be a member of *any* winning strategy; the proof is a straightforward induction on i, where h_i represents the ith projective homomorphism that was eliminated. Therefore, if the result of the procedure is an empty set, there is no winning strategy. Moreover, if the result of the procedure is a winning strategy, the result is in fact the maximal winning strategy. □

We have given a polynomial-time algorithm for deciding if there is a winning strategy for the quantified k-pebble game. However, we can show that our algorithm can be used to derive even stronger consequences. The given algorithm can also be used to compute the maximal winning strategy for the general quantified pebble game. In fact, we can observe that any projective homomorphism that can be eliminated in the computation of the maximal winning strategy H_k for the quantified k-pebble game, can also be eliminated in the computation of the maximal winning strategy for the general quantified pebble game! Thus, in the general quantified pebble game, any k-projective homomorphism in a winning strategy must fall into H_k, and indeed by the "subfunction" requirement, every projective homomorphism h in a winning strategy with $|\text{dom}(h)| > k$ must yield a k-projective homomorphism in H_k when restricted to a subdomain of size $\leq k$. After having computed the maximal winning strategy H_k for a QCSP instance, we can therefore add constraints of the form $R(a_1, \ldots, a_k)$ with

$$R^{\mathbf{B}} = \{(h(a_1), \ldots, h(a_k)) : h \in H_k, \text{dom}(h) = \{a_1, \ldots, a_k\}\}.$$

(Technically, the way we have formalized quantified constraint satisfaction, we need to expand the signature to add these constraints.) After adding these constraints, we have not changed the truth of the QCSP instance (since we have not changed whether or not there is a winning strategy for the general game, see Proposition 1), but in general we obtain a more constrained instance. In fact, in the CSP setting, adding these extra constraints corresponds exactly to the process of establishing k-consistency [20].

The overall algorithm we have described, namely, of computing H_k and then imposing new constraints, is thus a QCSP generalization of the CSP notion of establishing k-consistency. This algorithm preserves the truth of the QCSP instance and thus can be applied in general, *to any arbitrary QCSP instance*. In the case that H_k is found to be empty, we can immediately conclude that the instance is false. In the next section, we will show that in some interesting cases, the converse holds, that is, some classes of QCSP instances have the following property: if on an instance there is a winning strategy for the Duplicator in the quantified k-pebble game, then the instance is true. When this property holds, we will say that *establishing k-consistency is a decision procedure* for the class of instances.

4   Tractability

4.1   The Left-Hand Side: Bounded Treewidth

We define a notion of treewidth for quantified relational structures. Let $(p, \mathbf{A})$ be a quantified relational structure. The *Gaifman graph* of $\mathbf{A}$ is the graph with vertex set equal to the universe A of $\mathbf{A}$ and with an edge $\{a, a'\}$ for every pair of different elements $a, a' \in A$ that occur together in a $\mathbf{A}$-tuple, by which we mean an element of $R^{\mathbf{A}}$ for some relation symbol R. A *scheme* for $(p, \mathbf{A})$ is a supergraph (A, E) of the Gaifman graph of $\mathbf{A}$ along with an ordering $a_1, \ldots, a_n$ of the elements of A such that

- the ordering $a_1, \ldots, a_n$ preserves the "after" relation, that is, if $i < j$, then a_j comes after a_i in p, and
- for any a_k, its lower numbered neighbors form a clique, that is, for all k, if $i < k$, $j < k$, $\{a_i, a_k\} \in E$, and $\{a_j, a_k\} \in E$, then $\{a_i, a_j\} \in E$.

The *width* of a scheme is the maximum, over all vertices a_k, of the size of the set $\{i : i < k, \{a_i, a_k\} \in E\}$, that is, the set containing all lower numbered neighbors of a_k. The *treewidth* of a quantified relational structure $(p, \mathbf{A})$ is the minimum width over all schemes for $(p, \mathbf{A})$.

Our definition of treewidth strictly generalizes the definition of treewidth in the study of the CSP: the treewidth of a relational structure $\mathbf{A}$, as defined (for instance) in [9], is equivalent to the treewidth of $(p, \mathbf{A})$ where p existentially quantifies all elements of $\mathbf{A}$ (in any order). Note that having only existential quantifiers in p amounts to dropping the first requirement in the definition of scheme above, since in that case, *every* ordering preserves the "after" relation.

Let QCSP[treewidth $< k$] be the restriction of the QCSP problem to all instances $((p, \mathbf{A}), \mathbf{B})$ where $(p, \mathbf{A})$ has treewidth strictly less than k.

Theorem 4. *For all $k \geq 2$, establishing k-consistency is a decision procedure for* QCSP[treewidth $< k$].

Proof. Let $(p, \mathbf{A}), \mathbf{B}$ be an arbitrary instance of QCSP[treewidth $< k$], and suppose that there is a winning strategy H for the quantified k-pebble game. We show that there is a homomorphism from $(p, \mathbf{A})$ to $\mathbf{B}$. Let the graph (A, E) and the ordering q_s be a minimum width scheme for $(p, \mathbf{A})$, and let q be the quantifier prefix equal to q_s with the corresponding quantifiers from p placed in front of each variable. Now, $(q, \mathbf{A})$ is semantically equivalent to $(p, \mathbf{A})$, since q is equal to p up to interchanging quantifier-variable pairs *within* quantifier blocks. So it suffices to show that there is a homomorphism from $(q, \mathbf{A})$ to $\mathbf{B}$. We let n denote the size of the universe A of $\mathbf{A}$, and denote q by $Q_1 a_1 \ldots Q_n a_n$. By Proposition 1, it suffices to show that Duplicator can win the truth quantified pebble game.

The strategy for the Duplicator is to play so that in the rth round the partial function $a_i \to b_i$ determined by the pebbles, when restricted to the set containing a_r and its lower numbered neighbors (in the graph (A, E) and ordering q_s), falls into H.

We prove by induction that the Duplicator can play in this way. Assume that the Spoiler has initiated the rth round by placing a pebble on a_r. Let N_r denote all lower numbered neighbors of a_r. The partial function $a_i \to b_i$ determined by the pebbles, when restricted to N_r, gives a projective homomorphism h_r in H by the induction hypothesis. Since the scheme $((A, E), q_s)$ has width $\leq k$, the projective homomorphism h_r has domain size $|N_r| = |\mathrm{dom}(h_r)| < k$.

- Suppose that a_k is existentially quantified. By part (1) of the definition of winning strategy, there is a homomorphism h'_r extending h_r to a_k. The Duplicator can thus set b_k to be $h'_r(a_k)$.

– Suppose that a_k is universally quantified. In this case, the Spoiler sets b_k. By part (2) of the definition of winning strategy, there is a homomorphism h'_r extending h_r that maps a_k to b_k.

It remains to verify that after all n rounds have passed, the Duplicator has won the truth quantified pebble game, that is, the function h mapping a_i to b_i (the function determined by the pebbles) is a homomorphism from $\mathbf{A}$ to $\mathbf{B}$. Let $(c_1, \ldots, c_k) \in R^{\mathbf{A}}$ be a tuple of $\mathbf{A}$. Let a_r be the highest numbered element of this tuple with respect to q_s. In the rth round, it was ensured that the restriction of h to $\{a_r\} \cup N_r$ was equal to a projective homomorphism h'_r falling in H. Since $\{c_1, \ldots, c_k\} \subseteq \{a_r\} \cup N_r$, we have $(h(c_1), \ldots, h(c_k)) = (h'_r(c_1), \ldots, h'_r(c_k)) \in R^{\mathbf{B}}$.

$\square$

4.2   The Right-Hand Side: Relational Restrictions

For each relational structure $\mathbf{B}$, let $\mathrm{QCSP}(\mathbf{B})$ denote the problem of deciding, given a quantified constraint formula ϕ, whether or not $\mathbf{B} \models \phi$. That is, $\mathrm{QCSP}(\mathbf{B})$ is the restricted case of the QCSP where the relational structure is fixed as $\mathbf{B}$.

Near-Unanimity Polymorphisms. Let $\mathbf{B}$ be a relational structure with universe B. An operation $f : B^k \to B$ is a *polymorphism* of a relational structure $\mathbf{B}$ if it is a homomorphism from $\mathbf{B}^k$ to $\mathbf{B}$ [1]. A *near-unanimity operation* is an operation $f : B^k \to B$ of arity ≥ 3 satisfying the identities

$$d = f(e, d, d, \ldots, d) = f(d, e, d, \ldots, d) = \cdots.$$

These identities state that if all but (at most) one of the arguments to f are equal to d, then f outputs d.

Theorem 5. *Let $\mathbf{B}$ be a relational structure having a near-unanimity polymorphism of arity k. Establishing k-consistency is a decision procedure for $\mathrm{QCSP}(\mathbf{B})$.*

Proof. We assume that the Duplicator has a winning strategy H for the quantified k-pebble game on $(p, \mathbf{A})$, $\mathbf{B}$, and aim to show that there is a homomorphism from $(p, \mathbf{A})$ to $\mathbf{B}$. We in fact prove the following claim.

Claim. Let h be a projective homomorphism from $\mathbf{A}$ to $\mathbf{B}$ such that all restrictions of h onto a domain of size $\leq k$ is in H; and, let $a \notin \mathrm{dom}(h)$ be any variable coming after all elements of $\mathrm{dom}(h)$.

– If a is existentially quantified, then there exists a projective homomorphism h' extending h with $a \in \mathrm{dom}(h')$.
– If a is universally quantified, then for all $\mathbf{B}$-elements b there exists a projective homomorphism h' extending h with $h'(a) = b$.

---

[1] Recall that when $\mathbf{A}$ and $\mathbf{B}$ are relational structures over the same vocabulary with A and B, respectively, the relational structure $\mathbf{A} \times \mathbf{B}$ is defined to have universe $A \times B$ and so that for each relation symbol R, it holds that $R^{\mathbf{A} \times \mathbf{B}} = \{((a_1, b_1), \ldots, (a_k, b_k)) : (a_1, \ldots, a_k) \in R^{\mathbf{A}}, (b_1, \ldots, b_k) \in R^{\mathbf{B}}\}$.

Proving this claim suffices, since it follows immediately that the truth quantified pebble game can be won by the Duplicator, which by Proposition 1 implies that there is a homomorphism from $(p, \mathbf{A})$ to $\mathbf{B}$.

We prove the claim by induction on the size of $\mathsf{dom}(h)$. By the definition of winning strategy, the claim holds when $|\mathsf{dom}(h)| < k$. So, let us suppose that $|\mathsf{dom}(h)| \geq k$, and that $a \notin \mathsf{dom}(h)$ comes after all elements of $\mathsf{dom}(h)$. Pick any k distinct elements $a_1, \ldots, a_k \in \mathsf{dom}(h)$, and define h_i to be the restriction of h to $\mathsf{dom}(h) \setminus \{a_i\}$. We now consider two cases.

Case 1: Suppose that a is existentially quantified. By induction, each of the mappings h_i can be extended to a projective homomorphism h'_i with $a \in \mathsf{dom}(h'_i)$. Let $\mu : D^k \to D$ denote the near-unanimity polymorphism of $\mathbf{B}$. Define b to be $\mu(h'_1(a), \ldots, h'_k(a))$. We claim that the extension h' of h mapping a to b is a projective homomorphism. Let $(c_1, \ldots, c_m) \in R^{\mathbf{A}}$ be any $\mathbf{A}$-tuple. We wish to show that there is a tuple $(b_1, \ldots, b_m) \in R^{\mathbf{B}}$ such that $h'(c_j) = b_j$ for all $c_j \in \mathsf{dom}(h')$. Each h'_i is a projective homomorphism, and so for each $i = 1, \ldots, k$ there is a tuple $(b^i_1, \ldots, b^i_m) \in R^{\mathbf{B}}$ such that $h'_i(c_j) = b_j$ for all $c_j \in \mathsf{dom}(h'_i)$. We have $((b^1_1, \ldots, b^k_1), \ldots, (b^1_m, \ldots, b^k_m)) \in R^{\mathbf{B}^k}$; we claim that the desired tuple $(b_1, \ldots, b_m)$ can be obtained as $\mu((b^1_1, \ldots, b^k_1), \ldots, (b^1_m, \ldots, b^k_m))$. Let us consider any $c_j \in \mathsf{dom}(h')$.

- If $c_j = a$, then by our definition of h' we have $b_j = \mu(b^1_j, \ldots, b^k_j)$.
- If $c_j \in \{a_1, \ldots, a_k\}$ and $c_j = a_l$, then all elements $b^1_j, \ldots, b^k_j$ are equal to $h'(c_j)$ except possibly b^l_j, so $\mu(b^1_j, \ldots, b^k_j)$ is equal to $h'(c_j)$.
- If $c_j \in \mathsf{dom}(h) \setminus \{a_1, \ldots, a_k\}$, then all elements $b^1_j, \ldots, b^k_j$ are equal to $h'(c_j)$, and so $\mu(b^1_j, \ldots, b^k_j)$ is equal to $h'(c_j)$.

Case 2: Suppose that a is universally quantified and that b is a $\mathbf{B}$-element. By induction, each of the mappings h_i can be extended to a projective homomorphism h'_i with $h'_i(a) = b$. Let h' be the extension of h mapping a to b. Notice that, as in the previous case, b equals $\mu(h'_1(a), \ldots, h'_k(a))$. By reasoning as in the previous case, it can be established that h' is a projective homomorphism.   □

Set Function Polymorphisms. When $\mathbf{B}$ is a relational structure with universe B, define $\wp(\mathbf{B})$ to be the relational structure having universe $\wp(B) \setminus \{\emptyset\}$ (where $\wp(B)$ denotes the power set of B) and such that for every relation symbol R, it holds that

$$R^{\wp(\mathbf{B})} = \{(\mathsf{pr}_1 S, \ldots, \mathsf{pr}_k S) : S \subseteq R^{\mathbf{B}}, S \neq \emptyset\}$$

where k denotes the arity of R, and $\mathsf{pr}_i S$ denotes the set $\{s_i : (s_1, \ldots, s_k) \in S\}$, that is, the projection onto the ith coordinate. We say that $f : \wp(B) \setminus \{\emptyset\} \to B$ is a *set function polymorphism* of $\mathbf{B}$ if f is a homomorphism from $\wp(\mathbf{B})$ to $\mathbf{B}$. Alternatively, one can define $f : \wp(B) \setminus \{\emptyset\} \to B$ to be a set function polymorphism of $\mathbf{B}$ if and only if all of the mappings $f_k : B^k \to B$ defined by $f_k(b_1, \ldots, b_k) = f(\{b_1, \ldots, b_k\})$ (with $k \geq 1$) are polymorphisms of $\mathbf{B}$. (It is straightforward to verify that these two definitions are equivalent.) We show that a variant of consistency called default k-consistency solves any *QCSP-tractable set function polymorphisms*, that is, a set function polymorphism giving rise to a tractable case of the QCSP [8]. Note that default k-consistency is a stronger concept than k-consistency, so establishing default k-consistency gives a unified algorithm for all tractable cases discussed in this paper. (Details are left to the full version of this paper.)

Theorem 6. *Let* **B** *be a relational structure having a QCSP-tractable set function polymorphism. Establishing default k-consistency is a decision procedure for* QCSP(**B**).

5   Homomorphisms Between Quantified Relational Structures

We now define a notion of homomorphism called *Q-homomorphism* that allows us to compare quantified relational structures. Let $(p, \mathbf{A})$ and $(p', \mathbf{A}')$ be quantified relational structures. Let X (X') denote the existentially quantified variables of **A** (respectively, **A**$'$), and let Y (Y') denote the universally quantified variables of **A** (respectively, **A**$'$).

A pair of functions $(f : X \rightarrow X', g : Y' \rightarrow Y)$ is a *Q-homomorphism* from $(p, \mathbf{A})$ to $(p', \mathbf{A}')$ when the following conditions hold:

- For all relation symbols R and all tuples $(a_1, \ldots, a_k) \in R^{\mathbf{A}}$, there exists a tuple $(a'_1, \ldots, a'_k) \in R^{\mathbf{A}'}$ such that, for all i, either
  - both a_i and a'_i are existentially quantified and $f(a_i) = a'_i$, or
  - both a_i and a'_i are universally quantified and $a_i = g(a'_i)$.
- The mappings f, g preserve the "after" relation in that if

$$(a_1, a'_1), (a_2, a'_2) \in \{(a, f(a)) : a \in X\} \cup \{(g(a'), a') : a' \in Y'\}$$

  and a_2 comes after a_1, then a'_2 comes after a'_1.

Q-homomorphisms and homomorphisms (from quantified relational structures to relational structures) interact well in that a natural transitivity-like property holds.

Theorem 7. *Let $(p, \mathbf{A})$ and $(p', \mathbf{A}')$ be quantified relational structures, and let* **B** *be a relational structure. If there is a Q-homomorphism from $(p, \mathbf{A})$ to $(p', \mathbf{A}')$ and there is a homomorphism from $(p', \mathbf{A}')$ to* **B***, then there is a homomorphism from $(p, \mathbf{A})$ to* **B**.

Proof. (idea) Let (f, g) be a Q-homomorphism from $(p, \mathbf{A})$ to $(p', \mathbf{A}')$, and assume that the Duplicator wins the truth quantified pebble game on $(p', \mathbf{A}')$ and **B**. By Proposition 1, it suffices to show that the Duplicator wins the truth quantified pebble game on $(p, \mathbf{A})$ and **B**; we give a winning strategy for this game where the Duplicator simulates the truth quantified pebble game on $(p', \mathbf{A}')$ and **B**.

– When the Spoiler places a pebble on an existentially quantified **A**-element a, the Duplicator places a pebble on the **A**′-element $f(a)$ in the simulated game, and then uses the response of the Duplicator in the simulated game to determine where on **B** to place a pebble.
– When the Spoiler places a pebble on a universally quantified **A**-element a and a corresponding pebble on a **B**-element b, there are two cases. If a does not occur in any **A**-tuple, the Duplicator simply ignores the pebble on a, as well as the corresponding pebble on b. If a does occur in an **A**-tuple, then by definition of Q-homomorphism, there exists a **B**-element a' such that $a = g(a')$ and so the set $g^{-1}(a)$ is non-empty. The Duplicator then places pebbles on all elements of $g^{-1}(a)$ in the simulated game, and places down the $|g^{-1}(a)|$ corresponding pebbles all on b.

□

We say that two quantified relational structures $(p, \mathbf{A})$ and $(p', \mathbf{A}')$ are Q-*homomorphically equivalent* if there exists a Q-homomorphism from $(p, \mathbf{A})$ to $(p', \mathbf{A}')$, and there exists a Q-homomorphism from $(p', \mathbf{A}')$ to $(p, \mathbf{A})$. The following corollary is immediate from Theorem 7.

Corollary 8. *Let $(p, \mathbf{A})$ and $(p', \mathbf{A}')$ be Q-homomorphically equivalent quantified relational structures, and let $\mathbf{B}$ be a relational structure. There is a homomorphism from $(p, \mathbf{A})$ to $\mathbf{B}$ if and only if there is a homomorphism from $(p', \mathbf{A}')$ to $\mathbf{B}$.*

Theorem 7, as we have noted, generalizes Corollary 8. However, Corollary 8 can be generalized in a different direction. In particular, we can show that two Q-homomorphically equivalent quantified relational structures behave identically as regards the quantified k-pebble game.

Theorem 9. *Let $(p, \mathbf{A})$ and $(p', \mathbf{A}')$ be two Q-homomorphically equivalent quantified relational structures, and let $\mathbf{B}$ be a relational structure. For all $k \geq 2$, the Duplicator wins the quantified k-pebble game on $(p, \mathbf{A})$ and $\mathbf{B}$ if and only if the Duplicator wins the quantified k-pebble game on $(p', \mathbf{A}')$ and $\mathbf{B}$.*

Proof. (idea) Assume that the Duplicator wins the quantified k-pebble game on $(p', \mathbf{A}')$ and $\mathbf{B}$. We want to show that the Duplicator wins the quantified k-pebble game on $(p, \mathbf{A})$ and $\mathbf{B}$. We may assume without loss of generality that all universally quantified variables in $(p', \mathbf{A}')$ appear in a tuple of $\mathbf{A}'$, and likewise for $(p, \mathbf{A})$. Let (f, g) be a Q-homomorphism from $(p, \mathbf{A})$ to $(p', \mathbf{A}')$. As $(p, \mathbf{A})$ and $(p', \mathbf{A}')$ are Q-homomorphically equivalent, it can be verified that g is a bijection between the universally quantified variables of $(p', \mathbf{A}')$ and those of $(p, \mathbf{A})$. We can show that the Duplicator wins the quantified k-pebble game on $(p, \mathbf{A})$ and $\mathbf{B}$ using the same simulation strategy as in the proof of Theorem 7. Note that because g is a bijection, no more than k pebbles are ever placed on $\mathbf{A}'$ in the simulated game (which is played on $(p', \mathbf{A}')$ and $\mathbf{B}$). □

We remark that Corollary 8 can be derived from Theorem 9 (along with Proposition 1).

The results obtained in this section allow us to expand the tractability result concerning instances with bounded treewidth (Theorem 4). This result generalizes an analogous expansion that was carried out in the CSP setting [9]. Let $\mathsf{QCSP}[\mathcal{H}(\text{treewidth} < k)]$ be the restriction of the QCSP problem to all instances $((p, \mathbf{A}), \mathbf{B})$ where $(p, \mathbf{A})$ is Q-homomorphically equivalent to a quantified relational structure with treewidth strictly less than k.

Theorem 10. *For all $k \geq 2$, establishing k-consistency is a decision procedure for $\mathsf{QCSP}[\mathcal{H}(\text{treewidth} < k)]$.*

Proof. Immediate from Theorem 4, Corollary 8, and Theorem 9.   □

References

1. Andrei Bulatov. Combinatorial problems raised from 2-semilattices. Manuscript.
2. Andrei Bulatov. A dichotomy theorem for constraints on a three-element set. In *Proceedings of 43rd IEEE Symposium on Foundations of Computer Science*, pages 649–658, 2002.
3. Andrei Bulatov. Tractable conservative constraint satisfaction problems. In *Proceedings of 18th IEEE Symposium on Logic in Computer Science (LICS '03)*, pages 321–330, 2003. Extended version appears as Oxford University technical report PRG-RR–03-01.
4. Andrei Bulatov. A graph of a relational structure and constraint satisfaction problems. In *Proceedings of 19th IEEE Annual Symposium on Logic in Computer Science (LICS'04)*, 2004.
5. Ashok Chandra and Philip Merlin. Optimal implementation of conjunctive queries in relational data bases. In *STOC*, 1977.
6. Hubie Chen. *The Computational Complexity of Quantified Constraint Satisfaction*. PhD thesis, Cornell University, August 2004.
7. Hubie Chen. Quantified constraint satisfaction and bounded treewidth. In *ECAI*, 2004.
8. Hubie Chen. Quantified constraint satisfaction, maximal constraint languages, and symmetric polymorphisms. In *STACS*, 2005.
9. Victor Dalmau, Phokion G. Kolaitis, and Moshe Y. Vardi. Constraint satisfaction, bounded treewidth, and finite-variable logics. In *Constraint Programming '02*, LNCS, 2002.
10. Victor Dalmau and Justin Pearson. Closure functions and width 1 problems. In *CP 1999*, pages 159–173, 1999.
11. Rina Dechter and Judea Pearl. Tree clustering for constraint networks. *Artificial Intelligence*, pages 353–366, 1989.
12. J. Flum, M. Frick, and M. Grohe. Query evaluation via tree-decompositions. *JACM*, 2002.
13. Eugene Freuder. Complexity of k-tree structured constraint satisfaction problems. In *AAAI-90*, 1990.
14. Georg Gottlob, Gianluigi Greco, and Francesco Scarcello. The complexity of quantified constraint satisfaction problems under structural restrictions. In *IJCAI*, 2005.
15. Georg Gottlob, Nicola Leone, and Francesco Scarcello. A comparison of structural csp decomposition methods. *Artif. Intell.*, 124(2):243–282, 2000.

16. Martin Grohe. The complexity of homomorphism and constraint satisfaction problems seen from the other side. In *FOCS 2003*, pages 552–561, 2003.
17. Peter Jeavons, David Cohen, and Martin Cooper. Constraints, consistency, and closure. *Articial Intelligence*, 101(1-2):251–265, 1998.
18. Ph.G. Kolaitis and M.Y. Vardi. On the expressive power of Datalog: tools and a case study. *Journal of Computer and System Sciences*, 51(1):110–134, 1995.
19. Ph.G. Kolaitis and M.Y. Vardi. Conjunctive-query containment and constraint satisfaction. *Journal of Computer and System Sciences*, 61:302–332, 2000.
20. Ph.G. Kolaitis and M.Y. Vardi. A game-theoretic approach to constraint satisfaction. In *Proceedings 17th National (US) Conference on Artificial Intellignece, AAAI'00*, pages 175–181, 2000.
21. Thomas J. Schaefer. The complexity of satisfiability problems. In *Proceedings of the ACM Symposium on Theory of Computing (STOC)*, pages 216–226, 1978.

An Algebraic Approach for the Unsatisfiability of Nonlinear Constraints*

Ashish Tiwari

SRI International
333 Ravenswood Ave
Menlo Park, CA, USA
tiwari@csl.sri.com

Abstract. We describe a simple algebraic semi-decision procedure for detecting unsatisfiability of a (quantifier-free) conjunction of nonlinear equalities and inequalities. The procedure consists of Gröbner basis computation plus extension rules that introduce new definitions, and hence it can be described as a critical-pair completion-based logical procedure. This procedure is shown to be sound and refutationally complete. When projected onto the linear case, our procedure reduces to the Simplex method for solving linear constraints. If only finitely many new definitions are introduced, then the procedure is also terminating. Such terminating, but potentially incomplete, procedures are used in "incompleteness-tolerant" applications.

1 Introduction

The ability to solve nonlinear constraints is central to the task of developing and automating analysis technology for several classes of systems. Nonlinear constraints arise in robotics, control theory, hybrid system models of physical and embedded control systems and biological systems, and in solving games [8, 15, 17]. Fortunately, the full first-order theory of the reals is known to be decidable [22]. Unfortunately, it has a double exponential lower-bound and most of the decision procedures for this theory are complex, nonlogical, and involve considerable splitting (causing blowups) [5, 7, 18, 24]. Available implementations of the decision procedure [12, 13] can only solve very small-sized examples.

We are particularly interested in the verification of hybrid systems. Our methodology for verification is based on abstraction and model-checking [23]. Automation of this technique requires sound and fast implementations of a procedure for testing unsatisfiability of (a conjunction of) nonlinear constraints. The same is also needed in the lazy approach of extending constraint solvers to handle boolean combination of constraints. Tools such as ICS [9], CVC [21], and MathSat [1], which are used in bounded model-checking of discrete systems,

---

* Research supported in part by the National Science Foundation under grants CCR-0311348 and ITR-CCR-0326540.

also implement some form of incomplete nonlinear constraint solving. Fast and sound, but potentially incomplete, implementations that can solve large problem instances are useful in several "incompleteness-tolerant" applications such as the process of creating abstractions, where incompleteness only causes creation of a coarser abstraction.

This paper considers the problem of developing fast reasoners for (quantifier-free conjunction of) nonlinear constraints over the theory of reals. Our goal was to develop a method that efficiently detected the "easy" unsatisfiable instances. For instance, we do not want to compute a full cylindrical algebraic decomposition of the n-dimensional space based on the polynomial p to decide if $p > 0 \land p < 0$ is satisfiable. Our goal was to give a logical procedure that can be described using simple inference rules. Moreover, the procedure should be simple and easy to implement and incremental, that is, new constraints can be added without redoing everything.

In this paper, we describe a critical-pair completion approach to nonlinear constraint solving. The main ingredient is the Gröbner basis computation method. Apart from it, we only need some *extension* rules that introduce new definitions. Surprisingly, this is all that is needed for obtaining a sound and refutationally complete procedure for testing unsatisfiability of nonlinear constraints–a consequence of the Positivstellensatz theorem from real algebraic geometry.

Our approach is based on eliminating inequality constraints by introducing slack variables and then constructing a Gröbner basis of the polynomials in the equality constraints. For example, suppose we want to prove unsatisfiability of $\{u_1 + u_2 - 1 \approx 0, -u_2 + 2 \approx 0, u_1 \geq 0\}$. If we construct a (fully reduced) Gröbner basis of the polynomials that appear in the two equations, we get $\{u_1 + 1, -u_2 + 2\}$. The first polynomial, $u_1 + 1$, is a *witness* for unsatisfiability of the original constraints, since $u_1 \geq 0$ implies that $u_1 + 1$ should be strictly greater-than 0, but the equational constraints require that $u_1 + 1 \approx 0$. Unfortunately, it is not the case that whenever the original constraints are unsatisfiable, the corresponding Gröbner basis will necessarily contain such a witness. For example, if we change the constraints slightly to $\{u_1 + u_2 - 1 \approx 0, u_2 u_3 - u_2 + 2 \approx 0, u_1 \geq 0, u_2 \geq 0, u_3 \geq 0\}$, then the Gröbner basis computation does not yield anything new and we fail to detect the witness. The witness here is $u_2 u_3 + u_1 + 1$, which is obtained by adding the two equations. The reason why the witness is not explicitly present in the Gröbner basis is that it is not "small-enough" in the lexicographic ordering chosen to construct the Gröbner basis.

The basic idea in our paper is that new definitions that introduce new constants allow greater flexibility in choosing orderings. For example, we can make the witness polynomial $u_2 u_3 + u_1 + 1$ smaller by introducing a definition $u_2 u_3 \approx u_4$ and giving u_4 the lowest precedence. As a result, we now compute the Gröbner basis for $\{u_1 + u_2 - 1, u_2 u_3 - u_2 + 2, u_2 u_3 - u_4\}$, and we get $\{u_1 + u_4 + 1, u_2 - u_4 - 2, u_2 u_3 - u_4\}$. The first polynomial in this set, $u_1 + u_4 + 1$, is a witness for unsatisfiability of the original set of constraints.

In the linear case, our method would introduce definitions of the form $u_1 \approx u_2$, making the new variable u_2 smaller than all other variables. This has the effect of lowering the precedence of the old variable u_1. This is similar to the Simplex method, where the pivot steps transform a Gröbner basis with respect to a given precedence $\succ_1$ to a Gröbner basis with respect to a different precedence $\succ_2$, doing this until a witness to unsatisfiability is detected, or we have (implicitly) exhausted all possibilities.

The inference rules are nonterminating because of the possibility of introducing infinitely many new definitions. Our procedure can be made terminating by limiting the introduction of new definitions. We could still guarantee completeness if there were known degree bounds for Positivstellensatz, whence we could introduce enough new definitions to cover all polynomials upto the given degree bound. Obtaining such degree bounds is an active area of research [19].

The presentation of the procedure in this paper is incremental. We first present a simple and incomplete procedure in Section 3. Thereafter, we describe the version of Positivstellensatz we use in this paper in Section 4. Using this result, in Section 5 we develop a sound procedure that is refutationally complete relative to an oracle (that provides the new definitions). Finally, we present the complete set of inference rules in Section 6 and show how the job of the oracle can be performed using static analysis of the polynomials.

2   Term Rewriting and Polynomials

Let $\{x_1, \ldots, x_n\}$ be a set of indeterminates, often denoted using vector notation as $\boldsymbol{x}$. The set of power-products over $\boldsymbol{x}$ is the free commutative monoid $[\boldsymbol{x}]$ generated by $\boldsymbol{x}$. Elements of $[\boldsymbol{x}]$, such as $x_1 x_2^2 x_3$, are denoted by μ with possible subscripts. The polynomial ring over the field of rational numbers $\mathbb{Q}$ is the $\mathbb{Q}$ vector space generated by $[\boldsymbol{x}]$, denoted by $\mathbb{Q}[\boldsymbol{x}]$. Elements from $\mathbb{Q}[\boldsymbol{x}]$ are denoted by p, q with possible subscripts. Atomic formulas are given as $p \approx 0$, $p \geq 0$, and $p > 0$. Since we deal with quantifier-free conjunctions of atomic formulas, the indeterminates $\boldsymbol{x}$ are logically constants, but we call them variables. Positive variables are denoted by v and nonnegative by u, w. Elements from $\mathbb{Q}$ will be denoted by c, and hence a polynomial p can be written as $c_0\mu_0 + c_1\mu_1 + \ldots + c_k\mu_k$.

Orderings on Polynomials. Let $\langle \boldsymbol{c_1}, \boldsymbol{c_2}, \ldots, \boldsymbol{c_m} \rangle$ be a sequence of m non-negative vectors in $\mathbb{Q}^{+^n}$ such that $m \geq n$ and $\{\boldsymbol{c_1}, \boldsymbol{c_2}, \ldots, \boldsymbol{c_m}\}$ spans the n-dimensional vector space $\mathbb{Q}^n$. We define an ordering on power-products as follows: $x_1^{d_1} \ldots x_n^{d_n} \succ x_1^{d_1'} \ldots x_n^{d_n'}$ if there is a k such that $1 \leq k \leq m$ and (a) $\boldsymbol{c_k} \cdot \boldsymbol{d} > \boldsymbol{c_k} \cdot \boldsymbol{d'}$, and (b) $\boldsymbol{c_i} \cdot \boldsymbol{d} = \boldsymbol{c_i} \cdot \boldsymbol{d'}$ for all $i < k$. If $\boldsymbol{e_i}$ is a unit vector in i-th direction (that is, only the i-th component is nonzero), then choosing $\langle \boldsymbol{e_1}, \boldsymbol{e_2}, \ldots, \boldsymbol{e_n} \rangle$ results in the pure lexicographic ordering. Note that if $\boldsymbol{e_0}$ contains all 1's, then choosing $\langle \boldsymbol{e_0}, \boldsymbol{e_1}, \boldsymbol{e_2}, \ldots, \boldsymbol{e_n} \rangle$ results in *total-degree* lexicographic ordering. For other choices of $\boldsymbol{c_i}$'s, we can get certain "combinations" of these two orderings.

The ordering $\succ$ on power-products can be extended to monomials by just ignoring the coefficient (if it is nonzero). The ordering on monomials can be extended to polynomials by using the multiset extension of $\succ$ (and viewing a polynomial as a multiset of monomials) [10].

Term Rewriting Systems. Term rewriting systems are sets containing directed equations, $l \to r$, where the orientation is usually chosen so that $l \succ r$ for some reduction ordering on the set of terms. If R is a rewrite system, the binary relation on terms $\to_R$ is defined as the closure of R under contexts and substitution. We use the usual notation for symmetric ($\leftrightarrow$) and transitive ($\to^*$) closures.

A rewrite system R is said to be *convergent* if the relation $\to_R$ is well-founded and the relation $\to_R^*$ is confluent, that is, $\leftrightarrow_R^* \subseteq \to_R^* \circ \leftarrow_R^*$. A rewrite system R is *fully reduced* if for every rule $l \to r \in R$, the term r is not reducible by R and the term l is not reducible by $R - \{l \to r\}$.

A (finite) fully reduced convergent R has several nice properties. It can be used to decide the relation $\leftrightarrow_R^*$. In fact, if $s \leftrightarrow_R^* t$ and $s \succeq t$, then we actually have $s \to_{R|_{\nsucc s}}^* \circ \leftarrow_{R|_{\nsucc s}}^* t$, where $R|_{\nsucc s}$ contains only those rules in R that contain terms no bigger than s. Furthermore, if r is a $\succ$-minimal term in the R-equivalence class $[[r]]_R$ and l is $\succ$-minimal in $[[r]]_R - \{r\}$, then $l \to_R r$. In other words, the fully reduced convergent R will either contain the rule $l \to r$ explicitly, or some other rule that can be used to rewrite l to r in one step.

Polynomials as Rewrite Rules. A polynomial expression can be normalized into a sum of monomials form $c_0\mu_0 + \cdots + c_k\mu_k$—intuitively using the distributivity rules and formally using a convergent rewrite system for the theory of polynomial rings [2, 3]. We work modulo this theory of polynomial rings in this paper. As a result, we assume that all terms are automatically converted into sum of monomial form. If we assume that $\mu_0 \succ \mu_i$ for all $1 \le i \le k$, then the polynomial equation $c_0\mu_0 + \cdots + c_k\mu_k \approx 0$ can be oriented into a rewrite rule $c_0\mu_0 \to -c_1\mu_1 + \cdots + -c_k\mu_k$. This is a ground rewrite rule (that is, it contains no variables), but we use its *AC*-extension, $c_0\mu_0\nu \to -c_1\mu_1\nu + \cdots + -c_k\mu_k\nu$, for purposes of rewriting polynomials. Here ν is an extension variable (that can be instantiated by monomials). For example, $-u_2 + 2 \approx 0$ can be used as $u_2 \to 2$ to rewrite $u_1u_2 + u_1$ to $2u_1 + u_1$, which normalizes to $3u_1$. This is denoted by $u_1u_2 + u_1 \to_{u_2 \to 2} 3u_1$. Thus, the rewrite relation $\to_P$ induced by P is defined modulo the theory of the coefficient domain $\mathbb{Q}$ and polynomial ring axioms.

Given a set P, the Gröbner basis for P can now be constructed using standard critical-pair completion [3]. A Gröbner basis is a convergent rewrite system and we can even make it fully reduced. We will denote by $GB_\succ(P)$ the fully reduced Gröbner basis for P computed using the ordering $\succ$.

Given a set $P \subset \mathbb{Q}[\boldsymbol{x}]$, the ideal generated by P (in $\mathbb{Q}[\boldsymbol{x}]$) is defined by

$$Ideal(P) = \{q : q = \Sigma_i\, q_ip_i,\ p_i \in P,\ q_i \in \mathbb{Q}[\boldsymbol{x}]\} = \{q : q \leftrightarrow_P^* 0\}$$

Thus, an ideal of P is the equivalence class of 0, when P is viewed as a set of equations, in the theory of polynomial rings [3]. Elimination ideal consists of the projection of the ideal onto polynomials over a subset of variables (that is, it eliminates the other variables). If P is a set of polynomials in $\mathbb{Q}[\boldsymbol{x}, \boldsymbol{u}]$, then we can eliminate the $\boldsymbol{x}$ variables and define $Elim(P, \boldsymbol{x}) = Ideal(P) \cap \mathbb{Q}[\boldsymbol{u}]$.

The above-mentioned property of fully reduced convergent rewrite systems implies that Gröbner basis can be used to compute elimination ideals. In particular, if $\succ$ is an ordering such that $\mu \succ \nu$ for any $\mu \in [\boldsymbol{x}, \boldsymbol{u}] - [\boldsymbol{u}]$ and $\nu \in [\boldsymbol{u}]$, then $Elim(P, \boldsymbol{x}) = Ideal(GB_{\succ}(P) \cap \mathbb{Q}[\boldsymbol{u}])$. In fact, $GB_{\succ}(P) \cap \mathbb{Q}[\boldsymbol{u}]$ will be a fully reduced Gröbner basis for the elimination ideal $Elim(P, \boldsymbol{x})$. The pure lexicographic ordering with precedence $\boldsymbol{x} \succ \boldsymbol{u}$ satisfies this property. On the other hand, if $\succ$ is a total-degree lexicographic ordering with precedence $\boldsymbol{x} \succ \boldsymbol{u}$, then $Ideal(GB_{\succ}(P) \cap \mathbb{Q}[\boldsymbol{u}])$ will contain all *linear* polynomials over $\boldsymbol{u}$ in $Ideal(P)$.

3   Introducing New and Eliminating Old Variables

Let $E = \{p_i \approx 0 : i \in I_1\}$, $F_1 = \{q_i > 0 : i \in I_2\}$, and $F_2 = \{q_i \geq 0 : i \in I_3\}$, where $p_i, q_i \in \mathbb{Q}[\boldsymbol{x}]$. Here I_1, I_2, I_3 are mutually disjoint, finite sets of indices. As in the Simplex method, we introduce new *slack* variables to convert the inequality constraints into equational constraints. Specifically, we introduce the variables v_i, $i \in I_2$ and w_i, $i \in I_3$ and replace the sets F_1 and F_2 by $E_1 = \{q_i - v_i \approx 0 : i \in I_2\}$ and $E_2 = \{q_i - w_i \approx 0 : i \in I_3\}$.

The set $E \cup E_1 \cup E_2$ of equations now contains polynomials from $\mathbb{Q}[\boldsymbol{x}, \boldsymbol{v}, \boldsymbol{w}]$. We also have the implicit constraints $\boldsymbol{v} > 0$ and $\boldsymbol{w} \geq 0$. It is obvious that the set $E \cup E_1 \cup E_2 \cup \{\boldsymbol{v} > 0, \boldsymbol{w} \geq 0\}$ is satisfiable over the reals if and only if the set $E \cup F_1 \cup F_2$ is satisfiable over the reals (all variables are assumed to be existentially quantified).

Example 1. Let $E = \{x_1^3 \approx x_1\}$ and $F = \{x_1 x_2 > 1, -x_2^2 > -1/2\}$. The constraints $E \cup F$ are transformed into the constraints $E' \cup F'$, where $E' = \{x_1^3 - x_1 \approx 0, x_1 x_2 - 1 - v_1 \approx 0, -x_2^2 + 1/2 - v_2 \approx 0\}$, and $F' = \{v_1 > 0, v_2 > 0\}$.

3.1   Elimination Ideal

Let E denote a set of polynomial equations over $\mathbb{Q}[\boldsymbol{x}, \boldsymbol{v}, \boldsymbol{w}]$. We assume the implicit constraints $\boldsymbol{v} > 0$ and $\boldsymbol{w} \geq 0$. Our goal is to detect unsatisfiability of E. Toward this end, we compute the Gröbner basis for the polynomials in E. Since the witnesses are likely to be in terms of $\boldsymbol{v}, \boldsymbol{w}$, we use an ordering with precedence $\boldsymbol{x} \succ \boldsymbol{v}, \boldsymbol{w}$ (that is, we are eliminating $\boldsymbol{x}$). If we are lucky, the computed Gröbner basis may already contain a witness for unsatisfiability of E.

Example 2. Consider the set $E = \{x_1^3 - x_1 \approx 0, x_1 x_2 - 1 - v_1 \approx 0, -x_2^2 + 1/2 - v_2 \approx 0\}$ and $F' = \{v_1 > 0, v_2 > 0\}$ from Example 1.

Computing a Gröbner basis for the polynomials in E (using a lexicographic ordering with precedence $x_1 \succ x_2 \succ v_1 \succ v_2$) and then removing all polynomials that contain variables x_1 and x_2, we are left with $\{v_1^3 + 3v_1^2 + 1/2 v_1 v_2 + 5/2 v_1 + 1/2 v_2 + 1/2\}$. This set is a basis for the ideal $Elim(Poly(E), \{x_1, x_2\})$.

The equation $v_1^3 + 3v_1^2 + 1/2 v_1 v_2 + 5/2 v_1 + 1/2 v_2 + 1/2 \approx 0$ is a witness for unsatisfiability: since the constraints $v_1 > 0, v_2 > 0$ imply that $v_1^3 + 3v_1^2 + 1/2 v_1 v_2 + 5/2 v_1 + 1/2 v_2 + 1/2 > 0$ necessarily, whereas Gröbner basis computation shows that it is necessarily zero. We can conclude that the original set of equations and inequalities (from Example 1) is also unsatisfiable.

The method of introducing slack variables and computing Gröbner basis with respect to an ordering that makes the slack variables smallest is not complete.

Example 3. If $E = \{x^2 - 2x + 2 \approx 0\}$, then the procedure described above would not introduce any new "slack" variables. The elimination ideal contains only the 0 polynomial, which results in a set of consistent equations $(0 \approx 0)$. However, E is unsatisfiable over the reals.

We wish to make the procedure complete using the positivstellensatz characterization of unsatisfiability over the reals.

4   Positivstellensatz

The following result in real algebraic geometry characterizes the unsatisfiability of a conjunction of nonlinear equations and inequalities. Given a set Q of polynomials, the monoid $[Q]$ generated by Q is the set consisting of all finite products of polynomials in Q, and the cone generated by Q is the smallest set containing $[Q]$ and closed under addition and multiplication by "perfect-square polynomials", that is,

$$[Q] = \{\Pi_{i \in I} \, q_i : q_i \in Q \text{ for all } i \in I\}$$
$$Cone[Q] = \{\Sigma_{i \in I} \, p_i^2 q_i : q_i \in [Q], p_i \in \mathbb{Q}[\boldsymbol{x}] \text{ for all } i \in I\}$$

Note that $1 \in [Q]$ for any set Q and $c^2 \in Cone[\emptyset]$ for all $c \in \mathbb{Q}$.

Theorem 1. *[Positivstellensatz [6, 14, 20]] Let P, Q, and R be sets of polynomials over $\mathbb{Q}[\boldsymbol{x}]$. The constraint*

$$\{p \approx 0 : p \in P\} \cup \{q \geq 0 : q \in Q\} \cup \{r \not\approx 0 : r \in R\}$$

is unsatisfiable (over the reals) iff there exist polynomials p, q, and r such that $p \in Ideal(P)$, $q \in Cone[Q]$, and $r \in [R]$ and $p + q + r^2 = 0$.

The theorem is difficult to use in its above form. However, we can replace the inequality constraints by equality constraints using slack variables and use the following corollary.

Corollary 1. *Let P be a set of polynomials from $\mathbb{Q}[\boldsymbol{x}, \boldsymbol{v}, \boldsymbol{w}]$. The constraint*

$$\{p \approx 0 : p \in P\} \cup \{v > 0 : v \in \boldsymbol{v}\} \cup \{w \geq 0 : w \in \boldsymbol{w}\}$$

is unsatisfiable iff there is a polynomial p' such that $p' \in Ideal(P) \cap Cone[\boldsymbol{v}, \boldsymbol{w}]$ and there is at least one monomial $c\mu$ in p' such that $c > 0$ and $\mu \in [\boldsymbol{v}]$.

Proof. The $\Leftarrow$ direction is obvious. For the $\Rightarrow$ direction, we use the Positivstellensatz to conclude that there exist polynomials p, q, and r such that $p \in Ideal(P)$, $q \in Cone[\boldsymbol{v}, \boldsymbol{w}]$, and $r \in [\boldsymbol{v}]$ and $p + q + r^2 = 0$. Note that $r^2 \in Cone[\boldsymbol{v}, \boldsymbol{w}]$ and hence the polynomial $q + r^2 \in Cone[\boldsymbol{v}, \boldsymbol{w}] \cap Ideal(P)$.

To prove that the polynomial $q + r^2$, equivalently $-p$, is the required p', we need to show that the polynomial $q + r^2$ contains a monomial $c\mu$ such that $c > 0$ and $\mu \in [\boldsymbol{v}]$. (Note that r^2 is such a monomial, but it could get canceled when added to q.) Suppose $p' = q + r^2$ and p' contains no such monomial $c\mu$. But then, if we set all x, w to 0 and all v to 1 (or any positive number), then q will evaluate to something greater-than or equal to 0 (by definition of $Cone$), r^2 will evaluate to something strictly greater-than 0, and hence $q + r^2$ will evaluate to something strictly positive, whereas each monomial in p' will evaluate to either 0 or something negative (since every monomial $c\mu$ in p' has either $c < 0$ or a factor from x, w). This contradiction concludes the proof. ∎

We have now reduced the problem of testing unsatisfiability of nonlinear constraints to deciding if, given a finite set P of polynomials over $\mathbb{Q}[x, v, w]$, does there exist a polynomial $p \in Ideal(P) \cap Cone[v, w]$ that also contains a monomial $c\mu$ with $c > 0$ and $\mu \in [v]$. The polynomial p is the witness to unsatisfiability. We need to search for the existence of such a p.

5   Searching a Witness, Searching an Ordering

It would be nice if we could establish that if such a witness p (to unsatisfiability) exists, then it would be present in the Gröbner basis of P. Note that this was indeed the case in Example 2. But this may not be true always. Fortunately, the property of fully reduced convergent rewrite systems discussed in Section 2 guarantees that this will be true *if* p were the minimal nonzero polynomial in $Ideal(P)$ with respect to the ordering $\succ$ used to construct the Gröbner basis for P. However, standard restrictions on term-orderings, such as monotonicity $(xy \succ x)$, could mean that under no admissible ordering p were minimal.

Example 4. Consider $P = \{v + w_1 - 1, w_1 w_2 - w_1 + 1\}$. Note that we implicitly assume that $v > 0$ and $w_1, w_2 \geq 0$. The set P is a Gröbner basis for the ideal generated by P with respect to the lexicographic ordering with $v \succ w_1 \succ w_2$.

There is a witness polynomial, $v + w_1 w_2$, in $Ideal(P)$, but P itself does not contain any witness polynomial. In fact, none of the fully reduced canonical Gröbner basis computed using *any* lexicographic ordering contains a witness polynomial for this example.

The problem here is that the witness polynomial $v + w_1 w_2 \in Ideal(P)$ is not a minimal nonzero polynomial in $Ideal(P)$ under any ordering. However, if we could have $w_1 \succ w_1 w_2$ (contrary to the requirements of term orderings), then Gröbner basis computation "could" eliminate w_1 by adding the two polynomials in P and obtain the witness.

We know from Corollary 1 that the witness polynomial p is in $Ideal(P) \cap Cone[v, w]$ and hence it is of the form $p_1^2 \nu_1 + p_2^2 \nu_2 + \cdots + p_k^2 \nu_k$ where, for all i, $\nu_i \in [v, w]$ and p_i is an arbitrary polynomial. There are two issues with making this minimal:

(i) The power-products ν_i can not be smaller than the individual variables contained in them. This was illustrated in Example 4.

(ii) The squares p_i^2 can not be smaller than any of the monomials or variables contained in them.

We solve both these problems using the idea of introducing new definitions and new variables. The new variables will be forced to be smaller than the other variables.

Example 5. Consider $P = \{v + w_1 - 1, w_1 w_2 - w_1 + 1\}$ from Example 4. If we introduce a new definition $D = \{w_1 w_2 \approx w_3\}$, and we choose an ordering in which $v \succ w_1 \succ w_2 \succ w_3$, then $GB_\succ(P \cup \{w_1 w_3 - w_3\}) = \{v + w_3, w_1 - w_3 - 1, w_2 w_3 + w_2 - w_3\}$. The witness $v + w_3$ is present in the above fully reduced Gröbner basis.

Next consider $P = \{w_1^2 - 2w_1 w_2 + w_2^2 + 1\}$. There is no polynomial with all coefficients positive in $Ideal(P)$ [11]. But there is a witness containing perfect squares: $(w_1 - w_2)^2 + 1$. If we introduce the definition $D = \{(w_1 - w_2)^2 \approx w_3\}$ and compute the Gröbner basis for $P \cup \{(w_1 - w_2)^2 - w_3\}$, we get $\{w_3 + 1, w_1^2 - 2w_1 w_2 + w_2^2 - w_3 a\}$. The witness $w_3 + 1$ is present in the above fully reduced Gröbner basis.

5.1 Completeness Relative to an Oracle

If an oracle can identify the monomials $p_i^2 \nu_i$ that are present in the witness, then the introduction of definitions and computing Gröbner basis is a sound and complete method for detecting unsatisfiability of nonlinear constraints.

If all coefficients in a polynomial are positive (negative) when it is written in its sum of monomials normal form representation, then we call it a positive (negative) polynomial.

Theorem 2 (Relative Completeness). *Let P be a set of nonlinear equations over $\mathbb{Q}[\boldsymbol{x}, \boldsymbol{v}, \boldsymbol{w}]$. Let $\Sigma_{i=1}^k p_i^2 \nu_i$ be a witness for unsatisfiability of $\{p \approx 0 : p \in P\} \cup \{v > 0 : v \in \boldsymbol{v}\} \cup \{w \geq 0 : v \in \boldsymbol{w}\}$. Let D be the set of definitions $\{p_i^2 \nu_i - w_i' : i = 1, \ldots, k\}$, where w_i' are new constants.*
Then, there exists a precedence $\succ'$ on $\boldsymbol{w}'$ such that $GB_\succ(P \cup D)$ will contain a positive or negative polynomial over $[\boldsymbol{w}']$, where $\succ$ extends $\succ'$ such that the only power-products smaller than any w' are possibly other variables in $\boldsymbol{w}'$.

Proof. By Corollary 1, the polynomial $\Sigma_{i=1}^k p_i^2 \nu_i$ is in the ideal generated by P. Therefore, the linear polynomial $w_1' + \cdots + w_k'$ is in the ideal generated by $P \cup D$. Since the ordering $\succ$ guarantees that linear polynomials over $\boldsymbol{w}'$ are smaller than other polynomials, it follows that the polynomial $w_1' + \cdots + w_k'$ is in the ideal generated by $GB' = GB \cap \mathbb{Q}[\boldsymbol{w}']$ (property of fully reduced convergent systems). If this polynomial is in GB', we are done.

If not, let $p' = c_1 w_1' + \cdots + c_k w_k'$ be the minimal size (that is, with least cardinality of $\{i : c_i \neq 0\}$) linear positive polynomial in the ideal generated by GB'. We claim that p' will be in GB' if we choose $\succ$ so that each constant in $\{w_i' : c_i \neq 0\}$ has lower precedence than other variables. Suppose p' is not in GB'. Then p' is reducible by some polynomial q' in GB'. The polynomial $q' = d_1 w_1' + \cdots + d_k w_k'$ is necessarily linear. Wlog assume that $c_1 > 0$ and $d_1 > 0$. (i) If there is a j s.t. $d_j \neq 0$, but $c_j = 0$, then w_j' is greater than all constants in p', and hence q' can not reduce p'. (ii) Consider $p' - c_j q'/d_j$, where

$j = min\{c_l/d_l : d_l > 0, l = 1, \ldots, k\}$. Note that if q' is positive/negative, then we are done. Hence, we assume that q' is not positive, and consequently j is well-defined. Now, clearly $p' - c_j q'/d_j$ is positive, and smaller than p' in size, a contradiction. This completes the proof. ∎

As we have seen in Section 2, there are several orderings $\succ$ that can extend $\succ'$ in the required way. One example is the total degree lexicographic ordering with precedence $x \succ v \succ w \succ w'$.

6   The Inference Rules

Following the presentation of Gröbner basis computation as a critical-pair completion procedure [2, 3], we present the inference rules that describe a procedure for testing unsatisfiability of nonlinear constraints. It consists of rules that compute Gröbner basis and rules that insert new definitions, which are required for identifying witnesses.

The inference rules operate on states. A state (V, P) consists of a set P of polynomials and a set V of variables occurring in P. We also implicitly maintain subsets $V_{>0}$ and $V_{\geq 0}$ of V. The initial state is $(\{x, v, w\}, P)$, where P is the set of polynomials obtained by adding slack variables to the original nonlinear constraints as described in Section 3 before and $V_{>0} = \{v\}$ and $V_{\geq 0} = \{v, w\}$.

We use an ordering $\succ$ on polynomials. As we observed in Section 3, it is a good heuristic to use a precedence $x \succ v, w$; more generally, $V - V_{\geq 0} \succ V_{\geq 0}$. We also assume that the ordering guarantees that *only linear polynomials are smaller than a linear polynomial*, cf. Theorem 2. When we write a polynomial as $c_0\mu_0 + p$, then we implicitly mean that μ_0 is the largest power-product, that is, $\mu_0 \succ p$ and $c_0 \neq 0$. As we add new definitions, such as $p - w'$, where $w' \in V^{new}$ is a new variables, we need to extend the ordering. We can choose any extension that guarantees that $p \succ w'$. Note that the new variable w' can be added to either $V - V_{\geq 0}$ or $V_{\geq 0}$. In most cases, we can extend the ordering without violating the invariant that $V - V_{\geq 0} \succ V_{\geq 0}$.

The inference rules are presented in Table 1. The inference rules *Simplify*, *Deduce*, and *Delete* are used for constructing a Gröbner basis of the polynomials in the set P. Note that the rules for normalizing an arbitrary polynomial expression into a sum of monomial form (with the largest monomial moved to the left and its coefficient normalized to 1) are left implicit in the presentation here; they have been formalized in previous presentations of Gröbner basis algorithm as completion [2, 3]. The collapse rule is subsumed by the *Simplify* rule.

The novelty in the inference rules in Table 1 comes from the rules that add new definitions. We use the largest monomials in P to determine the terms to be named by new variables. The notation $|\mu|$ denotes the total degree of the power-product μ. The notation $[V]^{0,1}$ denotes power-products in which every variable in V occurs with degree at most one.

The *Extend1* rule introduces a new nonnegative variable w' as a name for leading power-product μ_0 that is known to be nonnegative, that is, $\mu_0 \in [V_{\geq 0}]$. The *Extend2* rule introduces a new name for $\nu_0 + \alpha\nu_1$, in the hope that some

Table 1. Inference rules for detecting unsatisfiability of nonlinear constraints

Simplify:
$$\frac{(V, P \cup \{c_0\mu_0 + p, q\})}{(V, P \cup \{c_0\mu_0 + p, q'\})} \qquad \text{if } q \rightarrow_{c_0\mu_0 \rightarrow -p} q'$$

Deduce:
$$\frac{(V, P' = P \cup \{c_0\mu_0 + p, d_0\nu_0 + q\})}{(V, P' \cup \{c_0\mu'q - d_0\nu'p\})} \quad \begin{array}{l} \text{if } \mu_0\mu' = \nu_0\nu' = lcm(\mu_0, \nu_0) \neq \\ \mu_0\nu_0 \end{array}$$

Delete:
$$\frac{(V, P \cup \{0\})}{(V, P)}$$

Extend1:
$$\frac{(V, P' = P \cup \{\mu_0 + p\})}{(V \cup \{w'\}, P' \cup \{\mu_0 - w'\})} \qquad \text{if } \mu_0 \in [V_{\geq 0}], \ w' \in V_{\geq 0}^{new}$$

Extend2:
$$\frac{(V, P)}{(V \cup \{x'\}, P \cup \{\nu_0 + \alpha\nu_1 - x'\})} \qquad \text{if } \langle \nu_0, \nu_1 \rangle \text{ occurs in } P, \ x' \in V^{new}$$

Extend3:
$$\frac{(V, P' = P \cup \{\mu_0 + p\})}{(V \cup \{x'\}, P' \cup \{\nu_0 - x'\})} \qquad \begin{array}{l} \text{if } \nu_0^2\nu_0' = \mu_0\mu_0', \ \nu_0' \in [V_{\geq 0}]^{0,1}, \\ x' \in V^{new}, \ |\nu_0| > 1 \end{array}$$

Detect:
$$\frac{(V, P' = P \cup \{c_0\mu_0 + p\})}{(V, P \cup \{c_0\mu_0, p\})} \qquad \begin{array}{l} \text{if } c_0\mu_0 + p \text{ is a positive/negative} \\ \text{polynomial over } [V_{\geq 0}] \end{array}$$

Witness:
$$\frac{(V, P \cup \{c\mu\})}{\perp} \qquad \text{if } \mu \in [V_{>0}], \ c \neq 0$$

polynomial of the form $(\nu_0 + \alpha\nu_1 + p)^2$ appears in the unsatisfiability witness. We say that a power-product ν *occurs directly in* P if there is polynomial in P which contains a monomial with power-product ν. We generalize this notion and say that a power-product ν *occurs in* P with factor $\nu_0' \in [V_{\geq 0}]^{0,1}$ if there exists $\mu_0 \in [V]$ such that $\mu_0|\nu\nu_0'$ and μ_0 occurs directly in P. (As a heuristic rule, we prefer cases when $\mu_0 = \nu\nu_0'$.) Finally, we say that a pair of power-products $\langle \nu_0, \nu_1 \rangle$ *occurs in* P if (i) $\nu_0\nu_1$ occurs in P with factor ν_0', and (ii) either $\nu_0^2\nu_0'$ occurs in P with factor 1 or $\nu_0^3\nu_0'/w$ occurs in P with factor 1 for some $w \in V_{\geq 0}$, and (iii) either $\nu_1^2\nu_0'$ occurs in P with factor 1 or $\nu_1^3\nu_0'/w$ occurs in P with factor 1 for some $w \in V_{\geq 0}$.

In the *Extend2* rule, the symbol α denotes a (real) rigid variable that needs to be instantiated by a constant in $\mathbb{Q}$. We use symbolic α here and continue application of the inference rules by working over the field $\mathbb{Q}(\alpha)$ (instead of just $\mathbb{Q}$). As soon as we obtain a nontrivial expression in $\mathbb{Q}(\alpha)$, we instantiate α by the zero of that expression. The *Extend3* rule says that we need not bother about finding ν_1 (and α) if total degree of ν_0 is greater-than one.

We have not shown that the new variables introduced in *Extend* rules are pushed appropriately into the sets $V_{\geq 0}$ or $V_{>0}$.

Example 6. Consider the set $P = \{v + w_1 - 1, w_1w_2 - w_1 + 1\}$ from Example 4. Assuming $v > 0, w_1 \geq 0, w_2 \geq 0$, one possible derivation to $\perp$ is shown below. To illustrate the other extension rules, we also show the derivation with a new set $P = \{x_1^2 - 2x_2 + 3, x_1x_2 - x_2^2\}$ below.

i	Polynomials P_i	Transition Rule
0	$\{v + w_1 - 1, w_1 w_2 - w_1 + 1\}$	
1	$\{v + w_1 - 1, w_1 w_2 - w_1 + 1, w_1 w_2 - w'\}$	**Extend1**
2	$\{v + w_1 - 1, -w_1 + w' + 1, w_1 w_2 - w'\}$	**Simplify**
3	$\{v + w', -w_1 + w' + 1, w_1 w_2 - w'\}$	**Simplify**
4	$\{v, w', -w_1 + w' + 1, w_1 w_2 - w'\}$	**Detect**
5	$\perp$	**Witness**

i	Polynomials P_i	Rule
0	$P = P_0 = \{x_1^2 - 2x_2 + 3, x_1 x_2 - x_2^2\}$	
1	$P_0 \cup \{x_1 + \alpha x_2 - y_1\}$	**Extend2**
2	$\{x_2^2 - 2\alpha x_2 y_1 + y_1^2 - 2x_2 + 3, -(\alpha + 1)x_2^2 + y_1 x_2, x_1 + \alpha x_2 - y_1\}$	**Simplify**
3	$\{x_2^2 + 2x_2 y_1 + y_1^2 - 2x_2 + 3, y_1 x_2, x_1 - x_2 - y_1\}$	$\alpha \mapsto -1$
4	$\{x_2^2 + y_1^2 - 2x_2 + 3, y_1 x_2, x_1 - x_2 - y_1\}$	**Simplify**
5	$P_4 \cup \{x_2 + \beta - y_2\}$	**Extend2**
6	$\{y_1^2 + y_2^2 - (2\beta + 2)y_2 + (\beta^2 + 2\beta + 3), y_1 y_2 - \beta y_1, \ldots\}$	**Simplify**
7	$\{y_1^2 + y_2^2 + 2, y_1 y_2 + y_1, \ldots\}$	$\beta \mapsto -1$
8	$\{y_1^2, y_2^2, 2, y_1 y_2 + y_1, \ldots\}$	**Detect**
9	$\perp$	**Witness**

Lemma 1. *Suppose $(V, P) \vdash (V', P')$ is a one-step application of the inference rules. Then, P is satisfiable over the reals iff P' is satisfiable over the reals.*

Theorem 3 (Refutational completeness). *Suppose P_0 is unsatisfiable and $(V_0, P_0) \vdash^* (V, P)$ is a derivation such that $P \neq \perp$. Then, there exists a derivation from (V, P) to $\perp$.*

Proof. By Lemma 1 we conclude that P is unsatisfiable. Therefore, by Corollary 1 we know that there exist several witness polynomials $wp = \Sigma_i p_i^2 \nu_i \in Ideal(P) \cap Cone[V_{\geq 0}]$ for unsatisfiability of P. Assume that $p_0 \succeq p_1 \succeq p_2 \succeq \cdots$; and whenever $p_i \not\succ p_{i+1}$ then $\nu_i \succeq \nu_{i+1}$. Let μ be the leading power-product (LPP) of the largest polynomial in P that divides the leading power-product of $p_0^2 \nu_0$. If no such μ exists, then μ is set to 1. Now, we say that the witness polynomials wp (and the corresponding μ) is *bigger than* wp' (and the corresponding μ') if either the multiset $\{|p_0|, |p_1|, \ldots\}$ of the sizes of the p_i's is greater-than the multiset of the sizes $\{|p_0'|, |p_1'|, \ldots\}$; or these are equal and the size of μ' is greater-than the size of μ. (Note here that the size of a polynomial is just the multiset of the sizes of its monomials and the size of a monomial is the total-degree of its power-product.) This ordering on witnesses is clearly well-founded and hence a minimal is well-defined. Let $wp = \Sigma_i p_i^2 \nu_i'$ be such a minimal witness.

Note that none of the inference steps can increase the size of the minimal witness. We will show below that either we can always reduce the size of the minimal witness by applying an appropriate inference rule, or reach $\perp$.

Since we have the inference rules for constructing a Gröbner basis, we can assume that the polynomials in P form a Gröbner basis. Hence, there exists a polynomial $\mu_0 - p \in P$ such that $\mu_0 | LPP(p_0^2 \nu_0')$. If $\nu_0 = LPP(p_0)$, then we should have $\mu_0 | \nu_0^2 \nu_0'$, or equivalently, $\mu_0 \mu_0' = \nu_0^2 \nu_0'$ for some μ_0'.

Case 1. $|\nu_0| > 1$. In this case, the *Extend3* rule is applicable. Using this rule, we can introduce a variable equivalent to ν_0. In the minimal witness, if we replace ν_0 by this variable, then we get a smaller witness.

Case 2. $|\nu_0| = 0$. In this case, $\mu_0|\nu_0'$ and hence $\mu_0 \in [V_{\geq 0}]$. Hence the *Extend1* rule is applicable. If $|\mu_0| > 1$, we can again get a smaller witness as in Case 1. If $|\mu_0| = 1$, then the rule $\mu_0 - p$ is necessarily linear (because the ordering guarantees that only linear polynomials are smaller than linear polynomials). Also, each p_i is a constant. Let $c_i = p_i^2$. Consider two cases now.

Case 2.1. There is a rewrite step using a nonlinear polynomial in the derivation $\Sigma_i c_i \nu_i' \to_P^* 0$. Wlog assume ν_0' is rewritten to $c''\nu_0'' + \ldots$ by linear rules, and ν_0'' is reducible by a nonlinear rule. Using *Extend1* rules, we make ν'' bigger than ν_0'. As a result, in the new system, the nonlinear rule directly reduces the witness. Hence, the size of the witness remains unchanged, but the the size of the corresponding μ_0 increases (see the example following the proof).

Case 2.2. There is a no rewrite step using a nonlinear polynomial in the derivation $\Sigma_i c_i \nu_i' \to_P^* 0$. In this case, the linear polynomials in P are unsatisfiable. Therefore, there exists a smallest linear witness $\Sigma_i c_i w_i$ s.t. $c_i > 0$ and $w_i \in V_{\geq 0}$ (and there is some j s.t. $w_j \in V_{>0}$). Again, using the *Extend1* rule, we can make the variables w_i appearing in this witness smaller. As a result, the polynomial $\Sigma_i c_i w_i$ will appear in the set P and we would detect inconsistency (as in the proof of Theorem 2).

Case 3. $|\nu_0| = 1$. Our assumption on the ordering guarantees that all p_i's in the witness $wp = \Sigma_{i \geq 0} p_i^2 \nu_i'$, where $\nu_i' \in [V_{\geq 0}]^{0,1}$ are linear polynomials. Suppose $p_i = c_{i0} w_{i0} + \cdots + c_{il} w_{il}$. In the monomial expansion of $p_i^2 \nu_i'$, we distinguish between the *cross-product terms*, which are of the form $2c_{ij}d_{ik}w_{ij}w_{ik}\nu_i'$ (for $j \neq k$), and the *square terms*, which are of the form $c_{ij}^2 w_{ij}^2 \nu_i'$. We wish to identify w_{ij} and w_{ik} and apply the *Extend2* rule. The problem is that the cross-product terms can cancel out in the summation $\Sigma_{i \geq 0} p_i^2 \nu_i'$ and hence the witness wp may not contain any monomial whose power-product is, for instance, $w_{ij}w_{ik}\nu_i'$ (and hence the polynomials in P also may not contain this power-product).

Case 3.1. There is no monomial in wp whose power-product comes from a cross-product term. In this case the polynomial wp is itself of the form $\Sigma_i q_i^2 \nu_i'$, where q_i's are all monomials now. We conclude that the original p_i's are necessarily monomials: if not, then the new witness would be a smaller witness. If $|q_0^2 \nu_0'| > 1$, we can use *Extend1* on the leading monomial $q_0^2 \nu_0'$ and reduce the size of the witness. If $|q_0^2 \nu_0'| = 1$, the witness polynomial wp is linear. We make the variables that occur in wp minimal using *Extend1*. This will cause the witness polynomial to appear explicitly in P, whence we can use detect and witness to reach the $\bot$ state.

Case 3.2. There is a monomial in wp whose power-product comes from a cross-product term. Let the power-product be $w_{ij}w_{ik}\nu_i'$. If both $w_{ij}^2\nu_i'$ and $w_{ik}^2\nu_i'$ are also present in wp, then they also necessarily occur in P, and hence, the *Extend2* rule would be applicable and it would introduce the polynomial $(w_{ij} + \alpha w_{ik}) - w'$ for some new variable w'. If α is appropriately instantiated, this reduces the witness size.

Suppose $w_{ij}^2 \nu_i'$ is not present in wp. This implies that it was canceled in the summation. It can only be canceled by a cross-product term. That cross-product term can only come from something of the form $(\cdots + w_{ij} + w + \cdots)^2 w_{ij}(\nu_i'/w)$ (ignoring coefficients). But this implies that there will be a square term of the form $w_{ij}^3(\nu_i'/w)$. This term can not be canceled by any other cross-product term. Hence we can detect w_{ij} by searching for the occurrence of either $w_{ij}^2 \nu_i'$ or $w_{ij}^3(\nu_i'/w)$. In the latter case, note that $w, w_{ij} \in V_{\geq 0}$. This completes the proof. ∎

To illustrate the second case of the above proof, consider $P = \{v - v_1 + v_2, v_1 w - v_2 w + 1\}$. The witness for unsatisfiability is $vw + 1$. We notice that $vw + 1 \rightarrow v_1 w - v_2 w + 1 \rightarrow 0$ by P. Here the nonlinear polynomial in P is used after reducing the witness using the linear rules. Hence, we apply *Extend1* to make v smaller than v_1 by adding $v - v'$. After closing under the Gröbner basis rules, the result is $P = \{v_1 - v_2 - v', v'w + 1\}$ and the new witness is $v'w + 1$. Now, $\mu_0 = v'w$, which divides the leading power-product $v'w$ of the witness. The size of μ_0 that reduces $LPP(wp)$ has increased from 1 to 2.

6.1 Other Remarks and Future Work

The *Extend* rules can potentially introduce infinitely many new definitions in a derivation, thus leading to nontermination. Specifically, there are infinitely many choices in the application of the *Extend3* rule. If effective degree bounds on the witness polynomial (obtained using the Positivstellensatz) are known, then the application of the *Extend* rules (*Extend3* in particular) can be restricted to enforce termination, resulting in a decision procedure. The problem of obtaining effective degree bounds for the Positivstellensatz is still open [19]. We conjecture that the *Extend3* inference rule can be restricted to use only the minimal instance of ν_0 (that is, only the most-general unifier of $\nu_0^2 \nu_0' = \mu_0 \mu_0'$) and that this could be used to obtain a terminating set of inference rules that are also complete. This could provide an alternate approach to obtaining degree bounds for the Positivstellensatz.

The process of searching for the witnesses can be understood in the context of the Gröbner basis P as follows. The monomials in a polynomial can be colored by *pos* and *unknown* based on whether we know if they are necessarily nonnegative or not. For example, in $x^2 - 2x + 2$, the monomials x^2 and 1 are *pos*, while $-2x$ is *unknown*. The goal is to construct a polynomial in the ideal of P in which the *unknown* monomials have been eliminated. There are two ways to eliminate the *unknown* monomials. First, they can be captured as a cross-product term in a perfect-square polynomial. For example, $(x - 1)^2$ captures $-2x$. The inference rule *Extend2* aims to achieve this. Second, the monomials can be canceled when constructing a polynomial in the ideal of P. For example, consider the polynomials $v^2 - w_1 w_2 + 1$ and $w_1 w_2 + w_3 - 1$. The monomial $-w_1 w_2$ can be canceled by adding the two polynomials. This is reflected in the "critical-pair" overlap between $-w_1 w_2$ and $w_1 w_2$. However, since $-w_1 w_2$ is not the largest monomial in the first polynomial, Gröbner basis computation will not perform this operation. The *Extend1* rule aims to make the leading monomials smaller, so that the inner monomials such as $-w_1 w_2$ are exposed for critical-pair computation. This is clearly a generalization of the pivoting step in Simplex. The colored monomials can also be used to restrict the choices in the application of the *Extend* rules.

The value of the proposed approach for unsatisfiability testing of nonlinear constraints arises from the fact that it successfully solves the "easy" instances cheaply. A lot of the easy unsatisfiable instances are detected just by adding slack variables and then projecting the polynomial ideal onto the slack variables. This was illustrated in Example 2. In most of the other instances, we noticed that we need to apply the *Extend* rules at most one or two times to detect inconsistency. We also remark here that several other decision procedures for nonlinear constraints tend to do expensive computations on relatively simple instances. For example, if we have two constraints, $p > 0$ and $p < 0$ over $\mathbb{Q}[x_1, \ldots, x_n]$, then a naive procedure based on cylindrical algebraic decomposition, for instance, would attempt to create a decomposition of $\mathbb{R}^n$ based on the polynomial p. For sufficiently large p, this process could fail (run out of memory or time). It is easy to see that our procedure will generate the unsatisfiability witness $v_1 + v_2$, where v_1 and v_2 are the two slack variables, in just one inference step.

We believe that fast implementations for unsatisfiability testing of nonlinear constraints can be obtained by implementing (an extension or variant of) the inference rules presented here. One missing aspect is detecting satisfiable instances quickly. However, simple ideas to detect satisfiable instances can be integrated. In particular, we can use the fact that *every odd degree polynomial has a zero* to eliminate an old variable when we introduce a new variable. This can be done if the new variable names an odd-degree polynomial over the old variable.

Example 7. Consider $P = \{x^2 + 2x - 1\}$. We introduce a new variable y for $x+1$ to get $P_1 = \{y^2 - 2, x + 1 - y\}$. Now, we can eliminate x from this set and just have $P_2 = \{y^2 - 2\}$. The reason is that if P_2 is satisfiable, then we are guaranteed that there will exist an assignment for x (since $x + 1 - y$ has an odd-degree in x). Since P_2 can be detected to be satisfiable, we can conclude that P is satisfiable.

7   Conclusion

We have presented an algebraic semi-decision procedure, based on Gröbner basis computation and extension rules, for detecting unsatisfiability of nonlinear constraints. The procedure is given as a set of inference rules that are sound and refutationally complete. Our approach has the potential of resulting in fast solvers for testing unsatisfiability of nonlinear constraints. This is especially significant in the context of satisfiability testing tools [1, 9, 21] that are being increasingly used for program analysis. There is also much recent progress in computational aspects of real algebraic geometry and computational tools for building a sums-of-squares representation using semi-definite programming [4, 15, 16], which indicates that our work will be actively refined and developed further in the future. We are presently exploring the effectiveness of the new procedure for improving the implementation of the abstraction algorithm for hybrid systems [23].

References

1. G. Audemard, P. Bertoli, A. Cimatti, A. Kornilowicz, and R. Sebastiani. A SAT based approach for solving formulas over boolean and linear mathematical propositions. In *CADE*, volume 2392 of *LNCS*, pages 195–210. Springer, 2002.
2. L. Bachmair and H. Ganzinger. Buchberger's algorithm: A constraint-based completion procedure. In *CCL*, volume 845 of *LNCS*. Springer, 1994.
3. L. Bachmair and A. Tiwari. D-bases for polynomial ideals over commutative noetherian rings. In *RTA*, volume 1103 of *LNCS*, pages 113–127. Springer, 1997.
4. S. Basu and L. Gonzalez-Vega, editors. *Algorithmic and Quantitative Real Algebraic Geometry*, volume 60 of *DIMACS Series in DMTCS*, 2003.
5. S. Basu, R. Pollack, and M.-F. Roy. On the combinatorial and algebraic complexity of quantifier elimination. *J. of the ACM*, 43(6):1002–1045, 1996.
6. J. Bochnak, M. Coste, and M.-F. Roy. *Real Algebraic Geometry*. Springer, 1998.
7. G. E. Collins. Quantifier elimination for the elementary theory of real closed fields by cylindrical algebraic decomposition. In *Proc. 2nd GI Conf. Automata Theory and Formal Languages*, volume 33 of *LNCS*, pages 134–183. Springer, 1975.
8. R. S. Datta. Using computer algebra to compute Nash equilibria. In *Intl. Symp. on Symbolic and Algebraic Computation, ISSAC 2003*, pages 74–79, 2003.
9. L. de Moura, S. Owre, H. Rueß, J. Rushby, and N. Shankar. The ICS decision procedures for embedded deduction. In *IJCAR*, volume 3097 of *LNAI*. Springer, 2004.
10. N. Dershowitz and Z. Manna. Proving termination with multiset orderings. *Communications of the ACM*, 22(8):465–476, 1979.
11. M. Einsiedler and H. Tuncel. When does a polynomial ideal contain a positive polynomial? *J. Pure Appl. Algebra*, 164(1-2):149–152, 2001.
12. J. Harrison. *Theorem proving with the real numbers*. Springer-Verlag, 1998.
13. H. Hong. Quantifier elimination in elementary algebra and geometry by partial cylindrical algebraic decomposition version 13, 1995.
    http://www.gwdg.de/~cais/systeme/saclib,www.eecis.udel.edu/~saclib/.
14. J. L. Krivine. Anneaux preordonnes. *J. Anal. Math.*, 12:307–326, 1964.
15. P. A. Parrilo. SOS methods for semi-algebraic games and optimization. In *HSCC 2005*, volume 3414 of *LNCS*, page 54. Springer, 2005.
16. S. Prajna, A. Papachristodoulou, and P. A. Parrilo. SOSTOOLS: Sum of Square Optimization Toolbox, 2002. http://www.cds.caltech.edu/sostools.
17. S. Ratschan. Applications of quantified constraint solving over the reals: Bibliography, 2004. http://http://www.mpi-sb.mpg.de/~ratschan/appqcs.html.
18. J. Renegar. On the computational complexity and geometry of the first order theory of the reals. *J. of Symbolic Computation*, 13(3):255–352, 1992.
19. M.-F. Roy. Degree bounds for Stengle's Positivstellensatz, 2003. Network workshop on real algebra. http://ihp-raag.org/index.php.
20. G. Stengle. A Nullstellensatz and a Positivstellensatz in semialgebraic geometry. *Math. Ann.*, 207:87–97, 1974.
21. A. Stump, C. W. Barrett, and D. L. Dill. CVC: A cooperating validity checker. In *CAV*, volume 2404 of *LNCS*, pages 500–504. Springer, 2002.
22. A. Tarski. *A Decision Method for Elementary Algebra and Geometry*. University of California Press, 1948. Second edition.
23. A. Tiwari and G. Khanna. Series of abstractions for hybrid automata. In *HSCC*, volume 2289 of *LNCS*, pages 465–478. Springer, 2002.
24. V. Weispfenning. The complexity of linear problems in fields. *J. of Symbolic Computation*, 5, 1988.

Coprimality in Finite Models

Marcin Mostowski[1] and Konrad Zdanowski[2]

[1] Department of Logic, Institute of Philosophy, Warsaw University
m.mostowski@uw.edu.pl
[2] Institute of Mathematics, Polish Academy of Science
konrad.zdanowski@wp.pl

Abstract. We investigate properties of the coprimality relation within the family of finite models being initial segments of the standard model for coprimality, denoted by $\mathrm{FM}((\omega, \bot))$.

Within $\mathrm{FM}((\omega, \bot))$ we construct an interpretation of addition and multiplication on indices of prime numbers. Consequently, the first order theory of $\mathrm{FM}((\omega, \bot))$ is Π_1^0–complete (in contrast to the decidability of the theory of multiplication in the standard model). This result strengthens an analogous theorem of Marcin Mostowski and Anna Wasilewska, 2004, for the divisibility relation.

As a byproduct we obtain definitions of addition and multiplication on indices of primes in the model $(\omega, \bot, \leq_{P_2})$, where P_2 is the set of primes and products of two different primes and $\leq_X$ is the ordering relation restricted to the set X. This can be compared to the decidability of the first order theory of $(\omega, \bot, \leq_P)$, for P being the set of primes (Maurin, 1997) and to the interpretation of addition and multiplication in $(\omega, \bot, \leq_{P^2})$, for P^2 being the set of primes and squares of primes, given by Bès and Richard, 1998.

Keywords: finite models, arithmetic, finite arithmetic, coprimality, interpretations, complete sets, FM–representability.

1 Introduction

This paper is devoted to the study of finite arithmetics, the research area concentrated on semantical and computational properties of arithmetical notions restricted to finite interpretations. Almost all computational applications of logic or arithmetic consider arithmetical notions in essentially finite framework. Therefore it is surprising that so little attention is directed to this area. This is particularly surprising when we observe that a few of classical papers in computer science (see e.g. Hoare [3], Gurevich [2]) postulate this research direction as particularly important.

Discussing the problem of analyzing algorithms in an implementation-independent way, Hoare essentially postulates proving their properties in appropriate axiomatic versions of finite arithmetic. We know that particular implementations of integers or unsigned integers (natural numbers) are essentially finite arithmetics with a distinguished upper bound.

By Trachtenbrot's theorem [17] we know that first order theories of nontrivial finite arithmetics cannot be axiomatizable. We know that the first order logic allowing arbitrary interpretations is axiomatizable. However, restricted to finite models it is axiomatizable only for poor vocabularies – for which it is recursive. Probably this was

L. Ong (Ed.): CSL 2005, LNCS 3634, pp. 263–275, 2005.

one of the reasons why Hoare's postulate did not motivate logicians to study the case of arithmetics with a finite bound. Nevertheless, let us observe that working in the standard infinite model of natural numbers is not easier in any way. The first order theory of this model is not arithmetical. On the other hand, the first order theory of any finite arithmetic is at most co-recursively enumerable, that is Π_1^0. Therefore we can expect much better axiomatic approximations in the finite case.

For this reason, we should firstly consider properties of finite arithmetics from the logical point of view. Only recently a few papers devoted mainly to this area have appeared, see [10], [14], [8], [12], [7][1]. Probably one of the reasons for the lack of interest in finite arithmetics in the past was the expectation that nothing surprising can be found under the restriction to a finite framework. Presently, we know that finite arithmetics have a lot of unexpected semantical and computational properties. Exponentiation is easier than multiplication [7], divisibility itself is as complicated as addition and multiplication [12].

In this paper we give a solution of a problem presented at the Finite Model Theory Workshop Będlewo 2003. The problem is to determine the strength of coprimality in finite models. We show that, although semantically essentially weaker than the full arithmetic, it is recursively equally complicated.

The other source of our inspiration was the method of truth definitions in finite models proposed in [10] and further investigated in [11], [5] and [6]. The crucial problem there was finding a way of representing some nontrivial infinite relations in finite models. This motivated the notion of FM–representability[2]. It is known that a large class of arithmetical relations can be FM–represented. One of the motivating problems of our investigation is the question how much built-in arithmetic we need to apply the method of truth definitions. We characterize FM–representability for the finite arithmetic of coprimality. Our characterization – surprisingly – means that coprimality is sufficiently strong for the application of the truth definitions method in finite models.

Finally, as a byproduct of our research, we obtain an improvement of some theorems by Bès and Richard [1] characterizing the expressive power of coprimality in the standard infinite model equipped with some weak fragments of the standard ordering.

2 Basic Notions

We start with the crucial definition of FM–domain.

Definition 1 *Let* $\mathcal{R} = (R_1, \ldots, R_k)$ *be a finite sequence of arithmetical relations on* ω *and let* $\mathcal{A} = (\omega, \mathcal{R})$. *We consider finite initial fragments of this model. Namely, for* $n \geq 1$, *by* $\mathcal{A}_n$ *we denote the following structure*

$$\mathcal{A}_n = (\{0, \ldots, n-1\}, R_1^n, \ldots, R_s^n),$$

where, for $i = 1, \ldots, k$, *the relation* R_i^n *is the restriction of* R_i *to the set* $\{0, \ldots, n-1\}$.
The FM–domain *of* $\mathcal{A}$, *denoted by* $\mathrm{FM}(\mathcal{A})$, *is the family* $\{\mathcal{A}_n : n > 0\}$.

---

[1] We do not claim that finite arithmetics were not considered in older papers at all, but not as the main topic.

[2] This notion was first considered in [10]. ("FM" stands for "Finite Models".) The paper [13] discusses some variants of the notion of FM–representability.

We assume that all considered models are in relational vocabularies. Thus, we think of addition or multiplication as ternary relations which describe graphs of corresponding functions. Nevertheless, we will write, e.g. $\varphi(x + y)$ with the intended meaning $\exists z\,(+(x, y, z) \land \varphi(z))$. Thus, the formula $\varphi(f(x))$ means that there exists z which is the value for $f(x)$ and φ is true about this z.

Definition 2 *We say that φ is true of $a_1, \ldots, a_r \in \omega$ in all sufficiently large finite models from $\mathrm{FM}(\mathcal{A})$ (shortly $\mathrm{FM}(\mathcal{A}) \models_{\mathrm{sl}} \varphi[a_1, \ldots, a_r]$) if and only if*

$$\exists k \forall n \geq k\; \mathcal{A}_n \models \varphi[a_1, \ldots, a_r].$$

Sometimes we also say that φ is true of $a_1, \ldots, a_r$ in almost all finite models from $\mathrm{FM}(\mathcal{A})$.

Of course, k as above should be chosen in such a way that $k > \max\{a_1, \ldots, a_r\}$.

Definition 3 *We say that $R \subseteq \omega^r$ is FM–represented in $\mathrm{FM}(\mathcal{A})$ by a formula $\varphi(x_1, \ldots, x_r)$ if and only if for each $a_1, \ldots, a_r \in \omega$ the following conditions hold:*

(i) $\mathrm{FM}(\mathcal{A}) \models_{sl} \varphi[a_1, \ldots, a_r]$ *if and only if $R(a_1, \ldots, a_r)$,*
(ii) $\mathrm{FM}(\mathcal{A}) \models_{sl} \neg\varphi[a_1, \ldots, a_r]$ *if and only if $\neg R(a_1, \ldots, a_r)$.*

The main characterization of the notion of FM–representability in $FM(\mathbb{N})$, for $\mathbb{N} = (\omega, +, \times)$, is given by the following theorem (see [10]).

Theorem 4 (FM–representability theorem) *Let $R \subseteq \omega^n$. R is FM–representable in $\mathrm{FM}(\mathbb{N})$ if and only if R is decidable with a recursively enumerable oracle.*

The first question related to FM–representability is the following: How weak arithmetical notions are sufficient for the FM–representability theorem? In [10] the theorem has been proven for addition, multiplication and concatenation. It is a straightforward observation that concatenation is superfluous. A few less trivial results in this direction were obtained in [7] and [12]. In particular, in the last paper it was proven that:

Theorem 5 *For each $R \subseteq \omega^r$, R is FM–representable in $\mathrm{FM}(\mathbb{N})$ if and only if R is FM–representable in FM–domain of divisibility, $\mathrm{FM}((\omega, |))$, where $a|b \equiv \exists x\; ax = b$.*

It is surprising that such a weak relation as divisibility is sufficient here. So, the following natural problem appears. Can this theorem be improved by replacing divisibility by some weaker notions? For example, coprimality, where the coprimality relation, $\perp$, is defined by the following equivalence:

$$a \perp b \equiv \forall x((x|a \land x|b) \Rightarrow \forall y\; x|y).$$

The answer is obviously negative. Let us consider the function f defined as

$$f(x) = \begin{cases} 4 & \text{if } x = 2, \\ 2 & \text{if } x = 4, \\ x & \text{otherwise.} \end{cases}$$

f is an automorphism of $(\omega, \perp)$. Moreover, f also preserves coprimality when it is restricted to initial segments $\{0, \ldots, n\}$, for $n \geq 4$,. Therefore, the set $\{2\}$ is not FM–representable in $\mathrm{FM}((\omega, \perp))$. However, surprisingly, in a weaker sense coprimality is as difficult as addition and multiplication, see Theorems 10, 18, and 19.

Let us observe that in the standard model coprimality, and even multiplication, are relatively weak relations. Indeed, the first order theory of $(\omega, \times, \leq_P)$ is decidable, see [9], where P is the set of prime numbers and $\leq_P$ is the ordering relation restricted to this set.

We use the notion, $\leq_X$, for various sets $X \subseteq \omega$, with the analogous meaning. The complement of the predicate $\perp$ is denoted by $\not\perp$.

In our work, we use the notion of a first order interpretation. For details, see the paper by Szczerba [16], where the method was codified for the first time in the model-theoretic framework. We recall shortly the main ideas.

Let τ and σ be vocabularies and, for simplicity, let σ contain only one n-ary predicate R. A sequence $\bar{\varphi} = (\varphi_U, \varphi_\approx, \varphi_R)$ of formulae in the vocabulary τ is a first order interpretation of models of the vocabulary σ if the free variables of φ_U are $x_1, \ldots, x_r$, the free variables of $\varphi_\approx$ are $x_1, \ldots, x_{2r}$ and the free variables of φ_R are $x_1, \ldots, x_{rn}$. The sequence $\bar{\varphi}$ defines in a model $\mathcal{A}$ of the vocabulary τ a model of the vocabulary σ in the following sense. A universe U, defined by φ_U, is the set of n–tuples from $\mathcal{A}$:

$$U = \{(a_1, \ldots, a_r) : \mathcal{A} \models \varphi_U[a_1, \ldots, a_r]\}.$$

The equality relation is given by $\varphi_\approx$ which should define an equivalence relation on U. The interpretation of R is defined by

$$R(\mathbf{a}_1, \ldots, \mathbf{a}_n) \text{ if and only if}$$

$$\exists \bar{a}_1 \in \mathbf{a}_1 \ldots \exists \bar{a}_n \in \mathbf{a}_n \ \mathcal{A} \models \varphi_R[\bar{a}_1, \ldots, \bar{a}_n],$$

where $\mathbf{a}_1, \ldots \mathbf{a}_n$ are equivalence classes of the relation defined by $\varphi_\approx$ in U. The number r is called the width of the interpretation.

We write $I_{\bar{\varphi}}(\mathcal{A})$ for the model defined by $\bar{\varphi}$ in $\mathcal{A}$.

Definition 6 *We say that $\bar{\varphi}$ is an interpretation of* $\mathrm{FM}(\mathcal{A})$ *in* $\mathrm{FM}(\mathcal{B})$ *if there is a monotone, unbounded function $f : \omega \longrightarrow \omega$ such that for each $n \geq 1$,*

$$I_{\bar{\varphi}}(\mathcal{B}_n) \cong \mathcal{A}_{f(n)}.$$

If $\bar{\varphi}$ is of width 1, φ_U defines an initial segment in each model from $\mathrm{FM}(\mathcal{B})$ and the isomorphism between $\mathcal{A}_{f(n)}$ and $\mathcal{B}_n$ is just identity then we say that $\bar{\varphi}$ is an IS–interpretation.

An IS–interpretation was used in [12] for proving Theorem 5. In our interpretation of $\mathrm{FM}(\mathbb{N})$ in $\mathrm{FM}((\omega, \perp))$ we define arithmetic on indices of prime numbers.

3   The Main Theorem

In what follows models of the form $(\omega, \perp)$ or $\mathrm{FM}((\omega, \perp))$ are called coprimality models.

Let $\{p_i : i \in \omega\}$ be the enumeration of primes, that is $p_0 = 2, p_1 = 3, \ldots$ For a natural number a we use the notion of the support of a, defined as $\mathrm{Supp}(a) = \{p_i : p_i | a\}$. We define the equivalence relation $\approx$ as follows:

$$a \approx b \iff \mathrm{Supp}(a) = \mathrm{Supp}(b).$$

For each a, the equivalence class of a is denoted by $[a]$. Let us observe, that in each model from $\mathrm{FM}((\omega, \perp))$ as well as in $(\omega, \perp)$ we cannot distinguish between elements being in the same equivalence class of $\approx$.

Definition 7 *A relation $R \subseteq \omega^r$ is coprimality invariant if $\approx$ is a congruence relation for R. This means that for all tuples $a_1, \ldots, a_r$ and $b_1, \ldots, b_r$ such that $a_i \approx b_i$, for $i = 1, \ldots, r$,*

$$(a_1, \ldots, a_r) \in R \iff (b_1, \ldots, b_r) \in R.$$

We define relations R_+ and $R_\times$ by the following conditions:

$$R_+([p_i], [p_k], [p_m]) \text{ if and only if } i + k = m,$$

$$R_\times([p_i], [p_k], [p_m]) \text{ if and only if } ik = m.$$

We identify these relations with their coprimality invariant versions on elements of ω, instead of $\omega/_\approx$. R_+ and $R_\times$ give an interpretation of addition and multiplication on indices of prime numbers. Our main result is that they are interpretable in $\mathrm{FM}((\omega, \perp))$

For the proof of our main theorem we need some facts about the distribution of prime numbers.

Let $\pi(x)$ be a function defined as

$$\pi(x) = \sum_{\substack{p \le x \\ p - \text{prime}}} 1.$$

The prime number theorem states that the limit $\pi(x)/(x/\ln(x))$ converges to 1 for x going to infinity. We need the following consequences of the prime number theorem.

Proposition 8 *For each $b \in \omega$ there is K such that for each $n \ge K$ and for each $i < b$ there is a prime q such that*

$$in \le q < (i+1)n.$$

Sierpiński has observed in [15] that $K = e^b$ suffices.

Proposition 9 *Let $0 < \varepsilon < 1$. There is N such that for all $x \ge N$ the interval $(x, x(1+\varepsilon))$ contains a prime.*

Essentially, Proposition 9 is one of the corollaries of the prime number theorem mentioned in [4].

The main theorem of this section is the following.

Theorem 10 *There is an interpretation $\bar{\varphi}$ of width 1 of $\mathrm{FM}(\mathbb{N})$ in $\mathrm{FM}((\omega, \perp))$ such that for each k there is n such that $\bar{\varphi}$ defines in the model $(\{0, \ldots, n-1\}, \perp)$ the relations R_+ and $R_\times$ on an initial segment of $\{0, \ldots, n-1\}$ of size at least k.*
   Moreover, the equality predicate is not used in the formulae from $\bar{\varphi}$.

Proof. We will prove the theorem through a sequence of lemmas.
   Firstly, we define some auxiliary notions. Let $\varphi_\approx(x, y)$ be the formula

$$\forall z(z \perp x \equiv z \perp y).$$

Obviously, this formula defines the relation $\approx$ in $(\omega, \perp)$. Ambiguously, we denote relations defined by $\varphi_\approx(x, y)$ in models $(\omega, \perp)$ and $\mathrm{FM}((\omega, \perp))$ by $\approx$. In all these models $\approx$ is a congruence relation. (It means that $\approx$ is an equivalence and for all $a, b, a', b' \in \omega$ such that $a \approx a'$ and $b \approx b'$ we have $a \perp b$ if and only if $a' \perp b'$.) Therefore, in all considered models we cannot differentiate elements which are in the relation $\approx$. So, we can consider models $M_{/\approx}$ instead of M. The equivalence class of $a \in |M|$ with respect to $\approx$ is denoted by $[a]$. The elements of $M_{/\approx}$ which are of the form $[a]$ for $a \in |M|$, can be identified with finite sets of primes, $\mathrm{Supp}(a)$.
   We define some useful predicates.

- $P(x) := \forall z, y(z \not\perp x \wedge y \not\perp x \Rightarrow z \not\perp y) - x$ is a power of prime,
- $x \in y := P(x) \wedge x \not\perp y - x$ is a power of prime dividing y.
- $\{p, q\}$ – a function denoting, for a pair of primes p, q, an element of an equivalence class of pq. We have no multiplication but elements a such that $a \approx pq$ are defined by the formula $\forall z(z \perp a \equiv (z \perp p \wedge z \perp q))$. Of course we cannot define the unique a with this property. Nevertheless, this element is unique up to $\approx$. So, when considering models of the form $M_{/\approx}$, it is simply unique.

We have some operations definable on the equivalence classes of $\approx$.

Lemma 11 *There are formulae in the coprimality language $\varphi_\cup(x, y, z)$, $\varphi_\cap(x, y, z)$, $\varphi_-(x, y, z)$ such that in each coprimality model M, the following conditions hold for each $a, b, c \in |M|$:*

- $M \models \varphi_\cup[a, b, c]$ *if and only if* $\mathrm{Supp}(a) \cup \mathrm{Supp}(b) = \mathrm{Supp}(c)$,
- $M \models \varphi_-[a, b, c]$ *if and only if* $\mathrm{Supp}(a) \setminus \mathrm{Supp}(b) = \mathrm{Supp}(c)$,
- $M \models \varphi_\cap[a, b, c]$ *if and only if* $\mathrm{Supp}(a) \cap \mathrm{Supp}(b) = \mathrm{Supp}(c)$.

Proof. As $\varphi_\cup(x, y, z)$ we can take

$$\forall w(w \perp z \equiv (w \perp x \wedge w \perp y)).$$

$\varphi_-(x, y, z)$ can be written as

$$\forall w(P(w) \Rightarrow (w \not\perp z \equiv (w \not\perp x \wedge w \perp y))).$$

$\varphi_\cap$ is expressible in terms of $\varphi_\cup$ and φ_-.   □

   It follows that in all coprimality models we can reconstruct a partial lattice of finite sets of primes. However, the operation $\cup$ is total only in the infinite model $(\omega, \perp)$.

The crucial fact is that in finite models from $FM((\omega, \bot))$ we can compare small elements of a given model by the following formula $\varphi_\prec(x, y) :=$

$$\exists z (P(z) \wedge z \bot x \wedge z \bot y \wedge \exists w\, \varphi_\cup(x, z, w) \wedge \neg \exists w\, \varphi_\cup(y, z, w)).$$

By $\varphi_\preceq(x, y)$ we mean the formula $\varphi_\prec(x, y) \vee \varphi_\approx(x, y)$.

For a finite set $X \subseteq \omega$, we write ΠX for the product of all numbers in X.

Lemma 12 *For each c there is N such that for all $n \geq N$ and for all a, b with $1 \leq a, b \leq n$ and $\max\{\Pi\text{Supp}(a), \Pi\text{Supp}(b)\} \leq c$ the following holds*

$$(\{0, \ldots, n-1\}, \bot) \models \varphi_\prec[a, b] \ \textit{if and only if} \ \Pi\text{Supp}(a) < \Pi\text{Supp}(b)$$

Proof. Let $\mathcal{A} = (\{0, \ldots, n-1\}, \bot)$. The direction from left to right is simple. If $\mathcal{A} \models \varphi_\prec[a, b]$ then there is a prime $d \in |\mathcal{A}|$ such that $d\Pi\text{Supp}(a) \leq n-1$ and $d\Pi\text{Supp}(b) > n-1$. So, $\text{Supp}(a) < \text{Supp}(b)$.

To prove the other direction let us set $a_1 = \Pi\text{Supp}(a)$ and $b_1 = \Pi\text{Supp}(b)$ and let $a_1 < b_1$. Then, $\varphi_\prec$ is satisfied by a and b if and only if $(\frac{n-1}{b_1}, \frac{n-1}{a_1}]$ contains a prime. In the worst case $b_1 = a_1 + 1$ and in this case $(\frac{n-1}{b_1}, \frac{n-1}{b_1}(1 + \frac{1}{a_1})]$ should contain a prime. Thus it suffices to take N from Proposition 9 for $\varepsilon = 1/a_1$. $\square$

Now, our aim is to define in models from $FM(\omega, \bot))$ the relations R_+, $R_\times$. We define these relations on an initial segment of the model $(\{0, \ldots, n-1\}, \bot)$.

Firstly, we introduce a tool for coding pairs of primes.

$\text{Code}(p, x, y, q) \Longleftrightarrow_{Def}$

$P(p) \wedge P(q) \wedge P(x) \wedge P(y) \wedge$ " q is the $\prec$-greatest prime less then $\{p, x, y\}$".

The statement in quotation marks can be written down as

$$\forall z \forall w [(\varphi_\cup(x, y, z) \wedge \varphi_\cup(p, z, w)) \Rightarrow \varphi_\prec(q, w)] \wedge$$

$$\forall r [(P(r) \wedge \varphi_\prec(q, r)) \Rightarrow \exists z \exists w (\varphi_\cup(x, y, z) \wedge \varphi_\cup(p, z, w) \wedge \varphi_\prec(w, r))].$$

In the above formula, the variable w plays the role of the set $\{p, x, y\}$. Then, with the help of $\varphi_\prec$ we easily express the maximality of q.

The intended meaning of the formula $\text{Code}(p, x, y, q)$ is that q is a code of an unordered pair consisting of x and y. The prime q is determined uniquely up to the equivalence $\approx$. The prime p is called a base of a coding. Now, we define a formula which states that coding with the base p is injective below x.

$\text{GoodBase}(p, x) :=$

$$P(p) \wedge \forall q_1 \ldots \forall q_4 \{ [\bigwedge_{i \leq 4} (P(q_i) \wedge \varphi_\preceq(q_i, x)) \wedge \neg\varphi_\approx(\{q_1, q_2\}, \{q_3, q_4\})] \Rightarrow$$

$$\exists c_1 \exists c_2 (\text{Code}(p, q_1, q_2, c_1) \wedge \text{Code}(p, q_3, q_4, c_2) \wedge \neg\varphi_\approx(c_1, c_2) \}.$$

The above formula states that p is a good base for our coding for primes which are less than x. Namely, for each pair of primes below x we obtain a different code q taking p as a base. The existence of a good base for each given x is guaranteed by Proposition 8. We subsume the above consideration in the following lemma.

Lemma 13 *For each k there is N and $p \leq N$ such that For all $n \geq N$, $\mathrm{Code}(p, x_1, x_2, z)$ defines an injective coding of pairs of primes less than k in each model $(\{0, \ldots, n-1\}, \perp)$.*

Proof. Let k be given and let K be chosen from Proposition 8 for $b = k^2$. Next, let p be a prime greater than K. By Proposition 8 p is a good base for our coding in all models $(\{0, \ldots, n-1\}, \perp)$, for $n \geq N = k^2 p$. □

When the exact base for our coding of pairs of primes is inessential we write simply $\langle x, y \rangle$ for a prime coding a pair x, y. Of course, in such a case a proper base for our coding should be assured to exist. Nevertheless, since we always will be interested in coding pairs of primes from a given initial segment, the existence of a proper base follows in this case by Lemma 13.

The last lemma allows to turn recursive definitions of addition and multiplication on indices of primes into explicit ones. The first needed relation is the successor relation on indices of primes. It is defined as

$$S_\prec(x) = y \iff_{Def} \varphi_\prec(x, y) \wedge P(x) \wedge P(y) \wedge$$
$$\forall z \, (P(z) \Rightarrow \neg(\varphi_\prec(x, z) \wedge \varphi_\prec(z, y))).$$

Let us observe that if $S_\prec(p_z)$ is defined in a given finite model then it is the case that $S_\prec(p_z) = p_{z+1}$. We have the following.

Lemma 14 *Partial functions on indices of primes FM–representable in coprimality models equipped with the relation $\prec$ are closed under the scheme of primitive recursion.*

Proof. Let $g : \omega^n \longrightarrow \omega$ and $h : \omega^{n+2} \longrightarrow \omega$ be functions on indices of primes FM–representable in coprimality models. We need to show that the function $f : \omega^{n+1} \longrightarrow \omega$ defined as

$$f(0, \bar{x}) = g(\bar{x}),$$
$$f(i+1, \bar{x}) = h(i+1, \bar{x}, f(i, \bar{x})).$$

is FM–representable in coprimality models with $\prec$. For simplicity we assume that $n = 1$. Since we have $\prec$ and $\perp$, we can define, by Lemma 13, a function $\langle x, y \rangle$ coding pairs of primes as primes. The formula defining $f(p_i, p_x) = p_t$ states that there is a set which describes a recursive computation of $f(p_i, p_x)$ with the output p_t. It can be written as

$$\exists X \{\langle p_0, g(p_x) \rangle \in X \wedge$$
$$\forall p_z \forall p_w [\varphi_\prec(p_z, p_i) \Rightarrow (\langle p_{z+1}, p_w \rangle \in X \iff$$
$$\exists p_v (\langle p_z, p_v \rangle \in X \wedge p_w \approx h(p_{z+1}, p_x, p_v)))] \wedge$$
$$\langle p_i, p_t \rangle \in X \}.$$

Let us observe that quantification over a set of primes X can be interpreted as first order quantification over numbers. Instead of X we can take a such that $X = \mathrm{Supp}(a)$. Thus, if we have formulas defining g and h, all the other notions can be defined in models for coprimality and $\prec$. □

Now, let φ_+ and $\varphi_\times$ be formulae, provided by means of Lemma 14, which define addition and multiplication on indices of primes. They define R_+ and $R_\times$ only on some initial segment of primes from a given finite model, but this segment grows with the size of a model.

We define the universe of our interpretation by the formula $\varphi_U(x_1)$ which states that φ_+ and $\varphi_\times$ define addition and multiplication on the set

$$\{y : P(y) \wedge (y \approx x_1 \vee y \prec x_1)\}.$$

Such a formula exists because there is a finite axiomatization of $\mathrm{FM}((\omega, +, \times))$ within the class of all finite models given explicitly in [11]. Thus, we have shown that $\mathrm{FM}((\omega, +, \times))$ is interpretable in finite models of coprimality even without equality. This ends the proof of Theorem 10. □

4   Some Applications in Finite Models

As a corollary of Theorem 10, we obtain a partial characterization of relations which are FM–representable in $\mathrm{FM}((\omega, \perp))$.

Definition 15 *Let $R \subseteq \omega^r$. We define R^* as*

$$R^* = \{(x_1, \ldots, x_r) : \exists a_1 \ldots \exists a_r (\bigwedge_{i \leq r} (x_i \approx p_{a_i}) \wedge (a_1, \ldots, a_r) \in R\}.$$

Corollary 16 *Let $R \subseteq \omega^r$. R is FM–representable in $\mathrm{FM}(\mathbb{N})$ if and only if R^* is FM–representable in $\mathrm{FM}((\omega, \perp))$.*

Now we are going to characterize the complexity of the first order theory of $\mathrm{FM}((\omega, \perp))$ and of relations which are FM–represented in $\mathrm{FM}((\omega, \perp))$. Firstly, we need a partial result in this direction.

Let us define the relation $S \subseteq \omega^2$ such that

$$(x, y) \in S \text{ if and only if } \exists z(z \approx x \wedge y \approx p_z).$$

Lemma 17 *The relation S is FM–representable in $\mathrm{FM}((\omega, \perp))$.*

Proof. To simplify the exposition we consider all the equivalences between formulae in the sense of being true in all sufficiently large models from $\mathrm{FM}((\omega, \perp))$. They will be justified for fixed parameters a, b for which we want to decide whether $(a, b) \in S$. Thus, we may safely assume that $b \approx p$, for some prime p.

Let $a_0, a_1, \ldots$ be the enumeration of all consecutive products of different primes ordered according to $<$. This enumeration lists $\approx$–representatives of all $\approx$–equivalence classes. For $x \in \omega$ we define $\mathrm{ind}(x)$ as the unique i such that $x \approx x_i$. We define an auxiliary relation W such that

$$(x, y) \in W \iff y \approx p_{\mathrm{ind}(x)}.$$

Now, we show how to write a formula $\varphi_W(x, y)$ which, for any pair of fixed parameters as x and y, holds in almost all finite models from $\mathrm{FM}((\omega, \perp))$ exactly when

$(x, y) \in W$. We take $n = \text{ind}(x)$ and let $a_0, \ldots, a_n$ be an initial segment of the above enumeration. By Proposition 8, there is a prime t such that each interval (ta_i, ta_{i+1}), for $i < n$, contains a prime. Let $q_0, \ldots, q_n$ be a sequence of primes such that

$$q_i = \min\{s : P(s) \wedge ta_i \prec q_i\}$$

and let $B = \Pi_{i \leq n} q_i$. Then, let $p_0, \ldots, p_k$ be a sequence of consecutive primes such that $p_k \approx y$ and let $C = \Pi_{i \leq k} p_i$. Let us observe that B and C are definable from x, t and y in terms of $\prec$ and $\perp$. Moreover, any t which allows this definition is good for our purpose. Thus, we can use B and C in our formulae.

The formula $\varphi_W(x, y)$ expresses the fact that sets coded by B and C, constructed as above, are equicardinal. This can be witnessed by a set X which is a set of pairs of primes from B and C determining a bijection between B and C. In the formula φ_W below we use $\exists^{=1} z$ for the quantifier "there exists exactly one z".

$$\exists X \{\forall q \in B \, \exists^{=1} p \in C \, \langle q, p \rangle \in X \wedge \forall p \in C \, \exists^{=1} q \in B \, \langle q, p \rangle \in X\}.$$

Of course, the existence of such an X proves that B and C are equicardinal. By the same argument as in the proof of Lemma 14 we can replace quantifying over X by first order quantification.

Now, we show how to define S from W. Let T be the following relation. For all $x, y \in \omega$,

$$(x, y) \in T \iff \text{ind}(x) = y.$$

This relation is recursive, thus also FM–representable in $\text{FM}((\omega, +, \times))$ and, by Corollary 16, the starred version of T is FM–representable in $\text{FM}((\omega, \perp))$. T^* satisfies the following condition: for all x, y,

$$(x, y) \in T^* \iff \exists z (p_z \approx x \wedge p_{\text{ind}(z)} \approx y).$$

So, let $\varphi_{T^*}(x, y)$ FM–represent T^*.

Let us also recall the definitions of S and W:

$$(x, y) \in S \iff \exists(z \approx x \wedge p_z \approx y),$$

$$(x, y) \in W \iff y \approx p_{\text{ind}(x)}.$$

Let us observe that in all sufficiently large finite models an element w such that $\varphi_W(x, w)$ is just $p_{\text{ind}(x)}$.

Now, the formula $\varphi_S(x, y)$ which FM–represents S can be written as

$$\exists w (\varphi_W(x, w) \wedge \varphi_{T^*}(y, w)).$$

Then, for all fixed parameters a and b, and for almost all finite models M from $\text{FM}((\omega, \perp))$, the following equivalence holds:

$$M \models \varphi_S(a, b) \iff (a, b) \in S.$$

For the direction from left to right let us assume that for some t we have $\varphi_W(a, t)$ and $\varphi_{T^*}(b, t)$. This means that

$$t \approx p_{\text{ind}(a)}$$

and that for some s we have

$$p_s \approx b \text{ and } p_{\text{ind}(s)} \approx t.$$

This gives $p_{\text{ind}(a)} \approx p_{\text{ind}(s)}$ and $\text{ind}(a) = \text{ind}(s)$. Therefore, $s \approx a$ and $p_s \approx b$, which gives $(a, b) \in S$.

Now let us assume that $(a, b) \in S$. Then for some z we have

$$z \approx a \text{ and } p_z \approx b.$$

This gives that $\text{ind}(z) = \text{ind}(a)$, $p_z \approx b$ and $p_{\text{ind}(z)} \approx t$, for $t = p_{\text{ind}(z)}$. Then $\varphi_{T^*}(b, t)$. Additionally, $t \approx p_{\text{ind}(a)}$ and $\varphi_W(a, t)$. Therefore, $\varphi_S(x, b)$.   □

Theorem 18 *Let $R \subseteq \omega^r$. R is FM–representable in $\text{FM}((\omega, \perp))$ if and only if R is FM–representable in $\text{FM}(\mathbb{N})$ and R is coprimality invariant.*

Proof. All relations which are FM–representable in $\text{FM}((\omega, \perp))$ are coprimality invariant. Therefore, the implication from left to right is obvious. So, we prove the converse.

For the sake of readability we consider only unary relations. Let us fix a coprimality invariant relation $R \subseteq \omega$ which is FM–representable in $\text{FM}(\mathbb{N})$. By Corollary 16, let us take a formula $\xi(x)$ FM–representing R^* in the FM–domain of coprimality.

By Lemma 17, there is a formula $\psi(x, y)$, with coprimality as the only predicate, such that $\psi(x, y)$ FM–represents S in the FM–domain of coprimality. Then the formula $\varphi(x)$ defined as

$$\exists y(\psi(x, y) \wedge \forall z(\varphi_{\prec}(z, y) \Rightarrow \neg\psi(x, z)) \wedge \xi(y))$$

FM–represents R.   □

Finally, let us consider the recursive complexity of the elementary theory of $FM((\omega, \perp))$. The classical Trachtenbrot theorem says that we can reduce the halting problem to the problem of satisfiability in finite models. By our interpretation, it suffices to consider only finite models for coprimality.

Theorem 19 (Trachtenbrot's theorem for coprimality FM–domain) *The first order theory of* $\text{FM}((\omega, \perp))$ *is Π_1^0–complete. Moreover, the theorem remains valid even if we do not have equality in the language.*

5   An Application in the Standard Model

Maurin has shown in [9] that the first order theory of $(\omega, \times, \leq_P)$, where $\leq_P$ is the standard ordering restricted to primes, is decidable. On the other hand, Bès and Richard have shown in [1] that adding the ordering on primes and squares of primes to coprimality allows an interpretation of addition and multiplication. In what follows, we prove a similar result for the structure $(\omega, \perp, \leq_{P_2})$, where P_2 is the set of primes and products of two different primes. Namely, we show that the relations R_+ and $R_\times$ are definable in $(\omega, \perp, \leq_{P_2})$. It follows that the first order theory of this model is as hard as the theory of $(\omega, +, \times)$. (Let us mention that it is not known whether R_+ and $R_\times$ are definable in the structure considered by Bès and Richard.)

Below, we show how to develop a coding for pairs of prime numbers below a given prime k. Then, the rest of the argument is the same as in the case of finite models. However, we cannot use coding of pairs of primes from the preceding sections since it uses a comparison of primes with products of three different primes. We defined such a coding there since it gives a simpler construction. Moreover, if one wants to estimate a fragment of a finite model on which we have definitions of R_+ and $R_\times$ then such a coding gives a better bound than the coding which we are going to present now. On the other hand, in the infinite model, we want to add to coprimality a relation as weak as possible to obtain our definability result.

Theorem 20 R_+ and $R_\times$ are definable in $(\omega, \perp, \leq_{P_2})$, where $\leq_{P_2}$ is the ordering relation restricted to primes and products of two different primes.

Proof. We only show how to define coding of pairs of primes by one prime, while the rest of the proof remains the same as in the finite case.

Let a prime k be given. We show how to code pairs of primes less or equal to k. Let ε be such that

$$(1+\varepsilon)^3 < k^2/(k^2-1), \tag{*}$$

and let p be a prime such that for all $n \geq p$, the interval

$$(n, n(1+\varepsilon))$$

contains a prime number. Then, our new formula $\text{Code}(p, x, y, r)$ is the following:

$$P(p) \wedge P(x) \wedge P(y) \wedge P(r)\wedge$$

$$\exists r_1 \exists r_2 (\text{``}r_1 \text{ is the smallest prime greater than } px\text{''}\wedge$$

$\text{``}r_2 \text{ is the smallest prime greater than } py\text{''} \wedge \text{``}r \text{ is the greatest prime less than } r_1 r_2\text{''}).$

All the notions needed in the above formula are definable in $(\omega, \perp, \leq_{P_2})$. Now, we only argue that the coding with p chosen as above is injective below k.

Let q, q' be two primes less or equal to k. By the choice of ε and p, there is a code r for this pair with the property

$$p^2(qq'-1)(1+\varepsilon)^2 < r < p^2 qq'(1+\varepsilon)^2.$$

The first inequality follows from the fact that $pq < r_1$ and $pq' < r_2$. Thus, r is greater than any z such that $z(1+\varepsilon) < p^2 qq'$. The maximal z with this property is greater than $p^2(qq'-1)(1+\varepsilon)^2$. Indeed,

$$p^2(qq'-1)(1+\varepsilon)^2(1+\varepsilon) \leq p^2 qq'(1-1/qq')(1+\varepsilon)^3$$
$$\leq p^2 qq'(1-1/k^2)(1+\varepsilon)^3$$
$$< p^2 qq',$$

where the last strict inequality follows by (*).

The second inequality follows from the fact that $r_1 < pq(1+\varepsilon)$, $r_2 < pq'(1+\varepsilon)$, and $r < r_1 r_2$.

Therefore, for any pair of primes $q, q' \leq k$, the code r for this pair is in the interval $(p^2(qq'-1)(1+\varepsilon)^2, p^2 qq'(1+\varepsilon)^2)$. However, since for any other pair of primes $t, t' \leq k$, qq' differs from tt' by at least one, these intervals are disjoint for different pairs of primes. This proves that our coding method with p as a base is injective below k. □

Acknowledgments

We are very greatful to our collegues Leszek Kołodziejczyk and Michał Krynicki for discussions and their comments which were very helpful in our work.

References

1. Bès, A. and Richard, D., *Undecidable extensions of Skolem arithmetic*, Journal of Symbolic Logic, 63(1998), pp. 379–401.
2. Gurevich, Y., *Logic and the Challenge of Computer Science*, in Current Trends in Theoretical Computer Science (ed. E. Börger) Computer Science Press, 1988, 1–7.p
3. Hoare, C. A. R., *An axiomatic basis for computer programming*, Communications of the ACM, 12(1969), pp. 576–583.
4. Jameson, G. J. O., The prime number theorem, Cambridge University Press, 2003.
5. Kołodziejczyk, L. A. *Truth definitions in finite models*, Journal of Symbolic Logic, 69(2004), pp. 183–200.
6. Kołodziejczyk, L. A. *A finite model-theoretical proof of a property of bounded query classes within PH*, Journal of Symbolic Logic, 69(2004), pp. 1105–1116.
7. Krynicki, M. and Zdanowski, K., *Theories of arithmetics in finite models*, Journal of Symbolic Logic, 70(2005), pp. 1–28.
8. Lee, T., *Arithmetical definability over finite structures*, in Mathematical Logic Quarterly, 49(2003), pp. 385–393.
9. Maurin, F. *The theory of integer multiplication with order restricted to primes is decidable*, Journal of Symbolic Logic, 62(1997), pp. 123–130.
10. Mostowski, M., *On representing concepts in finite models*, in Mathematical Logic Quarterly 47(2001), pp. 513–523.
11. Mostowski, M., *On representing semantics in finite models*, in Rojszczak, A., Cachro, J., Kurczewski, G. (ed.) Philosophical Dimensions of Logic and Science, Kluwer Academic Publishers, 2003, pp. 15–28.
12. Mostowski, M. and Wasilewska, A., *Arithmetic of divisibility in finite models*, Mathematical Logic Quarterly, 50(2004), pp. 169–174.
13. Mostowski, M. and Zdanowski, K., FM–*representability and beyond*, Cooper, S. B., Löwe, B. and Torenvliet, L. (eds.), CiE 2005, LNCS 3526, Springer, pp. 358–367, 2005.
14. Schweikardt, N., *Arithmetic, First-Order Logic, and Counting Quantifiers*, ACM Transactions on Computational Logic, 5(2004), pp. 1–35.
15. Sierpiński, W., Elementary Theory of Numbers, PWN (Polish Scientific Publishers) – North Holland, 1964.
16. Szczerba, L. W., *Interpretability of elementary theories*, in Proceedings 15th ICALP 88 Logic, foundations of mathematics and computability theory, eds. Butts, Hintikka, Reidel Publishing, 1977, pp. 129–145.
17. Trachtenbrot, B., *The impossibility of an algorithm for the decision problem for finite domains*, in Doklady Akademii Nauk SSSR, 70(1950), pp. 569–572, in russian.

Towards a Characterization of Order-Invariant Queries over Tame Structures

Michael Benedikt[1] and Luc Segoufin[2]

[1] Bell Laboratories, 2701 Lucent Lane, Lisle, IL 60532, USA
benedikt@research.bell-labs.com
[2] INRIA-Futurs, Parc-Club Orsay university
4 rue Jacques Monod, 91893 Orsay Cedex, France
http://www-rocq.inria.fr/~segoufin

Abstract. This work deals with the expressive power of logics on finite structures with access to an additional "arbitrary" linear order. The queries that can be expressed this way are the *order-invariant queries* for the logic. For the standard logics used in computer science, such as first-order logic, it is known that access to an arbitrary linear order increases the expressiveness of the logic. However, when we look at the separating examples, we find that they have satisfying models whose Gaifman Graph is complex – unbounded in valence and in treewidth. We thus explore the expressiveness of order-invariant queries over graph-theoretically well-behaved structures. We prove that first-order order-invariant queries over strings and trees have no additional expressiveness over first-order logic in the original signature. We also prove new upper bounds on order-invariant queries over bounded treewidth and bounded valence graphs. Our results make use of a new technique of independent interest: the application of algebraic characterizations of definability to show collapse results.

1 Introduction

In classical finite model theory, a logic $\mathcal{L}(\sigma)$ for models over the relational signature σ, associates words of a grammar (the syntax) to relations of the model (the semantics). One generally requires that the logic is *closed under isomorphisms*: that is if A and B are finite models over σ and h is an isomorphism between A and B then for all $q \in \mathcal{L}(\sigma)$, $q \circ h$ and $h \circ q$ give the same answer. This is the case for all standard logics: first-order logic, monadic second-order logic, fixed point logic etc.

In practice, for instance in the database context where logics correspond to query languages, one can refer in the syntax to a predicate which is not necessarily in the signature σ of the input: a linear order which corresponds to the order in which the elements of the universe are stored on disk. Sentences in the logic can then make use of this predicate to perform operations over all elements of the universe separately. If we denote by $\sigma_<$ the extension of σ with an additional binary relation symbol $<$ which is assumed to be interpreted by a linear order over the domain, then one actually has $\mathcal{L}(\sigma_<)$ available instead of $\mathcal{L}(\sigma)$.

L. Ong (Ed.): CSL 2005, LNCS 3634, pp. 276–291, 2005.

Of course it is preferable to restrict the use that a query language can make of the extra predicate $<$. One would not wish to allow queries that use $<$ in order to return the smallest, according to $<$, element of the universe; the answer would depend on how the universe is stored on disk, which in turn may vary with time (depending, for instance, on the presence of indexes). To be *meaningful* a formula in $\mathcal{L}(\sigma_<)$ should be closed under isomorphism. A sentence $\phi \in \mathcal{L}(\sigma_<)$ is closed under isomorphisms iff it is *order-invariant*: For every finite σ-structure A, for every two expansions A_1 and A_2 of A to an $\sigma_<$-structure, $A_1 \models \phi \leftrightarrow A_2 \models \phi$. We denote by Inv-$\mathcal{L}(\sigma_<)$ (Inv-$\mathcal{L}(<)$ if σ is understood) the fragment of $\mathcal{L}(\sigma_<)$ ($\mathcal{L}(<)$ if σ is understood) containing all order-invariant sentences.

There are two questions that immediately arise. The first one is whether there exists an effective syntax for Inv-$\mathcal{L}(<)$. That is whether there exists a logic $\mathcal{L}'$ with an effective syntax and the same expressive power as Inv-$\mathcal{L}(<)$. The second one is finding the expressive power of Inv-$\mathcal{L}(<)$, in particular whether it is strictly more that $\mathcal{L}$ or not (the converse inclusion being obvious).

These questions were first considered in the case of fixed-point logics. Indeed a rephrasing of the Immerman-Vardi Theorem (see e.g. [Lib04]) says that Inv-IFP$(<)$ is PTIME and that Inv-FP$(<)$ is PSPACE. Here IFP stands for the inflationary fixed-point semantics while FP is the non-inflationary semantics. Note that this immediately implies that Inv-IFP$(<)$ (resp. Inv-FP$(<)$) is strictly more expressive than IFP (FP) as IFP (FP) fails to express all of PTIME (PSPACE). The question of the existence of a logic with effective syntax for Inv-IFP$(<)$ is open: From the result above this is identical to the question whether there is a logic for PTIME, a longstanding open question in finite model theory [FmtOpen].

Another example is Monadic Second Order Logic (MSO). It is easy to see that Inv-MSO$(<)$ allows one to express every query in the extension of MSO with counting quantifiers (CMSO), which is strictly more expressive than MSO.

The example above shows that access to an arbitrary ordering increases expressiveness when one deals with powerful logics that can express recursive operators. What about weaker logics, such as first-order logic (FO)? A famous example due to Gurevich [AHV95] shows that Inv-FO$(<)$ is more expressive than FO for any σ including at least one binary predicate. Extensions of this result due to Otto [Otto00] give examples of Inv-FO$(<)$ sentences that are quite complex: in particular, [Otto00] shows that there are Inv-FO$(<)$ sentences not expressible in infinitary logic formed over first-order logic with a bounded number of variables and quantifiers of the form $\exists^{!i} x\ \phi(x, \boldsymbol{y})$. However, the example queries of Gurevich and Otto each have satisfying models that include a binary predicate which becomes graph-theoretically very complex as the models vary. Thus it is natural to conjecture that if one restricts the structures to be well-behaved, Inv-FO$(<)$ cannot express complex queries.

In this paper we investigate the expressiveness of Inv-FO$(<)$ sentences over graph-theoretically well-behaved structures, specifically structures of bounded treewidth and structures of bounded valence. We will show that Inv-FO$(<)$ collapses to FO on trees. We then show that Inv-FO$(<)$ collapses to MSO on structures of bounded valence and on structures of bounded treewidth.

One of our main contributions is a new proof technique for analyzing order-invariant queries. Suppose one wants to show bound of the form Inv-$\mathcal{L}(<) \subseteq \mathcal{L}'$ for some logic $\mathcal{L}'$. The most common method (e.g. [GS00]) is to show that two sufficiently $\mathcal{L}'$-equivalent structures can be ordered so that they agree on $\mathcal{L}$ sentences of a given quantifier rank. But ordering two such arbitrary equivalent structures is difficult. In addition, since ordering structures is a priori stronger than what is required, this technique may not be sufficient to show tight bounds. We show that algebraic characterizations of first-order definability, such as those in [BP89, BS05], can be utilized to give new results on collapse of order-invariant queries over logics on restricted structures. These characterization theorems show that if a query is not definable in the logic without the order, then there are witness structures that are similar and of a very restricted form which cannot be distinguished by the query. Only these special witness structures need to be ordered, and the restricted form of these structures makes the ordering arguments much more tractable.

Related Work. Following the initial example of Gurevich, order-invariant queries over first-order logic were investigated in [Otto00, GS00, Ross03]. It is clear that Inv-FO$(<)$ queries are expressible in both existential second order (ESO) and universal second order (USO) logic, and also that they are computable in NLOGSPACE. To our knowledge, there is no result giving containment of Inv-FO$(<)$ queries in a sublogic of ESO $\cap$ USO. While [Otto00] shows that there are Inv-FO$(<)$ queries that are not in infinitary logic with counting quantifiers, [GS00] shows that all Inv-FO$(<)$ open formulas are local (cannot distinguish points with similar local neighborhoods). [Ross03] shows there are Inv-FO$(<)$ queries that are not first-order which only make use of the successor relation in the order. All of the published examples of Inv-FO$(<)$ sentences are expressible in CMSO.

[Cou96] studies order-invariant MSO queries, showing that over trees, Inv-MSO$(<)$ has exactly the same expressiveness as CMSO. The results of [Lap98] combined with those of [Cou90, Cou91] show that the same equality holds for graphs of bounded treewidth. It is an open question whether the inclusion of Inv-MSO$(<)$ queries in CMSO holds over arbitrary structures (the conjecture is that it does not, see [FmtOpen]).

The algebraic proof technique we give here derives from [BS05], and we believe it gives a uniform method of treating invariant query questions. However, the bounds we give on Inv-FO$(<)$ queries on trees have been announced independently in the forthcoming [Nie05]. The result of [Nie05] relies on the locality of Inv-FO$(<)$ proved in [GS00], while the technique given here may be applicable to logics that are not local.

Note that the classical Craig interpolation theorem [CK90] implies that a first-order query that is invariant over *all* structures must be in FO. The interpolation theorem is known not to hold over finite structures (even for trees) [EF95]. Our results can be seen as showing that one can reclaim some consequences of interpolation by restricting to well-behaved classes of structures.

Organization: Section 2 gives the basic definitions of the paper and reviews results about regular languages that will be used here. Section 3 gives characterizations of invariant FO queries on strings. Section 4 extends these characterizations to trees. Section 5 gives bounds for graphs of bounded valence and bounded treewidth. Section 6 gives conclusions.

2   Background

By a *query*, we refer to any boolean function on finite relational structures in some vocabulary σ. We will generally have σ consist of at least a binary relation S, and Σ a finite set of unary predicates. We will refer to structures for such a σ as colored graphs. Given some class C of colored graphs (strings, trees, etc.) a query ϕ over the signature $\sigma_< = \sigma \cup \{<\}$ is *order-invariant over C* if for every finite σ-structure $G \in C$, for every two expansions G_1 and G_2 of G to a $\sigma_<$ structure, $G_1 \models \phi \leftrightarrow G_2 \models \phi$. Such queries can clearly also be considered as queries over σ. The queries we consider here will always be defined by logics (FO, Inv-FO($<$), etc.). Two logics are said to have the same expressiveness (over class C) if the set of queries they define (resp. set of restrictions of queries to domain C) are the same.

Our definition of first-order logic (FO) and Monadic Second Order Logic (MSO) over a vocabulary σ is standard. We will sometimes abuse notation and refer to an "Inv-FO($<$) query over C", to mean an FO($\sigma_<$) query that is order-invariant over C, and similarly for other logics. If P is a set of integers, the logic FO$_{mod(P)}$ extends FO by allowing formula to be built up by the rule $\psi(\boldsymbol{y}) = \exists^{q,r} x\, \phi(x, \boldsymbol{y})$, where $q < r$ are integers and $r \in P$. This holds in a structure $(G, \boldsymbol{y})$ iff the number of x such that $(G, \boldsymbol{y}, x)$ holds is equal to q modulo r. The logic FO$_{mod}$ is FO$_{mod(P)}$ with $P = \mathbb{N}$. Similarly, the logic CMSO is formed by allowing the above formation rule for first-order variables on top of the formation rules of MSO. For an integer k an FO(σ) k-type is a maximal consistent collection of first-order sentence of quantifier rank at most k. An MSO type is defined similarly. For $\mathcal{L}$ any one of MSO, FO, FO$_{mod}$ we write $G \equiv^{\mathcal{L}}_r G'$ if G and G' agree on $\mathcal{L}$ sentences of quantifier rank at most r. We assume familiarity with Ehrenfeucht-Fraïsse games (see e.g. [Lib04]), which characterize $\equiv^{FO}_r$.

The most restricted set of structures we consider are strings. Let σ^S consist of exactly $\Sigma \cup \{S\}$. We will use the terms string and word interchangeably to mean any σ^S-structure in which the domain with S is isomorphic to an initial segment of the integers with the successor relation. We will also assume (as part of the definition of string and word) that every element in the structure satisfies exactly one of the predicates in Σ: this assumption is only to simplify the presentation. The set of strings over a fixed Σ as above will be denoted Σ^*.

By a tree we mean a connected directed graph S that is acyclic and where every element has at most one predecessor. We will also sometimes use "tree" to mean an expansion of a tree in the above sense by a set of unary predicates in Σ. Let $\sigma^{S,S'}$ be the signature extending σ^S with a new binary predicate S'. A sibling ordering on a tree is any binary relation that compares only elements with a comment parent, and which is a linear order on the set of children of any node. By a "siblinged tree" we mean a $\sigma^{S,S'}$-structure, where S is a tree and S' is the successor relation corresponding to some sibling ordering. We distinguish between the set of unordered ranked trees RT_n for $n \in N$, (where n is the bound on the number of children of any node), sibling ranked trees SRT_n (which we consider as siblinged trees where there is a bound on the number of children) , unranked trees UT, and sibling unranked trees SUT. For any of these domains D, a collection $C \subseteq D$ is regular if it is MSO definable over the appropriate vocabulary. Queries over strings and trees will also be referred to as *languages*, and will often be identified with the set of elements that map to true under the query.

Our main technique is based on the use of algebraic machinery for analyzing logics on strings and trees. We thus review some of the known connections between definability and algebraic properties of string languages. All the results below can be found in [Str94]. The $\mathrm{MSO}(\sigma^S)$ sentences define exactly regular languages. With any language L, one can associate the equivalence relation $\equiv_L$ on Σ^*: $x \equiv_L y \leftrightarrow (\forall u \in \Sigma^* \ \forall v \in \Sigma^* \ uxv \in L \leftrightarrow uyv \in L)$. Regularity of L is equivalent to the fact that the set of equivalence classes is finite. The set of classes of $\equiv_L$ equipped with the concatenation operation forms a monoid, called the *syntactic monoid* of L, denoted η_L. An element e of the syntactic monoid is an *idempotent* if $e^2 = e$. We can consider a word to be idempotent if its class is; translating the above, we have that a word e is idempotent iff $uev \in L \leftrightarrow ue^2v \in L$. The theorems of MacNaughton and Papert and of Schützenberger characterize when a regular language is definable in first-order logic over σ^S augmented with an additional binary predicate $\prec$, where $\prec$ is interpreted as the transitive closure of S: this occurs exactly when the syntactic monoid of L is *aperiodic* (e.g. see [Str94] Theorem VI.1.1): there is l such that the monoid satisfies $\forall m \in \eta_L \ m^l = m^{l+1}$. Translated back to words, this means that $\forall u, v, w \in \Sigma^* \ uv^l w \in L \leftrightarrow uv^{l+1}w \in L$.

3   Order-Invariant Queries on Strings

We first deal with the decidability of membership in Inv-FO($<$). If one could decide membership in Inv-FO($<$), one would immediately have an effective syntax for Inv-FO($<$) queries. However, it is well-known [AHV95] that one cannot decide whether or not an FO($<$) query is order-invariant: this follows easily from the undecidability of the satisfiability problem for first-order logic. Over strings, satisfiability is decidable, hence it is a priori feasible that membership in Inv-FO($<$) over strings is decidable. We show that this is not the case:

Proposition 1. *The problem of deciding, given a sentence $\phi \in$ FO($<$), whether or not it is order-invariant over Σ^*, is undecidable.*

Proof. Consider the function that takes a $\phi \in$ FO($<$) and returns the conjunction of ϕ with a fixed sentence ϕ_0 that is not order-invariant. This function reduces non-order invariance to satisfiability of satisfiability of an FO($<$) sentence over expansions of structures in Σ^* by a linear order, and hence it suffices to show that this satisfiability problem is undecidable. To do this, from an input-free Turing machine M we construct a FO($<$) formula φ_M such that M halts iff φ_M has a model in Σ^*. Assume wlog that all Turing machines work over the binary alphabet $\{0, 1\}$. Let Σ consist of the unary predicates $P_0, P_1, P_\sharp, P_\square$. We consider strings from this alphabet.

A configuration c of a Turing machine using memory of size k can be described using a string of length k, where unused cells are colored with $\square$. Therefore a string of the form $\sharp c_1 \sharp c_2 \cdots \sharp c_n$ can code a set of configurations of a Turing Machine. We want to use S and $<$ in order to show that such words can code a run of a Turing Machine.

To do this we restrict the linear orders considered as follows. Let $\text{succ}_<$ be the successor relation corresponding to $<$ (note that $\text{succ}_<$ is definable in FO from $<$).

- All nodes labeled by $\sharp$ are ordered first by $<$.
- The remaining nodes are ordered in order to verify the following property:
  $$\forall x, y, u, v \ \neg P_\sharp(x) \wedge S(y, x) \wedge \text{succ}_<(y, u) \wedge S(u, v) \longrightarrow \text{succ}_<(x, v).$$

In words this says that once the order on the symbols $\sharp$ is fixed, then $<$ is completely defined by induction on strings of the form $\sharp c_1 \sharp c_2 \cdots \sharp c_n$: the order on $\sharp$ symbols induces an order $\prec$ on the $(c_i)_{1 \leq i \leq n}$ and, based on this, $<$ order the remaining symbols lexicographically based first on their position in one of the c_i and using $\prec$ for breaking ties. Note that the property given above is definable in FO($<$) by a formula that we denote by $\psi_<$. One can verify that all strings that are models of $\psi_<$ are of the form $\sharp c_1 \sharp c_2 \cdots \sharp c_n$ where the size of each c_i is the same. Given a model of $\psi_<$, let $\alpha_<$ be the bijection of $[n]$ such that $\alpha_<(i) = j$ where c_j is the string following the ith symbol $\sharp$ according to $<$. Each model of $\psi_<$ can thus be seen as a sequence of configurations $c_{\alpha(1)} \cdots c_{\alpha(n)}$.

We now fix M and construct a formula ψ_M which, assuming $\psi_<$ checks that the sequence $c_{\alpha(1)} \cdots c_{\alpha(n)}$ is an accepting run of M. For this notice that $\text{succ}_<$ associates cells located at the same place on the tape of M and at two successive steps of M. Using this relation it is a classical technique (see e.g. Chap. 9 in [Lib04]) to code in first-order logic the fact that two successive configurations are valid according to M.

From the discussion above it is now easy to see that the formula $\varphi_M = \psi_M \wedge \psi_<$ has the desired property. □

The proof technique can easily be modified to show undecidability for the other classes considered in this paper. We now turn to the expressiveness of Inv-FO($<$) over strings. We show that this is as low as it can possibly be (recall that every FO query is in Inv-FO($<$)). We prove this directly, in order to exhibit the main technique of the paper. It will also follow from our results on trees in Section 4.

Theorem 1. Inv-FO($<$) $=$ FO *over strings.*

Proof. The proof is based on an algebraic characterization of FO within the set of regular languages. The following theorem of Beauquier and Pin characterizes when a regular language L is FO definable (see also [Str94], Thm VI.3.1):

Theorem 2 ([BP89]). *A regular language L is FO definable iff its syntactic monoid is aperiodic, and additionally for any $e, f, u, v, w \in \eta_L$ with e, f idempotent, $eufvewf = ewfveuf$.*

Let $\phi \in$ Inv-FO($<$). Let M be a string model of ϕ. Then M is of the form $(\omega_M, <_M)$ where ω_M is a string and $<_M$ a linear order on the universe of ω_M. Let $L(\phi) = \{\omega_M \mid M \models \phi\}$. A model M is *obvious* if $<_M$ is exactly the transitive closure of the successor relation in ω_M. Let $L'(\phi) = \{\omega_M \mid M \models \phi$ and M is obvious$\}$. Because $L'(\phi)$ is definable in FO($<$), it is regular and its syntactic monoid is aperiodic (see Section 2). By order-invariance $L(\phi) = L'(\phi)$ thus $L(\phi)$ is aperiodic. To show that $L(\phi)$ is definable in FO, by Theorem 2, it suffices to show that for any $e, f, u, v, w \in \eta_{L(\phi)}$ with e, f idempotent, $eufvewf = ewfveuf$.

Recall that elements of $\eta_{L(\phi)}$ are equivalence classes of words. Replacing each of e, f, u, v, w with a word which is a representative of the corresponding equivalence class, and recalling what it means for the word $eufvewf$ to be equivalent to $ewfveuf$, we have the following equivalent condition: For all words e, f which represent idempotents in $\eta_{L(\phi)}$, for all words u, v, w and all words a, b:

$$aeufvewfb \in L(\phi) \leftrightarrow aewfveufb \in L(\phi)$$

Note that if e and f are idempotent, then $e^n = e$ and $f^n = f$ for any integer n. Therefore for any word c, d and any n we have $ced \in L(\phi)$ iff $ce^n d \in L(\phi)$. Hence $L(\phi)$ is FO-definable iff for some n, m, we have the following holding for all words a, e, u, f, w, v, b:

$$ae^n uf^n ve^n wf^n b \in L(\phi) \quad \text{iff} \quad ae^m wf^m ve^m uf^m b \in L(\phi)$$

Let n be 3^{2k}, where k is the quantifier rank of ϕ, and $m = 3^k + n$. To prove the theorem it is therefore sufficient to prove the following Lemma:

Lemma 1. $ae^n uf^n ve^n wf^n b \in L(\phi)$ *iff* $ae^m wf^m ve^m uf^m b \in L(\phi)$

Proof (of Lemma). Let a, e, u, v, w, f, b be arbitrary words, let I be the σ^S-structure $ae^n uf^n ve^n wf^n b$ and J be the σ^S-structure $ae^m wf^m ve^m uf^m b$. We construct two orders $<_I$ and $<_J$ such that the expansions I' and J' of I and J with these orders are equivalent up to quantifier rank k (recall that k is the quantifier rank of ϕ). This implies that I' and J' agree on ϕ and, by order-invariance of ϕ, that any linear-ordered expansion of I over $\sigma^S_<$ agrees on ϕ with any linear-ordered expansion of J over $\sigma^S_<$ and therefore $ae^n uf^n ve^n wf^n b \in L(\phi)$ iff $ae^m wf^m ve^m uf^m b \in L(\phi)$.

The orders $<_I$ and $<_J$ we construct assume a fixed order on the strings a, e, u, v, w, f, b and always order a, v, b before the rest of the strings. Therefore, wlog, we can ignore a, v, b and assume that each of the symbols e, f, u, w represents a single letter. So assume $I = e^n uf^n e^n wf^n$ and $J = e^m wf^m e^m uf^m$.

We will now create our ordering using the technique of [GS00] (in particular, see the discussion in the beginning of Lemma 2 of [GS00]). In both orderings we start with u and then w. We order the remaining nodes based on their distance from the set $\{u, w\}$ (which is at most n), with the nearer ones coming first in the ordering. It remains to describe how to order the four nodes at fixed distance i from $\{u, w\}$. We call this set in model M, $W_i(M)$. For M being either I or J, $W_i(M)$ consists of four elements called u-left$_i(M)$, u-right$_i(M)$, w-left$_i(M)$ and w-right$_i(M)$ with the obvious meanings. The ordering will alternate between "u-first ordering" and "w-first ordering" of $W_i(M)$. In the "u-first ordering" elements of $W_i(M)$ are ordered in the sequence u-left$_i$, u-right$_i$, w-left$_i$, w-right$_i$. In the "w-first ordering" elements of $W_i(M)$ are ordered in the sequence w-left$_i$, w-right$_i$, u-left$_i$, u-right$_i$. In model I, $W_i(I)$ is ordered with the u-first ordering for $1 \leq i \leq f(k)$ where $f(k)$ is 3^k). We then switch to a w-first ordering for $W_i(I)$ such that $f(k) < i \leq 2f(k)$, and continue switching this way, ending with the u-first ordering. In model J we do the same, and by the choice of m which contains one extra block of size $f(k)$, we end with a w-first ordering.

This completes the description of the ordering $<_I$ and $<_J$. Note that by construction $<_I$ and $<_J$ are completely identical close to u and w and, by the presence of this extra *switch*, close to the beginning and ending of I and J. We now give a sketch of the proof that duplicator can win the k-round Ehrenfeucht-Fraïsse game on the expansions I' and J'. Intuitively, the reason is that any of the switches in the ordering can be detected but there are so many of them that the parity of switches cannot be detected. An element x of one of the models is classified as u-first or w-first based on whether the ordering for the set W_i containing x is u-first or w-first. We refer to this as the *orientation* of x. Inside each W_i, the *rank* of x is a number between 1 and 4 which corresponds to the order of x relative to W_i. Each $W_i(M)$ is called a *segment* of M. Each model M is divided into *sections*, where a section is a set of the form $\bigcup_{i \in s} W_i(M)$, where s is an interval of $\mathbb{N}$ maximal with respect to the orientation of $W_i(M)$ being constant. The distance moving outward from $\{u, w\}$ gives us two orderings: the quotient ordering $<_{s,M}$ on sections, and within any section a partial order $<_{d,M}$.

Given a play $(x_1...x_j, y_1...y_j)$ our inductive invariant is that the relations in the signature are preserved, and the following distances are preserved (from $\boldsymbol{x}$ to $\boldsymbol{y}$) up to the threshold 3^{k-j}: the distances $<_{d,M}$ between the x_i's, if they lie within the same section, the distances between the section of x_i and the section of x_j, the distance between an x_i and the edges of its section, and the distance of the section of x_i to the final and initial section. The invariant also includes that $\text{rank}(x_i)=\text{rank}(y_i)$ and $d(x_i, \{u, w\}) < d(x_j, \{u, w\}) \leftrightarrow d(y_i, \{u, w\}) < d(y_j, \{u, w\})$.

Handling a play by the spoiler is done as follows. Assume WLOG that the x_i already include representatives from the final and initial section. Now say the spoiler plays $x = x_{j+1}$ in I. We denote by $s(x)$ the section of x. Consider the position of $s(x)$ among the sections already marked by $\boldsymbol{x}$. By the inductive hypothesis it is easy to find a section s' in J which respects all the inductive distance requirements up to 3^{k-j-1}. Consider now the set $W_i(I)$ which contains

x and its position relative to the other segments marked by x in $s(x)$. Again, using the inductive hypothesis, on can find a set $W_j(J)$ which respects the inductive requirements up to threshold 3^{k-j-1}. Inside $W_j(J)$, pick y such that rank$(x)=$rank(y).   □

    □

4   Order-Invariant Queries over Trees

We now deal with extending the above results to trees. We have done this for ranked trees, with or without a sibling ordering, and for unranked trees with no sibling ordering. For sibling unranked trees, we have only an MSO upper-bound on the expressiveness. To generalize the results for strings to trees, we use analogous algebraic machinery to characterize definability over ranked trees. As with strings, MSO definable sets of (siblinged or unordered) trees are exactly the regular sets of trees. Every such set is thus the acceptance set of a bottom-up tree automaton, which can be taken to be deterministic.

A *pointed tree* is a tree with a designated leaf which acts as a port. The concatenation of two pointed trees Δ and Δ' is denoted by $\Delta \cdot \Delta'$ and is the pointed tree constructed from Δ by plugging Δ' to its port. The set of pointed trees is denoted by T^1. A *k-pointed tree* is a tree with k designated leaves. For any k-pointed tree Δ and any (pointed or not) trees $t_1 \cdots t_k$, $\Delta[t_1 \cdots t_k]$ denotes the tree constructed from Δ by plugging t_1 in its first port and t_2 in its second port, etc. The set of k-pointed trees is denoted by T^k. Given a deterministic automaton A and a k-pointed tree Δ, we get a function Δ_A from Q^k to Q which, given $q_1 \ldots q_k$, gives the state obtained when running A starting at states q_i at the ith port and q_0 at all other leaves. When A is fixed, a pointed tree Δ is said to be *idempotent* if the function it defines is idempotent ($\Delta_A \circ \Delta_A = \Delta_A$).

Let L be a regular tree language. We say that L satisfies (†) if the following holds:

1. For any $\Delta \in T^2$, $e \in T^1$ idempotent and any $t, t' \in T^0$
$$\Delta[e \cdot t, e \cdot t'] \in L \quad \text{iff} \quad \Delta[e \cdot t', e \cdot t] \in L$$
2. For any $s, s', u, v \in T^1$, $e, f \in T^1$ idempotents, and $t \in T^0$
$$s \cdot e \cdot u \cdot f \cdot s' \cdot e \cdot v \cdot f \cdot t \in L \quad \text{iff} \quad s \cdot e \cdot v \cdot f \cdot s' \cdot e \cdot u \cdot f \cdot t \in L$$
3. there exists a l such that for any $s, u \in T^1$ and any $t \in T^0$
$$s \cdot u^l \cdot t \in L \quad \text{iff} \quad s \cdot u^{l+1} \cdot t \in L$$

In [BS05], the following algebraic characterization is proved:

Theorem 3 ([BS05]). *Let L be a regular tree language over* SRT, RT, *or* UT. *Then L is definable in* FO *iff L satisfies* (†).

Using this, we prove the following:

Theorem 4. Inv-FO$(<)$ = FO *over* RT, SRT, *and* UT.

The first step of the proof is to show that:

Theorem 5. Inv-FO$(<) \subseteq$ MSO *over* RT, SRT, SUT, *and* UT.

Proof. This is rather immediate for SRT, SUT and RT, as a linear order can be defined in MSO (with parameters in the case of RT). In the case of UT, it is no longer possible to define a linear order in MSO, we use Ehrenfeucht-Fraïsse techniques. Let ϕ be an Inv-FO($<$) sentence, and $<_s$ be a sibling ordering. Then there is a canonical linear order on t which is constructed by a lexicographical process from $<_s$. Linear orders obtained this way are called *natural* in the rest of the paper. Let $L'(\phi) = \{t \mid$ there exists a natural linear order $<$ such that $\langle t, < \rangle \models \phi\}$. Because ϕ is order-invariant we have $L'(\phi) = L(\phi)$. Therefore the next lemma, which shows that sibling-ordering-invariant FO collapses to MSO, concludes the proof of the theorem.

Lemma 2. $L'(\phi)$ *is definable in* MSO.

Proof. Recall that σ^S is our signature for unordered trees and $\sigma^{S,S'}$ is the extension of σ^S with an extra binary predicate S' which is interpreted as a successor among siblings. From the SUT case we know that $L'(\phi)$ is definable in MSO ($\sigma^{S,S'}$). We want to show that it is definable in MSO (σ^S).

Because $L'(\phi)$ is definable in MSO ($\sigma^{S,S'}$), there is a deterministic unranked bottom-up tree automaton $A = \langle \Sigma, Q, \delta, q_0, F \rangle$ that computes $L'(\phi)$ [BMW01]. Using classical minimization and completion techniques we can further assume that A satisfies the following properties:

- For every $q \in Q$ there is tree t^q such that when A runs on t^q it reaches state q at the root.
- For every $q \neq q' \in Q$ there exists a pointed tree $\Delta^{q,q'}$ such that the set $\{\Delta_A^{q,q'}(q), \Delta_A^{q,q'}(q')\}$ contains one accepting and one non-accepting state.

Recall from [BMW01] that for each $q \in Q$ and each $a \in \Sigma$ $\delta(q, a)$ is a regular expression over Q with the meaning that a node label a gets state q if the sequence of its children according to S' forms a word in $\delta(q, a)$. We first show that the regular expression, which could be expressed by a formula in MSO, is actually expressible in FO.

Claim. For each $q \in Q$ and each $a \in \Sigma$, $\delta(q, a)$ is definable by a formula of FO (Q), that is a formula using only unary predicates from Q.

Proof (of the claim). Take an arbitrary $q \in Q$ and $a \in \Sigma$, and let $L = \delta(q, a)$. Let $k = |Q|$ and fix r such that for all $n, m \geq r$ the linearly ordered sequence $1^n \equiv_k^{FO(<)} 1^m$. Fix also an arbitrary order $<_Q$ on Q. Take two words w and w' in Q^* such that $w \equiv_r^{FO(Q)} w'$. We show that $w \in L$ iff $w' \in L$. This immediately implies that L is in FO(Q).

Assume by way of contradiction that $w \in L$ but $w' \notin L$. Reorder w and w' according to $<_Q$, with an arbitrary order for ties. This yields two new strings $\bar{w} = q_1 \cdots q_u$ and $\bar{w}' = q_1' \cdots q_v'$. We first claim that $w \in L$ iff $\bar{w} \in L$. Assume, for a contradiction, that $w \in L$ but $\bar{w} \notin L$. By assumption on A we have two trees $t_1 = a[t^{p_1} \cdots t^{p_u}]$ and $t_2 = a[t^{q_1} \cdots t^{q_u}]$ such that w (resp. $\bar{w}$) is $p_1 \cdots p_n$ ($q_1 \cdots q_n$). Now, as $w \in L$, q is the state reached by A on t_1. Let q' be the state reached by A on t_2. By determinism of A and because $\bar{w} \notin L$ we have $q' \neq q$. Let $t = \Delta^{q,q'} \cdot t_1$ and $t' = \Delta^{q,q'} \cdot t_2$. By construction t is accepted by A but t' is not. This contradicts sibling invariance of ϕ.

By symmetry the above argument shows that $\bar{w} \in L$ but $\bar{w}' \notin L$. Consider now the trees $t_1 = a[t^{q_1} \cdots t^{q_u}]$ and $t_2 = a[t^{q'_1} \cdots t^{q'_v}]$. Again q is the state reached by A on t_1 while $q' \neq q$ is the state reached by A on t_2. Let $t = \Delta^{q,q'} \cdot t_1$ and $t' = \Delta^{q,q'} \cdot t_2$. By construction t is accepted by A but t' is not. We now claim that $t \equiv_k t'$, which implies that t and t' agree on ϕ, the desired contradiction.

We show this by giving a winning strategy for the corresponding Ehrenfeucht-Fraïssé game. On $\Delta^{q,q'}$ and the roots of t_1 and t_2, Duplicator plays using the identity map. On a move where Spoiler plays a node x in tree t^{q_i}, Duplicator always responds by an identical y in a tree t^{p_j} such that $q_i = p_j$. There might be several possible choices of p_j. Let n be the number of occurrences of state q_i in w and m be the number of occurrences of the same state in w'. By assumption on w and w' we have $n = m$ or $n, m \geq r$. In both cases Duplicator picks one appropriate p_j using its strategy in the $\equiv_k^{FO(<)}$ game between 1^n and 1^m. It is easy to verify that this strategy works.

Now by the claim, each transition is given by a $FO(Q)$ query. Hence, we know that A is given by an automaton that uses unordered first-order transitions. It is now immediate to see that such an automaton can be simulated in MSO. This completes the proof of Lemma 2.   □

□

From Theorem 3 and Theorem 5 it suffices now to show that $L(\phi)$ satisfies (†). Part 2) and 3) of (†) work exactly as in the proof of Theorem 1 as we can view wlog each pointed tree as a single character. We now check 1) of (†) and give the proof for SRT and RT together as it makes no difference.

Fix $\Delta \in T^2$, $e \in T^1$ idempotent and any $t, t' \in T^0$. Our goal is to prove that $\Delta[e \cdot t, e \cdot t'] \in L$ iff $\Delta[e \cdot t', e \cdot t] \in L$. We proceed as in the proof of Theorem 1 and show this it is enough to find n, m and linear orders $<_I$ and $<_J$ such that $\langle \Delta[e^n \cdot t, e^n \cdot t'], <_I \rangle \equiv_k \langle \Delta[e^m \cdot t', e^m \cdot t], <_J \rangle$, where k is the quantifier rank of ϕ.

We choose $n = 3^{2k}$ and $m = n + 3^k$ and we compute $<_I$ and $<_J$ section by section as in the proof of Theorem 1 with t and t' playing the role of u and v. It starts from $\{t, t'\}$ and move towards Δ, each section having 3^k segments, each segment $(W_i(M))$ having now only two elements instead of four, and being oriented t-first or t'-first depending on the section. We conclude elementary equivalence up to depth k as in the proof of Theorem 1   □

5   Order Invariance on Tree-Like and Bounded Valence Structures

We now consider how to extend the bounds given in the previous sections to graphs that may have cycles, but which are still well-behaved. We concentrate on two well-behaved classes: the bounded treewidth structures [RS84], and the bounded valence structures. We show that over such structures Inv-FO($<$) $\subseteq$ MSO. Note that this inclusion is not true in general as Gurevich's example is not definable in MSO.

A *tree decomposition* of a graph G consists of a tree T and a function d mapping nodes of T to sets of vertices of G, satisfying:

- For every edge $(v_1, v_2) \in G$, there is $t \in T$ with $v_1, v_2 \in d(t)$.
- For every vertex v of G, $\{n \in T : v \in d(n)\}$ is a connected subset of T.

The *width* of a decomposition (T, d) is $max_{t \in T} |d(t)| - 1$. The *treewidth* of a graph G is the minimal width of a tree decomposition of G. Let TW(b) be the set of colored graphs with treewidth at most b.

Theorem 6. *For every b,* Inv-FO($<$) $\subseteq$ MSO *over* TW(b).

Proof (sketch). Assume that ϕ has quantifier rank j, and let $L(\phi) = \{G \mid$ there exists a linear order $<$ such that $\langle G, < \rangle \models \phi\}$ be the set of graphs defined by ϕ.

We introduce some terminology based on the work of Courcelle and Lapoire (e.g. see [Lap98], page 34). We describe how a tree decomposition $D = (T, d)$ of G can be considered as a colored tree T_D. The underlying tree of T_D is T and predicates for the colors are defined as follows: Fix for each node x of T a graph G_x with vertices in $\{1 \ldots b+1\}$ that is isomorphic to the restriction of G to $d(x)$, and an isomorphism μ_x taking G_x onto this restriction. Let ν_x be the partial function from $\{1 \ldots b+1\}$ to $\{1 \ldots b+1\}$ that maps i to j exactly when $\mu_x(i) = \mu_y(j)$, where y is the parent node of x. For each graph τ with vertices in $\{1, b+1\}$, we have a predicate P_τ that holds at a node x of T_D iff $G_x = \tau$, and for each h from $\{1 \ldots b+1\}$ to $\{1 \ldots b+1\}$, we have a a predicate P_h that holds at x iff $\nu_x = h$. That is, the predicates specify which graphs are associated with a node, and how the graph at a node is linked to the graph of its parent. Let T_b be the set of trees for this signature, and let f be the evaluation map taking a colored tree in T_b to the corresponding graph.

Let $T(\phi) = f^{-1}(L(\phi))$. It follows from results of Courcelle [Cou90, Cou91] that $T(\phi)$ is CMSO definable. By a result of Lapoire [Lap98], there exists a MSO transduction g, which, given a graph G of treewidth b, computes a structure $D \in T_b$ such that $f(D) = G$. The exact definition of MSO transduction will not be needed here; the key fact about such transductions is that the pre-image of an MSO definable set under a transduction is again MSO definable [Cou91].

Let $T'(\phi) = \{(t, \prec_s) \mid t \in T(\phi), \prec_s$ a sibling ordering on $T\}$. Since $T(\phi)$ is CMSO definable, $T'(\phi)$ is definable in MSO($\sigma^{S,S'}$). We claim that $T'(\phi)$ (hence $T(\phi)$) is MSO(σ^S) definable. The theorem would follow from this, using $L(\phi) = g^{-1}(T)$ and the closure of MSO definability under inverse MSO transduction mentioned above.

As discussed above we view $T'(\phi)$ as a set of unranked trees labeled with a finite alphabet Σ. Because $T'(\phi)$ is definable in MSO($\sigma^{S,S'}$), let A be an unranked tree automaton for $T'(\phi)$ with state set Q and transition function δ. Recall that δ maps every pair $(q, a) \in Q \times \Sigma$ to a regular expression over Q. We also assume that A is minimal (see the assumptions on the automaton A listed in the proof of Lemma 2). For strings s, s' in Q^*, say $s \simeq_k s'$ if they have the same number of occurrences of each q up to threshold k.

Claim. There exists a k such that, for every $s, s' \in Q^*$, if $s \simeq_k s'$ then for every $(q, a) \in Q \times \Sigma$, $s \in \delta(q, a)$ iff $s' \in \delta(q, a)$.

From this claim, it is clear that T' (hence T) is $\mathrm{MSO}(\sigma^S)$ definable.

Proof (of the claim). Take $k = 3^j$ (recall that j is the quantifier rank of ϕ) and suppose s and s' are such that $s \simeq_k s'$. Assume by contradiction that s and s' disagree on $\delta(q_0, a)$: that is $s \in \delta(q_0, a)$ while $s' \in \delta(q_1, a)$ with $q_0 \neq q_1$. Fix an order on Q and let ω and ω' be the strings computed from s and s' by making the letters appear in the order of Q and let W and W' be the string structures computed from ω and ω' by adding the obvious linear order on ω and ω'. By the choice of k it follows from $s \simeq_k s'$ that $W \equiv_j^{\mathrm{FO}} W'$. For each state q choose a tree t_q such that A reaches state q when running on t_q. Let Δ be the pointed tree $\Delta_A^{q_0, q_1}$ that witnesses that $q_0 \neq q_1$ (for the precise definition, see the definition of $\Delta_A^{q, q'}$ in the proof of Lemma 2). Let $T^0 = \Delta \cdot a[t_{\omega_1} \cdots t_{\omega_n}]$ and $T^1 = \Delta \cdot a[t_{\omega'_1} \dots t_{\omega'_m}]$ where $\omega_1 \cdots \omega_n$ ($\omega'_1 \cdots \omega'_m$) is the sequence of letters of ω (ω'). By construction A accepts T^0 while it rejects T^1, therefore $f(T^0) \models \phi$ but $f(T^1) \models \neg \phi$.

We now create orderings $<_0$ on $G^0 = f(T^0)$ and $<_1$ on $G^1 = f(T^1)$ such that the resulting ordered graphs are equivalent on $\mathrm{FO}(<)$ sentences of quantifier rank j, a contradiction. For any subtree t of T^0, we let $\nu(t)$ be the union of all nodes of $f(t)$. We first choose fixed orderings on $\nu(\Delta)$, on $\nu(t_q)$ for all $q \in Q$. In G^0 we begin with the fixed ordering on $\nu(\Delta)$, then proceed with the fixed ordering for the nodes $\nu(t_{\omega_1}) - \nu(\Delta), \cdots, \nu(t_{\omega_n}) - \nu(\Delta)$. Note that the fact that t_{ω_i} are all subtrees below Δ in the tree decomposition T^0 implies that the sets $\nu(t_{\omega_i}) - \nu(t_0)$ are pairwise disjoint. In G^1 we proceed similarly, using $\nu(t_{\omega'_1}) - \nu(\Delta)$. Let $<_0$ and $<_1$ be these orderings. We show how to play the Ehrenfeucht-Fraïssé game between $\langle G^0, <_0 \rangle$ and $\langle G^1, <_1 \rangle$. Given a play of the game, let H be the function taking each pebble x in $G^0 - \nu(\Delta)$ to the unique ω_l such that $x \in \nu(t_{\omega_l}) - \nu(\Delta)$, and let H' be the similar function on G^1. By induction, one can show that Duplicator can play maintaining the following properties on the pebbles $x_1 \cdots x_i$ and $y_1 \cdots y_i$ at any step i: *(i)* the play is the identity for moves in $\nu(\Delta)$, *(ii)* $H(\boldsymbol{x}) \equiv_{k-i}^{\mathrm{FO}(<)} H'(\boldsymbol{y})$, *(iii)* $\langle t_{\omega_{H(x_i)}}, x_i \rangle$ is isomorphic to $\langle t_{\omega'_{H'(y_i)}}, y_i \rangle$.

From the claim, Theorem 6 follows. $\square$

We now turn to proving the analogous result for the set $\mathrm{BV}(b)$ of colored graphs of valence less than b.

Theorem 7. *For every b Inv-FO$(<) \subseteq$ MSO over $\mathrm{BV}(b)$.*

Proof. (sketch) Let $\phi \in$ Inv-FO$(<)$, and let l be the quantifier rank of ϕ.

For connected graphs of bounded valence there exists integer k, and MSO formulas $\psi(z, x, y, S_1 \dots S_k)$ and $\gamma(S_1 \dots S_k)$ such that, given a connected undirected graph G of valence b, and a distinguished point p, we have: *(i)* G has at least one expansion to $S_1 \ldots S_k$ such that $\gamma(S_1 \ldots S_k)$ holds, *(ii)* for any $S_1 \ldots S_k$ satisfying γ, $\rho(x, y) = \psi(p, x, y, S_1 \ldots S_k)$ defines a linear-ordering on G. The formula γ says that S_i suffices to give a local ordering of G, and ψ says that x comes before y in a depth-first traversal of G starting from p using the local ordering definable from the S_i (see [Cou96] for the detailed argument on the construction of γ and ψ). Let r be the maximum quantifier rank of ψ and γ above. Then any two connected graphs G and G' of valence b that agree on MSO sentences of rank $r + l + k + 3$ will have the property that: For any $p, S_1 \ldots S_k$ in G there are $p', S_1' \ldots S_k'$ in G' such that the corresponding expansions have the same FO $l + 2$-types. This implies that we can choose for each MSO $(r + l + k + 3)$-type τ, a distinguished FO l-type $\nu(\tau)$ of expanded structures such that τ implies the existence of $(S_i)_{1 \le i \le k}$ and p that give an expansion ordering satisfying $\nu(\tau)$.

Let $<_\tau$ be some fixed ordering of MSO $(r+k+l+3)$-types of graphs. Suppose there are two graphs G and G' that agree on MSO sentences of quantifier rank at most l', where l' is big enough that the cardinality of the number of components in G with a given $(r + k + l + 3)$-type in G agrees with the cardinality of the number of components in G' with that type, up to threshold 3^l (since being a component is MSO definable, one can guarantee this with a suitably large l'). Order G by the ordering $<_G$ as follows: use $<_\tau$ to order the components of G according to their $(r + k + l + 3)$-type, breaking ties arbitrarily; within any component of type τ choose a point p and an $S_1 \ldots S_k$ such that the expansion by the ordering corresponding to $p, (S_i)_{1 \le i \le k}$ realizes $\nu(\tau)$. Form the analogous ordering on G'.

We can now verify that $(G, <_G)$ and $(G', <_{G'})$ agree on first-order formulas of quantifier rank l, hence on ϕ. Inside a component of type τ, we play according to the strategy given by $\nu(\tau)$. For dealing with the arbitrary number of connected components, note that with l moves, in the presence of a linear order, one cannot count more than up to 2^l. This implies that any two BV(b) graphs with the same MSO l'-type must agree on ϕ. It follows that ϕ is MSO definable over BV(b).   □

The technique in the theorem above can be used to show that over arbitrary graphs in which a local order (i.e. on the successors of any given node) is definable, Inv-FO($<$) sentences are in MSO_2, Monadic Second Order Logic where quantification is over edges rather than nodes. One uses the fact that within a component, an ordering can always be defined in MSO_2 with parameters (see [Cou96]).

Proposition 2. *Over locally-ordered structures we have* Inv-FO($<$) $\subseteq MSO_2$.

We do not know yet whether Inv-FO($<$) is contained in CMSO over arbitrary structures.

6   Conclusions

Our aim is to show that over well-behaved classes of structures, order-invariant queries over first-order logic in any given signature σ collapse to first-order over the signature without the order. Thus far we have shown this for strings and trees. One method of extending this to first-order logic over bounded treewidth structures is to prove that if two graphs of treewidth b agree on first-order sentences of sufficiently large quantifier-rank, one can find tree decompositions of each graph that agree on fixed quantifier-rank; this would allow the characterizations of definability to be pushed from trees to graphs. We do not yet know of interesting classes for which a transfer of first-order equivalence from graphs to trees can be performed.

The technique presented here is applicable to logics other than first-order logic. For example, we have obtained an algebraic characterization of the logic FO_{mod} on trees, by replacing the aperiodicity condition in (†) by: there exists a l such that for any $s, u \in T^1$ and any $t \in T^0 s \cdot u^l \cdot t \in L$ iff $s \cdot u^{l+q} \cdot t \in L$. This extends the characterization of FO_{mod} on strings of Straubing (VII.3.1 of [Str94]). We believe that this can be used to prove analogous bounds on Inv-$FO_{mod}(<)$ definability on trees.

References

[AHV95]   S. Abiteboul, R. Hull and V. Vianu. *Foundations of Databases*. Addison-Wesley, 1995.

[BP89]   D. Beauquier and J.-E. Pin. Factors of words. In *ICALP*, 1989.

[BS05]   M. Benedikt and L. Segoufin. Regular Tree Languages Definable in FO. Available from the authors. An abstract has appeared in *STACS*, 2005.

[BMW01]   A. Brüggemann-Klein, M. Murata, and D. Wood. *Regular Tree and Regular Hedge Languages over Unranked Alphabets* Available at ftp://ftp11.informatik.tu-muenchen.de/pub/misc/caterpillars/, 2001.

[Büc60]   J. Büchi. Weak second-order logic and finite automata. *S. Math. Logik Grunlagen Math.*, 6:66–92, 1960.

[CK90]   C.C. Chang and H.J. Keisler. *Model Theory*. North-Holland Elsevier, 1990.

[Cou90]   B. Courcelle. *The Monadic Second Order Logic of Graphs I: Recognizable Sets of Finite Graphs Information and Computation*, 1990.

[Cou91]   B. Courcelle. *The Monadic Second Order Logic of Graphs V: On Closing the Gap Between Definability and Recognizability Theor. Comput. Sci*, 1991.

[Cou96]   B. Courcelle. *The Monadic Second Order Logic of Graphs X: Linear Orders Theoretical Computer Science* 160:87–143, 1996.

[Don70]   J. Doner. Tree acceptors and some of their applications. *Journal of Computer and System Sciences*, 4:406–451, 1970.

[EF95]   H.-D. Ebbinghaus and J. Flum. *Finite Model Theory*. Springer Verlag, 1995.

[Nie05]   H. Niemisto. On Locality and Uniform Reduction. To appear in *LICS*, 2005.

[FmtOpen] http://www-mgi.informatik.rwth-aachen.de/FMT Open Problems in Finite Model Theory

[GS00]   M. Grohe and T. Schwentick. *Locality of Order-Invariant First-Order Queries. ACM TOCL*, 2000.

[Lap98]   D. Lapoire. *Recognizability Equals Monadic Second-Order Definability for Sets of Graphs of Bounded Tree-Width*. In *STACS*, 1998.

[Lib04]   L. Libkin. *Elements of finite model theory.* springer, 2004.

[Otto00]   M. Otto. Epsilon-logic is more expressive then first-order logic on finite structures. *Journal of Symbolic Logic*, 65(4), pp. 1749-757, 2000.

[RS84]   N. Robertson and P. Seymour. Graph Minors III: planar tree-width. *J. Combin. Theory Ser. B*, 36:49-64,1984.

[Ross03]   B. Rossman. Successor-invariance in the finite. In *LICS*, 2003.

[Str94]   H. Straubing. *Finite Automata, Formal Logic, and Circuit Complexity.* Birkhäuser, 1994.

[TATA]   H. Comon et al. *Tree Automata: Techniques and Applications*. Available at **www.grappa.univ-lille3.fr/tata**.

[Tho97]   W. Thomas. *Handbook of formal languages*, volume 3, chapter 7. Springer, 1997.

[TW68]   J.W. Thatcher and J.B. Wright. Generalized finite automata woth an application to a decision problem of second order logic. *Math. Syst. Theory*, 2:57–82, 1968.

Decidability of Term Algebras
Extending Partial Algebras

Bakhadyr Khoussainov and Sasha Rubin

Department of Computer Science, University of Auckland, New Zealand

Abstract. Let $\mathcal{A}$ be a partial algebra on a finite signature. We say that $\mathcal{A}$ has decidable query evaluation problem if there exists an algorithm that given a first order formula $\phi(\bar{x})$ and a tuple $\bar{a}$ from the domain of $\mathcal{A}$ decides whether or not $\phi(\bar{a})$ holds in $\mathcal{A}$. Denote by $E(\mathcal{A})$ the total algebra freely generated by $\mathcal{A}$. We prove that if $\mathcal{A}$ has a decidable query evaluation problem then so does $E(\mathcal{A})$. In particular, the first order theory of $E(\mathcal{A})$ is decidable. In addition, if $\mathcal{A}$ has elimination of quantifiers then so does $E(\mathcal{A})$ extended by finitely many definable selector functions and tester predicates. Our proof is a refinement of the quantifier elimination procedure for free term algebras. As an application we show that any finitely presented term algebra has a decidable query evaluation problem. This extends the known result that the word problem for finitely presented term algebras is decidable.

1 Introduction

The (free) algebra of terms plays an important role in many areas of computer science and algebra. It is the unique universal object that can be mapped homomorphically onto any given algebra (over a fixed signature). This provides a bijection between congruences of the term algebra and the class of all algebras (over that signature). Since finite trees can be represented as terms, the algebra of terms appears in computer science. For instance, in automata theory regular languages of trees can be identified with congruences of finite index of the algebra of terms. In logic programming, terms are used as basic objects in unification algorithms. In modern object oriented programming many data structures are stored and manipulated as terms. Other applications are in constraint databases, pattern matching, type theory and the theory of algebraic specification.

In logic and computability, the term algebra attracts much attention due to the fact that its first order theory is decidable. This was first proved by Mal'cev in [15]. His proof uses the method of elimination of quantifiers. This result has been reproved and extended by others in different settings. Rybina and Voronkov in [17] applied the method of elimination of quantifiers to show that the term algebra with queues has a decidable first order theory. Korovin and Voronkov in [8] prove that the existential fragment of the term algebra with the Knuth-Bendix ordering is decidable. Manna, Zhang and Sipma in [16] prove that the term algebra with the length function for terms has a decidable theory using the elimination of quantifiers. They also prove that the theory of the term algebra

L. Ong (Ed.): CSL 2005, LNCS 3634, pp. 292–308, 2005.

that involves k-alternations of quantifiers, regardless of the total number of the quantifiers, is at most k-fold exponential. Vorobyov in [18] and Compton and Henson in [4] prove that that the decision problem for the first order theory of the term algebra has a non-elementary lower bound.

In this paper, we extend the decidability result for term algebras to a much more general setting. A *partial algebra* (or simply an *algebra*) $\mathcal{A}$ is a structure whose basic operations are partial functions on A. In case all the functions are total on A, we may stress this and call $\mathcal{A}$ a *total algebra*. Partial algebras naturally occur when one restricts the basic operations of a total algebra $\mathcal{A}$ to some $B \subset A$ that is not closed under the basic operations. In this case the value of a term in B may be undefined, in which case B is a partial algebra and not a total algebra.

The algebra freely generated by $\mathcal{A}$, called the *free total extension* of $\mathcal{A}$ and written as $E(\mathcal{A})$, is the total algebra generated by $\mathcal{A}$ with a sort of universal mapping property: every homomorphism from $\mathcal{A}$ into any total algebra $\mathcal{B}$ can be extended to a homomorphism from $E(\mathcal{A})$ into $\mathcal{B}$. We remark that $E(\mathcal{A})$ is a natural object in universal algebra (see [9, Section 28]) as well as in computer science. For instance in the theory of algebraic specification partial algebras are a natural way of treating errors such as division by zero (see [1]); here $E(\mathcal{A})$ are used as models of specifications. The algebra $E(\mathcal{A})$ is also used in providing non-standard models of Clarke's Axioms (see [14]). Also [11] uses $E(\mathcal{A})$ to extend the Myhill-Nerode theorem with 'finitely generated congruence' instead of 'congruence of finite index'.

We say that an algebra $\mathcal{A}$ has a *decidable query evaluation problem* if there exists an algorithm that given a first order formula $\phi(\bar{x})$ and a tuple $\bar{a}$ from the domain of $\mathcal{A}$ decides whether or not $\phi(\bar{a})$ holds in $\mathcal{A}$. In particular if $\mathcal{A}$ has a decidable query evaluation problem then its first order theory is decidable. Our main result states:

If $\mathcal{A}$ has a decidable query evaluation problem then so does $E(\mathcal{A})$.

In particular, the first order theory of $E(\mathcal{A})$ is decidable. In addition, if $\mathcal{A}$ has elimination of quantifiers then so does $E(\mathcal{A})$ extended by finitely many definable selector functions and tester predicates. These results are proved by rewriting a formula Φ of $E(\mathcal{A})$ into a formula Φ' that can be evaluated in A so that $E(\mathcal{A}) \models \Phi$ if and only if $\mathcal{A} \models \Phi'$. The technique is a refinement of the quantifier elimination procedure in [16].

A *finitely presented term algebra* is the quotient of a free term algebra by finitely many ground term equations. The word problem for finitely presented term algebras is decidable [10]. As a corollary to our main result we have, we feel, a clean proof that the first order theory of a finitely presented algebra is decidable [3].

We place our result, that decidability is preserved by the free total extension of $\mathcal{A}$, in the realm of other constructions that preserve decidability. Direct product, disjoint union and under suitable conditions the ω-product preserve decidability of the query evaluation problem. Another example is the construction of the tree-like unfolding from [19] that preserves the monadic second order theory. Finally we mention a construction from [12] that resembles the one in this paper. There a purely relational structure $\mathcal{A}$ is lifted by first extending the signature by adding new function symbols from a functional signature Σ. One then considers the Σ-term algebra generated by constants from A. Finally, the relations of the algebra $\mathcal{A}$ are lifted in a natural way to the domain of the terms. The resulting structure is called the Σ-term power of $\mathcal{A}$. It is proved that if $\mathcal{A}$ has decidable first order theory then so does the Σ-term power of $\mathcal{A}$.

Here is a brief outline of this paper. The next section gives basic definitions, examples, and facts about structures with decidable query evaluation problem. Section 3 presents a formal definition of the free total extensions for partial algebras and some of the properties of these extensions. Section 4 is devoted to proving the main result of this paper. The final section applies the main result to show that each finitely presented term algebra has decidable query evaluation problem.

2   Structures with Decidable Query Evaluation Problem

A *structure* consists of a domain A of elements, and basic operations $f^A, g^A \ldots$ on A and relations $P^A, Q^A, \ldots$ on A. The *signature* of $\mathcal{A}$ is $(f, g, \ldots, P, Q, \ldots)$. We view constants as operations of arity 0. In general, the operations f^A may be partial functions.

Convention: For the purpose of this paper, the domain A is a decidable set from a decidable domain such as the strings over a finite alphabet, or ground terms over a finite ranked alphabet, or natural numbers. In particular all the structures we consider are countable.

Definition 1. The query evaluation problem *for the structure $\mathcal{A}$ is the set $QEP(\mathcal{A})$ of all pairs $(\phi(\bar{x}), \bar{a})$ such that $\mathcal{A} \models \phi(\bar{a})$ where $\phi(\bar{x})$ is a first order formula of $\mathcal{A}$ and $\bar{a}$ is a tuple of elements from A. If there is an algorithm deciding $QEP(\mathcal{A})$ then we say that $\mathcal{A}$* **has decidable query evaluation problem**.

We will abbreviate the phrase 'query evaluation problem' as QEP.

In other words, $\mathcal{A}$ has decidable QEP means that its elementary diagram with constants naming every element in A is decidable. We remark that this definition can be extended to other logics besides first order. Here are several examples of structures with decidable QEP.

Example 1. Every finite structure has decidable QEP.

Recall that a theory T is a set of sentences closed under deduction. A theory T admits effective quantifier elimination if there is an effective procedure that transforms a formula into an equivalent (in T) quantifier free formula. Say that a structure $\mathcal{A}$ admits effective quantifier elimination if its first order theory does. Specific examples include algebraically closed fields, vector spaces over finite fields, term algebras extended with selector functions, etc (see [5] for examples).

Example 2. If $\mathcal{A}$ admits effective elimination of quantifiers and the domain and basic operations of $\mathcal{A}$ are decidable, then $\mathcal{A}$ has decidable QEP.

Automatic structures are relational structures whose predicates are recognised by synchronous finite automata. For precise definitions see [7].

Example 3. Automatic structures have decidable QEP.

For issues on complexity of query evaluation problems for automatic structures see [2], [13], and for examples of automatic structures see [6], [7]. For these structures definable relations are, in fact, recognised by finite automata.

Example 4. Every decidable consistent theory T has a model with decidable QEP.

Indeed, the classical Henkin construction can be made effective; and the constructed model has decidable QEP.

Example 5. If $\mathcal{A}$ has decidable QEP then every structure that is first order definable in $\mathcal{A}$ also has decidable QEP.

Example 6. Let $\mathcal{A}$ and $\mathcal{B}$ be structures of a signature Σ with decidable QEP. Then the following structures have decidable QEP: the product $\mathcal{A} \times \mathcal{B}$ (Feferman, Vaught, 59); the disjoint union $\mathcal{A} \oplus \mathcal{B}$, where the domain of $\mathcal{A} \oplus \mathcal{B}$ is the union of A and B and for each $P \in \Sigma$ the predicate $P^{\mathcal{A} \oplus \mathcal{B}}$ is the union $P^{\mathcal{A}} \cup P^{\mathcal{B}}$ and there is a unary predicate for A.

3   Free Total Extensions of Partial Algebras

Partial Algebras: A *partial algebra*, or simply an *algebra*, is a structure $\mathcal{A} = (A, f^A, g^A, \ldots, h^A)$ consisting of a domain of elements A and finitely many partial functions on A. Constants are viewed as functions of arity 0. Incase all the functions are total we call the structure a *total algebra*. The signature of $\mathcal{A}$ is $(f, g, \ldots, h)$ and is called a *functional signature* since it does not contains symbols for relations. Note that a term, such as $f(g(a, b))$, may not have a value in A, in which case we say that *(the value of) the term is undefined in $\mathcal{A}$*. All structures implicitly have the symbol $=$ for equality. In a partial algebra $\mathcal{A}$ two terms s and t are defined to be equal, written $s = t$, if they are both defined in $\mathcal{A}$ and have the same value in $\mathcal{A}$ (so called existential equality). Tuples of elements a_i of A and tuples of variables x_i are denoted by $\bar{a}$ and $\bar{x}$ respectively.

Term Algebras: Let Σ be a functional signature and C a non-empty domain. Then the set of *ground terms* over C, written $GT_\Sigma(C)$ or simply $GT(C)$, is defined inductively as follows:

- Every element of C is a ground term.
- If f is an n-ary symbol from Σ, and $\bar{t}$ is an n-tuple of ground terms, then $f(\bar{t})$ is a ground term.

The *free term algebra generated by* C is $(GT(C), (f)_{f \in \Sigma})$ where the value of f on $\bar{t}$ is defined as the ground term $f(\bar{t})$. It is a total algebra.

The *depth* of ground terms is defined as follows: terms in C have depth 0; if $f(t_1, \cdots, t_k)$ is not in C, then its depth is 1 more than the maximum of the depths of the t_i.

Free Total Extensions: Let $\mathcal{A}$ be a partial algebra on signature Σ. We define what it means to extend $\mathcal{A}$ to a total algebra $E(\mathcal{A})$ in the free-est possible way. First we give an explicit construction and then the usual one in terms of homomorphisms.

For any ground term $t \in GT(A)$ define its *canonical form* with respect to the structure $\mathcal{A}$, written $t_{\mathcal{A}}^{(c)}$ or simply $t^{(c)}$, by induction as follows.

- If $t = a$ and $a \in A$ then define $t^{(c)}$ as a.
- Assume $t = f(t_1, \ldots, t_n)$, and $t_1^{(c)}, t_2^{(c)}, \ldots, t_n^{(c)}$ have been defined. If each $t_i^{(c)}$ is in A and the value $f(t_1^{(c)}, t_2^{(c)}, \ldots t_n^{(c)})$ is defined in $\mathcal{A}$ and equals $b \in A$ then define $t^{(c)}$ as b. Otherwise $t^{(c)}$ is defined as $f(t_1^{(c)}, t_2^{(c)}, \ldots t_n^{(c)})$.

The *algebra of canonical terms with respect to* $\mathcal{A}$ is the total algebra over signature Σ, whose domain is the set of canonical terms $t^{(c)}$ for $t \in GT(A)$, and for which the value of f on $\bar{t}$, with t_i canonical, is simply $f(\bar{t})^{(c)}$.

As we will see in a moment, this algebra is isomorphic to the free total extension of $\mathcal{A}$. First we need some definitions (see [9][section 13]).

Let $\mathcal{C}$ and $\mathcal{B}$ be partial algebras over the same signature. A *homomorphism from* $\mathcal{C}$ *into* $\mathcal{B}$ is a total mapping $h : \mathcal{C} \to \mathcal{B}$ so that whenever $\bar{a} \in C$, $f \in \Sigma$, and $f^{\mathcal{C}}(\bar{a})$ is defined then $f^{\mathcal{B}}(h(\bar{a}))$ is defined and is equal to $h f^{\mathcal{C}}(\bar{a})$. For $B \subset C$, say that $\mathcal{C}$ *extends* $\mathcal{B}$ if for every $\bar{b}$, $f^{\mathcal{B}}(\bar{b})$ is defined and equal to $b \in B$ if and only if $f^{\mathcal{C}}(\bar{b})$ is defined and equal to $b \in B$. Note that this allows the possibility that $f^{\mathcal{C}}(\bar{b})$ is defined (and not in B) while $f^{\mathcal{B}}(\bar{b})$ is undefined. Finally recall that a total algebra $\mathcal{C}$ is *generated* by a set $X \subset C$ if $\mathcal{C}$ is the smallest (under $\subset$) total algebra containing every total subalgebra $\mathcal{B}$ with $X \subset B$. In this case every element of $\mathcal{C}$ is equal to $t(\bar{b})$ for some term t and some tuple of elements $\bar{b}$ from X.

Define a *free total extension of* $\mathcal{A}$ (compare [11]), written $E(\mathcal{A})$, as satisfying the following properties:

1. $E(\mathcal{A})$ is a total algebra extending $\mathcal{A}$.
2. $E(\mathcal{A})$ is generated by the elements of A.
3. Every homomorphism h from $\mathcal{A}$ into a total algebra $\mathcal{B}$ can be extended to a homomorphism of $E(\mathcal{A})$ into $\mathcal{B}$.

Note that $E(\mathcal{A})$ is unique. Here are some examples.

Example 7. Let $\mathcal{C}$ be the partial algebra $(C, (f)_{f \in \Sigma})$ where the f's are undefined everywhere. Then $E(\mathcal{C})$ is the free term algebra generated by C.

Example 8. Let $\mathcal{A}$ be a finite partial algebra. Then $E(\mathcal{A})$ is isomorphic to the quotient of $GT(A)$ by finitely many ground term equations [11].

Proposition 1. *The algebra of canonical terms with respect to $\mathcal{A}$ is isomorphic to $E(\mathcal{A})$.*

Convention. In what follows we will work on this algebra of canonical terms directly, instead of using the abstract definition of $E(\mathcal{A})$. Moreover, in order to distinguish the original domain A of $E(\mathcal{A})$ we introduce a unary predicate A to the language.

4   Main Theorem

The main result is the following theorem.

Theorem 1. *If a partial algebra $\mathcal{A}$ has decidable QEP then so does its free total extension $E(\mathcal{A})$.*

The proof is a mixture of the quantifier elimination procedure for free term algebras and the decision procedure of $\mathcal{A}$. The basic idea is to inductively remove existential quantifiers over variables specified to be outside of A.

We extend the signature $\Sigma \cup \{A\}$ of $E(\mathcal{A})$ to include new operations and relations. To avoid confusion, the operations of Σ are called *constructors*. The new operations consist of:

- unary *selector functions* f_i for every constructor f and $i \leq arity(f)$, and
- unary *tester predicates* IS_f for every constructor f.

From now on we work in this extended signature unless specified otherwise. Sequences of selector functions will be denoted by $L, M, \ldots$ and $L_i, M_i, \ldots$ for $i \in \mathbb{N}$.

Semantics: These new operations have the following semantics. Let t be a canonical term. If $t \notin A$ then there is a unique constructor f and canonical terms $\bar{s}$ so that $t = f(\bar{s})$. In this case, define $f_i(t)$ as s_i (for $i \leq arity(f)$) and define $\mathrm{IS}_f(t)$ as $\top$. Also define $g_i(t) = t$ and $\mathrm{IS}_g(t)$ as $\bot$ for every constructor $g \neq f$ (for $i \leq arity(g)$). On the other hand, if $t \in A$, then define $g_i(t)$ as t and $\mathrm{IS}_g(t)$ as $\bot$ for every constructor g (for $i \leq arity(g)$. Finally $A(t)$ holds if and only if $t \in A$. Note the selector functions f_i and the tester predicates IS_f are definable in the language of $E(\mathcal{A})$.

Terms: A term t over the extended signature of $E(\mathcal{A})$ with free variables amongst $\bar{x}$ will be written $t(\bar{x})$. The expression

$$t[x_1/s_1(\bar{v}), \cdots, x_k/s_k(\bar{v})]$$

denotes the term t where each of the mentioned x_i from $\bar{x}$ has been replaced with the corresponding term $s_i(\bar{v})$. For instance, if $t = f(h_2(x_1), x_2)$, then $t[x_1/g(v_1)]$ is the term $f(h_2(g(v_1)), x_2)$.

Literals: To be sure, every literal in the language of $E(\mathcal{A})$ is either an equation of terms $t = s$, a disequation of terms $t \neq s$, or a tester predicate applied to a term $\mathrm{IS}_f(t)$.

We list some basic properties of $E(\mathcal{A})$ that will be used implicitly in showing the correctness of the the algorithm provided in the next section. Here and unless specified otherwise, equivalence is in $E(\mathcal{A})$.

Lemma 1. *The algebra $E(\mathcal{A})$ satisfies the following properties:*

1. $t_1 = t_2$ implies that $t_1 \in A$ if and only if $t_2 \in A$.
2. $f(\bar{t}) \notin A \wedge f(\bar{s}) \notin A$ implies that $f(\bar{t}) = f(\bar{s})$ is equivalent to $\bigwedge t_i = s_i$.
3. $f(\bar{t}) \notin A \wedge f(\bar{s}) \notin A$ implies that $f(\bar{t}) \neq f(\bar{s})$ is equivalent to $\bigvee t_i \neq s_i$.
4. for $f \neq g$, $[f(\bar{t}) \notin A \wedge g(\bar{s}) \notin A]$ implies that $f(\bar{t}) = g(\bar{s})$ is equivalent to $\perp$.
5. for $f \neq g$, $[f(\bar{t}) \notin A \wedge g(\bar{s}) \notin A]$ implies that $f(\bar{t}) \neq g(\bar{s})$ is equivalent to $\top$.
6. $s \notin A \wedge f(\bar{t}) \notin A$ implies that the equation $s = f(\bar{t})$ is equivalent to $\bigwedge f_i s = t_i \wedge IS_f(s)$. Similarly it implies that the disequation $s \neq f(\bar{t})$ is equivalent to $\bigvee f_i s \neq t_i \vee \neg IS_f(s)$.

4.1   Quantifier Elimination

(Un)limited Quantification: An *unlimited quantification* is one of the form $Qz \notin A$ and a *limited quantification* is one of the form $Qz \in A$, where $Q \in \{\exists, \forall\}$. As a shorthand we write $Q^u z$ for unlimited quantification and $Q^l z$ for limited quantification. Also we write $\exists^u \bar{x}$ for $\exists^u x_1 \cdots \exists^u x_m$ where $\bar{x} = (x_1, \cdots, x_m)$. Note that a quantification may be neither limited nor unlimited.

From now on, let $\bar{x}$ denote existentially quantified unlimited variables and let $\bar{y}$ denote unlimited parameters. Similarly let $\bar{x}'$ denote quantified limited variables and let $\bar{y}'$ denote limited parameters.

The following technical lemma shows how to remove unlimited quantification. Its proof will be the focus of this subsection.

Lemma 2. *Every formula of the form $\exists^u \bar{x} \chi(\bar{x}, \bar{y}, \bar{y}')$ where all quantifications in χ are limited, is equivalent in $E(\mathcal{A})$ to a formula of the form $\chi'(\bar{y}, \bar{y}')$, where all quantifications in χ' are also limited.*

Proof. We may assume $\mathcal{A}$ is not a total algebra for otherwise $E(\mathcal{A})$ is isomorphic to $\mathcal{A}$ in which case the lemma is trivial.

Consider a formula of the form

$$\exists^u \bar{x} \chi(\bar{x}, \bar{y}, \bar{y}'),$$

where all the quantifications in χ are limited. We will describe a procedure that transforms this formula into an equivalent formula where all the quantifications are limited.

The procedure will use the following techniques implicitly. We can always assume that all quantified variables are distinct by renaming if neccessary.

disjunctive splitting: The process of replacing a formula of the form $\exists x(B \vee C)$ by its logical equivalent $(\exists x\, B) \vee (\exists x\, C)$.

disjunctive normal form: Every quantifier free formula can be written as $\bigvee (\bigwedge B_{i,j})$ where the $B_{i,j}$ are literals.

formula normal form: Every formula can be expressed as

$$Q_k v_k \cdots Q_1 v_1 \psi,$$

where the Q_i are blocks of quantifiers of the same type (namely $\exists$ or $\forall$) and ψ is quantifier free in disjunctive normal form.

We now explain the concepts of type and type completion that will be used throughout the algorithm.

Types and Type Completions: A *type* of a term s is one of the following formulae:

- $s \in A$, or
- $s \notin A \wedge \mathrm{IS}_g(s) \wedge \bigwedge_{f \neq g} \neg \mathrm{IS}_f(s)$, where g is some constructor.

Note that a term has finitely many types. This definition is used in the following important concept [16].

Say that a conjunction of literals B is *type-completed* if for every subterm s in B, exactly one type of s is expressed in B. Note that a conjunction of literals can be extended to finitely many non-equivalent in $E(\mathcal{A})$ type-completed formulae. For example, a type completion of the formula $f(x) = y$ is the conjunction of $f(x) = y$ and

$$[x \in A] \wedge [f(x) \notin A \wedge \mathrm{IS}_f(f(x)) \wedge \bigwedge_{g \neq f} \neg \mathrm{IS}_g(f(x))] \wedge [y \notin A \wedge \mathrm{IS}_h(y) \wedge \bigwedge_{g \neq h} \neg \mathrm{IS}_g(y)].$$

We remark that a type completion may not be satisfiable (for instance if $f \neq h$ in the example above) Indeed in the algorithm such type completions are identified and replaced with $\bot$.

The *type completion of a quantifier free formula* $\phi = \bigvee \psi_i$, where each ψ_i is a conjunction of literals, is the equivalent quantifier free formula

$$\bigvee_{i,k} \psi'_{i,k},$$

where $\{\psi'_{i,k} \mid k\}$ consist of all the non-equivalent type completions of ψ_i. Here $\bigvee_{i,k} \psi'_{i,k}$ is *the type completion of* ϕ and is called *type-completed*. Note that each $\psi'_{i,k}$ is a conjunction of literals.

The *type completion of a formula* Φ is defined by the following procedure. First put Φ into formula-normal-form. So Φ is of the form

$$Q_p v_p \cdots Q_1 v_1 \, \psi,$$

where each Q_i is either $\forall$ or $\exists$. Now ensure that every quantifier is either limited or unlimited as follows. We proceed by induction on p. Suppose by induction that we have transformed the formula Φ into an equivalent formula Ψ with the property that Ψ is in formula-normal-form and all its quantifiers are either limited or unlimited. Now $\exists v_{p+1} \Psi$ is replaced by

$$\left[\exists^l v_{p+1} \Psi \vee \exists^u v_{p+1} \Psi \right].$$

Similarly, $\forall v_{p+1} \Psi$ is replaced by

$$\left[\forall^l v_{p+1} \Psi \wedge \forall^u v_{p+1} \Psi\right].$$

Now put the result into formula-normal-form. This completes the inductive step. Finally type-complete the quantifier free part. So the formula is now of the form

$$Q_k^{\alpha_k} v_k \cdots Q_1^{\alpha_1} v_1 \psi,$$

where ψ is $\bigvee_i \psi_i$, and each ψ_i is a type-completed conjunction of literals, and $\alpha_i \in \{l, u\}$. Note that it is equivalent to the original formula Φ.

Limited and unlimited literals: Suppose that ψ_i is a (not necessarily type-completed) conjunction of literals with the property that there is a *unique* (up to equivalence in $E(A)$) type-completed conjunction of literals ψ_i' equivalent in $E(\mathcal{A})$ to ψ_i. For example if ψ_i is $x \in A \wedge f(x) \notin A$ then ψ_i' is $x \in A \wedge f(x) \notin A \wedge IS_f(f(x)) \wedge_{g \neq f} \neg IS_g(f(x))$. In most cases, $\psi_i = \psi_i'$ will be type-completed itself.

Call a term t *limited (with respect to ψ_i)* if ψ_i' contains the literal $t \in A$, and *unlimited (with respect to ψ_i)* if ψ_i' contains the literal $t \notin A$. An equation or disequation is (un)limited in ψ_i if both sides of it are (un)limited in ψ_i. A tester predicate $IS_g(t)$ is (un)limited (with respect to ψ_i) if t is (un)limited (with respect to ψ_i). For the rest of the proof, when a term or (dis)equation is called limited, implicitly it is with respect to the type-completed ψ_i' in which it occurs. Recall a quantification Qv is called limited if it is of the form $(Qv \in A)$, also written $Q^l v$. If it is of the form $Qv \notin A$, also written $Q^u v$, it is unlimited.

The Algorithm: We are now ready to describe the algorithm. It takes as input a formula Φ of $E(\mathcal{A})$ of the form $\exists^u \overline{x} \chi(\overline{x}, \overline{y}, \overline{y}')$, where every quantification in χ is limited. So we may write Φ as $\exists^u \overline{x} Q_k^l x_k' \cdots Q_1^l x_1' \psi(\overline{x}, \overline{x}', \overline{y}, \overline{y}')$. Also, recall our notation for variables: $\overline{x}$ denotes existentially quantified unlimited variables and $\overline{y}$ unlimited parameters; similarly, $\overline{x}'$ denote quantified limited variables and let $\overline{y}'$ denote limited parameters.

The algorithm proceeds in steps that transform the input formula to an equivalent output formula with additional syntactic properties. After describing each step we prove, unless obvious, termination and correctness of that step.

Step 1. Type Completion.

   Input: An $E(\mathcal{A})$-formula.

   Output: An equivalent type completed formula.

   So the formula is now of the form

$$\exists^u \overline{x} \, Q_k^l x_k' \cdots Q_1^l x_1' \bigvee_i \psi_i(\overline{x}, \overline{x}', \overline{y}, \overline{y}'), \qquad (\star)$$

where each ψ_i is a type-completed conjunction of literals.

Step 2. Put Every Term into Term-Normal-Form.

   Input: A type-completed formula in form $(\star)$.

   Output: An equivalent type-completed formula in form $(\star)$ for which every term is in term-normal-form.

A term is in *term-normal-form* if it is of the form

$$t(v_1, \cdots, v_{j+k})[v_1/L_1 w_1, \cdots, v_j/L_j w_j]$$

where $t(\overline{v})$ is a term built from constructors only. Here L_i is a sequence of selectors applied to the variable w_i.

This step proceeds by pushing selectors past constructors as follows.

For each ψ_i do the following until no more apply. Pick some t occurring as a subterm in ψ_i. There are two cases.

Case 1: t Is Limited. Then for every constructor f, replace $f_j(t)$ in ψ_i by t, and

Case 2: t Is Unlimited. Let g be the unique constructor for which $\mathrm{IS}_g(t)$ is a conjunct of ψ_i.

- For every $f \neq g$, replace $f_j(t)$ in ψ_i by t, and
- Say t is of the form $g(\overline{s})$ for some $\overline{s}$. Then replace $g_j(t)$ in ψ_i by s_j.

Termination: After applying each case the number of selectors in the formula decreases. Hence this step can be iterated only a finite number of times.

Correctness: By Lemma 1, the transformation preserves the equivalence of the formulas. Since only terms have changed, the resulting formula is still in formula-normal-form. Also, since every term in the resulting formula is already a term in the original formula, the result is also type-completed.

Step 3. Remove Selectors from All Unlimited Quantified Variables.

Input: A type-completed formula $\varPhi$ in form $(\star)$ for which every term is in term-normal-form.

Output: An equivalent type-completed formula for which every term is in term-normal-form and there are no selectors infront of unlimited quantified variables. That is one of the form

$$\exists^u \overline{x}\, Q_k^l x_m' \cdots Q_1^l x_1' \bigvee_i \psi_i(\overline{x}, \overline{x}', \overline{y}, \overline{y}'), \qquad (\dagger)$$

where each ψ_i is type-completed and no unlimited quantified variable x in ψ_i has a selector in front of it.

Recall $\overline{x}$ is the collection of unlimited quantified variables. For each $x \in \overline{x}$ do the following until no more apply:

- Pick an $x \in \overline{x}$ for which Lx occurs in $\varPhi$ where L is some non-empty block of selectors.
- Replace $\varPhi$ by

$$\exists \overline{v} \bigvee_f \left[f(\overline{v}) \notin A \wedge (\exists^u \overline{x}\, Q_k^l x_k' \cdots Q_1^l x_1' \bigvee_i \psi_i)[x/f(\overline{v})] \right],$$

where f varies over all constructors. Now remove the existential quantifier $\exists^u x$.
- Apply step 1 and then step 2.

Termination: The result of substituting $f(\overline{v})$ for x and then putting the terms in normal form results in every selector of the form Lx being transformed into one of the form $L'v_i$ where the length of L' is smaller than the length of L. And since steps 1 and 2 do not introduce selectors, the size of the largest block of selectors in front of unlimited quantified variables strictly decreases with each iteration.

Correctness: Once the procedure terminates no unlimited quantified variable x has a selector infront of it. That the output formula is equivalent follows from the fact that the following are equivalent in $E(\mathcal{A})$ for every formula Θ:

- $\exists^u x \, \Theta$.
- $\exists^u x \bigvee_f (\mathrm{IS}_f(x) \wedge x \notin A \wedge \Theta)$.
- $\exists \overline{v} \exists^u x \bigvee_f (x = f(\overline{v}) \wedge \mathrm{IS}_f(x) \wedge x \notin A \wedge \Theta)$.

Step 4a. Put Every (Dis)equation into Literal-Normal-Form.

Input: A formula of the form (†).

Output: An equivalent formula of the form (†) in which every equation and disequation is in literal-normal-form.

An *unlimited (dis)equation is in literal-normal-form* if it is of the form

$$L_1 v_1 \Delta L_2 v_2$$

for some (possibly empty) L_i, and $\Delta \in \{=, \neq\}$. Here the v_i and $L_i v_i$ are unlimited.

A *limited (dis)equation is in literal-normal-form* if each term is in term-normal-form

$$t(v_1, \cdots, v_{j+k})[v_1/L_1 y_1, \cdots, v_j/L_j y_j],$$

with the additional property that the $v_{j+1}, \cdots, v_{j+k}, L_1 y_1, \cdots, L_j y_j$ are limited (that is, stated in ψ_i to be in A).

Throughout this step ensure that every literal $t \Delta s$ satisfies $t \in A$ if and only if $s \in A$ by applying the following steps whenever possible.

- An equation between terms t and s with $t \in A$ and $s \notin A$ is replaced with $\bot$.
- A disequation between terms t and s with $t \in A$ and $s \notin A$ is replaced with $\top$.

Hence every (dis)equation is either limited or unlimited.

For the unlimited (dis)equations repeat the following steps in each ψ_i until none can be applied; and then finally put the result into formula-normal-form, and type complete it.

- An unlimited equation of the form $f(\overline{t}) = g(\overline{s})$ is replaced with $\bot$ if $f \neq g$ and with $\bigwedge_i t_i = s_i$ if $f = g$.
- An unlimited disequation of the form $f(\overline{t}) \neq g(\overline{s})$ is replaced with $\top$ if $f \neq g$ and with $\bigvee_i t_i \neq s_i$ if $f = g$.
- An unlimited equation of the form $Ly = f(\overline{t})$, *with y unquantified*, is replaced with $\bigwedge_i f_i Ly = t_i$.

– An unlimited disequation of the form $Ly \neq f(\bar{t})$, *with y unquantified*, is replaced with $\bigvee_i f_i Ly \neq t_i$.

Termination: In general each item removes a literal $t \Delta s$ and replaces it with (a boolean combination of) a set of literals $\{t_i \Delta' s_i\}$. The relevant property here is that the largest number of constructors appearing in a term from $t \Delta s$ is strictly greater than the largest number of constructors appearing in a term from any of the $t_i \Delta' s_i$.

Correctness: The procedure produces a formula that is equivalent to the original one as seen from Lemma 1. The formula is of the form (†) since the only introduced selectors are in front of unquantified unlimited variables from $\bar{y}$. Now if $t \Delta s$ is a resulting unlimited literal then it does not contain constructors. Also both t and s are in term-normal-form since each operation preserves being in term-normal-form. Hence $t \Delta s$ is in literal-normal-form.

Now we deal with the easier case of limited literals. Suppose t is limited (with respect to some ψ_i of the input formula). Note that t is already in term-normal-form. Now if some v_j or $L_j y_j$ occurring in t were unlimited (with respect to ψ_i) then t would also be unlimited, and so ψ_i is replaced with $\bot$. Hence every limited (dis)equation in the formula obtained is in literal-normal-form.

Step 4b. Put Tester Predicates into Literal-Normal-Form.

Input: A formula of the form (†) in which every equation and disequation is in literal-normal-form.

Output: An equivalent formula in formula-normal-form, for which every literal is in literal-normal-form, and there are no selectors infront of unlimited quantified variables. Also it has the property that no literal mentions both a quantified unlimited variable from $\bar{x}$ and a limited variable.

A *tester predicate is in literal-normal-form* if it is of the form $IS_g(Lv)$ for some g, where L is a possibly empty block of selectors, v is a variable and Lv is unlimited (that is, stated in ψ_i not to be in A).

We put every tester predicate into literal normal form by applying the following steps to every term t in ψ_i. Say t is of the form $f(\bar{s})$ for some $\bar{s}$ and f.

– If t is limited then replace $IS_g(t)$ by $\bot$ for every g.
– If t is unlimited then replace $IS_g(t)$ by $\top$ if $f = g$ and by $\bot$ otherwise.

Termination is clear in this case.

Correctness: By Lemma 1 the resulting formula is equivalent to the input formula. Every tester predicate is in normal form since each term in the input was in term-normal-form. Since in this step only tester predicates are removed the (dis)equations are still in literal-normal-form, and there are no selectors infront of unlimited quantified variables x from $\bar{x}$. We remark that although Φ is no longer necessarily type-completed, each disjunct ψ_i has a unique type-completion ψ_i' up to equivalence in $E(\mathcal{A})$. Recall that we call a term (un)limited in ψ_i if it is (un)limited in ψ_i'. So if x occurs in a term, that term is unlimited. Moreover if x occurs in an equation or disequation, then it is of the form $x \Delta Ly$, or $x \Delta x_1$, where x_1 is also from $\bar{x}$ and y is unlimited and unquantified. Similarly if x occurs in a tester predicate, it is of the form $\mathrm{IS}_g(x)$ for some g. Hence no literal mentions both x and a limited variable.

Step 5. Separate the Unlimited Quantifiers from the Limited Quantifiers.

Input: A formula Φ of the form

$$\exists^u \overline{x}\, Q_k^l x_k' \cdots Q_1^l x_1'\, \psi(x_k', \cdots, x_1', \overline{x}, \overline{y}, \overline{y}'),$$

and with the properties resulting from the previous step.

Output: An equivalent formula in disjunctive normal form where each conjunction consists of formulae of the form

$$Q_k^l x_k' \cdots Q_1^l x_1'\, \mu(x_k', \cdots, x_1', \overline{y}, \overline{y}'), \quad \exists^u \overline{x}\, \delta(\overline{x}, \overline{y}) \ \text{ and } \ \epsilon(\overline{y}, \overline{y}').$$

Here

- μ is a possibly empty quantifier free formula with the property that every literal in μ is limited and mentions some variable from $\overline{x}'$.
- δ is a possibly empty conjunct of literals, and every literal in it is unlimited and mentions some variable from $\overline{x}$.
- ϵ is a possibly empty conjunct of literals.

In other words literals not in the scope of some quantifier are separated out. This can be done since by the previous step no literal mentions both some x and some limited variable x_j' or y'.

Step 6a. Remove Equations from δ That Mention x.

Input: The formula resulting from the previous step.

Output: An equivalent formula of the same form with the additional property that there are no equations in any of the δ.

Repeat the following in every δ until no more apply.

- Replace $x = x$ by $\top$ and $x \neq x$ by $\bot$.
- If an equation $x = Ly$ is a conjunct in δ, then replace δ by $\delta[x/Ly]$.

Some literals may be transformed into literals of the form $L_1 y_1 \Delta L_2 y_2$. So put these new literals into the corresponding ϵ.

Termination: Since each stage removes the variable x from δ, this process eventually stops.

Correctness: After termination it is still the case that every literal in δ is unlimited and mentions some variable from $\overline{x}$. And the corresponding ϵ may have gained more literals. So the output formula has the same form but there are no equations left in δ since every equation mentioned some variable from $\overline{x}$.

Step 6b. Replace Each $\exists^u \overline{x}\, \delta$ with $\top$.

Input: The formula resulting from the previous substep.

Output: An equivalent formula without any unlimited quantification.

From the previous step there are no equations in any of the δ; the disequations in δ are of the form $x \neq Ly$ or $x \neq x_i$ for some other unlimited quantified variable x_i from $\overline{x}$.

Let $\bar{b}$ be an arbitrary instantiation of canonical terms for the parameters $\bar{y}$. We need to show that

$$E(\mathcal{A}) \models \exists^u \bar{x}\, \delta(\bar{x}, \bar{b})$$

First evaluate all selectors Lb by applying the definition of the selectors to the canonical terms from $\bar{b}$. Let $s \in \mathbb{N}$ be the maximum of the depth of the subterms (of the sentence) that do not mention variables from $\bar{x}$. For each f let $n_f \in \mathbb{N}$ be the number of distinct $x \in \bar{x}$ so that $\mathrm{IS}_f(x)$ is a conjunct of δ. Choose $k \in \mathbb{N}$ larger than s and with the property that for every constructor $f \in \Sigma$ there are at least n_f distinct canonical terms of depth k that start with an f. This can be done since $\mathcal{A}$ is not a total algebra.

Now let $\bar{a}$ be distinct canonical terms of depth k that satisfy the type data in δ. Then $E(\mathcal{A}) \models \delta(\bar{a}, \bar{b})$. Indeed an unlimited literal $a_i \neq a_j$ holds in $E(\mathcal{A})$ by assumption that the elements of $\bar{a}$ are distinct. An unlimited literal $a \neq b$ holds in $E(\mathcal{A})$ since the depth k of the term a is at least s which is greater than the depth of the term b. This completes the description and correctness of the algorithm.

Tracing through the proof, we see that we have transformed a formula of $E(\mathcal{A})$,

$$\exists^u \bar{x}\, Q_k^l x_k' \cdots Q_1^l x_1' \bigvee \psi_i(\bar{x}, \bar{x}', \bar{y}, \bar{y}'),$$

into one of the form

$$\bigvee_j [\epsilon_j(\bar{y}, \bar{y}') \wedge Q_k^l x_k' \cdots Q_1^l x_1' \, \mu_j(\bar{x}', \bar{y}', \bar{y})],$$

with only limited quantification, and where every literal in μ_j is limited and mentions some variable of $\bar{x}'$. This completes the proof of Lemma 2.

4.2   Corollaries and the Main Result

We can also give a characterisation of the definable relations of $E(\mathcal{A})$.

Theorem 2. *There is a procedure that given a formula $\Phi(\bar{y}, \bar{y}')$ of $E(\mathcal{A})$ returns an equivalent formula $\Phi'(\bar{y}, \bar{y}')$ with the property that every quantification is limited. In particular Φ' has the form*

$$\bigvee_j [\epsilon_j(\bar{y}, \bar{y}') \wedge Q_m^l x_m' \cdots Q_1^l x_1' \, \mu_j(\bar{x}', \bar{y}', \bar{y})]$$

where ϵ_j is a type-completed conjunct of literals, μ_j is a type-completed quantifier free formula, and every literal in μ_j is limited and mentions some variable from $\bar{x}' = \cup_i \bar{x}'_i$.

Proof. Given a formula Φ of $E(\mathcal{A})$, first replace the formula with its type-completion. So it is now of the form

$$Q_m v_m \cdots Q_1 v_1 \psi,$$

where every quantifier is either limited or unlimited. Now pick an innermost formula of the form

$$Q^u \overline{x} \, Q^l_k x'_k \cdots Q^l_1 x'_1 \, \psi(\overline{x}, \overline{x}', \overline{y}, \overline{y}'),$$

where the quantifiers Q^l_i are limited, ψ is in disjunctive normal form $\bigvee_i \psi_i$, where each ψ_i, a conjunction of literals, is type-completed. Note that k may be 0. Also, we may assume that $Q^u \overline{x}$ is $\exists^u \overline{x}$, for if it were $\forall^u \overline{x}$ then replace it with $\neg \exists^u \overline{x} \neg$, and push the second $\neg$ inward as usual. Applying the lemma results in a formula with no unlimited quantification. Now repeat this process until there are no more unlimited quantifiers in Φ. Finally although the lemma may result in a formula containing a conjunct of literals B that is not type-completed, it is the case that B is equivalent in $E(\mathcal{A})$ to a type-completed conjunction of literals B'. So replace B by B'. This completes the proof.

A formula $\Phi(\overline{y}, \overline{y}')$ of $E(A)$ is called an $\mathcal{A}$-*formula* if it is type-completed, and of the form

$$Q^l_k x'_k \cdots Q^l_1 x'_1 \, \mu(\overline{x}', \overline{y}', \overline{y}),$$

where every literal in μ is limited. Recall this means that each term is limited and of the form

$$t(v_1, \cdots, v_{j+k})[v_1/L_1 y_1, \cdots, v_j/L_j y_j],$$

where each of $v_{j+1}, \cdots, v_{j+k}, L_1 y_1, \cdots, L_j y_j$ is limited, and $t(\overline{v})$ consists of constructors only.

Observe that if χ_1 and χ_2 are $\mathcal{A}$-formulae, then they are equivalent in $\mathcal{A}$ if and only if they are equivalent in $E(\mathcal{A})$. Also the subformulae from Theorem 2 of the form

$$Q^l_k x'_k \cdots Q^l_1 x'_1 \, \mu(\overline{x}', \overline{y}', \overline{y}),$$

are $\mathcal{A}$-formulae. Hence we have the next corollary.

Corollary 1. *If $\mathcal{A}$ admits elimination of quantifiers, then so does $E(\mathcal{A})$.*

We now restate, and are ready to prove the main theorem.

Theorem 3. *If a partial algebra $\mathcal{A}$ has decidable QEP then so does its free total extension $E(\mathcal{A})$.*

Proof. Given a formula $\Phi(\overline{v})$ and a tuple of elements $\overline{w}$ from $E(\mathcal{A})$. Apply Theorem 2 and transform Φ into Φ'. Now form the sentence $\Phi'(\overline{w})$. This sentence consists of quantifier free sentences $\epsilon(\overline{a}, \overline{a}')$, and sentences of the form

$$Q^l_k v'_k \cdots Q^l_1 v'_1 \bigvee_i \psi_i(\overline{v}', \overline{a}', \overline{a}),$$

where each ψ_i consists of limited literals and type data. Here $\overline{a}'$ and $\overline{a}$ are amongst $\overline{w}$, and moreover $\overline{a}' \subset A$ and $\overline{a} \cap A = \emptyset$. Also $\overline{v}' = \cup_j v'_j$.

Now evaluate $\epsilon(\overline{a}, \overline{a}')$. This is done by first applying the definition of the selector functions and tester predicate and then applying the fact that $t_1 = t_2$, where the t_i are canonical terms not in A, if and only if t_1 and t_2 are syntactically equal.

This leaves an $\mathcal{A}$-sentence that is evaluated using the algorithm for the theory of $\mathcal{A}$.

5   Application

Recall that $GT_\Sigma(C)$ denotes the (total) algebra of ground terms generated by the non-empty set C of constants from Σ. We start with the following definition.

Definition 2. *Let E be a set of ground equations; that is equations of the form $t = s$ where t and s are ground terms. Consider the quotient of GT_Σ by the smallest congruence generated by E. This is a (total) algebra (over Σ) that we will denote by $\mathcal{A}_E$. Call a total algebra* **finitely presented** *if it is of the form $\mathcal{A}_E$ for some* finite *set E of ground equations.*

Easy examples include GT_Σ itself and every finite algebra. Decision problems for finitely presented algebras in a given variety of algebras have received much attention. For instance, the word problem in finitely presented semigroups or groups (in the variety of semigroups or groups) is, in general, undecidable. However in the variety of *all* algebras, the situation is different. For instance Kozen in [10] considers the uniform word problem, the finiteness problem, the subalgebra membership problem and the triviality problem: all are decidable in polynomial time. Comon in [3] proves that the first order theory of any finitely presented term algebra is decidable. The proof uses algebraic techniques combined with quantifier elimination methods. Our main theorem can now be applied to give another and, we think, simpler proof to decide the first order theory for finitely presented term algebras.

The relationship between finitely presented algebras and free total extensions is described in the next theorem (implicit in [11]).

Theorem 4. *A total algebra is a finitely presented term algebra if and only if it is the free total extension of a finite partial algebra.*

So we immediately have the following application of Theorem 1.

Theorem 5. *Let $\mathcal{A}_E$ be a finitely presented algebra. Then it has decidable QEP. In particular its first order theory is decidable.*

Acknowledgement

We thank the referees for their helpful comments used in improving the presentation and proof of the results.

References

1. E. Astesiano, M. Bidoit, H. Kirchner, B. Krieg-Brückner, P.D. Mosses, D. Sannella, A. Tarlecki: CASL: The Common Algebraic Specification Language. *Theoretical Computer Science* 286:2 (2002) 153–196
2. A. Blumensath, E. Gradel: Automatic structures. *Proceedings of the 15th Annual IEEE Symposium on Logic in Computer Science (LICS) 2000* 51–62
3. H. Comon: Complete axiomatization of some quotient term algebras: *Theoretical Computer Science* 122:1-2 (1993) 165–200
4. K. Compton, C. Henson: A uniform method for proving lower bounds on computational complexity of logical theories. *Annals of Pure and Applied Logic* 48 (1990) 1–79
5. W. Hodges: *Model theory.* Cambridge University press (1993)
6. B. Khoussainov, S. Rubin, F. Stephan: Automatic Linear Orders and Trees. *Transactions on Computational Logic (TOCL) special issue (selected papers from the LICS 2003 conference)*, editor Phokion G. Kolaitis (accepted)
7. B. Khoussainov, A. Nerode: Automatic Presentations of Structures. D. Lievant (editor) *Proceedings of the conference on Logic and Computational Complexity (1994)* vol. 960 of LNCS (1995) 367–393
8. K. Korovin, A. Voronkov: A decision procedure for the existential theory of term algebra with the Knuth-Bendix ordering. In *Proceedings IEEE Conference on Logic in Computer Science* (2000) 291–302
9. G. Grätzer: *Universal Algebra.* D. Van Nostrand Co., Inc., Princeton, N.J.-Toronto, Ont.-London (1968)
10. D. Kozen: Complexity of finitely presented algebras. In *Proceedings of the 9th ACM symposium on theory of computing* (1977) 164–177
11. D. Kozen: Partial automata and finitely generated congruences: an extension of Nerode's theorem. In J. Crossley, J. Remmel, R. Shore, M. Sweedler editors. Logical methods: in honour of Anil Nerode's Siztieth Birthday, Birkhauser (1993) 490–511
12. V. Kuncak, M. C. Rinard: Structural sybtyping of non-recursive types is decidable. In *Proceedings IEEE Conference on Logic in Computer Science* (2003) 96–107
13. M. Lohrey: Automatic structures of bounded degree. *In proceedings of 10th International conference Logic for programming, artificial intelligence and reasoning,* editors M. Vardi and A. Voronkov, vol. 2850 of LNCS (2003) 346–360
14. M. Maher: A CLP view of logic programming. In *Algebraic and Logic Programming* (1992) 364–383
15. A. Mal'cev: Axiomatizable classes of locally free algebras of various types. In *The Metamathimatics of Algebraic Systems. Anatolii Ivanovič Mal'cev. Collected papers: 1936-1967,* B. Wells III, Ed. Vol. 66. North Holland. Chapter 23 (1971) 262–281
16. T. Zhang, H. Sipma, Z. Manna: Term algebras with length function and bounded quantifier alternation. *the 17th International Conference on Theorem Proving in Higher Order Logics (TPHOLs'04)* Volume 3223 of LNCS (2004) 321–336
17. T. Rybina, A. Voronkov: *A Decision procedure for term algebras with queues.* ACM Transaction on Computational Logic 2:2 (2001) 155–181
18. S. Vorobyov: An improved lower bound for the elementary theories of trees. In *Proceedings of the 13th Intl. Conference ib automated deduction,* Volume 1104 of LNCS (1996) 275–287
19. I. Walukiewicz: Monadic second order logic on tree-like structures. *Theoretical computer science* 275:1-2 (2002) 311–346

Results on the Guarded Fragment
with Equivalence or Transitive Relations

Emanuel Kieroński

Institute of Computer Science, University of Wrocław

Abstract. We study the problem of the satisfiability of guarded formulas in models in which some distinguished binary symbols are interpreted as equivalence relations or as transitive relations. We sharpen the undecidability result for the two-variable guarded fragment with transitive relations by reducing the number of transitive relations to two. We prove that the satisfiability problem for the two-variable guarded fragment with two equivalence relations is 2EXPTIME-complete. We consider the guarded fragment with equivalence relations in guards and show that this variant is easily reducible to the variant with transitive relations in guards. However, in the case of two variables, the version with equivalence relations is easier: NEXPTIME-complete. Finally we show that the decidability results for the guarded fragment with either equivalence relations or transitive relations in guards cannot be generalized to the loosely guarded fragment.

1   Introduction

The *guarded fragment* GF is a fragment of first-order logic[1] FO in which the usage of quantifiers is restricted. GF is defined inductively: all atomic formulas belong to GF; GF is closed under Boolean connectives; if $\varphi(\mathbf{x}, \mathbf{y}) \in$ GF then $\forall \mathbf{x}(\alpha(\mathbf{x}, \mathbf{y}) \to \varphi(\mathbf{x}, \mathbf{y}))$ and $\exists \mathbf{x}(\alpha(\mathbf{x}, \mathbf{y}) \wedge \varphi(\mathbf{x}, \mathbf{y}))$ belong to GF, where $\alpha(\mathbf{x}, \mathbf{y})$ is an atomic formula containing all the free variables of φ. In particular, α may be of the form $x = x$. Atoms α are called *guards*.

The guarded fragment was introduced by Andréka, van Benthem and Németi [1] as a generalization of propositional modal logic. Since then various variants and extensions of GF have been investigated. It appeared that GF retains a lot of good properties of modal logics. In particular its satisfiability problem is decidable [1, 4], it has the finite model property and (a kind of) tree model property [4], remains decidable when augmented with fixed-point operators [8].

Since the transitivity of a binary relation is not expressible in GF, to obtain counterparts of some modal logics with axioms of transitivity or equivalence, like K4 or S5, it is natural to study the satisfiability of the guarded fragment in restricted classes of models, in which some binary relations are interpreted as transitive or, respectively, equivalence relations. In this paper we follow the convention, that when considering the guarded fragment, this requirement is expressed syntactically: we "extend" the guarded fragment by the possibility of stating that some binary relations have to be transitive

---

[1] In this paper we consider first-order logic without constants and function symbols.

L. Ong (Ed.): CSL 2005, LNCS 3634, pp. 309–324, 2005.

(equivalence) relations. Now, a formula consists of its purely guarded part and a list $trans[T_1, \ldots, T_l]$ ($equiv[T_1, \ldots, T_l]$) of transitive (equivalence) relations[2].

Unfortunately, it appeared that the guarded fragment with transitive relations GF+$trans$ is undecidable. The first proof of this fact was given in [4], where undecidability of GF^3+$trans[T_1, T_2]$, the guarded fragment with three variables and two transitive relations, was established. Since modal logic can be embedded into the two-variable guarded fragment GF^2, it was very interesting what happens on the level of two variables. This question was partially answered by Ganzinger, Meyer and Veanes [3] who showed undecidability of GF^2+$trans[T_1, \ldots, T_4]$ and GF^2+$trans[T_1, \ldots, T_5]$ without equality. In this paper we sharpen this result and show undecidability of GF^2+$trans[T_1, T_2]$ without equality and with T_1, T_2 being the only non-unary symbols. Moreover, we believe, that in spite of restricting the language, our proof is simpler than the proof in [3]. The result is optimal with respect to the number of transitive relations, since, as we argue, GF^2+$trans[T]$ is decidable, and in fact 2EXPTIME-complete.

The undecidability of GF^2+$trans[T_1, T_2]$ becomes even more interesting if compared to the results on the two-variable guarded fragment with equivalence relations. In [11] we proved undecidability of GF^2+$equiv[E_1, E_2, E_3]$ without equality, but it appeared that even the whole FO^2+$equiv[E_1, E_2]$ is decidable. Because of the high level of technical complication we were not able to give precise complexity bounds for the last variant. On the level of the guarded fragment the situation is easier and in this paper we show that GF^2+$equiv[E_1, E_2]$ is 2EXPTIME-complete. This reveals, that reasoning about two equivalence relations is more difficult than about one, since even FO^2+$equiv[E]$ is in NEXPTIME [11].

The undecidability of the guarded fragment with several transitive or equivalence relations seems to be quite surprising in the context of connections to modal logics: modal logics K4 and S5 are decidable even in their multi-modal variants involving more than one transitive or equivalence relation. To explain this, we should note that in the natural translation of modal logic into the guarded fragment all these special relations appear only in guards. And indeed, the guarded fragment with *transitive guards* GF+TG, the version of GF+$trans$ in which only the relations appearing only in guards can be required to be transitive, is decidable in 2EXPTIME [15]. See [3, 9, 10, 15, 16] for more discussions and results on this variant.

Here we consider the guarded fragment with equivalence relations in guards GF+EG and show that it is easily reducible to GF+TG. For two variables, however, GF^2+EG is easier than GF^2+TG. We show that it is NEXPTIME-complete, while the latter was shown to be 2EXPTIME-hard [10].

In the Table 1 we summarize the results on the satisfiability of the two-variable guarded fragment with transitive or equivalence relations.

In the final part of this paper we look at a more liberal extension of the guarded fragment, the so-called *loosely* guarded fragment, LGF. In LGF [2], the notion of guard is relaxed. A guard $\alpha(\mathbf{x}, \mathbf{y})$ may be a conjunction of atoms provided that each pair of variables from $\mathbf{x} \cup \mathbf{y}$ appears together in an atom. In [3] it was observed that LGF^3 with one transitive relation is undecidable. The proof went by a reduction from the inter-

---

[2] Note that the set of formulas obtained in this way is not really what we usually call "logic"; for example it is not closed under negation.

Table 1. Satisfiability of GF^2 with transitive or equivalence relations

type of special relations	only in guards	# of equivalence/transitive relations		
		1	2	3 or more
equivalence	NEXPTIME	NEXPTIME	2EXPTIME	undecidable
transitive	2EXPTIME	2EXPTIME	undecidable	undecidable

section emptiness problem for context-free languages. Here we give a straightforward, very easy proof of the slightly stronger result: undecidability of LGF^3 without equality with one transitive relation which is used only in guards. Similar result can be also obtained for LGF^3 with one equivalence relation in guards. Moreover, in the case of one transitive relation, we can make an effort to use the transitive symbol as the only non-unary symbol in our proof. These results show that the distinction between *relation in guards only* and *relations everywhere*, which is important for transitive and equivalence relations in GF, disappears in LGF.

Besides obvious connections to modal logics, the study of the satisfiability of the guarded fragment in restricted classes of models can be also motivated by similar study for FO^2. See [6, 7, 11, 13].

2 Plan of the Paper

In Section 3 we adapt the normal form theorems for the guarded fragment to our purposes. Section 4 contains results on $GF^2+trans$. In Section 5 we study GF+EG, in particular GF^2+EG, and prove that the $GF^2+equiv[E_1, E_2]$ is 2EXPTIME-complete. In Section 6 we show undecidability of $LGF^3+TG[T]$ and $LGF^3+EG[T]$.

A note for the reader. We assume that the reader knows basic concepts from model theory and theoretical computer science, in particular notions of (atomic) 1-types and 2-types and results on alternating Turing machines.

The paper is closely related to [15] and [11]. In a few places we refer not only to results from these papers but also to techniques used in proofs. In such parts we do not present whole constructions in detail, but rather sketch main ideas.

3 Normal Forms

We want to adapt the normal form theorems for the guarded fragment to make them more convenient to our purposes. Such theorems were proved by Grädel in [4] and Szwast and Tendera in [15]. Let us review the results from [15].

Definition 1. *We say that a GF^2 sentence is in normal form if it is a conjunction of sentences of the following form:*

- $\exists x \, (\alpha(x) \wedge \chi(x))$,
- $\forall x \, (\alpha(x) \rightarrow \exists y \, ((\beta(x, y) \wedge \chi(x, y))))$,
- $\forall x \forall y \, (\beta(x, y) \rightarrow \chi(x, y))$,

where $\alpha(x)$, $\beta(x, y)$ are atomic formulas and χ is quantifier-free.

It can be shown:

Theorem 1 (Szwast, Tendera). *With every GF^2 sentence φ of the length n one can effectively associate a set Δ of GF^2 sentences in normal form (over the extended vocabulary), $\Delta = \{\delta_1, \ldots, \delta_d\}$ such that*
(1) φ *is satisfiable if and only if $\bigvee_{i \leq d} \delta_i$ is satisfiable,*
(2) $d \leq O(2^n)$ *and for every $i \leq d$, $\lceil \delta_i \rceil = O(n \log n)$,*
(3) Δ *can be computed deterministically in exponential time and every sentence δ_i can be computed in time polynomial with respect to n,*
(4) $\bigvee_{i \leq d} \delta_i \models \varphi$ *and every model of φ can be expanded to a model of $\bigvee_{i \leq d} \delta_i$.*

The additional property of normal form of Szwast and Tendera is that if a binary symbol T appears in φ only in guards then it is also the case in Δ. It is important when working with either the guarded fragment with transitive guards GF+TG or the guarded fragment with equivalence relations in guards GF+EG. We use this property in Section 5.1.

Let us introduce the new definition:

Definition 2. *A $GF^2 + equiv[E_1, E_2]$ sentence φ is in $[E_1, E_2]$-guarded normal form if it is a conjunction of sentences of the following form:*

- $\exists x \, (\alpha(x) \wedge \chi(x))$,
- $\forall x \, (\alpha(x) \rightarrow \exists y \, (E_1 xy \wedge E_2 xy \wedge \chi(x,y)))$ *($\forall \exists^{++}$ form),*
- $\forall x \, (\alpha(x) \rightarrow \exists y \, (E_1 xy \wedge \neg E_2 xy \wedge \chi(x,y)))$ *($\forall \exists^{+-}$ form),*
- $\forall x \, (\alpha(x) \rightarrow \exists y \, (E_2 xy \wedge \neg E_1 xy \wedge \chi(x,y)))$ *($\forall \exists^{-+}$ form),*
- $\forall x \, (\alpha(x) \rightarrow \exists y \, (\beta(x,y) \wedge \neg E_1 xy \wedge \neg E_2 xy \wedge \chi(x,y)))$ *($\forall \exists^{--}$ form),*
- $\forall x \forall y \, (\beta(x,y) \rightarrow \chi(x,y))$ *($\forall \forall$ form),*

where $\alpha(x)$, $\beta(x,y)$ are atomic formulas, and χ is quantifier-free.

The polynomial transformation from normal form from Definition 1 to normal form from Definition 2 is standard and we skip it here. The existence of such a transformation allows us to substitute the words *normal form* in Theorem 1 with $[E_1, E_2]$-*guarded normal form*.

4   Guarded Fragment with Transitive Relations

4.1   A Note on the Case of One Transitive Relation

Essentially, the decidability of $GF^2 + trans[T]$ can be shown by applying the construction of Szwast and Tendera for GF^2+TG [15]. The only problem is that sometimes a conjunct of the form $\forall x \, (\alpha(x) \rightarrow \exists y \, (\beta(x,y) \wedge \chi(x,y)))$ without T in the guard β may say in χ that x and y have to be connected by T. To avoid this problem we first transform a formula into $[T]$-guarded normal form, which is similar to $[E_1, E_2]$-guarded normal form, with symbol T treated in a special way. The construction from [15] gives now the 2EXPTIME upper bound. The lower bound follows from [10]. We have:

Theorem 2. *The satisfiability problem for $GF^2 + trans[T]$ is 2EXPTIME-complete.*

4.2   Two Transitive Relations

In this subsection and in Section 6 we use the technique from [6, 13]. It was originally designed for extensions of FO^2 but adaptations for extensions of GF^2 without equality are straightforward. Due to the space limit we give only the statement of the lemma we require, without a proof.

We say that an infinite structure $\mathfrak{G} = (G, H, V)$ is *grid-like* if the standard grid $\mathfrak{G}_\mathbb{N}$ is homomorphically embeddable into $\mathfrak{G}$; a finite $\mathfrak{G}$ is grid-like if some $\mathfrak{G}_m$ is homomorphically embeddable into $\mathfrak{G}$, where $\mathfrak{G}_m$ is the standard grid on $m \times m$ torus.

Lemma 1. *Let $\mathcal{L}$ be an extension of GF^2 without equality. Let N be a distinguished unary symbol of the vocabulary. If there exists an $\mathcal{L}$-formula φ such that:*
(1) $\mathfrak{G}_\mathbb{N}$ can be expanded to a model of φ,
(2) for every $k \in \mathbb{N}$ there exists $m > k$ such that $\mathfrak{G}_m$ can be expanded to a model of φ,
(3) if $\mathfrak{A} \models \varphi$ is infinite then $\mathfrak{A}{\restriction}N$ is grid-like,
(4) if $\mathfrak{A} \models \varphi$ is finite then $\mathfrak{A}{\restriction}N$ is grid-like,
then $\mathcal{L}$ forms a conservative reduction class. If at least conditions (1) and (3) hold then the satisfiability problem for $\mathcal{L}$ is undecidable.

We improve the undecidability result from [3] by showing that the two-variable guarded fragment with two transitive relations is undecidable. Actually we prove the stronger result, implying also the undecidability of the finite satisfiability problem:

Theorem 3. *$GF^2 + trans[T_1, T_2]$ without equality is a conservative reduction class.*

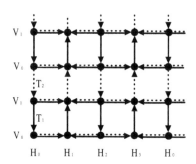

Fig. 1. Grid structure for $GF^2 + trans[T_1, T_2]$

Proof. Let us expand the standard grid to the structure $\mathfrak{G}_\mathbb{N}$ illustrated in the Fig. 1. Additionally, for the unary symbol N, Na is true for every element a of the structure. We capture some properties of $\mathfrak{G}_\mathbb{N}$ by the sentence φ, then observe that every $\mathfrak{G}_{4m}$ can also be expanded to a model of φ and prove that the restriction to N of every model of φ is grid-like.

The sentence φ is the conjunction of the following formulas:
(A) The initial formulas:

$$\exists x\, (H_0 x \wedge V_0 x \wedge N x), \tag{1}$$
$$\forall x\, (N x \rightarrow \exists y\, (H x y \wedge N y)), \tag{2}$$
$$\forall x\, (N x \rightarrow \exists y\, (V x y \wedge N y)). \tag{3}$$

(B) A formula axiomatising H, which has the following shape:

$$\forall xy \; (Hxy \rightarrow (\varphi_1 \vee \varphi_2 \vee \ldots \vee \varphi_8)), \tag{4}$$

where each φ_i describes one of the eight possible cases of values of unary predicates on H-connected vertices and transitive connections between them. For example:

$$\varphi_1 \equiv T_1 xy \wedge T_2 xy \wedge H_0 x \wedge V_0 x \wedge H_1 y \wedge V_0 y \wedge Nx \wedge Ny, \tag{5}$$

$$\varphi_2 \equiv T_1 yx \wedge T_2 yx \wedge H_1 x \wedge V_0 x \wedge H_2 y \wedge V_0 y \wedge Nx \wedge Ny. \tag{6}$$

(C) A formula axiomatising V, which is built similarly to the one for H.

(D) A group of formulas stating that some horizontally adjacent elements that are connected by T are also linked by H. Example formulas from this group are:

$$\forall xy \; (T_1 xy \rightarrow ((H_0 x \wedge V_1 x \wedge H_1 y \wedge V_1 y) \rightarrow Hxy)), \tag{7}$$

$$\forall xy \; (T_2 xy \rightarrow ((H_2 x \wedge V_0 x \wedge H_1 y \wedge V_0 y) \rightarrow Hyx)). \tag{8}$$

It is not hard to see that every $\mathfrak{G}_{4m}$ can be expanded to a model of φ. It is enough to take a natural quotient.

We sketch the argument for grid-likeness of $\mathfrak{A}' = \mathfrak{A} {\upharpoonright} N$, for $\mathfrak{A} \models \varphi$. It is enough to show that H is complete over V in $\mathfrak{A}'$, i.e. $\mathfrak{A}' \models \forall xyx'y'((Hxy \wedge Vxx' \wedge Vyy') \rightarrow Hx'y')$ (cf. [13]). Assume that for a, a', b, b':

$$\mathfrak{A}' \models Hab \wedge Vaa' \wedge Vbb'.$$

We show that then $\mathfrak{A}' \models Ha'b'$. One should consider several cases distinguished by values of the H_i and V_i on a. Let us go through one of them, for instance $\mathfrak{A}' \models H_1 a \wedge V_1 a$. By (B) we have:

$$\mathfrak{A}' \models H_2 b \wedge V_1 b \wedge T_2 ba.$$

Similarly (C) implies:

$$\mathfrak{A}' \models H_1 a' \wedge V_0 a' \wedge T_2 aa'$$

and

$$\mathfrak{A}' \models H_2 b' \wedge V_0 b' \wedge T_2 b'b.$$

From transitivity of T_2 it follows that $\mathfrak{A}' \models T_2 b'a'$. Now an appropriate formula of the form (D) guarantees $\mathfrak{A}' \models Ha'b'$ which finishes the proof for this case. The remaining seven cases can be treated in the similar way. $\qquad \square$

Remark. The binary relation symbols H and V do not play the crucial role in the above proof. They can be simulated by T_1, T_2 and values of unary predicates H_i and V_i. Thus we obtain undecidability of $GF^2+trans[T_1, T_2]$ without equality, even if T_1 and T_2 are the only non-unary symbols allowed.

5   Guarded Fragment with Equivalence Relations

5.1   Equivalence Relations in Guards

In [15] Szwast and Tendera study the guarded fragment with transitive guards GF+TG, in which merely relations appearing only in guards may be required to be transitive. It is also natural to consider the guarded fragment with equivalence relations in guards GF+EG. We observe that there is a simple reduction from GF+EG to GF+TG [3].

Lemma 2. *There is a polynomial time reduction which transforms a GF+EG sentence φ into a GF+TG sentence φ' such that φ is satisfiable if and only if φ' is satisfiable.*

Proof. Let $\varphi = \varphi' + equiv[E_1, \ldots, E_k]$ be a GF+EG sentence. We construct a GF+TG sentence ψ, in an extended vocabulary, which is satisfiable if and only if φ is satisfiable. For every $1 \le i \le k$ we introduce a new unary predicate P_i and put the formulas:

$$\forall xy \; (E_i xy \rightarrow ((P_i x \wedge P_i y) \rightarrow x = y)), \tag{9}$$

$$\forall x \; (x = x \rightarrow \exists y \; (E_i xy \wedge P_i y)) \wedge \forall x \; (x = x \rightarrow \exists y \; (E_i yx \wedge P_i y)). \tag{10}$$

We define ψ as the conjunction of φ', formulas (9)-(10) for $1 \le i \le k$, and the transitivity requirement: $trans[E_1, \ldots, E_k]$.

Now, every model of φ can be expanded to a model of ψ by choosing for every i exactly one element in each E_i-class, and marking it with P_i. On the other hand in every model of ψ each E_i has to be symmetric and reflexive and thus, because of transitivity of E_i, it has to be an equivalence relation.   □

Since the satisfiability of GF+TG was shown in [15] to be in 2EXPTIME, and the satisfiability of the pure guarded fragment GF is 2EXPTIME-hard we have:

Corollary 1. *The satisfiability problem for GF+EG is 2EXPTIME-complete.*

Let us turn our attention to the case of the two-variable fragment and observe that GF^2+EG is easier than GF^2+TG which is complete for 2EXPTIME [10, 15].

Theorem 4. *The satisfiability problem for GF^2+EG is NEXPTIME-complete.*

Proof. **The Upper Bound.** The proof of the upper bound is similar to the proof of the upper bound for GF^2+TG given by Szwast and Tendera [15]. In their proof it is shown how to construct a *ramified* model for a satisfiable formula. The most important properties of a ramified model are:

- it is *tree-controlled*,
- each pair of elements is connected by at most one transitive relation (possibly in both directions),

---

[3] Notice that the straightforward idea of expressing symmetry of a transitive relation E: $\forall xy \; (Exy \rightarrow Eyx)$ leads to the formula in which E appears outside guards.

– the size of every T_i-clique[4], for a transitive T_i, is at most exponential in the size of the input formula.

The similar properties may be obtained in our case. We sketch the construction. Let φ be a GF^2+EG sentence in normal form from Definition 1, with equivalence relations $E_1, \ldots E_k$. Let $\mathfrak{A} \models \varphi$. We build a new model $\mathfrak{B} \models \varphi$ which consists of (possibly) infinite number of layers. The Layer 0 consists of elements satisfying conjuncts of the form $\exists x(\alpha(x) \wedge \chi(x))$. The Layer 1 contains E_i-classes for $1 \leq i \leq k$ and free witnesses (witnesses for formulas without equivalence symbols in guards) for elements from the Layer 0. The E_i-classes of an element b may be constructed in such a way that for $i \neq j$, E_i-class of b and E_j-class of b have exactly one common element: b, and each E_i-class has size at most exponential in $|\varphi|$. The construction of these classes is the same as the construction of T_i-cliques of an element in the proof of Szwast and Tendera for GF^2+TG. We say that an E_i-class of b is its *petal*. All the free witnesses of b we put outside the petals of b. They are called *leaves*. Petals and leaves of b form the *flower* of b.

We proceed recursively. For an element b from the Layer i we build its flower and put its petals (with the exception of at most one which was built in the previous step) and leaves into the Layer $(i+1)$. We define all non-specified 2-types in $\mathfrak{B}$ as negative types consistent with 1-types, where by a negative 2-type we mean a type without positive occurrences of binary predicates.

This way we obtain a ramified model $\mathfrak{B} \models \varphi$. For every 1-type t_i which is realized in $\mathfrak{B}$ we choose an element b_i of type t_i. We argue that a list L of such elements and its flowers is sufficient for construction of another ramified model $\mathfrak{B}' \models \varphi$. Observe that the information we require is of size exponential in $|\varphi|$. To construct $\mathfrak{B}'$ we put for every t_i in L a new copy of the flower of b_i into the Layer 0. Now, we proceed recursively again. Let b be an element in the Layer i whose flower is not defined. There may be at most one petal built for b, say E_r-petal. In our list we find b' of the same type as b and we put a new copy of every petal of b', except its E_r-petal, to the Layer $(i+1)$. Similarly, we add all the required leaves for b.

The procedure deciding the satisfiability problem for GF^2+EG may be the following:

input: a GF^2+EG sentence φ
begin procedure
- compute the set Δ from Theorem 1,
- *guess* a sentence δ_i in normal form from Definition 1 from Δ,
- *guess* a list of 1-types realized in a ramified model of δ_i, and shapes of its flowers,
- check if a ramified model may be constructed basis of this list.
end procedure

Clearly, it works in nondeterministic exponential time.

The Lower Bound. The lower bound can be easily obtained without equality, even in the presence of only one equivalence relation E. Using two sets of unary predicates

---

[4] A T_i-clique is a maximal set X of elements such that for every $a, b \in X$ both $T_i(a, b)$ and $T_i(b, a)$ are true.

$H_1, \ldots H_n$ and $V_1, \ldots, V_n$ we can write a formula such that its model has to contain an E-class which can be viewed as a square of size $2^n \times 2^n$ such that for every element, H_i and V_i code in binary its horizontal and vertical position. Having such a class it is easy to simulate a nondeterministic Turing machine working in time exponential in the size of its input. $\qquad\square$

5.2   GF^2 with Two Equivalence Relations

In [11] we considered the whole FO^2 with equivalence relations. We proved that FO^2+ $equiv[E]$ has exponential model property. From this result and from the proof of the lower bound in Theorem 4 it follows that GF^2+$equiv[E]$ is NEXPTIME-complete. In [11] we proved also that the satisfiability problem for whole two-variable first order logic with two equivalence relations is decidable in 3NEXPTIME. Because of the high level of technical complication of our construction we were not able to give exact complexity bounds. In the case of the two-variable guarded fragment the situation is easier. However, we still have to take into account infinite models. The following example of an infinity axiom comes from [11]. If E_1 and E_2 are interpreted as equivalence relations then the conjunction φ of the formulas (11)-(16) is satisfiable only in infinite models.

$$\exists x \, (Px \wedge Sx) \tag{11}$$

$$\forall x \, (Px \rightarrow \exists y \, (E_1 xy \wedge x \neq y \wedge Qy)) \tag{12}$$

$$\forall x \, (Qx \rightarrow \exists y \, (E_2 xy \wedge x \neq y \wedge Py)) \tag{13}$$

$$\forall xy \, (E_1 xy \rightarrow ((Px \wedge Py) \rightarrow x = y)) \tag{14}$$

$$\forall xy \, (E_2 xy \rightarrow ((Qx \wedge Qy) \rightarrow x = y)) \tag{15}$$

$$\forall x \, (Sx \rightarrow \neg \exists y \, (E_2 xy \wedge x \neq y)) \tag{16}$$

An example model is illustrated in the Fig. 2.

Fig. 2. A model of φ

Here we show:

Theorem 5. *The satisfiability problem for GF^2+$equiv[E_1, E_2]$ is 2EXPTIME-complete.*

Proof. **The Lower Bound.** The proof of the lower bound is similar in spirit to the proof of 2EXPTIME-lower bound for GF^2+TG from [10].

It suffices to show that every problem in AEXPSPACE can be reduced in polynomial time to the satisfiability problem for GF^2+$equiv[E_1, E_2]$. Let M be an alternating Turing machine working in space bounded by 2^{n^k}. Let w be an input for M. We construct a GF^2+$equiv[E_1, E_2]$ sentence φ which is satisfiable if and only if M accepts w.

Without any loss of generality we can assume that in every configuration M has exactly two possible transitions and that it enters an accepting or rejecting state in exactly $2^{2^{n^k}}$ - th step. Moreover, to simplify the proof, we assume that after this step M does not stop, but works infinitely. More precisely, the accepting and rejecting states are universal. In each of such states M has two identical transitions: it does not write any symbol on the tape and it does not change its state. In other words, after accepting or rejecting M stays infinitely in the same configuration. Such assumption about M allows us not to bother about the numbering of configurations.

Every configuration is represented by a set of 2^{n^k} elements, each of them corresponding to a single cell of the tape. To encode a position of an element in a configuration, i.e. a consecutive number of a tape cell it represents, we use the unary relation symbols $P_1, \ldots, P_{n^k}$. Formally, $P_i a$ is true if the i-th bit of the position of a is set to 1. We use the abbreviation $\overline{P}(a)$ to describe this position ($0 \le \overline{P}(a) < 2^{n^k}$).

The following properties (for fixed l, $0 \le l < 2^{n^k}$) can be easily expressed by quantifier-free formulas of the length polynomial in n:

$$\overline{P}(x) = l, \quad \overline{P}(x) \ge l, \quad \overline{P}(x) = \overline{P}(y), \quad \overline{P}(x) = \overline{P}(y) + 1.$$

For example, the last property can be expressed as follows:

$$\bigvee_{0 \le i < n^k} (P_i x \wedge \neg P_i y \wedge \bigwedge_{j > i} (\neg P_j x \wedge P_j y) \wedge \bigwedge_{j < i} (P_j x \leftrightarrow P_j y)).$$

We connect each pair of elements a, b belonging to the representation of a same configuration with both E_1 and E_2, i.e. we want $E_1 ab \wedge E_2 ab$ to be satisfied in our model.

We use the standard description of a configuration: for each symbol a_i in the alphabet of M (including **blank**) we use the unary relation symbol A_i, for each state q_i - the unary symbol Q_i. We also have the unary symbol H describing the head position. An element x represents a tape cell scanned by the head if Hx and $Q_i x$, for some i, are true.

We begin our construction by enforcing that every model of φ contains a substructure that can be viewed as an infinite binary tree. The set of 2^{n^k} elements describing a single configuration of M is treated as a "node" of this tree.

We organize the structure in such a way that elements belonging to an even configuration, i.e. a configuration whose depth in the tree is even, are connected to the elements from the successor configurations by E_1, and elements belonging to an odd configuration are connected to the elements from the successor configurations by E_2. We do not impose any relations between elements that do not belong to a same configuration or to two consecutive configurations. Additionally, we introduce two unary symbols: D_0 true for elements belonging to even configurations, and D_1 true for elements belonging to odd configurations. One more unary symbol L indicates that the element belongs to the left son of some node.

Let us start the construction with stating the existence of the node representing the initial configuration of M on w. For elements of this node the special unary symbol I is true. We assume that for this configuration L is true:

$$\exists x \, (Ix \wedge D_0 x \wedge Lx \wedge \overline{P}(x) = 0, \tag{17}$$

$$\forall x (Ix \rightarrow (\overline{P}(x) \neq 2^{n^k} - 1 \rightarrow \exists y\ (E_1 xy \wedge E_2 xy \wedge Iy \wedge D_0 y \wedge Ly \wedge \overline{P}(y) = \overline{P}(x)+1))). \tag{18}$$

The next formula expresses that for every element, except the last one, belonging to a description of a configuration there exists a successor in this configuration:

$$\bigwedge_{i=0,1} \forall x\ (D_i x \rightarrow (\overline{P}(x) \neq 2^{n^k} - 1$$

$$\rightarrow (\exists y\ (E_1 xy \wedge E_2 xy \wedge D_i x \wedge (Lx \leftrightarrow Ly) \wedge \overline{P}(y) = \overline{P}(x) + 1)))). \tag{19}$$

The existence of successor nodes is implied by the following formulas, in which $n \oplus m \equiv n + m \pmod 2$:

$$\bigwedge_{i=0,1} \forall x\ (D_i x \rightarrow \exists y\ (E_{i+1} xy \wedge D_{i\oplus 1} y \wedge Ly \wedge \overline{P}(y) = 0)), \tag{20}$$

$$\bigwedge_{i=0,1} \forall x\ (D_i x \rightarrow \exists y\ (E_{i+1} xy \wedge D_{i\oplus 1} y \wedge \neg Ly \wedge \overline{P}(y) = 0)). \tag{21}$$

Now we say that a model of our formula satisfies several basic properties of a computation tree. We can say that there is exactly one alphabet symbol in every tape cell:

$$\bigwedge_{k=0,1} \forall x\ (D_k x \rightarrow \bigvee_i A_i x)\ \wedge\ \bigwedge_j \forall x\ (A_j x \rightarrow \bigwedge_{i\neq j} \neg A_i x). \tag{22}$$

In each configuration at most one element is scanned by the head:

$$\forall xy\ (E_1 xy \rightarrow (E_2 xy \rightarrow ((Hx \wedge Hy) \rightarrow x = y))). \tag{23}$$

Exactly the elements which represent tape cells observed by the head store information about state.

$$\bigwedge_i \forall x\ (Q_i x \rightarrow Hx)\ \wedge\ \forall x\ (H(x) \rightarrow \bigvee_i Q_i x). \tag{24}$$

The next formulas ensure that the root of the tree describes the initial configuration in the state Q_0 of M on the input $w = a_0 a_1 \ldots a_{n-1}$.

$$\forall x\ (Ix \rightarrow (\overline{P}(x) = 0 \rightarrow (Hx \wedge Q_0 x))), \tag{25}$$

$$\bigwedge_{i<n} (\forall x\ (Ix \rightarrow (\overline{P}(x) = i \wedge \rightarrow W_i x)))\ \wedge\ \forall x\ (Ix \rightarrow (\overline{P}(x) \geq n \rightarrow Bx)), \tag{26}$$

where B is the symbol representing **blank**.

The following formula says that if a tape cell of a configuration is not scanned by the head then in the same cell of both successor configurations the alphabet symbol does not change.

$$\bigwedge_{i=0,1} \forall xy\ (E_{i+1} xy \rightarrow ((D_i x \wedge D_{i\oplus 1} y \wedge \neg Hx \wedge \overline{P}(x) = \overline{P}(y)) \rightarrow \bigwedge_i (A_i x \leftrightarrow A_i y))). \tag{27}$$

Consider now a node t of the tree and the configuration c that is described by this node. There are two cases: the state of the machine in this configuration is existential or it is universal.

In the first case we enforce that the configuration represented by the left son of t is created by applying one of the two possible transitions on c. Assume that for an existential state q and a letter a there are two possible transitions: $(q,a) \to (q',a',\to)$ and $(q,a) \to (q'',a'',\leftarrow)$. We put:

$$\bigwedge_{i=0,1} \forall xy \quad \left(E_{i+1}xy \to \left(((D_ix \wedge D_{i\oplus1}y \wedge Qx \wedge Ax \wedge Ly \wedge \overline{P}(x) = \overline{P}(y))\right.\right.$$
$$\to ((A'y \wedge \forall x\, (E_1xy \to (Next(y,x) \to Hx \wedge Q'x)))$$
$$\left.\left.\vee (A''y \wedge \forall x\, (E_1xy \to (Next(x,y) \to Hx \wedge Q''x)))))))\right), \quad (28)$$

where the formula $Next(x,y)$ is the abbreviation stating that x and y are two consecutive elements in the representation of a configuration: $Next(x,y) \equiv E_1xy \wedge E_2xy \wedge (D_0x \wedge D_0y \vee D_1x \wedge D_1y) \wedge \overline{P}(y) = \overline{P}(x) + 1$. Other possible situations, when both transitions move the head forward or both transitions move the head backward, can be handled similarly.

Consider now the case of a universal configuration. We enforce that the left son of t is created by applying the first transition and the right son by applying the second one. For a universal state q, a letter a and transitions $(q,a) \to (q',a',\to)$ and $(q,a) \to (q'',a'',\leftarrow)$ we put:

$$\bigwedge_{i=1,2} \forall xy \quad \left(E_{i+1}xy \to \left(((D_ix \wedge D_{i\oplus1}y) \wedge Qx \wedge Ax \wedge \overline{P}(x) = \overline{P}(y))\right.\right.$$
$$\to (\neg Ly \to (A''x \wedge \forall x\, (E_1xy \to (Next(x,y) \to (Q''x \wedge Hx)))))$$
$$\left.\left.\wedge\ Ly \to (A'x \wedge \forall x\, (E_1xy \to (Next(y,x) \to (Q'x \wedge Hx))))))\right). \quad (29)$$

To finish our construction we give a formula stating that the machine never enters the only rejecting state q_r.

$$\forall x\, (Q_rx \to \textbf{false}). \quad (30)$$

Now, let φ be a conjunction of (17)–(30) (plus formulas symmetric to (28) and (29)). Observe that the number of conjuncts and the size of each of them are polynomial in the size of M and w.

We claim that φ is satisfiable if and only if M accepts w. Indeed, if M accepts w then an accepting computation tree can be transformed into a model $\mathfrak{M}$ of φ in the following way. The root of the computation tree is transformed into the root of $\mathfrak{M}$. Then we proceed recursively. Let c be a configuration in the computation tree and let c' be its code in $\mathfrak{M}$. If c is universal then we transform its left subtree into the left subtree of c' and its right subtree into the right subtree of c'. If c is existential then we transform its accepting subtree into the left subtree of c'. Since we want to have a complete binary tree, we have to define somehow also the right subtree of c'. We can for example construct all nodes of this subtree in such a way that they agree with c' in predicates denoting alphabet symbols and for each element a from these nodes $\mathfrak{M} \models \neg Ha \wedge \bigwedge_i \neg Q_i a$. It is easy to see that $\mathfrak{M}$ is indeed a model of our formula.

For the proof of the opposite direction we want to check that the existence of an accepting computation tree is implied by the existence of a model $\mathfrak{M}$ of φ. The set of elements of $\mathfrak{M}$ whose existence is enforced by (17), (18) is translated into the root of computation tree. It describes the initial configuration of M on w because of (25), (26). The construction of the further parts of the computation tree is recursive. Let us assume that we have constructed a configuration c, which is encoded in $\mathfrak{M}$ by the node consisting of $c_1, \ldots, c_{2^{nk}}$. Let us go for example through the case when c is an even configuration, i.e. for all i we have $\mathfrak{M} \models D_0 c_i$. Let a_0 and b_0 be the elements whose existence is ensured by (20), (21). We have $\mathfrak{M} \models D_1 a_0 \wedge D_1 b_0 \wedge L a_0 \wedge \neg L b_0$. By (19) there exist two nodes consisting of $a_0, \ldots a_{2^{nk}}$ and $b_0, \ldots b_{2^{nk}}$ connected to the node with c_i by E_1 such that:

- for all i: $\mathfrak{M} \models \bar{P}(a_i) = \bar{P}(b_i) = i$,
- for all i: $\mathfrak{M} \models L a_i \wedge \neg L b_i$,
- for $i < 2^{n^k} - 1$: $\mathfrak{M} \models a_i E_1 a_{i+1} \wedge a_i E_2 a_{i+1} \wedge b_i E_1 b_{i+1} \wedge b_i E_2 b_{i+1}$.

If c is universal then we translate the node $a_0, \ldots, a_{2^{nk}}$ into the left successor of c and $b_0, \ldots, b_{2^{nk}}$ into the right successor of c. Consider now the case of existential c. Because of the appropriate formula of the form (28) we can translate $a_0, \ldots, a_{2^{nk}}$ into one of the successors of c. At this moment we leave the second successor undefined.

At the end, for formal conformity, we substitute undefined subtrees of existential nodes with subtrees which agree with transition function of M. This is not essential since in existential nodes we demand only one accepting successor. The construction of φ, in particular its conjuncts of type (28) and (29), implies that the tree obtained in the described way is indeed an accepting computation tree of $\mathcal{M}$ on w.

Remark. Note that the usage of equivalence relation symbols outside guards in our proof is very limited. Actually, we only need guards which are conjunctions of two atoms: $E_1 xy \wedge E_2 xy$, instead of just single atomic formulas.

The Upper Bound. In this proof we use the notion of *intersection*. An intersection (in a relational structure with E_1, E_2 equivalence relations) is an equivalence class of the relation $E_1 \cap E_2$. To denote intersections we use letters $\mathcal{I}$, $\mathcal{J}$. By $\Theta^{\mathfrak{A}}(\mathcal{I})$ we denote the isomorphism type of $\mathcal{I}$ in $\mathfrak{A}$. Equivalence classes of E_1 and E_2 are called E_1-classes and E_2-classes, respectively. Let φ be a satisfiable formula in $[E_1, E_2]$-guarded normal form. From an arbitrary model of φ we may obtain a model with all intersections of size bounded exponentially in $|\varphi|$. It was shown for $FO^2 + equiv[E_1, E_2]$ in [11]. Let $\mathfrak{A} \models \varphi$ be a model of φ with so exponentially bounded intersections. We construct a new, forest-like model $\mathfrak{B} \models \varphi$. $\mathfrak{B}$ is built from isomorphic copies of intersections from $\mathfrak{A}$, thus we do not have to bother about witnesses for conjuncts of the form $\forall\exists^{++}$.

Several times in the construction we have the following task: for a given E_i-class $\mathcal{C}$ of $\mathfrak{A}$, with a distinguished intersection $\mathcal{I}$, build an E_i-class $\mathcal{D}$ of $\mathfrak{B}$, of size exponential in $|\varphi|$, such that $\mathcal{D}$ contains an isomorphic copy of $\mathcal{I}$ as an intersection. The desired $\mathcal{D}$ may be obtained in the same way as an exponential model $\mathfrak{M}$ of a satisfiable $FO^2 + equiv[E]$ sentence ψ in [11]. The class $\mathcal{C}$ in our case corresponds to $\mathfrak{M}$ in the construction in [11], intersections of $\mathcal{C}$ correspond to E-classes of $\mathfrak{M}$. We skip details here.

The construction of $\mathfrak{B}$ is divided into (possibly) infinite number of stages:

Layer 0. For every conjunct of φ of the form $\exists x(\alpha(x) \wedge \chi(x))$ we find an element $a \in \mathfrak{A}$ such that $\mathfrak{A} \models \alpha(x) \wedge \chi(x)$. Let $a \in \mathcal{I} \subseteq \mathcal{C}$, where $\mathcal{C}$ is the E_1-class of a. We put a new intersection $\mathcal{J}$, such that $\Theta^{\mathfrak{A}}(\mathcal{I}) = \Theta^{\mathfrak{B}}(\mathcal{J})$, into the Layer 0 of $\mathfrak{B}$. We construct a regular, exponential-size version $\mathcal{D}$ of $\mathcal{C}$, containing $\mathcal{J}$ as an intersection. After building $\mathcal{D}$ we may assume that all its elements have all the required witnesses for conjuncts of the form $\forall\exists^{+-}$.

The Next Layers. We proceed recursively. Having constructed the Layer i we construct the Layer $i + 1$. Let $\mathcal{J}$ be an intersection belonging to the i-th Layer. If i is even then the whole E_1-class of $\mathcal{J}$ is constructed. Similarly if i is odd then the whole E_2-class of $\mathcal{J}$ is constructed. Assume that i is even. The case of i odd is symmetric. We want to construct E_2-class of $\mathcal{J}$, thus providing all the witnesses for formulas of the type $\forall\exists^{-+}$. Let $\mathcal{I} \subseteq \mathfrak{A}$ be such that $\Theta^{\mathfrak{A}}(\mathcal{I}) = \Theta^{\mathfrak{B}}(\mathcal{J})$. Let $\mathcal{C}$ be the E_2-class containing $\mathcal{I}$. We construct a regular, exponential-size version $\mathcal{D}$ of $\mathcal{C}$, containing $\mathcal{J}$ as an intersection. We put all the intersections of $\mathcal{D}$, with the exception of $\mathcal{J}$ into the Layer $i + 1$ of $\mathfrak{B}$. Note that it may happen that $\mathcal{D}$ consists only of $\mathcal{J}$ and then there is nothing to put into the next layer.

Now, let us explain how to provide witnesses for conjuncts of the type $\forall\exists^{--}$ for elements from $\mathcal{J}$. Let $b \in \mathcal{J}$. Consider a conjunct of φ of the form $\forall x \, (\alpha(x) \rightarrow \exists y \, (\beta(x,y) \wedge \neg E_1 xy \wedge \neg E_2 xy \wedge \chi(x,y)))$. Let $a \in \mathfrak{A}$ be such that $type^{\mathfrak{A}}(a) = type^{\mathfrak{B}}(b)$ [5]. If $\mathfrak{A} \models \alpha(a)$ then find an element $a' \in \mathcal{I}' \subseteq \mathfrak{A}$ such that $\mathfrak{A} \models \beta(a,a') \wedge \neg E_1 aa' \wedge \neg E_2 aa' \wedge \chi(a,a)$. Put a new intersection $\mathcal{J}'$ into the Layer $i + 1$. Set $\Theta^{\mathfrak{B}}(\mathcal{J}') = \Theta^{\mathfrak{A}}(\mathcal{I}')$. Let b' be an element of type $type^{\mathfrak{A}}(a')$ in $\mathcal{J}'$. Set $type^{\mathfrak{B}}(b,b') = type^{\mathfrak{A}}(a,a')$. Set all the other 2-types between elements from $\mathcal{J}'$ and the elements from the Layer i in such a way that they are *negative*, i.e. contain no positive occurrences of binary predicates. Extend $\mathcal{J}'$ to its E_2-class.

For all the pairs of elements for which we have not set their 2-type yet, we set them in such a way that they are negative. We can do so safely, without violating the conjuncts of φ of the form $\forall\forall$, because all of such conjuncts say something only about elements connected by binary relations.

This finishes the construction of a regular model. Observe that to check the existence of a regular model constructed in the described way it suffices to build doubly exponential number of levels of a model. It is enough because the number of isomorphism types of intersections which may appear is at most doubly exponential. So later we may just repeat the structure of trees. An alternating procedure working in exponential space, checking whether a $GF^2+EQ[E_1, E_2]$ sentence in $[E_1, E_2]$-normal form has a regular model, can be naturally derived from the construction. $\qquad\square$

6 Loosely Guarded Fragment with One Equivalence/Transitive Relation

The guarded fragment with equivalence relations GF+$equiv$ and the guarded fragment with transitive relations GF+$trans$ are undecidable. We know, however, that if we restrict the usage of equivalence/transitive relations to guards only then the obtained versions: GF+EG and GF+TG become decidable. In this section we show that this result cannot be generalized to the loosely guarded fragment LGF. We prove that allowing

---

[5] $type^{\mathfrak{A}}(a)$ denotes the 1-type realized by a in $\mathfrak{A}$.

just one equivalence or transitive relation makes LGF undecidable, even if this special relation is used only in guards.

First we give an easy proof of:

Theorem 6. $LGF^3+TG[T]$ *forms a conservative reduction class.*

Proof. Let φ be the conjunction of (31)-(34) below and the transitivity axiom $trans[T]$.

$$\exists x\, Nx, \tag{31}$$

$$\forall x\ (Nx \rightarrow \exists y\ (Txy \wedge Hxy \wedge Ny)), \tag{32}$$

$$\forall x\ (Nx \rightarrow \exists y\ (Tyx \wedge Vxy \wedge Ny)), \tag{33}$$

$$\forall xyz((Vyx \wedge Hyz \wedge Txz) \rightarrow \exists y(Txz \wedge Txy \wedge Tyz \wedge Hxy \wedge Vzy \wedge Ny)). \tag{34}$$

It is very easy to expand $\mathfrak{G}_\mathbb{N}$ to a model of φ. We make N true for all elements and set T to be the transitive closure of $\{((x_1, y_1), (x_2, y_2)) : x_2 = x_1 + 1 \wedge y_1 = y_2 \vee y_1 = y_2 + 1 \wedge x_1 = x_2\}$. Similarly, an expansion of $\mathfrak{G}_m$ to a model of φ is obtained by setting T to be the transitive closure of $\{((x_1, y_1), (x_2, y_2)) : 0 \leq x_i, y_i < m, x_2 = x_1 + 1 \ (\mathrm{mod}\ m) \wedge y_1 = y_2 \vee y_1 = y_2 + 1 \ (\mathrm{mod}\ m) \ \wedge x_1 = x_2\}$.

Let us consider an infinite model $\mathfrak{A} \models \varphi$. We argue that $\mathfrak{A}' = \mathfrak{A}{\upharpoonright}N$ is grid-like, i.e. we construct a homomorphism $h : \mathfrak{G}_\mathbb{N} \rightarrow \mathfrak{A}'$. We start with embedding $\{(x, 0) : x \in \mathbb{N}\}$ into $\mathfrak{A}'$. Choose an element a_0 in $\mathfrak{A}'$ satisfying (31) and set $h(0, 0) = a_0$. Now if $h(x, 0) = a_x$ then let $h(x + 1, 0)$ be the witness for a_x and (32). Assume now that we have defined $h(x, y)$ for all $x \in \mathbb{N}$ and $y < k$. We say how to extend h to the $k - th$ row of $\mathfrak{G}_\mathbb{N}$. Choose $h(0, k)$ as a witness for $h(0, k - 1)$ and (33). Now, let $h(1, k)$ be a witness for $h(0, k), h(0, k-1)$ and $h(1, k-1)$ and (34). We define $h(x, k)$ in sequence as witnesses for $h(x - 1, k)$, $h(x - 1, k - 1)$, $h(x, k - 1)$ and (34).

We skip here the argumentation for grid-likeness of finite models. It goes the same way as the proof of Lemma 2.4 in [13]. □

Observe that if we exchange the transitivity axiom $trans[T]$ in the above proof by $equiv[T]$ then the proof still works correctly. Thus we have:

Theorem 7. $LGF^3+EG[T]$ *forms a conservative reduction class.*

Remark. If we want to obtain only undecidability of $LGF^3+TG[T]$ rather then the fact that it forms a conservative reduction class, we can make it even in the case when T is the only non-unary symbol allowed. In this case we use the formula stating that between two elements connected by T there is no other element:

$$NB(x, y) \equiv \neg \exists z\ (Txy \wedge Txz \wedge Tzy).$$

Now we can simulate Hxy and Vxy by $NB(x, y)$ and values of some additional unary predicates (in particular the predicate D which is true exactly for elements in even rows). For example the formula (34) should be substituted by

$$\forall xyz\ ((Txy \wedge Tyz \wedge Txz) \rightarrow (((NB(x, y) \wedge NB(y, z)$$
$$\wedge (Dx \wedge \neg Dy \wedge \neg Dz \vee \neg Dx \wedge Dy \wedge Dz))$$
$$\rightarrow \exists y(Txy \wedge Tyz \wedge Txz \wedge NB(x, y) \wedge NB(y, z) \wedge (Dx \leftrightarrow Dy)))) \tag{35}$$

Note that due to the lack of the orientation of E this trick does not work for $LGF^3+EG[E]$.

References

1. H. Andréka, J. van Benthem, I. Németi, *Modal Languages and Bounded Fragments of Predicate Logic*, J. Philos. Logic, 27 , No. 3, 217-274, 1998.
2. J. van Benthem, *Dynamic bits and pieces*, ILLC Research Report, 1997.
3. H. Ganzinger, C. Meyer, M. Veanes, *The Two-Variable Guarded Fragment with Transitive Relations*, Proc. of 14-th LICS, pages 24-34, 1999.
4. E. Grädel, *On the Restraining Power of Guards*, J. Symbolic Logic 64:1719-1742, 1999.
5. E. Grädel, P. Kolaitis, M. Vardi, *On the Decision Problem for Two-Variable First Order Logic*, Bull. of Symbolic Logic, 3(1):53-96, 1997.
6. E. Grädel, M. Otto, *On Logics with Two Variables*, TCS 224:77-113, 1999.
7. E. Grädel, M. Otto, E. Rosen, *Undecidability Results on Two-Variable First-Order Logic*, Archive of Mathematical Logic, volume 38, pp. 313-354, 1999.
8. E. Grädel, I. Walukiewicz, *Guarded Fixpoint Logic*, Proc. of 14-th LICS, pages 45-54, 1999.
9. E. Kieroński, *EXPSPACE-Complete Variant of Guarded Fragment with Transitivity*, Proc. of 19-th STACS, pages 608-619, 2002.
10. E. Kieroński, *The Two-Variable Guarded Fragment with Transitive Guards is 2EXPTIME-Hard*, Proc. of 6th FOSSACS, LNCS 2620, pages 299-312, 2003.
11. E. Kieroński, M. Otto, *Small Substructures and Decidability Issues for First-Order Logic with Two Variables*, accepted for LICS 2005.
12. H. R. Lewis, *Complexity Results for Classes of Quantificational Formulas*, J. Comp. and System Sci. 21, pages 317-353, 1980.
13. M. Otto, *Two Variable First-Order Logic Over Ordered Domains*, J. Symb. Log. 66, 2001.
14. W. Szwast, L. Tendera, *On the Decision Problem for the Guarded Fragment with Transitivity*, Proc. of 16-th LICS, pages 147-156, 2001.
15. W. Szwast, L. Tendera, *The Guarded Fragment with Transitive Guards*, Annals of Pure and Applied Logic, 128, pages 227-276, 2004.
16. L. Tendera, *Counting in the two variable guarded logic with transitivity*, Proc. 22nd STACS, LNCS 3404, pages 83-96, 2005.

The Modular Decomposition
of Countable Graphs: Constructions
in Monadic Second-Order Logic

Bruno Courcelle[1] and Christian Delhommé[2]

[1] Université Bordeaux 1, LaBRI (CNRS)
courcell@labri.fr
[2] Université de La Réunion, ERMIT
delhomme@univ-reunion.fr

Abstract. We show that the modular decomposition of a countable graph can be defined from this graph, given with an enumeration of its set of vertices, by formulas of Monadic Second-Order logic. A second main result is the definition of a representation of modular decompositions by a low degree relational structures, also constructible by Monadic Second-Order formulas.

1 Introduction

The present article investigates the *modular decomposition of countable graphs* and more precisely, its construction by *Monadic Second-Order (MS* in short) formulas. The notion of modular decomposition of a finite graph has been studied extensively in many articles, and under various names. Möhring and Radermacher give in [17] a survey of this frequently rediscovered notion. It is important, not only for algorithmic purposes, but also for establishing structural properties, in particular of partial orders and their *comparability graphs* (see Kelly [15] who discusses finite and infinite comparability graphs and their *modules*); for instance, one can determine the transitive orientations of a comparability graph from its modular decomposition.

The modular decomposition of a finite graph is the finite tree of its *strong modules*, with inclusion as ancestor relation, together with some structure attached to the nodes of the tree. Each node is a *graph operation*, either the *disjoint union*, the *complete product*, the *sequential product* or the *substitution* to the vertices of a *prime* graph, i.e., a graph that is not expressible in terms of these operations. The strong modules of an infinite graph can be defined in the very same way as for finite graphs. They are either pairwise disjoint or comparable for inclusion, but they do not form a tree, defined as a connected and directed graph without circuits. They form a *generalized tree* defined as a partial order such that the set of elements larger than any element is linearly ordered (and called below simply a tree). Such trees may have no root. The linearly ordered set $\mathbf{Q}$ of rational numbers is a tree in this sense. For defining the modular decomposition of a countable graph, we *do not take all strong modules*, but only some of them called *robust*. Doing so we obtain a countable tree associated with a countable graph. The basic definitions are reviewed in Section 2.

L. Ong (Ed.): CSL 2005, LNCS 3634, pp. 325–338, 2005.

Our goals here are to represent modular decompositions of countable graphs by relational structures, and to use *MS logic* to construct from a graph its modular decomposition. Another concern is to describe *dense* graphs (i.e., graphs having "lots of edges") by relational structures (actually vertex- and edge-labelled graphs) which are as sparse as possible. The linearly ordered set $\mathbf{Q}$ has an empty Hasse diagram. One may think that one *must* represent it by a *complete infinite graph*. However, it can be defined as a certain ordering of the nodes of the complete infinite binary tree (here in the usual sense). A similar binary tree can be constructed from any linearly ordered set A by first-order formulas using an auxiliary enumeration of A (i.e., an ordering of A isomorphic to the ordinal ω). By using these facts, we can represent the modular decomposition of a countable graph G by a countable graph of maximum degree $m + 3$ where m is the least upper bound of the degrees of the prime induced subgraphs of G. It may happen that m is finite, even if G has vertices of infinite degree. This is the case for countable cographs, defined as the graphs without induced P_4 (i.e., without induced path of length 3).

Because of space limitations, proofs are sketched or omitted. Complete proofs can be found in [11].

2   Robust Modules and Modular Decomposition

Unless otherwise specified, trees, forests, graphs and relational structures are countably infinite. A linear order isomorphic to ω is called an *ω-order*. An *ω-ordering* of a set is equivalent to an enumeration $x_0, x_1, ..., x_n, ...$ of this set.

Definitions 1. (*Trees, $\vee$-trees and leafy trees.*) A *forest* is a partial order $(T, \leq)$ such that for every element x, called a *node*, the set $T^x = \{y \in T \mid x \leq y\}$ is linearly ordered. A *tree* is a forest that is *directed*, i.e., such that every two nodes have an upper bound. A forest is the disjoint union of the trees which are the connected components of its comparability graph.

A tree is a *$\vee$-tree* (read a "sup-tree") if any two nodes x and y have a least upper bound denoted by $x \vee y$. A *sub-$\vee$-tree* must preserve the function $\vee$.

A *leaf* is a minimal node, a *root* is a maximal one. An *internal node* is one that is not a leaf. A forest may have one or several roots, or no root at all. It may have no leaf. A tree has at most one root. We say that a tree is *leafy* if it is a $\vee$-tree and every internal node is the least upper bound of two leaves. A finite tree is a finite rooted tree in the usual sense, and its root is the unique maximal element.

If $x \leq y$, we say that the node y is an *ancestor of* x. We say that y is the *father* of x if it the (unique) minimal node among those $> x$. We say in this case that x is a *son* of y.

For a partial order $(P, \leq)$ we use the notations $P^x = \{y \in P \mid x \leq y\}$, $P^{>x} = \{y \in P \mid y > x\}$, $P_x = \{y \in P \mid y \leq x\}$ and $P_{<x} = \{y \in P \mid y < x\}$. We let $HD(P)$ denote its *Hasse diagram*, i.e. the directed graph with set of vertices P and edges $x \to y$ whenever $x < y$ and there is no z with $x < z < y$. We say that P *is diagram-connected* if P is the transitive closure of $HD(P)$ and $HD(P)$ is connected.

Thus a tree is diagram-connected iff every node which is not the root has a father, and the graph of the father relation is connected. Any two nodes are thus at finite distance in the graph $HD(P)$. A diagram-connected tree may have no root.

Definitions 2. (*Directions in* $\vee$-*trees.*) Let T be a $\vee$-tree. For every x, $T_{<x}$ ordered by the induced ordering is a forest, hence a union of trees. Each of these trees D is called a *direction relative to* x. If $y \in D$, we say that D is the *direction of* y *relative to* x. We denote it by $dir_x(y)$. We denote by $Dir(x, T)$ the set of directions relative to x. The *degree* of a node x is the cardinality of $Dir(x, T)$. A tree is *binary* if every node has degree at most 2. If a node is $y \vee z$ where y and z are incomparable, it has degree at least 2. If T is finite, this definition of the degree of a node yields the number of its sons.

A $\vee$-tree $(T, \leq)$ is *ordered* if it is equipped with a linear order $\trianglelefteq_x$ on each set $Dir(x, T)$.

Definitions 3. (*Graph substitutions.*) Graphs are simple, directed, loop-free. Undirected graphs are those where each edge has an opposite edge. We denote by $x \to y$ the existence of an edge from x to y. Although a forest is a graph or can be considered as a graph, we will use the special term "nodes" for the vertices of a tree or a forest. We denote by V_G the set of vertices of a graph G.

If G is a graph and $X \subseteq V_G$, we denote by $G[X]$ the *induced* subgraph of G consisting of X and all the edges, the two ends of which are in X. If E is a set of edges of G, we denote by $G[E]$ the subgraph of G consisting the edges of E and all their end vertices.

If G and H are graphs with disjoint sets of vertices, and u is a vertex of G, we denote by $G[H/u]$ the graph resulting of the *substitution of H for u in G*. Its set of vertices is $V_G \cup V_H - \{u\}$, its edges are those of H, those of G that are not incident with u, the edges $x \to y$ whenever $x \in V_G - \{u\}$, $x \to u$ in G, $y \in V_H$, and the edges $y \to x$ whenever $x \in V_G - \{u\}$, $u \to x$ in G, $y \in V_H$.

If G and H are not disjoint, we replace H by an isomorphic copy disjoint with G. If $u_1, ..., u_n$ are vertices of G and $H_1, ..., H_n$ are graphs, we define $G[H_1/u_1, ..., H_n/u_n]$ as $G[H_1/u_1]...[H_n/u_n]$. The order in which substitutions are done is irrelevant, hence we can consider they are done simultaneously.

If H_v is a graph associated with each $v \in V_G$, we denote by $G[H_v/v, v \in V_G]$ the graph resulting from *the simultaneous substitution in G of H_v for $v \in V_G$*. It can be defined as the graph with vertices (v, w) for $v \in V_G$ and $w \in V_{H_v}$ and edges $(v, w) \to (v', w')$ iff either $v = v'$ and $w \to w'$ (in H_v), or $v \to v'$ (in G).

We will also use the graph operations $\oplus, \otimes$ and $\vec{\otimes}$: $G \oplus H$ is the disjoint union of G and H, $G \vec{\otimes} H$ is $G \oplus H$ augmented with edges from each vertex of G to each vertex of H, and $G \otimes H$ is $G \vec{\otimes} H$ augmented with edges from each vertex of H to each vertex of G. The graphs $G \oplus H$, $G \vec{\otimes} H$ and $G \otimes H$ can be defined as $K[G/u, H/v]$ for graphs K with two vertices u and v, and, respectively, no edge, an edge from u to v, edges between u and v in both directions. They are associative. We will consider them as operations of variable arity in the usual way. The operations $\oplus, \otimes$ are also commutative.

More generally, every graph can be turned into a graph operation. With a finite graph G, with vertices $v_1, ..., v_n$ we associate an n-ary graph operation, denoted by σ_G (where σ stands for *substitution*) defined by $\sigma_G(H_1, ..., H_n) = G[H_1/v_1, ..., H_n/v_n]$. If G is infinite, then σ_G is defined similarly as an operation of countably infinite arity.

Definition 4. (*Modules.*) Let G be a graph. A *module* of G is a subset M of its set of vertices V_G such that for every vertices x, y in M and every vertex z not in M: $x \rightarrow z$ implies $y \rightarrow z$ and: $z \rightarrow x$ implies $z \rightarrow y$. In words this means that every vertex not in M "sees" all vertices of M in the same way. (This notion is studied in several works in particular [15], [17] and [13], in different formal frameworks and using different terminologies). If M is a module then $G = H[G[M]/v]$ for some H and v. Hence the notion of a module identifies a way of expressing a graph as the result of a substitution.

A module is *strong* if it is non empty and does not overlap any module. (Two sets *meet* if they have a nonempty intersection. They *overlap* if they meet and are incomparable for inclusion.) The singletons and the set V_G are strong modules. We will identify frequently a module M and the subgraph $G[M]$ it induces. A graph G is *prime* if it has no trivial module, where the *trivial modules* are $\varnothing$, the singletons and V_G.

The smallest prime undirected graph is the path P_4 with 3 edges and 4 vertices. The smallest prime directed graphs have 3 vertices. The graph H is a module of $G[H/u]$ and the graphs G and H are modules of $G \oplus H$, $G \overrightarrow{\otimes} H$ and $G \otimes H$.

Fact. *A countable graph may have uncountably many strong modules.*

Given two vertices x, y in a graph G, we let $M(x, y)$ be the intersection of all strong modules containing x and y. It is a strong module. We call $M(x, y)$ a *robust module.* It may be the set of all vertices or $\{x\} = M(x, x)$. A countable graph has countably many robust modules. The *maximal proper strong (mps in short) modules of a graph* G are the maximal proper strong submodules of its robust modules.

Proposition 1. *For every graph G, the robust modules form a leafy tree, denoted by $rdec(G)$ which is a sub-$\vee$-tree of the tree of strong modules. Every strong module (in particular G) is the union of the directed set of robust modules included in it. A strong module that is not a singleton is robust iff it is the father of some strong module.*

Proposition 2. [12] *Let G be a graph.*

1. *For every non singleton robust module M, we have one and only one of the following possibilities:*
   (I) either $G[M]$ is the disjoint union (denoted $\oplus$) of a family of connected graphs $C_i, i \in I$,
   (II) or $G[M]$ is the complete product (denoted $\otimes$) of a family of graphs $C_i, i \in I$, where no C_i is of type II,

(III) or $G[M]$ is the linear product (denoted $\overrightarrow{\otimes}$) of a linearly ordered family of graphs $C_i, i \in I$, where no C_i is of type III,
(IV) or $G[M] = P[C_i/u_i, i \in I]$ where P is a prime graph ; this prime graph is unique up to isomorphism.

2. *The graphs C_i are the mps submodules of M. They are not necessarly robust. Their common father in the tree of strong modules of G is M.*
3. *The graphs P of Case IV are induced subgraphs of G.*

The graphs P of case IV are called the *prime factors* of G. If G is given as $Q[P[H_v/v, v \in V_P]/u]$, with P prime, then P is one of its a prime factors. In order to decompose G it suffices to decompose separately Q and the graphs H_v.

Definition 5. (*Modular decomposition.*) By decomposing all robust modules (using Proposition 2), we obtain a hierarchical structure yielding the modular decomposition. The *modular decomposition* of a (countable) graph G is defined formally as the *countable* tree $mdec(G)$ of its robust and mps modules. For finite graphs, the notions of a strong and of a robust module coincide. Hence this notion of modular decomposition is equivalent for finite graphs to the usual one which is the finite rooted tree of the strong modules.

We now analyze the structure of the trees $mdec(G)$. A *limit node* in a tree is a node which is the least upper bound of a directed set of strictly smaller elements. A *father node* is a node that has at least one son. In a ∨-tree, a father node may be also a limit node.

A ∨-tree is said to be *modular* if it satisfies the following conditions:

1. No father node is a limit node.
2. Every father node is the least upper bound of two leaves.
3. A limit node has degree one. Every limit node is the least upper bound of a directed set of non-limit nodes.

Proposition 3. *1. The tree of the modular decomposition of a graph is a modular tree.*
2. Every modular tree is the tree of the modular decomposition of some graph.

Definition 6. (*Embeddings of trees.*)
Let $(T, \leq, \trianglelefteq)$ and $(U, \leq', \trianglelefteq')$ be *ordered* trees. (We denote by $\trianglelefteq$ the family of linear orders $\trianglelefteq_x$ associated with nodes). A ∨-*embedding* of T *into* U is an injective mapping $h : T \to U$, such that for all x, y in T: $h(x) \leq' h(y)$ iff $x \leq y$, $h(x \vee y) = h(x) \vee' h(y)$, and if $D, E \in Dir(x, T)$ and $D \trianglelefteq_x E$, then $h(D) \trianglelefteq'_{h(x)} h(E)$, where $h(D)$ is the unique direction in $Dir(h(x), U)$ that contains $\{h(u) \mid u \in D\}$ (it is not the set extension of h on the set D).

If furthermore, $T \subseteq U$ and h is the inclusion mapping, we say that T *is a sub-∨-tree of U*. For trees which are not ordered, the definitions are the same without the conditions on the ordering of directions.

Proposition 4. *1. Every leafy tree T ∨−embeds into a unique (up to isomorphism) minimal (for ∨−embedding) modular tree denoted by $\widehat{T}$.*
2. T is the sub-∨-tree of $\widehat{T}$ induced by the non-limit nodes.

Proof (Sketch). This construction is a *completion*, where we add only the elements needed as greatest elements of certain directions. (Similar but different completions are used in semantics of recursive program schemes, see [5]). We let $\widehat{T}$ consist of the following sets: the set of all directions (i.e., the union of the sets $Dir(x, T)$) and the sets of the form T_u $(= \{w \in T \mid w \leq u\})$ for all nodes u, (a direction can be of the form T_u) ordered by inclusion. The "new" elements in $\widehat{T}$ are the directions which have no greatest element in T.

Proposition 5. *For every graph G, we have $mdec(G) = \widehat{rdec(G)}$, where $rdec(G)$ is the leafy tree of robust modules of G.*

We wish to have a representation of the modular decomposition of a graph G by a relational structure from which G can be defined in a unique way. Hence, it is not enough to know the "abstract" tree $mdec(G)$, we need also represent in a way or another the information attached to each node, that describes which of cases I-IV of Proposition 2 does apply.

The tree $mdec(G)$ can be seen as the syntactic tree of an algebraic expression denoting G, built with substitution operations, possibly of infinite arity. We do not develop here this algebraic aspect (see [11]), but we define a relational structure, somehow equivalent to these algebraic expressions and suitable for expressing properties of modular decomposition in MS logic. Hence we construct from $mdec(G)$ and the five types of nodes (of modules) a binary relational structure $Gdec(G)$, equivalently a vertex- and edge-labelled directed graph, from which G can be defined. We call it the *graph representation of the modular decomposition of G*.

Definition 7. (*Graph representations of modular decompositions.*)
The structure $Gdec(G)$ consists of the tree $mdec(G) = (T, \leq)$, augmented with edges between the sons of the nodes M of T (which are modules of G), in order to represent the edges of G between the submodules corresponding to these sons. It is a straightforward generalization of the similar notion defined in [8]. Formally, we define $Gdec(G)$ from $mdec(G)$ as follows:

For each node M of $mdec(G)$ which is neither a limit node nor a leaf, whence has at least two sons, we do the following according to its type (cf. Proposition 2):

- if M corresponds to a robust module of type I, we label it by $\oplus$,
- if M corresponds to a robust module of type II, we label it by $\otimes$,
- if M corresponds to a robust module of type III, we label it by $\overrightarrow{\otimes}$, and we define a linear order on the sons of M (which corresponds to the linear order of the strong modules C_i, cf. Proposition 2),
- if M corresponds to a robust module of type IV, we create edges between the sons of M corresponding to the edges of P in an obvious way.

We obtain thus the structure $Gdec(G)$ defined as:

$$(T, \leq, lab_\oplus, lab_\otimes, lab_{\overrightarrow{\otimes}}, edg, order)$$

where $(T, \leq)$ is the tree $mdec(G)$, $lab_\oplus, lab_\otimes, lab_{\overrightarrow{\otimes}}$ are unary predicates defining the labels $\oplus, \otimes, \overrightarrow{\otimes}$ of the nodes of types I,II,III, edg is a binary relation representing the edges created between sons of nodes of type IV, and *order* is the binary relation such that $order(x, y)$ iff $x \unlhd_{x \vee y} y$ and x, y are sons of $x \vee y$, which implies that $x \vee y$ is labelled by $\overrightarrow{\otimes}$. If G is undirected then *order* and $lab_{\overrightarrow{\otimes}}$ are empty and can be omitted. We can consider $Gdec(G)$ as a graph with three types of edges, corresponding to the binary relations $\leq, edg, order$, and labelled by $\leq, edg, order$. The symbols $\oplus, \otimes, \overrightarrow{\otimes}$ are thus vertex labels.

The objectives are to prove that $Gdec(G)$ and G can be defined from each other by transformations of relational structures specified by monadic second-order formulas, and thus to obtain that the monadic second-order properties of the modular decomposition of a graph G are monadic second-order expressible in G and vice-versa.

Monadic second-order logic and monadic second-order transformations of structures (called *MS transductions*) are presented in many works by the first author ([6], [8], [7]). Lacking of space, we only recall that an MS transduction (also called sometimes an *MS interpretation*, but this term conflicts with its use in semantics, cf. [5]) is a transformation of relational structures that is specified by MS formulas forming its *definition scheme*. It transforms a structure S into a structure T (possibly over a different set of relations) such that the domain D_T of T is a subset of $D_S \times \{1, ..., k\}$. (The numbers $1, ..., k$ are just a convenience for the formal definition ; we are actually interested by relational structures up to isomorphism). In many cases, this transformation involves a bijection of D_S onto a subset of D_T, and the definition scheme can be constructed in such a way that this bijection is the mapping: $x \mapsto (x, 1)$. Hence, in this case D_T contains $D_S \times \{1\}$, an isomorphic copy of D_S, and we will say that the MS transduction is *domain extending*, because it defines the domain of T as an extension of that of S. (This does not imply that the relations of T extend those of S). An *FO transduction* is a transduction defined by a first-order definition scheme.

Proposition 6. *A graph G can be defined by an FO transduction from $Gdec(G)$ as a graph, the vertices of which are the leaves of $mdec(G)$.*

Theorem 1. *There is a domain extending MS transduction, let γ, constructing $Gdec(G)$ from $(G, \preccurlyeq)$ where $\preccurlyeq$ is an ω-order. There is an FO transduction δ such that $\delta(Gdec(G)) = G$ for every graph G.*

Proof (Sketch). We describe the main steps of the construction of γ.

Step 1: The notions of a module, of a strong module and of a robust module are MS expressible. The types I, ..., IV of robust modules can also be identified by MS formulas. Moreover, there exist MS formulas $\varphi_1(X, Y)$ (resp. $\varphi_2(X, Y)$) such that for all sets of vertices M, M' of a graph G, $\varphi_1(M, M')$ (resp. $\varphi_2(M, M')$) holds in G iff M is a robust module of type I, (resp. of type II), and M' is one of the corresponding modules C_i. There exists an MS formula $\varphi_3(X, Y, Z)$ such that for all sets of vertices M, M', M'', $\varphi_3(M, M', M'')$ holds iff M is a robust module of type III, M' is a module C_i, M'' is a module C_j and $i < j$. Finally, there exists an MS formula $\varphi_4(X, Y, Z)$ such that for all sets of vertices M, M', M'', $\varphi_4(M, M', M'')$ holds iff M is a robust module of type IV, M' is a module C_i,

M" is a module C_j and $u_i \to u_j$ in the graph P. All these formulas can be constructed by straightforward translations from the definitions.

Step 2: Given a graph G, we construct the leafy tree $rdec(G)$ of its robust modules. The leaves of $rdec(G)$ are the vertices of the graph. We must define the internal nodes of $rdec(G)$ which correspond to the robust modules. The ω-order $\preccurlyeq$ on vertices is here useful. Each robust module M has at least 2 sons (they are also modules). We let $fl(X)$ be the $\preccurlyeq$-smallest vertex in $X \subseteq V_G$. We let N be the son of M containing $fl(M)$, and we take $fl(M - N)$ as the representative of M. Two robust modules are represented by different vertices. (This would not be the case if we decided to represent M by $fl(M)$). We can thus construct $rdec(G)$, by an MS transduction, as a tree with set of nodes $V_G \times \{1\} \cup R_G \times \{2\}$, where R_G is the set of vertices which represent some module.

Step 3: We know from Proposition 5 that $mdec(G)$ is the completion of $rdec(G)$. This completion is a domain extending MS-transduction, using again an auxiliary ω-ordering. The technique is similar to the one used in the previous step. We represent by some leaf, in a well-defined way, each direction to be completed: using $\preccurlyeq$ we define a linear order on directions ; we represent a direction D by $fl(D')$ where D' is the next one in this order (among directions relative to the same node x) ; a maximal direction (in the case where x has finite degree) is represented by x. Hence if N is the set of nodes of $rdec(G)$, the completed tree $mdec(G)$ has set of nodes $N_{mdec(G)} = N \times \{1\} \cup R \times \{2\} \cup S \times \{3\}$, where R (S) is the set of nodes representing non-maximal (maximal) directions to be completed.

Step 4: An MS transduction transforms $(V_G \cup N_{mdec(G)}, edg_G, \leq_{mdec(G)})$ into $Gdec(G)$. Its definition is a straightforward translation from the definition using Step 1.

Since the composition of several domain extending MS transductions is a domain extending MS transduction, we get a domain extending MS transduction γ that maps $(V_G, edg_G, \preccurlyeq)$ into $Gdec(G)$.

The inverse of γ: We define δ. The vertices of G are the leaves of the tree underlying $Gdec(G)$, hence can be identified by FO formulas. Given two vertices x and y of G, whether there is in G an edge $x \to y$ can be determined from the label of $x \vee y$ in $Gdec(G)$ and, when $x \vee y$ satisfies case III, by the ordering the directions of x and y relative to $x \vee y$ (by using the condition "there exist sons u, v of $x \vee y$ such that $order(u, v)$, $x \leq u$, and $y \leq v$"), and when $x \vee y$ satisfies case IV, the existence of an edge in P between the submodules containing x and y (by the condition "there exist sons u, v of $x \vee y$ such that $edg(u, v)$, $x \leq u$ and $y \leq v$").

3   Universal $\vee$−Trees

It is well-known that the linearly ordered set $\mathbf{Q}$ is *universal for finite and countable linear orders*: it embeds each of them and is itself countable. We will construct a *universal ordered tree*, where universality is relative to $\vee$-embeddings.

Definition 8. (*Ordered trees constructed from linear orders.*) We let S and D be two nonempty linearly ordered sets. We let $Aseq(S, D)$ denote the set of alternating sequences of the form: $s_1 d_1 s_2 d_2 ... d_n s_{n+1}$, for $n \geq 0$; they have at least one occurrence of an element in S. We order them by $\leq_{S,D}$ defined by: $w \leq_{S,D} u$ iff $u = u's, w = u's'w'$ for some u' in $(SD)^*$, some w' in $(DS)^*$, some s, s' in S with $s' \leq_S s$. In particular, $w \leq_{S,D} u$ if $u \leq_{pref} w$ (where $\leq_{pref}$ denotes the prefix order on sequences).

Lemma 1.

1. *The ordered set* $(Aseq(S, D), \leq_{S,D})$ *is a* $\vee$-*tree, denoted by* $T(S, D)$.
2. *The directions in* $T(S, D)$ *relative to a node* us *(for* $u \in (SD)^*$, $s \in S$*) are the nonempty sets of the following forms:*

$$D(0, us) = \{us'w \mid s' \in S, s' <_S s, w \in (DS)^*\} \text{ and:}$$
$$D(d, us) = \{usdw \mid w \in S(DS)^*\} \text{ for } d \in D.$$

$D(0, us)$ is called the *main direction* relative to us. We let $\mathbf{Q}_-$ be the set of negative rational numbers and $\mathbf{Q}_+$ be the set of positive ones. Of course they are both order-isomorphic to $\mathbf{Q}$, but it is more convenient to distinguish them. We let $D = \mathbf{Q}_- + \mathbf{Q}_+$ (i.e., we concatenate as ordered sets $\mathbf{Q}_-$ and $\mathbf{Q}_+$), we let $S = \mathbf{Q}$ and we make the $\vee$-tree $T(S, D)$ into an ordered tree by ordering directions as follows:

$$D(d, us) \trianglelefteq_{us} D(0, us) \trianglelefteq_{us} D(d', us) \text{ for } d \in \mathbf{Q}_-, d' \in \mathbf{Q}_+, \text{ and}$$
$$D(d, us) \trianglelefteq_{us} D(d', us) \text{ for } d, d' \in D, d < d'.$$

We denote this tree by $UT(\mathbf{Q}, \mathbf{Q}_-, \mathbf{Q}_+)$ (read *Universal Tree*).

Proposition 7. *Every ordered tree* $\vee$-*embeds into* $UT(\mathbf{Q}, \mathbf{Q}_-, \mathbf{Q}_+)$.

Proof (Sketch). We consider an ordered tree $(T, \leq, \trianglelefteq)$, ω-ordered by $\preccurlyeq$ with corresponding enumeration denoted by $t_0, ..., t_n, ...$ We define a structuring of T that depends on this enumeration and associates a finite depth with each node. This structuring will be the basis of a representation of ordered trees by "usual" binary trees that will be considered in Section 4.

Step 1: We associate with every $x \in T$ a unique subset $U(x)$ characterized as follows:

1. $U(x)$ is a maximal chain containing x,
2. it is lexicographically minimal with this property, which means that for every maximal chain W containing x and different from $U(x)$, the $\preccurlyeq$-smallest element of $(U(x) - W) \cup (W - U(x))$ is in $U(x)$.

We note for later use that this set is MS definable. Some facts: if $y < x$ and $y \in U(x)$, then $U(y) = U(x)$. If $U(x) \neq U(y)$, then $U(x) \cap U(y) = T^z$ for some z, and if x and y are incomparable, then $z = x \vee y$.

Step 2 : We define a sequence of chains $W_0, ..., W_n, ...$ by:
$W_0 = U(t_0)$, w_1 is the $\preccurlyeq$-smallest node not in W_0,
$W_1 = U(w_1) - W_0$, ...
w_n is the $\preccurlyeq$-smallest node not in $W_0 \cup W_1 \cup ... \cup W_{n-1}$,

$W_n = U(w_n) - (W_0 \cup W_1 \cup ... \cup W_{n-1}), ...$

Every node has a finite *depth*, $d(x) = 0$ if $x \in W_0$, $d(x) = 1 + d(p(x))$ if $x \in W_{n+1}$ and $p(x)$ is the $\leq$ −smallest node strictly above all nodes in W_{n+1}. (Note that $W_{n+1} = U(w_{n+1}) - T^{p(x)}$.)

Hence x of depth n has a sequence of ancestors $p(x), p^2(x), ..., p^n(x)$ of depths $n-1, n-2, ..., 0$.

Step 3: For each m we fix an (order preserving) embedding $h_m : W_m \to \mathbf{Q}$, and we let $h(x) = h_m(x)$ if $x \in W_m$.

Step 4: We associate with x as in Step 2 the sequence of rational numbers $h(p^n(x))...h(p(x))h(x)$. We have in this way the elements $s_1, s_2, ..., s_{n+1}$ of a sequence $s_1 d_1 s_2 d_2 ... d_n s_{n+1}$ in $ASeq(\mathbf{Q}, \mathbf{Q}_- \cup \mathbf{Q}_+)$, but the d_i's which encode directions are still missing.

Step 5: The main direction relative to x (cf. the lemma) is the one that meets the maximal chain $U(x)$. The set of directions $\trianglelefteq$-smaller than the main direction is linearly ordered by $\trianglelefteq$. We embed it into $\mathbf{Q}_-$ and we do the same into $\mathbf{Q}_+$ for the directions which are $\trianglelefteq$-larger than the main direction. Hence x is finally represented by: $h(p^n(x))f_{n-1}...h(p^2(x))f_1 h(p(x))f_0 h(x)$, where f_i represents the direction of $p^i(x)$ relative to $p^{i+1}(x)$.

We have defined in this way a $\vee$-embedding of $(T, \leq, \trianglelefteq)$ into the tree $UT(\mathbf{Q}, \mathbf{Q}_-, \mathbf{Q}_+)$.

Remarks. 1. Using an obvious extension of the notation, $UT(\mathbf{1}, \varnothing, \mathbf{N})$ is a tree where each infinite tree in the sense of [4] $\vee$-embeds.

2. Fraïssé defines in [14] (Theorem 6.2 of Chapter 10) a (countable) tree $\mathbf{W}$, which is actually the unordered tree underlying $UT(\mathbf{Q}, \varnothing, \mathbf{1})$. All finite or countable trees embed in $\mathbf{W}$. His theorem concerns trees and embeddings, and not ordered trees and $\vee$-embeddings as does our Proposition 7. The tree $\mathbf{W}$ is a binary $\vee$-tree and $\vee$-embeds all binary $\vee$-trees, but only binary $\vee$-trees since it is binary and least upper bounds of pairs of nodes are preserved.

4   Representing Modular Decompositions by Low Degree Relational Structures

Our objective is to represent ordered trees and modular decompositions by relational structures of *lowest possible degree* (the notion of degree is as for graphs) by generalizing the observation that the *dense* structure $(\mathbf{Q}, \leq)$ is isomorphic to the set of nodes of the complete infinite binary tree, a graph of degree 3, ordered appropriately. Let us give some motivations for this investigation. For finite objects like graphs and partial orders, space efficient representations are of interest. For an example, every finite partial order P can be represented by its Hasse diagram, which may contain $O(m^{1/2})$ edges whereas the directed graph of P has m edges. The same ratio holds for certain dense cographs represented by their modular decompositions. In both cases the original partial order (or graph) can be determined from its Hasse diagram (or its modular decomposition) by computations of transitive closures, hence by MS transductions.

This motivation does not apply to infinite graphs, but bounds on degrees of infinite structures are nevertheless interesting because they yield structural or logical properties. For examples, every *equational graph* of bounded degree is *prefix recognizable*, see Caucal [3], or Barthelmann [1] for a similar result. MS logic with edge set quantifications is as powerful as MS logic without them for expressing properties of sparse graphs, see [9].

We achieve this goal and we define mutual transformations of relational structures that are MS transductions.

Definition 9. (*Standard binary trees.*) By a *standard binary tree*, we mean a simple directed edge-labelled graph $T = (N_T, lson_T, rson_T)$ where N_T is the finite or countable set of nodes, $lson_T$ and $rson_T$ are two binary functional relations defining for each node its *left son* and its *right son*. A node may have no son, two sons, or just a right son or a left son. The *root* is the unique node of indegree 0 and every node is reachable from it by a unique directed path. For a standard binary tree T, and $x, y \in N_T$, we write $x \to_l y$ if y is the left son of x, $x \to_r y$ if y is the right son of x, and $x \to y$ if y is the left or the right son of x.

A linear order, the *in-order*, on N_T is defined by: $x \leq_{in,T} y$ iff $x = y$ or $x \to_r z \to^* y$ or $y \to_l z \to^* x$ for some z, or $t \to_l z \to^* x$ and $t \to_r z' \to^* y$ for some t, z, z'.

We let $\Omega(T)$ denote the linearly ordered set $(N_T, \leq_{in,T})$. The mapping Ω is an MS-transduction because the transitive closure of a given binary relation is expressible by an MS formula. Our objective is to construct T from $\Omega(T)$ by an MS transduction.

Proposition 8. *1. There exist first-order formulas $\lambda(x, y)$ and $\rho(x, y)$ that define in every structure $(N, \sqsubseteq, \preccurlyeq)$ such that $\sqsubseteq$ is a linear order and $\preccurlyeq$ is an ω-order, binary relations $lson$ and $rson$ such that $(N, lson, rson)$ is a standard binary tree T such that $\Omega(T) = (N, \sqsubseteq)$. This tree T is defined from $(N, \sqsubseteq, \preccurlyeq)$ by an FO transduction.*
2. There exists a domain extending FO transduction that transforms a standard binary tree T into a standard binary tree U such that $(Leaves(U), \leq_{in,U})$ is isomorphic to $(N_T, \leq_{in,T})$.

By combining the two constructions, one can represent, using an FO-transduction, the ordered set N as the in-ordered set of *leaves* of a standard binary tree whereas the first one represents it as the in-ordered set of *nodes*.

Proof (Sketch).

1. See [10].
2. This is a classical transformation: for an example, using the notation of trees by terms, $a(b, c)$ is replaced by $*(b, *(a, c))$.

Thus we can represent the universal tree $UT(\mathbf{Q}, \mathbf{Q}_-, \mathbf{Q}_+)$ and whence to all trees via Proposition 7, by standard binary trees with appropriate node labels.

Proposition 9. *There exists a domain extending MS transduction α that associates with every ordered tree T that is also ω-ordered, a node-labelled standard binary ω-ordered tree $W = (N_W, node_T, lson_W, rson_W)$ and an MS-transduction β that defines T from W.*

Intuitively, α encodes T into a binary tree and β is its inverse, the decoding transduction.

Proof (Sketch). We first describe the idea for a tree that is embedded into $UT(\mathbf{Q}, \mathbf{Q}_-, \mathbf{Q}_+)$, by Proposition 7. A node x is described by a sequence of rational numbers $s_1 d_1 s_2 d_2 ... d_n s_{n+1}$ such that s_1 is a node on the chain of nodes of depth 0, d_1 is a direction relative to s_1, saying "in which direction to go next below s_1". This direction indicates a chain of nodes of depth 1, in which s_2 is selected. Then d_2 indicates where to go next, etc... until one reaches s_{n+1}.

By Proposition 8, every rational number can be represented by a path in a standard binary tree, i.e. a word in $\{left, right\}^*$. We concatenate the words representing $s_1, d_1, s_2, d_2, ..., d_n, s_{n+1}$ in this order, and we obtain a path in a standard binary tree. The edges of this path are colored, say in *blue* for those encoding the positions $s_1, s_2, ..., s_{n+1}$ and in *red* for those encoding the directions $d_1, d_2, ..., d_n$. So we can distinguish in a path the portions encoding positions and those encoding directions. It follows that all trees, and in particular $UT(\mathbf{Q}, \mathbf{Q}_-, \mathbf{Q}_+)$ can be represented as subtrees of the complete standard binary tree with colored edges. Actually, coloring an edge is equivalent to coloring its target. So node labels are sufficient and we can use a single unary relation $node_T$.

Proposition 8 says also that for a linear order given with an auxiliary ω-order, the encoding of its elements by paths of the binary tree is definable by an MS transduction. By combining the transductions associated at each depth with the chains W_i and with the sets of directions (cf. the proof of Proposition 7), one obtains the desired one.

We now apply this result to the representation of modular decompositions.

Definition 10. (*Sparse representations of modular decompositions.*) Assuming that, by Proposition 9, $(T, \leq)$ is represented by a node-labelled standard binary tree $(W, node_T, lson_W, rson_W)$, then we define a *sparse representation* of the modular decomposition of G as a structure:

$$Sdec(G) = (W, node_T, lson_W, rson_W, lab_\oplus, lab_\otimes, lab_{\overrightarrow{\otimes}}, edg).$$

The relation *order* is no longer necessary because the linear order on directions in T is handled by the inorder on W derived from the left and right types of sons.

Theorem 2. *There exists a domain extending MS transduction that associates with an ω-ordered graph G a sparse representation $Sdec(G)$ of its modular decomposition. The structure $Sdec(G)$ is a vertex- and edge-labelled graph of degree $m+3$ where m is the maximum degree of a vertex in a prime factor of G (cf case IV of Proposition 2). There exists an MS-transduction that defines G from $Sdec(G)$.*

Proof. It suffices to combine the MS transductions of Proposition 1 and Proposition 9. The tree $(T, \leq)$ underlying $Gdec(G)$ is not ordered: only the sons of the nodes of type III are linearly ordered, whereas Proposition 9 uses ordered trees. But since an ω-order is available in G whence in T, we can use it to make T into an ordered tree, just by defining a linear order on the directions relative to the nodes of types I,II and IV. The bound on the degree of $Sdec(G)$ follows from the definitions.

5   Concluding Remarks and Questions

We have proved that the graph $Sdec(G)$ representing the modular decomposition of a countable graph G can be defined from G and any ω-order of its vertices by an MS transduction, and that, conversely, G is definable from $Sdec(G)$ also by an MS transduction.

Finite presentations of countable graphs of several types are studied by Blumensath and Graedel in [2]. One can thus ask whether a finite presentation of G yields one of same type of $Sdec(G)$. A graph G is *VR-equational* (i.e. is the canonical solution of a finite system of equations over so-called *VR operations*) iff it is the image of the standard binary tree $\mathbf{B} = (\{0,1\}^*, lson_\mathbf{B}, rson_\mathbf{B})$ under an MS transduction (Proposition 2.2 of [2]). If G is VR-equational, and if an ω-order of V_G is MS definable, then by Proposition 2, $Sdec(G)$ is also the image of $\mathbf{B}$ under an MS transduction, hence is VR-equational. (Since no ω-order on $\mathbf{B}$ is MS definable, the second assumption cannot be deleted). Conversely, if $Sdec(G)$ is VR-equational, so is G.

Question 1. Is the former assertion true without the hypothesis that an ω-order of V_G is MS definable?

It is possible that something weaker than an ω-order (e.g., a partial order of some kind) is sufficient for Theorems 1 and 2 to hold.

The article [2] studies in detail *automatic structures* (also considered in [16] ; they contain the VR-equational graphs, characterized also as *prefix-recognizable graphs*). These structures have domains defined as regular languages and relations defined by multihead synchronized automata. The tree $\mathbf{B}$ ordered by inorder is an automatic structure. So is the universal tree $UT(\mathbf{Q}, \mathbf{Q}_-, \mathbf{Q}_+)$ with domain defined as $(L_\mathbf{Q}.L_{\mathbf{Q}^*})^* L_\mathbf{Q}$ where $L_\mathbf{Q} = (0 \cup 11)^* 10$ represents $\mathbf{B}$ and $L_{\mathbf{Q}^*} = (0 \cup 1)(0 \cup 11)^* 10$ represents the linear order $\mathbf{Q}_- + \mathbf{Q}_+$.

If in the structure $Sdec(G)$ we replace $lson_W$ and $rson_W$ by $ldes_W$ and $rdes_W$ such that $ldes_W(x, y)$ holds iff $x \leq_T u$ where $lson_W(u, y)$ holds, and similarly for $rdes_W$, then we obtain a binary structure $Fdec(G)$ (that is no longer sparse) from which G can be constructed by an FO transduction. It follows that G is automatic if $Fdec(G)$ is, because the image of an automatic structure under an FO transduction is automatic ([2] Proposition 4.3).

Question 2. For which graphs G is it true that the binary structure $Fdec(G)$ is automatic?

References

1. K. Barthelmann, When can an equational simple graph be generated by hyperedge replacement? *Mathematical Foundations of Computer Science*, Lec. Notes Comput. Sci. 960 (1998) 543-552.
2. A. Blumensath, E. Grädel, Finite presentations of infinite structures: Automata and interpretations, *Theory of Computing Systems* 37 (2004) 641-674.
3. D. Caucal, On the regular structure of prefix rewriting. *Theoretical Computer Science* 106 (1992) 61 - 86.
4. B. Courcelle, Fundamental properties of infinite trees, *Theoretical Computer Science* 25 (1983) 95-169.
5. B. Courcelle, Recursive applicative program schemes, in *Handbook of Theoretical Computer Science vol. B*, J. Van Leeuwen ed., Elsevier, 1990, pp.459-492.
6. B. Courcelle, Monadic second-order graph transductions: A survey. *Theoretical Computer Science*, 126 (1994)53-75.
7. B. Courcelle, The expression of graph properties and graph transformations in monadic second-order logic. In G. Rozenberg, editor, *Handbook of graph grammars and computing by graph transformations, Vol. 1: Foundations*, World Scientific, 1997, pp. 313-400.
8. B. Courcelle, The monadic second-order logic of graphs X: Linear orderings, *Theoretical Computer Science* 160 (1996) 87-143.
9. B. Courcelle, The monadic second-order logic of graphs XIV: Uniformly sparse graphs and edge set quantifications. *Theoretical Computer Science* 299 (2003) 1-36.
10. B. Courcelle, The monadic second-order logic of graphs XV: On a Conjecture by D. Seese, to appear in *Journal of Applied Logic*, see:
   http://www.labri.fr/~courcell/ActSci.html
11. B. Courcelle, C. Delhommé, The modular decomposition of countable graphs, 2005, see: http://www.labri.fr/ ~courcell/ActSci.html.
12. A. Ehrenfeucht, T. Harju, G. Rozenberg, Decomposition of infinite labeled 2-structures, Lec. Notes Comput. Sci. 812 (1994) 145-158.
13. A. Ehrenfeucht, T. Harju, G. Rozenberg, The theory of 2-structures. A framework for decomposition and transformation of graphs, World Scientific Publishing Co., River Edge, New-Jersey,1999.
14. R. Fraïssé, Theory of relations, *Studies in logic* vol. 118, North-Holland, 1986 (Second edition, Elsevier, 2000).
15. D. Kelly, Comparability graphs, in *Graphs and order*, I. Rival ed., D. Reidel Pub. Co., 1985, pp. 3-40.
16. B. Khoussainov, A. Nerode, Automatic presentations of structures, in *Logic and Computational Complexity*, Lec. Notes Comput. Sci. 960 (1995) 367-392.
17. R. Möhring, R. Radermacher, Substitution decomposition of discrete structures and connections with combinatorial optimization, *Annals Discrete Maths* 19 (1984) 257-356.

On the Complexity of Hybrid Logics with Binders

Balder ten Cate[1] and Massimo Franceschet[1,2]

[1] Informatics Institute, University of Amsterdam
Kruislaan 403 – 1098 SJ Amsterdam, The Netherlands
[2] Department of Sciences, University "G. D'Annunzio"
Viale Pindaro, 42 – 65127 Pescara, Italy

Abstract. *Hybrid logic* refers to a group of logics lying between modal and first-order logic in which one can refer to individual states of the Kripke structure. In particular, the hybrid logic HL(@, ↓) is an appealing extension of modal logic that allows one to refer to a state by means of the given names and to dynamically create new names for a state. Unfortunately, as for the richer first-order logic, satisfiability for the hybrid logic HL(@, ↓) is undecidable and model checking for HL(@, ↓) is PSPACE-complete. We carefully analyze these results and we isolate large fragments of HL(@, ↓) for which satisfiability is decidable and model checking is below PSPACE.

1 Introduction

There is a general interest in well-behaved logical languages in-between the basic modal language and full first-order logic. Ideally, one would like such languages to combine the good properties of both: to be reasonably expressive, to be decidable, and to have other good properties, such as the interpolation property. Famous examples of fragments that have been studied are the *guarded fragment* [1, 2] and the *two variable fragment* [3, 4]. Both are decidable, reasonably expressive languages, but they lack interpolation.

The hybrid logic HL(@, ↓) is another example of a language in between the basic modal language and full first-order logic. It extends the basic modal language with three constructs: nominals, which act as names of states of the model, the satisfaction operator @, which allows one to express that a formula holds at the state named by a nominal, and the binder ↓, which allows one to give a name to the current state. Together, these three elements greatly increase the expressivity of the language. Moreover, like the basic modal language and full first-order logic, HL(@, ↓) has the interpolation property. Unfortunately, it is undecidable.

The language HL(@, ↓) is a natural fragment of first-order logic: it is the generated submodel invariant fragment [5], it is the least expressive extension of the basic hybrid language HL(@) with interpolation [6], and, finally, it has been characterized as the intersection of first-order logic with second-order propositional modal logic [7]. HL(@, ↓) has been used in the context of semistructured

L. Ong (Ed.): CSL 2005, LNCS 3634, pp. 339–354, 2005.

data. In particular, [8] gives an application of model checking in hybrid logic to the problems of query and constraint evaluation for semistructured data.

In this paper, we give an in-depth analysis of the undecidability of HL(@, ↓). We show how decidability can be regained by making a syntactic restriction on the formulas, or by restricting the class of models in a natural way. Moreover, we show how these and similar syntactic and semantic restrictions affect the complexity of the model checking problem for hybrid languages.

In Section 2 we introduce hybrid logic, and in Section 3 we revisit the undecidability result for HL(@, ↓). In Section 4 and 5, we show how decidability can be regained by restricting the language and the class of models, respectively. In Section 6 we investigate how these and similar restrictions affect the complexity of the model checking problem for hybrid logic. We conclude in Section 7.

2   Hybrid Logic

In its basic version, hybrid logic extends modal logic with devices for naming (individual) states and for accessing states by their names. The key idea is the use of *nominals*. Syntactically, nominals behave like ordinary propositions, but they have an important semantic property. A nominal is true at *exactly one state* of the model. In such a way, it gives a name to that point. Besides nominals, the hybrid language HL(@, ↓) also contains @-operators, that allow one to state that a formula is true at a state named by a nominal, and the ↓-binder, that allows one to introduce variables to name points. Formally, HL(@, ↓) is defined as follows.

Let PROP = $\{p, q, \ldots\}$ be a (countably) infinite set of proposition symbols, NOM = $\{i, j, \ldots\}$ be a (countably) infinite set of nominals, and SVAR = $\{x, y, \ldots\}$ be a (countably) infinite set of state variables. We assume that these sets are disjoint. The formulas of the hybrid language HL(@, ↓) are given by the following recursive definition.

$$\alpha := \top \mid p \mid t \mid \neg\alpha \mid \alpha \wedge \beta \mid \Diamond\alpha \mid @_t\alpha \mid \downarrow x.\alpha$$

where $p \in$ PROP, $t \in$ NOM $\cup$ SVAR and $x \in$ SVAR. We will use the familiar shorthand notations, such as $\Box\alpha$ for $\neg\Diamond\neg\alpha$. The notions of *free* and *bound* variables are defined similarly as in first-order logic. A hybrid *sentence* is a hybrid formula with no free variables. The *width* of a formula α is the maximum number of free variables of any subformula of α.

The binder ↓ binds a variable to the current state of evaluation. For instance, the formula $\downarrow x.\Diamond x$ says that the current state is reflexive. The @ operator combines naturally with the ↓ binder: while ↓ stores the current state of evaluation, @ enables us to retrieve the information stored by shifting the point of evaluation. As an example, the formula $\downarrow x.\Diamond\downarrow y.@_x\Box y$ states that the current point has exactly one successor.

Hybrid formulas are interpreted over *hybrid Kripke structures* (or *hybrid models*) of the form $M = (W, R, V)$ where W is a set of states, R is a binary relation over W called the accessibility relation, and $V :$ PROP$\cup$NOM $\to \wp(W)$

is a valuation function that assigns to each proposition letter or nominal a set of states, such that $V(i)$ is a singleton set for each nominals i. The pair $F = (W, R)$ is called the *frame* of M and M is said to be a model based on the frame F.

An assignment for M is a function $g : \text{SVAR} \to W$. Given such an assignment g, a variable $x \in \text{SVAR}$ and a state $w \in W$, we will use g_w^x to refer to the assignment that is identical to g except that maps x to w. Formally, $g_w^x(y) = x$ for $y = x$ and $g_w^x(y) = g(y)$ for $y \neq x$.

Let $M = (W, R, V)$ be a hybrid model, g an assignment for M, and let $w \in W$. For any nominal i, let $[i]^{M,g} = V(i)$, and for any state variable x, let $[x]^{M,g} = \{g(x)\}$. The semantics of $\text{HL}(@, \downarrow)$ is as follows:

$$M, g, w \Vdash \top$$
$$M, g, w \Vdash p \qquad \text{iff } w \in V(p)$$
$$M, g, w \Vdash t \qquad \text{iff } w \in [t]^{M,g} \text{ for } t \in \text{NOM} \cup \text{SVAR}$$
$$M, g, w \Vdash \neg\alpha \qquad \text{iff } M, g, w \not\Vdash \alpha$$
$$M, g, w \Vdash \alpha \wedge \beta \text{ iff } M, g, w \Vdash \alpha \text{ and } M, w \Vdash \beta$$
$$M, g, w \Vdash \Diamond\alpha \qquad \text{iff there is a } w' \in W \text{ such that } wRw' \text{ and } M, g, w' \Vdash \alpha$$
$$M, g, w \Vdash @_t\alpha \qquad \text{iff } M, g, w' \Vdash \alpha \text{ where } \{w'\} = [t]^{M,g}$$
$$M, g, w \Vdash {\downarrow}x.\alpha \text{ iff } M, g_w^x, w \Vdash \alpha$$

Define the *first-order correspondence language* to be the first-order language with equality that has one binary relation symbol R, a unary relation symbol p for each $p \in \text{PROP}$ and a constant i for each nominal $i \in \text{NOM}$. Every hybrid Kripke structure (W, R, V) can be viewed as a relational structure for the first-order correspondence language: the binary relation symbol R is interpreted by the accessibility relation R, the unary relation symbols p are interpreted by $V(p)$, and each constant i denotes the unique state w such that $V(i) = \{w\}$. Then, the following *Standard Translation*, defined by mutual recursion[1] between two functions ST_x and ST_y, embeds $\text{HL}(@, \downarrow)$ into the first-order correspondence language (where $p \in \text{PROP}$ and $t \in \text{NOM} \cup \text{SVAR}$)[2]:

$\text{ST}_x(\top)$	$= \top$		$\text{ST}_y(\top)$	$= \top$	
$\text{ST}_x(p)$	$= p(x)$		$\text{ST}_y(p)$	$= p(y)$	
$\text{ST}_x(t)$	$= x = t$		$\text{ST}_y(t)$	$= y = t$	
$\text{ST}_x(\neg\alpha)$	$= \neg\text{ST}_x(\alpha)$		$\text{ST}_y(\neg\alpha)$	$= \neg\text{ST}_y(\alpha)$	
$\text{ST}_x(\alpha \wedge \beta)$	$= \text{ST}_x(\alpha) \wedge \text{ST}_x(\beta)$		$\text{ST}_y(\alpha \wedge \beta)$	$= \text{ST}_y(\alpha) \wedge \text{ST}_y(\beta)$	
$\text{ST}_x(\Diamond\alpha)$	$= \exists y.(xRy \wedge \text{ST}_y(\alpha))$		$\text{ST}_y(\Diamond\alpha)$	$= \exists x.(yRx \wedge \text{ST}_x(\alpha))$	
$\text{ST}_x(@_t\alpha)$	$= \exists y.(y = t \wedge \text{ST}_y(\alpha))$		$\text{ST}_y(@_t\alpha)$	$= \exists x.(x = t \wedge \text{ST}_x(\alpha))$	
$\text{ST}_x({\downarrow}z.\alpha)$	$= \exists z.(z = x \wedge \text{ST}_x(\alpha))$		$\text{ST}_y({\downarrow}z.\alpha)$	$= \exists z.(z = y \wedge \text{ST}_y(\alpha))$	

---

[1] Mutual recursion is used in order to limit the number of variables occurring in the translation.

[2] As was pointed out by Guillaume Malod (personal communication), the clause for the $\downarrow$-binder in the Standard Translation for $\text{HL}(@, \downarrow)$ given in [5], i.e., $\text{ST}_x({\downarrow}z.\alpha) = \text{ST}_x(\alpha)[z/x]$ and $\text{ST}_y({\downarrow}z.\alpha) = \text{ST}_y(\alpha)[z/y]$, is incorrect. Indeed, consider the formula ${\downarrow}z.\Diamond\Diamond z$. The Standard Translation of this formula according to the definitions in [5] is $\exists y.(xRy \wedge \exists x.(yRx \wedge x = z))[z/x] = \exists y.(xRy \wedge \exists x.(yRx \wedge x = x))$, which clearly fails to capture the semantics of the hybrid formula.

Here, it is assumed that the variables x, y do not occur in α. For each HL($@, \downarrow$)-formula α with free variables $y_1, \ldots, y_n$, $ST_x(\alpha)$ is a first-order formula with free variables in $\{x, y_1, \ldots, y_n\}$. Moreover, it is easy to show that for any Kripke structure M, assignment g and world w, $M, g, w \Vdash \alpha$ if, and only if, $M, g^x_w \models ST_x(\alpha)$. It follows that HL($@, \downarrow$) is a fragment of the first-order correspondence language. In fact, this fragment admits several natural characterizations, as mentioned in the introduction.

In Section 4, we will consider a further extension of HL($@, \downarrow$), containing the global modality E and the converse operator $\Diamond^-$ (whose duals will be denoted by A and $\Box^-$, respectively). These have the following semantics:

$$M, g, w \Vdash E\alpha \quad \text{iff there is a } w' \in W \text{ such that } M, g, w' \Vdash \alpha$$
$$M, g, w \Vdash \Diamond^-\alpha \text{ iff there is a } w' \in W \text{ such that } w'Rw \text{ and } M, g, w' \Vdash \alpha$$

or, in terms of the Standard Translation:

$$
\begin{array}{l|l}
ST_x(E\alpha) = \exists y.(y = y \wedge ST_y(\alpha)) & ST_y(E\alpha) = \exists x.(x = x \wedge ST_x(\alpha)) \\
ST_x(\Diamond^-\alpha) = \exists y.(yRx \wedge ST_y(\alpha)) & ST_y(\Diamond^-\alpha) = \exists x.(xRy \wedge ST_x(\alpha))
\end{array}
$$

For $\theta_1, \ldots, \theta_n \in \{\downarrow, @, E, \Diamond^-\}$, we will use HL($\theta_1, \ldots, \theta_n$) to refer to the extension of the modal language with nominals and the operators $\theta_1, \ldots, \theta_n$ (if $\downarrow$ is among $\theta_1, \ldots, \theta_n$, then the language is understood to contain state variables as well). The language HL($@, \downarrow, E, \Diamond^-$), which we will also refer to as *the full hybrid language* (FHL), provides a natural upper bound on expressive power of hybrid languages: it is known to be expressively complete for first-order logic. In other words, every formula of the first-order correspondence language is equivalent to the standard translation of a FHL-formula.

So far, we have only introduced uni-modal HL($@, \downarrow$). This was only for convenience of exposition. It is straightforward to extend the above definitions to the multi-modal case. In fact, in the remainder of this paper, we will frequently make use of multi-modal formulas.

3   The Undecidability of HL($@, \downarrow$) Revisited

In this section, we revisit the negative result that is central to this paper: the undecidability of HL($@, \downarrow$) [9]. We present a new undecidability proof based on an encoding of the $\mathbb{N} \times \mathbb{N}$ tiling problem. It will help us identify the real source of the undecidability.

Let us first recall the $\mathbb{N} \times \mathbb{N}$ tiling problem. A tile type is a square, fixed in orientation, each side of which has a color. Formally, it can be identified with a 4-tuple of elements of some finite set of colors. To tile a space, we have to ensure that adjacent tiles have the same color on the matching sides. The $\mathbb{N} \times \mathbb{N}$ tiling problem is then: *given a finite set of tile types T, can the infinite grid $\mathbb{N} \times \mathbb{N}$ be tiled using only tiles of the types in T?* This problem is well known to be undecidable (see, e.g., [10]).

We will reduce this problem to the satisfiability problem for HL($@, \downarrow$) with three modalities: $\Diamond_1$ (to move one step up in the grid), $\Diamond_2$ (to move one step to the right in the grid), and $\Diamond$ (to reach all the points of the grid), interpreted by the accessibility relations R_1, R_2 and R, respectively. Let T be a finite set of tiles, and for each tile $t \in T$ let left(t), right(t), top(t), and bottom(t) denote the four colors of t. We will now give a hybrid formula π_T that describes a tiling of $\mathbb{N} \times \mathbb{N}$ using the tile types in T. Note that the formula π_T given below is not the simplest possible encoding of the tiling problem. The reason is that the specific syntactic shape of π_T will be further exploited later on in the paper.

Spypoint $\alpha = s \wedge \Diamond s \wedge \Box \Diamond s \wedge \Box\Box_1 \downarrow x.(\Diamond(s \wedge \Diamond x)) \wedge \Box\Box_2 \downarrow x.(\Diamond(s \wedge \Diamond x))$, where s is a nominal. This formula says that the current world is named s, that the set of its R-successors is closed under R_1 and R_2, and that each R-successor of s has s as an R-successor.

Functionality $\beta = \bigwedge_{i=1,2} \left(\Box\Diamond_i \top \wedge \Box \downarrow x.\Box(s \rightarrow \Box(\Box_i x \vee \Box_i \neg x))) \right)$. This formula, which is equivalent to $\bigwedge_{i=1,2} \left(\Box\Diamond_i \top \wedge \Box \downarrow x.\Box(s \rightarrow \Box(\Diamond_i x \rightarrow \Box_i x))) \right)$, says that, within the submodel consisting of all R-successors of s, the relations R_1 and R_2 are in fact total functions.

Grid $\gamma = \Box\downarrow x.\Box(s \rightarrow \Box(\Box_1\Box_2 \neg x \vee \Box_2\Box_1 x))$. This formula, which, in the presence of functionality is equivalent to $\Box\downarrow x.\Box(s \rightarrow \Box(\Diamond_1\Diamond_2 x \rightarrow \Diamond_2\Diamond_1 x))$, expresses that R_1 and R_2 commute.

Tiling $\delta = \Box(\delta_1 \wedge \delta_2 \wedge \delta_3)$, where

$$\delta_1 = \bigvee_{t \in T}(p_t \wedge \bigwedge_{t' \in T; t \neq t'} \neg p_{t'})$$
$$\delta_2 = \bigwedge_{t \in T}(p_t \rightarrow \Box_2 \bigvee_{t' \in T; \text{left}(t') = \text{right}(t)} p_{t'})$$
$$\delta_3 = \bigwedge_{t \in T}(p_t \rightarrow \Box_1 \bigvee_{t' \in T; \text{bottom}(t') = \text{top}(t)} p_{t'})$$

Formula δ_1 states that exactly one tile is placed at each node of the grid, δ_2 says that horizontally adjacent tiles must match, and δ_3 says that vertically adjacent tiles must match. Hence, δ states that the grid is well-tiled.

It is easy to prove that T tiles $\mathbb{N} \times \mathbb{N}$ iff the hybrid formula $\pi_T = \alpha \wedge \beta \wedge \gamma \wedge \delta$ is satisfiable.

Notice that the formula π_T does not contain any @-operators, it does not nest the $\downarrow$ binder, and it uses only one state variable. Hence, the source of undecidability for hybrid logic is neither the @-operator, nor the nesting degree of $\downarrow$, nor the number of state variables used the formulas. Instead, as we will show in the next section, the source of undecidability is the $\Box\downarrow\Box$-pattern of β and γ (i.e., a $\Box$-operator scoping over a $\downarrow$ that in turn has scope over a $\Box$-operator). For formulas not containing this pattern, the satisfiability problem is decidable.

We conclude this section by briefly surveying undecidability proofs for hybrid logic with $\downarrow$ binder. The first undecidability proofs appear in [9, 11]. Both the proofs reduce an undecidable tiling problem into the satisfiability for hybrid logic with $\downarrow$ binder. The reduction of Goranko [11] uses a global modality, whereas Blackburn and Seligman [9] eliminate the use of a global modality by means of a spy point construction (cf. the formula α above). The encoding of [9] uses nested occurrences of the $\downarrow$ binder. Areces, Blackburn, and Marx [12] give another undecidability proof by a reduction of the undecidable global satisfaction problem for K_{23} (the class of frames in which every state has at most 2 successors and at most 3 two-step successors). This proof has the advantage that it uses no nominals and no proposition letters. However, it does use nested occurrences of $\downarrow$. Finally, Marx [13] gives another proof of undecidability by a reduction from a tiling problem. The formulas used in this proof do not nest $\downarrow$ and contain only one state variable. Moreover, only one modality is used. However, the encoding is more involved than ours. Each of these proofs use formulas containing the $\Box\downarrow\Box$-pattern. The proof given above is reasonably simple, and it will help show the precise role of the $\Box\downarrow\Box$-pattern.

4 Syntactic Restrictions

In this section, we will show that the undecidability of $\mathrm{HL}(@, \downarrow)$ is caused by formulas containing the $\Box\downarrow\Box$-pattern. We show that without such formulas, the satisfiability problem is still decidable, even when the global modality and converse modalities are added to the language.

Consider the full hybrid language FHL. In what follows, it will be convenient to consider the universal operators $\Box$ and $\Box^-$, and the disjunction $\vee$, to be primitive operators (rather than shorthand notations). Moreover, we will restrict attention to sentences, i.e., formulas with no free state variables. This is not an essential limitation, since one can always replace free variables by nominals.

We say that a formula of FHL is in *negation normal form* (NNF) if the negation symbol appears only in front of atomic subformulas. Each hybrid formula is equivalent to a hybrid formula in NNF. For instance, $\neg\downarrow x.\Diamond(x\wedge\neg p)$ is equivalent to $\downarrow x.\Box(\neg x\vee p)$.

We call *universal operators* the modalities $\Box$, $\Box^-$ and A, and *existential operators* the modalities $\Diamond$, $\Diamond^-$ and E. We define a $\Box\downarrow$-formula (respectively, $\Diamond\downarrow$-formula) as a hybrid formula in NNF in which some occurrence of $\downarrow$ is in the scope of a universal (respectively, existential) operator. Moreover, we define a $\downarrow\Box$-formula (respectively, $\downarrow\Diamond$-formula) as a hybrid formula in NNF in which an occurrence of a universal (respectively, existential) operator is in the scope of a $\downarrow$. Similar definitions hold for different patterns. For example, $\Box\downarrow\Box$-formula is a formula in NNF containing a universal operator that contains in its scope a $\downarrow$ that contains in its scope a universal operator. A $\downarrow$-formula is simply a formula in NNF containing a $\downarrow$ binder. Given a pattern π, we define $\mathrm{FHL}\setminus\pi$ to be the fragment of FHL consisting of all formulas in NNF that are *not* of the form π. Notice that such fragments are not necessarily closed under negation.

Theorem 1. *There exists a polynomial satisfiability-preserving translation from* $\mathrm{FHL}\setminus\Box\downarrow$ *to* $\mathrm{HL}(@, \Diamond^-, E)$. *Moreover, the translation preserves satisfiability relative to any class of frames.*

Proof. It is convenient to introduce a new hybrid binder $\exists$. We add to the language formulas of the form $\exists x.\alpha$, where x is a state variable, interpreted as follows:

$$M, g, w \Vdash \exists x.\alpha \text{ iff } M, g_v^x, w \Vdash \alpha \text{ for some state } v$$

Notice that $\downarrow$ can be defined in terms of $\exists$ as follows: $\downarrow x.\alpha \equiv \exists x.(x \wedge \alpha)$.

Let us proceed with the proof. Let α_0 be a hybrid formula in $\text{FHL} \setminus \square\downarrow$. We show how to polynomially translate α_0 into a formula α_3 in $\text{HL}(@, \diamondsuit^-, E)$ such that α_0 is satisfiable if, and only if, α_3 is satisfiable. The translation consists of three steps:

1. Let α_1 be obtained from α_0 be replacing each subformula of the form $\downarrow x.\varphi$ by $\exists x(x \wedge \varphi)$. Since no occurrence of the $\downarrow$ binder in α_0 is in the scope of a universal operator, the same holds for the occurrences of the $\exists$ binder in α_1;
2. rewrite α_1 into quantifier prefix form (i.e., where all occurrences of $\exists$ are in front of the formula), using the following equivalences: $\diamondsuit\exists x.\varphi \equiv \exists x.\diamondsuit\varphi$, $\diamondsuit^-\exists x.\varphi \equiv \exists x.\diamondsuit^-\varphi$, $E\exists x.\varphi \equiv \exists x.E\varphi$, $@_t\exists x.\varphi \equiv \exists x.@_t\varphi$, $\psi \wedge \exists x.\varphi \equiv \exists x.(\psi \wedge \varphi)$, $\psi \vee \exists x.\varphi \equiv \exists x.(\psi \vee \varphi)$. Note that renaming of variables might be necessary. Let α_2 be the resulting formula;
3. Let α_3 be obtained from α_2 by replacing each state variable by a fresh nominal and removing the corresponding existential quantifiers.

The resulting formula α_3 is in $\text{HL}(@, \diamondsuit^-, E)$, the length of α_3 is linear in the length of α_0, and α_0 and α_3 are easily seen to be equi-satisfiable.   □

To illustrate the above proof, consider the formula $\downarrow x.\diamondsuit\downarrow y.@_x(\diamondsuit(y \wedge q) \wedge \square(\square\neg y \vee p))$, which does not contain the $\square\downarrow$-pattern. It can be rewritten as follows:

$$\begin{aligned}
\downarrow x.\diamondsuit\downarrow y.@_x(\diamondsuit(y \wedge q) \wedge \square(\square\neg y \vee p)) &\equiv \\
\exists x.(x \wedge \diamondsuit\exists y.(y \wedge @_x(\diamondsuit(y \wedge q) \wedge \square(\square\neg y \vee p)))) &\equiv \\
\exists x.\exists y.(x \wedge \diamondsuit(y \wedge @_x(\diamondsuit(y \wedge q) \wedge \square(\square\neg y \vee p)))) &\cong \\
i \wedge \diamondsuit(j \wedge @_i(\diamondsuit(j \wedge q) \wedge \square(\square\neg j \vee p)))
\end{aligned}$$

Corollary 1. *The satisfiability problem for* $\text{FHL} \setminus \square\downarrow$ *is* EXPTIME-*complete.*

Proof. The lower bound follows from the fact that $\text{FHL} \setminus \square\downarrow$ embeds the basic modal language with global modality, which is known to have an EXPTIME-complete satisfiability problem [14]. The upper bound follows from Theorem 1 since satisfiability of $\text{HL}(@, \diamondsuit^-, E)$-formulas can be decided in EXPTIME [15].

We now prove the mirror image of Theorem 1: satisfiability for $\text{FHL} \setminus \downarrow\square$ is decidable. The technique we use is similar to the one used by Marx [13]: we embed $\text{FHL} \setminus \downarrow\square$ into the $\forall$-guarded fragment. The $\forall$-guarded fragment of first-order logic consists of all formulas constructed from atomic formulas and their negations using conjunction, disjunction, existential quantification, and guarded universal quantification. Hence only the universal quantification is constrained. The satisfiability problem for $\forall$-guarded first-order formulas is 2EXPTIME-complete. It is EXPTIME-complete when there is a uniform bound on the width of the formula. For more details, cf. [16].

Theorem 2. *The satisfiability problem for* $\text{FHL} \setminus \downarrow\square$ *is in* 2EXPTIME. *The satisfiability problem for* $\text{FHL} \setminus \downarrow\square$-*formulas of bounded width is* EXPTIME-*complete.*

Proof. Let α be any FHL $\setminus \downarrow\Box$-sentence. We will show by induction on α that $ST_x(\alpha)$ is $\forall$-guarded. Since $ST_x(\alpha)$ can be obtained from α in polynomial time, this proved that the satisfiability problem for FHL $\setminus \downarrow\Box$ is in 2ExpTime.

To smoothen the induction, we will prove the result for any subformula α of a FHL $\setminus \downarrow\Box$-sentence. If α is a (negated) atomic formula, then $ST_x(\alpha)$ is quantifier-free, hence $\forall$-guarded. If α is of the form $\alpha_1 \wedge \alpha_2$ or $\alpha_1 \vee \alpha_2$, then by the induction hypothesis, $ST_x(\alpha)$ is the conjunction (respectively, disjunction) of two $\forall$-guarded formulas, and hence is $\forall$-guarded.

Next, suppose α is of the form $X\alpha_1$, where X is an existential operator or an @-operator. By the induction hypothesis, $ST_y(\alpha_1)$ is $\forall$-guarded. Inspection of the relevant clauses of the Standard Translation shows that $ST_x(\alpha)$ is also $\forall$-guarded.

Next, suppose α is of the form $X\alpha_1$, where X is a universal operator. Again, by induction hypothesis, $ST_y(\alpha_1)$ is $\forall$-guarded. Moreover, by assumption α is a subformula of a FHL $\setminus \downarrow\Box$-sentence. It follows α_1 cannot contain any free state variables (for, these would have to be bound higher up). It follows that $ST_y(\alpha_1)$ contains no free variables besides (possibly) y. Inspection of the relevant clauses of the Standard Translation shows that this variable y is appropriately guarded in $ST_x(\alpha)$, and hence $ST_x(\alpha)$ is $\forall$-guarded.

Finally, suppose α is of the form $\downarrow z.\alpha_1$. Then, $ST_x(\alpha) = \exists z.(z = x \wedge ST_x(\alpha_1))$. By induction hypothesis, $ST_x(\alpha_1)$ is $\forall$-guarded. It follows that $ST_x(\alpha)$ is also $\forall$-guarded.

It is easy to see that, if a hybrid formula α has width w, then the width of $ST_x(\alpha)$ is at most $w + 2$. Hence, a bound on the width of the FHL $\setminus \downarrow\Box$-formula implies a bound on the width of its $\forall$-guarded standard translation. Since the satisfiability problem for $\forall$-guarded formulas of bounded width is ExpTime-complete, this gives us an ExpTime upper bound. The lower bound follows from the ExpTime-hardness of the basic modal logic extended with the global modality [14]. □

Satisfiability for FHL $\setminus \downarrow\Box$ is ExpTime-hard, since satisfiability for modal logic with the global modality is already ExpTime-hard [14]. We don't know the exact complexity of FHL $\setminus \downarrow\Box$, but we conjecture that it is ExpTime-complete.

By combining the techniques used to prove Theorems 1 and 2, we have the main result of this section:

Theorem 3. *The satisfiability problem for* FHL $\setminus \Box\downarrow\Box$ *is in* 2ExpTime. *The satisfiability problem for* FHL $\setminus \Box\downarrow\Box$*-formulas of bounded width is* ExpTime-complete.

Proof. Let $\alpha \in$ FHL $\setminus \Box\downarrow\Box$. If $\alpha \in$ FHL $\setminus \downarrow\Box$, then the satisfiability of α can be decided in 2ExpTime by Theorem 2. Suppose therefore that $\alpha \notin$ FHL $\setminus \downarrow\Box$. Let β be a minimal $\downarrow\Box$-subformula of α. Since $\alpha \in$ FHL $\setminus \Box\downarrow\Box$, β cannot be in the scope of a universal operator in α. It follows that this occurrence of $\downarrow$ can be removed as in the proof of Theorem 1. Repeating this step for each minimal $\downarrow\Box$-subformula of α, we obtain a formula $\beta \in$ FHL $\setminus \downarrow\Box$ that is satisfiable iff α is satisfiable. By Theorem 2, satisfiability of β can be checked in 2ExpTime. The

ExpTime-completeness in the case of bounded width follows from the bounded width case of Theorem 2.   □

To illustrate the above proof, consider the formula $\alpha = \Diamond\downarrow x.\Box\downarrow y.@_y\Diamond x$. It contains both the $\downarrow\Box$- and the $\Box\downarrow$-pattern, hence neither Theorem 1 nor Theorem 2 can be applied. However, α does not contain the $\Box\downarrow\Box$-pattern, hence Theorem 3 can be invoked. There exists only one minimal $\downarrow\Box$-subformula of α, that is $\beta = \downarrow x.\Box\downarrow y.@_y\Diamond x$. The outermost occurrence of $\downarrow$ in β is not in the scope of any universal operator in α, hence it can be removed as done in Theorem 1. The resulting equi-satisfiable formula is $\alpha' = \Diamond(i \wedge \Box\downarrow y.@_y\Diamond i)$, which does not contain the $\downarrow\Box$-pattern anymore. Hence Theorem 2 can be applied to it.

Since the negation of an FHL$\setminus\Diamond\downarrow\Diamond$-formula is equivalent to an FHL$\setminus\Box\downarrow\Box$-formula, we have as a corollary the following dual result.

Corollary 2. *The validity problem for* FHL$\setminus\Diamond\downarrow\Diamond$ *is in* 2ExpTime. *The validity problem for* FHL$\setminus\Diamond\downarrow\Diamond$*-formulas of bounded width is* ExpTime-*complete.*

In particular, if a hybrid formula ϕ contains neither the $\Box\downarrow\Box$ pattern nor the $\Diamond\downarrow\Diamond$ pattern, then both satisfiability and validity of ϕ are decidable.

5   Semantic Restrictions

In this section, we restrict attention to uni-modal models of bounded width, i.e., models with only one binary relation R, in which each node is R-related only to a restricted number of points. More precisely, for any cardinal κ, let K_κ be the class of uni-modal models in which for every node d there are strictly less than κ nodes e such that $(d, e) \in R$. In particular, K_2 is the class of models in which every points has at most one R-successor, and K_ω is the class of models in which every node has only finitely many R-successors. We will refer to elements of K_κ as κ-models for short. In what follows we will consider the satisfiability problem of HL$(@, \downarrow)$ and of the first-order correspondence language on κ-models, for particular κ. Our results are summarized in Table 1. All results generalize to the case with multiple modalities, except for the decidability of the first-order correspondence language on K_2.

The terminology and results used in this section can be found in [17] and [10], or in other texts on computational complexity. In particular, we follow the usual terminology from recursion theory: the language of second-order arithmetic is the second-order language with constants 0, 1, function symbols + and ×, and equality. Formulas of second-order arithmetic are interpreted over the natural numbers. A Σ_1^1 formula of second order arithmetic is a formula of the form $\exists R_1 \ldots R_n.\phi$ where ϕ contains no second-order quantifiers. A set A of natural numbers is said to be in Σ_1^1 if it is defined by a Σ_1^1 formula that has one free first-order variable and no free second-order variables. A set A of natural numbers is Σ_1^1-hard if for every B in Σ_1^1 there is a computable function $f : \mathbb{N} \to \mathbb{N}$ such that for all $n \in \mathbb{N}$, $n \in B$ iff $f(n) \in A$. A set of natural numbers is Σ_1^1-complete if it is both in Σ_1^1 and Σ_1^1-hard. It is well known that Σ_1^1-hard sets are not recursively

Table 1. Complexity of the satisfiability problem on κ-models.

	HL($@, \downarrow$)	first-order correspondence language
$\kappa = 1$	NP-complete	NEXPTIME-complete
$\kappa = 2$	NP-complete	Decidable, not elementary recursive
$3 \leq \kappa < \omega$	NEXPTIME-complete	Π_1^0-complete (co-r.e., not decidable)
$\kappa = \omega$	Σ_1^0-complete (r.e., not decidable)	Σ_1^1-complete (highly undecidable)
$\kappa > \omega$	Π_1^0-complete (co-r.e., not decidable)	Π_1^0-complete (co-r.e., not decidable)

enumerable. When one speaks of an arbitrary decision problem as being in Σ_1^1 or Σ_1^1-hard, it is implicitly understood that the instances of the decision problem are coded into natural numbers (under some computable encoding).

Following [17], we call a decidable problem *elementary recursive* if the time complexity can be bounded by a constant number of iterations of the exponential function.

Theorem 4. *The satisfiability problem of* HL($@, \downarrow$) *on* K_κ *is:*

1. *NP-complete, for* $\kappa = 1, 2$.
2. *NEXPTIME-complete, for* $3 \leq \kappa < \omega$.
3. *Recursively enumerable but not decidable, for* $\kappa = \omega$.
4. *Co-recursively enumerable but not decidable, for* $\kappa > \omega$.

Proof. Point 1. The lower bound follows from the NP-hardness of propositional satisfiability. The upper bound is proved by establishing the polynomial size model property.

For $\kappa = 1, 2$, every κ-satisfiable HL($@, \downarrow$)-formula is satisfiable in a κ-model with at most $O(|\phi|^2)$ nodes. For, suppose $M, g, w \Vdash \phi$ for some κ-model $M = (W, R, V)$ and assignment g. Let $W' \subseteq W$ consist of all worlds that are reachable from w or from a world named by one of the nominals occurring in ϕ in at most $md(\phi)$ steps, where $md(\phi)$ is the modal depth of ϕ. Let M' be the submodel of M with domain W'. Clearly, M' is a κ-model and M' satisfies the cardinality requirements and a straightforward induction argument shows that $M', g, w \Vdash \phi$.

This leads to a non-deterministic polynomial time algorithm for testing satisfiability of an HL($@, \downarrow$)-formula ϕ on κ-models, for $\kappa = 1, 2$. The algorithm first non-deterministically chooses a candidate model (M, g, w) of size $O(|\phi|^2)$, and then it tests whether $M, g, w \Vdash \phi$ and $M \in \mathsf{K}_\kappa$. The latter tests can be performed in polynomial time using a top down model checking algorithm (cf. Theorem 6).

Point 2 (Upper bound). For $3 \leq \kappa < \omega$, every formula satisfiable on a κ-model is satisfiable on a κ-model with at most $O(|\phi| \cdot \kappa^{md(\phi)})$ nodes. For, suppose $M, g, w \Vdash \phi$ for some κ-model $M = (W, R, V)$ and assignment g. Let $W' \subseteq W$ consist of all worlds that are reachable from w or from a world named by one of the nominals occurring in ϕ in at most $md(\phi)$ steps. Let M' be the submodel of M with domain W'. Note that the cardinality of M' is $O(|\phi| \cdot \kappa^{|\phi|})$, and M' is still a κ-model. Furthermore, a straightforward induction argument shows that $M', g, w \Vdash \phi$.

This leads to a non-deterministic ExpTime algorithm for testing satisfiability of an $HL(@, \downarrow)$-formula ϕ on κ-models. The algorithm first non-deterministically chooses a candidate model (M, g, w) of size $O(|\phi| \cdot \kappa^{|\phi|})$, and then tests whether $M, g, w \Vdash \phi$. The latter test can be performed in time $O(|M|^{|\phi|})$ [8], which is $O((|\phi| \cdot \kappa^{|\phi|})^{|\phi|}) = O(|\phi|^{|\phi|} \cdot \kappa^{(|\phi|^2)})$.

Point 2 (Lower bound). Consider monadic first-order formulas without equality, i.e., first-order formulas containing unary predicates only, without equality. Any such satisfiable formula ϕ of length n has a model with at most 2^n nodes, and the satisfiability problem for such formulas is NExpTime-complete [17, Section 6.2.1]. We will reduce this problem to the satisfiability problem for $HL(@, \downarrow)$-formulas on κ-models (for $3 \le \kappa < \omega$), thus showing that the latter problem is NExpTime-hard.

Fix a nominal i, and for any monadic first-order formula ϕ without equality, define ϕ^+ inductively, such that $(x = y)^+ = @_x y$, $(Px)^+ = @_x p$, $(\cdot)^+$ commutes with the Boolean connectives and $(\exists x.\psi)^+ = @_i \Diamond^{|\phi|} \downarrow x.\psi$. In words, ϕ^+ states that ϕ holds in the submodel consisting of all points reachable from the point named i in exactly $|\phi|$ many steps. In general, there can be up to $(\kappa - 1)^{|\phi|}$ many points reachable from the point named i in exactly $|\phi|$ many steps (in particular, this will be the case if the submodel generated by i is a $(\kappa - 1)$-ary tree). Thus, ϕ is satisfiable iff ϕ is satisfiable in a model with at most $2^{|\phi|}$ nodes iff ϕ^+ is satisfiable in a κ-model, for $\kappa \ge 3$.

Point 3. We will provide polynomial reductions between this problem and the finite satisfiability problem for first-order logic. The satisfiability problem for first-order logic on finite models is Σ_1^0-complete, even in the case with only a single, binary relation [17, Section 3.2].

Trivially, if an $HL(@, \downarrow)$-formula is satisfiable in a finite model, it is satisfiable in an ω-model. Conversely, if an $HL(@, \downarrow)$-formula is satisfiable in an ω-model then is satisfiable in a finite model, since the modal depth of the formula provides a bound on the depth of the model. Hence, the satisfiability problem of $HL(@, \downarrow)$ on ω-models reduces (by the Standard Translation) to the satisfiability problem for first-order logic on finite models.

Conversely, the finite satisfiability problem for first-order logic can be reduced to satisfiability of $HL(@, \downarrow)$ on ω-models. Fix a nominal i, and for any first-order formula ϕ, define ϕ^+ inductively, such that $(x = y)^+ = @_x y$, $(Rxy)^+ = @_x \Diamond y$, $(\cdot)^+$ commutes with the Boolean connectives and $(\exists x.\psi)^+ = @_i \Diamond \downarrow x.\psi^+$. In words, ϕ^+ states that ϕ holds in the submodel consisting of the successors of the point named i. It follows that ϕ is satisfiable in a finite model iff the $HL(@, \downarrow)$-formula ϕ^+ is satisfiable on an finitely branching ω-model.

Point 4. By the Löwenheim-Skolem theorem, a first-order formula is satisfiable if and only if it is satisfiable on a finite or countably infinite model. Since $HL(@, \downarrow)$ is a fragment of first-order logic, the Löwenheim-Skolem theorem also applies to $HL(@, \downarrow)$-formulas. It follows that the satisfiability problem for $HL(@, \downarrow)$ on countably branching models coincides with the general satisfiability problem of $HL(@, \downarrow)$, which is in Π_1^0 by the Standard Translation and Π_1^0-hard by the tiling argument from Section 3.   □

As the following theorem shows, the first-order correspondence language performs much worse.

Theorem 5. *The satisfiability problem of first-order sentences of the correspondence language on* K_κ *is:*

1. NExpTime *complete, for* $\kappa = 1$.
2. *decidable but not elementary recursive, for* $\kappa = 2$.
3. *Co-recursively enumerable but not decidable, for* $3 \le \kappa < \omega$.
4. Σ_1^1*-hard, and hence neither recursively enumerable nor co-recursively enumerable, for* $\kappa = \omega$.
5. *Co-recursively enumerable but not decidable, for* $\kappa > \omega$.

Proof. We prove here only the decidable cases (points 1 and 2). The reader is referred to the full version of this paper [18] for a full proof of the theorem.

Point 1. This case coincides with the satisfiability problem for monadic first-order logic (on 1-models, every formula of the form Rst is equivalent to $\bot$), which is known to be NExpTime complete [17].

Point 2. Consider the satisfiability problem for first-order logic with one unary function symbol, an arbitrary number of unary relation symbols and equality ("the Rabin class"). This problem is decidable, but not elementary recursive [17]. We will provide reductions between this problem and the satisfiability problem for first-order logic on 2-models.

Let ϕ be any first-order formula containing one unary function symbol f and any number of unary relation symbols and equality. Let R be a binary relation symbol, and let ϕ_R be obtained from ϕ by repeatedly applying the rewrite rules

- replace atomic formulas of the form $Pf(t)$ by $\exists x.(Rtx \wedge Px)$
- replace atomic formulas of the form $f(s) = t$ or $t = f(s)$ by $\exists x.(Rsx \wedge x = t)$

until the function symbol f does not occur in the formula anymore (in case of nested function symbols, the above rules might need to be applied several times). It is not hard to see that ϕ is satisfiable iff $\phi_R \wedge \forall x \exists y.Rxy$ is satisfiable on a 2-model.

Let ϕ be any first-order formula with one binary relation symbol R and any number of unary relation symbols. Let f be a unary function symbol and let P be a new unary relation, and let ϕ_f be the result of replacing all subformulas of ϕ of the form Rst by $Ps \wedge (t = fs)$. Intuitively, the unary predicate P represents the existence of a successor, and the unary function f encodes the successor of a node, if it exists. One can easily see that ϕ is satisfiable on a 2-model iff ϕ_f is satisfiable (simply let R denote the graph of f, or viceversa).

It follows that the satisfiability problem of first-order logic on 2-models is decidable but not elementary recursive. $\qquad\Box$

6 Model Checking

So far, we only studied the satisfiability and the validity problems. It is natural to ask how our syntactic and semantic restrictions affect the complexity of the model checking problem.

Given a hybrid model M, an assignment g, a state w, and a hybrid formula α, the *model checking problem* is to check whether $M, g, w \Vdash \alpha$. We will restrict ourselves to hybrid sentences. This is not a limitation, since one can always replace the free variables by fresh nominals, expanding the model accordingly.

In [8], the authors give a polynomial time model checker for $\mathrm{HL}(@, \Diamond^-, E)$. Moreover, they prove that the model checking problem for $\mathrm{HL}(@, \downarrow)$ is PSPACE-complete (as it is for full first-order logic), even for formulas without nominals, @-operators and proposition letters.

Theorem 6. *The model checking problem for $\mathrm{HL}(@, \downarrow)$ on κ-models can be solved in polynomial time for $\kappa \leq 2$, and is PSPACE-complete for $\kappa \geq 3$.*

Proof. The first part of the theorem can be proved using a straightforward top-down model checking algorithm. Since each state in the model has at most one successor, the algorithm takes time linear in the length of the input formula. As for the second part, the proof of PSPACE-hardness of model checking for $\mathrm{HL}(@, \downarrow)$ given in [8] uses a model with out-degree 2. It follows that the model checking problem for $\mathrm{HL}(@, \downarrow)$ on κ-models, with $\kappa \geq 3$, is PSPACE-complete. □

For $\mathrm{HL}(@, E, \downarrow)$ and first-order logic, on the other hand, the model checking problem is PSPACE-complete even on 1-models [8].

In the following, we investigate how restrictions on the *syntax* of hybrid formulas affect the complexity of model checking. Our first result is that, if formulas containing the $\downarrow\Box\downarrow$ pattern are excluded, then the model checking problem drops from PSPACE to NP.

Theorem 7. *The model checking problem for $\mathrm{FHL} \setminus \downarrow\Box\downarrow$ is NP-complete.*

Proof. To prove NP-hardness, we embed the satisfiability problem for propositional formulas (SAT) into the model checking problem for $\mathrm{FHL} \setminus \downarrow\Box\downarrow$. Let $\phi(p_1, \ldots, p_n)$ be any propositional formula, and let $M = (W, R, V)$, where $W = \{0, 1\}$ and $R = W \times W$ (the valuation V is irrelevant). For each p_k occurring in ϕ, pick a corresponding state variable x_k. Furthermore, let y be a state variable distinct from all $x_1, \ldots, x_n$. Let ϕ' be obtained from ϕ by replacing each occurrence of a proposition letter p_k by $\Diamond(x_k \wedge y)$. Intuitively, the two states of M represent truth and falsity, and among these two states the variable y denotes the truth state. It is easily seen that the propositional formula ϕ is satisfiable iff $\Diamond\downarrow y \Diamond\downarrow x_1 \Diamond\downarrow x_2 \ldots \Diamond\downarrow x_n.\phi'$ is true in M (at any of the nodes $0, 1$). The latter formula contains no universal operators, and hence belongs to $\mathrm{FHL} \setminus \downarrow\Box\downarrow$.

To show that the problem is in NP, we give a nondeterministic algorithm that solves the model checking problem in polynomial time. Let α be an $\mathrm{FHL} \setminus \downarrow\Box\downarrow$ sentence, $M = (W, R, V)$ be a model, $v \in W$ and g be an assignment. Replace each subformula of α of the form $\downarrow x.\varphi$ by $\exists x.(x \wedge \varphi)$, and apply the equivalences given in the proof of Theorem 1 in order to move the existential quantifiers out of the scope of as many connectives as possible. The resulting sentence α' is equivalent to α and has the following properties:

1. α' is built up from literals (i.e., formulas of the form $(\neg)p$, $(\neg)i$ or $(\neg)x$) using conjunction, disjunction, existential operators $(\Diamond, \Diamond^-, E)$, universal operators $(\Box, \Box^-, A)$, and existential quantifiers.
2. All existential quantifiers in α' either immediately follow a universal operator (e.g., as in $\Box \exists x_1 \ldots x_n \gamma$) or occur at the start of the formula.
3. For all subformulas of α' of the form $X \exists x_1 \ldots x_n \gamma$, with X a universal operator, γ contains no free variables besides $x_1, \ldots, x_n$.

List all subformulas of α' of the form $X\beta$, with X a universal operator and $\beta = \exists x_1 \ldots \exists x_m . \gamma(x_1 \ldots x_m)$, in order of increasing length. For each such β do the following: create a new proposition symbol p_β and replace β by p_β in α'. For each state $w \in W$, check whether $M, g, w \Vdash \beta$, and, if the answer is positive, then insert the state w in $V(p_\beta)$.

The nondeterminism is hidden in the test $M, g, w \Vdash \beta$. Indeed, to check whether $M, g, w \Vdash \exists x_1 \ldots \exists x_m . \gamma(x_1 \ldots x_m)$, the algorithm guesses an assignment $w_1, \ldots, w_m$ for the variables $x_1, \ldots, x_m$, respectively, and then it checks whether $M, g^{x_1, \ldots, x_m}_{w_1, \ldots, w_m}, w \Vdash \gamma(x_1 \ldots x_m)$. Since γ does not contain any existential quantifiers (the subformulas were processed in order of increasing length), it belongs to $HL(@, \Diamond^-, E)$ and hence the model checking can be performed in polynomial time.

The resulting formula is in $HL(@, \Diamond^-, E)$ and thus it can be model checked in polynomial time. $\qquad\Box$

Notice that the NP-hardness holds even for formulas without proposition letters, nominals and @-operators. A typical example of a formula to which Theorem 7 does not apply is $\downarrow x . \Box\Box\downarrow y . @_x \Diamond y$, which expresses a local form of transitivity.

In Section 4, we saw that $FHL \setminus \Box\downarrow\Box$ has a decidable satisfiability problem. We leave it as an open question whether the model checking complexity of that fragment is also below PSPACE (since the SAT problem can be embedded into the model checking problem for $FHL \setminus \Box\downarrow\Box$ as done in the proof of Theorem 7, the problem is at least NP-hard). Conversely, the fragment $FHL \setminus \downarrow\Box\downarrow$, for which we have just proved that the model checking problem is NP-complete, has an undecidable satisfiability problem: it suffices to note that the encoding of the tiling problem given in Section 3 does not make use of $\downarrow\Box\downarrow$-formulas.

We conclude this section with a hierarchy of fragments of the full hybrid language with $\downarrow$ binder that admits polynomial time model checking. If a hybrid formula α has width w, then $ST_x(\alpha)$ has width at most $w+2$. Hence, a bound on the width of the hybrid formulas implies a bound on the width of the standard translations. Moreover, model checking for first-order formulas using a bounded number of variables can be performed in polynomial time [19]. It is known that first-order formulas of a bounded width can be rewritten using a bounded number of variables (cf. [20] for an explicit proof). Thus, we obtain the following.

Theorem 8. *The model checking problem for formulas of the full hybrid language of bounded width can be solved in polynomial time.*

7   Conclusion

In this paper, we described two ways to tame the power of hybrid logic with binders. These are: (i) restricting the syntax by excluding formulas containing the pattern $\Box\!\downarrow\!\Box$, and (ii) restricting the class of models by assuming a bound on the branching degree of the models. Furthermore, we showed that similar restrictions can be used to lower the complexity of the model checking task.

Our decidability result for $\mathrm{FHL} \setminus \Box\!\downarrow\!\Box$ may be seen from a more general perspective: one could consider any sequence $\pi \subseteq \{\Box, \Diamond, \downarrow, @\}^*$, where $\Box$ stands for "a sequence of universal modalities", and $\Diamond$ stands for "a sequence of existential modalities", and ask whether the satisfiability problem for $\mathrm{FHL} \setminus \pi$ is decidable. In particular, one could ask if there is such a sequence π that contains $\Box\!\downarrow\!\Box$ as a subsequence and such that the satisfiability problem for $\mathrm{FHL} \setminus \pi$ is still decidable. Our undecidability proof in Section 3 (and more in particular the shape of the formulas β and γ used there) shows that the answer is negative, and hence Theorem 3 is tight.

Some results in this paper show that, under certain natural conditions, the language $\mathrm{HL}(@, \downarrow)$ behaves better than the first-order correspondence language, computationally speaking. Incidentally, the full hybrid language FHL has the same expressive power as the first-order correspondence language, as was shown in [21] by means of a translation HT mapping formulas of the first-order correspondence to FHL-formulas. The most interesting clause of this translation says $\mathrm{HT}(\exists x.\phi) = E\!\downarrow\!x.\mathrm{HT}(\phi)$. It shows that, in some sense, the first-order quantifier $\exists x$ consist of two parts, namely the *picking a state of the model* part, which is captured by the global modality, and the *variable binding* part, which is captured by the $\downarrow$. The syntax of $\mathrm{HL}(@, E, \downarrow, \Diamond^-)$ allows us to distinguish these two parts. One could say that our results identify computationally tractable fragments of first-order logic that can only be distinguished once these two parts of the quantifiers are split. In this sense, our paper can be seen as a fine study of the structure of first-order quantifiers.

Finally, the outcomes of our investigation show once more that, from a computational point of view, the satisfiability problem and the model checking problem for a logic are sensitive to different *sources* of complexity. Restricting the model width makes satisfiability easier, but it does not lower the complexity of model checking. On the other hand, restricting the formula width makes model checking more tractable, but it does not affect the undecidability of satisfiability.

Acknowledgements

This paper has benefited from discussions with Maarten Marx, and from the comments of the anonymous reviewers. The authors were supported by NWO grants 612.069.006 and 612.000.207, respectively.

References

1. Andréka, H., van Benthem, J., Németi, I.: Modal logics and bounded fragments of predicate logic. Journal of Philosophical Logic **27** (1998) 217–274
2. Grädel, E.: On the restraining power of guards. Journal of Symbolic Logic **64** (1999) 1719–1742
3. Mortimer, M.: On languages with two variables. Zeitschrift für mathematische Logik und Grundlagen der Mathematik **21** (1975) 135–140
4. Grädel, E., Otto, M.: On logics with two variables. Theoretical computer science **224** (1999) 73–113
5. Areces, C., Blackburn, P., Marx, M.: Hybrid logics: Characterization, interpolation, and complexity. Journal of Symbolic Logic **66** (2001) 977–1010
6. ten Cate, B.: Interpolation for extended modal languages. Journal of Symbolic Logic **70** (2005) 223–234
7. ten Cate, B.: Model theory for extended modal languages. PhD thesis, University of Amsterdam (2005) ILLC Dissertation Series DS-2005-01.
8. Franceschet, M., de Rijke, M.: Model checking for hybrid logics (with an application to semistructured data). Journal of Applied Logics (2005) To appear.
9. Blackburn, P., Seligman, J.: Hybrid languages. Journal of Logic, Language and Information **4** (1995) 251–272
10. Harel, D.: Recurring dominoes: making the highly undecidable highly understandable. Annals of Discrete Mathematics **24** (1985) 51–72
11. Goranko, V.: Hierarchies of modal and temporal logics with reference pointers. Journal of Logic, Language, and Information **5** (1996) 1–24
12. Areces, C., Blackburn, P., Marx, M.: A road-map on complexity for hybrid logics. In Flum, J., Guez-Artalejo, M.R., eds.: Proceedings of the 8th Annual Conference of the EACSL, Madrid (1999)
13. Marx, M.: Narcissists, stepmothers and spies. In: Proceedings of the International Workshop on Description Logics. (2002)
14. Fisher, M., Ladner, R.: Propositional dynamic logic of regular programs. Journal of Computer and System Sciences **18** (1979) 194–211
15. Areces, C., Blackburn, P., Marx, M.: The computational complexity of hybrid temporal logics. Logic Journal of the IGPL **8** (2000) 653–679
16. ten Cate, B., Franceschet, M.: Guarded fragments with constants. Journal of Logic, Language, and Information (2005) To appear.
17. Börger, E., Grädel, E., Gurevich, Y.: The Classical Decision Problem. Springer, Berlin (1997)
18. ten Cate, B., Franceschet, M.: On the complexity of hybrid logics with binders. Technical Report PP-2005-02, ILLC, University of Amsterdam (2005)
19. Vardi, M.Y.: On the complexity of bounded-variable queries. In: Proceedings of the ACM SIGACT-SIGMOD-SIGART Symposium on Principles of Database Systems. (1995) 266–276
20. ten Cate, B., Franceschet, M.: Guarded fragments with constants. Technical Report PP-2004-32, ILLC, University of Amsterdam (2004)
21. Blackburn, P.: Representation, reasoning, and relational structures: A hybrid logic manifesto. Logic Journal of the IGPL **8** (2000) 339–365

The Complexity
of Independence-Friendly Fixpoint Logic

Julian Bradfield[1] and Stephan Kreutzer[2]

[1] Laboratory for Foundations of Computer Science, University of Edinburgh
jcb@inf.ed.ac.uk
[2] Institut für Informatik, Humboldt Universität, 10099 Berlin, Germany
kreutzer@informatik.hu-berlin.de

Abstract. We study the complexity of model-checking for the fixpoint extension of Hintikka and Sandu's independence-friendly logic. We show that this logic captures ExpTime; and by embedding PFP, we show that its combined complexity is ExpSpace-hard, and moreover the logic includes second order logic (on finite structures).

1 Introduction

In everyday life we often have to make choices in ignorance of the choices made by others that might have affected our choice. With the popularity of the agent paradigm, there is much theoretical and practical work on logics of knowledge and belief in which such factors can be explicitly expressed in designing multi-agent systems. However, ignorance is not the only reason for making independent choices: in mathematical writing, it is not uncommon to assert the existence of a value for some parameter uniformly in some earlier mentioned parameter.

Hintikka and Sandu [HiS96] introduced a logic, called Independence-friendly (IF) logic, in which such independent choices can be formalized by independent quantification. Some of the ideas go back some decades, for IF logic can also be viewed as an alternative account of branching quantifiers (Henkin quantifiers) in terms of games of imperfect information. Independent quantification is a subtle concept, with many pitfalls for the unwary. It is also quite powerful: it has long been known that it has existential second-order power. In previous work [BrF02], the first author and Fröschle applied the idea of independent quantification to modal logics, where it has natural links with the theory of true concurrency; this prompted some consideration of fixpoint versions of IF modal logics, since adding fixpoint operators is the easiest way to get a powerful temporal logic from a simple modal logic. This led the first author to an initial investigation [Bra03] of the fixpoint extension of first-order IF logic, which we call IF-LFP. It turned out that fixpoint IF logic is not trivial to define, and appears to be very expressive, with the interaction between fixpoints and independent quantification giving a dramatic increase in expressive power. In [Bra03], only some fairly simple complexity results were obtained; in this paper, we obtain much stronger results about the model-checking complexity of IF-LFP. For the data complexity, we

L. Ong (Ed.): CSL 2005, LNCS 3634, pp. 355–368, 2005.

show that not only is IF-LFP EXPTIME-complete, but it captures EXPTIME; and for the combined complexity, we obtain an EXPSPACEhardness result. This latter result is obtained by an embedding of partial fixpoint logic into IF-LFP, which shows that on finite structures IF-LFP even includes second-order logic, a much stronger result than the first author previously conjectured.

2   Independence-Friendly Fixpoint Logic

2.1   Syntax

First of all, we state one important **notational convention**: to minimize the number of parentheses, we take the scope of all quantifiers and fixpoint operators to extend as far to the right as possible.

Now we define the syntax of first-order IF logic. Here we use the version of Hodges [Hod97], and we confine the 'independence-friendly' operators to the quantifiers; in the full logic, one can also specify conjunctions and disjunctions that are independent, but these are not necessary for our purposes – their addition changes none of our results.

Definition 2.1. As for FOL, IF-FOL has proposition (P, Q etc.), relation (R, S etc.), function (f, g etc.) and constant (a, b etc.) symbols, with given arities. It also has individual *variables* v, x etc. We write $\boldsymbol{x}, \boldsymbol{v}$ etc. for tuples of variables, and similarly for tuples of other objects; we use concatenation of symbols to denote concatenation of tuples with tuples or objects.

For formulae φ and terms t, the (meta-level) notations $\varphi[\boldsymbol{x}]$ and $t[\boldsymbol{x}]$ mean that the free variables of φ or t are included in the variables $\boldsymbol{x}$, without repetition.

The notions of 'term' and 'free variable' are as for FOL.

We assume equality = is in the language, and atomic formulae are defined as usual by applying proposition or relation symbols to individual terms or tuples of terms. The free variables of the formula $R(\boldsymbol{t})$ are then those of $\boldsymbol{t}$.

The compound formulae are given as follows:

Conjunction and disjunction. If $\varphi[\boldsymbol{x}]$ and $\psi[\boldsymbol{y}]$ are formulae, then $(\varphi \vee \psi)[\boldsymbol{z}]$ and $(\varphi \wedge \psi)[\boldsymbol{z}]$ are formulae, where $\boldsymbol{z}$ is the union of $\boldsymbol{x}$ and $\boldsymbol{y}$.

Quantifiers. If $\varphi[\boldsymbol{y}, x]$ is a formula, x a variable, and W a finite set of variables, then $(\forall x/W.\, \varphi)[\boldsymbol{y}]$ and $(\exists x/W.\, \varphi)[\boldsymbol{y}]$ are formulae. If W is empty, we write just $\forall x.\, \varphi$ and $\exists x.\, \varphi$.

Game negation. If $\varphi[\boldsymbol{x}]$ is a formula, so is $(\sim\varphi)[\boldsymbol{x}]$.

Flattening. If $\varphi[\boldsymbol{x}]$ is a formula, so is $(\downarrow \varphi)[\boldsymbol{x}]$.

Negation. $\neg\varphi$ is an abbreviation for $\sim \downarrow \varphi$.

Definition 2.2. IF-FOL$^+$ is the logic in which $\sim$, $\downarrow$ and $\neg$ are applied only to atomic formulae.

In the rest of this paper, we shall be working with IF-FOL$^+$, in which $\sim$ and $\neg$ merge, and $\downarrow$ has no effect. We shall therefore omit $\sim$ and $\downarrow$ from future definitions, and take $\neg$ as primitive, which also allows a simpler semantics than that for the full IF-FOL. Since we are proving lower bounds, the results apply also to the logic with negations.

2.2   Traditional Semantics

In the independent quantifiers the intention is that W is the set of independent variables, whose values the player is not allowed to know at this choice point: thus the classical Henkin quantifier $\genfrac{}{}{0pt}{}{\forall x\, \exists y}{\forall u\, \exists v}$, where x and y are independent of u and v, can be written as $\forall x/\varnothing.\, \exists y/\varnothing.\, \forall u/\{x,y\}.\, \exists v/\{x,y\}$. This notion of independence is the reason for saying that IF logic is natural in mathematical English: statements such as "For every x, and for all $\epsilon > 0$, there exists δ, depending only on $\epsilon \ldots$" can be transparently written as $\forall x, \epsilon > 0.\, \exists \delta/x.\, \ldots$ in IF logic.

If one then plays the Hintikka evaluation game (otherwise known as the model-checking game) with this additional condition, which can be formalized by requiring strategies to be uniform in the 'unknown' variables, one gets a game semantics of imperfect information, and defines a formula to be true iff Eloise has a winning strategy.

These games are not determined, so it is *not* the case that Abelard has a winning strategy iff the formula is not true. For example, $\genfrac{}{}{0pt}{}{\forall x}{\exists y}.x = y$ (or $\forall x.\, \exists y/\{x\}.\, x = y$) is untrue in any structure with more than one element, but Abelard has no winning strategy.

An alternative interpretation of the logic, dating from the early work on branching quantifiers, and one that is easier to handle mathematically in straightforward cases, is via Skolem functions with limited arguments. In FOL, the first order sentence $\forall x.\, \exists y.\, x = y$, over some universe A, is converted via Skolemization to the existential second-order sentence $\exists f : A \to A.\, \forall x.\, x = f(x)$. In this procedure, the Skolem function always takes as arguments all the universal variables currently in scope. By allowing Skolem functions to take only some of the arguments, we get a similar translation of IF-FOL$^+$: for example, $\forall x.\, \exists y/\{x\}.\, x = y$ becomes $\exists f : 1 \to A.\, \forall x.\, x = f()$. It can be shown that these two semantics are equivalent, in that an IF-FOL$^+$ sentence is true in the game semantics iff its Skolemization is true.

It is also well known that IF-FOL$^+$ is equivalent to existential second-order logic (in the cases where this matters, 'second-order' here means function quantification rather than set quantification). This is because the Skolemization process can be inverted: given a Σ_1^1 sentence, it can be turned into an IF-FOL$^+$ sentence (or equivalently, a sentence with Henkin quantifiers). We shall make use of this procedure in later results. Details can be found in [Wal70, End70] or in the full version of the paper, but here let us illustrate it by a standard example that demonstrates the power of IF logic. Consider the sentence 'there is an injective endofunction that is not surjective'. This is true only in infinite domains, and therefore not first-order expressible. It can be expressed directly in Σ_1^1 as

$$\exists f.\, (\forall x_1, x_2.\, f(x_1) = f(x_2) \Rightarrow x_1 = x_2) \wedge (\exists c.\, \forall x.\, f(x) \neq c)$$

which for the sake of reducing complexity below we will simplify to

$$\exists f.\, \exists c.\, \forall x_1, x_2.\, (f(x_1) = f(x_2) \Rightarrow x_1 = x_2) \wedge f(x_1) \neq c.$$

The basic trick for talking about functions in IF-FOL is to replace $\exists f. \forall x$ by $\forall x. \exists y$, so that y plays the role of $f(x)$. In FOL, this works only if there is just one application of f; but in IF-FOL, we can do it for two (or more) applications of f: we write $\forall x_1. \exists y_1$, and then we write an independent $\forall x_2/\{x_1, y_1\}. \exists y_2/\{x_1, y_1\}$. Now in order to make sure that these two (x_i, y_i) pairings represent the same f, the body of the translated formula is given a clause $(x_1 = x_2) \Rightarrow (y_1 = y_2)$. Applying this procedure to the Σ_1^1 sentence above and optimizing a bit, we get

$$\forall x_1, x_2. \exists y_1/x_2. \exists y_2/x_1. \exists c/\{x_1, x_2\}. (y_1 = y_2 \Leftrightarrow x_1 = x_2) \wedge y_1 \neq c.$$

2.3 Trump Semantics

The game semantics is how Hintikka and Sandu originally interpreted IF logic. Later on, the trump semantics of Hodges [Hod97], with variants by others, gave a Tarski-style semantics, equivalent to the original. This semantics is as follows:

Definition 2.3. Let a structure A be given, with constants, propositions and relations interpreted in the usual way. A *deal* $\boldsymbol{a}$ for $\varphi[\boldsymbol{x}]$ or $\boldsymbol{t}[\boldsymbol{x}]$ is an assignment of an element of A to each variable in $\boldsymbol{x}$. Given a deal $\boldsymbol{a}$ for a tuple of terms $\boldsymbol{t}[\boldsymbol{x}]$, let $\boldsymbol{t}(\boldsymbol{a})$ denote the tuple of elements obtained by evaluating the terms under the deal $\boldsymbol{a}$.

If $\varphi[\boldsymbol{x}]$ is a formula and W is a subset of the variables in $\boldsymbol{x}$, two deals $\boldsymbol{a}$ and $\boldsymbol{b}$ for φ are $\simeq_W$-*equivalent* ($\boldsymbol{a} \simeq_W \boldsymbol{b}$) iff they agree on the variables not in W. A $\simeq_W$-*set* is a non-empty set of pairwise $\simeq_W$-equivalent deals.

The denotation $[\![\varphi]\!]$ of a formula is a set T of *trumps*. If φ has n free variables, then $T \in \wp(\wp(A^n))$ – that is, a trump is a set of deals.

- If $(R(\boldsymbol{t}))[\boldsymbol{x}]$ is atomic, then a non-empty set D of deals is a trump iff $\boldsymbol{t}(\boldsymbol{a}) \in R$ for every $\boldsymbol{a} \in D$.
- D is a trump for $(\varphi \wedge \psi)[\boldsymbol{x}]$ iff D is a trump for $\varphi[\boldsymbol{x}]$ and D is a trump for $\psi[\boldsymbol{x}]$.
- D is a trump for $(\varphi \vee \psi)[\boldsymbol{x}]$ iff it is non-empty and there are trumps E of φ and F of ψ such that every deal in D belongs either to E or F.
- D is a trump for $(\forall y/W. \psi)[\boldsymbol{x}]$ iff the set $\{ \boldsymbol{a}b \mid \boldsymbol{a} \in D, b \in A \}$ is a trump for $\psi[\boldsymbol{x}, y]$.
- D is a trump for $(\exists y/W. \psi)[\boldsymbol{x}]$ iff there is a trump E for $\psi[\boldsymbol{x}, y]$ such that for every $\simeq_W$-set $F \subseteq D$ there is a b such that $\{ \boldsymbol{a}b \mid \boldsymbol{a} \in F \} \subseteq E$.
- D is a trump for $(\neg R(\boldsymbol{t}))[\boldsymbol{x}]$ iff $\boldsymbol{t}(\boldsymbol{a}) \notin R$ for every $\boldsymbol{a} \in D$.

A trump for φ is essentially a set of winning positions for the model-checking game for φ, for a given *uniform* strategy, that is, a strategy where choices are uniform in the 'hidden' variables. The most intricate part of the above definition is the clause for $\exists y/W. \psi$: it says that a trump for $\exists y/W. \psi$ is got by adding a witness for y, uniform in the W-variables, to trumps for ψ.

It is easy to see that any subset of a trump is a trump. In the case of an ordinary first-order $\varphi(\boldsymbol{x})$, the set of trumps of φ is just the power set of the set of tuples satisfying φ. To see how a more complex set of trumps emerges, consider the following formula, which has x free: $\exists y/\{x\}. x = y$. Any singleton set of deals is a trump, but no other set of deals is a trump. Thus we obtain that $\forall x. \exists y/\{x\}. x = y$ has no trumps (unless the domain has only one element).

The strangeness of the trump definitions is partly to do with some more subtle features of IF logics, that we do not here have space to discuss, but which are considered in detail in Ahti-Veikko Pietarinen's thesis [Pie00]. However, to take one good example, raised by a referee, consider $\varphi = \exists x. \exists y/\{x\}. x = y$. What are its trumps? As above, the trumps of $\exists y/\{x\}. x = y$ are singleton sets of deals. The only potential trump for φ is the set containing the empty deal $D = \{\langle\rangle\}$. Applying the definition, D is a trump for φ iff there is a singleton deal set $\{a\}$ for x such that there is a b such that $\{b\} \subseteq \{a\}$. The right hand side is true – take $b = a$ – so D is a trump. How come, if there is more than one element in A? Surely we must choose y independently of x, and therefore φ can't be true? Not so: because the choices are both made by the same player (Eloise), she can, as it were, make a uniform choice of y that, by 'good luck' agrees with her previous choice of x. Since she is not in the business of making herself lose, she will always do so. In game-theoretic terms, this is the difference between requiring a strategy to make uniform moves, and requiring a player to choose a strategy uniformly. In fact Hintikka and Sandu avoided this problem by only allowing the syntax to express quantifications independent in the other player's variables, which is in practice all one wishes to use in any case. Hodges removed this restriction to make his semantics cleaner, exposing the curiosity we have just described.

A sentence is said to be true if $\{\langle\rangle\} \in T$ (the empty deal is a trump set), and false if $\{\langle\rangle\} \in C$; this corresponds to Eloise or Abelard having a uniform winning strategy. Otherwise, it is undetermined. Note that 'false' is reserved for a strong sense of falsehood – undetermined sentences are also not true, and in the simple cases where negation and flattening are not employed, an undetermined sentence is as good as false.

2.4   IF-LFP

We now describe the addition of fixpoint operators to IF-FOL. This is slightly intricate, although the normal intuitions for understanding fixpoint logics still apply.

Definition 2.4. IF-LFP extends the syntax of IF-FOL as follows:

- There is a set $\mathrm{Var} = \{X, Y, \ldots\}$ of fixpoint variables. Each variable X has an arity $\mathrm{ar}(X)$.
- If X is a fixpoint variable, and t an $\mathrm{ar}(X)$-vector of terms then $X(t)$ is a formula.
- Let φ be a formula with free fixpoint variable X. φ has free individual variables $x = \langle x_1, \ldots, x_{\mathrm{ar}(X)} \rangle$ for the elements of X, together with other free individual variables z; let $\mathrm{fv}_\varphi(X)$ be the length of z. Now if t is a sequence of $\mathrm{ar}(X)$ terms with free variables y, then $(\mu X(x).\varphi)(t)[z, y]$ is a

formula; **provided that** φ is IF-FOL$^+$. In this context, we write just fv(X) for fv$_\varphi(X)$.

— similarly for $\nu X(\boldsymbol{x}).\varphi$.

To give the semantics of IF-LFP, we first define valuations for free fixpoint variables, in the context of some IF-LFP formula.

Definition 2.5. A fixpoint valuation $\mathscr{V}$ maps each fixpoint variable X to a set $\mathscr{V}(X) \in \wp(\wp(A^{\mathrm{ar}(X)+\mathrm{fv}(X)}))$.

Let D be a non-empty set of deals for $X(\boldsymbol{t})[\boldsymbol{x}, \boldsymbol{z}, \boldsymbol{y}]$, where $\boldsymbol{y}$ are the free variables of $\boldsymbol{t}$ not already among $\boldsymbol{x}, \boldsymbol{z}$. A deal $d = \boldsymbol{acb} \in D$, where $\boldsymbol{a}, \boldsymbol{c}, \boldsymbol{b}$ are the deals for $\boldsymbol{x}, \boldsymbol{z}, \boldsymbol{y}$ respectively, determines a deal $d' = \boldsymbol{t}(d)\boldsymbol{c}$ for $X[\boldsymbol{x}, \boldsymbol{z}]$. Let $D' = \{\, d' \mid d \in D \,\}$. D is a trump for $X(\boldsymbol{t})$ iff $D' \in \mathscr{V}(X)$.

The intuition here is that a fixpoint variable needs to carry the trumps both for the elements of the fixpoint and for any free variables, as we shall see below. Then we define a suitable complete partial order on the range of valuations, which will also be the range of denotations for formulae; it is simply the inclusion order on trump sets.

Definition 2.6. If T_1 and T_2 are elements of $\wp(\wp(A^n))$, define $T_1 \preceq T_2$ iff $T_1 \subseteq T_2$.

This order gives the standard basic lemma for fixpoint logics:

Lemma 2.7. *If $\varphi(X)[\boldsymbol{x}, \boldsymbol{z}]$ is an IF-FOL$^+$ formula and $\mathscr{V}$ is a fixpoint valuation, the map on $\wp(\wp(A^{\mathrm{ar}(X)+\mathrm{fv}(X)}))$ given by*

$$T \mapsto [\![\varphi]\!]_{\mathscr{V}[X:=T]}$$

is monotone with respect to $\preceq$; hence it has least and greatest fixpoints, constructible by iteration from the bottom and top elements of the set of denotations.

Thus we have the familiar definition of the μ operator:

Definition 2.8. $[\![\mu X(x).\varphi(X)[\boldsymbol{x}, \boldsymbol{z}]]\!]$ is the least fixpoint of the map on $\wp(\wp(A^{\mathrm{ar}(X)+\mathrm{fv}(X)}))$ given by

$$T \mapsto [\![\varphi]\!]_{\mathscr{V}[X:=T]};$$

and $[\![\nu X(x).\varphi(x)[\boldsymbol{x}, \boldsymbol{z}]]\!]$ is the greatest fixpoint. $\mu^\zeta X(x).\varphi$ means the ζth approximant of $\mu X(x).\varphi$, defined recursively by $\mu^\zeta X(x).\varphi = \varphi(\bigcup_{\xi < \zeta} \mu^\xi X(x).\varphi)$.

A distinctive feature of the definition, compared to the normal LFP definition, is the way that free variables are explicitly mentioned. Normally, one can fix values for the free variables, and then compute the fixpoint, but because of independent quantification this is not possible in the IF setting. For example, consider the formula fragment

$$\forall z. \ldots \mu X(x). \ldots \vee \exists y/\{z\}. X(y)$$

The independent choice of y means that the trumps for the fixpoint depend on the possible deals for z, not just a single deal.

2.5 Examples of IF-LFP

In order to give some human-readable examples of IF-LFP, we here reproduce a section from [Bra03].

For convenience, we introduce the abbreviation $\varphi \Rightarrow \psi$ for $\psi \vee \neg\varphi$ provided that φ is atomic.

Let $G = (V, E)$ be a directed graph. The usual LFP formula $R(y, z) \overset{\text{def}}{=} (\mu X(x).z = x \vee \exists w.\, E(x, w) \wedge X(w))(y)$ asserts that the vertex z is reachable from y. Hence the formula $\forall y.\forall z.\, R(y, z)$ asserts that G is strongly connected. Now consider the IF-LFP formula

$$\forall y. \forall z. (\mu X(x).z = x \vee \exists w/\{y, z\}.\, E(x, w) \wedge X(w))(y).$$

At first sight, one might think this asserts not only that every z is reachable from every y, but that the path taken is independent of the choice of y and z. This is true exactly if G has a directed Hamiltonian cycle, a much harder property than being strongly connected.

Of course, the formula does not mean this, because the variable w is fresh each time the fixpoint is unfolded. In the trump semantics, the denotation of the fixpoint will include all the possible choice functions at each step, and hence all possible combinations of choice functions. Thus the formula reduces to strong connectivity.

It may be useful to look at the approximants of this formula in a little more detail, to get some intuitions about the trump semantics. Considering just

$$H \overset{\text{def}}{=} (\mu X(x).z = x \vee \exists w/\{y, z\}.\, E(x, w) \wedge X(w))[x, y, z],$$

we see that in computing each approximant, the calculation of $[\![\exists w/\{y, z\}.\, \ldots]\!]$ involves generating a trump for every possible value of a choice function $f: x \mapsto w$. This is a feature of the original trump semantics, and can be understood by viewing it as a second-order semantics: just as the compositional Tarskian semantics of $\exists x.\, \varphi(x)$ involves computing all the witnesses for $\varphi(x)$, so computing the trumps of $\exists x/\{y\}.\, \varphi$ involves computing all the Skolem functions; and unlike the first-order case, it is necessary to work with functions (as IF can express existential second-order logic). Consequently, the nth approximant includes all states such that $x \to f_1(x) \to f_2 f_1(x) \to \ldots \to f_n \ldots f_1(x) = z$ for any sequence of successor-choosing functions f_i. Thus we see that the cumulative effect is the same as for a normal $\exists w$, and the independent choice has indeed not bought us anything.

It is, however, possible to produce a slightly more involved formula expressing the Hamiltonian cycle property in this inductively defined way, by using the standard trick for expressing functions in Henkin quantifier logics. We replace the formula H by

$$\forall s. \exists t/\{y, z\}.\, E(s, t) \wedge (\mu X(x).x = z \vee$$
$$\forall u. \exists v/\{x, y, z, s, t\}.\, (s = u \Rightarrow t = v) \wedge (x = u \Rightarrow X(v)))(y).$$

This works because the actual function f selecting a successor for every node is made outside the fixpoint by $\forall s. \exists t/\{y, z\}.\, E(s, t) \wedge \ldots$; then inside the fixpoint, a new choice function g is made so that $X(g(x))$, and g is constrained to be the same as f by the clause $(s = u \Rightarrow t = v)$. (The reader who is not familiar with the IF/Henkin to existential second-order translation might wish to ponder why $\forall s.\, \exists t / \{y, z\}.\, E(s, t) \wedge \mu X(x).x = z \vee (x = s \Rightarrow X(t))$ does not work.)

3   Second-Order Inductions and Independence-Friendly Logics

It has been known from the early studies of Henkin quantifiers [Wal70, End70] that existential second-order sentences can be transformed into sentences with the Henkin quantifier, and thus into IF-FOL. A technique frequently used in our results is the translation of existential second order inductions into IF-LFP. For this we show that the translation of existential second-order logic into independence-friendly logic can be extended to a translation of positive existential second-order inductions into independence-friendly fixpoint logic. Throughout this paper we only consider finite structures. Therefore we only give the translation for finite structures here.

We first give a formal definition of positive Σ^1_1-inductive formulae.

Definition 3.1. An (n, k)-ary third-order variable $\mathscr{R}$ is a variable interpreted by a set whose members are n-tuples of k-ary functions. Let, for some $k, n < \omega$, $\mathscr{R}$ be a (n, k)-ary third-order variable. A formula $\varphi(\mathscr{R}, f_1, \ldots, f_n)$ is Σ^1_1-*inductive* if it is built up by the usual formula building rules for Σ^1_1 augmented by a rule that allows the use of atoms $\mathscr{R} f_1 \ldots f_n$, where the f_i are k-ary function symbols, provided that the variable $\mathscr{R}$ is only used positively in φ.

Σ^1_1-inductive formulae φ can be used to define least fixpoint inductions in the same way as first-order formulae with a free relation variable in which they are positive are used to define fixpoint inductions. So we can define the stages $\mathscr{R}^\alpha$, $\alpha < \omega$, of the fixpoint induction in φ which ultimately lead to the least fixpoint of the operator defined by the formula φ. We call a relation that is obtained as the least fixpoint of a Σ^1_1-inductive formula Σ^1_1-*inductive*. Note, that the Σ^1_1-inductive relations are third-order objects, i.e. sets of functions.

We show next that any Σ^1_1-inductive third-order relation $\mathscr{R}$ can be defined by an IF-LFP-formula in the sense that there is a formula $\varphi(R, \boldsymbol{x}, y)$, positive in the second-order variable R, such that the *maximal* trumps in the least fixpoint of the operator defined by φ are precisely the graphs of the functions in $\mathscr{R}$. For the sake of simplicity, we only consider the case of $(1, k)$-ary inductions, i.e. where the fixpoint is a set of functions.

An important concept used in the following proofs is the notion of *functional trumps*; and a technically useful concept is that of *maximal trumps*.

Definition 3.2. Let $\varphi(\boldsymbol{x}, y)$ be a formula. A trump T for φ is *functional in $\boldsymbol{x}$ and y*, if for all pairs $(\boldsymbol{a}, b), (\boldsymbol{a}', b')$ of deals in T we have $b = b'$ whenever $\boldsymbol{a} = \boldsymbol{a}'$. T is *maximal* if there is no $T' \supsetneq T$ that is a trump for φ.

Note that because any subset of a trump is a trump, the trumps of a formula are determined by its maximal trumps. Of course, any subset of a functional trump is functional.

Notation. In the following proofs we will frequently use a construction like

$$\forall \boldsymbol{x}/\{\boldsymbol{x}_1, y_1, \ldots, \boldsymbol{x}_n, y_n\}\exists y/\{\boldsymbol{x}_1, y_1, \ldots, \boldsymbol{x}_n, y_n\}\big((\boldsymbol{x} = \boldsymbol{x}_1 \rightarrow y = y_1) \wedge \varphi\big)$$

for some formula φ. We will abbreviate this by

$$\forall \boldsymbol{x}\exists y\,\mathrm{clone}(\boldsymbol{x}_1, y_1; \boldsymbol{x}_2, y_2 \ldots, \boldsymbol{x}_n, y_n)\varphi.$$

and we will usually omit the list $(\boldsymbol{x}_2, y_2 \ldots, \boldsymbol{x}_n, y_n)$ of other variables which appear in the independence sets of the quantifiers, assuming that all other variables than the clones and originals are in that list. Essentially, this formula says that the Skolem functions f_y and f_{y_1} chosen for y and y_1, respectively, are the same. The next lemma makes this precise and establishes some useful properties of the clone construction.

Lemma 3.3. *Let* $\mathfrak{A}$ *be a structure and let* $\boldsymbol{x}$ *be a* k*-tuple of variables.*

(i) *Let* ψ *be a formula defined as* $\psi(\boldsymbol{x}, y) := \forall \boldsymbol{x}'\exists y'\mathrm{clone}(\boldsymbol{x}, y)\psi'$. *Then the trumps for* ψ *are precisely the sets of deals functional in* $\boldsymbol{x}$ *and* y *with some Skolem function* f, *such that the deals* $(\ldots, \boldsymbol{x}, f(\boldsymbol{x}), \boldsymbol{x}', f(\boldsymbol{x}'))$ *form a trump for* ψ'. *In particular, if* ψ' *is* **true**, *then the trumps of* ψ *are just the deals functional in* $\boldsymbol{x}$ *and* y.

(ii) *Let* $\varphi(\boldsymbol{x}', y')$ *be a formula with only functional trumps and let* ψ *be defined as* $\psi(\boldsymbol{x}, y) := \forall \boldsymbol{x}'\exists y'\mathrm{clone}(\boldsymbol{x}, y)\, \varphi$. *Then the trumps for* ψ *and the trumps for* φ *are the same, in the sense that for every trump* $T' \subseteq A^{k+1}$ *of* φ *there is a trump* $T \subseteq A^{k+1}$ *of* ψ *such that an assignment of elements* $\boldsymbol{a}$ *to the variables* $\boldsymbol{x}'$ *and* b *to* y' *is a deal in* T' *if, and only if, the corresponding assignment of* $\boldsymbol{a}$ *to* $\boldsymbol{x}$ *and* b *to* y *is a deal in* T *and, conversely, for every trump* T *of* ψ *there is a corresponding trump* T' *for* φ.

Proof. We first prove Part (i) of the lemma. Following our notation, the formula ψ is an abbreviation for

$$\forall \boldsymbol{x}'/\{\boldsymbol{x}, y\}\exists y'/\{\boldsymbol{x}, y\}(\boldsymbol{x} = \boldsymbol{x}' \rightarrow y = y') \wedge \psi'.$$

Towards a contradiction, suppose there was a non-functional trump T for ψ, i.e. T contains deals $(\boldsymbol{a}, b)$ and $(\boldsymbol{a}, b')$ for some $\boldsymbol{a}$ and $b \neq b'$. By the semantics of universal quantifiers, this implies that there must be a trump for $\exists y_1/\{\boldsymbol{x}, y\}(\boldsymbol{x} = \boldsymbol{x}' \rightarrow y = y') \wedge \psi'$ containing $(\boldsymbol{a}, b, \boldsymbol{a})$ and $(\boldsymbol{a}, b', \boldsymbol{a})$. But then, the set $\{(\boldsymbol{a}, b, \boldsymbol{a}), (\boldsymbol{a}, b', \boldsymbol{a})\}$ is a $\{\boldsymbol{x}, y\}$-set (recall Definition 2.3). Hence, there must be trump T' for $(\boldsymbol{x} = \boldsymbol{x}' \rightarrow y = y') \wedge \psi'$ and an element c so that T' contains the deals $(\boldsymbol{a}, b, \boldsymbol{a}, c)$ and $(\boldsymbol{a}, b', \boldsymbol{a}, c)$. But this is impossible as not both $b = c$ and $b' = c$ can be true but obviously every deal $(\boldsymbol{d}, e, \boldsymbol{d}', e')$ in a trump for $(\boldsymbol{x} = \boldsymbol{x}' \rightarrow y = y')$ satisfies the condition that if $\boldsymbol{d} = \boldsymbol{d}'$ then also $e = e'$. Finally, if T is a functional trump, then the corresponding T' must be a trump for ψ', and so the deals $(\boldsymbol{a}, b, \boldsymbol{a}, b)$ must be a trump for ψ'.

Part (ii) of the lemma follows analogously.   □

The next lemma shows that every formula in Σ_1^1 is equivalent to a formula in IF-LFP. The proof of the lemma follows easily from the work on Henkin-quantifiers. However, some care has to be taken with free occurrences of function variables.

Lemma 3.4. *Let* $\varphi(f_1, \ldots, f_n)$ *be a* Σ_1^1*-formula with free function variables* $f_1, \ldots, f_k$. *Then there is a formula* $\hat{\varphi}(\boldsymbol{x}_{f_1}, y_{f_1}, \ldots, \boldsymbol{x}_{f_k}, y_{f_k}) \in$ IF-FOL *such that for every structure* $\mathfrak{A}$ *a set* T *is a maximal trump for* $\hat{\varphi}$ *if, and only if, there are functions* $F_1, \ldots, F_k$ *such that* $\mathfrak{A} \models \varphi(F_1, \ldots, F_k)$ *and*

$$T = \{(\boldsymbol{a}_1, b_1, \ldots, \boldsymbol{a}_k, b_k) : F_i(\boldsymbol{a}_i) = b_i \text{ for all } 1 \le i \le k\}.$$

We are now ready to prove the main theorem of this section.

Theorem 3.5. *Let* $\mathscr{R}$ *be a* $(1, k)$*-ary third-order variable and let* $\varphi(\mathscr{R}, f)$ *be a* Σ_1^1*-inductive formula where* f *is a* k*-ary function symbol. Then there is a formula* $\hat{\varphi}(R, \boldsymbol{x}, y) \in$ IF-LFP, *where* R *is a* $k + 1$*-ary second-order variable that only occurs positively in* φ *and* $\boldsymbol{x}$ *is a* k*-tuple of variables, such that the least fixpoint* R^∞ *of* φ *satisfies the following properties.*

1. *Every trump* T *in* R^∞ *is functional.*
2. *Every maximal trump encodes the graph of a function in* $\mathscr{R}^\infty$ *and, conversely,*
3. *for every function* $f \in \mathscr{R}^\infty$ *there is a trump* T *in* R^∞ *encoding the graph of* f.

Proof. Let $\varphi(\mathscr{R}, f)$ be as in the statement of the theorem. W.l.o.g. we assume that φ has the form $\varphi(\mathscr{R}, f_0) := \varphi_0(f_0) \vee \exists f_1 \ldots \exists f_n((\bigwedge_{i=1}^n \mathscr{R} f_i) \wedge \varphi_1)$ so that $\mathscr{R}$ does not occur in φ_0 or φ_1. (See [EF99] for a proof of this normal form for existential first-order inductions. The proof for this case is analogous.) The formula φ is translated into a formula $\hat{\varphi}(R, \boldsymbol{x}, y) \in$ IF-LFP defined as follows:

$$\hat{\varphi}(R, \boldsymbol{x}, y) := \forall \boldsymbol{x}_1. \exists y_1. \text{clone}(\boldsymbol{x}, y)(\psi_0(\boldsymbol{x}, y) \vee \psi_1(R, \boldsymbol{x}_1, y_1))$$

where

$$\psi_0(\boldsymbol{x}, y) := \forall \boldsymbol{x}_{f_0} \exists y_{f_0} \, \text{clone}(\boldsymbol{x}, y) \, \hat{\varphi}_0(\boldsymbol{x}_{f_0}, y_{f_0})$$

and

$$\psi_1(R, \boldsymbol{x}_1, y_1) := \forall \boldsymbol{x}_{f_0} \exists y_{f_0} \, \text{clone}(\boldsymbol{x}_1, y_1) \, \psi_1'(\boldsymbol{x}_{f_0}, y_{f_0})$$

and

$$\psi_1'(R, \boldsymbol{x}_{f_0}, y_{f_0}) := \forall \boldsymbol{x}_{f_1} \exists y_{f_1} \ldots \forall \boldsymbol{x}_{f_n} \exists y_{f_n} \bigwedge_{i=1}^n (\forall \boldsymbol{x}' \exists y' \text{clone}(\boldsymbol{x}_{f_i}, y_{f_i}) \, R\boldsymbol{x}' y') \wedge$$
$$\hat{\varphi}_1(\boldsymbol{x}_{f_0}, y_{f_0}, \boldsymbol{x}_{f_1}, y_{f_1}, \ldots, \boldsymbol{x}_{f_n}, y_{f_n}).$$

Here $\hat{\varphi}_0$ and $\hat{\varphi}_1$ are the formulae obtained from φ_0 and φ_1 by applying Lemma 3.4. We claim that the formula $\hat{\varphi}$ satisfies the properties stated in the theorem. Let $\mathfrak{A}$ be a structure with universe A. By Lemma 3.3(i), the trumps T for are functional in $\boldsymbol{x}$ and y, with Skolem function g such that g satisfies ψ_0 or ψ_1.

The theorem now follows by showing via an induction on the ordinals α that every maximal trump in R^α is the graph of a function in $\mathscr{R}^\alpha$ and, conversely, the graph of every function in $\mathscr{R}^\alpha$ is a trump in R^α. $\qquad\square$

4 Independence-Friendly vs. Partial Fixpoint Logic

By definition, independence-friendly fixpoint logic is a least fixpoint logic. However, contrary to the fixpoint logics usually considered in finite model theory, here the fixpoints are not sets of elements but sets of trumps and therefore essentially third-order objects. In particular, it is no longer guaranteed that any fixpoint induction closes in polynomially many steps in the size of the structure – to the contrary, it may take an exponential number of steps to close. We will see below, that this greatly increases the expressive power of IF-LFP compared to normal least fixpoint logics.

As a first step in this direction we relate independence-friendly fixpoint logic to partial fixpoint logic. Partial fixpoint logic is an important logic in finite model theory. Syntactically, PFP is defined as the extension of first-order logic by formulae $\psi := [\mathbf{pfp}_{R,\boldsymbol{x}} \, \varphi](\boldsymbol{t})$, where R is a second-order variable of arity k, $\boldsymbol{x}$ a k-tuple of variables, $\boldsymbol{t}$ a k-tuple of terms, and φ itself an arbitrary PFP-formula. In particular, R may occur positive and negative in φ. On any finite structure $\mathfrak{A}$ with universe A the formula φ defines a sequence R^α, $\alpha < \omega$, of sets defined as $R^0 := \varnothing$ and $R^{\alpha+1} := \{\boldsymbol{a} : (\mathfrak{A}, R^\alpha) \models \varphi(\boldsymbol{a})\}$. As there are no restrictions on φ, this sequence need not reach a fixpoint. In this case, ψ is equivalent on $\mathfrak{A}$ to false. Otherwise, if the sequence becomes stationary and reaches a fixpoint R^∞, then for any tuple $\boldsymbol{a} \in A^k$, $\mathfrak{A} \models [\mathbf{pfp}_{R,\boldsymbol{x}} \, \varphi](\boldsymbol{a})$ if, and only if, $\boldsymbol{a} \in R^\infty$.

Among the various fixpoint logics commonly considered in finite model theory, PFP is the most expressive subsuming logics such as LFP and IFP and, on ordered structures, even second-order logic SO.

A central issue in finite model theory is to relate the expressive power of logics to the computational complexity of classes of structures definable in the logic. Of particular interest are so-called *capturing results*: A logic $\mathscr{L}$ captures a complexity class $\mathfrak{C}$ if every class of finite structures definable in $\mathscr{L}$ can be decided in $\mathfrak{C}$ and conversely, for every class $\mathscr{C}$ of finite structures which can be decided in $\mathfrak{C}$ there is a sentence $\varphi \in \mathscr{L}$ such that for all structures $\mathfrak{A}$, $\mathfrak{A} \models \varphi$ if, and only if, $\mathfrak{A} \in \mathscr{C}$.

Capturing results are important as they provide logical characterisations of complexity classes, i.e. characterisations independent of machine models and time or space bounds. In particular, non-expressibility results on the logic transfer directly into non-definability results on the complexity class. As such results are notoriously hard to come by, capturing results provide an interesting alternative for proving non-definability of problems in a complexity class.

Much effort has been spent on capturing results and for all major complexity classes such results have been found (see [EF99] for a summary). However, in many cases it could only be shown that a logic captures a complexity class on the class of ordered structures. As for PFP, it has been shown by Abiteboul and Vianu [AV89], that PFP captures PSPACE on the class of finite ordered structures.

As every class of structures definable in second-order logic is decidable in the polynomial time hierarchy, it follows immediately that PFP contains SO on ordered structures. One feature that makes PFP so expressive is its ability to define fixpoint inductions of exponential length in the size of the structure. We show next that every formula of PFP is equivalent to one in IF-LFP. For this we show that every partial fixed-point induction can be translated into a Σ_1^1-inductive definition which, by Theorem 3.5, is equivalent to a formula in IF-LFP. Due to space restrictions we refrain from giving the full proof here and refer to the full version of the paper.

Theorem 4.1. *For every formula* $\varphi \in$ PFP *there is an equivalent formula* $\psi \in$ IF-LFP.

We have already mentioned that pure independence-friendly logic is equivalent to Σ_1^1 where we can use an existential second-order quantifier to state the existence of a linear order on the universe of a structure even on classes of otherwise unordered structures. Thus the theorem above implies that IF-LFP contains SO on all rather than just ordered structures.

Corollary 4.2. *On finite structures, every formula of* SO *is equivalent to a formula in* IF-LFP.

In the next section we will derive some further corollaries of this theorem concerning the model-checking complexity of IF-LFP.

5   Complexity of Independence-Friendly Fixpoint Logic

In this section we analyse the complexity of IF-LFP on finite structures, both with respect to data and model-checking complexity. By data-complexity we understand the complexity of deciding for a fixed formula $\varphi \in$ IF-LFP and a given structure $\mathfrak{A}$ whether $\mathfrak{A} \models \varphi$. In particular, the input only consists of the structure $\mathfrak{A}$. By model-checking we mean the problem of deciding for a given finite structure $\mathfrak{A}$ and formula $\varphi \in$ IF-LFP whether $\mathfrak{A} \models \varphi$. Here, both φ and $\mathfrak{A}$ are part of input.

We begin our analysis with data-complexity. In [Bra03], the first author already noticed that any given formula of IF-LFP can be evaluated in time exponential in the size of the structure. For, every fixpoint $\mu R(\boldsymbol{x}).\varphi$ can be evaluated in time linear in the number of trumps for φ and therefore exponential in the size of the structure.

Proposition 5.1. IF-LFP *has exponential time data-complexity.*

We aim at a much stronger result. Not only will we show that IF-LFP is EXPTIME-complete with respect to data-complexity but we will prove that it actually captures EXPTIME, i.e. every class of structures decidable by an exponential time Turing-machine can be defined in IF-LFP and vice versa every class of structures definable in IF-LFP can be decided in deterministic exponential time. Again we refrain from giving the full proof here and refer to the full paper.

Theorem 5.2. IF-LFP *captures* EXPTIME.

Clearly, if a logic $\mathscr{L}$ captures a complexity class $\mathfrak{C}$, then the evaluation problem of $\mathscr{L}$ must be $\mathfrak{C}$-complete with respect to data complexity. Thus we get the following simple corollary.

Corollary 5.3. *There exist formulae in* IF-LFP *with* ExpTime-*complete data-complexity.*

We continue our complexity analysis of IF-LFP with the study of its model-checking complexity. For an upper bound, it is easily seen that for any given structure $\mathfrak{A}$ and formula φ the formula can be evaluated in $\mathfrak{A}$ using space doubly exponential in $|\varphi|$ and exponential in $|\mathfrak{A}|$. For, every evaluation of a fixpoint only needs enough space to store all possible trumps, and the number of trumps is bounded by $O(2^{\mathfrak{A}^{|\varphi|}})$.

Theorem 5.4. *Every formula* $\varphi \in$ IF-LFP *can be evaluated in a structure* $\mathfrak{A}$ *in space doubly exponential in* $|\varphi|$ *and exponential in* $|\mathfrak{A}|$.

The theorem gives an upper bound on the model checking complexity of IF-LFP. We have seen in Section 4 above that every formula of PFP is equivalent to one of IF-LFP. Further, the translation is polynomial in the size of the PFP-formula. Consequently, model-checking for IF-LFP is at least as complex as it is for PFP. As model-checking for PFP is known to be hard for exponential space – in fact even complete for exponential space – we get the following theorem.

Theorem 5.5. *The model-checking problem for* IF-LFP *is hard for exponential space.*

6   Conclusion

In this paper we studied the computational complexity of various problems related to IF-LFP. As we have seen, adding independence to least fixpoint logic increases the expressive power and complexity significantly. Another indicator for this is the translation of formulae of PFP to formulae of IF-LFP. This showed that IF-LFP is even more expressive than second-order logic – unless, of course, PSpace = ExpTime.

Looking at the various proofs given for the results, it becomes clear that the common technique used in all proofs was to use independent quantification to define functions and then show that these functions can be passed through the fixpoint induction. This suggests that there might be a more general relation between independence-friendly logic and second-order logic, namely that the two logics are actually equivalent. Showing this, however, requires a careful analysis of the role of negation in independence friendly logics and is far from obvious. This is part of ongoing work.

Acknowledgements

Part of this work was done while the second author was a postdoctoral fellow in Edinburgh supported by the EU Research and Training Network GAMES (Games and Automata for Synthesis and Validation).

The authors thank anonymous referees for comments which have improved the paper.

References

[AV89]   S. Abiteboul and V. Vianu, Fixpoint extensions of first-order logic and Datalog-like languages. *Proc. 4th IEEE Symp. on Logic in Computer Science (LICS)*, 71–79 (1989).

[Bra99] J. C. Bradfield, Fixpoints in arithmetic, transition systems and trees. *Theoretical Informatics and Applications*, **33** 341–356 (1999).

[Bra00] J. C. Bradfield, Independence: logics and concurrency, *Proc. CSL 2000*, LNCS **1862** 247–261 (2000).

[Bra03] J. C. Bradfield, Parity of imperfection, *Proc. CSL 2003*, LNCS **2803** 72–85 (2003).

[BrF02] J. C. Bradfield and S. B. Fröschle, Independence-friendly modal logic and true concurrency, *Nordic J. Computing* **9** 102–117 (2002).

[EF99]   H.-D. Ebbinghaus and J. Flum, Finite Model Theory, 2nd edition, Springer, 1999.

[End70] H. B. Enderton, Finite partially ordered quantifiers, *Z. für Math. Logik u. Grundl. Math.* **16** 393–397 (1970).

[Gra03] E. Grädel, Finite Model Theory and Descriptive Complexity, in *Finite Model Theory and Its Applications*, Springer, 2005. See http://www-mgi.informatik.rwth-aachen.de/Publications/pub/graedel/Gr-FMTbook.ps

[HiS96] J. Hintikka and G. Sandu, A revolution in logic?, *Nordic J. Philos. Logic* **1**(2) 169–183 (1996).

[Hod97] W. Hodges, Compositional semantics for a language of imperfect information, *Int. J. IGPL* **5**(4), 539–563.

[Pie00] A. Pietarinen, Games logic plays. Informational independence in game-theoretic semantics. D.Phil. thesis, Univ Sussex (2000).

[Wal70] W. J. Walkoe, Jr, Finite partially-ordered quantification. *J. Symbolic Logic* **35** 535–555 (1970).

Closure Properties of Weak Systems
of Bounded Arithmetic

Antonina Kolokolova

Simon Fraser University & Mathematical Institute, Prague
kol@cs.toronto.edu

Abstract. In this paper we study the properties of systems of bounded arithmetic capturing small complexity classes and state conditions sufficient for such systems to capture the corresponding complexity class tightly. Our class of systems of bounded arithmetic is the class of second-order systems with comprehension axiom for a syntactically restricted class of formulas $\Phi \subset \Sigma_1^B$ based on a logic in the descriptive complexity setting. This work generalizes the results of [8] and [9][1].

We show that if the system 1) extends V_0 (second-order version of $I\Delta_0$), 2) Δ_1-defines all functions with bitgraphs from Φ, and 3) proves witnessing for all theorems from Φ, then the class of Σ_1^B-definable functions of the resulting system is exactly the class expressed by Φ in the descriptive complexity setting, provably in this system.

1 Introduction

There has been a lot of research in descriptive complexity and bounded arithmetic, as well as their connections with complexity theory. However the question of direct relationship between these two fields did not receive much attention. The language of bounded arithmetic is richer than that of many logics, but often logics capture complexity classes over languages that include some arithmetic predicates (order, plus and times, or, equivalently, BIT predicate).

Bounded arithmetic studies the complexity of proving properties of these classes of formulas, whereas descriptive complexity is concerned with their expressive power. The most important distinction between different systems of bounded arithmetic is the strength of their induction (or comprehension) axiom schemes. This leads to the following question: how does the expressive power of the class of formulas in the induction axioms of a system relate to the power of the resulting system? In which cases the formulas in the comprehension are more complex than the provably total functions of a system and under which conditions their complexity coincides?

In this paper, we discuss properties under which the complexity of formulas in comprehension axioms and of provably total functions of a system of arithmetic is the same. Our approach is geared towards feasible complexity classes, those

---

[1] More detailed presentation of most of this work can be found in my PhD thesis, [17], available on ECCC.

L. Ong (Ed.): CSL 2005, LNCS 3634, pp. 369–383, 2005.

between P and DLOGTIME (uniform AC⁰). Restricting our attention to small classes allows us to use definability by NP predicates (bounded Σ_1) for the definition of capture in the bounded arithmetic setting: we consider exactly the functions with bitgraphs represented by NP predicates that are provably total in our systems. By Fagin's theorem [12], NP predicates are representable by second-order existential formulas, so the formula classes we consider here are subsets of second-order existential formulas.

Traditionally, functions are introduced by their recursion-theoretic characterization (see [4] for the original such result or [26]), but since we are trying to relate the expressive power of the formulas in comprehension and complexity of functions, we introduce function symbols by setting their bitgraphs to be formulas from the comprehension scheme.

Let C be a complexity class. Suppose that Φ_C is a class of (existential second-order) formulas that captures C in the descriptive complexity setting. We define a theory of bounded arithmetic $V\text{-}\Phi_C$ to be Robinson's Q together with comprehension over bounded Φ_C. The following is an informal statement of our main result:

Claim: *Let* AC⁰ $\subseteq C \subseteq$ P. *Suppose that* Φ_C *is closed under first-order operations provably in* $V\text{-}\Phi_C$ *(1). Also, suppose that for every* $\phi(\bar{x}, \bar{Y}) \in \Phi_C$, *if* $V\text{-}\Phi_C \vdash \phi$ *then there is a function* F *on free variables of* ϕ *which is computable in* C *and witnesses existential quantifiers of* ϕ *(2). Then the class of provably total functions of* $V\text{-}\Phi_C$ *is the class of functions computable in* C.

It may seems that the second condition, that is witnessing for the Φ_C theorems, is almost a restatement of the result itself. However, the class Φ_C can be very small, with definition of one complete problem for the class (for example transitive closure). Then the second condition states that if this small set of theorems can be witnessed, then all functions from that complexity class are provably total in the system.

For conventional systems of bounded arithmetic, such as ones considered by Clote and Takeuti in [3], it was shown that the class of provably total functions of a system coincides with the function class in the complexity-theoretic sense. Under our conditions this is provable within the system itself, so more work is needed to prove the conditions, but the result is stronger. We hope that our framework can be useful for proving independence results for weak theories of arithmetic.

Examples of systems that provably capture complexity classes are V_1-Horn capturing P from [7, 8], V-Krom capturing NL from [9] and V^0 capturing AC⁰ from [6]. As an example of a similar system that captures a complexity class, but not (known to be) provably, we present a system of arithmetic V-SymKrom corresponding to symmetric logspace (SL), based on symmetric second-order 2-CNF formulas (with $\oplus$ instead of $\vee$ between literals). This system can prove that its class of provably total functions is the AC⁰ closure of SL functions. By the recent Reingold's result [22], SL = L and so symmetric 2-SAT is solvable in logspace; therefore, AC⁰(SL) = SL = L. However, this proof, and even the proof that SL is closed under complementation by Nisan and Ta-Shma [20], rely on algebraic properties on expander graphs. In their current form, these proofs are not formalizable using SL-reasoning: to talk about algebra, we need at least polynomial time. It is a very interesting open question whether there is a combinatorial version of Reingold's proof that is formalizable in a system for L, and whether our theory for SL is fully conservative over a system for L.

2   Descriptive Complexity Framework

The name "descriptive complexity" refers to the study of expressive power of logics: fixing a formula, we look at the complexity of evaluating this formula on different finite structures. It is more common to call this area "finite model theory"; however, here we stay with the term "descriptive complexity" to emphasize the complexity theory connection and the richness of the assumed vocabulary. Please see [11], [16], and [18] for the background.

Following [16], we consider logics over the vocabulary $\tau = \{\min, \max, +, \times, \leq \}$ (we do not include BIT operator since it can be defined from $+, \times$ in the weakest of our systems; see [6] for details). For many results it is sufficient to assume only the presence of order and successor relations in the vocabulary (these are the assumptions of [13, 14]); however it is more convenient to work with a vocabulary containing all basic arithmetic operations. We refer to structures where the arithmetic symbols of the vocabulary get the standard interpretation as "arithmetic structures". The way we connect logics with complexity classes is stated in this definition (following [18]):

Definition 1 (Capture by a logic). *Let C be a complexity class, L a logic and K a class of finite structures. Then L captures C on K if*

1. *For every L-sentence ϕ and every $\mathcal{A} \in K$, testing if $\mathcal{A} \models \phi$ with ϕ fixed and an encoding of $\mathcal{A}$ as an input can be done in C.*
2. *For every collection K' of structures closed under isomorphism, if this collection is decidable in C then there is a sentence $\phi_{K'}$ of L such that $\mathcal{A} \models \phi_{K'}$ iff $\mathcal{A} \in K'$, for every $\mathcal{A} \in K$.*

For our purposes, we fix K to be the arithmetic structures. In particular, the universe of a structure is always considered to be $\{0, \ldots, n-1\}$.

Many capture results are obtained by extending first-order logic with additional operators, such as fixed-point operators. We find it more convenient to work with restrictions of second-order logics rather than extensions of first-order. However, in many cases we can switch to the extended first-order logic framework by adding a defining axiom for a new operator, where the defining axiom is a second-order formula. We use this for theories of non-deterministic logspace and symmetric logspace (NL and SL), in order to introduce respective transitive closure operators.

Definition 2. *We will use the term* restricted $SO\exists$ *to refer to formulas of the form*

$$\exists P_1 \ldots P_k \forall x_1 \ldots x_l \psi(\bar{P}, \bar{x}, \bar{a}, \bar{Y}), \tag{1}$$

where k, l are constants, and ψ is a (sub)class of CNF closed under conjunction. Here, when defining a subclass of CNF we treat only the quantified second-order variables $\bar{P}$ as literals.

Note that there are no occurrences of existential first-order quantifiers in restricted $SO\exists$ formulas. This is because even when the class of ψ is restricted to 2CNF with at most one occurrence of a positive literal, with presence of an existential quantifier it is possible to capture all of $SO\exists$ [13, 14]. Universal first-order and quantifier-free formulas are restricted $SO\exists$.

Schaefer's theorem ([23]) presents several restrictions on CNF that correspond to different complexity classes. Grädel in [13, 14] described how to use some of them to capture complexity classes by restricted second-order formulas. Here we use systems based on the following restrictions of ψ:

Definition 3. *A formula $\psi(\bar{x}, \bar{P}, \bar{a}, \bar{Y})$ is* Horn *with respect to the second-order variables $P_1, ..., P_k$ if ψ is quantifier-free in conjunctive normal form and in every clause there is at most one positive literal of the form $P_i(\bar{x})$. It is* Krom *with respect to $\bar{P}$ if ψ is a CNF with at most two occurrences of a P-literal per clause. It is* SymKrom *if it is Krom with $\oplus$ instead of $\vee$ in every clause (so every clause is of the form $(\phi_i \rightarrow L_i \oplus L'_i)$, where the only P-literals are L_i and L'_i).*

Following Grädel, we can define classes $SO\exists$ Horn and $SO\exists$ Krom and $SO\exists$ SymKrom as restricted $SO\exists$, in which ψ is, respectively, Horn, Krom and SymKrom with respect to $\bar{P}$.

The following descriptive complexity characterizations provide classes of formulas on which our systems can be based. However, not all of them result in systems tightly capturing the corresponding complexity class.

Over arithmetic structures,

- First-order logic captures uniform AC^0 ([1, 15]).
- Second-order existential logic captures NP ([12]), and in general levels of SO hierarchy correspond to levels of PH ([24]).
- Second-order Horn, Krom and SymKrom capture P, NL and SL, respectively ([13, 14]).

In case of restricted second-order formulas, the formula evaluation direction of the capture proof consists of the following steps. First, the formula is brought into propositional form by making a copy of its quantifier-free part for every possible tuple of values of quantified first-order variables. Then first-order terms and free second-order terms are evaluated. Second-order terms of the form $P_i(t(\bar{x}))$, where P_i is quantified and $t(\bar{x})$ is a term, are assigned propositional variables so that $P_i(t(\bar{x}))$ and $P_i(t'(\bar{x}))$ are assigned to the same variable whenever $t(\bar{x})$ evaluates to the same value as $t'(\bar{x})$, on possibly different tuples $\bar{x}$. Now the problem is reduced to testing satisfiability of the resulting propositional formula.

3   Bounded Arithmetic Framework

In descriptive complexity, a language in the traditional complexity theory setting is thought of as interpretations of a unary predicate X (viewed as a binary string) in a set of structures. A class of recursively enumerable languages then naturally corresponds to a class of formulas: each language in the class corresponds to a formula which has, as its set of models, the structures with X interpreted as strings from the language. In the bounded arithmetic setting, the relationship with complexity classes is slightly different. Here, we consider representations of languages in the standard model of arithmetic $\mathbb{N}_2$ (two-sorted $\mathbb{N}$). So instead of a set of structures with one predicate getting different interpretation we are talking about one fixed structure and different (second-order) elements of it satisfying the formula.

Definition 4 (Representation). *A formula $A(X)$ represents a language L if $L = \{w(S)|\mathbb{N}_2 \models A(S)\}$, where w is some encoding of strings. More generally, $A(\bar{x}, \bar{Y})$ represents a relation $R(\bar{x}, \bar{Y})$ which holds on $\bar{x}, \bar{Y}$ iff $\mathbb{N}_2 \models A(\bar{x}, \bar{Y})$. A class of formulas Φ represents a complexity class $\mathbf{C}$ iff every relation R from $\mathbf{C}$ is representable by a formula from Φ, and every formula from Φ can be evaluated within $\mathbf{C}$.*

This notion is parallel to the notion of "capture" from descriptive complexity (see definition 1); essentially, they have the same meaning of describing the expressive power of formulas. But the notion of "capture" we will be using for systems of bounded arithmetic will be quite different.

The language of our systems of arithmetic is $\mathcal{L}_A^2 = \{0, 1, +, \cdot, | \ |; <, =, \in\}$, a natural second-order extension of the language of Peano Arithmetic $\mathcal{L}_A = \{0, 1, +, \cdot; <, =\}$. Let $\mathbb{N}_2$ be a standard structure with natural numbers and finite sets of natural numbers in the universe; our first-order objects (denoted by lower-case letters) are natural numbers; second-order objects (denoted by upper-case letters) are binary strings or, equivalently, (finite) sets of numbers. Treating a second-order variable X as a set, its upper bound ("length") $|X|$ is defined to be the largest element $y \in X$ plus one, or 0 if X is an empty set.

Arithmetic terms are constructed using $+$ and $\times$ from first-order variables, constants 0 and 1, and terms of the form $|X|$ where X is a second-order variable. The atomic formulas of $\mathcal{L}_A^2$ have one of the forms $s = t, s \leq t, t \in X$, where s and t are terms and X is a second-order variable. We usually write $X(t)$ instead of $t \in X$. Formulas are built from atomic formulas using the propositional connectives $\wedge, \vee, \neg$, the first-order quantifiers $\forall x, \exists x$ and the second-order quantifiers $\forall X, \exists X$.

Bounded first-order quantifiers get their usual meaning: $\forall x \leq t\phi$ stands for $\forall x(x \leq t \rightarrow \phi)$ and $\exists x \leq t\phi$ stands for $\exists x(x \leq t \wedge \phi)$. Second-order quantified variables are strings of bounded length; the notation $\exists Z \leq b$ corresponds to $\exists Z \ |Z| \leq b$.

Definition 5. *Σ_0^B and Π_0^B both denote the class of bounded formulas with no second-order quantifiers. We define inductively Σ_{i+1}^B as the least class of formulas containing Π_i^B and closed under disjunction, conjunction, and bounded existential second-order quantification. The class Π_{i+1}^B is defined dually. We use notation $\Sigma_0^B(\Phi)$ to refer to the closure of Φ under first-order operations: that is, under $\vee, \wedge, \neg$ and bounded first-order $\forall$ and $\exists$.*

3.1   Translation

Let Φ be a descriptive logic over a vocabulary τ. For every $\phi \in \Phi$, we can define a translation ϕ^* into $\mathcal{L}^2_A$ with the following properties:

1. Every interpreted symbol from τ that occurs in $\mathcal{L}^2_A$ gets the standard interpretation, e.g., successor becomes $+1$, min becomes 0, etc.
2. Translate max as n for a free variable n. For every quantified first-order variable, set $n+1$ (more generally, a polynomial of n) as a bound. Note that then $|X| = n + 1$ for a unary second-order predicate.
3. Translate uninterpreted relational symbols of τ occurring in ϕ as free second-order variables of ϕ^*. If a variable is k-ary, use a pairing function to encode the relational symbol as a unary second-order variable. Then any occurrence of $R(x_1, \ldots, x_k)$ becomes $R^*(\langle x_1, \ldots, x_k \rangle)$, where $\langle x_1, \ldots, x_k \rangle$ is a value obtained by applying the pairing function to $x_1, \ldots, x_k$.

Under this translation, a restricted second-order formula becomes a restricted Σ^B_1 formula with the same restriction on the quantifier-free part. The resulting Φ^* represents in the standard model the same complexity class as is captured by Φ in the descriptive complexity setting.

Table 1. The 2-BASIC axioms

B1: $x + 1 \neq 0$	B2: $x + 1 = y + 1 \rightarrow x = y$	B4: $x + (y + 1) = (x + y) + 1$
B3: $x + 0 = x$	B5: $x \cdot 0 = 0$	B6: $x \cdot (y + 1) = (x \cdot y) + x$
B7: $0 \leq x$	B9: $x \leq y \wedge y \leq z \rightarrow x \leq z$	B10: $(x \leq y \wedge y \leq x) \rightarrow x = y$
B8: $x \leq x + y$	B11: $x \leq y \vee y \leq x$	B12: $x \leq y \leftrightarrow x < y + 1$
L1: $X(y) \rightarrow y < |X|$	L2: $y + 1 = |X| \rightarrow X(y)$	B13: $x \neq 0 \rightarrow \exists y (y + 1 = x)$

3.2   Systems of Bounded Arithmetic

Now, for a set of formulas Φ, a system V-Φ is axiomatized by 2-BASIC axioms listed in table above together with a comprehension scheme of the form

$$\exists Z \leq b \forall i < b(Z(i) \leftrightarrow \phi(i, \bar{a}, \bar{X})), \qquad (\Phi\text{-comp})$$

where $\phi \in \Phi$.

To agree with the common notation, we abbreviate V-Σ^B_i as V^i, $i \geq 0$. These theories are axiomatized by the 2-BASIC together with a comprehension scheme for Σ^B_i formulas. For $i \geq 1$, V^i is equivalent to the first-order theory S^i_2 by RSUV isomorphism [21, 25]. The system V^0 corresponds to the complexity class uniform AC^0.

4   Definability in V-Φ

4.1   Basic Properties of V^0 and V-Φ

The system V^0 is robust enough to prove many natural properties. In particular, induction on the length of string (and thus on Σ^B_0 combinations of Φ) is a theorem of V-Φ extending V^0. Also, V^0 proves properties of the pairing function and simultaneous comprehension over several variables, resulting in an array (so several existential second-order quantifiers can be treated as one). We use $P^{[b]}$ to denote the "b-th row" when P is being used as a 2-dimensional array. If $\phi(P)$ is a formula with no occurrence of $|P|$, then $\phi(P^{[b]})$ is obtained from $\phi(P)$ by replacing every atomic formula $P(t)$ by $P(b, t)$.

The following property, Replacement, plays a major role in our definability proofs. It is a theorem for V^1 and stronger theories, however weaker theories do not prove full Σ_1^B replacement under cryptographic assumptions by [10]. For our purpose it is sufficient to prove it for restricted Σ_1^B formulas.

Lemma 1 (Replacement). *Let Φ be a class of restricted Σ_1^B formulas. Then for every formula $\exists \bar{P} \phi(y, \bar{P}) \in \Phi$, where ϕ can have additional free variables, V-Φ proves*

$$\forall y < t \exists \bar{P} \phi(y, \bar{P}) \leftrightarrow \exists \bar{P} \forall y < t \phi(y, \bar{P}^{[y]}) \qquad \text{(Replacement)}$$

where $\bar{P}^{[y]}$ is $P_1^{[y]}, ..., P_k^{[y]}$.

Proof. The proof is a generalization of a proof of Replacement in [8]. Here we are using the lack of existential first-order quantifiers and closure under conjunctions of the quantifier-free parts of Φ-formulas.

4.2   Function Classes

Complexity classes are defined as classes of relations. This is also the interpretation for the descriptive complexity setting. But in bounded arithmetic the measure of the power of a theory is the complexity of the corresponding functions. So we use relations as graphs to define number functions and as bit graphs to define string functions. The following definition is very general, but sometimes does not produce a robust function class: for example, there is nothing in this definition that would force the functions to be closed under composition. In order to make the function classes defined this way meaningful, we will need additional restrictions.

Definition 6. *Let C be a complexity class. We define the corresponding class FC of functions of C as follows: A string function $F : \mathbb{N}^k \times (\{0,1\}^*)^l \to \{0,1\}^*$ is in FC iff there is a relation R in C and a polynomial p such that $F(\bar{x}, \bar{Y})(i) \leftrightarrow i < p(\bar{x}, |\bar{Y}|) \wedge R(i, \bar{x}, \bar{Y})$ for all $i \in \mathbb{N}$. A number function $f(\bar{x}, \bar{Y})$ is in the class FC if there is a string function in $F(\bar{x}, \bar{Y}) \in FC$ such that $f(\bar{x}, \bar{Y}) = |F(\bar{x}, \bar{Y})|$. If formula class Φ represents C, then R can be replaced by a formula $\phi \in \Phi$ representing R.*

For string functions, we are only concerned with the bits with indices smaller than $p(\bar{x}, \bar{Y})$. Therefore, a string corresponding to the value of a function will be of length less than $p(\bar{x}, \bar{Y})$. In particular, by the length axioms, all bits with indices larger than $p(\bar{x}, \bar{Y})$ are 0.

This definition of FC does not directly impose any "robustness" conditions such as closure under function composition. To allow for that, we define an AC^0 closure of FC as follows.

Definition 7. *A (string) function $F(\bar{x}, \bar{Y})$ is AC^0 reducible to a set of function symbols $\mathcal{L}$ (denoted $F \in AC^0(\mathcal{L})$) iff there is a sequence $F_1 \ldots F_n$ of string functions such that $F_n = F$ and F_i is in $\Sigma_0^B(\mathcal{L} \cup \{F_1 \ldots F_{i-1}\})$ for $i = 1, \ldots, n$. If for any $F \in AC^0(\mathcal{L})$, $F \in \mathcal{L}$ we say that $\mathcal{L}$ is closed under AC^0 reductions.*

In case FC is definable by formulas from Φ, the definition naturally generalizes to $AC^0(\Phi)$.

Definition 8. *A relation $R(\bar{x}, \bar{Y})$ is Δ_1^B-definable in V-Φ iff there exist formulas $\phi, \tilde{\phi} \in \Sigma_1^B$ such that $R(\bar{x}, \bar{Y})$ is represented by $\phi(\bar{x}, \bar{Y})$ and V-$\Phi \vdash \phi(\bar{x}, \bar{Y}) \leftrightarrow \neg\tilde{\phi}(\bar{x}, \bar{Y})$. A string function F is Σ_1^B-definable in V-Φ if it has a defining axiom $Z = F(\bar{x}, \bar{Y}) \leftrightarrow \phi(Z, \bar{x}, \bar{Y}))$, with $\phi \in \Sigma_1^B$ such that V-$\Phi \vdash \forall\bar{x}\forall\bar{Y}\exists!Z\phi(Z, \bar{x}, \bar{Y}))$.*

By the second-order version of Parikh's theorem (see [6]), we can use Σ_1^B-definability and Σ_1-definability interchangeably. Also, Δ_1^B-definable relations and Σ_1^B-definable boolean functions are the same (consider characteristic functions of predicates).

Using definition 8, we can state the definition of "capture" in the bounded arithmetic setting. This gives us a way of measuring the power of a system of arithmetic.

Definition 9 (Capture in bounded arithmetic). *A system of arithmetic T captures a complexity class C if the class of Σ_1^B-definable functions of T is exactly FC. That is, FC is the class of functions representable by Σ_1^B formulas that are provably total in T.*

Note that this is quite different from the descriptive complexity notion of "capture" from definition 1: descriptive complexity "captures" is bounded arithmetic "representable". The reason we are using the same word is that in both cases we are relating a logic (system of arithmetic) and a complexity class; "capture" here is a generic name for such a connection.

4.3   Properties

The first property that we consider is (provable) closure under AC^0 reductions. We emphasize the provability part here: in the previous work, e.g., by Clote and Takeuti [2], the fact that the classes in question were closed under complementation was used but not proven within the system.

*Property 1 (**Closure**).* Let Φ represent a complexity class C and let FC be as in definition 6. Then the *closure property* holds if Φ is closed under AC^0 reductions. In particular, FC is closed under composition and substitution of a term for a variable. In addition, Φ is *strongly closed* if for every $\phi^* \in \Sigma_0^B(\Phi)$ there exists $\phi \in \Phi$ such that V-$\Phi \vdash \phi^* \leftrightarrow \phi$.

If this property holds, the corresponding C must be closed under complementation and Φ extends Σ_0^B (that is, defines all of first-order). For some Φ, notably restricted Σ_1^B, it is not syntactically true that $\Sigma_0^B \subseteq \Phi$, but it can be proved that for any Σ_0^B formula there is an equivalent formula of Φ.

In order for a logic to translate into a "nice" system of arithmetic, the logic has to be in some sense "natural". That is, its properties such as closure under composition and complementation have to be provable using only simple concepts. Moreover, it should be easy to verify whether a given formula holds on a structure. More formally, we need the following property:

Property 2 (**Constructiveness**). Let Φ be a class of restricted Σ_1^B formulas, and let Φ represent C. This Φ has the *constructiveness property* if the following two conditions hold. Firstly, every $\phi \in \Phi$ defines a relation R that is Δ_1^B-definable in V-Φ, with ϕ being its Σ_1^B definition. Secondly, there are witnessing functions $\bar{F}$ with bit graphs in $\Sigma_0^B(\Phi)$ such that $\bar{F}(\bar{a}, \bar{Y})$ witness the existential quantifiers of the prenex form of $\phi \vee \tilde{\phi}$.

If, additionally, Φ is strongly closed, that is, has property 1, then the conclusion of the constructiveness property can be stated simpler as follows.

Property 2' (**Strong constructiveness**) For every $\phi \equiv \exists \bar{P} \psi(\bar{P}, \bar{a}, \bar{Y}) \in \Phi$ such that V-$\Phi \vdash \phi$ there are functions $\bar{F}$ witnessing $\bar{P}$ such that bitgraphs of $\bar{F}$ are in Φ.

It is enough to consider ϕ-theorems of V-Φ because if Φ is closed, then $\tilde{\phi} \in \Phi$ and so is $\phi \vee \tilde{\phi}$. Also, the assumption that bitgraphs of $\bar{F}$ are in $\Sigma_0^B(\Phi)$ becomes bitgraphs $\in \Phi$.

Sometimes we use the term "weak constructiveness" to refer to the original constructiveness property, and "strong constructiveness" for the second version.

4.4   Main Results

Now we are ready to state the main theorem of this paper.

Theorem 1 (Definability theorem). *Suppose that Φ is restricted Σ_1^B or Σ_0^B, constructive, and represents a complexity class C. Then all functions from FC are Σ_1^B-definable in V-Φ and all Σ_1^B-definable functions of V-Φ are in $\mathrm{AC}^0(FC)$.*

Suppose, additionally, that Φ is strongly closed. In this case, the class of Σ_1^B-definable functions of V-Φ coincides with FC provably in V-Φ.

We will refer to the first statement as "weak definability" and the second statement as "strong definability".

The proof of this theorem consists of two parts. The part that FC is Σ_1^B-definable in V-Φ follows by the fact that we have comprehension for $\Sigma_0^B(\Phi)$-formulas, which gives us replacement for both ϕ and its Σ_1^B negation.

The second part, which we call the *generalized witnessing theorem*, is used to show that the class of witnessing functions for ϕ-formulas is $\mathrm{AC}^0(FC)$.

Theorem 2 (Generalized witnessing theorem). *Let Φ be a class of restricted Σ_1^B formulas representing C. Suppose that Φ is constructive. Then Σ_1^B-theorems of V-Φ can be witnessed by functions from $\mathrm{AC}^0(FC)$ provably in V-Φ. That is, if V-$\Phi \vdash \exists Z \phi(\bar{x}, \bar{Y}, Z)$, where $\phi \in \Sigma_1^B$, then there is a string function $F(\bar{x}, \bar{Y})$ in $\mathrm{AC}^0(FC)$ such that*

$$V\text{-}\Phi, AX(F) \vdash \phi(\bar{x}, \bar{Y}, F(\bar{x}, \bar{Y})),$$

where $AX(F)$ is a defining axiom for F. If Φ is strongly closed and constructive, then V-Φ proves that the defining axiom for F is equivalent to a formula from Φ.

The witnessing theorem looks similar to the constructiveness property, but they talk about different classes of formulas. Whereas constructiveness is concerned with witnessing an existential quantifier in a $\phi \in \Phi$ (or finding a counterexample to ϕ), the witnessing theorem describes the power of a system in terms of the strength of Σ_1^B-theorems that the system in question can prove.

The theorem 2 is a generalization of the witnessing theorem for V^0 as presented in [6] (hence the name "Generalized witnessing"). The proof uses proof-theoretic techniques. Taking a Σ_1^B theorem of V-Φ, we analyze its anchored proof in a second-order version of quantified Gentzen calculus LK^2 and prove, by induction on the structure of the proof, that in every line existential quantifiers can be witnessed by the functions of given complexity. To ensure that every line in the proof has only Σ_1^B formulas, we replace the comprehension axiom of V-Φ by a statement of the form $\exists Z < t \forall i \le t(\phi(i) \wedge Z(i)) \vee (\tilde{\phi}(i) \wedge \neg Z(i))$, $\phi \in \Phi$, where $\tilde{\phi}$ is a Σ_1^B formula equivalent to the negation of ϕ, provided by the constructiveness property. This gives us the base case (witnessing for the axioms). The witnesses in the rest of the cases are AC^0 combinations of witnesses in the previous steps.

Note that if the conditions do not hold, then the class of witnessing functions can be smaller than representable by formulas in the comprehension axiom. An example of that is the theory V^1, with comprehension over NP predicates. By the second-order version of Buss' witnessing theorem [6, 26], the class of Σ_1^B-definable functions of V^1 is P. But not every Σ_1^B formula is Δ_1^B-definable in V^1. Moreover, even if NP = coNP and for every Σ_1^B formula there is an equivalent Π_1^B formula, it might not be the case that these equivalences are provable in V^1.

5   Applications of the Definability Theorem

In this section we restate several previously known capture results in our framework. Three such examples when the strong case of Theorem 1 applies are V^0 itself, V_1-Horn and V-Krom. Below, we show that these theories are built on classes of formulas satisfying our two properties.

Example 1([5, 6, 26]) Functions bit-definable by Σ_0^B formulas in V^0 are AC^0 functions, and Σ_0^B formulas correspond to the first-order logic which captures AC^0 in the descriptive sense ([1]). The constructiveness property is satisfied trivially, since Σ_0^B is closed under complementation syntactically and there are no quantifiers to witness. It was shown in [5, 26] that AC^0 functions are closed under composition and thus under AC^0 reductions. Therefore, theorem 1 applies, so the class of Σ_1^B-definable functions of V^0 is FAC^0.

Example 2 ([7, 8]) The class of Σ_1^B-Horn formulas comes from $SO\exists$-Horn formulas capturing P in the descriptive setting. The resulting system V_1-Horn defines polynomial-time functions by Σ_1^B-Horn formulas, and is equivalent in power to Zambella's P-def (and thus PV). In this case, the properties hold with $\Phi = \Sigma_1^B$-Horn and $FC = FP$. So by the definability theorem Σ_1^B-definable functions of V_1-Horn are precisely polynomial-time functions. The bulk of work is a formalization of the satisfiability algorithm for propositional Horn formulas, which is needed already to prove closure of Σ_1^B-Horn formulas under complementation. This algorithm is constructive: a satisfying assignment (or, equivalently, values for quantified second-order variables) is obtained as part of the algorithm (the value $T^{[a]}$ in the description of RUN). This gives the constructiveness property.

Example 3 ([9]) Now take the class of Σ_1^B-Krom formulas, a translated version of Grädel's $SO\exists$-Krom (second-order 2CNF). It is possible to formalize Immerman-Szelepcsényi's proof of closure of NL under complementation in the resulting theory V-Krom ([9]). Also, proving that transitive closure function is Σ_1^B-definable in V-Krom results in a proof of constructiveness for V-Krom: values for quantified second-order variables are expressed as Σ_0^B combination of transitive closure function calls.

The next example, a system of arithmetic for SL, presents a case when we were not able to prove the strong version of the properties; this led to the formulation of the weaker properties.

6 Weak Case of the Definability Theorem

A class of Σ_1^B-SymKrom formulas is very similar to Σ_1^B-Krom, except it is based on symmetric 2CNF (that is, 2CNF with XOR instead of disjunctions). From the same Grädel's paper as before, [14], we know that $SO\exists$-SymKrom captures SL. We define V-SymKrom to be V-Φ with $\Phi \equiv \Sigma_1^B$-SymKrom.

It seems that showing that a system V-SymKrom would capture FSL should be straightforward. However, the methods used to prove closure of SL under complementation (Nisan and Ta-Shma, [20]), and, recently, that SL = L (Reingold, [22]) use properties of expander graphs and rely on algebraic methods for the proofs. But those are not known to be formalizable in less complexity than P. By Reingold's result, the class of Σ_1^B-definable functions of V-SymKrom is thus all logspace functions, but this is not known to be provable in V-SymKrom itself, as opposed to the cases of AC^0, NL and P. It might still be possible that such a theory for SL is not fully conservative over a theory for L.

6.1 Symmetric Transitive Closure

To simplify proofs, we introduce symmetric transitive closure operator by the following axiom:

$$STC_{x,y}\phi(x,y,\bar{a},\bar{Y})[a,b,n] \leftrightarrow \forall R(CondS(\phi,R,n) \rightarrow R(a,b)), \qquad \text{(AxSTC)}$$

where

$$CondS(\phi,R,n) \equiv \forall x,y,z < n(R(x,x) \wedge (\phi(x,y) \rightarrow (R(y,z) \leftrightarrow R(x,z))))$$

Note that if ϕ is quantifier-free except for bounded existential first-order quantifiers, then the negation of the $STC_{x,y}\phi(x,y)[a,b,n]$ defining axiom is equivalent to a Σ_1^B-SymKrom formula. Therefore, V-SymKrom proves induction on Σ_0^B combinations of STC functions.

By the same reasoning as for V-Krom in [9], STC defined in this manner is reflexive, transitive and robust against adding an edge on the left versus on the right (that is, conditions with $\phi(x,y) \rightarrow (R(x,z) \leftrightarrow R(y,z))$ and $\phi(y,z) \rightarrow (R(x,z) \leftrightarrow R(x,y))$ are equivalent). It is also provable in V-SymKrom that STC is symmetric: $STC(a,b,n) \leftrightarrow STC(b,a,n)$.

To see that $V^0 \subset V$-SymKrom, we encode a first-order formula as a graph and apply the STC operator to it. A first-order existential quantifier in $\exists z < n\psi(z)$ is simulated by STC applied to the graph with an edge relation defined by $E(x,y) \leftrightarrow \neg\psi(x) \wedge y = x + 1$. That is, a graph is a path from vertex 0 to vertex n with every edge $(z, z+1)$ labeled $\neg\psi(z)$; if $\psi(z)$ holds for some z_0 then the edge $(z_0, z_0 + 1)$ is absent so the start of the path and the end of it are disconnected. Similarly, a first-order universal quantifier is encoded by a graph with $E(x,y)$ such that $E(s,u) \leftrightarrow E(u,t) \leftrightarrow \neg\psi(u)$. This construction is applied for every block $\exists z < n\forall u < n\psi(z,u)$: such block is encoded as a path with every edge replaced by a "nested diamonds" gadget encoding a universal quantifier. A vertex $\langle n,n \rangle$ is reachable from the vertex $\langle 0,0 \rangle$ iff $\exists z < n\forall u < n\psi(z,u)$ holds.

Now we need to show the weak constructiveness property. First, we show how to witness formulas from Σ_1^B-SymKrom using $\Sigma_0^B(STC)$. Second, we give a Σ_1^B predicate equivalent to the negation of STC and show how to witness it: since the value of every formula can be expressed using STC, this is sufficient for Δ_1^B-definability of Σ_1^B-SymKrom.

6.2   Constructing a Witness for a Σ_1^B-SymKrom Formula

Given a Σ_1^B-SymKrom formula $\phi^* \equiv \exists P\forall\bar{x} < \bar{n}\psi(P,\bar{x})$, we create a formula $\phi'(u,s,v,s')$ encoding the structure of ψ; this encoding is similar to the encoding used in [9] for Σ_1^B-Krom formulas. For every clause, $\phi'(u,s,v,s')$ says that P-literals contain terms evaluating to u and v, with s and s' being 0 if the literal is negated and 1 otherwise. A propositional version of the formula is satisfiable if the corresponding graph is bipartite, that is, $\exists R\forall u,v < b\forall s,s' < 2(\phi'(u,s,v,s') \rightarrow \neg R(u,s) \leftrightarrow R(v,s'))$. Now, to use STC to test bipartiteness we use the standard technique of "doubling" the graph, with every vertex having "even" and "odd" version and every edge connecting the literals on opposite sides. There is an odd cycle in the original graph (and thus the formula evaluates to false) iff there is a path from a vertex on one side to the same numbered vertex on the other; this can be expressed using STC. From the witness to the negation of STC we construct a value for P (all literals on the same side as the constant $\top$ are set to true).

6.3 Δ_1^B-Definability of STC

Saying that a pair (a, b) is in the symmetric transitive closure of a graph is equivalent to the statement that b is reachable from a in an undirected graph. The following Σ_0^B predicate REACHCOND$(R, E, n + 1, a)$ states that $R(x, i)$ is true iff x is at most distance i from a:

$$\forall x \leq n \forall i \leq n (R(x, 0) \leftrightarrow x = a) \wedge$$
$$(R(x, i + 1) \leftrightarrow (\exists y \leq n R(y, i) \wedge (E(y, x) \vee y = x)))$$

Let ϕ be a formula defining an edge relation of a graph. Let

$$UDist_\phi(x, y, d) \equiv STC_{(u,c),(v,c')}\alpha[(x, 0), (y, d), (n, n)],$$

where $\alpha(u, c, v, c') \equiv (c' = c + 1 \wedge (\phi(u, v) \vee u = v))$. For simplicity, we assume that ϕ is represented by the corresponding graph E, and write $UDist(x, y, d)$ in that case. Then, $R(x, i) \equiv UDist(a, x, i)$ satisfies $\exists R$REACHCOND$(R, E, n+1, a)$, and V-SymKrom $\vdash STC(a, b, n) \leftrightarrow \exists R$REACHCOND$(R, E, n + 1, a) \wedge R(b, n)$.

Now, we showed that the weak constructiveness property holds. Therefore, every SL function is Σ_1^B-definable in V-SymKrom and every Σ_1^B-definable function of V-SymKrom is in $AC^0(FSL)$ provably in V-SymKrom. We know that $AC^0(FSL) = FL$, that is every $AC^0(SL)$ function is already computable in logspace, but this is not known to be provable in V-SymKrom. Also, just like V-Krom, V-SymKrom is finitely axiomatizable by finite set of axioms of V^0 together with comprehension over $\neg AxSTC$.

7   Conclusion

In this work we present a general framework for constructing systems of arithmetic with predefined power based on descriptive complexity results. The setback is that whereas for capture results in the descriptive complexity setting it is sufficient to have "some" proof of capture, in our bounded arithmetic framework we need an "easy" proof of capture, getting in return a "provable" capture result. It is interesting to see in which cases the complexity classes behave nicely, like P or NL, and in which cases, like SL, the proofs use concepts not (known to be) formalizable within the class itself.

A general witnessing theorem applying to slightly different types of theories was presented recently by Cook and Nguyen [19]. Their framework applies to theories equivalent to universal theories. They have a large number of applications, including different theories for NL, SL and P. However, they do not talk about provable capture.

Yet another property, uniqueness, that can be used instead of constructiveness was suggested to me by Sam Buss. This property states that for every formula from Φ there is an equivalent Σ_1^B formula with at most one witness to the quantifiers. The uniqueness property immediately implies constructiveness.

In general, it is interesting to explore the "robustness" of complexity classes such as provability of their properties. We hope that our framework provides a natural setting for such study.

Acknowledgments

I am very grateful to Stephen Cook for his supervision of my PhD thesis, which contained most of the ideas that appear in this paper. In particular, the main ideas used for the SL theory were suggested to me by him.

References

1. D. M. Barrington, N. Immerman, and H. Straubing. On uniformity within NC^1. *Journal of Computer and System Sciences*, 41(3):274 – 306, 1990.
2. P. Clote and G. Takeuti. Bounded arithmetic for NC, ALOGTIME, L and NL. *Annals of Pure and Applied Logic*, 56:73 – 117, 1992.
3. P. Clote and G. Takeuti. First order bounded arithmetic and small boolean circuit complexity classes. In *Feasible Mathematics*, volume II. Birkhäuser Inc., 1995.
4. A. Cobham. The intrinsic computational difficulty of functions. In Y. Bar-Hillel, editor, *Logic, Methodology and Philosophy of Science*, pages 24–30, Amsterdam, 1965. North-Holland.
5. S. Cook. Theories for complexity classes and their propositional translations. *submitted*, pages 1–36, 2004.
6. S. A. Cook. CSC 2429S: Proof Complexity and Bounded Arithmetic. Course notes, URL: "http://www.cs.toronto.edu/~sacook/csc2429h", Spring 1998-2002.
7. S.A. Cook and A. Kolokolova. A second-order system for polynomial-time reasoning based on Grädel's theorem. In *Proceedings of the Sixteens annual IEEE symposium on Logic in Computer Science*, pages 177–186, 2001.
8. S.A. Cook and A. Kolokolova. A second-order system for polytime reasoning based on Grädel's theorem. *Annals of Pure and Applied Logic*, 124:193–231, 2003.
9. S.A. Cook and A. Kolokolova. Bounded arithmetic of NL. In *Proceedings of the Nineteens annual IEEE symposium on Logic in Computer Science*, pages 398–407, 2004.
10. Stephen Cook and Neil Thapen. The strength of replacement in weak arithmetic. In *Proceedings of the Nineteens annual IEEE symposium on Logic in Computer Science*, pages 256–264, 2004.
11. H.-D. Ebbinghaus and J. Flum. *Finite model theory*. Springer Verlag, 1995.
12. R. Fagin. Generalized first-order spectra and polynomial-time recognizable sets. *Complexity of computation, SIAM-AMC proceedings*, 7:43–73, 1974.
13. E. Grädel. The Expressive Power of Second Order Horn Logic. In *Proceedings of 8th Symposium on Theoretical Aspects of Computer Science STACS '91, Hamburg 1991*, volume 480 of *LNCS*, pages 466–477. Springer-Verlag, 1991.
14. E. Grädel. Capturing Complexity Classes by Fragments of Second Order Logic. *Theoretical Computer Science*, 101:35–57, 1992.
15. N. Immerman. Relational queries computable in polytime. In *14th ACM Symp. on Theory of Computing, Springer Verlag (Heidelberg, FRG and NewYork NY, USA)-Verlag*, pages 147 –152, 1982.
16. N. Immerman. *Descriptive complexity*. Springer Verlag, New York, 1999.
17. A. Kolokolova. *Systems of bounded arithmetic from descriptive complexity*. PhD thesis, University of Toronto, October 2004.
18. L. Libkin. *Elements of Finite Model Theory*. Springer Verlag, 2004.
19. Phuong Nguyen and Stephen Cook. Theories for TC^0 and other small complexity classes. *submitted*, 2004.

20. Noam Nisan and Amnon Ta-Shma. Symmetric logspace is closed under complement. In *Proc. 27th Ann. ACM Symp. on Theory of Computing (STOC'95)*, pages 140–146, 1995.

21. A. Razborov. An equivalence between second-order bounded domain bounded arithmetic and first-order bounded arithmetic. In P. Clote and J. Krajiček, editors, *Arithmetic, proof theory and computational complexity*, pages 247–277. Clarendon Press, Oxford, 1993.

22. O. Reingold. Undirected ST-Connectivity in Log-Space. *Electronic Colloquium on Computational Complexity*, ECCC Report TR04-094, 2004.

23. T. J. Schaefer. The complexity of satisfiability problems. In *Proceedings of the Tenth Annual ACM Symposium on Theory of Computing*, pages 216–226, 1978.

24. L. J. Stockmeyer. The polynomial-time hierarchy. *Theoretical Computer Science*, 3:1–22, 1977.

25. G. Takeuti. RSUV isomorphism. In P. Clote and J. Krajiček, editors, *Arithmetic, proof theory and computational complexity*, pages 364–386. Clarendon Press, Oxford, 1993.

26. D. Zambella. Notes on polynomially bounded arithmetic. *The Journal of Symbolic Logic*, 61(3):942–966, 1996.

Transfinite Extension of the Mu-Calculus

Julian Bradfield[1], Jacques Duparc[2], and Sandra Quickert[1]

[1] Laboratory for Foundations of Computer Science, University of Edinburgh
{jcb,squicke1}@inf.ed.ac.uk
[2] Ecole des HEC, Université de Lausanne
jacques.duparc@unil.ch

Abstract. In [1] Bradfield found a link between finite differences formed by Σ_2^0 sets and the mu-arithmetic introduced by Lubarski [7]. We extend this approach into the transfinite: in allowing countable disjunctions we show that this kind of extended mu-calculus matches neatly to the transfinite difference hierarchy of Σ_2^0 sets. The difference hierarchy is intimately related to parity games. When passing to infinitely many priorities, it might not longer be true that there is a positional winning strategy. However, if such games are derived from the difference hierarchy, this property still holds true.

1 Introduction

Modal mu-calculus, the logic obtained by adding least and greatest fixpoint operators to modal logic, has long been of great practical and theoretical interest in systems verification. The problem of understanding alternating least and greatest fixpoints gave rise to a powerful and elegant theory relating them to alternating parity automata and to parity games, developed by many people including particularly Emerson, Lei, Jutla and Streett. Meanwhile, mu-arithmetic, the logic obtained by adding fixpoints to first-order arithmetic, made a brief appearance in the early 90s when Lubarsky studied its ordinal-defining capabilities – curiously, the logic had not previously been studied *per se* even by logicians. Then Bradfield used mu-arithmetic as a meta-language for modal mu-calculus, in which to prove a theorem on alternating fixpoints. Subsequently, Bradfield looked further into the analogies between mu-arithmetic and modal mu-calculus, and showed a natural equation between arithmetic fixpoints and the finite difference hierarchy over Σ_2^0, corresponding to the equation between modal fixpoints and parity games. Once in the world of arithmetic, it becomes natural to think about transfinite hierarchies. In this paper, we study the transfinite extension of the connection between mu-arithmetic and the difference hierarchy, and connect it to the Wadge hierarchy.

2 The Transfinite Mu-Calculus

2.1 Syntax and Semantics of the Transfinite Mu-Calculus

The logic we are considering is an extension of the usual mu-arithmetic, as introduced by Lubarski [7]. First, let us establish basic notation and conventions.

L. Ong (Ed.): CSL 2005, LNCS 3634, pp. 384–396, 2005.

ω is the set of non-negative integers; variables $i, j, \ldots, n$ range over ω. The set of finite sequences of integers is denoted ω^*; finite sequences are identified with integers via standard codings; the length of a sequence s is denoted $\mathrm{lh}(s)$. The set of infinite sequences of integers is $^\omega\omega$. For $\alpha \in \,^\omega\omega$, $\alpha(i)$ is the i'th element of α, and $\alpha(<i)$ is the finite sequence $\langle \alpha(0), \ldots, \alpha(i-1) \rangle$. Concatenation of finite and infinite sequences is written with concatenation of symbols or with $^\frown$, and extended to sets pointwise. The usual Kleene lightface hierarchy is defined on ω, $^\omega\omega$ and their products: $\Sigma_1^0 = \Sigma_0^1$ is the semi-recursive sets, $\Sigma_{n_1}^0 = \exists x \in \omega.\Pi_n^0$, $\Pi_n^i = \neg\Sigma_n^i$ and $\Sigma_{n+1}^1 = \exists \alpha \in \,^\omega\omega.\Pi_n^1$. The corresponding boldface hierarchy is similar, but starts with $\mathbf{\Sigma}_1^0 = \mathbf{\Sigma}_0^1$ being the open sets.

Mu-arithmetic has as basic symbols the following: function symbols f, g, h; predicate symbols P, Q, R; first-order variables x, y, z; set variables X, Y, Z; and the symbols $\vee, \wedge, \exists, \forall, \mu, \nu, \neg, \in$. The language has expressions of three kinds, individual terms, set terms, and formulae. The individual terms comprise the usual terms of first-order logic. The set terms comprise set variables and expressions $\mu(x, X).\phi$ and $\nu(x, X).\phi$, where X occurs positively in ϕ. Here μ binds both an individual variable and a set variable; henceforth we shall often write just $\mu X.\phi$, and assume that the individual variable is the lower-case of the set variable. We also use $\mu\nu$ to mean 'μ or ν as appropriate'. The formulae are built by the usual first-order construction, together with the rule that if τ is an individual term and Ξ is a set term, then $\tau \in \Xi$ is a formula.

The semantics of the first-order connectives is as usual; $\tau \in \Xi$ is interpreted naturally; and the set term $\mu X.\phi(x, X)$ is interpreted as the least fixpoint of the functional $X \mapsto \{ m \in \omega \mid \phi(m, X) \}$ (where $X \subseteq \omega$).

To produce a transfinite extension, we add the following symbols and formulae. If we have countably many recursively given Φ_i, $i \in \omega$, whose free set variables are contained in the same finite set of set variables, then we allow infinite countable disjunction $\bigvee_{i<\omega} \Phi_i$ and conjunction $\bigwedge_{i<\omega} \Phi_i$. The restriction on free variables means that we can transform any formula to a closed formula by adding finitely many fixpoint operators. The semantics is obvious.

Any formula in the mu-calculus can be rewritten in a prenex normal form:

$$\tau_n \in \mu X_n.\tau_{n-1} \in \nu X_{n-1}.\tau_{n-2} \in \mu X_{n-2} \ldots \tau_1 \in \mu\nu X_1.\Phi$$

For the transfinite mu-arithmetic we need an extension of this formulation.

Definition 1. By induction on the construction of the formula we say that a formula in the transfinite mu-calculus is written in extended prenex normal form

- if it is a formula in the finite mu-calculus and written in prenex normal form, or
- if the formula is an infinite disjunction or conjunction of extended prenex normal form formulae, or
- if it is some $\mu\nu X.\Phi$ where Φ is in extended prenex normal form.

Given formulae Φ_i for $i < \omega$ in the mu-arithmetic, we observe that the formula $\bigvee_{i<\omega} \Phi_i$ can be written in extended prenex normal form, simply by writing each Φ_i in prenex normal form. Given an arbitrary formula of the extended arithmetic mu-calculus, an easy proof on induction by the formula's construction shows that it can be written in extended prenex normal form. Furthermore, we can unfold its complexity and represent it by a wellfounded tree on ω^*.

2.2   A Hierarchy of the Transfinite Mu-Calculus

The fixpoint alternation hierarchy of of mu-arithmetic is thus: the first order formulae and all set variables form the class Σ_0^μ which is the same as Π_0^μ. For any natural number n let Σ_{n+1}^μ be generated from $\Sigma_n^\mu \cup \Pi_n^\mu$ by closing it under $\vee, \wedge$ and the operation $\mu X.\Phi$ for $\Phi \in \Sigma_{n+1}^\mu$. Π_{n+1}^μ contains all negations of formulae and set terms in Σ_{n+1}^μ. In order to extend the hierarchy we need to describe the limit step. We allow recursively countable disjunctions and conjunctions, but we want to stay in the lightface hierarchy. Therefore we extend the hierarchy to ω_1^{ck}, the first non-recursive ordinal. Let λ be a recursive limit ordinal. In Σ_λ^μ we collect all formulae of earlier stages and close it under $\bigvee_{i<\omega}, \vee$ and $\wedge$. Observe that a formula in Σ_λ^μ is equivalent to a formula $\bigvee_{i<\omega} \Phi_i$ where each $\Phi_i \in \Sigma_{\alpha_i}^\mu$ with $\alpha_i < \lambda$. Finally, we let $\Pi_\lambda^\mu = \neg\Sigma_\lambda^\mu$. The transfinite successor stages are built in the same way as the finite successor stages.

Later, this hierarchy will be linked to the effective version of the Hausdorff–Kuratowski difference hierarchy of Σ_2^0-sets: a set is in Σ_α^∂ iff it is of the form

$$\bigcup_{\xi \in \mathrm{Opp}(\alpha)} A_\xi \setminus \bigcup_{\zeta < \xi} A_\zeta$$

where $(A_\xi)_{\xi<\alpha}$ is an effective enumeration of a $\subseteq$-increasing sequence of Σ_2^0-sets, $\alpha < \omega_1^{ck}$, and $\mathrm{Opp}(\alpha)$ is the set of ordinals $< \alpha$ and of opposite parity to α, where the parity of a limit ordinal is even.

3   Model-Checking for the Transfinite Mu-Calculus

Fixpoints are often calculated by iteration, computing successive approximants until convergence. (Recall that the αth approximant of a least fixpoint $\mu X.\phi$ is defined as $\mu X^\alpha.\phi = \phi((\bigcup_{\beta<\alpha} \mu X^\beta.\phi)/X)$, and dually for greatest fixpoints.) In the finite case, this is the straightforward 'global' algorithm used for model-checking modal fixpoint logics.

The main focus of this paper is the relation between infinite parity games and fixpoint calculation. The games are infinite, and have somewhat complex payoff sets. This also has an analogue in the world of finite modal fixpoint logics, where it corresponds to the use of parity automata. Of course, in the finite world, it is well known that one does not have to play infinite games – repeats can be detected. It is perhaps of some interest to see that even in this infinite world of infinite formulae, it is possible to extend techniques well known from modal mu-calculus, and characterize truth of transfinite mu-arithmetic by a game in which all plays are finite (and therefore the payoff sets are clopen). Of course, there is a small catch – the moves involve playing ordinals, which amounts (for countable structures) to having to make second-order moves. The techniques being extended have somewhat intricate full definitions and proofs, and this is not the main focus of our paper, so given space restrictions we will just outline the game for those who have some familiarity with the modal mu-calculus work (as described for example in [2]).

A concept used in many basic theorems about modal mu-calculus, and in soundness proofs for techniques such as local model-checking with tableaus, is that of *(μ-)signature*. Consider a least fixpoint variable X_1 somewhere inside a formula, and suppose that $n \in X_1$ when X_1 has its actual value. We can consider at which approximant of X_1 the value n enters X_1; suppose this is α_1. However, X_1 itself may be defined in terms of some outer least fixpoint variable X_2; so we also need to know what approximant α_2 is currently being used as the value of X_2. Then a μ-signature of n at some subformula is an assignment of ordinals α_i to all the enclosing least fixpoint operators X_i that makes the given subformula true of n. When one is doing local model-checking, which effectively means exploring the proof tree that justifies $n \in \Phi$, signatures can be thought of as sets of 'clocks' for the verifier: every time verifier passes through a least fixpoint variable X, she has to decrement the clock for that variable, and if the clock ever hits zero, she loses. This ensures that she only passes through least fixpoints finitely often.

Dually, if (as done for example in completeness proofs) we are arguing about the falsity of a formula, we can consider the ν-signatures: now we look at the approximants of the greatest fixpoints at which a formula becomes false – and again, the signatures give a clock to bound the time by which refuter must establish the falsity.

It is possible to combine the clock intuitions for both μ-signatures and ν-signatures into a single game, which works as well for the transfinite logic as for the normal logic. Consider the usual rules for the model-checking game, with the obvious rule for the infinite disjunctions and conjunctions. Now extend it thus: whenever play enters a least fixpoint formula $\mu X.\phi$, verifier gets to choose an ordinal α_X. This amounts to a promise that she will 'bottom out' of the inductive definition in finite time, measured by α. Then any time play passes through a formula $\tau \in X$, verifier must decrease α_X, and if she can't she loses. Similarly, refuter chooses clocks for the greatest fixpoint formulae, and decreases them when passing through the variables.

This amendment to the model-checking game rules forces all plays to be finite, since it can be shown by standard arguments that signatures are well-ordered with respect to the game moves. Thus the payoff sets are simply sets of finite plays (i.e. clopen subsets of $^\omega\omega$) rather than parity conditions. Then one can show by minor extensions of the standard mu-calculus arguments that indeed verifier/refuter has a winning strategy iff the initial formula is true/false.

4 Parity Games

The use of parity games in model checking has been described by many authors. A very detailed survey is given by Niwiński [9]. Let us mention that we follow the convention that if the maximal priority seen infinitely often is odd, then player I

wins. When looking at a formula in the transfinite mu-calculus, we need to play a parity game with infinitely many priorities: for each set variable we need a distinct priority. If we take the binary tree and attach to each node a priority in an arbitrary fashion, then, when playing a parity game on this tree, we might end up having a "wild" payoff set for player I, and we might also lose the nice property of having a memoryless winning strategy [4]. Furthermore, it might be that there is no maximum among the priorities seen infinitely often, and infinite runs might even meet each priority only finitely many times. However, as we will see, a labelling derived from a model checking game of a transfinite mu-calculus formula avoids all these undesired effects. Moreover, such a labelling describes some set of the transfinite difference hierarchy $\bigcup_{\alpha < \omega_1^{ck}} \Sigma_\alpha^\partial$ and vice versa.

5   Connecting the Transfinite Difference Hierarchy and the Transfinite Mu-Calculus

Our aim is to extend Bradfield's following theorem [1]:

Theorem 2. *For every natural number n the equality $\partial \Sigma_n^\partial = \Sigma_{n+1}^\mu$ holds true.*

The extension into the transfinite is our main result.

Theorem 3. *For every recursive ordinal α the equality $\partial \Sigma_\alpha^\partial = \Sigma_{\alpha+1}^\mu$ holds true. Thus, $\bigcup_{\alpha < \omega_1^{ck}} \Sigma_\alpha^\mu = \partial \Delta_3^0$.*

Proof. Let α be a recursive limit ordinal, and let $\mu X_{\alpha+1}. \bigvee_{i<\omega} \Phi_i \in \Sigma_{\alpha+1}^\mu$, so in particular each Φ_i is in some Σ_β^μ for some $\beta < \alpha$. We need to find a game with payoff set in Σ_α^∂ whose winning positions for player I are calculated by this formula.

Assume that the formula $\mu X_{\alpha+1}. \bigvee_{i<\omega} \Phi_i$ describes a nonempty subset of ω, and choose some witness n for this nonemptyness. Now consider the game tree which results from the parity game played as a model checking game. We might think of it as a subtree in ω^*, each node labelled with the position in the model checking game. In extending the tree in an appropriate way we may assume that it does not contain finite maximal branches, and in further simplifying the tree we may assume that each node marks a loop-back, i.e. we see some X_β at each node. This can be done because any infinite branch must hit such nodes infinitely many times. In omitting which element $n' \in \omega$ is inspected in the model checking game, we get a tree which is simply labelled by the indices of the set variables, i.e. by countable ordinals up to $\alpha + 1$ without limit ordinals. Observe that the labelling has the structure of a set in $\Sigma_{\alpha+1}^\partial$. Let us describe the payoff set for player I.

Since the outmost variable is under the scope of a minimal fixpoint operator, player I wins the model checking game iff at some point, the game gets captured inside some subformula Φ_i and player I wins the subgame for Φ_i, where $X_{\alpha+1}$ is replaced by $\bot$. By the induction hypothesis, the payoff set of such a subgame has the complexity of some set in Σ_β^∂ with $\beta < \alpha$. Since the payoff set of the whole game is the effective union of all these subsets, we obtain some set of complexity Σ_α^∂, which shows $\Sigma_{\alpha+1}^\mu \subseteq \partial\Sigma_\alpha^\partial$ for α limit. The successor case is an induction over the construction of formulae as in [1].

It remains to show $\partial\Sigma_\alpha^\partial \subseteq \Sigma_{\alpha+1}^\mu$. As before, we only need to consider the limit step, the successor case is done as in [1].

Let α be a countable limit ordinal, and let $A \in \Sigma_\alpha^\partial$. We may assume that $A = \bigcup_{i<\omega} A_i$ with $A_i \in \Sigma_{\alpha_i}^\partial$, $\alpha_i < \alpha$. We let A be the payoff set for player I and calculate her winning positions. By induction hypothesis, for each A_i we have a formula $\Phi_i \in \Sigma_\beta^\mu$, $\beta < \alpha$, describing the winning positions for player I with the payoff set A_i. Let

$$H_0 = \{s \in \omega^* \mid \exists i \ s \in \Phi_i\}$$

be the set of all nodes s.t. player I wins if some A_i is the payoff set. By recursion we define H_β for $\beta < \omega_1$: If β is a limit ordinal, then let $H_\beta = \bigcup_{\gamma<\beta} H_\gamma$, and if $\beta = \gamma + 1$, then let

$$H_\beta = \{s \in \omega^* \mid \exists i \ \text{player } I \text{ wins with payoff set } A_i \text{ or she can reach } H_\gamma\}$$

To reach a certain set of nodes is an open condition, thus, H_β can be viewed as the set of winning conditions of a set of complexity less than Σ_α^∂. It is immediate from the definition that the H_β's describe an increasing sequence of subsets of a countable set. Therefore, at some countable stage ξ the process stabilizes, we have reached the minimal fixpoint H_ξ of this process. Thus, we can express the calculation of the winning positions within the transfinite mu-calculus, within complexity $\Sigma_{\alpha+1}^\mu$:

$$\mu X_{\alpha+1}. \quad (\exists n \exists m \ \mathrm{lh}(x_{\alpha+1}) = 2n \wedge x_{\alpha+1}{}^\frown m \in X_{\alpha+1})$$
$$\vee \ (\exists n \forall m \ \mathrm{lh}(x_{\alpha+1}) = 2n + 1 \wedge x_{\alpha+1}{}^\frown m \in X_{\alpha+1})$$
$$\vee \ (\bigvee_{i<\omega} \Phi_i)$$

By the same arguments used in [1], as a result that Borel games are determined [8] H_ξ describes exactly the winning positions for player I.

6   Nicely Behaving Labellings

When extending the mu-arithmetic into the transfinite we need to check whether we keep key properties, namely the existence of positional winning strategies. This leads to

Definition 4. Let P be a parity game with priorities in some $\alpha < \omega_1$. P is called max-closed iff for every infinite run the set of all labels seen infinitely often is non-empty and contains a maximum.

Clearly, the rules of the model checking game ensure that the parity game derived from a model checking game of a transfinite mu-calculus formula is max-closed.

Theorem 5. *Each max-closed parity game admits a positional winning strategy for one of the players.*

Proof. We proceed by induction on the set of labels α. Of course, any set of countably many labels can be relabelled by natural numbers, but max-closedness is not preserved in general. In the sequel l will always denote the labelling function, $l : \alpha \to V$ where V is the set of vertices in the considered game graph.

Let us first consider the easier case, i.e. α is a limit ordinal. Assume player I has a winning strategy f, we need to find a positional winning strategy.

Let T be the tree of all possible plays. We define

$$A_0 = \left\{ s \in T \mid \exists \beta < \alpha. \forall t \in T[s]. l(t) \leq \beta \right\}$$

i.e. the cone of T above s is labelled with values up to β. Observe that by max-closedness, A_0 is dense in T. Otherwise, we could select a cone $T[t]$ having an empty intersection with A_0, meaning that every subcone of this cone is labelled with values cofinal in α. Since α is countable, there is a sequence $(\alpha_i)_{i<\omega}$ with each $\alpha_i < \alpha$ and $\bigcup_{i<\omega} \alpha_i = \alpha$. In the cone $T[t]$ it is easy to construct an infinite path x s.t. for each i there is some n_i with $l(x(n_i)) > \alpha_i$, contradicting the max-closedness.

Although A_0 is dense, it might be that the complement still contains infinite paths. Thus, we define by recursion:

$$A_\beta = \left\{ s \in T \setminus \bigcup_{\alpha<\beta} A_\alpha \mid \exists \gamma < \alpha. \forall t \in T[s] \setminus \bigcup_{\alpha<\beta} A_\alpha. l(t) \leq \gamma \right\}$$

The process stops at some countable γ. From these sets we can easily determine the set of winning positions for player I. We let H_0 be the set of all elements in A_0 such that player I can win the game starting at that position. Since the labels in the cone of the game tree are bounded by some $\beta < \alpha$, by induction hypothesis player I has a positional winning strategy within H_0. In particular, the game stays within H_0. In general we let H_β be the subset of A_β such that player I has a winning strategy as long as the game stays within A_β, and as soon as the game leaves A_β, some $H_{\beta'}$ is entered with $\beta' < \beta$. Again, within H_β player I has a positional winning strategy. Analogously to Section 5 the process stabilizes at some countable γ, and $H_\gamma = \bigcup_{\beta<\gamma} H_\beta$ is the set of all winning positions of player I, and it can be described by a formula of the transfinite mu-arithmetic provided the set of labels does not exceed ω_1^{ck}. It is fairly easy to describe a positional winning strategy for player I: as long as the game takes place in some H_β, she follows the positional winning strategy within H_β. It might be that player I cannot force the game to stay inside H_β, but if this set is left, then some $H_{\beta'}$ is entered with $\beta' < \beta$, and from that moment on player I follows the positional winning strategy for $H_{\beta'}$. Since all the H_β are pairwise disjoint, the concatenation of all positional winning strategies for the H_β gives a positional winning strategy for all her winning positions.

Analogously, if player II has a winning strategy, then he has a positional winning strategy as well.

Now let us consider the successor case. Assume $\alpha = \beta + 1$ is odd, thus player I needs to make sure that β is seen only finitely many times. Assume player I has a winning strategy, we need to find a positional winning strategy.

Let H_0 consist of those vertices s such that, starting from s, player I has a winning strategy which never leads to any vertex labelled with β. Such vertices must exist, otherwise player II has a winning strategy. In particular, being at such a node player I can force the game to stay in H_0. We claim that within H_0 player I has a positional winning strategy. Consider the game tree starting from $s \in H_0$ and remove all nodes outside H_0 together with the cones above those nodes. The remaining tree is labelled with values smaller than β, and by induction hypothesis on this subtree (and the corresponding subgraph) player I has a positional winning strategy. This positional winning strategy is clearly a winning strategy for the whole game. Constructing H_γ analogously to the limit case yields a positional winning strategy for the whole game.

Now assume player II has a winning strategy. This means that he either can manage to see β infinitely often, or, if player I keeps the occurrence of β finite, he wins the induced subgame. A positional strategy is described as follows: if a vertex belongs to player II's winning region, and if he has a winning strategy which guarantees him to reach some vertex labelled with β, then he plays in a way that he never leaves his winning region and after finitely many steps he will reach β. Clearly, to reach some node within the winning region labelled with β is an open condition, thus there exists a positional strategy for achieving that goal. If, after reaching β, player II can still reach another β within his winning region, he goes for it. At some point it might be that he still has a winning strategy, but he cannot make sure that β is seen again. At this stage consider the subgraph S wich consists of all nodes in player II's winning region with labels smaller than β, the edge relation restricted to S stays the same. Observe that by being a winning region player I can only leave the subgraph S in moving to a vertex which is still in player II's winning region, but from where player II can reach β memoryless again while remaining in his winning region. As long as the run stays in S, by induction hypothesis player II has a positional winning strategy. Thus, in concatenating the positional winning strategies for the different regions we obtain a positional winning strategy for the whole game.

The remaining case, $\alpha = \beta + 1$ even, is handled similarly. We can construct a positional winning strategy for player I as in the odd case for player II and vice versa. □

Corollary 6. *For any formula in the transfinite mu-arithmetic, model checking with parity games admits positional winning strategies.*

7   A Descriptive Set Theoretical Approach

In this section we step into Descriptive Set Theory, and see that the effective version of the very refined Wadge Hierarchy of sets of infinite words unveils a clue to understanding what goes on in the result $\partial \Sigma_n^\partial = \Sigma_{n+1}^\mu$. The reader may find most of the basic material in [5]. The effective Wadge hierarchy is studied in [6].

7.1   The Difference Hierarchy of Σ_2^0 Sets

This result admits a first step: $\Sigma_1^\mu = \partial \Sigma_1^0$. Recall that Σ_1^0 is the class of all semi-recursive sets, Σ_n^∂ being the class of n differences of effectively countable unions of complements of semi-recursive sets. To be more precise, the space of infinite sequences over an alphabet Σ is equipped with the usual topology, that is the product topology of the discrete topology over the alphabet. So, Σ_1^0 is the class of all sets of the form $W\Sigma^\omega$ where $W \subseteq \Sigma^*$ is a recursively enumerable set of finite words (possibly empty in which case $W\Sigma^\omega = \varnothing$). And Σ_n^∂ stands for the the class of sets of the form $A = A_n \smallsetminus A_{n-1} \cup A_{n-2} \smallsetminus A_{n-3} \cup \ldots$, where $A_1 \subseteq A_2 \subseteq \ldots \subseteq A_n$ is a sequence of sets in Σ_2^0 - the class of effectively countable unions of complements of semi-recursive sets.

So, as we see $\Sigma_1^\mu = \partial\Sigma_1^0$, $\Sigma_2^\mu = \partial\Sigma_2^0$, but then, $\Sigma_n^\mu = \partial\Sigma_n^0$ fails for $n > 2$, and must be replaced with $\Sigma_n^\mu = \partial\Sigma_{n-1}^\partial$. At first glance it seems there is no logic behind this. However, the effective Wadge hierarchy, a refinement of the effective difference hierarchy, gives the solution.

The Wadge Ordering. A natural improvement of the Hausdorff–Kuratowski hierarchy was induced by Wadge's work based on a reduction relation defined in terms of continuous functions. This means, a natural way to compare the topological complexity of sets A and B was to say $A \leq_W B$ – intuitively meaning A is topologically less complicated than B – if the problem of knowing whether x belongs to A *reduces* to knowing whether $f(x)$ belongs to B for some *simple* function, where simple meant continuous. The effective version deals with recursive functions instead, and in the sequel we will concentrate only on the effective version:

$$A \leq_W B \quad \text{iff} \quad \exists \text{ recursive } f : \Sigma_A^\omega \to \Sigma_B^\omega . f^{-1}B = A$$

The Wadge ordering ($\leq_W$) induces the strict ordering ($<_W$) and the Wadge equivalence ($\equiv_W$):

$$A <_W B \quad \text{iff} \quad A \leq_W B \wedge B \not\leq_W A$$

$$A \equiv_W B \quad \text{iff} \quad A \leq_W B \leq_W A$$

When restricted to Kleene pointclasses, this ordering becomes a quasi-well-orderingd, i.e. it is well-founded, and has antichains of length at most two. Moreover, if A and B are incomparable, then $A \equiv_W B^{\complement}$. The reason for this is that all these properties derive from Borel Determinacy [8]. Indeed, Wadge defined the relation $A \leq_W B$ in terms of the existence of a winning strategy in a suitable game: the Wadge game.

Definition 7 (The Wadge game). Let $A \subseteq \Sigma_A^\omega$, $B \subseteq \Sigma_B^\omega$, $\mathbf{W}(A,B)$ is an infinite two-player game where players (I, and II) take turn playing letters in Σ_A for I, and in Σ_B for II. As opposed to I, player II is allowed to skip provided he plays infinitely many letters.

So that at the end of a run (in ω moves), I has produced an ω-word $x \in \Sigma_A^\omega$ and II has produced $y \in \Sigma_B^\omega$. The winning conditions are:

$$II \text{ wins } \mathbf{W}(A, B) \quad \text{iff} \quad (x \in A \Leftrightarrow y \in B)$$

Wadge designed the rules of the game $\mathbf{W}(A, B)$ so that a strategy for II induces a continuous mapping $x \mapsto y$, and conversely; and the winning condition so that

$$II \text{ has a w.s. in } \mathbf{W}(A, B) \quad \text{iff} \quad A \leq_W B.$$

Let us define the equivalence relation $\sim$ by

$$A \sim B \quad \text{iff} \quad A \equiv_W B \ \text{ or } \ A \equiv_W B^{\complement} \ \text{ or } \ A \equiv_W 0B \cup 1B^{\complement}$$

Quotiented by $\sim$, and using determinacy, the Wadge ordering $\leq_W$ turns into a well-ordering (denoted by $\leq_{/\sim}$) whose minimal elements are all clopen sets. This induces the notion of the Wadge degree defined inductively:

$$d^\circ A = 0 \quad \text{iff} \quad A \text{ is clopen}$$

$$d^\circ A = \sup\{d^\circ B + 1 : \ B <_{/\sim} A\}$$

where $<_{/\sim}$ stands for the strict Wadge ordering $<_W$ quotiented by $\sim$.

7.2 Multiplication by ω_1^{ck}

Now, given a topological class, that is a class closed under pre-image by recursive functions (such as $\Sigma_1^0, \Sigma_n^\partial$), a set A is complete for the class if it reduces all sets in it. As usual, a complete set is a set of maximal complexity, therefore of maximal Wadge degree. In other words, the Wadge degree of a complete set of a given class is a measure of the topological complexity of this class.

If we look at the sequence of Wadge degrees of complete sets for respectively $\Sigma_1^0, \Sigma_2^0 = \Sigma_1^\partial, \Sigma_2^\partial, \Sigma_3^\partial, \dots$ We find $1, \omega_1^{ck}, \omega_1^{ck^2}, \omega_1^{ck^3}, \dots$. Surprisingly, the progression is precisely multiplication by ω_1^{ck}. More surprisingly indeed, is that multiplication of a Wadge degree by ω_1^{ck} ($\alpha \mapsto \alpha \cdot \omega_1^{ck}$) corresponds to a simple set theoretical operation ($A \mapsto A \bullet \omega_1^{ck}$). Namely,

$$A \bullet \omega_1^{ck} = (\Sigma \cup \{a_+, a_-\})^* a_+ A \ \cup \ (\Sigma \cup \{a_+, a_-\})^* a_- A^{\complement}$$

for a_1, a_- two different letters not in Σ. For a better understanding, a player (either I or II) in charge of $A \bullet \omega_1^{ck}$ in a Wadge game is exactly like the same player being in charge of A with the extra possiblity to erase all his moves and decide to start all over again being in charge of $A^{\complement}$ instead of A, and erase everything again and switch from A to $A^\complement$, and so on. Playing a_+ or a_- takes care of both the initialization of the play, and the choice between A and $A^\complement$. A word containing infinitely many a_+ or a_- being not in $A \bullet \omega_1^{ck}$.

This operation preserves the Wadge ordering

$$A \leq_W B \Rightarrow A \bullet \omega_1^{ck} \leq_W B \bullet \omega_1^{ck}$$

and satisfies the required property:

$$d^\circ(A \bullet \omega_1^{ck}) = d^\circ(A) \cdot \omega_1^{ck}$$

7.3   Division by ω_1^{ck}

For inductive proofs on the degree of sets, an inverse operation ω_1^{ck} is needed. It must be a set theoretical counterpart of division by ω_1^{ck}: given any set A the set $\frac{A}{\omega_1^{ck}}$ must verify:

1. $A \leq_W B \Rightarrow \frac{A}{\omega_1^{ck}} \leq_W \frac{A}{\omega_1^{ck}}$
2. $d^\circ A = \alpha \cdot \omega_1^{ck} \Rightarrow d^\circ \frac{A}{\omega_1^{ck}} = \alpha$

Unfortunately, we do not know how to obtain condition 2 directly. So we weaken our expectations and ask for the following instead:

$$d^\circ A = \alpha \cdot \omega_1^{ck} \Rightarrow d^\circ \frac{A}{\omega_1^{ck}} = \alpha + 1$$

How to get precisely condition 2 can be deduced from the whole artillery Duparc developed in [3]. The idea to define $\frac{A}{\omega_1^{ck}}$ is that a player in a Wadge game in charge of $\frac{A}{\omega_1^{ck}}$ is like this player being in charge of $A \subseteq \Sigma^\omega$ but having his opponent asking him questions whether or not the infinite word x he is actually constructing step by step, will remain in a tree $T_i \subseteq \Sigma^*$. The opponent is allowed to ask questions about as many trees as he wants as long as the player answers no. Once the player answers yes, there is no more questioning allowed. The precise definition is as follows. Given $x \in \Sigma^\omega$, we write x_{even} for the word $x(0)x(2)x(4)\ldots\ldots$

Definition 8. Given an alphabet Σ, a $\mathscr{T}$ree $\mathscr{T}$ on Σ is some *non empty pruned tree* that satisfies for any $u \in \mathscr{T}$ and any integer $n < \mathrm{lh}(u)$:

if n is even: then $u(n) \in \Sigma$ (these are the nodes that correspond to the main run), and

if n is odd: then $u(n)$ is an auxiliary move with three different options:

    Option $\langle$no$\rangle$: in this case $u(n) = \langle$no$, v\rangle$ for some $v \in \Sigma^*$ and $u_{even} \subseteq v$. So v is some position in Σ^* that extends the position u_{even}. But then, we demand that any position w in $\mathscr{T}$ that extends u must verify $w_{even} \subseteq v$ or $v \subseteq w_{even}$. Moreover, we also require that $\mathscr{T}$ verifies the following condition: if $(u \upharpoonright n) \langle$no$, v'\rangle$ belongs to $\mathscr{T}$ with $v' \neq v$, then both $v' \subseteq v$ and $v \subseteq v'$ must fail. Or, to say it differently, T must satisfy the condition:

$$\left(u \in \mathcal{T} \; \wedge \; u \langle \mathbf{no}, v \rangle \in \mathcal{T} \wedge \; u \langle \mathbf{no}, v' \rangle \in \mathcal{T} \right) \Rightarrow v = v' \; \vee \; v \perp v'.$$

Option $\langle \mathbf{yes} \rangle$: in this case $u(n) = \langle \mathbf{yes} \rangle$. This must be regarded as the option to avoid all other positions of the form $(u \upharpoonright n) \langle \mathbf{no}, v \rangle$. Formally, this means that any w in $\mathcal{T}$ that extends $(u \upharpoonright n) \langle \mathbf{yes} \rangle$ must satisfy

$$v \subseteq w_{even} \text{ fails for any } v \text{ such that } (u \upharpoonright n) \langle \mathbf{no}, v \rangle \in \mathcal{T}.$$

Option $\langle - \rangle$: this case should be regarded as no question asked at all. We require $\mathcal{T}$ to satisfy:

if $(u \upharpoonright n) \langle - \rangle \in \mathcal{T}$
  then $(u \upharpoonright n) \langle \mathbf{yes} \rangle \notin \mathcal{T}$ and for any $v \in \Sigma^*$ $(u \upharpoonright n) \langle \mathbf{no}, v \rangle \notin \mathcal{T}$.

Finally, $\mathcal{T}$ must verify:

$$\forall k \; \forall n > k \; \; x(2k+1) = \langle \mathbf{yes} \rangle \; \Rightarrow \; x(2n+1) = \langle - \rangle$$

We remark that given any infinite word $x \in \Sigma^\omega$, there exists a unique infinite branch $y \in [\mathcal{T}]$ such that $y_{even} = x$.

Definition 9. Let $A \subseteq \Sigma^\omega$, and $\mathcal{T}$ be a $\mathcal{T}$ree on Σ,

$$A^{\mathcal{T}} = \{ x \in [\mathcal{T}] : x_{even} \in A \}$$

$$\frac{A}{\omega_1^{ck}} = \text{ a } \leq_{/\sim}\text{-minimal element in } \{ A^{\mathcal{T}} : \mathcal{T} \text{ a } \mathcal{T}\text{ree on } \Sigma \}$$

Remark 10.

1. If $A \leq_W B$ then for any $\mathcal{T}$ree $\mathcal{T}_B$, one can easily design a $\mathcal{T}$ree $\mathcal{T}_A$ such that $A^{\mathcal{T}_A} \leq_W B^{\mathcal{T}_B}$. By minimality, $\frac{A}{\omega_1^{ck}} \leq_{/\sim} \frac{B}{\omega_1^{ck}}$.

2. If $d^\circ A = \alpha \cdot \omega_1^{ck}$, then take any B with $d^\circ B = \alpha$, consider $B \bullet \omega_1^{ck} \subseteq (\Sigma \cup \{b_+, b_-\})^\omega$. Let $\mathcal{T}$ be the $\mathcal{T}$ree that asks questions about the trees $(\Sigma^* \{b_+, b_-\})^n \Sigma^*$ for each n as long as the opponent does not agree on restricting his moves to such a tree. Clearly

$$\frac{A}{\omega_1^{ck}} \leq_{/\sim} (B \bullet \omega_1^{ck})^{\mathcal{T}} \leq_W \bigcup_{n \in \mathbb{N}} 0^{2n} B \cup 0^{2n+1} B^{\complement} \leq_{/\sim} B$$

This shows

$$d^\circ A = \alpha \cdot \omega_1^{ck} \; \Rightarrow \; d^\circ \frac{A}{\omega_1^{ck}} \leq \alpha + 1.$$

The other inequality is by induction on $d^\circ B$ and requires the complete knowledge of the Wadge hierarchy [3].

8 Outlook

The transfinite mu-arithmetic gives us a class of parity games with infinite labellings, but still they are well behaving, there are positional winning strategies. On the other hand, there are examples of automata which induce parity games without positional winning strategies, even requiring an infinite memory for a winning strategy [4]. It is interesting to draw the line between parity games with or without positional strategies sharper. Clearly, if there is only one path in the game tree which is a counter-example of max-closedness, then there is still a positional winning strategy, since this one path can always be left. Thus, the question arises how many ill behaving paths a game tree can allow and still having a positional winning strategy. Which is the right notion of smallness of the set of all ill-behaving paths, maybe meagreness?

Acknowledgement

Quickert is supported by the EU Research and Training Network GAMES (Games and Automata for Synthesis and Validation).

References

1. J. Bradfield. Fixpoints, Games and the Difference Hierarchy. *Theoretical Informatics and Applications*, 37:1–15, 2003.
2. J. Bradfield and C. Stirling. Modal Logics and mu-calculi: An Introduction. In J. Bergstra, A. Ponse, and S. Smolka, editors, *Handbook of Process Algebra*, pages 293–329. Elsevier Science B.V., 2001.
3. J. Duparc. Wadge Hierarchy and Veblen Hierarchy, Part I: Borel sets of finite rank. *Journal of Symbolic Logic*, 66:56–86, 2001.
4. E. Grädel and I. Walukiewicz. Positional Determinacy of Games with Infinitely Many Priorities. *Manuscript*.
5. A. Kechris. *Classical descriptive set theory*. Springer Verlag, 1995.
6. A. Louveau. Some results in the Wadge hierarchy of Borel sets. In A. Kechris, D. Martin, and Y. Moschovakis, editors, *Cabal Seminar 79-81*, Lecture notes in Mathematics, pages 28–55. Springer, 1983.
7. R. S. Lubarski. μ-definable sets of integers. *Journal of Symbolic Logic*, 58:291–313, 1993.
8. D. Martin. Borel determinacy. *Annals of Mathematics*, 102:336–371, 1975.
9. D. Niwiński. Fixed point characterisation of infinite behavior of finite state systems. *Theoretical Computer Science*, 189:1–69, 1997.

Bounded Model Checking of Pointer Programs

Witold Charatonik[1], Lilia Georgieva[2,*], and Patrick Maier[3,**]

[1] Instytut Informatyki, Uniwersytet Wrocławski
http://www.ii.uni.wroc.pl/~wch/
[2] School of Math. and Comp. Sci., Heriot-Watt Univ.
http://www.macs.hw.ac.uk/~lilia/
[3] Max-Planck-Institut für Informatik
http://www.mpi-inf.mpg.de/~maier/

Abstract. We propose a bounded model checking procedure for programs manipulating dynamically allocated pointer structures. Our procedure checks whether a program execution of length n ends in an error (e. g., a NULL dereference) by testing if the weakest precondition of the error condition together with the initial condition of the program (e. g., program variable x points to a circular list) is satisfiable. We express error conditions as formulas in the 2-variable fragment of the Bernays-Schönfinkel class with equality. We show that this fragment is closed under computing weakest preconditions. We express the initial conditions by unary relations which are defined by monadic Datalog programs.

Our main contribution is a small model theorem for the 2-variable fragment of the Bernays-Schönfinkel class extended with least fixed points expressible by certain monadic Datalog programs. The decidability of this extension of first-order logic gives us a bounded model checking procedure for programs manipulating dynamically allocated pointer structures. In contrast to SAT-based bounded model checking, we do not bound the size of the heap a priori, but allow for pointer structures of arbitrary size. Thus, we are doing bounded model checking of infinite state transition systems.

1 Introduction

Automatic verification of programs that can manipulate pointers into dynamically allocated memory is a challenging task, even for simple safety properties such as "there is no NULL dereference". In general, the problem is undecidable as the reachable state space of programs with dynamic memory allocation is infinite. Decidability can be traded off for precision by over- or under-approximating the reachable state space. Over-approximation is used by techniques based on abstraction, e. g., the shape analysis framework [24]. In bounded model checking (BMC), the set of reachable states is under-approximated by limiting the runtime or the memory of a program to an a priori chosen bound [7, 14]. Thus, BMC cuts off all states that require more than the chosen amount of time or space to be reached. As a consequence of under-approximation, BMC can not prove safety properties (unless the diameter of the state space is less than the time bound), it can only detect their violation. Progressively increasing the bound

---

* Partially supported by a DAAD grant.
** Corresponding author.

L. Ong (Ed.): CSL 2005, LNCS 3634, pp. 397–412, 2005.
© Springer-Verlag Berlin Heidelberg 2005

actually yields a semi-algorithm for detecting errors, provided the BMC problem is decidable.

The decidability of the BMC problem may seem trivial at first view. Note, however, that although we assume an explicit bound on the length of the program execution, we do not have any bound on the size of the initial data structure. This implies that the explored model is an infinite and infinitely branching transition system: given an initial condition like "the variable x points to a structure of type T" we have to consider all transitions from the initial state to a state where x points to a structure of type T and size n, for infinitely many numbers n. Although it is obvious that in a finite execution a program may explore only a finite fragment of the initial data structure and it is relatively easy to compute a bound on the size of this explored part, this observation alone still does not give any bound on the size of the initial data structure as a whole. In other words, even if we are in a bounded-model-checking setup, we still have to deal with an infinite set of reachable states due to the infinite branching in the underlying transition system. To overcome this problem we prove a kind of pumping lemma that implies that it is enough to consider initial structures up to a certain size.

The worst-case complexity of our method (2-NEXPTIME) may also look discouraging. Again one has to note that the doubly-exponential blowup comes from the specification of the initial data structure. As we show in section 3.3, in common cases the complexity is doubly exponential in the specification, but not in the length of the execution. In particular, for non-branching pointer structures like lists, and for fixed programs, the complexity boils down to NPTIME. Note also that it is not appropriate to directly compare the general complexity of our method with other approaches where the initial structure is given explicitly or its size is a priori bounded – one should not expect that any method could explore a data structure of doubly exponential size in less than doubly exponential time.

Our method to decide the BMC problem relies on the following observations.

- Error conditions like "x is a dangling pointer" are expressible in the 2-variable fragment of the Bernays-Schönfinkel class with equality.
- The fragment is closed under weakest preconditions w. r. t. finite paths.
- Data structures like trees, singly or doubly linked lists and even circular lists are expressible in a fragment of monadic Datalog.
- The combination of the Datalog fragment and the above fragment of the Bernays-Schönfinkel class is decidable.

In section 2, we reduce the BMC problem to satisfiability in the Bernays-Schönfinkel class with Datalog. Section 3 presents our main technical contribution: decidability of satisfiability of the 2-variable fragment of the Bernays-Schönfinkel class with equality extended by a certain class of monadic Datalog programs. This result follows from a small model property of the logic, which we prove by a kind of pumping technique. Proofs which have been omitted here due to lack of space can be found in [6].

2   Pointer Programs

We investigate imperative programs that manipulate dynamic data structures on the heap. Given a program and a specification of its input, i. e., the heap contents upon start-up, we want to check safety properties, e. g., whether the program can crash by dereferencing a dangling pointer.

2.1 Syntax

A program P consists of two parts, a struct declaration specifying templates of heap cells and a control flow graph specifying the possible program executions.

A *struct declaration* is finite directed graph with uniquely labeled edges, i. e., there are no two edges with the same label. We call the vertices of this graph *templates*, the edge labels we call *fields*. An edge label r is a field of a template T, denoted by $r \in T$, if r labels one of T's outgoing edges; note that due to unique labeling, each r is a field of exactly one template T.

A *control flow graph (CFG)* is a finite directed graph whose edges are labeled by actions. We call the vertices of a control flow graph *(control) locations*, and we assume that there is a distinguished location $init$, which does not have incoming edges. The set of *actions* Act is defined by the following grammar, where T is a template, s is a field, x and y are program variables, e is a program variable or a constant (including $NULL$), and γ is a formula built from program variables and constants (including $NULL$), the equality symbol $\approx$ and the Boolean connectives.

$$
\begin{aligned}
Act ::=\ &assume(\gamma) &&\textit{Assume condition } \gamma. \\
\mid\ &y := e &&\textit{Assign the value e to the variable y.} \\
\mid\ &y := s(x) &&\textit{Read the s-field of the cell pointed to by x into y.} \\
\mid\ &s(x) := e &&\textit{Write e to the s-field of the cell pointed to by x.} \\
\mid\ &free_T(x) &&\textit{Deallocate the T-cell pointed to by x.} \\
\mid\ &y := new_T() &&\textit{Allocate a new T-cell and assign its address to y.}
\end{aligned}
$$

A *path* π (of length n) in the CFG is a sequence $\langle \ell_0, \alpha_1, \ell_1, \ldots, \alpha_n, \ell_n \rangle$ alternating between locations ℓ_i and actions α_j. Note that there is no action for procedure calls, yet in bounded model checking, calls can be handled by inlining procedure bodies.

Figure 1 shows a sample program. Upon start it expects that the variable e points to a doubly linked circular list (realized by `next`- and `prev`-pointers). The program deallocates the cell pointed to by e, allocates a new cell and inserts it in place of the old one (using the temporary variables ne and pe).

2.2 Semantics

Given a program, we will provide a transition system semantics, i. e., a directed graph whose vertices are states and whose edges are transitions. Informally, a state is the contents of the program variables, i. e., an assignment of the program variables to values, and the contents of the heap, i. e., an assignment of heap addresses to values; hereby, a value may again be a heap address. We represent both assignments by means of a relational first-order structure with constants. The interpretations of constants make up the contents of the program variables, whereas the interpretations of the relations make up the contents of the heap. More precisely, each field r of a template T is interpreted by a functional binary relation, so $r(u, v)$ means that an instance of T lives at address u

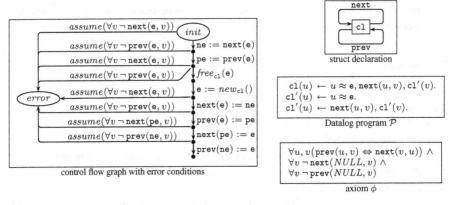

Fig. 1. Replacing an element in a doubly linked circular list by a new one; the initial condition that e points to a doubly linked circular list is expressed by $\phi \wedge \mathcal{P} \wedge \texttt{cl(e)}$.

in the heap, and v is the uniquely defined value of the r-field of that instance. Note that the universe of the first-order structures can be assumed finite since the number of heap addresses is finite (yet unbounded).

Formally, we associate a *vocabulary* σ to a program P, where $\sigma = \langle \bar{r}, \bar{c} \rangle$ declares a set of binary relation symbols $\bar{r}$ and a set of constants $\bar{c}$. Hereby, $\bar{c}$ is the set of program variables occurring in the control flow graph of P, and $\bar{r}$ is the set of fields occurring in the struct declaration of P. A σ-*structure* $\mathbf{A}$ is a tuple $\langle A, \bar{r}^{\mathbf{A}}, \bar{c}^{\mathbf{A}} \rangle$, where A is a non-empty universe, $\bar{r}^{\mathbf{A}} = \{r^{\mathbf{A}} \mid r \in \bar{r}\}$ is a set of binary relations on A interpreting the symbols in $\bar{r}$, and $\bar{c}^{\mathbf{A}} = \{c^{\mathbf{A}} \mid c \in \bar{c}\}$ is a set of elements of A interpreting the constants in $\bar{c}$. A *state* of the program P is a pair $\langle \ell, \mathbf{A} \rangle$ consisting of a location ℓ and a σ-structure $\mathbf{A}$. We require that the universe A is finite and that all relations $r^{\mathbf{A}}$ are functional, i.e., for all $a, b, b' \in A$, if $r^{\mathbf{A}}(a, b)$ and $r^{\mathbf{A}}(a, b')$ then $b = b'$. Additionally, we require that the interpretation of $NULL$ is a dangling pointer, i.e., $r^{\mathbf{A}}(NULL^{\mathbf{A}}, b)$ is false for all $b \in A$ and all relations $r^{\mathbf{A}}$.

Transitions are certain pairs of states. For specifying which pairs, we have to extend the vocabulary σ. We define $\sigma' = \langle \bar{r}', \bar{c}' \rangle$ to be a copy of σ, where $\bar{r}' = \{r' \mid r \in \bar{r}\}$ and $\bar{c}' = \{c' \mid c \in \bar{c}\}$. By $\sigma + \sigma'$, we denote the union of the vocabularies σ and σ'. Given a σ-structure $\mathbf{A}$, we denote the corresponding σ'-structure (with universe A) by $\mathbf{A}'$. Thus, given σ-structures $\mathbf{A}$ and $\mathbf{B}$ with $A = B$, $\mathbf{A} + \mathbf{B}'$ can be viewed as a $(\sigma + \sigma')$-structure. Likewise, a $(\sigma + \sigma')$-formula can be viewed as defining a binary relation on σ-structures. Figure 2 specifies the semantics of actions α as $(\sigma + \sigma')$-formulas $[\![\alpha]\!]$.

A *transition* of the program P is pair $\langle \langle \ell, \mathbf{A} \rangle, \langle \ell', \mathbf{B} \rangle \rangle$ of states such that $A = B$ and $\mathbf{A} + \mathbf{B}' \models [\![\alpha]\!]$, where the action α is the label of some edge from ℓ to ℓ' in the CFG of P. We call $\langle \ell, \mathbf{A} \rangle$ the *pre-state* and $\langle \ell', \mathbf{B} \rangle$ the *post-state* of the transition. Note that $[\![y := s(x)]\!]$ is false if x is dangling, i.e., the semantics models read dereferences of dangling pointers by deadlock. On the other hand, $[\![y := new_T()]\!]$ defines a total relation, so allocation never fails due to lack of memory[1].

---

[1] A slightly modified version of $[\![y := new_T()]\!]$ models failed allocation by returning $NULL$.

$$[\![assume(\gamma)]\!] \equiv \gamma \wedge \bigwedge_{c \in \bar{c}} c' \approx c \wedge \bigwedge_{r \in \bar{r}} r' = r$$

$$[\![y := e]\!] \equiv y' \approx e \wedge \bigwedge_{c \in \bar{c} \setminus \{y\}} c' \approx c \wedge \bigwedge_{r \in \bar{r}} r' = r$$

$$[\![y := s(x)]\!] \equiv s(x, y') \wedge \bigwedge_{c \in \bar{c} \setminus \{y\}} c' \approx c \wedge \bigwedge_{r \in \bar{r}} r' = r$$

$$[\![s(x) := e]\!] \equiv \forall u, v\big(s'(u, v) \Leftrightarrow u \approx x \wedge v \approx e \vee u \not\approx x \wedge s(u, v)\big) \wedge$$
$$\bigwedge_{c \in \bar{c}} c' \approx c \wedge \bigwedge_{r \in \bar{r} \setminus \{s\}} r' = r$$

$$[\![free_T(x)]\!] \equiv \bigwedge_{s \in T} \forall u, v\big(s'(u, v) \Leftrightarrow u \not\approx x \wedge s(u, v)\big) \wedge$$
$$\bigwedge_{c \in \bar{c}} c' \approx c \wedge \bigwedge_{r \in \bar{r} \setminus T} r' = r$$

$$[\![y := new_T()]\!] \equiv y' \not\approx NULL \wedge \bigwedge_{s \in T} \forall u, v(s(u, v) \Rightarrow u \not\approx y') \wedge$$
$$\bigwedge_{s \in T} \forall u, v\big(s'(u, v) \Leftrightarrow u \approx y' \wedge v \approx NULL \vee s(u, v)\big) \wedge$$
$$\bigwedge_{c \in \bar{c} \setminus \{y\}} c' \approx c \wedge \bigwedge_{r \in \bar{r} \setminus T} r' = r$$

Fig. 2. Semantics of actions. Note that the second-order equalities $r' = r$ are to be understood as abbreviations for first-order formulas $\forall u, v\big(r'(u, v) \Leftrightarrow r(u, v)\big)$.

Given a path $\pi = \langle \ell_0, \alpha_1, \ell_1, \ldots, \alpha_n, \ell_n \rangle$ in the CFG, we call a sequence of states $\langle \langle \ell_0, \mathbf{A}_0 \rangle, \langle \ell_1, \mathbf{A}_1 \rangle, \ldots, \langle \ell_n, \mathbf{A}_n \rangle \rangle$ a π-*execution* (of length n) if for $1 \leq i \leq n$, $\langle \langle \ell_{i-1}, \mathbf{A}_{i-1} \rangle, \langle \ell_i, \mathbf{A}_i \rangle \rangle$ is a transition. We call ℓ_n π-*reachable* from $\langle \ell_0, \mathbf{A}_0 \rangle$ if there is a π-execution $\langle \langle \ell_0, \mathbf{A}_0 \rangle, \ldots, \langle \ell_n, \mathbf{A}_n \rangle \rangle$. In general, we call a location ℓ' *reachable* from a state $\langle \ell, \mathbf{A} \rangle$ if there is a path π such that ℓ' is π-reachable from $\langle \ell, \mathbf{A} \rangle$.

2.3 Error Conditions

Run time errors in pointer programs occur when dereferencing dangling pointers or NULL pointers, or when freeing memory that is not allocated. We can check for such errors by introducing error conditions into the control flow graph just before the dangerous actions read, write and deallocate. As expressing the error conditions requires quantifiers, we have to allow more complex conditions in assume-actions.

Let P be a program and let σ be its associated vocabulary. We say that a σ-formula φ (of first-order logic with equality) is in the *Bernays-Schönfinkel* class [3, 21] with n variables[2], denoted by $\varphi \in BS_n$, if φ is equivalent to a sentence $\exists u_1, \ldots, u_m \forall v_1, \ldots, v_n \, \psi$, where ψ is quantifier-free. For expressing error conditions, we admit actions $assume(\gamma)$ where $\gamma \in BS_n$ for $n \geq 0$.

We extend the CFG of the program P to a *CFG with error conditions (ECFG)* in the following way. We introduce a new distinguished location *error*. For every edge from ℓ to ℓ' that is labeled by $y := s(x)$ or $s(x) := e$, we add an edge from ℓ to *error* labeled by $assume(\forall v \neg s(x, v))$. And for every edge from ℓ to ℓ' labeled by $free_T(x)$ and every $r \in T$, we add an edge from ℓ to *error* labeled by $assume(\forall v \neg r(x, v))$. Note that the error condition $\forall v \neg s(x, v)$ for the read- and write-actions is true if and only if the pointer x is dangling, i.e., there is no value for the s-field at address x. This condition also captures *NULL* dereferences since we assume the special address *NULL*

---

[2] We count only universally quantified variables; the other variables can be viewed as constants.

to be a dangling pointer. For deallocation, the error conditions $\forall v \neg r(x, v)$ are true if and only if there is no instance of T at address x, e.g., because it has been deallocated earlier. Note that all error conditions are in BS_1.

2.4   Weakest Preconditions

Let P be a program and let σ be the associated vocabulary. Given an action α and a σ-formula φ, informally the weakest precondition of φ w.r.t. α captures those states which upon execution of α may lead to a state satisfying φ. Formally, the *weakest precondition* $pre(\alpha; \varphi)$ is defined as $pre(\alpha; \varphi) \equiv \exists \bar{r}' \exists \bar{c}' (\llbracket \alpha \rrbracket \wedge \varphi[\bar{r}'/\bar{r}][\bar{c}'/\bar{c}])$. Here, $\varphi[\bar{r}'/\bar{r}][\bar{c}'/\bar{c}]$ is a short hand for $\varphi[r_1'/r_1, \ldots, r_m'/r_m][c_1'/c_1, \ldots, c_n'/c_n]$ where $\bar{r} = \{r_1, \ldots, r_m\}$ and $\bar{c} = \{c_1, \ldots, c_n\}$, i.e., $\varphi[\bar{r}'/\bar{r}][\bar{c}'/\bar{c}]$ is the formula obtained from φ by replacing every relation symbol r by r' and every constant c by c'. Further, $\exists \bar{r}' \exists \bar{c}'$ denotes the existential quantification of all relation symbols r' and all constants[3] c'. Note that $pre(\alpha; \varphi)$ is a second-order formula due to quantification over relations. Given a path π in the CFG of P and a σ-formula φ, we define the weakest precondition $pre(\pi; \varphi)$ of φ w.r.t. π in the usual way by induction on the length of π.

Depending on the actions α, we can rewrite the second-order formula $pre(\alpha; \varphi)$ to an equivalent first-order formula. We need to extended our notion of substitution to allow for the substitution of atomic formulas. Given two formulas φ and ψ and an atomic formula $r(u_1, u_2)$ with free variables u_i, we write $\varphi[\psi/r(u_1, u_2)]$ for the formula obtained from φ by replacing every atomic formula $r(t_1, t_2)$ by the formula $\psi[t_1/u_1, t_2/u_2]$, where the t_i are arbitrary terms. Note that the variables u_i just function as parameters for the terms t_i.

Lemma 1. *Given a σ-formula φ and an action α, we have the following characterization of $pre(\alpha; \varphi)$, where T is a template, $\bar{s} = \{s_1, \ldots, s_n\}$ are the fields of T, s is an arbitrary field, x and y are program variables, e is a program variable or constant (including NULL), and γ is a σ-formula.*

$$pre(assume(\gamma); \varphi) \equiv \gamma \wedge \varphi$$
$$pre(y := e; \varphi) \equiv \varphi[e/y]$$
$$pre(y := s(x); \varphi) \equiv \exists y'(s(x, y') \wedge \varphi[y'/y])$$
$$pre(s(x) := e; \varphi) \equiv \varphi[u \approx x \wedge v \approx e \vee u \not\approx x \wedge s(u, v)/s(u, v)]$$
$$pre(free_T(x); \varphi) \equiv \varphi[u \not\approx x \wedge \bar{s}(u, v)/\bar{s}(u, v)]^4$$
$$pre(y := new_T(); \varphi) \equiv \exists y'(\, y' \not\approx NULL \wedge \bigwedge_{s \in \bar{s}} \forall u, v(s(u, v) \Rightarrow u \not\approx y') \wedge$$
$$\varphi[y'/y][u \approx y' \wedge v \approx NULL \vee \bar{s}(u, v)/\bar{s}(u, v)])$$

Lemma 2. *Let P be a program, π a path in the ECFG and φ a σ-formula. If $\varphi \in BS_n$, $n \geq 2$, then $pre(\pi; \varphi) \in BS_n$. Moreover, the size of $pre(\pi; \varphi)$ is in $O(|\pi|^2 \cdot |\varphi|)$, and the length of the quantifier prefix of $pre(\pi; \varphi)$ is in $O(|\pi| + |\varphi|)$.*

---

[3] We quantify over constants as if they were free variables.

[4] We consider a formula like $\varphi[u \not\approx x \wedge \bar{s}(u, v)/\bar{s}(u, v)]$ to be an abbreviation for the formula $\varphi[u \not\approx x \wedge s_1(u, v)/s_1(u, v)] \ldots [u \not\approx x \wedge s_n(u, v)/s_n(u, v)]$.

2.5 Initial Conditions

Using weakest preconditions, we can propagate an error condition backwards along a given error path $\pi = \langle init, \ldots, error \rangle$ and obtain a condition $pre(\pi; \top)$ expressing precisely when the *error* location is π-reachable from *init*. Due to Lemma 2 and the decidability of the Bernays-Schönfinkel class, we can thus decide whether there is some state $\langle init, \mathbf{A} \rangle$ from which *error* is π-reachable. However, there may be many such states which are irrelevant since they do not satisfy certain initial conditions that we want to impose, e. g., that program variable e points to a doubly linked circular list. Such properties are not expressible in first-order logic, however, list or tree structures, even certain circular ones, can be specified by logic programs. We will express initial conditions as conjunctions $\phi \wedge \mathcal{P} \wedge Q$, where ϕ is a σ-formula in the Bernays-Schönfinkel class with 2 variables, $\mathcal{P}$ is a (restricted) monadic Datalog program and Q is a (ground) query. The use of Datalog allows us, without losing decidability, to express initial data structures of arbitrary size (expressing reachability of all memory cells in such structures usually requires the use of some kind of transitive closure, which often leads to undecidability.) The Datalog program $\mathcal{P}$ will be interpreted over models of ϕ and will extend them with unary relations, hence the *extensional database (EDB)* vocabulary is $\sigma = \langle \bar{r}, \bar{c} \rangle$, and the *intensional database (IDB)* vocabulary $\sigma_I = \langle \bar{p} \rangle$ declares a set of unary predicates $\bar{p}$.

A *monadic Datalog program* $\mathcal{P}$ is a finite set of clauses $A_0 \leftarrow A_1, \ldots, A_k, k \geq 0$, where the *head* A_0 is a unary IDB atom and the *body* $A_1, \ldots, A_k$ is a conjunction of IDB atoms and EDB literals (i. e., possibly negated EDB atoms). A *query* Q is a conjunction $A_1, \ldots, A_k, k \geq 0$, of ground IDB atoms (i. e., IDB atoms of the form $p(c)$). Note that we consider the order of the atoms in queries and clause bodies irrelevant. In section 3, we will impose further restrictions on monadic Datalog programs in order to prove a decidability result.

Let $\mathcal{P}$ be a monadic Datalog program. Given an EDB (i. e., a σ-structure) $\mathbf{A}$, the *least model* of $\mathcal{P}$ over $\mathbf{A}$ is the least extension of $\mathbf{A}$ to a $(\sigma + \sigma_I)$-structure $\mathbf{B}$ such that $\mathbf{B}$ is a model of the clause set $\mathcal{P}$. Given a σ-formula ϕ and a query Q, we say that $\mathbf{A}$ *satisfies* $\phi \wedge \mathcal{P} \wedge Q$, denoted by $\mathbf{A} \models \phi \wedge \mathcal{P} \wedge Q$, if $\mathbf{A}$ is a model of ϕ and the least model of $\mathcal{P}$ over $\mathbf{A}$ is a model of Q. In section 3, we will show that satisfiability of $\phi \wedge \mathcal{P} \wedge Q$ is decidable, provided that ϕ is in BS_2 and $\mathcal{P}$ is a restricted monadic Datalog program.

Figure 1 shows the initial condition $\phi \wedge \mathcal{P} \wedge$ cl(e) for a program operating on a circular doubly linked list. The monadic Datalog program $\mathcal{P}$ together with the query cl(e) ensure that e points to a circular list linked via the next-fields. The axiom ϕ ensures that the binary relation prev is the converse of next, so the circular list is doubly linked. Furthermore, ϕ ensures that the address *NULL* is a dangling pointer. Note that ϕ can not ensure functionality of next and prev, because functionality is not expressible in BS_2. Therefore, we need a semantic restriction on σ-structures, i. e., we will only consider σ-structures where all binary relations are functional.

2.6 The Bounded Model Checking Problem

Let P be a program. Informally, we call P pointer-safe if there is no initial state from which execution of P can result in a runtime error by dereferencing *NULL* or a dangling pointer. To define pointer-safety formally, we assume that the initial conditions of P are given as a conjunction $\phi \wedge \mathcal{P} \wedge Q$, where ϕ is a BS_2-formula, $\mathcal{P}$ a monadic Datalog program and Q a query. We call P *pointer-safe* if for all states $\langle init, \mathbf{A} \rangle$ such that $\mathbf{A} \models \phi \wedge \mathcal{P} \wedge Q$, the location *error* is unreachable from $\langle init, \mathbf{A} \rangle$ in the ECFG. In bounded model checking, we do not solve the full reachability problem but restrict to paths of an a priori bounded length. Since there are only finitely many such bounded paths in the ECFG, for showing decidability of the bounded model checking problem it suffices to restrict to one path. We call an ECFG-path $\pi = \langle init, \dots, error \rangle$ *pointer-safe* if for all $\mathbf{A}$ with $\mathbf{A} \models \phi \wedge \mathcal{P} \wedge Q$, *error* is not π-reachable from $\langle init, \mathbf{A} \rangle$.

Theorem 3. *Let P be a program with initial condition $\phi \wedge \mathcal{P} \wedge Q$, where ϕ is a BS_2-formula, $\mathcal{P}$ a Datalog program and Q a query. Let $\pi = \langle init, \dots, error \rangle$ be an ECFG-path of P. It is decidable whether π is pointer-safe.*

Proof. The path π is pointer-safe if and only if $pre(\pi; \top) \wedge \phi \wedge \mathcal{P} \wedge Q$ is unsatisfiable. As $pre(\pi; \top) \wedge \phi \in BS_2$ by assumption and Lemma 2, the decidability follows from Theorem 8 in section 3. $\square$

For the complexity of bounded model checking, we refer to section 3.3.

It turns out that the program from figure 1 is not pointer-safe. Consider the path $\pi = \langle init, \mathtt{ne} := \mathtt{next(e)}, \ell_1, \dots, \ell_6, assume(\forall v \neg \mathtt{next(pe}, v)), error \rangle$, then the formula $pre(\pi; \top) \wedge \phi \wedge \mathcal{P} \wedge Q$ is satisfiable, which means that the program can crash when executing the last but one action $\mathtt{next(pe)} := \mathtt{e}$. An analysis of the models of $pre(\pi; \top) \wedge \phi \wedge \mathcal{P} \wedge Q$ reveals the reason: If e points to a circular list of length 1 then pe $\approx$ e after the second action, so pe is dangling after $free_{c1}(\mathtt{e})$.

3 Deciding the Bernays-Schönfinkel Class with Datalog

In this section, we develop our main result, the decidability of the 2-variable fragment of the Bernays-Schönfinkel class extended by a class of monadic Datalog programs.

3.1 Syntax and Semantics of Bernays-Schönfinkel with Datalog

We are interested in satisfiability of formulas of the form $\phi \wedge \mathcal{P} \wedge Q$, where

- ϕ is a universal σ-formula in BS_2 (see section 2.3), i.e., ϕ is of the form $\forall u, v\, \psi$ with ψ quantifier-free[5],
- $\mathcal{P}$ is a monadic Datalog program with EDB vocabulary σ and IDB vocabulary σ_I (see section 2.5), and
- Q a query (see section 2.5).

We are not interested in general satisfiability of $\phi \wedge \mathcal{P} \wedge Q$ but we impose two additional restrictions on models $\mathbf{A}$. One restriction is motivated by the fact that we model pointer structures (see section 2.2), the other is used in our decidability proof.

---

[5] We handle non-universal formulas $\phi = \exists z_1, \dots, z_n\, \forall x, y\, \psi \in BS_2$ by extending the vocabulary $\sigma = \langle \bar{r}, \bar{c} \rangle$, adding the variables z_i as constants.

Functionality. We require that all binary relations are functional, i.e., for all EDB predicates r, $\mathbf{A}$ must be a model of $\forall u, v_1, v_2 (r(u, v_1) \land r(u, v_2) \Rightarrow v_1 \approx v_2)$. This ensures that $\mathbf{A}$ represents a pointer structure, which is functional graph since every pointer at any given moment points to at most one heap cell.

Non-Sharing. We require that the binary relations occurring in $\mathcal{P}$ represent pointers in data structures that do not share memory with other data structures defined by $\mathcal{P}$. That is, $\mathbf{A}$ must be a model of all sentences of the form $\forall u_1, u_2, v (s_1(u_1, v) \land s_1(u_2, v) \land u_1 \not\approx u_2 \Rightarrow const(v))$ and $\forall u_1, u_2, v (s_1(u_1, v) \land s_2(u_2, v) \Rightarrow const(v))$, where s_1 and s_2 are two distinct EDB predicates occurring in $\mathcal{P}$ and $const(v)$ is a shorthand for the disjunction $\bigvee_{c \in \bar{c}} v \approx c$. Note that the non-sharing restriction is not imposed on all binary predicates but just on the ones occurring in the Datalog program $\mathcal{P}$.

Obviously, the functionality and non-sharing restrictions are expressible in the Bernays-Schönfinkel class with equality. However, they require more than two variables, so we cannot add them to the formula ϕ but must deal with them on the semantic level.

Besides restrictions on the class of models, we need to impose two restrictions on monadic Datalog programs $\mathcal{P}$. We call $\mathcal{P}$ *tree-automaton-like* (*TA-like* for short) if all clauses are of the form

$$p(u) \leftarrow B_1, \ldots, B_l, r_1(u, v_1), q_1(v_1), \ldots, r_k(u, v_k), q_k(v_k) \qquad (1)$$

for some $k, l \geq 0$, where $u, v_1, \ldots, v_k$ are $k + 1$ distinct variables, $r_1, \ldots, r_k$ are k distinct relation symbols, and the B_i are EDB literals containing at most the variable u. We define the *degree* of a clause of the form (1) to be the natural number k, i.e., the number of IDB atoms in the body. The *degree* of a TA-like monadic Datalog program is the maximal degree of its clauses. We call $\mathcal{P}$ *intersection-free* if for all EDBs $\mathbf{A}$, the least model of $\mathcal{P}$ over $\mathbf{A}$ satisfies all sentences of the form $\forall v (p(v) \land q(v) \Rightarrow const(v))$, where p and q are two distinct IDB predicate symbols. Note that the EDBs are σ-structures satisfying the above functionality and non-sharing restrictions. Viewing the IDB predicates as shape types for heap cells, an intersection-free Datalog program associates to most heap cells only one shape type; i.e., there is no intersection of shape types (except for cells pointed to by program variables).

Besides the model-theoretic semantics from section 2.5 there is a proof theoretic semantics for Datalog programs. For simplicity, we will define the proof-theoretic semantics only for TA-like monadic Datalog programs $\mathcal{P}$. Let $\mathbf{A}$ be an EDB, i.e., a σ-structure. A *fact* $p(a)$ consists of an IDB predicate symbol p and an element $a \in A$. We say that a list of facts $q_1(a_1), \ldots, q_k(a_k)$, $k \geq 0$, *produces* a fact $p(a)$ (w.r.t. $\mathcal{P}$ and $\mathbf{A}$) if $\mathcal{P}$ contains a clause $p(u) \leftarrow B_1, \ldots, B_l, r_1(u, v_1), q_1(v_1), \ldots, r_k(u, v_k), q_k(v_k)$ such that $\mathbf{A}$ satisfies all $r_i(u, v_i)$ and all B_j when interpreting the variables $u, v_1, \ldots, v_k$ by $a, a_1, \ldots, a_k$, respectively. A *proof tree* T for $\mathcal{P}$ w.r.t. $\mathbf{A}$ is an ordered tree where each node is labeled by a fact. Depending on the situation, we call a node n which is labeled by $p(a)$ an *a-node*, a *p-node* or a *$p(a)$-node*. For each $p(a)$-node n in T with k sons $n_1, \ldots, n_k$, $k \geq 0$, labeled by $q_i(a_i)$, we require that the list of facts $q_1(a_1), \ldots, q_k(a_k)$ produces the fact $p(a)$ w.r.t. $\mathcal{P}$ and $\mathbf{A}$. The proof tree T *proves* a ground IDB atom $p(c)$ if its root is labeled by the fact $p(c^\mathbf{A})$. The proof- resp. model-theoretic semantics are linked in that the least model of $\mathcal{P}$ over $\mathbf{A}$ satisfies a ground IDB atom $p(c)$ if and only if $p(c)$ is proven by some proof tree T. Note that for all facts

$$\begin{array}{ll}
\texttt{list}(u) \leftarrow u \approx NULL. & \texttt{tree}(u) \leftarrow u \approx NULL. \\
\texttt{list}(u) \leftarrow \texttt{next}(u,v), \texttt{list}(v). & \texttt{tree}(u) \leftarrow \texttt{left}(u,v), \texttt{right}(u,w), \\
& \qquad \texttt{tree}(v), \texttt{tree}(w). \\
\texttt{dll}(u) \leftarrow u \approx \texttt{1st}, \texttt{next}(u, NULL). & \\
\texttt{dll}(u) \leftarrow \texttt{next}(u,v), \texttt{dll}(v). & \texttt{gtree}(u) \leftarrow u \approx NULL. \\
& \texttt{gtree}(u) \leftarrow \texttt{sons}(u,v), \texttt{gtrees}(v). \\
\texttt{ring}(u) \leftarrow u \approx \texttt{p}, \texttt{next}(u,v), \texttt{ring}'(v). & \texttt{gtrees}(u) \leftarrow u \approx NULL. \\
\texttt{ring}'(u) \leftarrow u \approx \texttt{p}. & \texttt{gtrees}(u) \leftarrow \texttt{tree}(u,v), \texttt{gtree}(v), \\
\texttt{ring}'(u) \leftarrow \texttt{next}(u,v), \texttt{ring}'(v). & \qquad \texttt{next}(u,w), \texttt{gtrees}(w).
\end{array}$$

$$\texttt{prev}(\texttt{fst}, NULL) \wedge \forall u,v\big(u \approx NULL \vee v \approx NULL \vee (\texttt{prev}(u,v) \Leftrightarrow \texttt{next}(v,u))\big)$$
$$\forall u,v\big(u \approx NULL \vee v \approx NULL \vee (\texttt{up}(u,v) \Leftrightarrow \texttt{left}(v,u) \vee \texttt{right}(v,u))\big)$$

Fig. 3. Datalog programs $\mathcal{P}$ and axioms ϕ representing initial conditions.

$p(a)$, we can w. l. o. g. assume that all $p(a)$-subtrees of a proof tree (i. e., all subtrees rooted at $p(a)$-nodes) are isomorphic.

Figure 3 shows examples of monadic Datalog programs $\mathcal{P}$ and BS_2 axioms ϕ that represent several initial conditions. The initial condition "the program variable p points to a list" can be expressed by the formula $\mathcal{P} \wedge \texttt{list}(\texttt{p})$ where $\mathcal{P}$ is the first of the programs on figure 3. The condition "the program variables fst and lst point to the first and the last elements of a doubly linked list" can be expressed by the formula $\phi \wedge \mathcal{P} \wedge \texttt{dll}(\texttt{fst})$ where $\mathcal{P}$ is the second program and ϕ is the first of the axioms. Here, an atom $\texttt{dll}(u)$ expresses that u points to a doubly linked list whose last element is pointed by lst; note that lst (unlike u) is a logical constant. The condition "p points to a binary tree" can be expressed by $\mathcal{P} \wedge \texttt{tree}(\texttt{p})$. In the same way as for doubly linked lists, one may add an axiom (the second one on figure 3) defining a predicate up to obtain a doubly linked binary tree. The condition "p points to a singly linked circular list" can be expressed by $\mathcal{P} \wedge \texttt{ring}(\texttt{p})$ where the Datalog program $\mathcal{P}$ defines two IDB predicates ring and ring'. The corresponding doubly linked circular list can also be defined, as was shown in figure 1. In the last example, general (i. e., arbitrarily branching) singly linked trees can be handled by representing them as trees of lists, i. e., every tree node points (via sons) to a list of sons (singly linked via next), each node of which points to a tree node (via tree). The condition "p points to a arbitrarily branching tree" can be expressed by $\mathcal{P} \wedge \texttt{gtree}(\texttt{p})$.

Note that all these monadic Datalog programs are TA-like and intersection-free, even if they are appear together in one initial condition (provided that the predicate next is suitably renamed to list_next, dll_next, ring_next and gtrees_next).

3.2 Decidability of Bernays-Schönfinkel with Datalog

In this section we prove that satisfiability of formulas of the form $\phi \wedge \mathcal{P} \wedge Q$ is decidable, where ϕ is a universal σ-formula in BS_2, $\mathcal{P}$ is a TA-like intersection-free monadic Datalog program and Q is a query. This is done by showing the small model property for these formulas: Every satisfiable formula has a model of size bounded by a function depending only on the formula. Before we prove this we recall some definitions and lemmas.

Finite Model Property. The proof of the well-known lemma below can be found in [8].

Lemma 4. *Let ϕ be a universal σ-formula in the Bernays-Schönfinkel class with equality. If $\mathbf{A} \models \phi$ and $\mathbf{B}$ is obtained from $\mathbf{A}$ by removing from its universe any number of elements not interpreting constants then $\mathbf{B} \models \phi$.*

The above lemma immediately gives us a finite model property for formulas of the form $\phi \wedge \mathcal{P} \wedge \mathcal{Q}$ with $\mathcal{Q} = q_1(c_1), \ldots, q_k(c_k)$. It is enough to take any model $\mathbf{A}$ of ϕ and any k proof trees T_i proving $q_i(c_i)$ and restrict $\mathbf{A}$ to the interpretations of constants and to the elements that occur in the proof trees T_i.

Corollary 5. *Every satisfiable formula of the form $\phi \wedge \mathcal{P} \wedge \mathcal{Q}$ has a finite model.*

Types. Recall the EDB vocabulary $\sigma = \langle \bar{r}, \bar{c} \rangle$. Let u and v be variables. A 1-atom $\alpha(u)$ is an atomic σ-formula containing at most the variable u, i.e., $\alpha(u)$ is a ground atom or it is of the form $u \approx u$, $u \approx c$, $r(u, u)$, $r(u, c)$ or $r(c, u)$ for $c \in \bar{c}$ and $r \in \bar{r}$; likewise we define 1-atoms $\alpha(v)$. A 2-atom $\alpha(u, v)$ is an atomic σ-formula containing at most the variables u and v, i.e., $\alpha(u, v)$ is a 1-atom or it is of the form $u \approx v$, $r(u, v)$ or $r(v, u)$ for $r \in \bar{r}$. A 1-literal (resp. 2-literal) is a possibly negated 1-atom (resp. 2-atom). A 1-type $\tau(u)$ (resp. $\tau(v)$) is a maximal propositionally consistent conjunction of 1-literals, i.e., all possible 1-atoms $\alpha(u)$ (resp. $\alpha(v)$) occur exactly once in the conjunction $\tau(u)$ (resp. $\tau(v)$). Similarly, a 2-type $\tau(u, v)$ is a maximal propositionally consistent conjunction of 2-literals. Note that there are only finitely many different types: the number of 1-types is $2^{(|\bar{c}|+1)^2(|\bar{r}|+1)}$ and the number of 2-types is $2^{(|\bar{c}|+2)^2(|\bar{r}|+1)}$.

In a given σ-structure $\mathbf{A}$, for every element $a \in A$ there is exactly one 1-type $\tau(u)$ that is true when one assigns a to the variable u; we denote this type $\tau(u)$ by $\tau_{\mathbf{A}}(a)$. Likewise, for every two elements $a, b \in A$ there is exactly one 2-type $\tau(u, v)$, denoted by $\tau_{\mathbf{A}}(a, b)$, that is true when one assigns a to u and b to v. Note that for $a, a', b, b' \in A$, the 2-type equality $\tau_{\mathbf{A}}(a, b) = \tau_{\mathbf{A}}(a', b')$ implies the 1-type equalities $\tau_{\mathbf{A}}(a) = \tau_{\mathbf{A}}(a')$ and $\tau_{\mathbf{A}}(b) = \tau_{\mathbf{A}}(b')$. A type $\tau(u)$ resp. $\tau(u, v)$ is *inhabited* in $\mathbf{A}$ if we have $\tau(u) = \tau_{\mathbf{A}}(a)$ resp. $\tau(u, v) = \tau_{\mathbf{A}}(a, b)$ for some $a, b \in A$. In general, not all types are inhabited in a fixed σ-structure $\mathbf{A}$, e.g., if $\mathbf{A}$ interprets the constants c_1 and c_2 by different elements, no type containing the conjuncts $u \approx c_1$ and $u \approx c_2$ can be inhabited. Therefore, $(|\bar{c}| + 1) \cdot 2^{(|\bar{c}|+1)^2|\bar{r}|}$ resp. $(|\bar{c}|^2 + 2|\bar{c}| + 2) \cdot 2^{(|\bar{c}|+2)^2|\bar{r}|}$ is an upper bound on the number of inhabited 1- resp. 2-types.

In bounded branching structures, we can get a tighter bound on the number of inhabited types. We say that a σ-structure $\mathbf{A}$ is *k-branching*, $k \geq 0$, if for all $r \in \bar{r}$, every $a \in A$ which is not interpreting a constant has at most k predecessors and k successors w.r.t. the relation $r^{\mathbf{A}}$. If $\mathbf{A}$ is k-branching then the number of inhabited 1-types is bounded by $|\bar{c}| + (((|\bar{c}| + 1)^{2k} + 1)^{|\bar{r}|}$, because there are $|\bar{c}|$ types inhabited by elements interpreting constants, and for the types inhabited by the other elements each binary predicate r contributes at most $(|\bar{c}| + 1)^k$ possibilities due to predecessors, $(|\bar{c}| + 1)^k$ due to successors and one due to the self loop. Similarly, the number of inhabited 2-types is bounded by $|\bar{c}|^2 + 2|\bar{c}|(((|\bar{c}| + 1)^{2k} + 3)^{|\bar{r}|} + (((|\bar{c}| + 1)^{4k} + 4)^{|\bar{r}|}$.

Lemma 6. *The number of 2-types inhabited in a σ-structure $\mathbf{A}$ is bounded by a function singly exponential in the size of the vocabulary σ. If $\mathbf{A}$ is k-branching, $k \geq 0$, then* *the bound on the number of inhabited 2-types is exponential in k and in the number of binary relations but polynomial in the number of constants.*

Pumping Lemma. The core of our decidability result is the following technical lemma which "pumps down" big proof trees by compressing long paths, thus bounding the depth of proof trees. With the bound, the small model theorem follows easily. Recall the EDB vocabulary $\sigma = \langle \bar{r}, \bar{c} \rangle$ and the IDB vocabulary $\sigma_I = \langle \bar{p} \rangle$.

Lemma 7. *If a formula of the form $\phi \wedge \mathcal{P} \wedge Q$ is satisfiable then it is satisfiable by a σ-structure $\mathbf{A}$ such that all IDB atoms $q(c)$ in the query Q have proof trees of depth bounded by*

$$|\bar{p}||\bar{c}| + |\bar{p}||\bar{c}| \cdot (\theta \cdot |\bar{p}| \cdot \delta + 1), \tag{2}$$

where δ is the degree of $\mathcal{P}$, and θ is the maximal number of inhabited 2-types in models of $\phi \wedge \mathcal{P} \wedge Q$.

Proof sketch. Let $\mathbf{A}$ be a model of $\phi \wedge \mathcal{P} \wedge Q$ and let T be a proof tree for $\mathcal{P}$ w.r.t. $\mathbf{A}$. Suppose T contains a path of length greater than (2). Then there exist two pairs of nodes $\langle m, n \rangle$ and $\langle m', n' \rangle$ on the path, labeled by $d, e, d', e' \in A$ respectively, such that (among other properties) the 2-types of the pairs $\langle d, e \rangle$ and $\langle d', e' \rangle$ coincide and the nodes n and n' are labeled by the same IDB-predicate q. Because n and n' are q-nodes, we can construct a smaller tree T' by replacing the subtree rooted at n with the subtree rooted at n'. Furthermore, we update the EDB such that in the updated EDB $\mathbf{B}$ a binary relation r is true between the elements d and e' if and only if r is true between d and e in $\mathbf{A}$. By construction of $\mathbf{B}$ and the type equality $\tau_\mathbf{A}(d, e) = \tau_\mathbf{A}(d', e')$, formulas in BS_2 cannot distinguish $\mathbf{A}$ and $\mathbf{B}$. Using this fact, we can show that $\mathbf{B}$ is a model of $\phi \wedge \mathcal{P} \wedge Q$, i.e., T' is a proof tree for $\mathcal{P}$ w.r.t. $\mathbf{B}$, $\mathbf{B}$ satisfies the functionality and non-sharing restrictions, and $\mathbf{B}$ is a model of ϕ. For a detailed proof see [6]. $\square$

Theorem 8. *Let ϕ be a universal σ-formula with 2 variables, let $\mathcal{P}$ be an intersection-free TA-like monadic Datalog program and let Q be a query. If the formula $\phi \wedge \mathcal{P} \wedge Q$ is satisfiable then it has a model of cardinality at most doubly exponential in the size of the formula. Deciding satisfiability of such formulas is in 2-NEXPTIME.*

Proof. By the lemmas 7 and 6, a satisfiable formula has a model $\mathbf{A}$ where the proof trees for the query atoms are bounded by a function singly exponential in the size of the formula, so their size is at most doubly exponential. By the observation following Lemma 4, the model $\mathbf{A}$ can be reduced to a model $\mathbf{B}$ consisting only of interpretations of the constants and elements occurring in the proof trees. $\square$

3.3   Complexity of Bounded Model Checking

It follows from Theorem 8, that the bounded model checking problem from Theorem 3 (see section 2.6) is in 2-NEXPTIME, because the size of the formula $pre(\pi; \top) \wedge \phi \wedge \mathcal{P} \wedge Q$ is polynomial in $|\pi|$ by Lemma 2. The double exponential complexity originates from two sources, the exponential bound on the number of inhabited 2-types and the (linear) degree of the Datalog program, leading to proof trees of exponential depth and double exponential size. In common situations, however, the complexity of bounded model checking can be improved significantly.

In the following, we consider bounded model checking problems for a fixed program P with a fixed initial condition $\phi \wedge \mathcal{P} \wedge Q$. We say that the formula $\phi \wedge \mathcal{P} \wedge Q$ is a *bounded branching formula* if there is $k \geq 0$ such that all models of $\phi \wedge \mathcal{P} \wedge Q$ are k-branching.

Theorem 9. *Let* $\pi = \langle init, \ldots, error \rangle$ *be an ECFG-path of a program P. If the initial condition* $\phi \wedge \mathcal{P} \wedge Q$ *is a bounded branching formula then (for fixed program and fixed initial condition) deciding whether* π *is pointer-safe is in* NExpTime. *If additionally the degree of* $\mathcal{P}$ *is 1 then the problem is in* NPTime.

Proof. Assume that all models of $\phi \wedge \mathcal{P} \wedge Q$ are k-branching for some fixed $k \geq 0$. By Lemma 2, the size of the precondition formula $pre(\pi; \top)$ is polynomial in $|\pi|$, in particular the length of the existential quantifier prefix is linear in $|\pi|$. For checking satisfiability of $pre(\pi; \top) \wedge \phi \wedge \mathcal{P} \wedge Q$, we convert $pre(\pi; \top)$ into a universal formula, which requires extending the EDB vocabulary σ by adding a number (linear in $|\pi|$) of new constants. By Lemma 6, the number of inhabited 2-types in models of $pre(\pi; \top) \wedge \phi \wedge \mathcal{P} \wedge Q$ is polynomial in the number of constants[6], hence polynomial in $|\pi|$. Thus, Lemma 7 yields a polynomial depth bound for proof trees, which implies a singly exponential bound on the size of the models. If the degree of $\mathcal{P}$ is 1, the polynomial depth bound implies a polynomial size bound. $\square$

Functionality and non-sharing restrictions ensure that all initial conditions in our examples are bounded branching formulas. The models of the initial conditions for lists (singly or doubly linked, circular or not) and singly linked trees (binary or general) are all 1-branching, whereas the models of the initial conditions for doubly linked binary trees are 2-branching. Thus for all these data structures, bounded model checking can be done in NExpTime, even if the program manipulates all these data structures simultaneously. Moreover, if a program works on list data structures only then bounded model checking can be done in NPTime, which is the optimal worst-case complexity for BMC of list manipulating programs.

4   Related Work

Automatic verification of pointer programs has received quite some attention recently. Dynamically allocated heap memory and properties such as sharing, cyclicity, and reachability in the heap have been formalized in various logical languages.

Abstraction from possibly unbounded state space to a finite model has been studied in [16, 24, 26]. These approaches use the framework of abstract interpretation to over-approximate the set of reachable states. This is achieved by interpreting program statements and properties in a 3-valued first-order logic with transitive closure (TC). Recently there have been attempts to increase the precision of the approximation by incorporating automated theorem for classical 2-valued first-order logic into the 3-valued setting [26].

---

[6] That the number of 2-types is exponential in k and the number of binary relations is irrelevant because those parameters are fixed.

Other approaches for shape analysis use decidable extensions of first-order fragments to reason about shape graphs. [15] proves decidable the $\exists\forall$-fragment with restricted occurrences of TC and deterministic TC. Unfortunately, without severe restrictions on transitive closure, most decidable fragments of first-order logic become undecidable [15]. In [25], the decidable guarded fixed-point logic μGF [11] is used for shape analysis. In μGF, one can express reachability from specified points along specified paths, but full transitive closure (i. e., reachability between a pair of variables) is inexpressible. Moreover, μGF lacks the finite model property [11] and becomes undecidable when functionality restrictions are added [9].

Special syntactically defined logics for expressing reachability have been designed. The reachability logic $\mathcal{RL}$ defined in [1] is a fragment of 2-variable first-order logic with transitive closure and additional Boolean variables. Expressive logics like PDL and CTL* can be embedded into it. Model checking for $\mathcal{RL}$ is efficient, but decidability of satisfiability has not been investigated.

In order to employ decision procedures for monadic second order logic over trees, Schwartzbach et al. model linked data structures using graph types [18, 20]. Graph types are logical representations of sets of graphs, where each graph has a tree backbone which uniquely defines the other (tree-violating) edges. The approach is similar to ours because the Datalog proof trees can be seen as tree backbones. However, the two approaches differ in how to specify the tree-violating edges. We can (but do not have to) specify global restrictions on the tree-violating edges in a fragment of first-order logic whereas graph types specify their tree-violating edges in a dynamic logic.

Graphs as models for software systems that contain pointers have been studied in [19, 22] where graph logics based on C^2, the 2-variable first-order logic with counting quantifiers have been defined. These logics can also be seen as variants of description logics [2] without fixed-points or transitive closure, hence they can model graphs but cannot express reachability. Via translations to C^2, the logics in [19] and [22] inherit decidability [10].

The functional modal fragment of first-order logic as defined by Herzig [13] is a target logic for mapping basic modal and description logics into the framework of first-order logic. The good computational properties of these logic, i. e., PSPACE-decidability and the finite model property, carry over to the functional modal fragment of first-order logic. In this fragment universal and existential quantification can be permuted [12] (i. e., $\forall\exists$ can be exchanged by $\exists\forall$), hence deciding satisfiability can be reduced to deciding the Bernays-Schönfinkel class.

Reynolds and O'Hearn introduced separation logic [23] and a Hoare-style proof system for local reasoning about pointer programs. In this approach, verification requires manual construction of proofs for Hoare-triplets because the logic is undecidable. Towards more automation, in [5] a decidable fragment of separation logic is studied. To obtain decidability expressiveness has to be sacrificed: the fragment can only specify singly linked lists. On the other hand, besides satisfiability also entailment is decidable, which is crucial for verification.

In hardware verification, bounded model checking using propositional SAT solvers was adopted as a standard technique almost immediately after its introduction by Biere et al. [4] in 1999. Jackson and Vaziri [17] extended SAT-based BMC from hardware circuits to Java-like imperative programs with heap references. Essentially, they translate the program specification and bounded executions of the program to a formula in first-order logic with transitive closure, which they check for satisfiability in small models using a SAT-solver. The use of first-order logic with TC as a specification language is very convenient, however, SAT-checking is only feasible on very small models, i. e., the size of the heap must be bounded a priori to only few cells. Similar approaches to SAT-based BMC of pointer programs are pursued in [7] and [14].

5   Conclusion and Future Work

We proposed a bounded model checking procedure for programs manipulating dynamically allocated pointer structures of arbitrary size. The worst-case complexity of our method is 2-NEXPTIME, but in common cases it goes down to NEXPTIME or even to NPTIME. Our approach is based on a combination of two logics, both of which are efficiently decidable in practice. Therefore, we hope that our algorithm can be implemented (e. g., by integrating a Datalog inference engine into a Bernays-Schönfinkel decision procedure) quite efficiently.

There are several possible directions for the future work. One of them is implementation via combination of decision procedures for Bernays-Schönfinkel class and Datalog. Another one is investigation of further applications of our method, e. g., in the analysis whether counterexamples generated by an abstraction-refinement model checker for pointer programs are spurious. Still another direction is to extend the applicability of the method, e. g., by releasing the non-sharing restriction (currently we are not able to express structures like DAG representation of trees). Further possibility is to extend our bounded model checking (which is a debugging method) to a verification method – this requires the ability to express the negation of initial conditions which leads to Datalog programs with greatest fixed point semantics (as opposed to the least fixed point semantics considered here). Finally, we consider the use of other decidable fragments of the first-order logic whose combination with Datalog could lead to a decidable logic expressive enough for reasoning about pointer structures. A good candidate is C^2, the 2-variable fragment of first-order logic with counting quantifiers, in which restrictions like functionality or non-sharing are easily expressible. As C^2 is closed under negation, a Datalog extension with greatest fixed points should also be well suited for verification of invariants.

References

1. N. Alechina and N. Immerman. Reachability logic: An efficient fragment of transitive closure logic. *Logic Journal of IGPL*, 8:325–337, 2000.
2. F. Baader, D. Calvanese, D. L. McGuinness, D. Nardi, and P. F. Patel-Schneider, editors. *The Description Logic Handbook*. Cambridge University Press, 2003.
3. P. Bernays and M. Schönfinkel. Zum Entscheidungsproblem der Mathematischen Logik. *Math. Annalen*, 99:342–372, 1928.
4. A. Biere, A. Cimatti, E. Clarke, and Y. Zhu. Symbolic model checking without BDD. In *Proc. TACAS'99*, LNCS 1579, pages 193–207. Springer, 1999.

5. R. Bornat, C. Calcagno, and P. O'Hearn. A decidable fragment of separation logic. In *Proc. FSTTCS'04*, LNCS 3328, pages 97–109. Springer, 2004.
6. W. Charatonik, L. Georgieva, and P. Maier. Bounded model checking of pointer programs. Technical Report MPI-I-2005-2-002, Max-Planck-Institut für Informatik, 2005.
7. E. Clarke, D. Kroening, and F. Lerda. A tool for checking ANSI-C programs. In *Proc. TACAS'04*, LNCS 2988, pages 168–176. Springer, 2004.
8. B. Dreben and W. D. Goldfarb. *The Decision Problem. Solvable Classes of Quantificational Formulas*. Addison-Wesley, 1979.
9. E. Grädel. On the restraining power of guards. *J. Symbolic Logic*, 64:1719–1742, 1999.
10. E. Grädel, M. Otto, and E. Rosen. Two-variable logic with counting is decidable. In *Proc. LICS'97*, pages 306–317. IEEE Computer Society Press, 1997.
11. E. Grädel and I. Walukiewicz. Guarded fixed point logic. In *Proc. LICS'99*, pages 45–54. IEEE Computer Society Press, 1999.
12. A. Herzig. *Raisonnement automatique en logique modale et algorithmes d'unification*. PhD thesis, Université Paul Sabatier, Toulouse, 1989.
13. A. Herzig. A new decidable fragment of first order logic, June 1990. In Abstracts of the 3rd Logical Biennial, Summer School & Conference in honour of S. C. Kleene, Varna, Bulgaria.
14. M. Huth and S. Pradhan. Consistent partial model checking. *Electronic Notes in Theoretical Computer Science*, 23, 2003.
15. N. Immerman, A. Rabinovich, T. Reps, M. Sagiv, and G. Yorsh. The boundary between decidability and undecidability for transitive-closure logics. In *Proc. CSL'04*, LNCS 3210, pages 160–174. Springer, 2004.
16. N. Immerman, A. Rabinovich, T. Reps, M. Sagiv, and G. Yorsh. Verification via structure simulation. In *Proc. CAV'04*, LNCS 3114, pages 281–294. Springer, 2004.
17. D. Jackson and M. Vaziri. Finding bugs with a constraint solver. In *Proc. ISSTA'00*, pages 14–25, 2000.
18. N. Klarlund and M. I. Schwartzbach. Graph types. In *Proc. POPL'93*, pages 196–205, 1993.
19. V. Kuncak and M. Rinard. On role logic. Technical Report 925, MIT Computer Science and Artificial Intelligence Laboratory, 2003.
20. A. Møller and M. I. Schwartzbach. The pointer assertion logic engine. In *Proc. PLDI'01*, pages 221–231, 2001.
21. F. Ramsey. On a problem of formal logic. In *Proc. London Mathematical Society*, pages 338–384, 1928.
22. A. Rensink. Canonical graph shapes. In *Proc. ESOP'04*, LNCS 2986, pages 401–415. Springer, 2004.
23. J. Reynolds. Intuitionistic reasoning about shared mutable data structure, 1999. Proc. of the 1999 Oxford-Microsoft Symposium in Honour of Sir Tony Hoare.
24. M. Sagiv, T. Reps, and R. Wilhelm. Parametric shape-analysis problems via 3-valued logic. *ACM TOPLAS*, 24(2):217–298, 2002.
25. T. Wies. Symbolic shape analysis. Master's thesis, MPI Informatik, Saarbrücken, Germany, 2004.
26. G. Yorsh, T. Reps, and S. Sagiv. Symbolically computing most-precise abstract operations for shape analysis. In *Proc. TACAS'04*, LNCS 2988, pages 530–545. Springer, 2004.

PDL with Intersection
and Converse Is Decidable

Carsten Lutz

Institute for Theoretical Computer Science
TU Dresden, Germany
lutz@tcs.inf.tu-dresden.de

Abstract. In its many guises and variations, propositional dynamic logic (PDL) plays an important role in various areas of computer science such as databases, artificial intelligence, and computer linguistics. One relevant and powerful variation is ICPDL, the extension of PDL with intersection and converse. Although ICPDL has several interesting applications, its computational properties have never been investigated. In this paper, we prove that ICPDL is decidable by developing a translation to the monadic second order logic of infinite trees. Our result has applications in information logic, description logic, and epistemic logic. In particular, we solve a long-standing open problem in information logic. Another virtue of our approach is that it provides a decidability proof that is more transparent than existing ones for PDL with intersection (but without converse).

1 Introduction

Propositional Dynamic Logic (PDL) has originally been proposed as a modal logic for reasoning about the behaviour of programs [12, 13, 22]. Since then, the adaptation of PDL to a growing number of applications has led to many modifications and extensions. Nowadays, these additional applications have become the main driving force behind the continuing interest in the PDL family of logics, see e.g. [1, 2, 5, 8, 14]. An important family of variations of PDL is obtained by adding an intersection operator on programs, and possibly additional program operators. Alas, the extension of PDL with intersection (IPDL) is notorious for being "theoretically difficult". This is mostly due to an intricate model theory: in contrast to most other extensions of PDL, the addition of intersection destroys the tree model property in a rather dramatic way. In particular, original PDL and many of its extensions can be decided by using automata on infinite trees [24] or embedding into the alternation-free fragment of Kozen's μ-calculus [16]. By adding intersection to PDL and destroying the tree model property, we leave this framework and thus lose the toolkit of results and techniques that have been established over the last twenty years. Consequently, the results obtained for IPDL are quickly summarized: the first result about the computational properties of PDL with intersection is due to Harel, who proved that satisfiability in IPDL with deterministic programs is undecidable [15]. In

L. Ong (Ed.): CSL 2005, LNCS 3634, pp. 413–427, 2005.

1984, Danecki showed that dropping determinism regains decidability [7]. He also establishes a 2-ExpTime upper bound. It was long unknown whether this upper bound is tight: only in 2004, the ExpTime lower bound for IPDL stemming from original PDL was improved to an ExpSpace one and then even to a tight 2-ExpTime one [17, 18]. An axiomatization for IPDL is long sought, but until now only the axiomatization of relatively weak fragments has been successfully accomplished [4].

It appears that virtually nothing is known about extensions of IPDL. Most strikingly, the natural extension of IPDL with converse programs (*ICPDL*) has never been investigated. The aim of this paper is to perform a first investigation of the computational properties of ICPDL: we show that satisfiability in ICPDL is decidable by developing a (satisfiability preserving) translation into the monadic second order logic of infinite trees (from now on simply called MSO). This result has several interesting consequences:

First, decidability of ICPDL implies decidability of the information logic DAL (*Data Analysis Logic*), a problem that has been open since DAL was proposed in 1985 [11]. The purpose of DAL is to aggregate data into sets that can be characterized using given properties, and, dually, to determine properties that best characterize a given set of data. Technically, DAL may be viewed as the variant of IPDL obtained by requiring all relations to be equivalence relations and admitting only the program operators $\cap$ and $\cup^*$, where the latter is a combination of PDL's operators $\cup$ and $\cdot^*$. In ICPDL, equivalence relations can be simulated using $(a \cup a^-)^*$ for some atomic program a. Thus, DAL can be viewed as a fragment of ICPDL.

Second, there is a close correspondence between variants of PDL and description logics (DLs). In particular, the description logic $\mathcal{ALC}_{\mathrm{reg}}$ [3, 14] is a syntactic variant of PDL without the test operator [23], and the intersection operator of IPDL corresponds to the intersection role constructor in description logics. The latter is a traditional constructor that is present in many DL formalisms, see e.g. [6, 9, 20, 21]. Decidability and complexity results play a central role in the area of description logic, but have never been obtained for the natural extension $\mathcal{ALC}_{\mathrm{reg}}^{\cap}$ of $\mathcal{ALC}_{\mathrm{reg}}$ with role intersection. Clearly, $\mathcal{ALC}_{\mathrm{reg}}^{\cap}$ is a syntactic variant of test-free ICPDL, and thus our decidability result carries over.

Third, ICPDL can be applied to obtain results in epistemic logic [10]. The basic observation is as in the case of DAL: ICPDL can simulate equivalence relations by writing $(a \cup a^-)^*$. Since union and transitive closure of programs can be combined to express the common knowledge operator of epistemic logic, and intersection of programs corresponds to the distributed knowledge operator, decidability of ICPDL can be used to obtain decidability for epistemic logic with both common knowledge and distributed knowledge. We should admit, however, that this approach is rather brute force: since the common knowledge and distributed knowledge operators of epistemic logic cannot be nested to build up more complex operations on relations, epistemic logic lacks much of the complexity of ICPDL. Therefore and as noted in [10], decidability can also be obtained using more standard techniques.

Apart from the applications just mentioned, we believe that there is an additional virtue of the MSO translation exhibited in this paper: without intending to derogate the admirable work of Danecki that provided the basic ideas for the tree encoding of ICPDL models developed in this paper [7], it seems fair to claim that Danecki's decidability proof for IPDL is rather intricate and difficult to understand. Moreover, the correctness is hard to verify since the only available presentation (a conference paper) lacks many non-trivial details. Although the MSO translation presented in the current paper also involves some non-trivial encodings, in our opinion it is the easiest proof of the decidability of IPDL that has been obtained so far. Together with the technical report accompanying this paper [19], the proofs are fully rigorous and readily checked in detail.

This paper is organized as follows. In Section 2, we introduce ICPDL. Section 3 prepares for the MSO translation by discussing, on an intuitive level, how ICPDL models can be abstracted into trees. The translation itself is exhibited in Section 4 which also contains a correctness proof. We discuss future work and conclude in Section 5.

2 The Language

Let Var and Prog be countably infinite sets of propositional variables and atomic programs, respectively. The sets of *ICPDL programs* and *ICPDL formulas* are defined by simultaneous induction as follows:

- each atomic program is a program;
- each propositional variable is a formula;
- if α and β are programs and φ is a formula, then the following are also programs:

$$\alpha^-, \ \alpha \cap \beta, \ \alpha \cup \beta, \ \alpha;\beta, \ \alpha^*, \ \varphi?$$

- if φ and ψ are formulas and α is a program, then the following are also formulas:

$$\neg\varphi, \ \langle\alpha\rangle\varphi$$

We use $\varphi_1 \wedge \varphi_2$ as an abbreviation for $\langle\varphi_1?\rangle\varphi_2$, $\varphi_1 \vee \varphi_2$ for $\neg(\neg\varphi_1 \wedge \neg\varphi_2)$, and $[\alpha]\varphi$ for $\neg\langle\alpha\rangle\neg\varphi$. Moreover, we use $\top$ to abbreviate an arbitrary (but fixed) propositional tautology, and $\bot$ for $\neg\top$.

The semantics of ICPDL is defined in the usual way through Kripke structures. A *Kripke structure* is a triple $K = (W, R, L)$, where

- W is a set of points,
- R assigns to each atomic program $a \in$ Prog a binary relation $R(a)$ on W,
- L assigns to each atomic proposition $p \in$ Var the set of points $L(p)$ in which it holds.

The extension of R to complex programs and the definition of the consequence relation $\models$ for ICPDL are, again, by simultaneous induction:

$$\begin{aligned}
R(\alpha^-) &\quad \text{is the converse of } R(\alpha) \\
R(\alpha_1 \cap \alpha_2) &= R(\alpha_1) \cap R(\alpha_2), \\
R(\alpha_1 \cup \alpha_2) &= R(\alpha_1) \cup R(\alpha_2), \\
R(\alpha_1; \alpha_2) &= R(\alpha_1) \circ R(\alpha_2). \\
R(\alpha^*) &\quad \text{is the reflexive-transitive closure of } R(\alpha) \\
R(\varphi?) &= \{(w,w) \in W^2 \mid K, w \models \varphi\}
\end{aligned}$$

$$\begin{aligned}
K, w \models p &\quad \text{iff } w \in L(p) \text{ for } p \in \mathsf{Var} \\
K, w \models \neg\varphi &\quad \text{iff } K, w \not\models \varphi \\
K, w \models \langle \alpha \rangle \varphi &\quad \text{iff there is } w' : (w, w') \in R(\alpha) \text{ and } K, w' \models \varphi
\end{aligned}$$

Let φ be a formula and $K = (W, R, L)$ a Kripke structure. Then K is a *model* of φ if there is a $w \in W$ with $K, w \models \varphi$. The formula φ is called *satisfiable* if it has a model.

3   ICPDL Models

Our aim is to devise a satisfiability preserving translation from ICPDL to MSO over infinite trees. The main difficulty is posed by the fact that ICPDL does not have the tree model property. This is witnessed e.g. by the formulas

$$\neg p \wedge \langle a \cap a^- \rangle p \quad \text{and} \quad \neg p \wedge [b]\bot \wedge \langle (a; p?; a) \cap b^* \rangle \top$$

which both enforce a cycle of length 2 [1]. To carry out the translation to MSO, it is important to develop a tree-shaped abstraction of ICPDL models. Such an abstraction is described in the current section. Although it provides the guiding intuitions for developing the translation to MSO, there is no need to formally establish the correctness of the abstraction beforehand. Therefore, our discussion will remain on an intuitive level.

Intersection

ICPDL's lack of the tree model property is clearly due to the intersection operator on relations. Even the simple formula $\langle a \cap b \rangle \top$ does not have a tree model: it enforces a Kripke structure K as shown on the left-hand side of Figure 1. For the MSO translation, we represent K using the tree displayed on the right-hand side of the same figure. In this tree, the left son represents the substructure of K that is obtained by dropping the b edge, and the right son describes the substructure obtained by dropping the a edge. The symbol "$\cap$" labelling the root node indicates that a parallelization operation is required to construct K from these two substructures: simply identify their roots and sinks. Intuitively, the root node represents the whole structure K.

The tree representation does not only encode the relational structure of K, but also records satisfaction of relevant formulas by states of K. The following

---

[1] It is easy to modify these formulas such that they enforce a cycle whose length is exponential in the length of the formula.

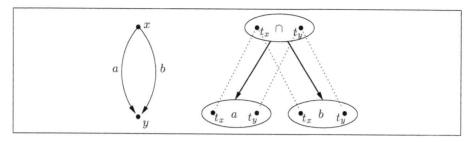

Fig. 1. Tree for intersection.

definition fixes the set of formulas relevant for deciding satisfiability of an ICPDL formula φ: the (Fischer-Ladner) closure of φ.

Definition 1 (Closure). *The set of subprograms* $\mathsf{subp}(\alpha)$ *of ICPDL programs* α *and the set of subformulas* $\mathsf{subf}(\varphi)$ *of ICPDL formulas* φ *is defined simultaneously as follows:*

- $\mathsf{subp}(a) = \{a\}$ *if* a *is atomic;*
- $\mathsf{subp}(\alpha) = \{\alpha\} \cup \mathsf{subp}(\beta) \cup \mathsf{subp}(\gamma)$ *if* $\alpha = \beta \cap \gamma$ *or* $\alpha = \beta; \gamma$;
- $\mathsf{subp}(\alpha) = \{\alpha\} \cup \mathsf{subp}(\beta)$ *if* $\alpha = \beta^*$ *or* $\alpha = \beta^-$;
- $\mathsf{subp}(\varphi?) = \{\varphi?\} \cup \bigcup_{\langle\beta\rangle\psi\in\mathsf{subf}(\varphi)} \mathsf{subp}(\beta)$;
- $\mathsf{subf}(p) = \{p\}$ *if* $p \in \mathsf{Var}$;
- $\mathsf{subf}(\neg\varphi) = \{\neg\varphi\} \cup \mathsf{subf}(\varphi)$;
- $\mathsf{subf}(\langle\alpha\rangle\varphi) = \{\langle\alpha\rangle\varphi\} \cup \mathsf{subf}(\varphi) \cup \bigcup_{\psi?\in\mathsf{subp}(\alpha)} \mathsf{subf}(\psi)$.

Finally, we define the closure *of an ICPDL formula* φ *as*

$$\mathsf{cl}(\varphi) := \{\psi, \neg\psi \mid \psi \in \mathsf{cl}(\varphi)\}.$$

For x a state in a Kripke structure, the *type* of x is the set of formulas $\{\varphi \in \mathsf{cl}(\varphi_0) \mid K, x \models \varphi\}$, where φ_0 is the formula whose satisfiability is to be decided. In the tree representation of a model, each node stores the type of the root state and of the sink state of the substructure that this node represents. In the case of Figure 1, all three tree nodes store the type t_x of x and t_y of y since they all describe a substructure of K with root x and sink y. We say that t_x is stored *in the first place* of each node, and t_y is stored *in the second place*. Observe that distinct places in tree nodes may represent identical states in the model. This induces an equivalence relation on places, whose skeleton is given as dotted lines in Figure 1. This relation will play a central role in the translation to MSO.

Composition

Now consider a formula $\langle a; b\rangle\top$. It enforces the model on the left-hand side of Figure 2. Again, the right-hand side displays the corresponding tree abstraction with the dotted edges providing a skeleton for the equivalence relation on places. The symbol ";" of the root nodes indicates that the structure represented by the

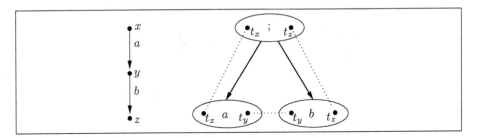

Fig. 2. Tree for composition.

root node is obtained from the structures represented by the leaves through a composition operation: identify the sink of the left son with the root of the right son.

Kleene Star

Formulas $\langle a^* \rangle \top$ enforce an a-path of arbitrary length. To represent a path of length zero (i.e., a single state), we use a tree consisting of a single node labelled "=". The two places of this node are equivalent, i.e., represent the same state. To represent longer paths, we may repeatedly apply the composition operation to nodes labelled "a" and "=". A tree representation of a path of length two can be found in Figure 3.

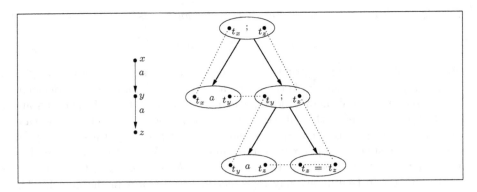

Fig. 3. Tree for Kleene star.

Observe the dotted edge connecting the two places of the "=" node. It should be clear that, by combining the representation schemata given in Figures 1 and 2 and by using "=" nodes, we can construct a tree representation of models enforced by any formula $\langle \alpha \rangle \top$, with α composed from the operators $\{\cup, \cap, \varphi?, ;, {}^*\}$ in an arbitrary way: the operator "$\cup$" requires no explicit representation in the tree structure and the operator "$\varphi?$" can be treated via a node labelled "=".

Converse

To deal with the converse operator, we take an approach that may not be what one would expect on first sight. As discussed later, the seemingly complicated treatment of converse allows to simplify other parts of the MSO translation. Consider a formula $\langle a^- \rangle \top$ and the enforced model given on the right-hand side of Figure 4.

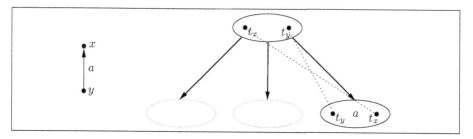

Fig. 4. Tree for converse programs.

Until now, all considered models have been abstracted into *binary* trees. For dealing with converse, we switch to *ternary* trees. The Kripke structure from Figure 4 is represented by the tree given on the right-hand side of the same figure. The third son represents the structure in which there is an a-edge from root y to sink x, i.e., the horizontal mirror image of the Kripke structure on the left. In contrast, the root represents the original structure, where there is an a-edge from *sink* y to *root* x. Observe that the equivalence relation induced by the pointed edges swaps the places of the root and the third son as expected. Also observe that the root node does not have a particular type such as "$\cap$" or ";". We need not introduce a dedicated type for converse since, for technical reasons discussed below, *every* node in the tree has a third son whose places are obtained by swapping the places of the original node. Finally, note that the first and second son of the root are simply dummies. Although they will be required to exist for technical reasons, intuitively they carry no meaningful information.

Multiple Diamonds

So far, we have mostly concentrated on tree abstractions of models for simple formulas of the form $\langle \alpha \rangle \varphi$. Tree abstractions of models for arbitrarily shaped formulas can be obtained by joining, in a suitable way, the tree abstractions of models for such simple formulas. Consider the formula $\langle a; b \rangle \top \wedge \langle c \rangle \top$, which enforces the structure shown on the left-hand side of Figure 5. As usual, the tree abstraction is shown on the right-hand side. The root together with the first two sons are the tree abstraction of the substructure witnessing $\langle a; b \rangle \top$, where the dotted edges are as in Figure 2 but omitted for simplicity. The third son exists because every node is required to have a third son. The dotted edges

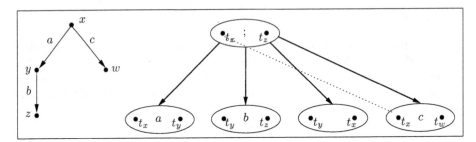

Fig. 5. Tree for multiple diamonds.

connecting the root and the third son are as in Figure 4, but again omitted. Finally, the fourth son by itself (i.e., without the root) is the tree abstraction of the substructure witnessing $\langle c \rangle \top$.

The ratio of this representation is as follows: suppose that a state x in a Kripke structure satifies multiple diamonds $\langle \alpha_1 \rangle \varphi_1, \ldots, \langle \alpha_k \rangle \varphi_k$. For $1 \leq i \leq k$, we take the representation of the model enforced by $\langle \alpha_i \rangle \varphi_i$ as a ternary tree as described above. Let these trees be $T_1, \ldots, T_k$. To join them into a single tree, we attach the roots of $T_2, \ldots, T_k$ as sons number 4 to $k+3$ to the root of T_1. Observe that, in the resulting tree, the first place of the root node is equivalent to the first place of sons number 4 to $k+3$. This is indicated by the dotted edge in Figure 5.

Using this method, we can deal with the problem that a state represented by the *left*-hand place of a tree node may have to satisfy more than a single diamond. What will we do if a state x represented by a right-hand place of a tree node has to satisfy diamonds $\langle \alpha_1 \rangle \varphi_1, \ldots, \langle \alpha_k \rangle \varphi_k$? We simply exploit the fact that every node has a third son swapping the places: we attach the trees $T_1, \ldots, T_k$ representing the models enforced by the diamonds $\langle \alpha_1 \rangle \varphi_1, \ldots, \langle \alpha_k \rangle \varphi_k$ as sons number 4 to $k+4$ to the third son of the node whose right-hand place represents x. By composing the dotted edges displayed in Figures 4 and 5, it is easily verified that, then, the second place of the root of T_1 is equivalent to the first place of the root of T_2 as required.

4   Translation to MSO

We now put the ideas developed in the previous section to work. The goal is to prove the main result of this paper:

Theorem 1. *Satisfiability in PDL with intersection and converse is decidable.*

Let φ_0 be an ICPDL formula whose satisfiability is to be decided. Moreover, let k be the number of diamond formulas $\langle \alpha \rangle \varphi$ in $\mathsf{cl}(\varphi_0)$. We translate φ_0 into an eqi-satisfiable formula φ_0^* of monadic second-order logic of the infinite $k+3$-ary tree. More precisely, we assume MSO models to have domain $\{1, \ldots, k+3\}^*$, which from now on we abbreviate with $[k+3]^*$. There are $k+3$ unary functions s_i mapping each node to it's i-th son.

Intuitively, the formula φ_0^* is constructed such that the models of φ_0^* are precisely the tree abstractions of models of φ_0. In particular, the intuition behind the $k + 3$ successors is as explained in the previous section. The assembly of φ_0^* involves several steps. First, we fix the MSO signature used:

- unary predicates F_φ^1 and F_φ^2 for every $\varphi \in \mathsf{cl}(\varphi_0)$;
- unary predicates $T_=$, $T_\cap$, $T_;$, and $T_\perp$;
- a unary predicate T_a for each atomic program a.

The predicates F_φ^i are used to store types in the first and second place of tree nodes (c.f. previous section): if $\mathfrak{M}$ is an MSO model and $x \in [k + 3]^*$, then $\{\varphi \mid \mathfrak{M} \models F_\varphi^1(x)\}$ is the type stored in the first place of x and $\{\varphi \mid \mathfrak{M} \models F_\varphi^2(x)\}$ is the type stored in the second place of x.

The predicates T_a, $T_=$, $T_\cap$, $T_;$, and $T_\perp$ are markers for the different kinds of nodes in trees. The only kind of node that was not discussed in the previous section is $T_\perp$. This kind of node is used when the i-th son is not needed, for some i with $3 < i \leq k + 3$. For example, assume that $\mathfrak{M} \not\models F_\varphi^1(x)$ for some node $x \in [k + 3]^*$ and all formulas $\varphi \in \mathsf{cl}(\varphi_0)$ of the form $\langle a \rangle \varphi$. Then the sons $x4, \ldots, x(k + 3)$ of x are not needed. Since our MSO models should be full $k + 3$-ary trees, we simply mark such sons with $T_\perp$.

To ensure that the sets $\{\varphi \mid \mathfrak{M} \models F_\varphi^1(x)\}$ describe valid types, we have to describe the semantics of negation and of diamonds—recall that all other operators are merely abbreviations. Dealing with negation is easy:

$$\psi_1^* := \bigwedge_{\neg\varphi \in \mathsf{cl}(\varphi_0)} \forall x \, . \, F_{\neg\varphi}^1(x) \leftrightarrow \neg F_\varphi^1(x) \wedge$$

$$F_{\neg\varphi}^2(x) \leftrightarrow \neg F_\varphi^2(x)$$

To treat diamonds, we need some preliminaries. First, we define a formula with two free variables that characterizes the identitiy of places as discussed in the previous section. More precisely, it is convenient to define four such formulas $\chi_{i,j}$, $i, j \in \{1, 2\}$, as shown in Figure 6. Intuitively, we have $\mathfrak{M} \models \chi_{i,j}[x, y]$ iff the i'th place of x is equivalent to the j'th place of y. According to the idea of place equivalence, all equivalent places should have the same type:

$$\psi_2^* := \bigwedge_{i,j \in \{1,2\}} \forall x, y \, . \, \chi_{i,j}(x, y) \rightarrow (\bigwedge_{\varphi \in \mathsf{cl}(\varphi_0)} F_\varphi^i(x) \leftrightarrow F_\varphi^j(y))$$

We now define, for each program $\alpha \in \mathsf{subp}(\varphi_0)$, a formula σ_α that relates the *first* place of a node x to the *second* place of a node y iff the states represented by these two places are related via the program α: for each $\alpha \in \mathsf{subp}(\varphi_0)$, set:

- $\sigma_a(x, y) := \exists z . \chi_{1,1}(x, z) \wedge T_a(z) \wedge \chi_{2,2}(y, z)$;
- $\sigma_{\varphi?}(x, y) := \chi_{1,2}(x, y) \wedge F_\varphi^1(x)$;
- $\sigma_{\alpha \cup \beta}(x, y) := \sigma_\alpha(x, y) \vee \sigma_\beta(x, y)$;

$$\vartheta(P_1, P_2) := \forall z.(T_=(z) \rightarrow (P_1(z) \leftrightarrow P_2(z))) \wedge \tag{1}$$

$$\forall z.(T_\sqcap(z) \rightarrow (P_1(s) \leftrightarrow P_1(s_1(z)))) \wedge \tag{2}$$

$$\forall z.(T_\sqcap(z) \rightarrow (P_1(s) \leftrightarrow P_1(s_2(z)))) \wedge \tag{3}$$

$$\forall z.(T_\sqcap(z) \rightarrow (P_2(s) \leftrightarrow P_2(s_1(z)))) \wedge \tag{4}$$

$$\forall z.(T_\sqcap(z) \rightarrow (P_2(s) \leftrightarrow P_2(s_2(z)))) \wedge \tag{5}$$

$$\forall z.(T_;(z) \rightarrow (P_1(z) \leftrightarrow P_1(s_1(z)))) \wedge \tag{6}$$

$$\forall z.(T_;(z) \rightarrow (P_2(z) \leftrightarrow P_2(s_2(z)))) \wedge \tag{7}$$

$$\forall z.(T_;(z) \rightarrow (P_2(s_1(z)) \leftrightarrow P_1(s_2(z)))) \wedge \tag{8}$$

$$\forall z.(P_1(z) \leftrightarrow P_2(s_3(z))) \wedge \tag{9}$$

$$\forall z.(P_2(z) \leftrightarrow P_1(s_3(z))) \wedge \tag{10}$$

$$\bigwedge_{3 < \ell \leq k+3} \forall z.(\neg T_\perp(s_\ell(z)) \rightarrow (P_1(z) \leftrightarrow P_1(s_\ell(z)))) \tag{11}$$

$$\chi_{i,j}(x, y) := \forall P_1, P_2.(P_i(x) \wedge \vartheta(P_1, P_2)) \rightarrow P_j(y)$$

Fig. 6. The formulas $\chi_{i,j}(x, y)$.

- $\sigma_{\alpha \cap \beta}(x, y) := \sigma_\alpha(x, y) \wedge \sigma_\beta(x, y)$;
- $\sigma_{\alpha;\beta}(x, y) := \exists z, z'.\sigma_\alpha(x, z) \wedge \chi_{2,1}(z, z') \wedge \sigma_\beta(z', y)$;
- $\sigma_{\alpha^-}(x, y) := \sigma_\alpha(s_3(y), s_3(x))$;
- $\sigma_{\alpha^*}(x, y) := \chi_{1,2}[x, x] \vee \forall P.\big((P(s_3(x)) \wedge \vartheta'_\alpha(P)) \rightarrow P(y) \big)$

with

$$\vartheta'_\alpha(P) := \forall x, y, z.\big((P(x) \wedge \chi_{2,1}(x, y) \wedge \sigma_\alpha(y, z)) \rightarrow P(z) \big)$$

Some remarks are in order. To see why σ_a does not simply read $x = y \wedge T_a(x)$, consider Figure 1: the left place of the root node is clearly related to the right place of the root node via the program a although the root is not labelled "a". In $\sigma_{\alpha;\beta}$, the middle conjunct is necessary since we only relate first places to second places. The formula σ_{α^-} is easily understood by considering the equivalence of places indicated in Figure 4. Finally, consider σ_{α^*}. The first disjunct reflects the fact that, in Kripke structures, α^* relates every state to itself. The formula $\vartheta'_\alpha(P)$ states that the set of nodes P is closed under making α-steps from second places of nodes in P: if $x \in P$, the second place of x is equivalent to the first place of some y, and y is related to some z via σ_α, then the second place of z can be reached from the second place of x by making an α transition and we add z to P. Note that, in the definition of σ_{α^*}, we put $s_3(x)$ into P as the initial element rather than x. This is necessary since σ_{α^*} relates first places to second places, but $\vartheta'_\alpha(P)$ closes off under making α-steps from *second* places of nodes in P. Moreover, the second place of $s_3(x)$ is clearly equivalent to the first place of x.

Using the formulas σ_α, we can now describe the semantics of diamonds:

$$\psi_3^* := \bigwedge_{\langle\alpha\rangle\varphi\in\mathsf{cl}(\varphi_0)} \forall x \,.\, F_{\langle\alpha\rangle\varphi}^1(x) \leftrightarrow \exists y.\sigma_\alpha(x,y) \wedge F_\varphi^2(y))$$

It pays off here that we require every node to have a third son with swapped places: due to this son, there is no need to explicitly describe the semantics of diamonds satisfied by second places, i.e., recorded via formulas $F_{\langle\alpha\rangle\varphi}^i(x)$ with $i = 2$. We thus save the definition of counterparts of the formulas σ_α that relate second places to first places. Also, there is no need to define counterparts of the formulas σ_α that relate first places to first places, or second places to second places: via the third son, such relationships can always be understood as a relationship from a first place to a second place.

Finally, we assemble φ_0^*:

$$\varphi_0^* := \psi_1^* \wedge \psi_2^* \wedge \psi_3^* \wedge \exists x.F_{\varphi_0}^1(x)$$

In [19], we prove correctness of the translation:

Lemma 1. φ_0 is satisfiable in ICPDL iff φ_0^* is satisfiable in MSO.

For the "if" direction, assume that φ_0^* is satisfiable in MSO, i.e. there is an MSO structure $\mathfrak{M}$ based on a tree of out-degree $k + 3$ such that φ_0^* is satisfied in $\mathfrak{M}$. Let $P := [k + 3]^* \times \{1, 2\}$ be the set of *places*. We define the relation $\sim$ on P by setting $(x, i) \sim (y, j)$ iff $\mathfrak{M} \models \chi_{i,j}[x, y]$. It is not hard to show that $\sim$ is an equivalence relation. Let $[x, i]$ denote the equivalence class of $(x, i) \in P$ w.r.t. $\sim$. We define a Kripke structure $K = (W, R, L)$ as follows:

- $W = \{[x, i] \mid (x, i) \in P\}$;
- $R(a) = \{([x, 1], [y, 2]) \mid \mathfrak{M} \models \sigma_a[x, y]\}$ for all atomic programs a;
- $L(p) = \{[x, 1] \mid x \in (F_p^1)^\mathfrak{M}\} \cup \{[x, 2] \mid x \in (F_p^2)^\mathfrak{M}\}$ for all $p \in \mathsf{Var}$.

Note that K is well-defined: due to φ_2^*, $(x, 1) \sim (y, 1)$ implies that $x \in (F_p^1)^\mathfrak{M}$ iff $y \in (F_p^1)^\mathfrak{M}$ for all $p \in \mathsf{Var}$, and likewise for F_p^2. Additionally, by definition of σ_a, $(x, 1) \sim (x', 1)$ and $(y, 2) \sim (y', 2)$ implies that $\mathfrak{M} \models \sigma_a[x, y]$ iff $\mathfrak{M} \models \sigma_a[x', y']$, for all atomic programs a.

In [19], we then prove the following, central claim.

Claim. For all $x, y \in [k + 3]^*$, $\varphi \in \mathsf{cl}(\varphi_0)$, and $\alpha \in \mathsf{subp}(\varphi_0)$, we have

1. $([x, 1], [y, 2]) \in R(\alpha)$ iff $\mathfrak{M} \models \sigma_\alpha[x, y]$;
2. $\mathfrak{M} \models F_\varphi^i[x]$ iff $K, [x, i] \models \varphi$

Since φ_0^* is satisfied in $\mathfrak{M}$, there is an $x \in [k + 3]^*$ such that $\mathfrak{M} \models F_{\varphi_0}^1[x]$. By Point 2 of the claim, this implies that K is a model of φ_0.

For the "only if" direction, let $K = (W, R, L)$ be a model of φ_0, and let $w_0 \in W$ such that $K, w_0 \models \varphi_0$. To construct an MSO model with domain $[k + 3]^*$ satisfying φ_0^* at the root, we inductively define three mappings

$$\tau_1 : [k + 3]^* \to W$$
$$p : [k + 3]^* \to \mathsf{subp}(\varphi_0) \cup \{\varepsilon, \bot\}$$
$$\tau_2 : [k + 3]^* \to W$$

such that the following condition is satisfied:

$$\text{for all } x \in [k + 3]^*, p(x) \neq \bot \text{ implies } (\tau_1(x), \tau_2(x)) \in R(p(x)), \qquad (\dagger)$$

where $R(\varepsilon)$ is defined as the identitiy relation on W. Intuitiy, $\tau_1(x)$ identifies the state described by the first place of x, $\tau_2(x)$ identifies the state described by the second place of x, and $p(x)$ is the program that we want to hold between these two places. The case $p(x) = \bot$ means that the mapping $p(\cdot)$ carries no relevant information for the node x. Before we can start the definition, we need some preliminaries. First, we assume that the diamond formulas in $\mathsf{cl}(\varphi_0)$ are linearly ordered, and that $\mathcal{E}_i$ yields the i-th such formula (the numbering starts with 0). Second, we call a program α *determined* if the top-level operator is not "$\cup$". We inductively fix a choice function ch that maps every triple $(w, \alpha, w') \subseteq W \times \mathsf{subp}(\varphi_0) \times W$ with $(w, w') \in R(\alpha)$ to a determined program $\mathsf{ch}(w, \alpha, w') \in \mathsf{subp}(\alpha)$ such that $R(\mathsf{ch}(w, \alpha, w')) \subseteq R(\alpha)$ and $(w, w') \in R(\mathsf{ch}(w, \alpha, w'))$: let $(w, w') \in R(\alpha)$.

- if α is determined, set $\mathsf{ch}(w, \alpha, w') := \alpha$.
- if α is not determined, then $\alpha = \beta \cup \gamma$. By the semantics, $(w, w') \in R(\alpha)$ implies $(w, w') \in R(\beta)$ or $(w, w') \in R(\gamma)$. In the first case, set $\mathsf{ch}(w, \alpha, w') := \beta$ if β is determined, and $\mathsf{ch}(w, \alpha, w') := \mathsf{ch}(w, \beta, w')$ otherwise. In the second case, set $\mathsf{ch}(w, \alpha, w') := \gamma$ if γ is determined, and $\mathsf{ch}(w, \alpha, w') := \mathsf{ch}(w, \gamma, w')$ otherwise.

Now, the three mappings are defined simultaneously by making a case distinction as follows. To understand this definition, it may help to recall the intuitions laid out in Section 3.

1. To start, set $\tau_1(\varepsilon) := w_0$, $p(\varepsilon) := \varepsilon$, and $\tau_2(\varepsilon) := w_0$. (The choice of $p(\varepsilon)$ and $\tau_2(\varepsilon)$ is not crucial).
2. Let $\tau_1(x)$ be defined, $\tau_1(s_1(x))$ undefined, and $p = \alpha_1 \cap \alpha_2$. Then set, for $i \in \{1, 2\}$: $\tau_1(s_i(x)) := \tau_1(x)$, $p(s_i(x)) := \mathsf{ch}(\tau_1(x), \alpha_i, \tau_2(x))$, and $\tau_2(s_i(x)) := \tau_2(x)$.
3. Let $\tau_1(x)$ be defined, $\tau_1(s_1(x))$ undefined, and $p = \alpha; \beta$. By $(\dagger)$ and the semantics, there is a $w \in W$ with $(\tau_1(x), w) \in R(\alpha)$ and $(w, \tau_2(x)) \in R(\beta)$. Set

$$\tau_1(s_1(x)) := \tau_1(x) \qquad p(s_1(x)) := \mathsf{ch}(\tau_1(x), \alpha, w) \qquad \tau_2(s_1(x)) := w$$
$$\tau_1(s_2(x)) := w \qquad p(s_2(x)) := \mathsf{ch}(w, \beta, \tau_2(x)) \qquad \tau_2(s_2(x)) := \tau_2(x)$$

4. Let $\tau_1(x)$ be defined, $\tau_1(s_1(x))$ undefined, $p = \alpha^*$, and $\tau_1(x) = \tau_2(x)$. Set, for $i \in \{1, 2\}$, $\tau_1(s_i(x)) := w_0$, $p(s_i(x)) := \varepsilon$, and $\tau_2(s_i(x)) := w_0$. Intuitively, the first and second successor of x are not needed. To nevertheless obtain a full $k + 3$-ary tree, we "restart" at w_0.

5. Let $\tau_1(x)$ be defined, $\tau_1(s_1(x))$ undefined, $p = \alpha^*$, and $\tau_1(x) \neq \tau_2(x)$. By (†) and the semantics, there is a sequence $w_0, \ldots, w_n \in W$ such that $\tau_1(x) = w_0$, $\tau_2(x) = w_n$, $(w_i, w_{i+1}) \in R(\alpha)$ for $i < n$, and $w_i \neq w_j$ for $i < j \leq n$. Let $w_0, \ldots, w_n \in W$ be the shortest such sequence. Set

$$\tau_1(s_1(x)) := \tau_1(x) \quad p(s_1(x)) := \mathsf{ch}(\tau_1(x), \alpha, w_1) \quad \tau_2(s_1(x)) := w_1$$
$$\tau_1(s_2(x)) := w_1 \quad p(s_2(x)) := \alpha^* \quad \tau_2(s_2(x)) := \tau_2(x)$$

6. Let $\tau_1(x)$ be defined, $\tau_1(s_1(x))$ undefined, and $p \in \mathsf{Prog}$ or p of the form α^-. Set, for $i \in \{1, 2\}$, $\tau_1(s_i(x)) := w_0$, $p(s_i(x)) := \varepsilon$, and $\tau_2(s_i(x)) := w_0$. Similar to Case 4, the first and second successor of x are not needed.

7. Let $\tau_1(x)$ be defined and $\tau_1(s_3(x))$ be undefined. Set $\tau_1(s_3(x)) := \tau_2(x)$, $\tau_2(s_3(x)) := \tau_1(x)$, and

$$p(s_3(x)) := \begin{cases} \mathsf{ch}(\tau_2(x), \alpha, \tau_1(x)) & \text{if } p(x) = \alpha^- \\ \bot & \text{if } p(x) \text{ is not of the form } \alpha^- \end{cases}$$

8. Let $\tau_1(x)$ be defined and $\tau_1(s_n(x))$ undefined for some n with $3 < n \leq k+3$, and $K, \tau_1(x) \models \mathcal{E}_{n-3} = \langle\alpha\rangle\varphi$. Then by the semantics there is a $w \in W$ with $(\tau_1(x), w) \in R(\alpha)$ and $K, w \models \varphi$. Set $\tau_1(s_n(x)) := \tau_1(x)$, $p(s_n(x)) := \mathsf{ch}(\tau_1(x), \alpha, w)$, and $\tau_2(s_n(x)) := w$.

9. Let $\tau_1(x)$ be defined and $\tau_1(s_n(x))$ undefined for some n with $3 < n \leq k+3$, and $K, \tau_1(x) \not\models \mathcal{E}_{n-3} = \langle\alpha\rangle\varphi$. Then set $\tau_1(s_n(x)) := w_0$, $p(s_n(x)) := \varepsilon$, and $\tau_2(s_n(x)) := w_0$. As in Cases 4 and 6, we restart at w_0 since the n-th successor of x is not needed.

Now we construct an MSO model $\mathfrak{M}$ as follows:

- for all $\varphi \in \mathsf{cl}(\varphi_0)$ and $i \in \{1, 2\}$, set $(F_\varphi^i)^{\mathfrak{M}} := \{x \in [k+3]^* \mid K, \tau_i(x) \models \varphi\}$
- $T_=^{\mathfrak{M}} := \{x \in [k+3]^* \mid p(x) = \varepsilon\}$
  $\cup \{x \in [k+3]^* \mid p(x) = \varphi?$ for some formula $\varphi\}$
  $\cup \{x \in [k+3]^* \mid p(x) = \alpha^*$ for some $\alpha \in \mathsf{subp}(\varphi_0)$ and $\tau_1(x) = \tau_2(x)\}$
  $T_\cap^{\mathfrak{M}} := \{x \in [k+3]^* \mid p(x) = \alpha \cap \beta$ for some $\alpha, \beta \in \mathsf{subp}(\varphi_0)\}$
  $T_;^{\mathfrak{M}} := \{x \in [k+3]^* \mid p(x) = \alpha; \beta$ for some $\alpha, \beta \in \mathsf{subp}(\varphi_0)\}$
  $\cup \{x \in [k+3]^* \mid p(x) = \alpha^*$ for some $\alpha \in \mathsf{subp}(\varphi_0)$ and $\tau_1(x) \neq \tau_2(x)\}$
  $T_\bot^{\mathfrak{M}} := \{s_n(x) \mid K, \tau_1(x) \not\models \mathcal{E}_{n-3}\}$
- for $a \in \mathsf{prog}$, set $T_a^{\mathfrak{M}} := \{x \in [k+3]^* \mid p(x) = a\}$.

In [19], we show that $\mathfrak{M} \models \varphi_0^*[\varepsilon]$.

5   Conclusion

In this paper, we have proved decidability of ICPDL, i.e. PDL extended with intersection and converse. As laid out in the introduction, this result that has several interesting applications. One additional virtue of the presented decidability proof is that, compared to existing proofs for PDL with intersection (but without converse), it is relatively simple and fully rigorous. There is, however, a price to be paid for this simplicity: our translation to MSO only yields a non-elementary upper bound. Indeed, when translating the following sequence $(\varphi_i)_{i \in \mathbb{N}}$ of ICPDL formulas, we obtain a sequence of MSO formulas with a strictly increasing quantifier alternation depth:

$$\varphi_i := [(\cdots ((a_0^*; a_1)^*; a_2)^*; \cdots; a_i)^*]p.$$

We believe that this upper bound is not tight. Indeed, it seems likely that satisfiability in ICPDL is 2-ExpTime-complete, just as satisfiability in IPDL. For proving this, however, it seems inevoidable to use the complex techniques of Danecki [7], in particular his "⊢" relation. Therefore, we believe that it is useful and illustrative to first prove only decidability in a more transparent way. Pinpointing the exact computational complexity of ICPDL is left for future work. Another interesting question is whether or not there are useful fragments of ICPDL that involve both intersection and Kleene star and for which reasoning is in ExpTime—thus not harder than in PDL. We suspect that the set of program operators $\{\cup, \cap, \cdot^*, \cdot^-, \varphi?\}$ induces such a fragment. Note that the mentioned fragment of ICPDL is still strong enough to capture the information logic DAL.

Acknowledgements

I am indebted to Ulrike Sattler, Lidia Tendera, and Martin Lange for many intense and fruitfull discussions about PDL with intersection.

References

1. L. Afanasiev, P. Blackburn, I. Dimitriou, B. Gaiffe, E. Goris, M. Marx, and M. de Rijke. PDL for ordered trees. *Journal of Applied Non-Classical Logic*, 2005. To appear.
2. N. Alechina, S. Demri, and M. de Rijke. A modal perspective on path constraints. *Journal of Logic and Computation*, 13(6):939–956, 2003.
3. F. Baader. Augmenting concept languages by transitive closure of roles: An alternative to terminological cycles. In *Proc. of IJCAI-91*, pages 446–451, Sydney, Australia, 1991.
4. P. Balbiani and D. Vakarelov. Iteration-free PDL with intersection: a complete axiomatization. In *Fundamenta Informaticae*, volume 45, pages 1–22. 2001.
5. D. Berardi, D. Calvanese, G. De Giacomo, M. Lenzerini, and M. Mecella. Automatic composition of e-services that export their behavior. In *Proc. of the 1st Int. Conf. on Service Oriented Computing (ICSOC 2003)*, volume 2910 of *LNCS*, pages 43–58. Springer, 2003.

6. D. Calvanese, G. De Giacomo, and M. Lenzerini. On the decidability of query containment under constraints. In *Proc. of PODS'98*, pages 149–158, 1998.
7. R. Danecki. Nondeterministic propositional dynamic logic with intersection is decidable. In *Proc. of the Fifth Symposium on Computation Theory*, volume 208 of *LNCS*, pages 34–53, Zaborów, Poland, Dec. 1984. Springer.
8. G. De Giacomo and M. Lenzerini. PDL-based framework for reasoning about actions. In *Proc. of AI*IA'95*, volume 992, pages 103–114. Springer, 1995.
9. F. M. Donini, M. Lenzerini, D. Nardi, and W. Nutt. The complexity of concept languages. *Information and Computation*, 134(1):1–58, 1997.
10. R. Fagin, J. Y. Halpern, Y. Moses, and M. Y. Vardi. *Reasoning About Knowledge*. MIT Press, 1995.
11. L. Farinas Del Cerro and E. Orlowska. DAL-a logic for data analysis. *Theoretical Computer Science*, 36(2-3):251–264, 1985.
12. M. J. Fischer and R. E. Ladner. Propositional modal logic of programs. In *Conference record of the ninth annual ACM Symposium on Theory of Computing*, pages 286–294. ACM Press, 1977.
13. M. J. Fischer and R. E. Ladner. Propositional dynamic logic of regular programs. *J. Comput. Syst. Sci.*, 18:194–211, 1979.
14. G. D. Giacomo and M. Lenzerini. Boosting the correspondence between description logics and propositional dynamic logics. In *Proc. of AAAI'94. Volume 1*, pages 205–212. AAAI Press, 1994.
15. D. Harel. Dynamic logic. In *Handbook of Philosophical Logic, Volume II*, pages 496–604. D. Reidel Publishers, 1984.
16. D. Kozen. Results on the propositional μ-calculus. In *Automata, Languages and Programming, 9th Colloquium*, volume 140 of *Lecture Notes in Computer Science*, pages 348–359. Springer-Verlag, 1982.
17. M. Lange. A lower complexity bound for propositional dynamic logic with intersection. In *Advances in Modal Logic Volume 5*. King's College Publications, 2005.
18. M. Lange and C. Lutz. 2-ExpTime lower bounds for propositional dynamic logics with intersection. 2005. Submitted.
19. C. Lutz. PDL with intersection and converse is decidable. LTCS-Report 05-05, Technical University Dresden, 2005. Available from http://lat.inf.tu-dresden.de/research/reports.html.
20. C. Lutz and U. Sattler. Mary likes all cats. In *Proc. of DL2000*, number 33 in CEUR-WS (http://ceur-ws.org/), pages 213–226, 2000.
21. F. Massacci. Decision procedures for expressive description logics with role intersection, composition and converse. In *Proc. of IJCAI-01*, pages 193–198, San Francisco, CA, Aug. 4–10 2001. Morgan Kaufmann Publishers, Inc.
22. V. Pratt. Considerations on floyd-hoare logic. In *FOCS: IEEE Symposium on Foundations of Computer Science (FOCS)*, 1976.
23. K. D. Schild. A correspondence theory for terminological logics: Preliminary report. In *Proc. of IJCAI-91*, pages 466–471. Morgan Kaufmann, 1991.
24. M.Y. Vardi and P. Wolper. Automata-theoretic techniques for modal logic of programs. Journal of Computer and System Sciences **32** (1986) 183–221

On Deciding Topological Classes of Deterministic Tree Languages

Filip Murlak*

Institute of Informatics, Warsaw University
ul. Banacha 2, 02–097 Warszawa, Poland
fmurlak@mimuw.edu.pl

Abstract. It has been proved by Niwiński and Walukiewicz that a deterministic tree language is either Π_1^1-complete or it is on the level Π_3^0 of the Borel hierarchy, and that it can be decided effectively which of the two takes place. In this paper we show how to decide if the language recognized by a given deterministic tree automaton is on the Π_2^0, the Σ_2^0, or the Σ_3^0 level. Together with the previous results it gives a procedure calculating the exact position of a deterministic tree language in the topological hierarchy.

Keywords: deterministic tree automata, index hierarchy, Borel hierarchy

1 Introduction

Tree automata, introduced by Rabin [14] in order to prove decidability of second order monadic logic of two successors, are today – together with μ-calculus – the basic tool in modeling and verification of concurrent systems. A tree represents all possible behaviours of an analysed system and an automaton is a coded correctness condition. An interesting measure of complexity of such a condition is the nesting depth of positive and negative constraints on the events occurring infinitely often. The formalization of that criterion gives the notion of the index of an automaton. The languages recognized by automata of different indices constitute an ascending hierarchy. This hierarchy was proved to be strict for the classes of deterministic [19], nondeterministic [9], alternating [1, 7] and weak alternating automata [8].

Another approach to estimating the complexity of a language is to calculate its level in the topological hierarchy. Skurczyński [16] proved that the finite part of the Borel hierarchy is strict for languages recognized by weak alternating automata. Deterministic tree languages surprisingly turned out to be either Π_1^1-complete (hence non-Borel) or on the Π_3^0 level of the Borel hierarchy. The paper by Niwiński and Walukiewicz [11] containing the proof of the above suggests also that two basic complexity criteria, combinatorial and topological, are closely related – at least for deterministic automata.

---

* Supported by KBN Grant 4 T11C 042 25.

L. Ong (Ed.): CSL 2005, LNCS 3634, pp. 428–441, 2005.

While the efficiency of verification methods depends on the brevity of the correctness conditions, they often result redundant when modeling real systems. Therefore, developing algorithmic methods for calculating actual automata's complexity would be interesting. So far, there have been presented procedures calculating the index of tree languages consisting of trees which have all the paths in a given regular ω-language [6, 10], deciding if a deterministic automaton is equivalent to a Büchi automaton [18], and calculating the (nondeterministic) index of deterministic automata [12]. The μ-calculus approach resulted in a procedure deciding if a given formula of modal μ-calculus is equivalent to a formula of modal logic [13].

In this paper we concentrate on algorithmic calculation of a language's position in the topological hierarchy and its connections with the deterministic index hierarchy. In Sect. 2 and Sect. 3 we remind the basic notions of automata theory and present simple criteria determining a language's position in the deterministic index hierarchy and weak deterministic index hierarchy. Section 4 recalls some necessary topological notions. In Sect. 5 we show how to decide if a deterministic language is in the classes Σ_2^0, Π_2^0 and Σ_3^0. When combined with the previous characteristics by Niwiński and Walukiewicz, it provides a complete procedure calculating the position of a deterministic language in the topological hierarchy.

2   Basic Notions

We shall use the symbol ω to denote the set of natural numbers $\{0, 1, 2, \ldots\}$. For an alphabet X, X^* is the set of finite words over X and X^ω is the set of infinite words over X. The concatenation of words $u \in X^*$ and $v \in X^* \cup X^\omega$ will be denoted by uv, and the empty word by ε. The concatenation is naturally generalized for infinite sequences of words $v_1 v_2 v_3 \ldots$. The concatenation of sets $A, B \subseteq X^*$ is $AB = \{uv : u \in A, v \in B\}$.

A *binary tree* is any subset of $\{0, 1\}^*$ closed under the prefix relation. An element of a tree is usually called a *node*. A *leaf* is any node of a tree which is not a (strict) prefix of some other node. We shall be dealing mainly with *labeled trees* over Σ which are functions $t : \mathrm{dom}\, t \to \Sigma$ such that $\mathrm{dom}\, t$ is a tree. The symbol T_Σ will denote the set of full infinite binary trees over Σ, i. e. functions $\{0, 1\}^* \to \Sigma$.

For any trees t, s and v, a node of t, the result of the *substitution* of v with s is a tree t' whose domain is the set $\mathrm{dom}\, t \cup v\mathrm{dom}\, s$ and

$$t'(u) = \begin{cases} s(u') \text{ if } u = vu' \text{ for some } u' \\ t(u) \text{ otherwise} \end{cases}.$$

Note that u' may be empty, so if $t(u) \neq s(\varepsilon)$, for the label of u in t' we choose $s(\varepsilon)$. We find it more convenient since the state of an automaton in a node depends on every predecessor of the node, but not on the node itself.

The concatenation of tree languages A, B is a tree language AB consisting of all trees obtained from any $t \in A$ by substituting every leaf u of t with any tree $s_u \in B$. The concatenation of infinite sequence of tree languages is a natural generalization of the above. A more precise definition requires an auxiliary notion of a limit. Let $t_0, t_1, \ldots$ be a sequence of trees such that

- $\operatorname{dom} t_0 \subseteq \operatorname{dom} t_1 \subseteq \cdots$,
- $\forall v \in \bigcup_{m \in \omega} \operatorname{dom} t_m \ \exists n_v \ \forall n \geq n_v \ t_n(v) = t_{n_v}(v)$.

The *limit* $t = \lim t_n$ is defined as follows:

- $\operatorname{dom} t = \bigcup_{m \in \omega} \operatorname{dom} t_m$,
- $t(v) = t_{n_v}(v)$.

An *infinite concatenation* of tree languages $L_0 L_1 \ldots$ consists of the limits of all sequences $t_0, t_1, \ldots$ such that $t_0 \in L_0$ and $t_{n+1} \in \{t_n\} L_{n+1}$ for all n.

The concatenation of trees s, t is the only element of the concatenation $\{s\}\{t\}$. Similarly, the concatenation of infinite sequence of trees $t = t_1 t_2 t_3 \ldots$ is the only element of $\{t_1\}\{t_2\}\{t_3\} \ldots$.

For $v \in \operatorname{dom} t$ we define $t.v$ as a subtree of t rooted in v, i. e. $\operatorname{dom}(t.v) = \{u : vu \in \operatorname{dom} t\}$, $t.v(u) = t(vu)$. A *segment* of a tree t between u and uv is the restriction of the function $t.u$ to the set $\{w \in \operatorname{dom}(t.u) : v$ is not a strict prefix of $w\}$.

A *(nondeterministic) automaton on words* is a tuple $A = \langle \Sigma, Q, \delta, q_0, \operatorname{rank} \rangle$, where Σ is a (finite) input alphabet, Q is the set of states, $\delta \subseteq Q \times \Sigma \times Q$ is the relation of transition and $q_0 \in Q$ is the initial state. The meaning of the function $\operatorname{rank} : Q \to \omega$ will be explained later. Instead of $(q, \sigma, q_1) \in \delta$ one usually writes $q \xrightarrow{\sigma} q_1$. A *run* of an automaton A on a word $w \in \Sigma^\omega$ is a word $\rho_w \in Q^\omega$ such that $\rho_w(0) = q_0$ and if $\rho_w(n) = q$, $\rho_w(n+1) = q_1$, and $w(n) = \sigma$, then $q \xrightarrow{\sigma} q_1$. A run ρ_w is *accepting* if the highest rank repeating infinitely often in ρ_w is even; otherwise ρ_w is *rejecting*. A word is *accepted* by A if there exist an accepting run on it. The language of words accepted by A is denoted by $L(A)$. One says that L is *recognized* by A if $L = L(A)$. An automaton is *deterministic* if its relation of transition is a function $Q \times \Sigma \to Q$. Note, that we do not let the transition relation be a partial function, and so there is a run – accepting or not – on every word. We call a language deterministic if it is recognized by a deterministic automaton.

An *(nondeterministic) automaton on trees* is a tuple $A = \langle \Sigma, Q, \delta, q_0, \operatorname{rank} \rangle$, the only difference being that $\delta \subseteq Q \times \Sigma \times Q \times Q$. Like before, $q \xrightarrow{\sigma} q_1, q_2$ means $(q, \sigma, q_1, q_2) \in \delta$. A *run* of A on a tree $t \in T_\Sigma$ is a tree $\rho_t \in T_Q$ such that $\rho_t(\varepsilon) = q_0$ and if $\rho_t(v) = q$, $\rho_t(v0) = q_1$, $\rho_t(v1) = q_2$ and $t(v) = \sigma$, then $q \xrightarrow{\sigma} q_1, q_2$. A path π of the run ρ_t is *accepting* if the highest rank repeating infinitely often in π is even; otherwise π is *rejecting*. A run is called accepting if all its paths are accepting. If at least one of them is rejecting, so is the whole run. An automaton is called deterministic if its transition relation is a function $Q \times \Sigma \to Q \times Q$.

An automaton is called *weak* if the rank of visited states does not increase during the run, i. e. whenever there is a transition $p \xrightarrow{\sigma} q_1, q_2$, then $\operatorname{rank} p \geq \operatorname{rank} q_1$ and $\operatorname{rank} p \geq \operatorname{rank} q_2$.

The symbol G_A denotes a directed edge-labeled graph representing the transition relation of A. The set of vertices is Q and whenever in A there is a transition

$p \xrightarrow{\sigma} q_1, q_2$, in G_A there is an edge between p and q_1 labeled with $(\sigma, 0)$ and an edge between p and q_2 labeled with $(\sigma, 1)$. For sake of brevity we shall write $p \xrightarrow{\sigma,0} q_1$ and $p \xrightarrow{\sigma,1} q_2$. A state is called *productive* if it is used in some accepting run. The *productive graph* G_A^+ is analogous to G_A, only now the set of vertices is restricted to productive states and when defining the edges we demand that *all* of the states p, q_1, q_2 are productive. We shall call a path in G_A *productive* if it is also a path in G_A^+.

A *partial run* of A is a segment of any run of A. A partial run ρ *realizes* a finite path π in the graph G_A^+ if it is a segment of an accepting run ρ' between two nodes x and y such that ρ' agrees with π between x and y. More precisely, if $\pi = p_0 \xrightarrow{\sigma_1, d_1} \ldots \xrightarrow{\sigma_m, d_m} p_m$, then $y = x d_1 d_1 \ldots d_m$, $\rho'(x) = p_0$ and $\rho'(x d_1 \ldots d_i) = p_i$ for all i. Note that, since ρ is a segment of an accepting run, all its infinite paths are accepting. A tree segment f *realizes* a path π if the corresponding partial run ρ_f realizes π.

When analysing the way an automaton works, one finds it useful to let the automaton begin its run in states other than initial. An automaton starting in the state q will be denoted by A_q.

The *index of an automaton* A is a pair $(\min \operatorname{rank} Q, \max \operatorname{rank} Q)$. Scaling down the rank function if necessary one may assume that $\min \operatorname{rank} Q \in \{0, 1\}$. For an index (i, j) we shall denote by $\overline{(i, j)}$ the dual index, i. e. $\overline{(0, j)} = (1, j+1)$, $\overline{(1, j)} = (0, j-1)$. The *index hierarchy* for a certain class of automata consists of (roughly speaking) ascending sets (levels) of languages recognized by (i, j)-automata, where $(i, j) \in \{0, 1\} \times \omega$. It is known that index hierarchies are strict for deterministic [19], nondeterministic [9], alternating [1, 7] and weak alternating automata [8].

3   Deterministic Index Hierarchies

Given a deterministic language, one may ask about its *deterministic index*, i. e. the exact position in the index hierarchy of deterministic automata. This question can be answered effectively. Here we follow the method introduced by Niwiński and Walukiewicz [10].

A sequence of loops $\lambda_i, \lambda_{i+1}, \ldots, \lambda_j$ in a graph of an automaton is called *an alternating chain* if the highest rank appearing on λ_k has the same parity as k and it is higher then the highest rank on λ_{k-1}. A (i, j)-*flower* is an alternating chain $\lambda_i, \lambda_{i+1}, \ldots, \lambda_j$ such that all loops start in the same state q. Let $\operatorname{Paths}(L) \subseteq \Sigma^\omega$ be the set of paths of trees from L and $\operatorname{Paths}'(L) \subseteq (\Sigma \times \{0, 1\})^\omega$ denote the language of generalized paths of L,

$$\operatorname{Paths}'(L) = \{\langle (\sigma_1, d_1), (\sigma_2, d_2), \ldots \rangle : \exists t \in L \ t(d_1 d_2 \ldots d_{i-1}) = \sigma_i \} \ .$$

Niwiński and Walukiewicz show that a language L is recognized by a (i, j)-automaton iff no deterministic automaton recognizing $\operatorname{Paths}(L)$ contains a $\overline{(i, j)}$-flower. As an intermediate pass they prove the following fact.

Theorem 1 (Niwiński, Walukiewicz [10]). *A deterministic automaton on words is equivalent to a deterministic (i, j)-automaton iff it does not contain a $\overline{(i, j)}$-flower.*

For a deterministic tree automaton A, the graph G_A^+ can be treated as a deterministic automaton recognizing Paths'$(L(A))$. Conversely, given a deterministic word automaton recognizing Paths'$(L(A))$, one may interpret it as a graph of a tree automaton, obtaining thus a deterministic automaton recognizing $L(A)$. Hence, applying Theorem 1 one gets the following result.

Proposition 1. *For a deterministic tree automaton A the language $L(A)$ is recognized by a deterministic (i, j)-automaton iff G_A^+ does not contain a $\overline{(i, j)}$-flower.*

Similarly, one can calculate the exact position of a deterministic language in the hierarchy of weak deterministic automata. A *weak (i, j)-flower* is a sequence of loops $\lambda_i, \lambda_{i+1} \ldots, \lambda_j$ such that λ_k is reachable in G_A^+ from λ_{k+1}, and λ_k is accepting iff k is even. Intuitively, the notion is to provide long enough alternation of rank parity. Therefore we have to extend it to cover the case when i is odd and instead of λ_i there is an unproductive state r reachable in G_A from λ_{i+1}.

Proposition 2. *For any deterministic tree automaton A the language $L(A)$ can be recognized by a weak deterministic (i, j)-automaton iff G_A^+ does not contain a weak $\overline{(i, j)}$-flower.*

Proof. ($\Rightarrow$) Let us suppose that G_A^+ contains a weak (i', j')-flower, $(i', j') = \overline{(i, j)}$. Let $g_{j'}$ be a tree segment realizing some path from the initial state q_0 to a state $r_{j'}$ on $\lambda_{j'}$. By induction, let g_k realize a path from the state $r_{k+1} \in \lambda_{k+1}$ to a state $r_k \in \lambda_k$ for $k = j' - 1, \ldots, i'$. Finally, let f_k realize the loop λ_k (from r_k to r_k) for all k. Let B be a weak deterministic automaton recognizing $L(A)$. Clearly, we can choose numbers $n_{i'}, \ldots, n_{j'}$ so that the run on $g_{j'}(f_{j'})^{n_{j'}} g_{j'-1}(f_{j'-1})^{n_{j'-1}} \ldots g_{i'}(f_{i'})^{n_{i'}}$ would need $j' - i'$ changes of rank parity and thus $j' - i' + 1$ different ranks. Consequently, the index of B cannot be (i, j).

($\Leftarrow$) A weak deterministic (i, j)-automaton is obtained by setting rank (q) equal to the highest number m such that there exists a weak (ι, m)-flower with a path from q to λ_m (recall that an unproductive state is a weak $(1, 1)$-flower). $\square$

Finally, for a deterministic language one may want to calculate its *nondeterministic index*, i.e. the position in the hierarchy of nondeterministic automata. This may be lower than the deterministic index, due to greater expressive power of nondeterministic automata. Consider for example the language $L_M^{0^\omega}$ consisting of trees whose leftmost paths are in a regular ω-language M. It can be recognised by a (nondeterministic) Büchi automaton, but its deterministic index is equal to the deterministic index of M, which can be arbitrarily high.

The problem transpired to be rather difficult and has only just been solved in [12]. The analogous problem for nondeterministic languages remains open.

4   Topological Hierarchy

We start with a short recollection of elementary notions of descriptive set theory. For further information see [5].

Let 2^ω be the set of infinite binary sequences with a metric given by the formula

$$d(u,v) = \begin{cases} 2^{-\min\{i\in\omega \,:\, u_i\neq v_i\}} & \text{iff } u \neq v \\ 0 & \text{iff } u = v \end{cases}$$

and T_Σ be the set of infinite binary trees over Σ with a metric

$$d(s,t) = \begin{cases} 2^{-\min\{|x| \,:\, x\in\{0,1\}^*,\, s(x)\neq t(x)\}} & \text{iff } s \neq t \\ 0 & \text{iff } s = t \end{cases} \;.$$

Both 2^ω and T_Σ, with the topologies induced by the above metrics, are Polish spaces (complete metric spaces with countable dense subsets). In fact, both of them are homeomorphic to the Cantor discontinuum.

The class of Borel sets of a topological space X is the closure of the class of open sets of X by countable sums and complementation. Within this class one builds so called *Borel hierarchy*. The initial (finite) levels of it are defined as follows:

Σ^0_1 – open relations, i. e. open subsets of X^n for some $n < \omega$,
Π^0_k – complements of relations from Σ^0_k,
Σ^0_{k+1} – countable sums of relations form Π^0_k.

For example, Π^0_1 are closed relations, Σ^0_2 are F_σ relations, and Π^0_2 are G_δ relations.

Even more general classes of sets form the *projective hierarchy*. We will need only its lowest level:

Σ^1_1 – *analytical* sets, i. e. projections of Borel relations,
Π^1_1 – complements of relations from Σ^1_1.

Let $\varphi : X \to Y$ be a continuous map of topological spaces. One says that φ *reduces* $A \subseteq X$ to $B \subseteq Y$, if $\forall x \in X \; x \in A \leftrightarrow \varphi(x) \in B$. Note that if B is in a certain class of the above hierarchies, so is A. For any class $\mathcal{C}$ a set B is $\mathcal{C}$-*hard*, if for any set $A \in \mathcal{C}$ there exists a reduction of A to B. The topological hierarchy is strict for Polish spaces, so if a set is $\mathcal{C}$-hard, it cannot be in any lower class. If a $\mathcal{C}$-hard set B is also an element of $\mathcal{C}$, then it is $\mathcal{C}$-*complete*.

For a deterministic automaton A one may define a function $\varphi_A : T_\Sigma \to T_{\text{im (rank)}}$ so that $\varphi_A(t)(v) = \operatorname{rank}(\rho_t(v))$, where ρ_t is the run of A on t. Note that φ_A is a continuous map that reduces $L(A)$ to the set P of all trees satisfying A's parity condition.

We shall continue with a handful of examples which will turn out useful later.

Example 1. Consider the set $P_{(1,2)} \subseteq T_{\{1,2\}}$ consisting of trees having infinitely many 2s on every path. For each $n < \omega$ let G_n be the set of all trees that have at least one 2 below the level n on every path. From König lemma it follows that each G_n is open. Clearly, $P_{(1,2)} = \bigcap_{n\in\omega} G_n$ and so it is a Π^0_2 set.

Example 2. Let $P_{(0,1)}^{fin} \subseteq T_{\{0,1\}}$ be the set of trees in which there are only finitely many 1s. For any $n < \omega$ a set $F_n \subseteq T_{\{0,1\}}$ consisting of trees having no 1s below the level n is closed. $P_{(0,1)}^{fin} = \bigcup_{n \in \omega} F_n$, hence $P_{(0,1)}^{fin} \in \Sigma_2^0$.

Example 3. Let $L_a^{0^*1^\omega} \subseteq T_{\{a,b\}}$ be the set of trees which have an a on every path from the set 0^*1^ω. Suppose that it is a Σ_2^0 set. Let $L_a^{0^*1^\omega} = \bigcup_{n \in \omega} F_n$, F_n is closed for all n. We claim that for every n there exists m_n such that in every tree from F_n the letter a occurs on the path 0^n1^ω above the level m_n. If there was no such number, then we could find a sequence t_k of trees having no letters a on the path 0^n1^ω above the level l_k, where $l_1 < l_2 < l_3 < \dots$. As $T_{\{a,b\}}$ is compact, there exists a subsequence t_{k_i} convergent in F_n. However the limit of t_{k_i} cannot be in F_n for it has no letter a on the path 0^n1^ω. Now, consider a tree t with a in nodes $0^n1^{m_n+1}$ and b in other nodes. Clearly, $t \in L_a^{0^*1^\omega}$, but $t \notin \bigcup_{n \in \omega} F_n$. This way we have shown that $L_a^{0^*1^\omega} \notin \Sigma_2^0$.

Example 4. In quite a similar way one proves that the set $Q = (0^*1)^*0^\omega$ is not in Π_2^0 (in fact, it is Σ_2^0-complete).

Example 5. Let $L_Q^{0^*1^\omega}$ denote the language of trees such that the rightmost path from every node of the form 0^* belongs to the language Q defined above. We shall see that it is Π_3^0-complete, and therefore it is not in Σ_3^0. Let us take any $M = \bigcup_{i<\omega} X_i$ with X_i in Σ_2^0. Since Q is Σ_2^0-complete, for each i there exists f_i reducing X_i to Q. One easily defines a continuous reduction of M to $L_Q^{0^*1^\omega}$ assigning to each t a tree having the word $f_i(t)$ on the path 0^i1^ω for all i, and 0s in all the other nodes.

5   Deciding Levels of Topological Hierarchy

The basic tool for investigating automata's properties is the technique of gadgets or difficult patterns in the graph of an automaton. In the topological context, the general recipe goes like this. For every class identify a gadget satisfying the following condition: if the gadget appears in an automaton A, then it provides a reduction of some difficult language to $L(A)$; otherwise $L(A)$ is in the class considered.

Wagner used this technique successfully in his solution of the general problem of continuous reductions between ω-languages [19]. For infinite words, the Borel hierarchy collapses at the level Δ_3^0 and below it is strict. The levels Π_1^0 and Σ_1^0 correspond to weak deterministic $(1, 2)$ and $(0, 1)$ automata; the class Δ_2^0 consists of all weak deterministic automata; Π_2^0 and Σ_2^0 are exactly deterministic Büchi and co-Büchi languages. We shall see that the situation for trees is slightly different.

We start with the gap property for deterministic tree languages. An automaton A admits a *split* if in G_A^+ there are two loops $q_0 \xrightarrow{\rho,0} q_1 \to \dots \to q_0$ and $q_0 \xrightarrow{\sigma,1} q_2 \to \dots \to q_0$ such that the highest ranks occurring on them are of different parity and the higher is odd.

Theorem 2 (Niwiński, Walukiewicz [11]). *For a deterministic automaton A, $L(A)$ is on the level Π_3^0 of the Borel hierarchy iff A does not admit split; otherwise $L(A)$ is Π_1^1-complete (hence non-Borel).*

Owing to this result, it is enough to decide if a language is on the levels Σ_1^0, Π_1^0, Σ_2^0, Π_2^0, Σ_3^0 and use the split criterion to get complete information on its position in the topological hierarchy. Before dealing with this task we shall see that it does not get any easier, and so, not only there exist non-Borel tree languages but even the Borel hierarchy for trees is higher than for words.

Proposition 3. *The Borel hierarchy for deterministic tree languages is strict below Π_3^0.*

Proof. The language $L_a^{0^*}$ consisting of trees having an a on the leftmost path is open, but obviously is not closed, $L_a^{0^*} \in \Sigma_1^0 \setminus \Pi_1^0$. An example of a language from $\Pi_1^0 \setminus \Sigma_1^0$ can be $\{t_0\}$, where $t_0(v) = 0$ for every node v. The set Q can be reduced to $P_{(0,1)}^{fin}$ by a map $\{0,1\}^\omega \ni w \mapsto t_w \in T_{0,1}$ where t_w is a tree whose leftmost path is w and having 0 in all the other nodes. Hence $P_{(0,1)}^{fin} \in \Sigma_2^0 \setminus \Pi_2^0$. It can also be easily seen that $L_a^{0^*1^\omega}$ is a Π_2^0 set and we have already proved that it is not a Σ_2^0. Finally, the language $L_Q^{0^*1^\omega}$ has been shown not to be in the Σ_3^0 class, but clearly is in Π_3^0.   □

Having seen the strictness of our confined hierarchy, we shall continue with the characterization of its levels. The description of the open and the closed languages is probably well known, so we state it here, together with a short proof, just for the sake of completeness.

Proposition 4. *For any deterministic tree automaton A*

    1. $L(A)$ is closed iff A is equivalent to a weak deterministic $(1,2)$-automaton[1],
    2. $L(A)$ is open iff A is equivalent to a weak deterministic $(0,1)$-automaton.

Proof. We will prove only (1). First, suppose that A is not equivalent to a weak deterministic $(1,2)$-automaton. It follows from Proposition 2 that in G_A^+ there must be an accepting loop λ_2 reachable from a rejecting loop λ_1. Let g_1 realize a path from q_0 to some $q_1 \in \lambda_1$, g_2 realize a path form q_1 to some $q_2 \in \lambda_2$, and f_1, f_2 realize loops λ_1 (from q_1 to q_1), λ_2 (from q_2 to q_2) respectively. Consider $t_n = g_1(f_1)^n g_2(f_2)^\omega$ and $t = g_1(f_1)^\omega$. Clearly, $t_n \in L(A)$ and $t_n \to t$ when $n \to \infty$, but $t \notin L(A)$. Hence $L(A)$ is not closed.

Now, if $L(A)$ is recognized by a weak deterministic automaton B, then it is the inverse image of a point under the continuous map φ_B and so it is closed.   □

The combinatorial characterization of Π_2^0-languages transpires to be equally elegant.

---

[1] Recall that we do not let an automaton stop. If we did, there should be $(0,0)$ instead of $(1,2)$.

Theorem 3. *For a deterministic tree automaton A, the language $L(A)$ is on the level Π_2^0 of the Borel hierarchy iff A is equivalent to a deterministic Büchi automaton.*

Proof. ($\Rightarrow$) Suppose that $L(A)$ is not recognized by a deterministic Büchi automaton. From Proposition 1 it follows that in G_A^+ there exist an accepting loop λ_0 and a rejecting loop λ_1 forming a $(0,1)$-flower. The loops λ_0 and λ_1 cannot be equal, so there is a node q lying on both loops, such that the next edges going out of q in λ_0 and λ_1 have different labels. Let us assume first that the edges are labeled with different letters $a \neq b$:

$$\lambda_0 : q \xrightarrow{a,d_0} r \longrightarrow \ldots \longrightarrow q \ ,$$

$$\lambda_1 : q \xrightarrow{b,d_1} s \longrightarrow \ldots \longrightarrow q \ .$$

Let f_0, f_1 be tree segments realizing the loops λ_0 and λ_1 respectively (both from q to q). Note that f_0 and f_1's roots are labeled with different letters a and b. Consider a map $\varphi : 2^\omega \to T_\Sigma$ defined by the formula

$$\varphi(x_0 x_1 x_2 \ldots) = f f_{x_0} f_{x_1} f_{x_2} \cdots \ ,$$

where f is a tree segment realizing a path from q_0 to q. The map φ is continuous, since $d(\varphi(x), \varphi(x')) \leq d(x, x')^{|\lambda_0|}$. Thus we have reduced $(0^*1)^*0^\omega$ to $L(A)$, which, by Example 4, implies that $L(A)$ is not a Π_2^0 language.

The second case is slightly more sophisticated. We have

$$\lambda_0 : q \xrightarrow{a,0} r \longrightarrow \ldots \longrightarrow q$$

$$\lambda_1 : q \xrightarrow{a,1} s \longrightarrow \ldots \longrightarrow q$$

(or dual). Consider a path in the graph G_A^+ along the edges of the loop λ_1 starting from q. We claim that it must reach a node q' such that there is an edge e from q' to q''' labeled with the same direction (0 or 1) as the edge in the loop but with a different letter, e.g.

$$\lambda_1 : q \xrightarrow{a,1} s \longrightarrow \ldots \longrightarrow q' \xrightarrow{b,0} q'' \longrightarrow \ldots \longrightarrow q$$
$$\searrow{\scriptstyle c,0}$$
$$q'''$$

Were there no such an edge, all the runs starting in q would loop on λ_1 and q would be unproductive. Let π_0 be the path $q \xrightarrow{a,1} s \longrightarrow \ldots \longrightarrow q' \xrightarrow{e} q'''$. The state q''' is productive, so we can extend π_0 to an infinite accepting path π in G_A^+. For f_0 we choose a tree segment realizing both λ_0 and π. This is possible owing to the fact, that λ_0 and π start with edges labeled with the same letter and different directions. As before, f_1 can be any tree segment realizing λ_1. Now we can continue like in the previous case.

($\Leftarrow$) $L(A)$ can be reduced to $P_{(1,2)}$, which is a Π_2^0-set. $\square$

We shall now continue with describing the Σ_2^0 languages. Recall the language $L_a^{0^*1^\omega} \subseteq T_{\{a,b\}}$ consisting of trees which have an a on every path from the set 0^*1^ω. Even though one may easily construct a deterministic $(0,1)$-automaton recognizing this language, it is not a Σ_2^0 set. Since a simple analog of the Π_2^0 case condition has shown insufficient, a more careful analysis of the automaton's graph is needed. We will say that a node $v \in G_A^+$ is *accessible with a split* if in G_A^+ there exist an accepting loop $u_1 \xrightarrow{\sigma, d_0} u_2 \longrightarrow \ldots \longrightarrow u_1$ and a path $u_1 \xrightarrow{\sigma, d_1} u_2' \longrightarrow \ldots \longrightarrow v$, where $d_0 \neq d_1$. We will say that a loop or a flower is accessible with a split, meaning that it contains a node accessible with a split.

Theorem 4. *For a deterministic tree automaton A, the language $L(A)$ is on the level Σ_2^0 of the Borel hierarchy iff A is equivalent to a deterministic $(0,1)$-automaton and G_A^+ does not contain a rejecting loop accessible with a split.*

Proof. ($\Rightarrow$) Let us suppose that $L(A)$ is a Σ_2^0 language. To prove the equivalence to a deterministic $(0,1)$-automaton follow the dual version of the method used in the previous theorem. There exist a rejecting loop λ_1 and an accepting loop λ_0 forming a $(1,2)$-flower. Find tree segments f_0, f_1 realizing λ_0, λ_1. Make sure they are different by finding an accepting path π leaving λ_1. The map φ defined in the previous proof reduces $(1^*0)^\omega$ to $\text{im}\,\varphi \cap L(A)$. Were $L(A)$ a Σ_2^0 set, so would $(1^*0)^\omega$, which, by Example 4, is not true. Let us now concentrate on the second part of the condition. Suppose that G_A^+ does contain a rejecting loop λ_1 accessible with a split from an accepting loop λ_0 along a path π. For $n \in \omega$ let π_n be an infinite accepting path having a prefix $\pi(\lambda_1)^n$ but no prefixes $\pi(\lambda_1)^m$ for $m > n$ (find an edge leaving the rejecting loop λ_1 just as it was done in the second case of the previous proof) and $\pi_\omega = \pi(\lambda_1)^\omega$. For each $\alpha \in \omega + 1 = \omega \cup \{\omega\}$ consider a tree segment f_α realizing both λ_0 and π_α, this being possible since the first edges of both paths are labeled with the same letter σ and different directions d_0, d_1. For any $x = (x_1, x_2, \ldots) \in (\omega+1)^\omega$ let $t_x = f f_{x_1} f_{x_2} \ldots$, where f is a tree segment realizing a path from the initial state q_0 to u_1. We shall define a continuous map $\varphi : T_{\{a,b\}} \to T_\Sigma$. For $s \in T_{\{a,b\}}$ let $y_i = \min(\{|w| : w \in 0^i1^*,\ s(w) = a\} \cup \omega)$. Let $z_i = \max(y_i - i, 0)$ if $y_i < \omega$ and $z_i = \omega$ if $y_i = \omega$. Let us now set $\varphi(s) = t_z$, where $z = (z_0, z_1, \ldots)$. The map φ reduces $L_a^{0^*1^\omega}$ to $L(A)$. However, we have already shown that $L_a^{0^*1^\omega}$ is not a Σ_2^0 language. Hence G_A^+ cannot contain a rejecting loop accessible with a split.

($\Leftarrow$) Investigating the proof of Theorem 1 one easily observes that the reduction is careful enough not to introduce any rejecting loops accessible with a split, provided there are no such loops in the original automaton. Therefore, we may assume that A is a $(0,1)$-automaton such that G_A^+ does not contain a rejecting loop accessible with a split. A state is called *relevant* if it has the highest rank on some productive loop. We may change the ranks of productive irrelevant states to 0, and assume from now on that the odd states are either relevant or unproductive. We claim that the odd states occur only finitely many times on accepting runs of A. Suppose that an odd state p occurs infinitely many times in an accepting run ρ. Then it appears in an infinite number of incomparable nodes $v_0, v_1, \ldots$ of ρ. Let π_i be a path of ρ going through the node v_i. Since

2^ω is compact, we may assume, passing to a subsequence, that the sequence π_i converges to a path π. As v_i are incomparable, at most one of them, say v_{i_0}, may lie on π. Let us remove π_{i_0} from the sequence π_i. Consider the node w_i in which p occurs for the first time on π_i after leaving π and let π_i^0 be the path from the last common node of π and π_i to w_i. Cutting the loops off if needed, we may assume that $|\pi_i^0| \leq |Q|$ for all $i \in \omega$. Subsequently, there exist a path π^0 repeating infinitely often in the sequence $\pi_0^0, \pi_1^0, \dots$. Moreover, the path π is accepting, so the starting node of π^0 must lay on an accepting productive loop. As p is productive, the assumption implies that it is relevant and, being odd, lies on some productive rejecting loop. Hence, G_A^+ contains a rejecting loop accessible with a split – a contradiction. This way we have shown that φ_A reduces $L(A)$ to $P_{0,1}^{fin}$, and so $L(A)$ is a Σ_2^0 language. $\square$

Let us now consider the class Σ_3^0. Every deterministic Σ_3^0 language is, due to Theorem 2, in the $\Delta_3^0 = \Pi_3^0 \cap \Sigma_3^0$ class. Below we present a combinatorial description of this class of languages.

Theorem 5. *For a deterministic tree automaton A, $L(A)$ is a Σ_3^0 set iff G_A^+ does not contain a $(0,1)$-flower accessible with a split.*

Proof. First let us suppose that G_A^+ contains a $(0,1)$-flower accessible with a split. Following the method used in Theorem 3 one easily finds a map reducing the language $L_Q^{0^*1^\omega}$ to $L(A)$. Subsequently, $L(A)$ is not a Σ_3^0 language.

Now, suppose that G_A^+ does not contain a $(0,1)$-flower accessible with a split. We shall find a Σ_3^0 representation of the set $R \subseteq T_Q$ of accepting runs of A. The theorem will follow since the map $T_\Sigma \ni t \mapsto \rho_t \in T_Q$ is continuous. Let us consider, then, the set $\mathcal{X}$ of strongly connected components of G_A^+. Recall that they form a directed acyclic graph, i. e. no path returns to a component it has left. The language R can be expressed by the following formula

$$R = \bigcap_{X \in \mathcal{X}} R_X \ ,$$

where R_X is the set of runs whose every path staying forever in X is accepting. Owing to this simple observation, it is enough to prove that the sets R_X are Σ_3^0.

Let Π_X denote the set of all paths from the initial state q_0 to some state in X containing only one state from X. Note that Π_X is countable for every X. Let us first suppose that X is accessible with a split. For $\pi \in \Pi_X$ let $R_{X,\pi}$ denote the set of runs whose every path going along π either leaves X or is accepting. By the hypothesis, X contains no $(0,1)$-flowers, and so it can be replaced by an equivalent component X' using only ranks 1 and 2. Therefore, given $q \in X'$, the set of runs of A_q, whose all paths are accepting or leave X', is a Π_2^0 set. Obviously, so is $R_{X,\pi}$. As it also holds that

$$R_X = \bigcap_{\pi \in \Pi_X} R_{X,\pi} \ ,$$

R_X is a Π_2^0 set.

The case of X not accessible with a split is slightly fastidious. Let ρ be an accepting run of A. Consider ρ_X, a subtree of ρ formed by the nodes which have a successor whose labeling state is in X. No state from X is accessible with a split, therefore it cannot appear in infinitely many incomparable nodes of ρ. Hence, ρ_X has only finitely many branches. Let ρ_X^0 denote the tree ρ_X restricted to the highest level below which there are no branching points. Let R_X^0 denote the set of all such trees; note that, although R_X may be uncountable, R_X^0 is countable. Obviously,

$$R_X = \bigcup_{s \in R_X^0} R_{X,s} \ ,$$

where $R_{X,s}$ is the set of runs from R_X coinciding with a tree $s \in R_X^0$ in its domain. Observe that $R_{X,s}$ is equal to the set of runs ρ' satisfying the following conditions:

(1) ρ' coincides with s in its domain,
(2) the states from X appear only in successors of leaves of s,
(3) in every subtree of ρ' rooted in a leaf of s the states from X appear infinitely often on at most one path,
(4) in every subtree of ρ' rooted in a leaf of s the highest rank of the states from X appearing infinitely often is even.

The condition (1) obviously defines an open set. The condition (2) defines a closed set and so does the condition (3), because it is equivalent to saying that no node of the subtree has both children in X. The condition (4) is of the $B(\Sigma_2^0)$ type. By $B(\Sigma_2^0)$ we mean the closure of Σ_2^0 by the finite Boolean operations; it is clearly a subclass of Σ_3^0. Hence $R_{X,s}$ is a Σ_3^0 set and so is R_X.   □

As a conclusion we obtain the main result of this paper.

Corollary 1. *The problem of calculating the exact position in the topological hierarchy of a language recognized by a deterministic tree automaton is decidable within the time of finding the productive states of a deterministic automaton.*

Proof. From Proposition 4 it follows that the language recognised by a deterministic automaton A is closed iff A is equivalent to a weak $(1,2)$-automaton. This, by Proposition 2, can be reformulated as follows: $L(A)$ is closed iff G_A^+ does not contain a weak $(0,1)$-flower. Now, to decide whether a deterministic automaton recognises a closed set, first determine its productive states, then build its productive graph and check for weak $(0,1)$-flowers. Note that two last steps can be easily done in polynomial time. The case of open languages is entirely dual.

For Π_2^0 and Σ_2^0 levels follow analogous argument only now using Theorem 3 and Theorem 4 respectively, and Proposition 1. For Π_3^0 and Σ_3^0 levels use the gap property and Theorem 5.

This way for a given deterministic language one obtains its exact level in the topological hierarchy.   □

In general, deciding the topological complexity of a deterministic tree language is as difficult as calculating the unproductive states of an automaton, the latter being equivalent to deciding a language's emptiness. In 1969 Rabin [14] showed that the emptiness problem is decidable, and in 1988 Emerson and Jutla [3] presented an algorithm with time complexity $\mathcal{O}((nd)^{3d})$, where n is the number of states and d is the number of ranks used. The emptiness problem can be easily reduced to solving parity games. The late nineties brought improved algorithms for this problem by Browne et al. [2] with complexity $\mathcal{O}(d^2 mn^{\frac{d}{2}})$ and by Seidl [15] with complexity $\mathcal{O}(dm(\frac{n+d}{d})^{\frac{d}{2}})$, where n, m, and d are the numbers of vertices, edges, and ranks in the game graph. The investigation of parity games resulted in polynomial algorithms in plenty of special cases, but so far it is not known if the original problem is polynomial. One of the last achievements in this field is the procedure by Jurdziński and Vöge [4] which is apparently quite efficient, however its complexity has not, at present, got any nontrivial upper bounds.

Acknowledgment

The author would like to thank Damian Niwiński for helpful comments and discussions, and the anonymous referees for helpful comments.

References

1. Bradfield, J. C.: The modal mu-calculus alternation hierarchy is strict. Theoret. Comput. Sci. **195** (1998) 133–153
2. Browne, A., Clarke, E. M., Jha, S., Long, D. E., Marrero, W.: An improved algorithm for the evaluation of fixpoint expressions. Theoret. Comput. Sci. **178** (1997) 237–255
3. Emerson, E. A., Jutla, C. S.: The complexity of tree automata and logics of programs. In: Proc. FoCS '88. IEEE Computer Society Press (1988) 328–337
4. Jurdziński, M., Vöge, J.: A discrete strategy improvement algorithm for solving parity games. In: Proc. CAV 2000. Lecture Notes in Computer Science, Vol. 1855. Springer-Verlag (2000) 202–215
5. Kechris, A. S.: Classical Descriptive Set Theory. Graduate Texts in Mathematics, Vol. 156. Springer-Verlag (1995)
6. Kupferman, O., Safra, S., Vardi, M.: Relating Word and Tree Automata. In: 11th IEEE Symp. on Logic in Comput. Sci. (1996) 322–332
7. Lenzi, G.: A hierarchy theorem for the mu-calculus. In: auf der Heide, F. M., Monien, B. (eds.): Proc. ICALP '96. Lecture Notes in Computer Science, Vol. 1099. Springer-Verlag (1996) 87–109
8. Mostowski, A. W.: Hierarchies of weak automata and weak monadic formulas. Theoret. Comput. Sci. **83** (1991) 323-335.
9. Niwiński, D.: On fixed point clones. In: Kott, L. (ed.): 13th ICALP '86. Lecture Notes in Computer Science, Vol. 226. Springer-Verlag (1986) 464–473
10. Niwiński, D., Walukiewicz, I.: Relating hierarchies of word and tree automata. In: STACS '98. Lecture Notes in Computer Science, Vol. 1373. Springer-Verlag (1998) 320–331
11. Niwiński, D., Walukiewicz, I.: A gap property of deterministic tree languages. Theoret. Comput. Sci. **303** (2003) 215–231

12. Niwiński, D., Walukiewicz, I.: Deciding nondeterministic hierarchy of deterministic tree automata. In: Proc. WoLLiC 2004 (to appear in Electronic Notes in Theoretical Computer Science)
13. Otto, M.: Eliminating recursion in μ-calculus. In: STACS'99. Lecture Notes in Computer Science, Vol. 1563. Springer-Verlag (1999) 531–540
14. Rabin, M. O.: Decidability of second-order theories and automata on infinite trees. Trans. Amer. Soc. **141** (1969) 1–35
15. Seidl, H.: Fast and simple nested fixpoints. Information Processing Letters **59** (1996) 303–308
16. Skurczyński, J.: The Borel hierarchy is infinite in the class of regular sets of trees. Theoret. Comput. Sci. **112** (1993) 413–418
17. Thomas, W.: Languages, automata, and logic. In: Rozenberg, G., Salomaa, A. (eds.): Handbook of Formal Languages, Vol. 3. Springer-Verlag (1997) 389–455
18. Urbański, T. F.: On deciding if deterministic Rabin language is in Büchi class. In: Montanari, J. R. U., Welzl, E. (eds.): Proc. ICALP 2000. Lecture Notes in Computer Science, Vol. 1853. Springer-Verlag (2000) 663–674
19. Wagner, K.: Eine topologische Charakterisierung einiger Klassen regulärer Folgenmengen. J. Inf. Process. Cybern. EIK **13** (1977) 473–487

Complexity and Intensionality in a Type-1 Framework for Computable Analysis

Branimir Lambov*

BRICS**
Department of Computer Science
University of Aarhus, Denmark
barnie@brics.dk

Abstract. This paper describes a type-1 framework for computable analysis designed to facilitate efficient implementations and discusses properties that have not been well studied before for type-1 approaches: the introduction of complexity measures for type-1 representations of real functions, and ways to define intensional functions, i.e. functions that may return different real numbers for the same real argument given in different representations.

This approach has been used in a recently developed package for exact real number computations, which achieves performance comparable to the performance of machine precision floating point.

1 Introduction

This paper presents an approach to computable analysis which corresponds to interval arithmetic supplied with a mechanism for increasing precision. The approach is designed to allow very efficient implementations of exact real arithmetic.

The main problem this paper addresses is the question of measuring complexity in this approach. As in other type-1 models for analysis, complexity cannot be simply introduced as a restriction on the class of functions used in the system, because the system contains an implicit unbounded search which makes all complexity classes that can define the minimization normal form, down to polynomials over the integers, equivalent in terms of expressive power. Suitable additional objects are introduced that can be used to characterize the complexity of numbers and functions.

In addition to this, the paper discusses the possibility to implement intensional representations of multi-valued real functions. This is not possible directly in the framework. However, using a modified definition of the representation of a real function that corresponds to access to a memory cell dedicated to the operation, intensional functions can be implemented, and their complexity can still be assessed using the methods described in this paper.

---

* Currently at FB Mathematik, TU-Darmstadt, Schlossgartenstrasse 7, D-64289 Darmstadt, Germany.
** Basic Research in Computer Science (www.brics.dk), funded by the Danish National Research Foundation.

L. Ong (Ed.): CSL 2005, LNCS 3634, pp. 442–461, 2005.

Previous implementations of exact real arithmetic [3, 7, 30], with the exception of Müller's *iRRAM* [26], implement real arithmetic by creating functional representations of real numbers via terms that describe the computation through which they were computed. This approach is very inefficient, leading to ratios in the computation time between real numbers and machine floating point greater than 100 to one even in cases where high precisions are not actually needed. The reason for this is the complexity of dealing with the history of the computation.

In contrast to this, in a type-1 approach such as the one described here, one can define real functions working on the level of approximations, effectively expressing the computation history in the code of a function, avoiding the overheads of creating a term representation for it and making it possible to make use of memory caches and even optimizations performed by a compiler.

The framework described here is used in an actual implementation (*RealLib*, [22]) of an exact real numbers package, whose most significant distinctive feature is the possibility to perform real number computations at a speed comparable to machine precision floating point when the computation does not require the precision to be raised (for details and performance comparison with floating point and other exact real number packages, see [21]).

To make use of this, the user has to express the computation as a partial approximation representation of a real function as described in this paper. Most details are handled automatically and such a representation rarely looks different from just expressing the computation as a series of operations applied to real numbers. In fact, a function working with machine precision numbers can usually be changed to a type-1 representation operating on real numbers just by changing the types and the function's signature.

With the help of this library, the user need not wonder whether to trust a quick machine precision implementation or invest time to redo it at higher precision and eventually in slow exact real number arithmetic. If the problem is easy and the machine precision can indeed be trusted, *RealLib*, unlike other exact real number systems, will return the result very quickly, typically after twice and in the worst case after no more than ten times the time is takes to perform the computation in machine precision. If it is hard, *RealLib* will be slow to compute the result, but a machine precision result in such a case can be completely wrong and lead to disastrous consequences.

2   Prerequisites

In this paper we will be referring to different levels of the finite type hierarchy. Type 0 is the natural numbers and every other finite type is a function that takes arguments of a lower type. Type 1 is the type of the functions over the natural numbers and Type 2 is the set of the functionals that take at least one type-1 function as argument.

Let $\mathbb{V}$ be an enumerable dense subset of the reals that contains the dyadic numbers and is closed under addition, substraction and division by 2. We will be using the following definitions as the established notion of computable real numbers and functions to which we will compare our model:

Definition 1. *A Cauchy function representation (CF-representation) of a real number α is a function $a : \mathbb{N} \to \mathbb{V}$, such that $\forall n \in \mathbb{N} \ \left(|a(n) - \alpha| < 2^{-\text{lth}(n)} \right)$*

(we use $\mathrm{lth}(n) = \lfloor \log_2(n+1) \rfloor$ instead of simply n in the exponent as the latter would not allow us to define the class of feasible real numbers as the ones having a poly-time CF-representation)

With variations in the rate of convergence and the specific representation of the numbers in $\mathbb{V}$, this definition appears in works by Grzegorczyk [11], Pour-El and Richards [24], Ko [17] and Weihrauch [29]. In the latter, Lemma 4.2.1 shows the equivalence of these variations.

Definition 2 ([29], P. 108). *A Cauchy function representation of a partial function ϕ :* $\mathbb{R} \to \mathbb{R}$ *is a partial functional* $\Phi : (\mathbb{N} \to \mathbb{V}) \times \mathbb{N} \to \mathbb{V}$, *such that*

$$\forall \alpha \in \mathrm{dom}\ \phi, \forall a - CF - \text{representations of } \alpha$$
$$\forall n \in \mathbb{N}\left((a, n) \in \mathrm{dom}\ \Phi \wedge |\Phi(a, n) - \phi(\alpha)| < 2^{-\mathrm{lth}(n)}\right)$$

In the case of total functions, this definition is equivalent to the Grzegorczyk's approach [11], used also by Pour-El and Richards in [24], which defines real functions as maps of rational approximations to infinite sequences of rational approximations supplied with a computable $\omega(k, n)$, a modulus of uniform continuity of the function on $[-k; k]$. For a most clear definition of the latter and proof of the equivalence the reader can refer to Cor. 2.16 of [17]. In contrast to that, the type-2 definition we are using works well with partial and multi-valued functions together with complexity measures.

Definition 3. *A real number or a real function is computable in a class C of computable functions, resp. functionals, iff there exists a representation in C for it in the sense of Definitions 1 or 2 respectively.*

To assess the complexity of real numbers we can use restrictions on the class of functions used in Definition 1. The complexity notions for real functions are somewhat more complicated, thus we have decided to compare our approach to two different complexity measures: Ko's notion of feasibility for Computable Analysis and type-2 complexity by restriction of the class of functionals used in Definition 2.

A well studied notion for feasibility of type-2 functionals is the class of the Basic Feasible Functionals (**BFF**, defined in [14]). They can also be defined as the type-2 restriction of the Basic Feasible Functionals of finite type (**BFF**$^\omega$, [6]), and the latter can be defined as the higher-type functions that can be written as terms containing only

- constants 0 for every finite type
- variables for every finite type
- constants for every poly-time function
- the Σ and Π combinators for every combination of finite types (i.e. typed lambda calculus)
- bounded recursion on notation R_{bn}:

$$R_{bn}(x, y, g, h) = \begin{cases} y, & \text{if } x = 0 \\ \min\left(g(x, R_{bn}(\lfloor x/2 \rfloor, y, g, h), h(x)\right), & \text{otherwise} \end{cases}$$

Two essential properties that we will be using in the complexity part of our paper are the facts that **BFF**$^\omega$ restricted to Type 1 coincides with the poly-time functions, and **BFF**$^\omega$ restricted to Type 2 (i.e. **BFF**) coincides with the functionals that are computable by an oracle Turing machine in time which is a second order polynomial to the lengths of the inputs ([14]), where the length of a function is taken to be $\mathrm{lth}(B) = \lambda n.\max_{\mathrm{lth}(k) \leq n} \mathrm{lth}(B(k))$. We will also use the fact that **BFF** can define the first normal form for type-2 functionals ([14]). Here, as well as in the following definition, an oracle Turing machine can use the oracle to retrieve arbitrarily good approximations of the real argument in one computational step.

We will use this definition for feasibility of real functions in the usual sense for Computable Analysis:

Definition 4 (Ko, [17], Def. 2.18). *A function ϕ is poly-time computable on $[a, b] \subseteq$ $\mathrm{dom}\,\phi$ in Ko's sense iff there is an oracle Turing machine computing it in the dyadic representation, which runs in time polynomial to the precision given in unary notation.*

This definition of complexity is equivalent to the signed digit one defined in [29], Def. 7.2.6, among others, and, using the next theorem, can be easily translated to the setting of **BFF**.

Theorem 1. *A partial real function is poly-time computable on $[a, b]$ in the sense of Ko iff it is **BFF**-computable on the same interval.*

Proof. A functional is in **BFF** if and only if there exists an oracle Turing machine computing it, running in time which is a second-order polynomial in the length of the inputs. In a compact interval, there exist CF-representations B of any real number that satisfy $\forall k(\mathrm{lth}(B(k)) \leq p(\mathrm{lth}(k)))$ for a polynomial p (using dyadic representations cut after the $\mathrm{lth}(k)$'th digit), and therefore a second order polynomial in $\mathrm{lth}(k), \mathrm{lth}(B)$ does not give more power than simply a polynomial in $\mathrm{lth}(k)$.   □

Our investigation of the connections between complexity in our model and type-2 and Computable Analysis complexity also require a very useful tool from Proof Theory, the concept of majorizability:

Definition 5 (W.A. Howard, [13]). *We define $x^* \,\mathrm{maj}_\rho\, x$ for a finite type ρ by induction on the type:*

$$x^* \,\mathrm{maj}_0\, x := x^* \geq x,$$

$$x^* \,\mathrm{maj}_{\tau \to \rho}\, x := \forall y^*, y \left(y^* \,\mathrm{maj}_\tau\, y \to x^*(y^*) \,\mathrm{maj}_\rho\, x(y) \right).$$

We will say that a class of function(al)s C is majorizable, if for every function(al) f in C there exists $f^* \in C$ with $f^* \,\mathrm{maj}\, f$, where the majorization relation is of the appropriate type.

The majorizability relation defines monotonicity in higher types, and for us the most interesting instantiation of the definition is in the case $\rho \equiv 0$ used in the form $x^* \,\mathrm{maj}_\tau\, x \wedge y^* \,\mathrm{maj}_\tau\, y \to x^*(y^*) \geq x(y)$, giving us the possibility to bound the result of the application of one higher-type functional to another as a number.

It is not hard to see that the poly-time functions and $\mathbf{BFF}^\omega$ are majorizable classes (detailed proof can be found in [20]). Other majorizable classes are the class of the primitive recursive functionals of finite type (in the sense of Kleene S1-S8 ([16]) as well as in the sense of Gödel ([10]) and any specific level in Gödel's primitive recursive hierarchy), any level of the Grzegorczyk hierarchy and many others, but not the class of all type-2 functionals (e.g. if the function

$$F(f) = \begin{cases} 0, & \text{if } \forall x(f(x) = 0) \\ \mu x[f(x) = 0], & \text{otherwise} \end{cases}$$

were majorized by F^*, then $F^*(\lambda x.1)$ would bound F applied to all zero/one functions which is not possible) or the class of the partial computable functionals if the notion is extended in a suitable way to accommodate partiality.

The definitions of representations of real functions used so far only admit extensional functions, i.e. ones that respect equality on real numbers and do not depend on the actual representation of the argument. Because dependence on the representation can be used to circumvent the impossibility to define discontinuous real functions, it is sometimes useful to consider *intensional* (i.e. non-extensional, functions that depend on the representation of the argument) functions. One cannot call such objects "functions on real numbers", because on the real numbers level they do not represent functions. Instead, the term *multi-valued functions* has been used in the literature ([29]) to denote the fact that they can return different results for arguments that are equal as real numbers.

We will be using the following definition of an intensional real function:

Definition 6. *A Cauchy function representation of a multi-valued function* $\phi : \mathbb{R} \to P(\mathbb{R})$ *is a partial functional* $\Phi : (\mathbb{N} \to \mathbb{V}) \times \mathbb{N} \to \mathbb{V}$, *such that*

$$\forall \alpha \in \text{ dom } \phi, \forall a - \text{CF} - \text{representations of } \alpha$$
$$\exists \beta \in \phi(\alpha) \forall n \in \mathbb{N} \left((a, n) \in \text{ dom } \Phi \wedge |\Phi(a, n) - \beta| < 2^{-\text{lth}(n)} \right)$$

Definition 7. *We will call a CF-representation* Φ *of a multi-valued function* ϕ *extensional iff*

$$\forall \alpha \in \text{ dom } \phi \forall x, y - \text{CF} - \text{representations of } \alpha \ \forall n (|\Phi(x, n) - \Phi(y, n)| < 2^{-n+1}).$$

Definition 8. *A CF-representation* Φ *of a multi-valued function* ϕ *is* intensional *if it is not extensional, i.e. iff*

$$\exists \alpha \in \text{ dom } \phi \exists x, y - \text{CF} - \text{representations of } \alpha \exists n (|\Phi(x, n) - \Phi(y, n)| \geq 2^{-n+1}).$$

Theorem 2. *There exists a multi-valued function* ϕ *with a computable intensional CF-representation such that no computable extensional CF-representation of* ϕ *exists.*

Proof. Take the function

$$\phi(x) = \begin{cases} \{0\}, & \text{if } x < 0 \\ \{0; 1\}, & \text{if } 0 \leq x \leq 1 \\ \{1\}, & \text{otherwise} \end{cases}$$

Then

$$\Phi(a, n) = \begin{cases} 0, & \text{if } a(1) \leq 0.5 \\ 1, & \text{otherwise} \end{cases}$$

is computable and a valid intensional CF-representation of Φ, but on the other hand any extensional representation would have a point of discontinuity and would thus be non-computable. $\square$

An extensional representation is a representation of one of the (single-valued) real functions that is possible in the multi-valued specification, and intensional functions do not have a corresponding real function. Nevertheless, the latter can be used in practice, especially in cases where a mathematical function is defined by forcing the result to one of a selection of equally valid choices. For example, the square root of a complex number as a function is discontinuous and thus its computable extensional implementations must be undefined at the line of discontinuity. However, if we define the square root as a multi-valued function, computable intensional implementations without holes in the domain can be given.

In this text extensionality and single-valued functions are assumed everywhere unless the alternative is explicitly specified.

3   The Partial Approximation Representations Approach

The basic objects we are going to use contain approximation information and an estimation of the amount of error in this approximation. To be able to define a class of real functions equivalent to the computable ones in the sense of Definition 2, a totally indeterminate value has to be allowed (otherwise e.g. division cannot be defined, see [28]). We do this by allowing an infinitely large value for the error.

Let $\mathbb{E}$ be a subset of the positive rational numbers which contains 1 and is closed under multiplication and division by 2, to which the special value ∞ is added, and which has a poly-time encoding and a poly-time comparison operator that respects ∞. It is possible to define encodings $\langle \cdot \rangle_{\mathbb{V}}$, $\langle \cdot \rangle_{\mathbb{E}}$ and $\langle \cdot, \cdot \rangle$ of, respectively, $\mathbb{V}$, $\mathbb{E}$ and pairs $\mathbb{V} \times \mathbb{E}$ with the following properties:

- $\mathrm{lth}(\langle a, b \rangle)$ is polynomial in $\max(\mathrm{lth}(\langle a \rangle_{\mathbb{V}}), \mathrm{lth}(\langle b \rangle_{\mathbb{E}}))$
- $\langle a, b \rangle \geq \langle a \rangle_{\mathbb{V}}$ and $\langle a, b \rangle \geq \langle b \rangle_{\mathbb{E}}$
- $\langle 2^{-n} \rangle_{\mathbb{E}} \geq 2^n$
- there exist poly-time functions that convert between the encodings of $\mathbb{V}$ and $\mathbb{E}$, rounding up if a number in $\mathbb{V}$ cannot be represented in $\mathbb{E}$
- multiplication and division by 2 are poly-time (and thus also multiplication by $2^{\pm \mathrm{lth}(k)}$) in both $\mathbb{V}$ and $\mathbb{E}$
- addition and the floor function $\lfloor \cdot \rfloor$ in $\mathbb{V}$ are poly-time
- there exists a function $\mathrm{dya}(n, d)$ that selects a code for the dyadic number $n2^{-d}$, such that whenever a, b, c, d are positive integers, $a \leq c \wedge b \leq d \rightarrow \mathrm{dya}(a, b) \leq \mathrm{dya}(c, d)$
- the absolute value operator on the codes is such that $\langle v \rangle_{\mathbb{V}} \leq \langle |v| \rangle_{\mathbb{V}}$ for any $v \in \mathbb{V}$

These properties can be satisfied e.g. for $\mathbb{V} = \mathbb{Q}$ and $\mathbb{E} = \mathbb{Q}_{\infty}^+$ by the Cantor pairing Π, the encoding of rational numbers q as $\Pi(n, d)$, such that

$$q = (-1)^n \frac{\lfloor (n+1)/2 \rfloor}{d},$$

and the encoding of ∞ as $\Pi(0, 0)$.

Having the distinction between the sets $\mathbb{V}$ and $\mathbb{E}$ is prompted by the need to include ∞, but also closely follows the choice one would often make when an actual implementation is developed as one might prefer a simpler (and thus more efficient) representation of the error information.

3.1 Definitions

Definition 9. *A partial approximation to a real number α is a pair (v, e) of type $\mathbb{V} \times \mathbb{E}$, such that $|v - \alpha| < e$. We will denote the class of partial approximations to α with $\mathbb{A}_\alpha$, and the class of partial approximations to any real with $\mathbb{A}_\mathbb{R} = \cup_{\alpha \in \mathbb{R}} \mathbb{A}_\alpha$. If $a \in \mathbb{A}_\mathbb{R}$ we will use a_v, a_e to denote respectively the value and error in a.*

Definition 10. *A partial approximation representation, p.a.r., of a real number α is a function $A : \mathbb{N} \to \mathbb{A}_\alpha$, for which $\forall k \exists n((A(n))_e \leq 2^{-k})$.*

If a real number is computable, then it certainly has a computable p.a.r.: if B is a representation of α, then $\lambda n.(B(n), 2^{-\text{lth}(n)})$ is one of its p.a.r.'s. Conversely, if a is a p.a.r. of α, then

$$\lambda k.(A(\mu n[(A(n))_e \leq 2^{-\text{lth}(k)}]))_v \tag{1}$$

is a valid CF-representation for it.

This equivalence does not hold for restrictions of the notion of computability. Because of the unbounded search in (1), it is possible to define all computable reals using p.a.r.'s in subrecursive classes such as primitive recursive, elementary or poly-time functions. For a proof of this, see [28].

For real functions, we want to have objects that operate on partial approximations instead of the full representations. They will have to convert approximations to an input to approximations to the result of the application of the function, and also we need to require that the precision of the output approximations gets arbitrarily good as the precision of the input increases. In other words,

Definition 11. *A partial approximation representation of a partial function $\phi : \mathbb{R} \to \mathbb{R}$ is a partial function $F : \mathbb{A}_\mathbb{R} \to \mathbb{A}_\mathbb{R}$, such that for any choice of $\alpha \in \text{dom } \phi$ and a partial approximation representation A of α, $\lambda n.F(A(n))$ is a partial approximation representation of $\phi(\alpha)$.*

Remark 1. This definition implies $a \in \mathbb{A}_\alpha \to F(a) \in \mathbb{A}_{\phi(\alpha)}$ for $\alpha \in \text{dom } \phi$.

3.2 Computability

We have severely restricted the information to which the function object has access; nevertheless, this does not restrict the class of real functions that are computable. The following theorem is a proof of this fact that uses a construction which we will later modify to use in our complexity and intensionality results:

Theorem 3. *A partial function $\phi : \mathbb{R} \to \mathbb{R}$ is computable if and only if it has a computable p.a.r.*

Proof. ($\leftarrow$) If we have a p.a.r. F of a function ϕ, and $\alpha \in \text{dom } \phi$, then the functional

$$\Phi(B, n) := (F(\langle B(m), 2^{-\text{lth}(m)} \rangle))_v, \text{ where} \tag{2}$$

$$m = \mu p \left[(F(\langle B(p), 2^{-\text{lth}(p)} \rangle))_e \leq 2^{-\text{lth}(n)} \right]$$

is total in n for any CF-representation B of α since from Definitions 11 and 10 the minimization will always stop, and Definition 9 together with Remark 1 ensures $|\Phi(a, n) - \phi(\alpha)| < 2^{-\mathrm{lth}(n)}$.   □

Proof. ($\rightarrow$) Fix a CF-representation Φ for ϕ.

For any $a \in \mathbb{A}_{\mathbb{R}}$ with $a_e < 1$, we can effectively find the largest natural number m with the property $2^m a_e < 1$. If $a_e \geq 1$, we take $m = 0$. Define the function

$$b := \lambda n.2^{-\mathrm{lth}(n)} \lfloor 2^{\mathrm{lth}(n)} a_v + 1/2 \rfloor. \tag{3}$$

For $0 \leq \mathrm{lth}(n) < m$ we have that if $\alpha \in \mathbb{A}_\alpha$

$$|b(n) - \alpha| \leq |a_v - \alpha| + 2^{-(\mathrm{lth}(n)+1)} \leq 2^{-m} + 2^{-(\mathrm{lth}(n)+1)} \leq 2^{-\mathrm{lth}(n)}$$

In the following we will use the language of exceptions[1]. Given the code of a computable functional Φ, we can construct an equivalent one $\Phi^\dagger$ that honors a new exception x. We can create a function

$$b\lceil m := \lambda n. \begin{cases} b(n), & \text{if } n < m \\ \mathbf{raise}\ x, & \text{otherwise} \end{cases}$$

and then define

$$\Phi^\ddagger(B, n) := \begin{cases} \langle 0, \infty \rangle, & \text{if } n = 0 \\ \mathbf{try}\ \langle \Phi^\dagger(B, \lfloor \frac{n}{2} \rfloor), 2^{-\mathrm{lth}(\lfloor \frac{n}{2} \rfloor)} \rangle \\ \mathbf{catch}(x)\ \Phi^\ddagger(B, \lfloor \frac{n}{2} \rfloor) \end{cases}, \text{ otherwise} \tag{4}$$

($\Phi^\ddagger(b\lceil m, n)$ finds the largest $l \leq n-1$ for which $\Phi(b, l)$ only refers to the first m values in b, or returns a completely undefined value if such an n cannot be found).

We will now prove that the function

$$F(a) := \Phi^\ddagger(b\lceil m, 2^{m+1}) \tag{5}$$

is the required p.a.r. of ϕ. To do this, we need to prove that $G = \lambda n.F(A(n))$ is a p.a.r. of $\phi(\alpha)$ for any p.a.r. A of α.

The first condition, $F(a) \in \mathbb{A}_{\phi(\alpha)}$ for any $a \in \mathbb{A}_\alpha$, follows from the requirement for Φ and the fact that there is a CF-representation for α that starts with $b(0), b(1), \ldots, b(m-1)$.

For the second condition, we need to prove the existence of 2^{-k}-approximations to $\phi(\alpha)$ among $G(n)$ for any k. The sequence defined by

$$c := \lambda n.2^{-\mathrm{lth}(n)} \lfloor 2^{\mathrm{lth}(n)} \alpha + 1/2 \rfloor$$

is a proper CF-name for α. If α is not a dyadic number, then for an arbitrary n, $|\alpha - c(n)| < 2^{-\mathrm{lth}(n)-1}$. There exists q depending on n, such that $|\alpha - c(n)| \leq$

---

[1] The reader can refer to a current book on semantics (e.g. [23]) for a proper definition of the concept and its implementation. Essentially the same approach (but explicitly specified and not identified as a case of using exceptions) is used e.g. in [18] and [1] and even in the definition of Kleene's associates (see Sec. 4). Through the use of exceptions we avoid the tedious explicit construction of the functional $\Phi^\ddagger$ from the code of Φ.

$2^{-\text{lth}(n)}(1/2 - 2^{-(q-\text{lth}(n))})$, and for all partial approximations a with $a_e < 2^{-q}$ we have $2^{\text{lth}(i)}|a_v - c(i)| < 1/2$ for all $0 \le i \le n$. But this implies that the sequence obtained by (3) coincides with c on the first $n+1$ elements.

Now, since Φ would look at finitely many elements of c to produce a value with any precision 2^{-k}, using that count in the procedure described above, we can come up with a q supplying a long enough sequence. Combining this with a requirement that m in (5) is sufficient for the target precision, we have $(F(a))_e \le 2^{-k}$ for all a's with $a_e \le 2^{-\max(q,k)}$, and since A has arbitrarily close approximations, this can be satisfied for $a = A(n)$ for some n.

If α is a dyadic number, i.e. $\exists n(c(n) = \alpha)$, then there are only finitely many variations of b that can exists, because they have to coincide after the first $n+1$ positions. Then there exists a maximum m for the number of lookups Φ can make to any of these b's in order to get a 2^{-k}-precise result. Hence $a_e \le 2^{-\max(m,k)}$ suffices to get the required precision for $F(a)$. $\qquad\square$

As in the case of real numbers, this equivalence does not hold for subclasses of the type-2 computable functions. To define all computable functions, it suffices to use severely restricted type-1 computability subclasses:

Theorem 4. *A partial real function is computable if and only if it has a p.a.r. in any subrecursive class C that contains the poly-time functions.*

Proof. ($\rightarrow$) It suffices to change the definition of $\Phi^{\ddagger}$ to a version bounded in execution time

$$\Phi^{\ddagger}(B,n,m) := \Phi^{\ddagger}(B,n,m) := \begin{cases} \langle 0, \infty \rangle, & \text{if } n = 0 \\ \textbf{try } \langle \Phi^m(B, \lfloor \frac{n}{2} \rfloor), 2^{-\text{lth}(\lfloor \frac{n}{2} \rfloor)} \rangle & \\ \textbf{catch}(x) \ \Phi^{\ddagger}(B, \lfloor \frac{n}{2} \rfloor, m) & \text{otherwise} \end{cases},$$

where by Φ^m we denote $\Phi^{\dagger}$ executed for m steps throwing the exception x if Φ did not halt, which can be done in a basic feasible functional (because **BFF** can define the first normal form for type-2 functionals), and modify F correspondingly to pass this additional argument: $F(a) := \Phi^{\ddagger}(b\lceil m, 2^{m+1}, m)$.

Since m is of the order of the length of the encoding of a it is possible to do all required steps in time polynomial to $\text{lth}(a)$. The proof of the existence of good approximations can be carried out here as well, the only difference being the need to satisfy a condition in the form $a_e \le 2^{-max(q,k,s)}$ for s being the number of steps it takes for Φ to complete its evaluation on b of length q.

The p.a.r. is type-1 basic feasible, therefore it is poly-time. $\qquad\square$

Proof. ($\leftarrow$) Follows from the previous theorem. $\qquad\square$

3.3   Complexity

Real Numbers. In order to be able to speak about different complexity classes of real numbers, we must make a definition which requests more from our functions in order to avoid the minimization in (1). This gives rise to the following definitions and equivalence property:

Definition 12. *A modulus for a p.a.r. A of a real number α is a function $m : \mathbb{N} \to \mathbb{N}$, such that for all k, $(A(m(k)))_e \le 2^{-\mathrm{lth}(k)}$.*

Definition 13. *We will say that a real number is p.a.r.-computable in a given class C of computable functions, if there exist both a p.a.r. and a modulus for it in C.*

Theorem 5. *A real number is computable in a subrecursive class C that contains the poly-time functions and is closed under composition if and only if it is p.a.r.-computable in C.*

Proof. If B is a CF-representation of the number, take the p.a.r. $A := \lambda n.$ $\langle B(n), 2^{-\mathrm{lth}(n)} \rangle$ and the modulus $m := \lambda n.n$.

For the other direction $B := \lambda k.(A(m(k)))_v$ is a CF-representation of the number if A and m are, respectively, its p.a.r. and modulus.   □

On the level of feasible functions, poly-time p.a.r. computability coincides with Ko's notion of poly-time computable real numbers [17] (Ko speaks about numbers given in unary notation, which is equivalent to the parameter $\mathrm{lth}(n)$ used in our definitions).

Type-2 Complexity for Functions. Again taking the p.a.r. of a real function we lose all complexity information about that function. To talk about complexity classes, we define a function that can replace the minimization in (2):

Definition 14. *A modulus for a p.a.r. F of a partial real function ϕ is a partial functional $M : (\mathbb{N} \to \mathbb{A}_{\mathbb{R}}) \times (\mathbb{N} \to \mathbb{N}) \times \mathbb{N} \longrightarrow \mathbb{N}$, such that for all $\alpha \in \mathrm{dom}\ \phi$, p.a.r. A of α, moduli m for A,*

$$\forall k((F(A(M(A, m, k))))_e \le 2^{-\mathrm{lth}(k)}). \tag{6}$$

Note that even though the actual function object is a type-1 object, we now introduce a type-2 operation to characterize it. However, some extra flexibility comes from the separation of these two objects: to implement e.g. a feasible real function one does not have to implement a feasible type-2 object, but only needs to prove that it exists. Moreover, if a CF-representation of a function needs extra information to be in a certain class (e.g. division needs evidence that the denominator is non-zero to be primitive recursive), it will in general only be needed for the modulus.

Definition 15. *We will say that a real function is p.a.r.-computable in a given class C of computable type-2 functionals, if both a computable p.a.r.[2] and its modulus can be found in C.*

Theorem 6. *If a function is p.a.r.-computable in a given class C that contains **BFF** and is closed under functional composition ([12], Def. 1.1) and substitution ([12], Def. 3.8), then it is computable in the same class.*

---

[2] Via the implicit embedding of Type 1 in Type 2.

Proof. For $\phi : \mathbb{R} \to \mathbb{R}$, $\alpha \in$ dom ϕ, F- p.a.r. of ϕ, M-modulus for F, and B - CF-representation of α, take

$$\Phi(B, n) := (F(A(M(A, \lambda p.p, n))))_v$$

where

$$A := \lambda p. \left\langle B(p), 2^{-\mathrm{lth}(p)} \right\rangle.$$

A is a p.a.r. for α with a modulus $\lambda p.p$, and hence from M being a modulus to F, we have $|\Phi(B, n) - \phi(\alpha)| < 2^{-\mathrm{lth}(n)}$. Φ is a basic feasible functional relative to F and M, therefore it is in C. $\qquad\square$

The other direction is more complicated. First we will verify that p.a.r.-computability coincides with CF-computability, i.e. that, in addition to the p.a.r., a modulus can be found for every computable function:

Theorem 7. *If a partial function $\phi : \mathbb{R} \to \mathbb{R}$ is computable, then it is p.a.r.-computable in the class of all partial computable functionals.*

Proof. We've already proved in Theorem 3 that there exists a computable p.a.r. to every computable real function. If it is F, then

$$M(A, m, n) := \mu p[(F(A(p)))_e \leq 2^{-\mathrm{lth}(n)}]$$

is a modulus for F. $\qquad\square$

This modulus does not even use the modulus for the real number. This is true, because in the presence of minimization brute force search makes the moduli redundant.

This is not the case for restricted complexity classes. To prove the equivalence between p.a.r. and CF-computability on some of them, we need the higher-type monotonicity we have in the majorizable classes and the following lemma:

Lemma 1. *Let b be defined as*

$$b(n) := \mathrm{dya}(\lfloor 2^{\mathrm{lth}(n)} a_v + 1/2 \rfloor, \mathrm{lth}(n)). \tag{7}$$

Then for all $\alpha \leq a_0, a \in \mathbb{A}_\alpha, b$, created by (7) for a with $a_e \leq 1$,

$$J(a_0) \, \mathrm{maj}_1 \, b$$

where

$$J(a_0) = \lambda n.\mathrm{dya}(1 + \lfloor 2^{\mathrm{lth}(n)} a_0 + 1/2 \rfloor, \mathrm{lth}(n)). \tag{8}$$

Proof. Since $a_e \leq 1$, we have $|a_v| < a_0 + 1$ and therefore by the properties of the encoding $J(a_0)(n) \geq b(n)$, and also, since when n is increased both the numerator and denominator in (8) do not decrease, we have $\forall k \leq n(J(a_0)(n) \geq J(a_0)(k) \geq b(k))$, which means $J(a_0) \, \mathrm{maj}_1 \, b$. $\qquad\square$

Theorem 8. *If a partial real function is computable in a majorizable class of type-2 functionals that contains **BFF** and is closed under functional composition and substitution, then it is p.a.r.-computable in that class.*

Proof. We will use the proof of Theorem 3, substituting the definition (3) of b with (7). All operations used in the generation of F can be done without leaving the class of Φ (this is true because m is of the order of lth(a)). Hence F is in the class. We now need to find a modulus for it.

In the class of Φ there exists a functional Ψ that does exactly the same job as Φ, but instead of returning the approximation it gives the largest k to which B was applied. Since the class contains this functional and is majorizable, it also contains a majorizer Ψ^* for it. The modulus for A gives us means to bound the absolute value of the real number described by it, therefore, with the previous lemma, there is a poly-time function $b^* := J(|A(m(0))| + 1)$ which majorizes all functions b generated by partial approximations with error less than 1.

Hence $l = \Psi^*(b^*, n) \geq \Psi(b, n)$ for all good b's, in particular for the one (call it b_0) generated by $a_0 = A(m(l))$, which means $\Phi^\dagger(b_0 \lceil l, n)$ will not raise an exception, and $F(a_0)$ will give a result with the required precision.

Hence $M(A, m, n) = \max(m(\Psi^*(J(|A(m(0))| + 1), n)), n)$ is a modulus for F.

$\square$

Real Number Complexity for Functions. In the previous subsection we found correspondence between complexities in this model and type-2 complexity. As Ko's approach, the complexity measure normally used for real functions, is different, we also define notions which are more closely related to the latter by defining type-1 moduli on closed subsets of the domain:

Definition 16. *A uniform modulus on* $[a, b] \subseteq \text{dom}\,\phi$ *of a p.a.r. F of a real function ϕ is a function $U : \mathbb{N} \to \mathbb{N}$, such that*

$$\forall \alpha \in [a, b]\forall a \in \mathbb{A}_\alpha \forall k \forall n(a_e \leq 2^{-\text{lth}(U(k))} \to (F(a))_e \leq 2^{-\text{lth}(k)})$$

Theorem 9. *A partial real function ϕ is computable in a majorizable class of type-2 functionals closed under functional composition and substitution on $[a, b] \subseteq \text{dom}\,\phi$ if and only if it has a p.a.r. and a uniform modulus in the same class.*

Proof. ($\to$) Use a and b to find an upper bound for the absolute value of α, then apply the same reasoning as in the previous proof.

$\square$

Proof. ($\leftarrow$) $M(A, m, k) = m(U(k))$ is a modulus for all A's representing reals in the interval, thus ϕ is p.a.r.-computable in the class.

$\square$

With this definition we're back at the type-1 level, and we also have a few important equivalences:

Corollary 1. *A partial real function ϕ is primitive recursive in the sense of Kleene ([16]) on $[a, b] \subseteq \text{dom}\,\phi$ if and only if it has a primitive recursive p.a.r. and a primitive recursive uniform modulus on $[a, b]$.*

Corollary 2. *A partial real function ϕ is **BFF**-computable on $[a, b] \subseteq \text{dom}\,\phi$ if and only if it has a poly-time p.a.r. and a poly-time uniform modulus on $[a, b]$.*

Combined with Theorem 1, the latter allows us to state the following equivalence property:

Corollary 3. *A partial real function ϕ is feasible in the sense of Ko on $[a, b] \subseteq \mathrm{dom}\ \phi$ if and only if it has a poly-time p.a.r. and a poly-time uniform modulus on $[a, b]$.*

3.4 Intensional Functions

Intensionality does not work well with the type-1 frameworks, because intensional functions rely on information that is not available in an approximation. If Φ is not extensional, Theorem 3 does not hold. More specifically, a partial function given by a p.a.r. is always extensional:

Theorem 10. *Let $F : \mathbb{A}_\mathbb{R} \rightarrow \mathbb{A}_\mathbb{R}$ and let $\alpha \in \mathbb{R}$ such that for all p.a.r. A of α, $\lambda n.F(A(n))$ is a p.a.r. of some $\beta \in \mathbb{R}$. Then β depends only on α and not on its representation A.*

Proof. Let X and Y be two p.a.r.'s of α. Then

$$Z(n) = \begin{cases} X(\frac{n}{2}), & \text{if } n \text{ is even} \\ Y(\frac{n-1}{2}), & \text{otherwise} \end{cases}$$

is also a representation of α. Then $\lambda n.F(Z(n))$ is a p.a.r. of a real number β and therefore $\lambda n.F(X(n))$ and $\lambda n.F(Y(n))$ are also p.a.r.'s to β as subsequences of $\lambda n. F(Z(n))$. $\qquad\square$

Still, intensional functions are interesting for us and we want to find a way to accommodate them. To do this, we have to pass additional information to the functions.

The most straightforward solution is to supply information about the history of the approximation as an argument to the p.a.r., i.e. essentially use Kleene's associate definition. We will not be treating this approach, because the amount of information that has to be passed to the associate in a direct application of Kleene's approach is too big and complexity reasoning would be very difficult if not impossible.

A different approach, carrying less information, is to give the function access to the previous value it has produced, i.e.

Definition 17. *A recursion-p.a.r. of a multi-valued function ϕ is a function $F : \mathbb{A}_\mathbb{R} \times \mathbb{A}_\mathbb{R} \rightarrow \mathbb{A}_\mathbb{R}$, such that for any choice of $\alpha \in \mathrm{dom}\ \phi$ and p.a.r. A of α, $\lambda n.F(A(n), F(A(n-1), F(A(n-2), \cdots F(A(0), 0) \cdots)))$ is a p.a.r. of a $\beta \in \phi(\alpha)$.*

Alternatively, one can extract the "history information" in a separate function:

Definition 18. *A storage-p.a.r. of a multi-valued function ϕ is a pair of functions $F : \mathbb{N} \rightarrow \mathbb{A}_\mathbb{R}$ and $H : \mathbb{A}_\mathbb{R} \times \mathbb{N} \rightarrow \mathbb{N}$, such that for any choice of $\alpha \in \mathrm{dom}\ \phi$ and p.a.r. A of α, $\lambda n.F(H(A(n), H(A(n-1), H(A(n-2), \cdots H(A(0), 0) \cdots)))$ is a p.a.r. of a $\beta \in \phi(\alpha)$.*

The idea behind this is that the function has access to a memory cell where it can store information about past calls and update at each call. This can be very efficient, especially in practical cases where one bit of external storage[3] can be sufficient.

We will be treating the storage-p.a.r. approach and in the end we will show that the two are equivalent.

Theorem 11. *A multi-valued function has a CF-representation if and only if it has a storage-p.a.r.*

Proof. ($\leftarrow$) Given a storage-p.a.r. pair F, H, the function

$$\Phi(a, n) = F(h(k))_v, \text{ where}$$
$$h(m) = \begin{cases} 0, & \text{if } m = 0 \\ H(\langle a(m-1), 2^{-\text{lth}(m-1)} \rangle, h(m-1)), & \text{otherwise} \end{cases}$$
$$k = \mu m. \left[F(h(m))_e \leq 2^{-\text{lth}(n)} \right]$$

is a CF-representation of the function ϕ: h builds a sequence of applications of H which is only lengthened when we move ahead in the approximation, and since the sequence $\langle a(i), 2^{-\text{lth}(i)} \rangle_{i \in \mathbb{N}}$ is a p.a.r. to the argument, the storage-p.a.r. of ϕ has to return approximations to one of the possible results, and the minimization for k always terminates. $\square$

Proof. ($\rightarrow$) We will define H that builds a signed digit representation of the real number and adds more information to it with consecutive calls. We will be storing the signed digit representation as a pair $\Pi(h_i, h_s)$, where h_i is an integer approximating the number with error 1, and h_s is a string of $\{-1; 0; 1\}$ encoded in base 4. The following function implements this using bounded recursion on the notation of a_e (the bound is not explicitly specified, but h_s only grows by two bits for every bit of precision in a_e and h_i is bounded by $\lfloor a_v + 1 \rfloor$ or the previous value of h_i):

$$H(\langle a_v, a_e \rangle, \Pi(h_i, h_s)) = \begin{cases} \Pi(h_i, h_s), & \text{if } a_e > 2^{-(\exp(\Pi(h_i, h_s)) + 1)} \\ \Pi(\lfloor a_v + \frac{1}{2} \rfloor, 0), & \text{if } \frac{1}{4} < a_e \leq \frac{1}{2} \wedge \Pi(h_i, h_s) = 0 \\ \Pi(g_i, 4g_s + 1), & \text{if } a_e \leq 2^{-(\exp(\Pi(h_i, h_s)) + 1)} \\ & \wedge \text{num}(\Pi(h_i, h_s)) - a_v 2^{\exp(\Pi(h_i, h_s))} > \frac{1}{2} \\ \Pi(g_i, 4g_s + 3), & \text{if } a_e \leq 2^{-(\exp(\Pi(h_i, h_s)) + 1)} \\ & \wedge a_v 2^{\exp(\Pi(h_i, h_s))} - \text{num}(\Pi(h_i, h_s)) > \frac{1}{2} \\ \Pi(g_i, 4g_s + 2), & \text{otherwise,} \end{cases}$$

where

$$\Pi(g_i, g_s) = H(\langle a_v, 2a_e \rangle, \Pi(h_i, h_s))$$
$$\text{num}(\Pi(h_i, h_s)) = \begin{cases} h_i, & \text{if } h_s = 0 \\ 2\text{num}(\Pi(h_i, \lfloor \frac{h_s}{4} \rfloor)) - 1, & \text{if } h_s \equiv 1 (\text{mod} 4) \\ 2\text{num}(\Pi(h_i, \lfloor \frac{h_s}{4} \rfloor)), & \text{if } h_s \equiv 2 (\text{mod} 4) \\ 2\text{num}(\Pi(h_i, \lfloor \frac{h_s}{4} \rfloor)) + 1, & \text{if } h_s \equiv 3 (\text{mod} 4) \end{cases}$$
$$\exp(\Pi(h_i, h_s)) = \left\lceil \frac{\text{lth}(h_s)}{2} \right\rceil.$$

---

[3] In practice, F will usually take the current approximation as an additional argument. This argument is not needed for the proofs that follow and does not interfere with them because H can encode it in its result.

We use the same construction as in Theorem 3 (changing only the definitions of m and b), to prove that the following is a storage-p.a.r. of ϕ if Φ is its CF-representation and $\Phi^\dagger$ is a version of it that honors a new exception x:

$$F(\Pi(h_i, h_s)) = \Phi^\dagger(b\lceil m, m+1)$$

for

$$\Phi^\dagger(B, n) = \begin{cases} \langle 0, \infty \rangle, & \text{if } n = 0 \\ \textbf{try } \langle \Phi^\dagger(B, \lfloor \frac{n}{2} \rfloor), 2^{-\text{lth}(\lfloor \frac{n}{2} \rfloor)} \rangle \\ \quad \textbf{catch}(x) \, \Phi^\dagger(B, \lfloor \frac{n}{2} \rfloor) \end{cases}, \text{ otherwise}$$

$$m = \exp(\Pi(h_i, h_s)) + 1$$

$$b\lceil m = \lambda n. \begin{cases} b(n), & \text{if } n < m \\ \textbf{raise } x, & \text{otherwise} \end{cases}$$

$$b(n) = \text{dya}(\text{num}(g(n)), \exp(g(n)))$$

$$g(n) = \Pi(h_i, \lfloor h_s 4^{n+1-m} \rfloor).$$

In this b decodes the information stored in h to a unary function which gives correct approximations to the argument up to its $m - 1$'st value and $\Phi^\dagger$ computes $\Phi(b, n)$ for the largest $n \leq m$ for which this information is sufficient.

Let A be a p.a.r. of an $\alpha \in \text{dom}\,\phi$ and h be a shorthand for $h(n) = H(A(n), H(A(n-1), \cdots H(A(0), 0) \cdots))$.

Since A contains approximations to α for any precision, the string built by h is has no limit for its length and encodes a CF-representation of α. Since Φ is a computable CF-representation of ϕ, by passing to it finite parts of this representation of α, we are getting finite parts of the representation of a number $\beta \in \phi(\alpha)$, and the construction of $\Phi^\dagger$ ensures $F(h(n)) \in \mathbb{A}_\beta$. To get arbitrarily precise approximations to β it suffices to be able to provide arbitrarily long finite parts of the CF-representation of α, which we can do.   □

Unlike in Theorem 3, where b can be different at consecutive calls to F with different approximations to the number, here the initial part of b does not change and this makes the proof simpler.

Complexity measures can be introduced similarly to the extensional case, but here we also want to make sure the history information does not grow too quickly:

Definition 19. A modulus *for a storage-p.a.r. pair* F, H *of a function* ϕ *is a pair of functions* $M, N : (\mathbb{N} \to \mathbb{A}_\mathbb{R}) \times (\mathbb{N} \to \mathbb{N}) \times \mathbb{N} \longrightarrow \mathbb{N}$, *such that for any p.a.r.* A *to* $\alpha \in \text{dom}\,\phi$ *with modulus* m,

$$\forall k \left(F(H(A(m(2^{\text{lth}(n)})), H(A(m(2^{\text{lth}(n)-1})), \cdots H(A(m(0)), 0) \cdots)))_e \leq 2^{-k} \right),$$

where $n = M(A, m, k)$ *and*

$$\forall n \left(H(A(m(2^{\text{lth}(n)})), H(A(m(2^{\text{lth}(n)-1})), \cdots H(A(m(0)), 0) \cdots) \leq N(A, m, n) \right)$$

Theorem 12. *A multi-valued function has a CF-representation in **BFF** if and only if it has a poly-time storage-p.a.r. with a modulus in **BFF**.*

Proof. ($\leftarrow$) This variation of what we did in the previous theorem is in **BFF** if F, H, M and N are in **BFF**:

$$\Phi(a, n) = F(h(k))_v, \text{ where}$$

$$h(m) = \begin{cases} 0, & \text{if } m = 0 \\ H(\langle a(2^{\text{lth}(m)-1}), 2^{-(\text{lth}(m)-1)}\rangle, h(2^{\text{lth}(m)-1})), & \text{otherwise} \end{cases}$$

$$k = M(\lambda p.(a(p), 2^{-\text{lth}(p)}), \lambda p.p, n),$$

because the recursion on notation h is bounded by $N(\lambda p.(a(p), 2^{-\text{lth}(p)}), \lambda p.p, n)$. $\square$

Proof. ($\rightarrow$) All the constructions used in the previous theorem can be done in **BFF**. The M part of the modulus can be constructed exactly as in Theorem 8, and the bound N is

$$N(A, m, n) = \Pi\left(\lfloor A(m(0)) + 2\tfrac{1}{2}\rfloor, 2^{2\text{lth}(\max_{i \leq \text{lth}(n)} A(m(2^{\text{lth}(n)-i})))}\right).$$

(the maximum can be computed in **BFF** as shown in [5]) Because of the properties of the encoding, if $A(\text{lth}(n) - i)_e \leq 2^{-k}$, $\text{lth}(A(\text{lth}(n) - i)) \geq 2^k$, and since in H we're adding two bits for every bit of precision in the approximation, N gives us a bound for the size of the history information. $\square$

Finally, it remains to show that the recursion-p.a.r. approach shares the same properties:

Theorem 13. *A multi-valued function ϕ has a recursion-p.a.r. if and only if it has a storage-p.a.r. Moreover, the conversion is poly-time.*

Proof. Let R be a recursion-p.a.r. of ϕ. Then

$$H(a, h) = R(a, h)$$
$$F(h) = h$$

is a storage-p.a.r. of ϕ. Conversely,

$$R(a, h) = \text{hide}(F(H(a, \text{extr}(h))), H(a, \text{extr}(h)))$$

$$\text{hide}(a, h) = \begin{cases} \langle h, \infty\rangle, & \text{if } a_e \geq \tfrac{1}{2} \\ \text{hh}(a, h), & \text{otherwise} \end{cases}$$

$$\text{hh}(a, h) = \left\langle 2^{-\text{ll}(a_e)}\left(\lfloor 2^{\text{ll}(a_e)} a_v\rfloor + 2^{-\text{lth}(h)}\left((2^{\text{lth}(h)} - 1) + 2^{-(\text{lth}(h)+1)} h\right)\right), 2a_e\right\rangle$$

$$\text{extr}(a) = \begin{cases} a_v, & \text{if } a_e \geq 1 \\ \text{ee}(2^{\text{ll}(a_e)+1} a_v - \lfloor 2^{\text{ll}(a_e)+1} a_v\rfloor), & \text{otherwise} \end{cases}$$

$$\text{ee}(z) = 2^{\text{count}(z)+1}(2^{\text{count}(z)} z - (2^{\text{count}(z)} - 1))$$

$$\text{ll}(e) = \begin{cases} 0, & \text{if } e \geq 1 \\ 1 + \text{ll}(2e), & \text{otherwise} \end{cases}$$

$$\text{count}(z) = \begin{cases} 0, & \text{if } z < \tfrac{1}{2} \\ 1 + \text{count}(2z - 1), & \text{otherwise} \end{cases}$$

does the translation in the other direction: R hides the values of h inside the results it returns by truncating a_v to the precision of a_e and adding to it a string of ones as long as the binary representation of h, followed by a zero and h itself. Because a_e is doubled, a is still a partial approximation to the result of the application of the function, and the value of h can be extracted by first removing the truncated a_v, counting the number of consecutive ones in the remainder and then recovering h as the string of this length that follows the separating zero.   □

Because the conversion between a storage-p.a.r. and a recursion-p.a.r. is poly-time, the complexity results also apply to recursion-p.a.r.'s.

4   Related Work

4.1   Domain Theory

The existing approach most closely related to this work is the domain theoretic approach. It relies on the monotonicity of the functions representing real numbers and has a built-in mechanism to treat intervals as equivalent to real numbers through the idea of "partial real numbers".

Our approach uses a significantly more relaxed requirement in place of the monotonicity as we feel the latter can be hard to ensure especially in the presence of innacurate operations on the approximations (e.g. via fixed or multiple precision floating point). In our approach, we do not provide proper treatment of intervals either but only request (crude) overestimations of the resulting intervals without a mechanism that allows for an improvement of these overestimations. We believe that interval arithmetic using exact reals is preferable to an internal mechanism to treat intervals and real numbers equivalently.

Escardo, Hofmann and Streicher [9] have shown that correct treatment of partial reals in a language for exact real number computations is a cause of serious inefficiencies, showing that even simple operations like addition are inherently parallel. In a later work by Marcial-Romero and Escardo [25] it is shown that this problem can be avoided using a multi-valued test to implement addition which is single-valued on real numbers, but multi-valued on real intervals. Unfortunately, with this the usefulness of the intrinsic mechanism to process intervals becomes somewhat questionable.

To the knowledge of the author, suitable complexity measures for numbers and functions in the domain theoretic approach are yet to be derived. The similarities of the present approach suggest that the notions of moduli for numbers and functions could be used for the domain theoretic framework as well.

4.2   Kleene's Associates

Kleene [15] and Kreisel [19] independantly describe a standard method to translate a type-2 functional into a type-1 object called *associate*. The associates can be seen as a very early use of exceptions, as the associate object either fails (formally returns a 0, which can be viewed as raising an exception) or returns the proper result (formally adding one to it, which can be viewed as honoring the exception) of the application of the type-2 functional to the type-1 argument that is being approximated in a number encoding a finite initial segment of it. The type-2 functional can be recovered from its associate α via the following application operation for an arbitrary type-1 argument β:

$$\alpha(\beta) = \alpha(\mu n.\alpha(\overline{\beta n} \neq 0)) - 1,$$

(in the language of exceptions this is a block that catches the exceptions raised by α and reiterates with higher n until α returns a value) where $\overline{\beta n}$ is an encoding of the initial n values of β.

In the context of total type-2 functionals, Buss and Kapron [4] show that preserving feasibility between type-2 functionals and their associates requires a feasible *modulus of continuity*, similar to our Def. 14 of a modulus of a p.a.r.

The partial approximation representations can be seen as a reformulation of the application of Kleene's approach to the signed-digit representation of real numbers modified to relax the requirements for the representations of real numbers without increasing the complexity of the objects that approximate them. As in [4], the definition of feasible function in our model requires a separate modulus.

4.3   Constructive Analysis with Witnesses

Schwichtenberg also describes a type-1 approach [27] to the computability of real functions using representations of real numbers as rational sequences with separate Cauchy moduli and real functions as rational maps with explicit moduli of uniform continuity (the approach also taken by Pour-El and Richards [24] but kept at Type 1 through the use of the separate Cauchy modulus). In this approach the Cauchy moduli are an integral part of the computation, since the rational approximations alone are not sufficient to extract properties of the real.

The modulus is a worst-case analysis which is useful for complexity reasoning and theoretical extraction of bounds, while error propagation analysis and unbounded search for a sufficient precision is easier and more efficient in practice, because the results usually come out earlier than the worst-case analysis predicts. Our approach makes use of the former for complexity reasoning but the implementations rely on the latter to achieve significantly better performance on average.

5   Conclusion

We have defined a new type-1 approach to computability of real numbers which uses very simple approximation objects. We have shown that the resulting p.a.r. objects do not admit complexity reasoning by themselves, but their complexity can be characterized for both partial functions via type-2 moduli and functions total on closed intervals via uniform type-1 moduli, and these characterizations coincide with the existing approaches for measuring real number complexity. The complexity reasoning is separate from the function or number representation which allows for theoretical reasoning about the complexity of existing implementations that rely on unbounded search to find a sufficient computational precision. As an example, a single implementation of the reciprocal function can be shown to be poly-time if the argument can be witnessed to be different from zero, and to require unbounded minimization if such a witness can not be found, by presenting different moduli which, however, do not need to be implemented at all.

We have also shown that the concept of intensional (or multi-valued) functions can be admitted in this approach with a simple modification of the object representing a function. Unlike a direct application of Kleene's translation of Type 2 to Type 1, our representations of intensional functions do not require vast complexity. Complexity reasoning for these functions is also possible and coincides with existing complexity measures for type-2 representations.

This paper only investigates the approach in the single argument case. In the case of binary functions, certain problems may arise if the two arguments do not produce arbitrarily good approximations at the same time. To enforce this, a modification of the definition of a real number can be used.

While in the unary case the condition $\forall k \exists n ((A(n))_e < 2^{-k})$ suffices, binary and multiple-argument functions would require the condition $\forall k \exists n \forall m \geq n((A(m))_e < 2^{-k})$. The theorems and proofs presented in this paper remain valid with this modification and corresponding changes of the definitions of moduli, with the exception of Theorem 7 which can be shown to be true using the fact that computable functions have computable moduli of continuity.

We are grateful to the suggestions made by the anonymous referees as well as Ulrich Kohlenbach which led to an improved presentation of the paper.

References

1. Berger, U., Oliva, P., *Modified Bar Recursion and Classical Dependent Choice.* Lecture Notes in Logic **20**, 89–107 (2005).
2. Brattka, V., *Recursive characterisation of computable real-valued functions and relations.* Theoret. Comput. Sci. **162**, 47–77 (1996).
3. Briggs, K., *Implementing exact real arithmetic in python, C++ and C,* to appear in Journal of theoretical computer science
   See also http://more.btexact.com/people/briggsk2/xrc.html
4. Buss, S.R., Kapron, B.M., *Resource-bounded continuity and sequentiality for type-two functionals.* ACM Transactions on Computational Logic, vol. **3**, no. 3, 402–417 (7/2002).
5. Cook, S.A., Kapron, B.M., *Characterizations of the basic feasible functionals of finite type.* Feasible Mathematics: A Mathematical Sciences Institute Workshop, Birkhauser, Eds. S. Buss, P. Scott, pp. 71-96 (1990).
6. Cook, Stephen A. *Computability and complexity of higher type functions.* Logic from computer science (Berkeley, CA, 1989), 51–72, Math. Sci. Res. Inst. Publ., **21**, Springer, New York (1992).
7. Edalat, A., *Exact Real Number Computation Using Linear Fractional Transformations.* Final Report on EPSRC grant GR/L43077/01.
   Available at http://www.doc.ic.ac.uk/~ae/exact-computation/exactarithmeticfinal.ps.gz
8. Edalat, Abbas; Sünderhauf, Philipp *A domain-theoretic approach to computability on the real line.* Theoret. Comput. Sci. **210**, no. 1, 73–98 (1999).
9. Escardo, M., Hofmann, M., Streicher, T., *On the non-sequential nature of the interval-domain model of real-number computation.* Math. Struct. in Comp. Science vol. **14** (2004), 803–814.

10. Gödel, K., *Über eine bisher noch nicht benützte Erweiterung des finiten Standpunktes*. Dialectica **12**, 280–287 (1958).

11. Grzegorczyk, A., *On the definitions of computable real continuous functions*. Fundamenta Matematicae **44**, 61–67 (1957).

12. Hinman, P.G., *Recursion-theoretic hierarchies.*, Springer (1978).

13. Howard, W.A., *Hereditarily majorizable functionals of finite type*. In: Troelstra (ed.), Metamathematical investigation of intuitionistic arithmetic and analysis, pp. 454-461. Springer LNM **344** (1973).

14. Kapron, B. M.; Cook, S. A. *A new characterization of type-2 feasibility*. SIAM J. Comput. **25**, no. 1, 117–132 (1996).

15. Kleene, S.C., *Countable Functionals*. In: A. Heyting (ed), Constructivity in Mathematics, North-Holland, Amsterdam, 81–100 (1959).

16. Kleene, S.C., *Recursive Functionals and Quantifiers of Finite Types I*. Trans. Amer. Math. Soc. **91**, 1–52 (1959).

17. Ko, K.-I., *Complexity theory of real functions*. Birkhäuser, Boston-Basel-Berlin (1991).

18. Kohlenbach, U., *Theory of majorizable and continuous functionals and their use for the extraction of bounds from non-constructive proofs: effective moduli of uniqueness for best approximations from ineffective proofs of uniqueness*(german). PhD Dissertation, Frankfurt (1990).

19. Kreisel, G., *Interpretation of analysis by means of functionals of finite type*. In: A. Heyting (ed), Constructivity in Mathematics, North-Holland, Amsterdam, 101–128 (1959).

20. Lambov, B., *A two-layer approach to the computability and complexity of real functions*. Computability and complexity in analysis (Cincinnati, 2003), 279–302, Informatik Berichte, **302**, Fernuniversität Hagen (8/2003).
    See also http://www.brics.dk/~barnie/RealLib/

21. Lambov, B., *RealLib, an Efficient Implementation of Exact Real Arithmetic*, submitted. Available at http://www.brics.dk/~barnie/RealPractical.pdf

22. Lambov, B., *RealLib3 Manual*.
    Available at http://www.brics.dk/~barnie/RealLib/

23. Mosses, P. D., *Action Semantics*. Cambridge Tracts in Theoretical Computer Science **26**, Cambridge University Press (1992).

24. Pour-El, M.B., Richards, J.I., *Computability in Analysis and Physics*. Springer (1989).

25. Marcial-Romero, J.R. and Escardo, M. *Semantics of a sequential language for exact real-number computation*. Proceedings of the 19th Annual IEEE Symposium on Logic in Computer Science, pp. 426-435, (7/2004).

26. Müller, N., *The iRRAM: Exact arithmetic in C++*. Computability and complexity in analysis. (Swansea, 2000). Lecture Notes in Computer Science **2064**. Springer (2001).
    See also http://www.informatik.uni-trier.de/iRRAM/

27. Schwichtenberg, H., *Constructive Analysis with Witnesses* (Marktoberdorf '03)
    Available at http://www.mathematik.uni-muenchen.de/~schwicht/papers/mod03/modart03.ps

28. Skordev, D., *Characterization of the computable real numbers by means of primitive recursive functions*. Computability and complexity in analysis (Swansea, 2000), 296–309, Lecture Notes in Computer Science **2064**, Springer (2001).

29. Weihrauch, K., *Computable Analysis*. Springer (2000).

30. Yap, Chee, *Towards Exact Geometric Computation*. Computational Geometry : Theory and application, **3-23** (9/1997).
    See also http://www.cs.nyu.edu/exact/core/

Computing with Sequences, Weak Topologies and the Axiom of Choice

Vasco Brattka[1] and Matthias Schröder[2]

[1] Department of Mathematics & Applied Mathematics
University of Cape Town, Rondebosch 7701, South Africa
BrattkaV@maths.uct.ac.za
[2] LFCS, School of Informatics
University of Edinburgh, Edinburgh, UK
mschrode@inf.ed.ac.uk

Abstract. We study computability on sequence spaces, as they are used in functional analysis. It is known that non-separable normed spaces cannot be admissibly represented on Turing machines. We prove that under the Axiom of Choice non-separable normed spaces cannot even be admissibly represented with respect to any compatible topology (a compatible topology is one which makes all bounded linear functionals continuous). Surprisingly, it turns out that when one replaces the Axiom of Choice by the Axiom of Dependent Choice and the Baire Property, then some non-separable normed spaces can be represented admissibly on Turing machines with respect to the weak topology (which is just the weakest compatible topology). Thus the ability to adequately handle sequence spaces on Turing machines sensitively relies on the underlying axiomatic setting.

1 Introduction

In this paper we study computability on certain normed spaces X and their dual spaces X'. The framework for this investigation is computable analysis [2, 3, 8], the Turing machine based theory of computability and complexity on real numbers and other topological spaces. We will, in particular, use the representation based approach to computable analysis [8].

Some of our results depend on the underlying axiomatic setting and we will use the following notations to indicate the axioms:

- **ZF** for Zermelo-Fraenkel set theory.
- **AC** for the Axiom of Choice.
- **DC** for the Axiom of Dependent Choice.
- **BP** for the Baire Property Axiom (which states that any subset of the reals can be represented as a symmetric difference of an open and a meager set).

We will not make any direct use of these axioms but we will use certain results which can either be proved in **ZF+AC** or in **ZF+DC+BP**. It is known that in

L. Ong (Ed.): CSL 2005, LNCS 3634, pp. 462–476, 2005.

ZF the Hahn-Banach Theorem can be considered as a weak version of the Axiom of Choice **AC** and the Axiom of Dependent Choice **DC** is equivalent to the Baire Category Theorem (see [5] for a general discussion of the role of these axioms in functional analysis). Some counterexamples in functional analysis do only exist in the setting **ZF+AC** whereas **ZF+DC+BP** allows to exclude the existence of the corresponding objects. Such pathological objects are called "intangibles" by Schechter [5] since their existence cannot be proved constructively. Here, it is important to notice that the consistency of **ZF** implies the consistency of **ZF+AC** (proved by Gödel) as well as the consistency of **ZF+DC+BP** (proved by Shelah, see 14.73 and 14.74 in [5]). If not mentioned otherwise, we will work throughout this paper in the setting **ZF+DC**. Only if the full Axiom of Choice is needed, we will explicitly mention that we are working in **ZF+AC** or in case that we need the Baire Property, we will explicitly mention that we are working in **ZF+DC+BP**.

In the following section we will discuss compatible representations of normed spaces and their dual spaces. Such representations are well-behaved in the sense that they make all bounded linear functionals continuous. Our results show that in **ZF+AC** non-separable normed spaces X and their duals X' do not admit compatible representations.

In Section 3 we will consider the sequence spaces ℓ_p, as they are well-known in functional analysis. The space ℓ_∞ is a typical example of a non-separable normed space and we will prove that this spaces admits a compatible representation in **ZF+DC+BP**, but not in **ZF+AC**.

In Section 4 we discuss a canonical representation which is admissible with respect to the so-called weak* topology. Such representations exist at least for dual spaces of spaces with compatible representations. For separable reflexive spaces we obtain a representation which is admissible with respect to the weak topology.

2   Compatible Representations

In this section we will prove that neither non-separable normed spaces nor their dual spaces admit compatible representations. We start with recalling some notions from computable analysis [8]. The basic idea of the representation based approach to computable analysis is to represent infinite objects like real numbers, functions or sets, by infinite strings over some alphabet Σ (which should at least contain the symbols 0 and 1). Thus, a *representation* of a set X is a surjective mapping $\delta :\subseteq \Sigma^\omega \to X$ and in this situation we will call (X, δ) a *represented space*. Here Σ^ω denotes the set of infinite sequences over Σ and the inclusion symbol is used to indicate that the mapping might be partial. If we have two represented spaces, then we can define the notion of a computable function.

Definition 1 (Computable function). Let (X, δ) and (Y, δ') be represented spaces. A function $f :\subseteq X \to Y$ is called (δ, δ')–*computable*, if there exists some computable function $F :\subseteq \Sigma^\omega \to \Sigma^\omega$ such that $\delta' F(p) = f\delta(p)$ for all $p \in \mathrm{dom}(f\delta)$.

Of course, we have to define computability of functions $F :\subseteq \Sigma^\omega \to \Sigma^\omega$ to make this definition complete, but this can be done via Turing machines: F is computable if there exists some Turing machine, which computes infinitely long and transforms each sequence p, written on the input tape, into the corresponding sequence $F(p)$, written on the one-way output tape. If the represented spaces are fixed or clear from the context, then we will simply call a function f *computable*.

For the comparison of representations it will be useful to have the notion of *reducibility* of representations. If δ, δ' are both representations of a set X, then δ is called *reducible* to δ', $\delta \leq \delta'$ in symbols, if there exists a computable function $F :\subseteq \Sigma^\omega \to \Sigma^\omega$ such that $\delta(p) = \delta'F(p)$ for all $p \in \mathrm{dom}(\delta)$. Obviously, $\delta \leq \delta'$ holds if and only if the identity id : $X \to X$ is (δ, δ')–computable. Moreover, δ and δ' are called *equivalent*, $\delta \equiv \delta'$ in symbols, if $\delta \leq \delta'$ and $\delta' \leq \delta$.

Analogously to the notion of computability we can define the notion of (δ, δ')–*continuity* by substituting a continuous function $F :\subseteq \Sigma^\omega \to \Sigma^\omega$ for the computable function F in the definition above. On Σ^ω we use the *Cantor topology*, which is simply the product topology of the discrete topology on Σ. The corresponding reducibility will be called *continuous reducibility* and we will use the symbols $\leq_t$ and $\equiv_t$ in this case. Again we will simply say that the corresponding function is *continuous*, if the representations are fixed or clear from the context. The category Rep of represented spaces and of continuous (w.r.t. the ambient representations) functions is cartesian-closed. There is a canonical function space representation $[\delta \to \delta']$ of the set $\mathcal{C}(\delta, \delta')$ of (δ, δ')–continuous functions. It has the property that the represented space $(\mathcal{C}(\delta, \delta'), [\delta \to \delta'])$ is the exponential of (X, δ) and (Y, δ') in the category Rep. Moreover, evaluation and currying are even computable (see [7, 8] for details).

If not mentioned otherwise, we will always assume that a represented space is endowed with the final topology induced by its representation. This will lead to no confusion with the ordinary topological notion of continuity, as long as we are dealing with *admissible* representations. A representation δ of a topological space X is called *admissible*, if δ is maximal among all continuous representations δ' of X, i.e. if $\delta' \leq_t \delta$ holds for all continuous representations δ' of X. If δ_X, δ_Y are admissible representations of topological spaces X, Y, then a function $f : X \to Y$ is (δ_X, δ_Y)–continuous if and only if it is sequentially continuous, cf. [6]. Moreover, $[\delta_X \to \delta_Y]$ is an admissible representation of the space of the sequentially continuous functions between X and Y. Hence the category of sequential topological spaces having an admissible representation and of sequentially continuous functions is cartesian closed as well.

Now we introduce compatible representations of normed spaces. Here we assume that by $\mathbb{F}$ the underlying field is denoted, which might either be the field $\mathbb{R}$ of real numbers or the field $\mathbb{C}$ of complex numbers, in each case endowed with the ordinary Euclidean norm and topology.

Definition 2. Let X be a normed space. Then a topology τ on X is called *compatible*, if any bounded linear functional $f : X \to \mathbb{F}$ is continuous with respect to τ. The smallest topology $\tau^{\mathrm{w}} = \sigma(X, X')$ with this property is called the *weak topology* on X.

As usual we will say that a topological space (X, τ) is *separable* if there exists a countable subset $D \subseteq X$ which is dense in X with respect to τ, i.e. such that the closure of D coincides with X.

Lemma 1. *Let* $(X, \| \ \|)$ *be a normed space and let* τ *be a compatible topology. In* **ZF+AC** *the space* $(X, \| \ \|)$ *is separable, if* (X, τ) *is separable.*

Proof. Let $(X, \| \ \|)$ be a normed vector space over $\mathbb{F}$ with a compatible topology τ and let $D = \{d_0, d_1, ...\}$ be a countable dense subset with respect to τ. By $\mathbb{Q}_{\mathbb{F}}$ we denote either $\mathbb{Q}$ or $\mathbb{Q}[i]$ depending on whether $\mathbb{F} = \mathbb{R}$ or $\mathbb{F} = \mathbb{C}$. Let us assume that $(X, \| \ \|)$ is not separable. Then the countable set

$$U := \left\{ \sum_{i=0}^{\infty} q_i \cdot d_i : (q_i)_{i \in \mathbb{N}} \in \mathbb{Q}_{\mathbb{F}}^{\mathbb{N}} \text{ and } q_j = 0 \text{ for almost all } j \right\}$$

is not dense in X with respect to the norm $\| \ \|$. Hence there is some $y \in X$ which is not in the closure $\overline{U}$ of U with respect to the norm $\| \ \|$. Thus $s := \mathrm{dist}(\overline{U}, y) := \inf_{u \in \overline{U}} \|y - u\| > 0$. One easily verifies that $\overline{U}$ and

$$V := \{c \cdot y + u : c \in \mathbb{F}, u \in \overline{U}\}$$

form linear subspaces of X. Since $y \notin \overline{U}$, we can unambiguously define a linear functional $f : V \to \mathbb{F}$ by $f(c \cdot y + u) := c$ for all $c \in \mathbb{F}$ and $u \in \overline{U}$. Since

$$\frac{|f(c \cdot y + u)|}{\|c \cdot y + u\|} = \frac{1}{\|y - \frac{-u}{c}\|} \leq \frac{1}{s}$$

for $c \neq 0$ it follows that f is bounded. By the Hahn-Banach Theorem f can be extended to a bounded linear functional $F : X \to \mathbb{F}$. Since τ is compatible, if follows that F is continuous with respect to τ and since $F(y) = 1$, it follows that $F^{-1}(B(1, 1/2)) \in \tau$ is an open set containing y. By density of D there is some $i \in \mathbb{N}$ with $d_i \in F^{-1}(B(1, 1/2))$ which contradicts $F(d_i) = 0$. $\square$

This lemma can also be obtained as a consequence of the result in functional analysis that a convex subset of a locally convex space X is dense if and only if it is dense with respect to the weak topology on X. However, we present a direct proof in order to pinpoint how **AC** is used, namely in the shape of the Hahn-Banach Theorem. In the setting **ZF+DC+BP**, the space ℓ_∞ turns out to be a counterexample to this lemma (cf. Section 3).

Now we extend the notion of compatibility to representations. Therefore, we assume that $\delta_{\mathbb{F}}$ denotes some standard representation of the field $\mathbb{F}$ which is admissible with respect to the Euclidean topology (e.g. its Cauchy representation, see [8]).

Definition 3. A representation δ of a normed space X is called *compatible*, if every bounded linear functional $f : X \to \mathbb{F}$ is $(\delta, \delta_{\mathbb{F}})$–continuous.

If δ is a compatible representation of X, then the function space representation $[\delta \to \delta_\mathbb{F}]$ can be considered as a representation of the dual space X' (which is the set of bounded linear functionals $f : X \to \mathbb{F}$ endowed with the operator norm $||f|| := \sup_{x \in B(0,1)} |f(x)|$). Here and in the following we will use for every $x \in X$ the canonical linear bounded evaluation functional

$$\iota_x : X' \to \mathbb{F}, f \mapsto f(x),$$

defined on the dual space X' of X. The maps ι_x induce a linear bounded map

$$\iota : X \to X'', x \mapsto \iota_x$$

and with the help of the Hahn-Banach Theorem one can prove that ι is injective and even an isometry (see Corollaries III.1.6 and III.1.7 in [9]). Those spaces for which ι is even bijective and thus isometric isomorphism, are called *reflexive*. For the moment we will use the embedding ι in order to transfer compatible representations of X' to compatible representations of X.

Proposition 1. *In* **ZF+AC** *a normed space X admits a compatible representation, if its dual space X' admits a compatible representation.*

Proof. Let δ' be a compatible representation of the dual space X'. In **ZF+AC** one can prove that ι is injective and since δ' is compatible, we can define a representation δ of X by

$$\delta(p) = x :\iff [\delta' \to \delta_\mathbb{F}](p) = \iota_x.$$

Since the evaluation

$$\mathrm{ev} : X' \times X \to \mathbb{F}, (f, x) \to f(x) = \iota_x(f)$$

is $([\delta', \delta], \delta_\mathbb{F})$–continuous, it follows that each bounded linear functional

$$f : X \to \mathbb{F}, x \mapsto f(x) = \mathrm{ev}(f, x)$$

is $(\delta, \delta_\mathbb{F})$–continuous. This means that δ is compatible. $\square$

Now we are prepared to prove the main result of this section from which we can conclude that the possibilities to introduce a computability theory on non-separable normed spaces which is well-behaved with respect to dual spaces are very limited (given the Axiom of Choice).

Theorem 1. *Let X be a non-separable normed space. In* **ZF+AC** *neither X nor its dual space X' admit a compatible representation.*

Proof. Assume δ is a compatible representation of X and let τ be the final topology of δ, viewed as a total function from the domain of δ endowed with the countably based subspace topology inherited from the Cantor space. Then every linear bounded functional $f : X \to \mathbb{F}$ is $(\delta, \delta_\mathbb{F})$–continuous and hence continuous with respect to τ. Therefore, τ is a compatible topology. But since (X, τ) is a quotient of a countably based space, it admits a countable dense subset. This contradicts Lemma 1. The statement on the dual space follows from Proposition 1. $\square$

3   Sequence Spaces

In this section we will study the sequence spaces

$$\ell_p := \{x \in \mathbb{F}^{\mathbb{N}} : ||x||_p < \infty\}$$

with the norms

$$||x||_p := \sqrt[p]{\sum_{i=0}^{\infty} |x_i|^p}$$

in case of $1 \leq p < \infty$ and

$$||x||_{\infty} := \sup_{i \in \mathbb{N}} |x_i|$$

in case of $p = \infty$ for all $x = (x_i)_{i \in \mathbb{N}}$, as they are known in functional analysis. One important duality property of these spaces is expressed by the following theorem (see, for instance, Theorem II.2.3 in [9]):

Theorem 2 (Landau). *Let* $p, q > 1$ *be real numbers such that* $\frac{1}{p} + \frac{1}{q} = 1$ *or* $p = 1$ *and* $q = \infty$. *Then the map* $\lambda : \ell_q \to \ell_p'$, $a \mapsto \lambda_a$ *with* $\lambda_a : \ell_p \to \mathbb{F}$, $(x_k)_{k \in \mathbb{N}} \mapsto \sum_{k=0}^{\infty} a_k x_k$ *is an isometric isomorphism. The map* λ *is also isometric in case of* $p = \infty$ *and* $q = 1$.

The proof is mainly based on Hölder's Inequality. It is known that the fact that λ is an isomorphism cannot be generalized to the case $p = \infty$ and $q = 1$ straightforwardly, since the result depends on the underlying axiomatic setting in this case. On the one hand, using the Hahn-Banach Theorem one can extend the limit functional lim : $c \to \mathbb{F}$ on the space of convergent sequences c to a functional $L : \ell_{\infty} \to \mathbb{F}$ with the same norm and it is easy to see that this functional cannot be represented as λ_a with some $a \in \ell_1$. Thus we obtain the following classical property of the map λ defined in Landau's Theorem (see, for instance, Theorem II.1.11 in [9]):

Theorem 3. *In* **ZF+AC** *the map* $\lambda : \ell_1 \to \ell_{\infty}'$ *is not surjective.*

Thus, one could say that ℓ_{∞}' is a proper superset of ℓ_1. On the other hand, Pincus proved a result, first stated by Solovay, which shows that the situation changes if we replace the Axiom of Choice by Dependent Choice and the Baire Property (see 29.37 in [5]):

Theorem 4 (Solovay, Pincus). *In* **ZF+DC+BP** *the map* $\lambda : \ell_1 \to \ell_{\infty}'$ *is an isometric isomorphism.*

The Theorem of Landau and its counterpart for the case $p = \infty$ and $q = 1$ have certain consequences concerning the existence of compatible representations of the sequence spaces. As a preparation we prove a characterization of weak convergence for these spaces. We recall that in functional analysis weak convergence means convergence with respect to the weak topology, i.e. a sequence $(x_n)_{n \in \mathbb{N}}$ in a normed space X is said to *converge weakly* to x, if $(f(x_n))_{n \in \mathbb{N}}$ converges to

$f(x)$ for any linear bounded functional $f : X \to \mathbb{F}$. The first part of the following lemma is a known fact. We include the proof in order to indicate how the second part follows from the previous theorem.

Lemma 2. *Let $1 < p < \infty$. A sequence $((x_{ij})_{j \in \mathbb{N}})_{i \in \mathbb{N}}$ in ℓ_p converges weakly to $(x_j)_{j \in \mathbb{N}}$, if and only if the sequence converges with respect to the product topology on $\mathbb{F}^{\mathbb{N}}$ to $(x_j)_{j \in \mathbb{N}}$ and if it is bounded in $\| \ \|_p$. For $p = \infty$, the equivalence holds in **ZF+DC+BP**, whereas in **ZF+AC** merely the only-if-part is true.*

Proof. Let $((x_{ij})_{j \in \mathbb{N}})_{i \in \mathbb{N}}$ be a sequence in ℓ_p which converges weakly to $(x_j)_{j \in \mathbb{N}} \in \ell_p$, i.e. $(f((x_{ij})_{j \in \mathbb{N}})_{i \in \mathbb{N}}$ converges for any linear bounded functional $f : \ell_p \to \mathbb{F}$ to $f((x_j)_{j \in \mathbb{N}})$. Since the canonical projections

$$\mathrm{pr}_j : \ell_p \to \mathbb{F}, (y_j)_{j \in \mathbb{N}} \to y_j$$

are linear bounded functionals, it follows that $(\mathrm{pr}_j((x_{ij})_{j \in \mathbb{N}}))_{i \in \mathbb{N}} = (x_{ij})_{i \in \mathbb{N}}$ converges for any fixed $j \in \mathbb{N}$ to $\mathrm{pr}_j((x_j)_{j \in \mathbb{N}}) = x_j$ and hence $((x_{ij})_{j \in \mathbb{N}})_{i \in \mathbb{N}}$ converges with respect to the product topology on $\mathbb{F}^{\mathbb{N}}$ to $(x_j)_{j \in \mathbb{N}}$. Moreover, it is known that any weakly convergent sequence in ℓ_p is bounded. (This is a consequence of the Uniform Boundedness Theorem, see for instance Korollar IV.2.3 in [9], and can be proven in **ZF+DC**.)

Now let us assume that $((x_{ij})_{j \in \mathbb{N}})_{i \in \mathbb{N}}$ is a sequence in ℓ_p which converges to $(x_j)_{j \in \mathbb{N}} \in \ell_p$ with respect to the product topology on $\mathbb{F}^{\mathbb{N}}$ and which is bounded in $\| \ \|_p$. We have to prove that the sequence $(f((x_{ij})_{j \in \mathbb{N}}))_{i \in \mathbb{N}}$ converges for any functional $f : \ell_p \to \mathbb{F}$ to $f((x_j)_{j \in \mathbb{N}})$. Let q be such that $1/p + 1/q = 1$ or $q = 1$ in case of $p = \infty$. If the map $\lambda : \ell_q \to \ell_p'$ from Landau's Theorem 2 is an isometric isomorphism, then it suffices to prove that $(\lambda_a((x_{ij})_{j \in \mathbb{N}}))_{i \in \mathbb{N}} = (\sum_{j=0}^{\infty} a_j x_{ij})_{i \in \mathbb{N}}$ converges for any $a = (a_j)_{j \in \mathbb{N}} \in \ell_q$ to $\lambda_a((x_j)_{j \in \mathbb{N}}) = \sum_{j=0}^{\infty} a_j x_j$. Therefore, let $a = (a_j)_{j \in \mathbb{N}} \in \ell_q$, i.e. $\|a\|_q = (\sum_{j=0}^{\infty} |a_j|^q)^{1/q} < \infty$. Since $((x_{ij})_{j \in \mathbb{N}})_{i \in \mathbb{N}}$ is bounded in ℓ_p, it follows that $S := \sup_{i \in \mathbb{N}} \|(x_{ij})_{j \in \mathbb{N}} - (x_j)_{j \in \mathbb{N}}\|_p + 1$ exists. Let $\varepsilon > 0$. There is some $J \in \mathbb{N}$ such that $(\sum_{j=J+1}^{\infty} |a_j|^q)^{1/q} < \varepsilon/(2S)$. Let $M := \max\{|a_0|, |a_1|, ..., |a_J|\}$. Since $((x_{ij})_{j \in \mathbb{N}})_{i \in \mathbb{N}}$ converges to $(x_j)_{j \in \mathbb{N}}$ with respect to the product topology on $\mathbb{F}^{\mathbb{N}}$ there is some $I \in \mathbb{N}$ such that $|x_{ij} - x_j| < \varepsilon/(2M(J+1))$ for all $i \geq I$ and $j = 0, ..., J$. Now we obtain by Hölder's Inequality for all $i \geq I$

$$\left| \sum_{j=0}^{\infty} a_j x_{ij} - \sum_{j=0}^{\infty} a_j x_j \right|$$

$$\leq \sum_{j=0}^{J} |a_j \cdot (x_{ij} - x_j)| + \sum_{j=J+1}^{\infty} |a_j \cdot (x_{ij} - x_j)|$$

$$\leq \sum_{j=0}^{J} |a_j| \cdot |x_{ij} - x_j| + \|(0, ..., 0, a_{J+1}, a_{J+2}, ...)\|_q \cdot \|(x_{ij} - x_j)_{j \in \mathbb{N}}\|_p$$

$$\leq (J+1)M \cdot \frac{\varepsilon}{2M(J+1)} + \frac{\varepsilon}{2S} \cdot S = \varepsilon.$$

This proves the desired convergence.

It remains to recall that by Landau's Theorem 2 $\lambda : \ell_q \to \ell'_p$ is an isometric isomorphism for $1 < p < \infty$ and by the Theorem of Solovay and Pincus 4 this also holds in $\mathbf{ZF+DC+BP}$ for the case $p = \infty$.   □

The reader should notice that the previous result does not capture the case $p = 1$. This is not an accidental omission, but the result cannot be extended to this case. This is due to the following well-known result (see 28.20 in [5]):

Lemma 3 (Schur). *A sequence in ℓ_1 converges weakly to a certain limit if and only if it converges to the same limit with respect to the norm $|| \; ||_1$.*

We recall that a topology τ is called *sequential*, if any sequentially open set is open. A set U is called *sequentially open*, if any sequence with limit in U is eventually in U. The *sequentialization* $\mathrm{seq}(\tau)$ is the set of all sequentially open sets or, in other words, the smallest sequential topology which contains τ. Two sequential topologies coincide, if their convergence relations on sequences are identical. Any topology induced by a norm is sequential.

By the Lemma of Schur, the sequentialization of the weak topology of ℓ_1 is just the norm topology induced by the norm $|| \; ||_1$. Since it is known that for infinite dimensional normed spaces $(X, || \; ||)$ the norm $|| \; || : X \to \mathbb{R}$ itself is not continuous with respect to the weak topology (see 28.18 in [5]) and, in particular, the norm topology is different from the weak topology, it follows that the weak topology on ℓ_1 is not a sequential topology.

Now we will discuss compatible representations of sequence spaces. In particular, we will exploit the characterization of weak convergence to show that under certain assumptions such representations exist. In particular, we are interested in the following representations (which have been introduced in the more general context of general computable normed spaces [1]; here $\delta_{\mathbb{N}}$ denotes some canonical representation of the natural numbers $\mathbb{N}$):

Definition 4. Let $1 \leq p \leq \infty$. We define three representations $\delta_p, \delta_p^=, \delta_p^\geq$ of ℓ_p as follows:

- $\delta_p(r) = x : \iff [\delta_{\mathbb{N}} \to \delta_{\mathbb{F}}](r) = x,$
- $\delta_p^=\langle r, s \rangle = x : \iff \delta_p(r) = x$ and $\delta_{\mathbb{R}}(s) = ||x||_p,$
- $\delta_p^\geq\langle r, s \rangle = x : \iff \delta_p(r) = x$ and $\delta_{\mathbb{R}}(s) \geq ||x||_p,$

for all $r, s \in \Sigma^\omega$.

The representation δ_p is nothing but the standard representation of $\mathbb{F}^{\mathbb{N}}$ restricted to ℓ_p and it is admissible with respect to the subtopology τ_p on ℓ_p of the product topology on $\mathbb{F}^{\mathbb{N}}$. In [1] it has been shown that $\delta_p^=$ is admissible with respect to the weakest topology $\tau_p^=$ on ℓ_p which contains the topology τ_p and which makes the norm $|| \; ||_p$ continuous. Finally, $\delta_p^\geq$ is admissible with respect to the inductive limit topology $\tau_p^\geq = \varinjlim \sigma_k$ of the subtopologies σ_k of τ_p on $X_k := \{x \in \ell_p : ||x||_p \leq k\}$. These results mainly rely on closure properties provided in [6]. Moreover, the product topology on $\mathbb{F}^{\mathbb{N}}$ is a sequential topology with a countable basis and thus it follows that τ_p is a sequential topology with a countable basis. The topology $\tau_p^=$ is obtained by an initial construction from sequential topologies with countable bases and $\tau_p^{\geq}$ is obtained by a final construction from sequential topologies. Using these properties, one can conclude that all three topologies τ_p, $\tau_p^=$ and $\tau_p^{\geq}$ are sequential topologies as well (see [6, 7, 10]).

Now the question occurs how these topologies are related to topologies considered in functional analysis. Firstly, we will characterize the topology $\tau_p^=$ for $1 \leq p < \infty$ which turns out to be just the norm topology $\tau_{\|\ \|_p}$ induced by the norm $\|\ \|_p$. This does not hold true in case of $p = \infty$, where the sequence $(e_1 + e_{2+i})_{i \in \mathbb{N}}$ built from the unit vectors e_i (which are zero except for the i-th position where they are one) is an obvious counterexample. The following lemma expresses a fact which is folklore in functional analysis. For completeness we include the proof.

Lemma 4. *Let* $1 \leq p < \infty$. *A sequence* $((x_{ij})_{j \in \mathbb{N}})_{i \in \mathbb{N}}$ *in* ℓ_p *converges to* $(x_j)_{j \in \mathbb{N}}$ *with respect to the norm* $\|\ \|_p$, *if and only if the sequence converges with respect to the product topology on* $\mathbb{F}^{\mathbb{N}}$ *to* $(x_j)_{j \in \mathbb{N}}$ *and if* $(\|(x_{ij})_{j \in \mathbb{N}}\|_p)_{i \in \mathbb{N}}$ *converges to* $\|(x_j)_{j \in \mathbb{N}}\|_p$.

Proof. If $((x_{ij})_{j \in \mathbb{N}})_{i \in \mathbb{N}}$ converges to $(x_j)_{j \in \mathbb{N}}$ with respect to the norm $\|\ \|_p$, then it converges weakly to the same limit. Literally the same proof as for the first part of Lemma 2 shows that it also converges to the same limit with respect to the product topology. Moreover, the norm $\|\ \|_p : \ell_p \to \mathbb{R}$ is continuous with respect to the norm topology, hence it is sequentially continuous which proves that the norm of the sequence converges to the norm of the limit.

For the other direction let us assume that $((x_{ij})_{j \in \mathbb{N}})_{i \in \mathbb{N}}$ converges to $(x_j)_{j \in \mathbb{N}}$ with respect to the product topology and that $(\|(x_{ij})_{j \in \mathbb{N}}\|_p)_{i \in \mathbb{N}}$ converges to $\|(x_j)_{j \in \mathbb{N}}\|_p$. Let $\varepsilon > 0$. There is some $J \in \mathbb{N}$ such that $\sum_{j=J+1}^{\infty} |x_j|^p < \varepsilon/8$ and there is some $I \in \mathbb{N}$ such that

$$\max\{|x_{ij} - x_j|^p, |x_j|^p - |x_{ij}|^p\} < \frac{\varepsilon}{4(J+1)}$$

for all $j = 0, ..., J$ and $i > I$ and such that

$$\|(x_{ij})_{j \in \mathbb{N}}\|_p^p - \|(x_j)_{j \in \mathbb{N}}\|_p^p < \frac{\varepsilon}{4}$$

for all $i > I$. We obtain

$$\|(x_{ij})_{j \in \mathbb{N}} - (x_j)_{j \in \mathbb{N}}\|_p^p$$
$$= \sum_{j=0}^{J} |x_{ij} - x_j|^p + \sum_{j=J+1}^{\infty} |x_{ij} - x_j|^p$$
$$\leq (J+1)\frac{\varepsilon}{4(J+1)} + \sum_{j=J+1}^{\infty} (|x_{ij}|^p + |x_j|^p)$$

$$= \frac{\varepsilon}{4} + \sum_{j=0}^{\infty} |x_{ij}|^p - \sum_{j=0}^{J} |x_{ij}|^p + 2 \sum_{j=J+1}^{\infty} |x_j|^p - \sum_{j=0}^{\infty} |x_j|^p + \sum_{j=0}^{J} |x_j|^p$$

$$< \frac{\varepsilon}{4} + \|(x_{ij})_{j \in \mathbb{N}}\|_p^p - \|(x_j)_{j \in \mathbb{N}}\|_p^p + 2\frac{\varepsilon}{8} + \sum_{j=0}^{J} (|x_j|^p - |x_{ij}|^p)$$

$$\leq \frac{\varepsilon}{4} + \frac{\varepsilon}{4} + \frac{\varepsilon}{4} + (J+1)\frac{\varepsilon}{4(J+1)} = \varepsilon.$$

This proves the desired convergence.   □

Using the same estimations one can prove the following slightly more general result, where δ_{ℓ_p} denotes the so-called Cauchy representation of the space $(\ell_p, \| \ \|_p)$ (which is a standard representation that is admissible with respect to the norm topology, see [8]).

Proposition 2. *Let $1 \leq p < \infty$. Then $\delta_{\ell_p} \equiv \delta_p^=$.*

In [1] it has also been proved that in general we obtain $\tau_{\| \ \|_p} \supseteq \tau_p^= \supseteq \tau_p^\geq \supseteq \tau_p$ for the corresponding topologies, where $\tau_{\| \ \|_p}$ denotes the norm topology again. This raises the question whether the weak topology $\tau_p^w = \sigma(\ell_p, \ell_p')$ can be included in this inclusion chain. Using Lemma 2 and Lemma 4 we can directly conclude the following corollary.

Corollary 1. *For the spaces ℓ_p with $1 < p < \infty$ we obtain $\tau_{\| \ \|_p} = \tau_p^= \supsetneq \tau_p^\geq = \text{seq}(\tau_p^w)$.*

For the space ℓ_1 the situation is different and we can conclude from the Lemma of Schur 3 and Lemma 4 the following result.

Corollary 2. *For the space ℓ_1 we obtain $\tau_{\| \ \|_1} = \tau_1^= = \text{seq}(\tau_1^w) \supsetneq \tau_1^\geq$.*

For the non-separable space ℓ_∞ the situation is yet different again and it depends on the underlying axiomatic setting.

Theorem 5. *For the space ℓ_∞ we obtain $\tau_{\| \ \|_\infty} \supsetneq \tau_\infty^= \supsetneq \tau_\infty^\geq$. Additionally,*

- *in ZF+AC, $\tau_{\| \ \|_\infty} \supsetneq \text{seq}(\tau_\infty^w) \supsetneq \tau_\infty^\geq$ and $\text{seq}(\tau_\infty^w)$ is incomparable with $\tau_\infty^=$, whereas*
- *in ZF+DC+BP, $\text{seq}(\tau_\infty^w) = \tau_\infty^\geq$.*

Proof. The first two strict inclusions have been proved in [1] (and they hold for non-separable general computable normed spaces in general). The fact that $\text{seq}(\tau_\infty^w) \supseteq \tau_\infty^\geq$ holds follows from the only-if-part of Lemma 2 (which does not require the Axiom of Choice). The inclusion has to be strict and $\tau_\infty^= \not\supseteq \text{seq}(\tau_\infty^w)$, both in ZF+AC, since the contrary would contradict Theorem 1 (this is because $\delta_\infty^=$ is admissible with respect to $\tau_\infty^=$ whereas by Theorem 1 no representation is admissible with respect to $\text{seq}(\tau_\infty^w)$).

Next we prove $\text{seq}(\tau_\infty^w) \not\supseteq \tau_\infty^=$. We consider the sequence $(e_i)_{i \in \mathbb{N}}$ of unit vectors (which are zero except for the i-th position where they are one; for simplicity we assume $e_0 = 0$). Let $f : \ell_\infty \to \mathbb{F}$ be some arbitrary linear bounded functional with $s := ||f||$. Let us assume that $(f(e_i))_{i \in \mathbb{N}}$ does not converge to 0. Then there is some $k \in \mathbb{N}$ and some strictly increasing $\varphi : \mathbb{N} \to \mathbb{N}$ with $|f(e_{\varphi(i)})| > 1/k$ for all $i \geq 1$ (in particular, $\varphi(i) \geq 1$ for all $i \geq 1$). Now we consider

$$z := \sum_{i=1}^{ks} e_{\varphi(i)} \frac{|f(e_{\varphi(i)})|}{f(e_{\varphi(i)})}$$

and we obtain $||z||_\infty = 1$ and $|f(z)| > s$ which is a contradiction! Thus, $(f(e_i))_{i \in \mathbb{N}}$ does converge to $f(0) = 0$ and hence $(e_i)_{i \in \mathbb{N}}$ converges to 0 with respect to τ_∞^w and hence also with respect to $\mathrm{seq}(\tau_\infty^w)$. On the other hand, it is obvious that $(e_i)_{i \in \mathbb{N}}$ does not converge to 0 with respect to $\tau_\infty^=$.

Since the weak topology is always contained in the norm topology and the norm topology is sequential, we obtain $\tau_{||\ ||_\infty} \supseteq \mathrm{seq}(\tau_\infty^w)$. The inclusion is strict, since otherwise $\mathrm{seq}(\tau_\infty^w) = \tau_{||\ ||_\infty} \supseteq \tau_\infty^=$ would follow.

Finally, in **ZF+DC+BP** $\mathrm{seq}(\tau_\infty^w) = \tau_\infty^\geq$ by Lemma 2.   □

We could also prove the previous theorem without reference to Theorem 1 by a direct usage of the following example.

Example 1. In **ZF+AC** one can apply the Hahn-Banach Theorem in order to prove that the limit functional $\lim : c \to \mathbb{F}$ has a linear bounded extension $L : \ell_\infty \to \mathbb{F}$. Any such extension L is not continuous with respect to $\tau_\infty^=$ and hence not with respect to $\tau_\infty^\geq$: the sequence $(x_n)_{n \in \mathbb{N}}$ with elements $x_n = (1, 0, ..., 0, 1, 1, ...) \in \ell_\infty$ (with n zeros) converges to $x = (1, 0, 0, ...) \in \ell_\infty$ with respect to $\tau_\infty^=$, but $L(x_n) = 1 \neq 0 = L(x)$ for all n. In particular, $\tau_\infty^= \not\supseteq \mathrm{seq}(\tau_\infty^w)$.

We can also combine our results on compatible representations of ℓ_∞ as follows.

Corollary 3. *In* **ZF+AC** *neither ℓ_∞ nor its dual space admit compatible representations, whereas in* **ZF+DC+BP** *the space ℓ_∞ as well as its dual space ℓ_1 admit compatible representations.*

In fact, in **ZF+DC+BP** the representation $\delta_\infty^\geq$ is a compatible representation of ℓ_∞ which is admissible with respect to the weak topology on ℓ_∞. In **ZF+AC**, $\delta_\infty^\geq$ has at least the property that a linear function $f : \ell_\infty \to \mathbb{F}$ is $(\delta_\infty^\geq, \delta_\mathbb{F})$-continuous if and only if there is some $a \in \ell_1$ such that $f = \lambda_a$. Moreover, $\delta_\infty^\geq$ is admissible with respect to the weakest topology on ℓ_∞ for which every function λ_a, $a \in \ell_1$, is continuous.

4   The Weak-Star Topology

In this section we investigate the so-called weak* topology in our context. This will also allow us to generalize some of the positive results of the previous section to a more general setting.

Definition 5. Let X be a normed space. The *weak* topology* on the dual space X' is the smallest topology $\tau^{w^*} = \sigma(X', X)$ which makes for all $x \in X$ the functionals $\iota_x : X' \to \mathbb{F}$, $f \mapsto f(x)$ continuous.

It is obvious that the weak* topology $\sigma(X', X)$ on X' is weaker or equal to the weak topology $\sigma(X', X'')$ on X', i.e. $\sigma(X', X'') \supseteq \sigma(X', X)$. By a Theorem of Banach, Smulian, James and others (see 28.41 in [5]) the topologies coincide exactly for reflexive spaces, i.e. such spaces for which the canonical embedding $\iota : X \to X''$, $x \mapsto \iota_x$ is bijective (however, this result requires the Axiom of Choice).

The next lemma shows that a sequence of functionals converges with respect to the weak* topology if and only if it converges with respect to the compact-open topology. Readers familar with topological vector spaces might derive this fact from Theorem 4.6 in Paragraph 5 of Chapter 3 in [4]. For completeness we include a direct proof.

Lemma 5. *Let X be a Banach space and let $(f_n)_{n \in \mathbb{N}}$ be a sequence of linear bounded functionals $f_n : X \to \mathbb{F}$ and let $f : X \to \mathbb{F}$ be another such functional. Then $(f_n)_{n \in \mathbb{N}}$ converges to f with respect to the weak* topology on X' if and only if it converges to f with respect to the compact-open topology on $\mathcal{C}(X, \mathbb{F})$.*

Proof. First of all, by definition of the weak* topology the sequence $(f_n)_{n \in \mathbb{N}}$ converges with respect to the weak* topology to f if and only if it converges pointwise to f.

Now if the sequence $(f_n)_{n \in \mathbb{N}}$ converges pointwise to f, then $\sup_{n \in \mathbb{N}} |f_n(x)|$ exists for each $x \in X$ and by the Uniform Boundedness Theorem $M := \sup_{n \in \mathbb{N}} ||f_n||$ also exists. Let us assume that $(f_n)_{n \in \mathbb{N}}$ does not converge to f with respect to the compact-open topology. Then there exists a non-empty compact subset $K \subseteq X$ and some $\varepsilon > 0$ such that for any $n \in \mathbb{N}$ there is some $k_n > n$ with $\sup_{x \in K} |f_{k_n}(x) - f(x)| > \varepsilon$. We can assume that $(k_n)_{n \in \mathbb{N}}$ is a strictly increasing sequence. Then for any $n \in \mathbb{N}$ there is some $x_n \in K$ such that $|f_{k_n}(x_n) - f(x_n)| > \varepsilon$ and since K is compact the sequence $(x_n)_{n \in \mathbb{N}}$ has a convergent subsequence $(x_{n_i})_{i \in \mathbb{N}}$ which converges to some $x \in K$. Since $(f_{k_{n_i}}(x))_{i \in \mathbb{N}}$ converges to $f(x)$, there is some $i \in \mathbb{N}$ such that $|f_{k_{n_i}}(x) - f(x)| < \varepsilon/2$ and $||x_{n_i} - x|| < \varepsilon/((M + ||f|| + 1)2)$. Now we obtain

$$
\begin{aligned}
\varepsilon &< |f_{k_{n_i}}(x_{n_i}) - f(x_{n_i})| \\
&\leq |f_{k_{n_i}}(x_{n_i}) - f_{k_{n_i}}(x)| + |f_{k_{n_i}}(x) - f(x)| + |f(x) - f(x_{n_i})| \\
&\leq (||f_{k_{n_i}}|| + ||f||) \cdot ||x_{n_i} - x|| + |f_{k_{n_i}}(x) - f(x)| \\
&< (M + ||f||) \cdot \frac{\varepsilon}{(M + ||f|| + 1)2} + \frac{\varepsilon}{2} \\
&< \varepsilon
\end{aligned}
$$

which is a contradiction. Thus, the assumption was wrong and $(f_n)_{n \in \mathbb{N}}$ converges to f with respect to the compact-open topology.

Finally, if $(f_n)_{n \in \mathbb{N}}$ converges to f with respect to the compact-open topology, then it also converges to f pointwise. □

For the compact-open topology on $X' \subseteq \mathcal{C}(X, \mathbb{F})$, we can conclude that $\mathrm{seq}(\tau^{co}) = \mathrm{seq}(\tau^{w^*})$. We do not know whether the two topologies agree themselves or whether they are sequential.

Now let δ be a compatible representation of X and let τ_δ be the final topology of δ. Analogously to the previous proof, one can show that a sequence of linear bounded functionals $(f_n)_n$ converges to a linear bounded functional f with respect to the weak* topology if and only if it converges with respect to sequentially-compact-open topology[1] on $\mathcal{C}((X, \tau_\delta), \mathbb{F})$. From [7], we know that the *dual representation* δ' of X', defined by

$$\delta'(p) = f : \iff [\delta \to \delta_\mathbb{F}](p) = f,$$

is admissible with respect to the sequentially-compact-open topology on X'. Thus we obtain the following corollary.

Corollary 4. *Let X be a Banach space with some compatible representation δ. Then the dual representation δ' of X' is admissible with respect to the weak* topology $\tau^{w^*} = \sigma(X', X)$ on X'.*

Let us denote by $\tau_p^{w^*} = \sigma(\ell_p, \ell_q)$ the weak* topology on ℓ_p induced by the corresponding conjugate space ℓ_q with $1/p + 1/q = 1$. Then we can formulate our results on the ℓ_p spaces as follows.

Theorem 6. *For the spaces ℓ_p with $1 < p \leq \infty$ we obtain $\tau_p^{\geqslant} = \mathrm{seq}(\tau_p^{w^*})$.*

Proof. In case of $1 < p \leq \infty$ it follows from the effective Theorem of Landau (see Theorem 7.2 in [1]) that $\delta_p^{\geqslant} \equiv_t \delta_q' = [\delta_{\ell_q} \to \delta_\mathbb{F}]$ for the conjugate q, but the latter representation is admissible with respect to the weak* topology $\tau_p^{w^*}$ by the previous corollary. $\square$

Note that in case of $1 < p < \infty$ we could also conclude the claim from Corollary 1 and the fact that for these p the spaces ℓ_p are reflexive. For reflexive spaces the weak and the weak* topologies coincide.

In general the previous corollary opens a possibility to define a canonical representation of a separable reflexive normed space which is admissible with respect to the weak topology.

Corollary 5. *Let X be a reflexive normed space with some compatible representation δ. Define a representation δ^w of X by*

$$\delta^w(p) = x : \iff \delta''(p) = \iota_x$$

Then δ^w is admissible with respect to the weak topology on X.

Similarly as in Proposition 1 one can prove that for a compatible representation δ of X, the representation δ' is a compatible representation of X' (now using reflexivity instead of **ZF+AC**). Since X' always is complete, we can now apply Corollary 4 to δ' in order to derive the previous corollary.

---

[1] A subbase is given by the sets $\{f \in \mathcal{C}((X, \tau_\delta), \mathbb{F}) \mid f[K] \subseteq O\}$, where $K \subseteq X$ is sequentially compact and $O \subseteq \mathbb{F}$ is open.

p	ZF+AC	ZF+DC+BP
$p=\infty$	$\tau_{\lVert\ \rVert_\infty} \supsetneqq \begin{smallmatrix} seq(\tau^w_\infty) \\[-2pt] \tau^=_\infty \end{smallmatrix} \supsetneqq seq(\tau^{w^*}_\infty) = \tau^\geq_\infty$	$\tau_{\lVert\ \rVert_\infty} \supsetneqq \tau^=_\infty \supsetneqq seq(\tau^w_\infty) = seq(\tau^{w^*}_\infty) = \tau^\geq_\infty$
$1<p<\infty$	$\tau_{\lVert\ \rVert_p} = \tau^=_p \supsetneqq seq(\tau^w_p) = seq(\tau^{w^*}_p) = \tau^\geq_p$	
$p=1$	$\tau_{\lVert\ \rVert_1} = \tau^=_1 = seq(\tau^w_1) \supsetneqq \tau^\geq_1$	

Fig. 1. Weak topologies on the ℓ_p spaces

5 Conclusions

In this paper we have proved that the possibilities to handle non-separable spaces on Turing machines sensitively rely on the underlying axiomatic setting. In **ZF+AC** non-separable normed spaces do not admit compatible representations whereas in **ZF+DC+BP** such representations do exist for certain spaces.

In particular we have studied the sequence spaces ℓ_p which can be handled in a uniform way and which include ℓ_∞ as a typical example of a non-separable normed space. The results for these spaces turned out to be surprisingly diverse and the table in Figure 1 summarizes the inclusions which we have obtained comparing different weak topologies for these spaces. The last two rows contain the results for $1 \leq p < \infty$ that do not depend on the axiomatic setting.

Our results suggest that the setting **ZF+DC+BP** is more natural from the point of view of computable analysis. However, functional analysis is classically developed in **ZF+AC** and a **ZF+DC+BP** version would be substantially different, even classically. Nevertheless, even in **ZF+DC+BP** the separable version of the Hahn-Banach Theorem is available (see [5]). Hence for a computable version of functional analysis the setting **ZF+DC+BP** might be sufficient.

References

1. Brattka, V.: Computability on non-separable Banach spaces and Landau's theorem. In Crosilla, L., Schuster, P., eds.: From Sets and Types to Topology and Analysis: Towards Practicable Foundations for Constructive Mathematics. Oxford University Press (to appear)
2. Ko, K.I.: Complexity Theory of Real Functions. Progress in Theoretical Computer Science. Birkhäuser, Boston (1991)
3. Pour-El, M.B., Richards, J.I.: Computability in Analysis and Physics. Perspectives in Mathematical Logic. Springer, Berlin (1989)
4. Schäfer, H.H.: Topological Vector Spaces. Macmillan, New York (1966)
5. Schechter, E.: Handbook of Analysis and Its Foundations. Academic Press, San Diego (1997)

6. Schröder, M.: Extended admissibility. Theoretical Computer Science **284** (2002) 519–538

7. Schröder, M.: Admissible Representations for Continuous Computations. PhD thesis, Fachbereich Informatik, FernUniversität Hagen (2002)

8. Weihrauch, K.: Computable Analysis. Springer, Berlin (2000)

9. Werner, D.: Funktionalanalysis. 4th edn. Springer, Berlin (2002)

10. Willard, S.: General Topology. Addison-Wesley, Reading (1970)

Light Functional Interpretation
An Optimization of Gödel's Technique
Towards the Extraction of (More) Efficient Programs from (Classical) Proofs

Mircea-Dan Hernest*

Laboratoire d'Informatique (LIX)
École Polytechnique, F-91128 Palaiseau, France
danher@lix.polytechnique.fr

Abstract. We give a Natural Deduction formulation of an adaptation of Gödel's functional (*Dialectica*) interpretation to the extraction of (more) efficient programs from (classical) proofs. We adapt Jørgensen's formulation of pure Dialectica translation by eliminating his "Contraction Lemma" and allowing free variables in the extracted terms (which is more suitable in a Natural Deduction setting). We also adapt Berger's *uniform* existential and universal quantifiers to the Dialectica-extraction context. The use of such quantifiers *without computational meaning* permits the identification and isolation of contraction formulas which would otherwise be redundantly included in the pure-Dialectica extracted terms. In the end we sketch the possible combination of our refinement of Gödel's Dialectica interpretation with its adaptation to the extraction of bounds due to Kohlenbach into a *light monotone* functional interpretation.

Keywords: Program extraction from (classical) proofs, Complexity of extracted programs, Berger's uniform quantifiers, Gödel's Functional interpretation, Proof-Carrying Code, Proof Mining.

1 Introduction

Important practical results have been obtained in recent years in the field of extractive Proof Theory (also dubbed *proof mining* [22]). The implemented algorithms coming from metamathematical research have yielded interesting and in many cases quite unexpected programs [7, 15, 31, 32]. Various approaches to program extraction from *classical*[1] proofs have been developed over years of research [1, 3, 6, 8, 9, 21, 24, 25, 27–30, 32]. The use of proof interpretations instead of the direct application of cut-elimination has opened the path to the obtention of better practical results by means of such modular techniques (see

---

* *Project LogiCal* – Pôle Commun de Recherche en Informatique du Plateau de Saclay, CNRS, École Polytechnique, INRIA et Université Paris-Sud – **FRANCE** and *Graduiertenkolleg Logik in der Informatik* (**GKLI**) – München, **GERMANY**. Partly financed by **Deutsche Forschungsgemeinschaft**.
[1] *Constructive* proofs have a more or less explicit computational content, see, e.g., [4].

L. Ong (Ed.): CSL 2005, LNCS 3634, pp. 477–492, 2005.
© Springer-Verlag Berlin Heidelberg 2005

[15] for a discussion on this). Gödel's functional (*Dialectica*) interpretation [2, 13] directly applies to (far) more complex proofs than (its weaker form) Kreisel's Modified Realizability [23]. Refinements [1, 6, 9] of the latter's combination with Friedman's A-translation [12] only partly repair this disparity (see [15] or [20] for discussions on this issue). However, the (rough) exact realizers yielded by the Dialectica interpretation are generally (far) more complex than those produced by, e.g., the technique of Berger, Buchholz and Schwichtenberg [6]. The main reason for such a situation seems to be the inclusion of *contraction formulas* in the rough Dialectica realizing terms. Even though in some cases the primary realizers extracted via the two techniques normalize to basically the same programs (see [15] for such an example), the normalization of programs extracted by proof interpretations is generally far more expensive than their synthesis (see [16] for a detailed complexity exposition). The simpler the rough extracted term is, the lower the overall cost of producing the final normalized realizer becomes. In order to handle this contraction problem of the Dialectica interpretation, Kohlenbach [21] devised the *monotone* functional interpretation, an adaptation of Gödel's technique to the extraction of uniform bounds for the exact realizers (see also [11] for the more recent *bounded* functional interpretation of Ferreira and Oliva). While so prolific (see [20] or [22]) in the context of mathematical Analysis, the monotone functional interpretation, used alone, is generally not as practically useful for the synthesis of exact realizers in discrete mathematics.

We propose in this paper a different kind of optimization of Gödel's functional interpretation by the elimination already from the primary extracted terms of a number of contraction formulas which are identified as *computationally redundant* by means of an adaptation of Berger's *uniform* quantifiers from [5] to the Dialectica-extraction context. These are called "quantifiers without computational content" in [34] and we will here call them quantifiers *without computational meaning* (abbreviated ncm, with n from *non*). We will here denote by $\overline{\forall}$ the ncm universal quantifier and by $\overline{\exists}$ the ncm existential quantifier. While our $\overline{\exists}$ is identical[2] to Berger's $\{\exists\}$, our $\overline{\forall}$ requires a further strengthening of the restriction set by Berger on his $\{\forall\}^+$ rule – we must take into account the inclusion of the *computationally relevant* contractions into the Dialectica realizing terms.

We build on top of Jørgensen's [19] Natural Deduction formulation of pure Dialectica interpretation which we transform by eliminating his "Contraction Lemma"[3] and by allowing free variables in the extracted terms[4]. We call *light*[5] functional interpretation this refinement of Gödel's Dialectica technique for the extraction of more efficient programs from (classical) proofs. We generally abbreviate (both regular and ncm) *quantifier free* by qfr. We will use expressions like "qfr formula", "ncm variable" or "ncm quantifiers" with the obvious meanings.

---

[2] Modulo the formulation as axioms like in [34] of Berger's rules for $\{\exists\}$ from [5].

[3] Which we find too complicated for a computer-implemented Dialectica extraction.

[4] Which is more suitable in a Natural Deduction context (see [16] for a discussion).

[5] Where "light" is to be understood as the opposite of "heavy" and not otherwise.

2   An Arithmetic for Gödel Functionals

We devise a weakly extensional variant WE–Z of the intuitionistic arithmetical system Z of Berger, Buchholz and Schwichtenberg [6] which restricts extensionality and adds the elements peculiar to Gödel's *Dialectica* interpretation [2, 13]. It moreover integrates the *non-computational-meaning* (abbreviated ncm) quantifiers mentioned in Section 1 above . System (WE–)Z is an extension of Gödel's **T** with the logical and arithmetical apparatus which renders it suitable to the applied program extraction from (classical) proofs (by means of Dialectica interpretation and its variants), see [6, 34, 36].

Finite types are inductively generated from base types by the rule that if σ and τ are types then $(\sigma\tau)$ is a type. For simplicity we take as *base* types only the type ι for natural numbers and o for booleans. We make the convention that concatenation is right associative and consequently omit unnecessary parenthesis, writing $\delta\,\sigma\,\tau$ instead of $(\delta(\sigma\tau))$. We denote tuples of types by $\underline{\sigma} :\equiv \sigma_1,\ldots,\sigma_n$. We abbreviate by $\underline{\sigma}\tau$ the type $\sigma_1\ldots\sigma_n\tau$. It is immediate that every type τ can be written as either $\tau \equiv \underline{\sigma}\iota$ or $\tau \equiv \underline{\sigma}o$.

The *term system* is a variant of Gödel's **T** formulated over the finite types with λ-abstraction as primitive. This is most appropriate in a Natural Deduction context. Terms are hence built from variables and term constants by λ-abstraction and application. We represent the latter as concatenation and we agree that it is left-associative in order to avoid excessive parenthesizing. All variables and constants have an *a priori* fixed type and terms have a type fixed by their formation. Written term expressions are always assumed to be well-formed in the sense that types match in all applications between sub-terms. As particular (term) *constants* we distinguish the following:

- $\mathtt{tt}^o$ and $\mathtt{ff}^o$ which denote boolean truth and falsity;
- for each type τ the *selector* $\mathtt{If}_\tau$ of type $o\,\tau\,\tau\,\tau$ which denotes choice according to a boolean condition with the usual *if-then-else* semantics;
- 0^ι (zero), S^ι (successor) and Gödel's recursor R_τ of type $\tau\,(\iota\,\tau\,\tau)\,\iota\,\tau$;
- equality $=^{\iota\,\iota\,o}$ – a functional constant and not predicate in our system.

Variables are denoted by $a, b, c, p, q, u, v, x, y, z, U, V, X, Y, Z$ such that, if not otherwise specified, a, b, c are free and u, v, x, y, z are bound variables of type ι. Also p, q denote variables of type o (be them free or bound) and U, V, X, Y, Z are *functional* variables (i.e., not of base type). We denote terms by r, s, t, S, T. We use sub- or super- scripts to enlarge the classes of symbols. We use underlined letters to denote tuples of corresponding objects. Tuples are just comma-separated lists of objects. If $\underline{t} \equiv t_1,\ldots,t_n$ we denote by $s(\underline{t})$ or even $s\underline{t}$ the term $st_1\ldots t_n$, i.e., $((st_1)\ldots)t_n$ by the left-associativity convention. Also $\underline{s}(\underline{t})$ and $\underline{s}\,\underline{t}$ denote the tuple $s_1(\underline{t}),\ldots,s_m(\underline{t})$. As particular *terms* we distinguish the following:

- Boolean conjunction and implication (with their usual semantics):

$$\mathtt{And}^{ooo} :\equiv \lambda p, q.\,\mathtt{If}_o\, p\, q\, \mathtt{ff}$$
$$\mathtt{Imp}^{ooo} :\equiv \lambda p, q.\,\mathtt{If}_o\, p\, q\, \mathtt{tt}$$

- For each higher-order type $\tau \equiv \underline{\sigma}\iota$ or $\tau \equiv \underline{\sigma}o$ we define the *zero* term 0_τ of type τ. We also let $0_\iota := 0$ and $0_o := \mathtt{ff}$ and thus every type is inhabited by a zero term (since we consider only ι and o as base types).
- For each positive integer n and type τ we define the *n-selector* $\mathtt{If}_\tau^n$ of type $\overbrace{o...o}^{n}\,\overbrace{\tau...\tau}^{n}\,\tau\tau$ by $\mathtt{If}_\tau^1 := \mathtt{If}_\tau$ and for $n \geq 2$ the definition of $\mathtt{If}_\tau^n$ is $\lambda p_1,\ldots,p_n,x_{n+1},x_n,\ldots,x_1.\,\mathtt{If}_\tau\,p_1\,(\mathtt{If}_\tau^{n-1}\,p_2\,\cdots\,p_n\,x_{n+1}\,x_n\,\cdots\,x_2)\,x_1$ s.t. $\mathtt{If}_\tau^n(r_1,\ldots,r_n,t_{n+1},t_n,\ldots,t_1)$ selects the first t_i with $i \in \overline{1,n}$ for which r_i is false, if it exists, otherwise t_{n+1} – if all $\{r_i\}_{i=1}^n$ are true.

The base logical system is a Natural Deduction formulation (see [6] and [34]) of Intuitionistic Logic[6]. We use $\wedge$ (logical conjunction), $\rightarrow$ (logical implication), $\forall$ (forall) and $\exists$ (*strong*, intuitionistic exists) as base logical constants. The only predicate symbol is the unary $\mathtt{at}$ which takes a single boolean argument. If t^o is a boolean term then $\mathtt{at}(t)$ is the *atomic* formula which (informally) denotes the fact that t is true. We *define* the logical falsum in terms of boolean falsity by $\perp := \mathtt{at}(\mathtt{ff})$. The *weak* (classical) existential quantifier $\exists^{\underline{\mathtt{cl}}}$ is defined in terms of $\forall$ by $\exists^{\underline{\mathtt{cl}}}x\,A(x) := (\forall x.\,A(x) \rightarrow \perp) \rightarrow \perp$. Negation $\neg$ and equivalence $\leftrightarrow$ are defined as usual, i.e., $\neg A := A \rightarrow \perp$ and $A \leftrightarrow B := ((A \rightarrow B) \wedge (B \rightarrow A))$. Disjunction is defined by $A \vee B := \exists p^o\,((\mathtt{at}(p) \rightarrow A) \wedge ((\neg\mathtt{at}(p)) \rightarrow B))$.

Predicate equality at base types is defined for boolean terms s and t by $s =_o t := \mathtt{at}(s) \leftrightarrow \mathtt{at}(t)$ and for natural terms s and t by $s =_\iota t := \mathtt{at}(= s\,t)$. Equality between terms s and t of type $\tau \equiv \sigma_1 \ldots \sigma_n \sigma$, with $\sigma \in \{o, \iota\}$, is extensionally defined as $s =_\tau t := \forall x_1^{\sigma_1}, \ldots, x_n^{\sigma_n}\,(s\,x_1 \ldots x_n =_\sigma t\,x_1 \ldots x_n)$. For any type τ we denote by $s \neq_\tau t := \neg(s =_\tau t)$ *non-equality* between the terms s^τ and t^τ. If non-ambiguous, we often omit to specify the type τ of (non-)equality.

We also introduce in our system an adaptation of Berger's [5] *uniform* quantifiers, here denoted $\bar{\forall}$ (forall ncm) and $\bar{\exists}$ (exists ncm) to the extraction of (more) efficient programs by Gödel's Dialectica interpretation. From a logic viewpoint $\bar{\forall}$ and $\bar{\exists}$ behave exactly like $\forall$ and $\exists$ – a theorem stating that "the (purely syntactic) replacement of $\bar{\forall}$ and $\bar{\exists}$ with their *computationally meaningful*[7] (or *regular*) correspondents in a proof $\mathcal{P}$ yields a(nother) proof in the corresponding system without ncm quantifiers" can easily be established. However, the converse to this (meta)theorem does not hold (in general) because of the (necessary) restriction which is set on the introduction rules for the ncm-universal quantifier and implication (see Sections 2.1 and 2.2 below). The special rôle of $\bar{\forall}$ and $\bar{\exists}$ is played in the program-extraction process only. There they act like some kind of labels for parts of the proof at input which are to be ignored since they are *a priory* (i.e., at the proof-building stage) distinguished as having no computational content. They also bring an important optimization with respect to the maximal type degree of programs extracted from those proofs for which the use of the com-

---

[6] The prominent place Johansson's Minimal Logic [18] has in the presentation of system Z in [6] or [34] is no longer needed for our (light) Dialectica-extraction exposition.

[7] Notice that $\forall$ is as computationally meaningful as $\exists$ in the context of program extraction by (light) Dialectica interpretation, see Definition 31 and Theorem 34.

putationally meaningful correspondents would have just brought an unjustified increase of this maximal type degree[8].

In order to avoid excessive parenthesizing we make the usual conventions that $\forall, \overline{\forall}, \exists^{\mathfrak{A}}, \exists, \overline{\exists}, \neg, \wedge, \vee, \rightarrow, \leftrightarrow$ is the decreasing order of precedence and $\rightarrow$ is right associative. For (more efficient) program-extraction purposes we impose that all axioms are closed formulas and for optimization purposes – at the example of Schwichtenberg's MINLOG system [34, 36] – their closure is ensured with $\overline{\forall}$ rather than $\forall$. The only exception[9] to this rule is the *Induction Axiom* IA, see Section 2.2 for the various definitions of induction within system WE–Z. Therefore it will be understood that even though a formula presented below as axiom is literally open, in fact the axiom it denotes is the $\overline{\forall}$ closure of the respective formula.

2.1   The Logical Axioms and Rules of System WE–Z

We begin by adapting the set of rules for Minimal Logic from [34] to the setting of program-extraction by the light Dialectica interpretation (defined in Section 3). First of all we define the following two *variable conditions* which will be used to constrain the rules concerning the (ncm-)universal quantifier:

- $\mathsf{VC}_1(z)$: the variable z does not occur free in any of the undischarged assumptions of the proof of the premise of the rule;
- $\mathsf{VC}_2(z, t)$: the term t is "free for" z in the conclusion, i.e., no free variable of t gets quantified after substituting $\{z \leftarrow t\}$ in the conclusion.

We also define an *ncm-formula condition* which is required to constrain the rule of implication introduction *with contraction* (see below) in order to attain the soundness theorem for the light Dialectica interpretation. For a formula A, the condition ncm-FC(A) says that if A contains (at least) a positive universal or a negative existential (regular) quantifier[10], then A *must not* contain any ncm quantifier[11]. The logical rules of our system WE–Z are then as follows:

- Deduction from an (arbitrary, undischarged) assumption: $A \vdash A$.
- Conjunction elimination left: $\dfrac{A \wedge B}{A}\ \wedge_l^-$, conjunction elimination right: $\dfrac{A \wedge B}{B}\ \wedge_r^-$ and conjunction introduction: $\dfrac{A,\, B}{A \wedge B}\ \wedge^+$.
- Implication elimination: $\dfrac{A,\, A \rightarrow B}{B}\ \rightarrow^-$ (Modus Ponens).

---

[8] Upon which the run-time complexity of the normalization algorithm directly depends, regardless of the reduction strategy, see Berger's paper [5] for more on this.

[9] This *notable* exception is necessary only in the context of Dialectica extraction because the Dialectica realizers of IA integrate the induction formula via an $\rightarrow^+$ *with contraction*, see Section 2 of [14], Section 2.2 and the proof of Theorem 34 below.

[10] Which means that A is *computationally relevant*, see "Implication introduction".

[11] So that the light D-interpretation of A has a qfr *base* formula A_D, see Definition 31.

– Implication introduction: $\dfrac{[A] \ldots /B}{A \to B} \to^+$, where some particular (possibly empty) class of instances of the formula A among the undischarged assumptions of the proof of B gets discharged. If at least two instances of A get discharged (i.e., the $\to^+$ is with contraction) then the $\mathtt{ncm\text{-}FC}(A)$ restriction applies. If moreover the premise of $\mathtt{ncm\text{-}FC}(A)$ holds, then A is named *computationally relevant*, otherwise A is a *computationally irrelevant* (*redundant for Dialectica*) contraction formula.

– [ncm-]ForAll elimination: $\left[\dfrac{\overline{\forall}z\, A(z)}{A(t)}\, \overline{\forall}_{z,t}^- \right] \dfrac{\forall z\, A(z)}{A(t)}\, \forall_{z,t}^-$, s.t. $\mathtt{VC_2}(z,t)$.

– ForAll introduction: $\dfrac{A(z)}{\forall z\, A(z)}\, \forall_z^+$, such that $\mathtt{VC_1}(z)$.

– ncm-ForAll introduction: $\dfrac{\mathcal{P}\colon A(z)}{\overline{\forall}z\, A(z)}\, \overline{\forall}_z^+$, such that $\mathtt{VC_1}(z)$ and $\mathtt{VC_3}(z,\mathcal{P})$. The latter (third) *variable condition* applies to the $\overline{\forall}$-quantified variables only. Although basically the same as the pre-condition set by Berger on his $\{\forall\}^+$ rule in [5], a strengthening peculiar to light-Dialectica extraction is necessary:

  – $\mathtt{VC_3}(z,\mathcal{P})$: the variable z does not occur free in any of the *instantiating terms* t involved by a $\forall_{\bullet,t}^-$ in the proof $\mathcal{P}$ (so far Berger's restriction) and z is also not free in the computationally relevant *contraction formulas* of $\mathcal{P}$ (defined above at the Implication Introduction item).

Intuitionistic Logic is then obtained by adding the axioms defining $\exists$ and $\overline{\exists}$:

$\mathtt{Ax}\exists^-$:	$\exists z_1\, A(z_1) \wedge \forall z_2\, [\, A(z_2) \to B\,] \to B$	($\exists$ elimination)
$\mathtt{Ax}\exists^+$:	$\forall z_1\, [\, A(z_1) \to \exists z_2\, A(z_2)\,]$	($\exists$ introduction)
$\mathtt{Ax}\overline{\exists}^-$:	$\overline{\exists} z_1\, A(z_1) \wedge \overline{\forall} z_2\, [\, A(z_2) \to B\,] \to B$	($\overline{\exists}$ elimination)
$\mathtt{Ax}\overline{\exists}^+$:	$\overline{\forall} z_1\, [\, A(z_1) \to \overline{\exists} z_2\, A(z_2)\,]$	($\overline{\exists}$ introduction)

with the usual restriction that z_2 is not free in B and, most important,

$\mathtt{AxEFQ}$: $\perp \to A$ (Ex-Falso-Quodlibet)

Notice that Intuitionistic Logic could have equally been formulated with rules for $\exists$ and $\vee$[12] (instead of axioms), see [5, 38]. We here follow [34] in choosing a formulation which is more suitable for computer-applied program-extraction.

2.2   Weakly Extensional Intuitionistic Arithmetic WE–Z

We now add the basic arithmetical apparatus to Intuitionistic Logic. We first introduce the axioms which give the (usual) behavior of (higher-order exten-

---

[12] The Boolean Induction axiom $\mathtt{AxBIA}$ from Section 2.2 is strictly required for attaining the usual logical behavior of $\vee$ in our formulation.

sional) equality, stressing from the beginning that *extensionality* must[13] be restricted in the context of Dialectica interpretation. The extensionality *axiom* $E_{\sigma,\tau} : \forall z^{\sigma\tau}, x^\sigma, y^\sigma. \ x =_\sigma y \to zx =_\tau zy$ must not be derivable in our system. We here deviate from system Z of [6, 34, 36] which derives $E_{\sigma,\tau}$ and therefore is fully extensional. We first present the more basic axioms of Reflexivity, Symmetry and Transitivity which we retain (modulo our definition of higher-order equality, hence they are no longer quantifier-free here) from system Z of [6, 34, 36] :

AxREF$_\tau$:	$x =_\tau x$	(Reflexivity)
AxSYM$_\tau$:	$x =_\tau y \to y =_\tau x$	(Symmetry)
AxTRZ$_\tau$:	$x =_\tau y \wedge y =_\tau z \to x =_\tau z$	(Transitivity)

Notice that although AxSYM$_\tau$ and AxTRZ$_\tau$ have no computational content under Modified Realizability, they must be provided with realizing terms under (light) Dialectica interpretation for higher-order τ[14]. We thus stay as close as possible to the axiomatic of system Z of [6, 34, 36], rendering easier the task of implementing program-extraction by (light) Dialectica interpretation in MIN-LOG [34, 36]. We do, however, have to deviate from system Z when it comes to the *Compatibility Axiom* (which implies $E_{\sigma,\tau}$): $x =_\sigma y \to B(x) \to B(y)$, which we replace by the following (strictly) weaker Compatibility Rule:

$$
\text{COMPAT}_\sigma : \quad
\begin{array}{c}
A_0 \\
\vdots \\
x =_\sigma y \\
\hline
B(x) \to B(y)
\end{array}
\qquad
\begin{array}{l}
\text{with the restriction that} \\[4pt]
\text{all undischarged assumptions used} \\[4pt]
\text{in the proof of } x =_\sigma y \text{ (here denoted } A_0) \\[4pt]
\text{are quantifier-free}
\end{array}
$$

Had the above restriction[15] not been present, the Compatibility Axiom would be directly deducible by $\to^+$, hence full extensionality would be derivable and (light) Dialectica interpretation would fail to interpret all proofs of our system[16].

---

[13] See, e.g., the chapter on Dialectica interpretation in [20] for detailed explanations. Howard's original counterexample to the Dialectica realizability of the extensionality axiom $E_{\iota\iota,\iota}$ by Gödel primitive recursive functionals is exposed in [17]. See also [35] for a counterexample to the Dialectica realizability of $E_{\iota\iota,\iota}$ by Van de Pol – Schwichtenberg monotone majorizable functionals (a class of functionals intersecting but independent of Gödel's **T**).

[14] We could have used only AxREF$_\iota$, AxSYM$_\iota$ and AxTRZ$_\iota$ as axioms since higher-type Reflexivity, Symmetry and Transitivity can be deduced in (pure) Minimal Logic from the Reflexivity, Symmetry and respectively Transitivity of natural numbers. The latter are quantifier-free and hence realizer-free under both Realizability and Dialectica interpretations. We however chose the above presentation for practical reasons – proofs are shorter and the realizers for the (light) Dialectica interpretation of higher-order Symmetry and Transitivity are immediate (see Section 3 of [14]).

[15] This restriction is not only sufficient but also necessary. Already by allowing purely universal undischarged assumptions in the proof of $x =_\sigma y$ we can deduce $E_{\iota\iota,\iota}$ in our system and we therefore become subject to Howard's counterexample [17].

[16] Details of the fact that Dialectica is valid for COMPAT are given in Section 3 of [14].

We now present the *boolean axioms*. While the desired behavior of ff is already given by AxEFQ, for tt we must introduce the Truth Axiom

AxTRH: $at(tt)$.

In order to attain the usual logical behavior of $\vee$ we must complement its definition with the following[17]

AxBIA: $A(tt) \wedge A(ff) \rightarrow \forall p^o A(p)$ (Boolean Induction Axiom)

and for the selectors If_τ we introduce the following expected axioms:

$$AxIf_\tau : \begin{cases} If_\tau \; tt \; x^\tau \; y^\tau =_\tau x \\ If_\tau \; ff \; x^\tau \; y^\tau =_\tau y \end{cases} .$$

Defining axioms for the constants specific to *natural numbers* are as usual

$$AxS : \begin{cases} Sz \neq_\iota 0 \\ Sx =_\iota Sy \rightarrow x =_\iota y \end{cases} \qquad AxR_\tau : \begin{cases} R_\tau \, x \, y \, 0 =_\tau x \\ R_\tau \, x \, y \, (Sz) =_\tau y(z, R_\tau \, x \, y \, z) \end{cases}$$

We finally arrive at integrating Induction for natural numbers in WE–Z. There are very simple realizing terms under Kreisel's Modified Realizability for the usual Induction Axiom IA : $A(0) \rightarrow \forall z \, (A(z) \rightarrow A(Sz)) \rightarrow \forall z \, A(z)$ – see [6, 34]. In contrast, the Dialectica realizing terms for IA are far more complicated because they include the Dialectica-translation of $\forall z(A(z) \rightarrow A(Sz))$ – see Section 2 of [14], Footnote 9 and the proof of Theorem 34 below. We will therefore use the following variant of the simpler induction rule employed by Jørgensen in [19]:

IR_0 : $\emptyset \cdots / A(0) \,$, $\, \emptyset \cdots / \forall z(A(z) \rightarrow A(Sz)) \vdash \forall z A(z)$

If one ignores the ncm quantifiers (and hence also the ncm–FC restriction), this system of intuitionistic arithmetic is in fact *the* weakly extensional version of system Z of [6, 34, 36], extended with $\bot \leftrightarrow at(ff)$. We denote the system with the ncm quantifiers by WE–Z and the system without the ncm quantifiers (hence without the ncm–FC restriction) by WE–Z⁻. In WE–Z⁻ there is a full equivalence between IA, IR_0 and the usual Induction Rule[18] IR : $A(0) \,$, $\, \forall z(A(z) \rightarrow A(Sz)) \vdash \forall z A(z)$. We denote deduction in systems WE–Z and WE–Z⁻ by $\vdash$ and respectively $\vdash_-$.

Remark 21 The following lemmas hold in WE–Z⁻:

LmAND: $\vdash_- \forall p^o, q^o \, (\, at(And \, p \, q) \, \leftrightarrow \, at(p) \wedge at(q) \,)$

LmIMP: $\vdash_- \forall p^o, q^o \, (\, at(Imp \, p \, q) \, \leftrightarrow \, at(p) \rightarrow at(q) \,)$

They establish the equivalence for terms of boolean and logical conjunction and implication, fact which permits the treatment of qfr formulas as prime (atomic) formulas, see Remark 22 and Section 1 of [14]. The WE–Z⁻ lemmas LmOι : $\vdash_- 0^{\underline{\sigma}\iota} \underline{z}^{\underline{\sigma}} =_\iota 0$ and LmOo : $\vdash_- 0^{\underline{\sigma}o} \underline{z}^{\underline{\sigma}} =_o ff$ describe the behavior of the

---

[17] Only the "disjunction introduction" theorems $\vdash A \rightarrow A \vee B$ and $\vdash B \rightarrow A \vee B$ are ensured by the definition of $\vee$. The "elimination of disjunction" requires AxBIA, see Remark 22 below.

[18] See Section 2 of [14] for details. The simulation of IR in terms of IR_0 fails in WE–Z.

zero terms. The expected behavior of the selector $\mathtt{If}^n_\tau$ is given by the following $n+1$ lemmas of $\mathtt{WE\text{–}Z}^-$ grouped under the name $\mathtt{LmIf}^n_\tau$:

$$\begin{cases} \{\ \vdash \bigwedge_{j=1}^{i-1}\mathtt{at}(p_j) \wedge \neg\mathtt{at}(p_i) \ \rightarrow\ \mathtt{If}^n_\tau(p_1,\ldots,p_n,x_{n+1},x_n,\ldots,x_1) =_\tau x_i\ \}_{i=1}^n \\ \ \ \vdash \bigwedge_{i=1}^n\mathtt{at}(p_i)\ \rightarrow\ \mathtt{If}^n_\tau(p_1,\ldots,p_n,x_{n+1},x_n,\ldots,x_1) =_\tau x_{n+1} \end{cases}$$

Remark 22 There exists a unique bijective association of boolean terms to $\mathtt{qfr}$ formulas $A_0 \mapsto \mathtt{t}_{A_0}$ such that $\vdash A_0 \leftrightarrow \mathtt{at}(\mathtt{t}_{A_0})$ for all A_0. In particular $\vdash (p =_o \mathtt{tt}) \leftrightarrow \mathtt{at}(p)$ and $\vdash (p =_o \mathtt{ff}) \leftrightarrow \neg\mathtt{at}(p)$. It then follows that all $\mathtt{qfr}$ formulas A_0 of system $\mathtt{WE\text{–}Z}^-$ are decidable in the sense that $\vdash A_0 \vee \neg A_0$. Using $\mathtt{AxBIA}$ it further follows that we can produce $\mathtt{WE\text{–}Z}^-$ proofs by *case distinction* over $\mathtt{qfr}$ formulas, i.e., $\vdash (A_0 \rightarrow A) \wedge (\neg A_0 \rightarrow A) \rightarrow A$. More general, the following schema of *disjunction elimination*[19] is deducible in system $\mathtt{WE\text{–}Z}^-$:

$$\vdash \bigwedge_{i=1}^n(A_i \rightarrow B)\ \rightarrow\ (\bigvee_{i=1}^n A_i \rightarrow B) \tag{1}$$

Stability for $\mathtt{qfr}$ formulas of $\mathtt{WE\text{–}Z}$, i.e., $\vdash \neg\neg A_0 \rightarrow A_0$ follows as an immediate application of case distinction over $\mathtt{qfr}$ formulas with $A := \equiv \neg\neg A_0 \rightarrow A_0$ [20].

2.3 The Semi-classical (Plus Choice) System $\mathtt{WE\text{–}Z}^+$

We present below the extension of $\mathtt{WE\text{–}Z}$ with three axioms which have simple realizers *directly* under the (light) Dialectica interpretation (defined in Section 3). The first two principles are (logically) deducible in the (fully) classical version of $\mathtt{WE\text{–}Z}$ (obtained by adding *full* stability $\neg\neg A \rightarrow A$) but not in $\mathtt{WE\text{–}Z}$. These *semi-classical* (logical) axioms are Markov's Principle[21]

$\mathtt{AxMK}$: $\exists^{\mathtt{1}} z\, A_0(z) \rightarrow \exists z\, A_0(z)$

and Independence of premises for universal premises

$\mathtt{AxIP}_\forall$: $[\forall x\, A_0(x) \rightarrow \exists y\, B(y)] \rightarrow \exists y\, [\forall x\, A_0(x) \rightarrow B(y)]$.

System $\mathtt{WE\text{–}Z}^+$ is obtained by further adding to $\mathtt{WE\text{–}Z}$ (besides $\mathtt{AxMK}$ and $\mathtt{AxIP}_\forall$) the non-logical (i.e., not logically deducible in $\mathtt{WE\text{–}Z}$) Axiom of Choice:

$\mathtt{AxAC}$: $\forall x\, \exists y\, B(x,y) \rightarrow \exists Y\, \forall x\, B(x,Y(x))$.

We will denote deduction in $\mathtt{WE\text{–}Z}^+$ by $\vdash_+$.

3   The Light Functional (*Light Dialectica*) Interpretation

By $\mathtt{LD}$-*interpretation* we call below our adaptation of Gödel's functional (*Dialectica*) interpretation [2, 13] to the extraction of (more) efficient programs

---

[19] Very useful to prove soundness for the (light) D-interpretation of $\rightarrow^+$, in Section 3.
[20] Section 1 of [14] contains complete proofs of all the results stated above.
[21] The usual formulation of Markov's principle as $\mathtt{AxMK_1}$: $\neg\neg \exists z\, A_0(z) \rightarrow \exists z\, A_0(z)$ is equivalent over $\mathtt{WE\text{–}Z}$ to $\mathtt{AxMK}$. Also the form $\mathtt{AxMK_2}$: $\neg\neg \exists z\, A_0(z) \rightarrow \exists z\, \neg\neg\, A_0(z)$, which was preferred by the authors of [16] because of complexity considerations. Our choice of $\mathtt{AxMK}$ is motivated by the particularity of $\exists^{\mathtt{1}}$ in [6, 34].

from (classical) proofs. It is a recursive syntactic translation from proofs in WE–Z$^+$ (or in WE–Z^{c+}, after a double-negation translation, see Section 3.2) to proofs in WE–Z$^-$ such that positive occurrences of $\exists$ and negative occurrences of $\forall$ in the proof's conclusion get actually realized by terms in Gödel's **T**. These *realizing terms* are also called *the programs extracted* by the LD-interpretation and (if only the extracted programs are wanted) the translation process is also referred to as *program-extraction*. The translated proof is also called the *verifying proof* since it verifies the fact that the extracted programs actually *realize* the LD-interpretation of the conclusion of the proof at input. Gödel's Dialectica interpretation (abbreviated D-*interpretation*) is relatively (far) more complicated when it has to face *contraction*, which in Natural Deduction amounts to discharging more than one copy of an assumption in an Implication Introduction $\rightarrow^+$. Kohlenbach presents in [21] an elegant way of simplifying the treatment of contraction when the goal is to extract Howard majorizing functionals for the Dialectica realizers. He named "monotone functional interpretation" this variant of the D-interpretation which we here abbreviate by MD-*interpretation*. The MD-interpretation has been used with great success over the last years for producing mathematical proofs to important new theorems in numerical functional analysis[22].

However, when the extraction of *exact* realizers is concerned, the monotone D-interpretation does not necessarily bring an efficient answer. A different kind of optimization of Gödel's D-interpretation appears to be necessary. We here propose the LD-interpretation as a refinement of Gödel's technique which allows for the extraction of more efficient exact realizers. Moreover, the same refinement equally applies to the extraction of more efficient bounds via what might be called the *light monotone*[23] functional interpretation (abbreviated LMD-interpretation).

The D-interpretation was first introduced in [13] for a Hilbert-style formulation of Arithmetic – see also [2, 10, 20, 26, 38] for other (more modern) formulations within Hilbert-style systems. Natural Deduction formulations of the Diller-Nahm [10] variant of D-interpretation were provided by Diller's students Rath [33] and Stein [37]. Only in the year 2001 Jørgensen [19] provided a first Natural Deduction formulation of the original Gödel's functional interpretation. In the Diller-Nahm setting all choices between the potential realizers of a contraction are postponed to the very end by collecting all candidates and making a single final global choice. In contrast, Jørgensen's formulation respects Gödel's original treatment of contraction by immediate (local) choices. Jørgensen devises a so-called "Contraction Lemma" in order to handle (in the given Natural Deduction context) the discharging of more than one copy of an assumption in an Implication Introduction $\rightarrow^+$. If $n + 1$ undischarged occurrences of an assumption are to be canceled in an $\rightarrow^+$, then Jørgensen uses his Contraction Lemma n times, shifting partial results n times back and forth over the "proof gate" $\vdash$. We consider this to be less efficient from the applied program-extraction perspec-

---

[22] Papers [22] and [20] contain comprehensive surveys of such to-day applications of MD-interpretation to concrete mathematical proofs from the literature.

[23] Or *light bounded* functional (*Dialectica*) interpretation, following [11] instead of [21].

tive. We also think that the soundness proof for the (L)D-interpretation is thus somewhat more complicated w.r.t. contraction. We will here use the n-selector $\mathtt{If}^n_\tau$ for equalizing in one single (composed) step all the (L)D-interpretations of the $n+1$ undischarged occurrences (see the proof of Theorem 34 below). While technically impossible to have a direct n-selector available for all $n \in \mathbb{N}$, in actual optimizing implementations $\mathtt{If}^n_\tau$ could be given a (more) direct definition for $n \leq N$ for a certain convenient[24] upper margin N instead of being simulated in terms of n times $\mathtt{If}^1_\tau$. The practical gain w.r.t. Jørgensen's solution is that the handling of contraction is directly moved from the proof level to the term level: back-and-forth shifting over $\vdash$ is no longer required when building the verifying proof. We also modify Jørgensen's variant of D-interpretation by allowing free variables in the extracted terms. This corresponds to the formulation of Gödel's $\mathbf{T}$ with λ-abstraction as primitive and is more natural in a Natural Deduction setting. In addition we include the treatment of our adaptation of Berger's uniform existential and universal quantifiers from [5] to the Dialectica-extraction context.

The light functional (*Dialectica*) interpretation/translation assigns a formula $A^D(\underline{a}) \equiv \exists \underline{x}\, \forall \underline{y}\, A_D(\underline{x}; \underline{y}; \underline{a})$ with A_D not necessarily quantifier-free and $\underline{x}, \underline{y}$ tuples of fresh variables of finite type (such that $\{\underline{x}, \underline{y}, \underline{a}\}$ are all free variables of A_D) to each instance of the formula $A(\underline{a})$ (with $\underline{a}$ all free variables of A). The types of $\underline{x}, \underline{y}$ depend only on the types of the regularly bound variables of A and on the logical structure of A. We will also denote by $B^D(\underline{b}) :\equiv \exists \underline{u}\, \forall \underline{v}\, B_D(\underline{u}; \underline{v}; \underline{b})$ the LD-interpretation of $B(\underline{b})$. If non-ambiguous, we sometimes omit to display some of the free variables of the LD-translated formulas.

Definition 31 (The light Dialectica interpretation of formulas)

$$A^D :\equiv (A_D :\equiv A) \text{ for prime formulas } A$$

$$(A \wedge B)^D :\equiv \exists \underline{x}, \underline{u}\, \forall \underline{y}, \underline{v}\, [\,(A \wedge B)_D :\equiv A_D(\underline{x}; \underline{y}; \underline{a}) \wedge B_D(\underline{u}; \underline{v}; \underline{b})\,]$$

$$(\exists z A(z, \underline{a}))^D :\equiv \exists z^\dagger, \underline{x}\, \forall \underline{y}\, [\,(\exists z A(z, \underline{a}))_D(z^\dagger, \underline{x}; \underline{y}; \underline{a}) :\equiv A_D(\underline{x}; \underline{y}; z^\dagger, \underline{a})\,]$$

$$(\forall z A(z, \underline{a}))^D :\equiv \exists \underline{X}\, \forall z^\dagger, \underline{y}\, [\,(\forall z A(z, \underline{a}))_D(\underline{X}; z^\dagger, \underline{y}; \underline{a}) :\equiv A_D(\underline{X}(z^\dagger); \underline{y}; z^\dagger, \underline{a})\,]$$

$$(\overline{\exists} z A(z, \underline{a}))^D :\equiv \exists \underline{x}\, \forall \underline{y}\, [\,(\overline{\exists} z A(z, \underline{a}))_D(\underline{x}; \underline{y}; \underline{a}) :\equiv \exists z\, A_D(\underline{x}; \underline{y}; z, \underline{a})\,]$$

$$(\overline{\forall} z A(z, \underline{a}))^D :\equiv \exists \underline{x}\, \forall \underline{y}\, [\,(\overline{\forall} z A(z, \underline{a}))_D(\underline{x}; \underline{y}; \underline{a}) :\equiv \forall z\, A_D(\underline{x}; \underline{y}; z, \underline{a})\,]$$

$$(A \rightarrow B)^D :\equiv \exists \underline{Y}, \underline{U}\, \forall \underline{x}, \underline{v}\, [\,(A \rightarrow B)_D :\equiv A_D(\underline{x}; \underline{Y}(\underline{x}, \underline{v})) \rightarrow B_D(\underline{U}(\underline{x}); \underline{v})\,]$$

Here $\cdot \mapsto \cdot^\dagger$ is a mapping which assigns to every given variable z a completely new variable $z^\dagger$ which has the same type of z. Different variables $z^\dagger$ are returned for different applications on the same argument variable z when processing a given formula A. This ensures that two nested quantifications of the same variable in A are correctly distinguished in A_D. The variables $\underline{X}, \underline{Y}, \underline{U}$ produced in the treatment of $\rightarrow$ and $\forall$ are also completely new but in contrast to the variables

---

[24] Only a certain limited number of n-selectors is needed for most practical applications.

produced by $\cdot^\dagger$, their type is strictly more complex than the type of the original variable. The free variables of A^D are exactly the free variables of A. If A is quantifier-free then $A^D = A_D = A$.

Remark 32 By abuse of notation from now on we use non-underlined letters also for tuples of objects (variables, terms, ...) and identify an individual object with the tuple containing only that object.

Definition 33 (Dialectica terms) For every ncm-quantifier free formula $A(a)$ we denote by $t_A^D[x; y; a]$ the boolean term associated to (the quantifier-free formula) $A_D(x; y; a)$ by the mapping from Remark 22. We call it "the Dialectica term associated to" $A(a)$. The following holds:

$$\vdash_- A_D(x; y; a) \leftrightarrow \mathtt{at}(t_A^D[x; y; a]) \tag{2}$$

Theorem 34 (Exact realizer synthesis by the LD-interpretation) There exists an algorithm which given at input a proof $\mathcal{P} : \{C^i\}_{i=1}^n \vdash_+ A$ produces at output the tuples of terms $\{T_i\}_{i=1}^n$ and T, the tuples of variables $\{x_i\}_{i=1}^n$ and y all together with the verifying proof $\mathcal{P}_D : \{C_D^i(x_i; T_i(\underline{x}, y))\}_{i=1}^n \vdash_- A_D(T(\underline{x}); y)$ — where $\underline{x} :\equiv x_1, \ldots, x_n$. Moreover,

1. the variables $\underline{x}$ and y do not occur in $\mathcal{P}$ (they are all completely new)
2. the free variables of T and $\{T_i\}_{i=1}^n$ are among the free variables of A and $\{C^i\}_{i=1}^n$ (we call this "the *free variable condition* (FVC) for programs extracted by the D-interpretation")

hence $\underline{x}$ and y also do not occur free in the *extracted* terms $\{T_i\}_{i=1}^n$ and T.

Proof: The algorithm proceeds by recursion on the structure of the input proof $\mathcal{P}$. Realizing terms must be presented for all the axioms and then realizing terms for the conclusion of a rule must be deduced from terms which realize the premise of that rule. Since x, y are produced by the LD-interpretation of formulas (see Definition 31) it is immediate that they do not occur in $\mathcal{P}$. We present below only the (sub)case of Implication Introduction $\rightarrow^+$ in which at least two copies of the implicative assumption get canceled. Since it involves *contraction*, this case is far more difficult than all the other axioms and rules[25].

$$\frac{[A] \ldots /B}{A \rightarrow B} \rightarrow^+$$ We are given, with $n \geq 1$, $\underline{z} \equiv \overbrace{z, \ldots, z}^{n+1}$ and $\underline{x} \equiv x_{n+2}, \ldots, x_m$, that :

$$\{A_D(z; T_i(\underline{z}, \underline{x}, y))\}_{i=1}^{n+1}, \{C_D^i(x_i; T_i(\underline{z}, \underline{x}, y))\}_{i=n+2}^m \vdash_- B_D(T(\underline{z}, \underline{x}); y) \tag{3}$$

It has been assumed that $n + 1 \leq m$, where $n + 1$ is the number of copies of the assumption A which get discharged in this $\rightarrow^+$. Each of these $n + 1$ instances of A produces the same tuple z of existential variables under the LD-interpretation, see Definition 31. The **ncm-FC(A)** constraint ensures that the tuples $\{T_i\}_{i=1}^{n+1}$ are

---

[25] See the comments in the beginning of this section. A full proof is in Section 3 of [14].

all of length 0 (denoted $\sqcup$), i.e., A is *computationally irrelevant*, orelse A_D is `qfr` (pre-condition for the association to A of *Dialectica terms*, see Definition 33). Only in the former case can we directly discharge in a single $\to^+$ all $n + 1 \geq 2$ copies of $A_D(z; \sqcup)$ in the LD-interpretation of the premise of this rule, see (3). In contrast, for the latter case we must first *equalize* the assumptions involving the $n + 1$ terms $\{T_i\}_{i=1}^{n+1}$. This is because the terms $\{T_i\}_{i=1}^{n+1}$ can be mutually different since they are extracted from different sub-proofs which involve the different copies of A. We hereafter treat this case. We achieve the *equalizing* of $\{T_i\}_{i=1}^{n+1}$ in one single step[26], using the n-selector $\mathtt{If}_\tau^n$. For all $i \in \overline{1,n}$ let T^i abbreviate $T_i(\underline{z}, \underline{x}, y)$ and let

$$\widetilde{S} :\equiv \lambda\underline{x}, z, y. \, \mathtt{If}_\tau^n(\mathtt{t}_A^D[z; T^1], \ldots, \mathtt{t}_A^D[z; T^n], T_{n+1}(\underline{z}, \underline{x}, y), T^n, \ldots, T^1) \qquad (4)$$

By n applications of $\overline{\forall}^-$ to $\mathtt{LmIf}_\tau^n$ with $\{p_i \leftarrow \mathtt{t}_A^D[z; T^i]\}_{i=1}^n$ and (2) we get $\vdash_- \wedge_{j=1}^{i-1} A_D(z; T^j) \wedge \neg A_D(z; T^i) \to \widetilde{S}(\underline{x}, z, y) = T_i(\underline{z}, \underline{x}, y)$ for all $i \in \overline{1,n}$ and $\vdash_- \wedge_{i=1}^n A_D(z; T^i) \to \widetilde{S}(\underline{x}, z, y) = T_{n+1}(\underline{z}, \underline{x}, y)$ from which we further obtain

$$\vdash_- [\wedge_{j=1}^{i-1} A_D(z; T^j) \wedge \neg A_D(z; T^i)] \to [A_D(z; \widetilde{S}(\underline{x}, z, y)) \to \wedge_{k=1}^{n+1} A_D(z; T^k)]$$

$$\vdash_- [\wedge_{j=1}^n A_D(z; T^j)] \to [A_D(z; \widetilde{S}(\underline{x}, z, y)) \to \wedge_{k=1}^{n+1} A_D(z; T^k)]$$

for all $i \in \overline{1,n}$. We used one $\mathtt{COMPAT}$[27] in each of the $n + 1$ above deductions and an $\mathtt{AxEFQ}$ in each of the the the first n of these. Due to the decidability of `qfr` formulas we have $\vdash_- \vee_{i=1}^n [\wedge_{j=1}^{i-1} A_D(z; T^j) \wedge \neg A_D(z; T^i)] \vee [\wedge_{j=1}^n A_D(z; T^j)]$. It then follows by (1) that $\vdash_- A_D(z; \widetilde{S}(\underline{x}, z, y)) \to \wedge_{i=1}^{n+1} A_D(z; T^i)$, hence for all $i \in \overline{1, n+1}$ we obtain $A_D(z; \widetilde{S}(\underline{x}, z, y)) \vdash_- A_D(z; T_i(\underline{z}, \underline{x}, y))$. We then discharge all assumptions A_D of (3) in $n + 1$ applications of $\to^+$ and combine with the previously obtained proofs of $A_D(z; T_i(\underline{z}, \underline{x}, y))$ in $n + 1$ applications of $\to^-$ to conclude, with $S :\equiv \lambda\underline{x}, z. \, T(\underline{z}, \underline{x})$ and $\{S_i :\equiv \lambda\underline{x}, z. \, T_i(\underline{z}, \underline{x})\}_{i=n+2}^m$, that

$$\{A_D(z; \widetilde{S}(\underline{x}, z, y))\}_{i=1}^{n+1}, \, \{C_D^i(x_i; S_i(\underline{x}, z, y))\}_{i=n+2}^m \vdash_- B_D(S(\underline{x}, z); y) \, .$$

We can now cancel all $\{A_D(z; \widetilde{S}(\underline{x}, z, y))\}_{i=1}^{n+1}$ in a single $\to^+$ and thus get

$$\{C_D^i(x_i; S_i(\underline{x}, z, y))\}_{i=n+2}^m \vdash_- A_D(z; \widetilde{S}(\underline{x}, z, y)) \to B_D(S(\underline{x}, z); y) \, .$$

Notice that $\mathtt{t}_A^D$ introduces in $\widetilde{S}$ new occurrences of the free variables of A. We finally obtain, with the `FVC` obviously satisfied, that

$$\{C_D^i(x_i; S_i(\underline{x}, z, y))\}_{i=n+2}^m \vdash_- (A \to B)_D(\widetilde{S}(\underline{x}), S(\underline{x}); z, y) \, .$$

$\square$

Corollary 35 There exists an algorithm which from a given proof $\vdash_+ A(a)$ produces exact realizing terms $T[a]$ with a verifying proof $\vdash_- \forall y A_D(T; y; a)$.

---

[26] Jørgensen's solution uses here n steps which correspond to the simulation of $\mathtt{If}_\tau^n$ in terms of n instances of $\mathtt{If}_\tau^1$, see also the comments in the preamble of Section 3.

[27] Since A_D is quantifier-free, the restriction on undischarged assumptions is respected.

3.1 The Light Monotone (or Light Bounded) Functional Interpretation

We sketch below the combination of our refinement of Goedel's Dialectica interpretation [2, 13, 19] with its optimization for the extraction of bounds[28] due to Kohlenbach [21]. We add to systems $\texttt{WE-Z}^-$, $\texttt{WE-Z}$ and $\texttt{WE-Z}^+$ a functional inequality constant for naturals (of type $\iota\iota o$), denoted $\geq$. We define *predicate inequality* between terms s^ι and t^ι as an abbreviation for $\texttt{at}(\geq s t)$, denoted $\geq_\iota$. Similar to equality, predicate *inequality* at higher types $\tau \equiv \sigma_1 \dots \sigma_n \iota$ is extensionally defined as $s \geq_\tau t :\equiv \forall x_1^{\sigma_1}, \dots, x_n^{\sigma_n} (s x_1 \dots x_n \geq_\iota t x_1 \dots x_n)$. We also define Howard's *majorization* relation $\succeq_\tau$ by $\succeq_\iota :\equiv \geq_\iota$ and

$$x \succeq_{\sigma\tau} y :\equiv \forall z_1^\sigma, z_2^\sigma (z_1 \geq_\sigma z_2 \rightarrow x z_1 \geq_\tau y z_2) \ .$$

The monotonic systems $\texttt{WE-Z}_\texttt{m}^-$, $\texttt{WE-Z}_\texttt{m}$ and $\texttt{WE-Z}_\texttt{m}^+$ are then obtained by further adding the axioms defining the usual behavior of predicate inequality on naturals: $z \geq_\iota 0$, $0 \geq_\iota \texttt{S}z \rightarrow \bot$ and $\texttt{S}x \geq_\iota \texttt{S}y \rightarrow x \geq_\iota y$. We denote by $\vdash_-^\texttt{m}$, $\vdash^\texttt{m}$ and $\vdash_+^\texttt{m}$ deductions in $\texttt{WE-Z}_\texttt{m}^-$, $\texttt{WE-Z}_\texttt{m}$ and respectively $\texttt{WE-Z}_\texttt{m}^+$.

Conjecture 36 (Uniform bound synthesis by a LMD-interpretation)
There exists an algorithm which from a given proof $\vdash_+^\texttt{m} A(a)$ produces *closed* uniform bounds T with a verifying proof $\vdash_-^\texttt{m} \exists x (T \succeq x \wedge \forall a, y A_\texttt{D}(x(a); y; a))$.

Notice that the optimization brought by the light MD-interpretation does not concern contraction (which is sufficiently handled by the pure MD-interpretation) as much as the diminishing of the maximal type degree of the extracted bounds.

3.2 Extensions of L(M)D-Interpretation to Extractions from Fully Classical Proofs. Systems $\texttt{WE-Z}^c$, $\texttt{WE-Z}^{c+}$ and $\texttt{WE-Z}_\texttt{m}^{c+}$

The system $\texttt{WE-Z}^c$ of weakly extensional Classical Arithmetic is obtained by adding to $\texttt{WE-Z}$ the Stability principle:

$$\texttt{AxSTAB} : \qquad \neg\neg A \rightarrow A \qquad\qquad\qquad \text{(Stability)}$$

In fact it would be sufficient that AxSTAB replaces AxEFQ since the latter is deducible from AxSTAB in Minimal Logic. However we keep AxEFQ as axiom of $\texttt{WE-Z}^c$ since it has simple Dialectica realizers[29]. The problem of AxSTAB is that it has no (direct) realizer under Dialectica interpretation – a preprocessing double negation translation, denoted $\cdot \mapsto \cdot^\texttt{N}$ will be necessary to interpret fully classical systems (see, e.g., [16, 20, 22]).

We will denote by $\texttt{WE-Z}^{c+}$ the system $\texttt{WE-Z}^c$ extended with $\texttt{AxAC}_0$ (the quantifier-free version of AxAC, obtained by restricting B to qfr formulas). We will denote deductions in $\texttt{WE-Z}^c$ and $\texttt{WE-Z}^{c+}$ by $\vdash_c$ and $\vdash_{c+}$ respectively.

Conjecture 37 There exists an algorithm which from a given proof $\vdash_{c+} A(a)$ produces exact realizing terms $T[a]$ with a verifying proof $\vdash_- \forall y (A^\texttt{N})_\texttt{D}(T; y; a)$.

---

[28] See also [11] for a more recent adaptation of Dialectica to the extraction of bounds.
[29] Input proofs to the Dialectica-extraction algorithm may thus become shorter.

We denote by $\mathtt{WE}\text{-}\mathcal{Z}_{\mathtt{m}}^{\mathtt{c+}}$, $\vdash_{\mathtt{c+}}^{\mathtt{m}}$ the monotonic correspondents of $\mathtt{WE}\text{-}\mathcal{Z}^{\mathtt{c+}}$ and $\vdash_{\mathtt{c+}}$.

Conjecture 38 There exists an algorithm which, given at input a proof $\vdash_{\mathtt{c+}}^{\mathtt{m}} A(a)$, produces at output *closed* uniform bounds T and the verifying proof

$$\vdash_{-}^{\mathtt{m}} \exists x (T \succeq x \wedge \forall a, y (A^{\mathtt{N}})_{\mathtt{D}}(x(a); y; a)).$$

Acknowledgements

We would like to thank Prof. Helmut Schwichtenberg for his guidance of our interaction with MINLOG and for many useful discussions.

References

1. J. Avigad. Interpreting classical theories in constructive ones. *The Journal of Symbolic Logic*, 65(4):1785–1812, 2000.
2. J. Avigad and S. Feferman. Gödel's functional ('Dialectica') interpretation. In S.R. Buss, editor, *Handbook of Proof Theory*, volume 137 of *Studies in Logic and the Foundations of Mathematics*, pages 337–405. Elsevier, 1998.
3. F. Barbanera and S. Berardi. Extracting constructive content from classical logic via control–like reductions. In M. Bezem and J.F. Groote, editors, *Typed Lambda Calculi and Applications*, pages 45–59. LNCS Vol. 664, 1993.
4. J.L. Bates and R.L. Constable. Proofs as programs. *ACM Transactions on Programming Languages and Systems*, 7(1):113–136, January 1985.
5. U. Berger. Uniform Heyting Arithmetic. *Annals of Pure and Applied Logic*, 133(1-3):125–148, May 2005. Festschrift for H. Schwichtenberg's 60th birthday.
6. U. Berger, W. Buchholz, and H. Schwichtenberg. Refined program extraction from classical proofs. *Annals of Pure and Applied Logic*, 114:3–25, 2002.
7. U. Berger, H. Schwichtenberg, and M. Seisenberger. The Warshall algorithm and Dickson's lemma: Two examples of realistic program extraction. *Journal of Automated Reasoning*, 26:205–221, 2001.
8. R.L. Constable and C. Murthy. Finding computational content in classical proofs. In G. Huet and G. Plotkin, editors, *Logical Frameworks*, pages 341–362. Cambridge University Press, 1991.
9. T. Coquand and M. Hofmann. A new method for establishing conservativity of classical systems over their intuitionistic version. *Mathematical Structures in Computer Science*, 9(4):323–333, 1999.
10. J. Diller and W. Nahm. Eine Variante zur Dialectica Interpretation der Heyting Arithmetik endlicher Typen. *Archiv für Mathematische Logik und Grundlagenforschung*, 16:49–66, 1974.
11. F. Ferreira and P. Oliva. Bounded Functional Interpretation. *Annals of Pure and Applied Logic*, 48 pp., To appear, see Elseviers's *Science Direct* on the Internet.
12. H. Friedman. Classical and intuitionistically provably recursive functions. In G.H. Müller and D.S. Scott, editors, *Higher Set Theory*, volume 669 of *Lecture Notes in Mathematics*, pages 21–27. Springer Verlag, 1978.
13. K. Gödel. Über eine bisher noch nicht benützte Erweiterung des finiten Standpunktes. *Dialectica*, 12:280–287, 1958.
14. M.-D. Hernest. Technical Appendix to this paper. See the author's web-page.
15. M.-D. Hernest. A comparison between two techniques of program extraction from classical proofs. In M. Baaz, J. Makovsky, and A. Voronkov, editors, *CSL 2003: Extended Posters*, vol. VIII of *Kurt Gödel Society's Collegium Logicum*, pp. 99–102. Springer Verlag, 2004.

16. M.-D. Hernest and U. Kohlenbach. A complexity analysis of functional interpretations. *Theoretical Computer Science*, 338(1-3):200–246, 10 June 2005.

17. W.A. Howard. Hereditarily majorizable functionals of finite type. [38], pp. 454–461.

18. I. Johansson. Der Minimalkalkül, ein reduzierter intuitionistischer Formalismus. *Compositio Matematica*, 4:119–136, 1936.

19. K.F. Jørgensen. Finite type arithmetic. Master's thesis, Departments of Mathematics and Philosophy, University of Roskilde, Roskilde, Denmark, 2001.

20. U. Kohlenbach. Proof Interpretations and the Computational Content of Proofs. *Lecture Course*, latest version in the author's web page.

21. U. Kohlenbach. Analysing proofs in Analysis. In W. Hodges, M. Hyland, C. Steinhorn, and J. Truss, editors, *Logic: from Foundations to Applications, Keele, 1993*, European Logic Colloquium, pages 225–260. Oxford University Press, 1996.

22. U. Kohlenbach and P. Oliva. Proof mining: a systematic way of analysing proofs in Mathematics. *Proc. of the Steklov Institute of Mathematics*, 242:136–164, 2003.

23. G. Kreisel. Interpretation of analysis by means of constructive functionals of finite types. In A. Heyting, editor, *Constructivity in Mathematics*, pages 101–128. North-Holland Publishing Company, 1959.

24. J.-L. Krivine. Classical logic, storage operators and second-order lambda-calculus. *Annals of Pure and Applied Logic*, 68:53–78, 1994.

25. D. Leivant. Syntactic translations and provably recursive functions. *The Journal of Symbolic Logic*, 50(3):682–688, September 1985.

26. H. Luckhardt. *Extensional Gödel Functional Interpretation*, volume 306 of *Lecture Notes in Mathematics*. Springer Verlag, 1973.

27. H. Luckhardt. Bounds extracted by Kreisel from ineffective proofs. In P. Odifreddi, editor, *Kreiseliana: About and around Georg Kreisel*, pp. 289–300. A.K. Peters, Wellesley, MA, 1996.

28. C. Murthy. Extracting constructive content from classical proofs. Tech. Report 90-1151, Dep.of Comp.Science, Cornell Univ., Ithaca, NY, U.S.A.,1990. PhD thesis.

29. G.E. Ostrin and S.S. Wainer. Elementary arithmetic. *Annals of Pure and Applied Logic*, 133(1-3):275–292, May 2005. Festschrift for H. Schwichtenberg's 60s.

30. M. Parigot. $\lambda\mu$–calculus: an algorithmic interpretation of classical natural deduction. In *Proc. of Log. Prog. and Automatic Reasoning, St. Petersburg*, volume 624 of *LNCS*, pages 190–201. Springer Verlag, Berlin, Heidelberg, New York, 1992.

31. C. Paulin-Mohring and B. Werner. Synthesis of ML programs in the system Coq. *Journal of Symbolic Computation*, 15(5/6):607–640, 1993.

32. C. Raffalli. Getting results from programs extracted from classical proofs. *Theoretical Computer Science*, 323(1-3):49–70, 2004.

33. P. Rath. *Eine verallgemeinerte Funktionalinterpretation der Heyting Arithmetik endlicher Typen*. PhD thesis, Universität Münster, Germany, 1978.

34. H. Schwichtenberg. *Minimal logic for computable functions*. Lecture course on program-extraction from (classical) proofs. Author's page or MINLOG distrib. [36].

35. H. Schwichtenberg. Monotone majorizable functionals. *Studia Logica*, 62:283–289, 1999.

36. H. Schwichtenberg and Others. Proof- and program-extraction system MINLOG. Free code and documentation at http://www.minlog-system.de.

37. M. Stein. *Interpretation der Heyting-Arithmetik endlicher Typen*. PhD thesis, Universität Münster, Germany, 1976.

38. A.S. Troelstra, editor. *Metamathematical investigation of intuitionistic Arithmetic and Analysis*, volume 344 of *Lecture Notes in Mathematics*. Springer-Verlag, 1973.

Feasible Proofs of Matrix Properties with Csanky's Algorithm

Michael Soltys

McMaster University, Computing and Software
1280 Main Street West, Hamilton, ON., Canada L8S 4K1
soltys@mcmaster.ca

Abstract. We show that Csanky's fast parallel algorithm for computing the characteristic polynomial of a matrix can be formalized in the logical theory **LAP**, and can be proved correct in **LAP** from the principle of linear independence. **LAP** is a natural theory for reasoning about linear algebra introduced in [8]. Further, we show that several principles of matrix algebra, such as linear independence or the Cayley-Hamilton Theorem, can be shown equivalent in the logical theory **QLA**. Applying the separation between complexity classes $\mathbf{AC}^0[2] \subsetneq \mathbf{DET}(GF(2))$, we show that these principles are in fact not provable in **QLA**. In a nutshell, we show that linear independence is "all there is" to elementary linear algebra (from a proof complexity point of view), and furthermore, linear independence cannot be proved trivially (again, from a proof complexity point of view).

Keywords: Proof complexity, Csanky's algorithm, matrix algebra.

1 Introduction

This paper makes the following claim: our intuition that the principle of linear independence is all that there is to elementary linear algebra is justified from a proof complexity point of view. This means that from the principle of linear independence we can prove other strong principles of linear algebra (for example, the Cayley-Hamilton Theorem) using concepts of very low computational complexity. Furthermore, we claim that linear independence itself cannot be proved using concepts of low computational complexity.

To argue this claim, we present a new feasible proof of the Cayley-Hamilton Theorem (CHT) from the principle of linear independence in a weak theory of linear algebra (**QLAP**). The proof is based on Csanky's algorithm for computing the characteristic polynomial of a matrix. Csanky's algorithm is a fast parallel algorithm that computes the characteristic polynomial of a matrix over fields of characteristic zero.

QLAP is a first order theory for reasoning about matrices. Our new proof of the CHT with Csanky's algorithm leads to **QLAP** proofs of equivalence of important principles of linear algebra (for example, linear independence and the CHT). We also show that these principles are independent of **QLAP**. To show

L. Ong (Ed.): CSL 2005, LNCS 3634, pp. 493–508, 2005.

this independence we use the previously known result that $\mathbf{AC}^0[2]$ is properly contained in $\mathbf{DET}(\mathrm{GF}(2))$.

The class $\mathbf{AC}^0[2]$ consists of problems solvable with polynomial size circuits (in the size of the input), bounded depth, where besides the usual gates $\{\wedge, \vee, \neg\}$ we are also allowed to use the parity gate $\oplus$. The class $\mathbf{DET}(\mathrm{GF}(2))$ consists of problems $\mathbf{AC}^0$ reducible to computing the determinant over the field of two elements. Another class which will make a frequent appearance in this paper is $\mathbf{NC}^2$, which consists of those problems which are solvable with polynomial size circuits of depth $O(\log^2)$ (in the size of the input).

It is known that $\mathbf{AC}^0[2] \subsetneq \mathbf{DET}(\mathrm{GF}(2)) \subseteq \mathbf{NC}^2 \subseteq \mathbf{PolyTime}$, and the separation between the first two complexity classes is the famous result of Razborov and Smolensky ([5, 6])). This separation will be instrumental in showing our independence result in the last section.

In this line of research we are motivated by a dual purpose: we want to understand the proof complexity of linear algebra, and we are also searching for good candidates for separating the Frege and extended Frege propositional proof systems. This separation is a central problem in theoretical computer science, and the theorems of universal linear algebra are considered to be good candidates to show such a separation – see [1] for more background on this quest.

In [8] we introduced the logical theory $\mathbf{LA} \subset \mathbf{LAP} \subset \exists\mathbf{LA}$ we gave the first feasible (i.e., using polynomial time concepts) proof of the CHT, a central theorem of matrix algebra from which many other universal theorems follow (in $\mathbf{LAP}$). Our proof was based on Berkowitz's algorithm, which is an efficient parallel algorithm for computing the characteristic polynomial of a matrix (and hence the inverse, adjoint, and determinant of a matrix). Berkowitz's algorithm is field independent (that is, it works over any field), and it can be formalized with $\mathbf{NC}^2$ circuits. Both Berkowitz's algorithm and Csanky's algorithm are $\mathbf{NC}^2$ algorithms, and have the following interesting relationship: if they could be shown to compute the same thing in $\mathbf{LAP}$, they could both be shown correct in $\mathbf{LAP}$. As things stand now, are best proofs of correctness for both are polytime.

In section 2 we describe the relevant theories, $\mathbf{LA}$, $\mathbf{LAP}$, $\mathbf{QLA}$, and $\exists\mathbf{LA}$. In section 3 we describe Csanky's and Berkowitz's algorithms, and show that they can be formalized in $\mathbf{LAP}$. In section 4 we show that the CHT follows in $\mathbf{LAP}$ from the principle of linear independence. This result is obtained using Csanky's algorithm, and so it requires fields of characteristic zero. In section 5 we show that five main principles of linear algebra can all be shown equivalent in $\mathbf{QLA}$, and furthermore, $\mathbf{QLA}$ does not prove any of them.

2   The Theories LA, LAP, ∃LA, and QLA

We define a quantifier-free theory of Linear Algebra (matrix algebra), and call it $\mathbf{LA}$. Our theory is strong enough to prove the ring properties of matrices such as $A(BC) = (AB)C$ and $A + B = B + A$ but weak enough so that all the theorems of $\mathbf{LA}$ (over finite fields or the field of rationals) translate into propositional tautologies with short Frege proofs.

Our theory has three sorts of object: *indices* (i.e., natural numbers), *field elements*, and *matrices*, where the corresponding variables are denoted $i, j, k, \ldots$; $a, b, c, \ldots$; and $A, B, C, \ldots$, respectively. The semantics assumes that objects of type field are from a fixed but arbitrary field, and objects of type matrix have entries from that field.

Terms and formulas are built from the function and predicate symbols:

$$0_{\text{index}}, 1_{\text{index}}, +_{\text{index}}, *_{\text{index}}, -_{\text{index}}, \text{div}, \text{rem}, 0_{\text{field}}, 1_{\text{field}},$$

$$+_{\text{field}}, *_{\text{field}}, -_{\text{field}}, {}^{-1}\text{r}, \text{c}, \text{e}, \Sigma, \leq_{\text{index}}, =_{\text{index}}, =_{\text{field}}, \tag{1}$$

$$=_{\text{matrix}}, \text{cond}_{\text{index}}, \text{cond}_{\text{field}}$$

The intended meanings should be clear, except for the following operations on a matrix A: $\text{r}(A), \text{c}(A)$ are the numbers of rows and columns in A, $\text{e}(A, i, j)$ is the field element A_{ij}, $\Sigma(A)$ is the sum of the elements in A. Also $\text{cond}(\alpha, t_1, t_2)$ is interpreted **if** α **then** t_1 **else** t_2, where α is a formula all of whose atomic subformulas have the form $m \leq n$ or $m = n$, where m, n are terms of type index, and t_1, t_2 are terms either both of type index or both of type field. The subscripts index and field are usually omitted, since they are clear from the context.

In addition to the usual rules for constructing terms we also allow the terms $\lambda ij\langle m, n, t\rangle$ of type matrix. Here i and j are variables of type index bound by the λ operator, intended to range over the rows and columns of the matrix. Here also m, n are terms of type index *not* containing i, j (representing the numbers of rows and columns of the matrix) and t is a term of type field (representing the matrix element in position (i, j)).

The λ terms allow us to construct the sum, product, transpose, etc., of matrices. For example, suppose first that A and B are $m \times n$ matrices. Then, $A + B$ can be defined as $\lambda ij\langle m, n, \text{e}(A, i, j) + \text{e}(B, i, j)\rangle$. Now suppose that A and B are $m \times p$ and $p \times n$ matrices, respectively. Then:

$$A * B := \lambda ij\langle m, n, \Sigma \lambda kl\langle p, 1, \text{e}(A, i, k) * \text{e}(B, k, j)\rangle\rangle$$

However, even if matrices are of incompatible size, their addition and product is well defined, since the "smaller" matrix is implicitly padded with zeros (as $\text{e}(A, i, j) = 0$ for i or j outside the range). Thus, all terms are well defined.

Atomic formulas and formulas are built in the usual manner, but in **LA** and **LAP** we only allow bounded index quantifiers (note that **LA**, respectively **LAP**, with bounded index quantifiers is conservative over **LA**, respectively **LAP**, without them).

We use Gentzen's sequent calculus LK (with quantifier rules omitted) for the underlying logic. We include 34 non-logical axioms in four groups: Axioms for equality, indices, field elements, and matrices (all quantifier-free). These specify the basic properties of the function and predicate symbols (1). By convention each instance of an axiom resulting from substituting terms for variables is also an axiom, so the axioms are really axiom schemes. All the axioms are given in [8].

We need an extra axiom to ensure that the underlying field is of characteristic zero. This can be stated with $\Sigma I_n \neq 0$, where I_n is the $n \times n$ identity matrix, which is given with a constructed term $\lambda ij \langle n, n, \text{cond}(i = j, 1, 0) \rangle$. This requirement is necessary for Csanky's algorithm which works only over fields of characteristic zero, as it performs divisions by integers.

We need just two non-logical rules: an equality rule for terms of type matrix, and the induction rule:

$$\frac{\Gamma, \alpha(i) \rightarrow \alpha(i+1), \Delta}{\Gamma, \alpha(0) \rightarrow \alpha(n), \Delta} \tag{2}$$

To formalize Newton's and Berkowitz's algorithms we extend the theory **LA** to the theory **LAP** by adding a new function symbol P, where $P(n, A)$ means A^n. We also add two new axioms, which give a recursive definition of P; namely, $P(0, A) = I$ and $P(n+1, A) = P(n, A) * A$. This is enough to formalize the coefficients of the characteristic polynomial of a matrix, as computed by either algorithm, as terms in the language of **LAP**. However, it seems that **LAP** is too weak to prove strong properties of the characteristic polynomial (such as the CHT or the multiplicativity of the determinant).

The theory $\exists$**LA** is an extension of **LA** where we allow induction over formulas of the form $(\exists X \leq t)\alpha$, where α has no quantifiers, and $\exists X \leq t$ is a bounded existential matrix quantifier ($X \leq t$ is just shorthand for $\mathbf{r}(X) \leq t \wedge \mathbf{c}(X) \leq t$). Note that the theory $\exists$**LAP**, defined analogously, is conservative over $\exists$**LA** because matrix powering (P) can be defined in $\exists$**LA**; so we don't really need to include P (see [10]).

Finally, **QLA** is **LA** with quantification over matrices, but induction restricted to formulas of **LA**.

This concludes a brief tour through the theories **LA**, **LAP**, $\exists$**LA**, and **QLA**. They are natural theories, in that they include what one would expect to formalize matrix algebra. **LA** is the weakest, and it can be thought off as the theory that proves the ring properties of matrices. **LAP** is **LA** together with the matrix powering function (and defining axioms), and it can formalize Csanky's and Berkowitz's algorithm, but it seems too weak to prove strong properties about them. $\exists$**LA** is **LA** together with an induction over formulas with bounded matrix quantifiers (which also allows it to simulate **LAP**).

3   Csanky's and Berkowitz's Algorithms

Both Csanky's and Berkowitz's algorithms compute the characteristic polynomial of a matrix, which is usually defined as $p_A(x) = \det(xI - A)$, for a given matrix A. Let p_A^{CSANKY} and p_A^{BERK} denote the coefficients of the characteristic polynomial of A given as column vectors, respectively. Let $p_A^{\text{CSANKY}}(x)$ and $p_A^{\text{BERK}}(x)$ denote the actual characteristic polynomials, with coefficients computed by the respective algorithms.

Newton's symmetric polynomials are defined as follows: $s_0 = 1$, and for $1 \leq k \leq n$, by:

$$s_k = \frac{1}{k} \sum_{i=1}^{k} (-1)^{i-1} s_{k-i} \text{tr}(A^i) \tag{3}$$

Then, $p_A^{\text{CSANKY}}(x) = s_0 x^n - s_1 x^{n-1} + s_2 x^{n-2} - \cdots \pm s_n x^0$. It is shown in the proof of lemma 1 how Csanky's algorithm computes the s_i's more efficiently (in $\mathbf{NC}^2$) than in the straightforward way suggested by the recurrence (3).

Lemma 1. p_A^{CSANKY} can be given as a term of $\mathbf{LAP}$.

Proof. We follow the ideas in [11, Section 13.4]. We restate (3) in matrix form: $s = Ts - b$ where s, T, b are given, respectively, as follows:

$$
\begin{pmatrix} s_1 \\ s_2 \\ \vdots \\ s_n \end{pmatrix}, \quad
\begin{pmatrix}
0 & 0 & 0 & \cdots \\
\frac{1}{2}\text{tr}(A) & 0 & 0 & \cdots \\
\frac{1}{3}\text{tr}(A^2) & \frac{1}{3}\text{tr}(A) & 0 & \cdots \\
\frac{1}{4}\text{tr}(A^3) & \frac{1}{4}\text{tr}(A^2) & \frac{1}{4}\text{tr}(A) & \cdots \\
\vdots & \vdots & \vdots & \ddots
\end{pmatrix}, \quad
\begin{pmatrix} \text{tr}(A) \\ \frac{1}{2}\text{tr}(A^2) \\ \vdots \\ \frac{1}{n}\text{tr}(A^n) \end{pmatrix}
$$

Then $s = -b(I - T)^{-1}$. Note that $(I - T)$ is an invertible matrix as it is lower triangular, with 1s on the main diagonal. The inverse of $(I-T)$ can be computed recursively using the following idea: if C is lower-triangular, with no zeros on the main diagonal, then

$$
C = \begin{pmatrix} C_1 & 0 \\ E & C_2 \end{pmatrix} \quad \Rightarrow \quad C^{-1} = \begin{pmatrix} C_1^{-1} & 0 \\ -C_2^{-1}EC_1^{-1} & C_2^{-1} \end{pmatrix}
$$

There are $O(\log(n))$ many steps and the whole procedure can be simulated with circuits of depth $O(\log^2(n))$ and size polynomial in n.

This, however, does not give us an $\mathbf{LAP}$-term, and it would be difficult to formalize the proof of correctness of this recursive inversion procedure in $\mathbf{LAP}$. Thus, instead of this recursive computation, we use the fact that the CHT can be proved correct in $\mathbf{LAP}$ for *triangular* matrices (see [7, Section 5.2]). From the characteristic polynomial of $(I - T)$ we obtain its inverse, and the inverse can be proved correct (i.e., $(I - T)(I - T)^{-1} = (I - T)^{-1}(I - T) = I$) using the the CHT for triangular matrices, and this can be formalized in $\mathbf{LAP}$.

Berkowitz's algorithm, just as Csanky's algorithm, allows us to reduce the computation of the characteristic polynomial to matrix powering. Its advantage is that it works over any field; however, certain properties (such as the fact that similar matrices have the same characteristic polynomial) have easy proofs in weak theories ($\mathbf{LAP}$) for Csanky's algorithm, but (seem to) require polytime theories ($\exists \mathbf{LA}$) for Berkowitz's algorithm.

Berkowitz's algorithm computes the characteristic polynomial of a matrix in terms of the characteristic polynomial of its principal minor:

$$
A = \begin{pmatrix} a_{11} & R \\ S & M \end{pmatrix} \tag{4}
$$

where R is an $1 \times (n-1)$ row matrix and S is a $(n-1) \times 1$ column matrix and M is $(n-1) \times (n-1)$. Let $p(x)$ and $q(x)$ be the characteristic polynomials of A and M respectively. Suppose that the coefficients of p form the column vector

$$p = \begin{pmatrix} p_n & p_{n-1} & \cdots & p_0 \end{pmatrix}^t \tag{5}$$

where p_i is the coefficient of x^i in $\det(xI - A)$, and similarly for q. Then:

$$p = C_1 q \tag{6}$$

where C_1 is an $(n+1) \times n$ Toeplitz lower triangular matrix (Toeplitz means that the values on each diagonal are constant) and where the entries in the first column are defined as follows: $c_{i1} = 1$ if $i = 1$, $c_{i1} = -a_{11}$ if $i = 2$, and $c_{i1} = -(RM^{i-3}S)$ if $i \geq 3$. Berkowitz's algorithm consists in repeating this for q, and continuing so that p is expressed as a product of matrices. Thus:

$$p_A^{\mathrm{BERK}} = C_1 C_2 \cdots C_n \tag{7}$$

where C_i is an $(n+2-i) \times (n+1-i)$ Toeplitz matrix defined as above except A is replaced by its i-th principal sub-matrix. Note that $C_n = (1 \quad -a_{nn})^t$.

Since each element of C_i can be explicitly defined in terms of A using matrix powering, and since the iterated matrix product can be reduced to matrix powering by a standard method, the entire product (7) can be expressed in terms of A using matrix powering. Thus the right-hand side of (7) can be expressed as a term in **LAP**.

Since we can define the characteristic polynomial in **LAP** (as p^{CSANKY} or p^{BERK}), it follows immediately that we can also define the determinant and the adjoint as terms of **LAP**.

4   Correctness of Csanky's Algorithm

The main result of this section, given as theorem 1, is the following:

$$\mathbf{QLAP} \vdash \text{Linear Independence} \supset \text{CHT} \tag{8}$$

where CHT (the Cayley-Hamilton Theorem) stands for $p_A(A) = p_A^{\mathrm{CSANKY}}(A) = 0$. Since $\exists \mathbf{LA}$ proves the principle of linear independence (see [10]), we have a new proof that $\exists \mathbf{LA}$ can prove the CHT. We assume that the characteristic polynomial of A, p_A, is computed with Csanky's algorithm, i.e., in this section $p_A = p_A^{\mathrm{CSANKY}}$.

Lemma 2. *LAP proves that similar matrices have the same characteristic polynomial; that is, if P is any invertible matrix, then $p_A = p_{PAP^{-1}}$.*

Proof. Observe that $\operatorname{tr}(AB) = \sum_i \sum_j a_{ij} b_{ji} = \sum_j \sum_i b_{ji} a_{ij} = \operatorname{tr}(BA)$, so using the associativity of matrix multiplication, $\operatorname{tr}(PA^iP^{-1}) = \operatorname{tr}(A^iPP^{-1}) = \operatorname{tr}(A^i)$. Inspecting (3), we see that a proof by induction on the s_i proves this lemma.

Lemma 3. *LAP proves that if A is a matrix of the form:*

$$\begin{pmatrix} B & 0 \\ C & D \end{pmatrix} \tag{9}$$

where B and D are square matrices (not necessarily of the same size), and the upper-right corner is zero, then $p_A(x) = p_B(x) \cdot p_D(x)$.

Proof. Let s_i^A, s_i^B, s_i^D be the coefficients of the characteristic polynomials (as given by (3)) of A, B, D, respectively. We want to show by induction on i that

$$s_i^A = \sum_{j+k=i} s_j^B s_k^D,$$

from which the claim of the lemma follows. The Basis Case: $s_0^A = s_0^B = s_0^D = 1$. For the Induction Step, by definition and by the induction hypothesis, we have that s_{i+1}^A equals

$$= \sum_{j=0}^{i} (-1)^j s_{i-j}^A \mathrm{tr}(A^{j+1}) = \sum_{j=0}^{i} (-1)^j \left[\sum_{p+q=i-j} s_p^B s_q^D \right] \mathrm{tr}(A^{j+1})$$

and by the form of A (i.e., (9)):

$$= \sum_{j=0}^{i} (-1)^j \left[\sum_{p+q=i-j} s_p^B s_q^D \right] (\mathrm{tr}(B^{j+1}) + \mathrm{tr}(D^{j+1}))$$

to see how this formula simplifies, we divide it into two parts:

$$= \sum_{j=0}^{i} (-1)^j \left[\sum_{p+q=i-j} s_p^B s_q^D \right] \mathrm{tr}(B^{j+1}) + \sum_{j=0}^{i} (-1)^j \left[\sum_{p+q=i-j} s_p^B s_q^D \right] \mathrm{tr}(D^{j+1}).$$

Consider first the left-hand side. When $q = 0$, p ranges over $\{i, i-1, \ldots, 0\}$, and $j+1$ ranges over $\{1, 2, \ldots, i+1\}$, and therefore, by definition, we obtain s_{i+1}^B. Similarly, when $q = 1$, we obtain s_i^B, and so on, until we obtain s_1^B. Hence we have:

$$= \sum_{j=0}^{i+1} s_{i-j}^B s_j^D + \sum_{j=0}^{i} (-1)^j \left[\sum_{p+q=i-j} s_p^B s_q^D \right] \mathrm{tr}(D^{j+1}).$$

The same reasoning, but fixing p instead of q on the right-hand side, gives us:

$$= \sum_{j=0}^{i+1} s_{i-j}^B s_j^D + \sum_{j=0}^{i+1} s_j^B s_{i-j}^D = \sum_{j+k=i+1} s_j^B s_k^D$$

which gives us the induction step and the proof of the lemma.

To show that $p_A(A) = 0$ it is sufficient to show that $p_A(A)e_i = 0$ for all vectors e_i in the standard basis $\{e_1, e_2, \ldots, e_n\}$. Let k be the largest integer such that

$$\{e_i, Ae_i, \ldots, A^{k-1}e_i\} \tag{10}$$

is linearly independent; we know that $k - 1 < n$, by the principle of linear independence (this is the first place where we use linear independence). Then, (10) is a basis for a subspace W of $\mathbb{F}^n$, and W is invariant under A, i.e., given any $w \in W$, $Aw \in W$.

Using Gaussian Elimination we write $A^k e_i$ as a linear combination of the vectors in (10). Using the coefficients of this linear combination we write a monic polynomial

$$g(x) = x^k + c_1 x^{k-1} + \cdots + c_k x^0 \tag{11}$$

such that $g(A)e_i = 0$.

Let A_W be A restricted to the basis (10), that is, A_W is a matrix representing the linear transformation $T_A : \mathbb{F}^n \longrightarrow \mathbb{F}^n$ induced by A, restricted to the subspace W. The matrix A_W^t has the following simple form:

$$\begin{pmatrix} 0 & 0 & 0 & \ldots & 0 & -c_k \\ 1 & 0 & 0 & \ldots & 0 & -c_{k-1} \\ 0 & 1 & 0 & \ldots & 0 & -c_{k-2} \\ \vdots & & & \ddots & & \vdots \\ 0 & 0 & 0 & \ldots & 1 & -c_1 \end{pmatrix} \tag{12}$$

i.e., it is the *companion matrix* of the polynomial $g(x)$. Since $p_A = p_{A^t}$, we consider the transpose of A_W, since A_W^t has the property that its principal submatrix is also a companion matrix, and that will be used in a proof by induction in the next lemma.

The proof of the next lemma is the crucial technical result of this section. The proof is given in the appendix.

Lemma 4. **LAP** *proves that the polynomial $g(x)$ is the characteristic polynomial of A_W, in other words, $g(x) = p_{A_W}(x)$.*

It is interesting to note that lemma 4 can also be proved (feasibly) for Berkowitz's algorithm instead, and the proof is in fact much simpler: consider again the matrix given by (12). We assume inductively that p_M^{BERK} (the characteristic polynomial of the principal submatrix of (12)) is given by $(1\ c_1\ c_2 \ldots c_{k-1})^t$. Since $R = (0 \ldots 0 -c_k)$ and $S = e_1$, $p_A^{\text{BERK}} = B \cdot p_M^{\text{BERK}}$, where B (the matrix given by Berkowitz's algorithm) is an $(n+1) \times n$ matrix with 1s on the main diagonal, 0s everywhere else, except for $+c_k$ in position $(n+1, 1)$. From this, it is easy to see that p_A^{BERK} is given by $(1\ c_1\ c_2 \ldots c_k)^t$.

As was pointed out in the introduction, if we managed to prove in **LAP** that Csanky's and Berkowitz's algorithms compute the same thing (i.e., $p^{\text{CSANKY}} = p^{\text{BERK}}$) we would have an **LAP** proof of the CHT for both. The reason is that the CHT follows for Berkowitz's algorithm from $\det(A) = \det(PAP^{-1})$, which is trivial to prove for Csanky's algorithm (see proof of Lemma 2).

Lemma 5. $\exists LA$ *proves that the polynomial $g(x)$ divides $p_A(x)$.*

Proof. Extend (10) to a full basis of $\mathbb{F}^n$:

$$B = \{e_i, Ae_i, \ldots, A^{k-1}e_i, e_{j_1}, e_{j_2}, \ldots, e_{j_{n-k}}\}.$$

This extension can be carried out easily with Gaussian Elimination, by checking which vectors from the standard basis ($\{e_1, e_2, \ldots, e_n\}$) are in the span consisting of (10) and those vectors that have already been added, and adding only those that are not. This is the only other place (besides the paragraph following the proof of lemma 3) where we need to use the principle of linear independence.

Let P be the change of basis for A from the standard basis to B. Then,

$$PAP^{-1} = \begin{pmatrix} A_W & 0 \\ * & E \end{pmatrix}$$

where A_W is a $k \times k$ block, and E is a $(n-k) \times (k-n)$ block (corresponding to the extension), and we have a block of zeros above E since W is invariant under A. By lemma 3 it follows that $p_A(x) = p_{PAP^{-1}}(x) = p_{A_W}(x) \cdot p_E(x)$. By lemma 4, $p_{A_W} = g(x)$, and so $g(x)$ divides $p_A(x)$.

Theorem 1. $QLAP$ *proves the Cayley-Hamilton Theorem (CHT) from the principle of linear independence, when the characteristic polynomial is computed by Csanky's algorithm.*

Proof. By lemma 5,

$$p_A(A)e_i = (p_{A_W}(A) \cdot p_E(A))e_i = (g(A) \cdot p_E(A))e_i = p_E(A) \cdot (g(A)e_i) = 0.$$

Since this is true for any e_i in the standard basis, it follows that $p_A(A) = 0$.

The proof of the multiplicativity of the determinant is a $\exists LA$ corollary of this theorem, as can be seen in [8]. Together, the CHT and the multiplicativity of the determinant, are two powerful universal principles of linear algebra from which many others follow directly. An important open question remains: are they provable in **LAP**?

5   Equivalence of Matrix Principles

Consider the following five central principles of linear algebra:

1. The Cayley-Hamilton Theorem
2. $(\exists B \neq 0)[AB = I \vee AB = 0]$
3. Linear Independence ($n+1$ vectors in $\mathbb{F}^n$ must be linearly dependent)
4. Weak Linear Independence (n^k vectors ($n, k > 1$) in $\mathbb{F}^n$ must be linearly dependent)
5. Every matrix has an annihilating polynomial

In this section we are going to show that **QLA** proves their equivalence. Furthermore, we show that these principles are *independent* of **QLA**. Thus, even though **QLA** is strong enough to show them equivalent, it is too weak to prove any of them.

Notice however that **QLA** does not have the matrix powering function, yet two of these principles, namely 1 and 5, require matrix powering to be stated. Let $\mathrm{POW}(A, n)$ be the formula:

$$\exists \langle X_0 X_1 \ldots X_n \rangle (\forall i \leq n)[X_0 = I \wedge (i < n \supset X_{i+1} = X_i * A)] \qquad (13)$$

The size of $\langle X_0 X_1 \ldots X_n \rangle$ can be bounded as it is a $\mathbf{r}(A) \times (\mathbf{r}(A) \cdot (n+1))$ matrix. (The abuse of notation in (13) is for better readability, but this formula can be stated formally as a bounded Σ_1 formula of **QLA**.)

Theorem 2. *The five principles of linear algebra can be proved equivalent in* **QLAP** *with* $POW(A, n)$.

Proof. 3 implies 1 because of the results of the previous section. Note that here we need fields of characteristic zero (because of Csanky's algorithm). It is an open question whether we can prove this over arbitrary fields – for example in the context of Berkowitz's algorithm.

1 implies 2 because B is just the adjoint, for which we have the desired properties from the Cayley-Hamilton Theorem.

2 implies 3, because suppose that we have $(n + 1)$ vectors in $\mathbb{F}^n$, and that they are linearly independent. Let A be the $n \times (n + 1)$ matrix whose columns are these $n + 1$ vectors. Let A' be the matrix resulting by appending a row of zeros to A. Since the vectors are linearly independent, there is no B such that $A'B = 0$, so by 2 there must be a B such that $A'B = I$; but that is not possible, given that the last row of A' is zero.

3 obviously implies 4.

4 implies 5 because we can look at $\{I, A, A^2, \ldots, A^{n^k}\}$, where A is $n \times n$, and k as large as we want, and as vectors these matrices are linearly dependent by 4.

5 implies 2, because if $p(A) = 0$, we can choose the largest s such that $p(A) = q(A)A^s$. If $q(A) \neq 0$, we choose the largest $k \leq s$ so that $q(A)A^k \neq 0$, and this is our zero divisor for A. If $q(A) = 0$, then it has a non-zero constant coefficient, and hence we can obtain from $q(A)$ the inverse for A.

Recall that the **Steinitz Exchange Theorem (SET)** says the following: if T is a (finite) total set for a vector space V, i.e., $\mathrm{span}(T) = V$, and E is a *linearly independent* set, then there exists an $F \subseteq T$, such that $|F| = |E|$, and $(T - F) \cup E$ is total. (Note that in general, SET is stated for any T, not necessarily finite, but here we assume that T is finite.)

We can state SET in the language of **QLA** as follows: associate the finite set T of m vectors in $\mathbb{F}^n$ with a $n \times m$ matrix T, and we can state that T is total with $(\exists A)[TA = I]$. Let E be a $n \times k$ matrix representing the k vectors in E. We want to find k column in T, and replace them by E. We can state that there exists a permutation matrix so that TP has those k columns as the last k

columns. Using the λ-constructor, we can "chop of" those last k columns, and replace them by E, and then state that the result is also total. Thus, SET can be stated in **QLA**.

Lemma 6. **QLA** *proves that the Steinitz Exchange Theorem implies the five principles listed at the beginning of this section.*

Proof. We show that SET implies (in **QLA**) the existence of an annihilating polynomial. Consider the set $E = \{I, A, A^2, A^3, \ldots, A^{n^2-1}\}$, where A is an $n \times n$ matrix. If E is linearly dependent, we are done: we have an annihilating polynomial. Otherwise, suppose that E is linearly independent.

Let $V = M_{n \times n}(\mathbb{F})$, that is V is the vector space of $n \times n$ matrices, over some field $\mathbb{F}$ (note that our argument is field independent). Let $T = \{e_{ij}\}_{1 \le i, j \le n}$, that is, T is the set of all elementary matrices e_{ij}, which are matrices with 1 in position (i, j), but zeros everywhere else. Note that $|T| = |E| = n^2$, and T is clearly total.

Therefore, by the Steinitz Exchange Theorem, $(T - F) \cup E$ is total for some $|F| = |E|$, and so E is total since $T = F$ if $|T| = |E| = n^2$. If E is total, then $A^{n^2} \in \text{span}(E)$, and hence $E \cup \{A^{n^2}\}$ is linearly dependent, and so we have an annihilating polynomial once again.

Can we show that the five principles, listed at the beginning of this section, prove (in **QLA**) the SET? Here is an obvious proof of SET: pick E_1 in E, and since T is total, we can write it as a linear combination of elements in T, say $E_1 = a_1 T_1 + a_2 T_2 + \cdots a_n T_n$, all $a_i \ne 0$. So, T_1 can be written as a sum of elements in $T - \{T_1\} \cup \{E_1\}$. So, put T_1 in F. Note that $T - \{T_1\} \cup \{E_1\}$ remains total. Now pick E_2, and write it as a linear combination of a finite subset of elements in $T - \{T_1\} \cup \{E_1\}$. By the assumed linear independence of E, E_2 cannot be written in terms of E_1 alone, so like before, we can pick some T_2 and put it in F. We proceed inductively, at each step putting some T_i in F.

The problem with the proof outlined above is that it requires induction over formulas with matrix quantifiers, which we do not have in **QLA** (on the other hand, this proof could be easily formalized in $\exists$**LA**). Thus we propose the following open problem: can SET be proved in **QLA** from the five principles? More generally: can Gaussian Elimination, properly stated, be shown correct in **QLA** from the five principles?

We conjecture that the answer is "yes" to those two questions, and that they are not too hard to prove.

Lemma 7. $\mathbf{QLA} \vdash (\exists B \ne 0)[AB = I \vee AB = 0] \supset POW(A, n).$

Proof. We use reduction of matrix powering to matrix inverse described in [3]. Let N be the $n^2 \times n^2$ matrix consisting of $n \times n$ blocks which are all zero except for $(n-1)$ copies of A above the diagonal zero blocks. Then $N^n = 0$, and $(I - N)^{-1} = I + N + N^2 + \ldots + N^{n-1} =$

$$\begin{pmatrix} I & A & A^2 & \dots & A^{n-1} \\ 0 & I & A & \dots & A^{n-2} \\ \vdots & & \ddots & & \vdots \\ 0 & 0 & 0 & \dots & I \end{pmatrix}.$$

Set $C = I - N$. Show that if $CB = 0$, then $B = 0$, using induction on the rows of B, starting with the bottom row. Using $(\exists B \neq 0)[CB = I \lor CB = 0]$, conclude that there is a B such that $CB = I$. Next, show that $B = I + N + N^2 + \cdots + N^{n-1}$, again, by induction on the rows of B, starting with the bottom row. Thus, B contains $I, A, A^2, \dots, A^{n-1}$ in its top rows, and $\mathrm{POW}(A, n)$ follows.

Thus, not every implication in theorem 2 requires $\mathrm{POW}(A, n)$. In particular, $2 \Leftrightarrow 3$ and $3 \Rightarrow 4$ can be shown in **QLA** (for $2 \Leftrightarrow 3$ see proof of corollary below). It is an open question whether 4 implies 3 in **QLA**.

Lemma 8. *QLA $\nvdash$ $\mathrm{POW}(A, n)$.*

Proof. We can turn **QLA** into a three-sorted universal theory in the style of **QPV** ([2]), by introducing function symbols for all the λ-terms, so we have number-valued functions, field-valued functions, and matrix valued-functions. Further, if the underlying field is $\mathrm{GF}(2)$, then all these functions are in the complexity class $\mathbf{AC}^0[2]$ (by translations given in [8]). Hence, by the Herbrand Theorem, every existential theorem of **QLA** can be witnessed by an $\mathbf{AC}^0[2]$ function.

Let $\mathbf{DET}(\mathrm{GF}(2))$ be the complexity class of functions $\mathbf{NC}^1$ reducible to the determinant over $\mathrm{GF}(2)$. This class is equal to the class $\mathbf{POW}(\mathrm{GF}(2))$, by results in [3]. On the other hand, $\mathbf{AC}^0[2]$ is properly contained in $\mathbf{DET}(\mathrm{GF}(2))$, since $\mathbf{L} \subseteq \mathbf{DET}(\mathrm{GF}(2))$ (see [4]), while $\mathrm{MAJORITY} \in \mathbf{L}$ but it is *not* in $\mathbf{AC}^0[2]$ (see [5, 6]).

Corollary 1. *QLA does not prove the principles 2 and 3 (while it can show them equivalent without $\mathrm{POW}(A, n)$).*

Proof. By lemmas 7 and 8 we see that **QLA** does not prove 2. Now, 3 implies 2 by the following argument: take A and add e_i (the elementary column vector with 1 in the i-th entry, and zeros everywhere else) as the last column. By linear independence, we know that there exist $b_{1i}, b_{2i}, \dots, b_{(n+1)i}$, not all zero, such that $b_{1i}A_1 + b_{2i}A_2 + \cdots b_{ni}A_n + b_{(n+1)i}e_i = 0$, where A_i is the i-th column of A. If for all i, $b_{(n+1)i}$ is not zero, we found B such that $AB = I$. If, on the other hand, some $b_{(n+1)i} = 0$, then B consisting of columns given by $[b_{1i}b_{2i} \dots b_{ni}]^t$ is a zero divisor of A, i.e., $AB = 0$.

6   Conclusions and Open Problems

We gave a new feasible proof of the Cayley-Hamilton Theorem via Csanky's algorithm. The new proof requires fields of characteristic zero, but it shows that the CHT follows in **LAP** from the principle of linear independence. It is an open question whether the CHT follows in **LAP** from the principle of linear independence over general fields.

We showed that five important principles of linear algebra can be shown equivalent in **QLA**, and using a previously known separation of complexity classes (namely $\mathbf{AC}^0[2] \subsetneq \mathbf{DET}(\mathrm{GF}(2))$) we showed that none of these principles is provable in **QLA**.

It is an interesting open problem whether the principles listed in theorem 2 can be proved in $\mathbf{QLA} + \mathrm{POW}(A, n)$. Likewise, it is an open problem whether Berkowitz's and Csanky's algorithm are provable correct in **LAP** (they can be stated in **LAP**, and weak properties of correctness are provable in **LAP**).

Acknowledgments

The author would like to thank Stephen Cook for pointing out the proof of the Cayley-Hamilton Theorem in [9], which is the basis for the proof in section 4. The material in section 5 came from discussions with Mark Braverman and Stephen Cook. Finally, the author is grateful to the anonymous referees, especially to the referee who succinctly and elegantly expressed the contribution of this paper (see the first sentence of the introduction).

References

1. M. Bonet, S. Buss, and T. Pitassi. Are there hard examples for Frege systems? *Feasible Mathematics*, II:30–56, 1994.
2. Stephen Cook and Alasdair Urquhart. Functional interpretations of feasible constructive arithmetic. *Annals of Pure and Applied Logic*, 63:103–200, 1993.
3. Stephen A. Cook. A taxonomy of problems with fast parallel algorithms. *Information and Computation*, 64(13):2–22, 1985.
4. Erich Grädel. Capturing complexity classes by fragments of second-order logic. *Theoretical Computer Science*, 101(1):35–57, 1992.
5. A. A. Razborov. Lower bounds on the size of bounded depth networks over a complete basis with logical addition. *Mathematicheskie Zametki*, 41:598–607, 1987. English translation in Mathematical Notes of the Academy of Sciences of the USSR 41:333-338, 1987.
6. R. Smolensky. Algebraic methods in the theory of lower bounds for Boolean circuit complexity. In *Proceedings, 19th ACM Symposium on Theory of Computing*, pages 77–82, 1987.
7. Michael Soltys. *The Complexity of Derivations of Matrix Identities*. PhD thesis, University of Toronto, 2001.
8. Michael Soltys and Stephen Cook. The complexity of derivations of matrix identities. *Annals of Pure and Applied Logic*, 130:277–323, 2004.
9. Lawrence E. Spence, Arnold J. Insel, and Stephen H. Friedberg. *Elementary Linear Algebra: A Matrix Approach*. Prentice Hall, 1999.
10. Neil Thapen and Michael Soltys. Weak theories of linear algebra. *Archive for Mathematical Logic*, 44(2):195–208, 2005.
11. Joachim von zur Gathen. Parallel linear algebra. In John H. Reif, editor, *Synthesis of Parallel Algorithms*, pages 574–617. Morgan and Kaufman, 1993.

Appendix

Proof (lemma 4). We will drop the W from A_W as there is no danger of confusion (the original matrix A does not appear in the proof); thus, A is a $k \times k$ matrix, with 1s below the main diagonal, and zeros everywhere else except (possibly) in the last column where it has the negations of the coefficients of $g(x)$.

As was noted above, A is divided into four quadrants, with the upper-left containing just 0. Let $R = (0 \ldots 0 \ -c_k)$ be the row vector in the upper-right quadrant. Let $S = e_1$ be the column vector in the lower-left quadrant, i.e., the first column of A without the top entry. Finally, let M be the principal submatrix of A, $M = A[1|1]$; the lower-right quadrant.

Let $s_0, s_1, \ldots, s_k$ be the Newton's symmetric polynomials of A.

To prove that $g(x) = p_{A_{T_W}}(x)$ we prove something stronger: we show that (i) for all $0 \le i \le k$ $(-1)^i s_i = c_i$, and (ii) $p_A(A) = 0$.

We show this by induction on the size of the matrix A. Since the principal submatrix of A (i.e., M) is *also* a companion matrix, we assume that for $i < k$, the coefficients of the symmetric polynomial of M are equal to the c_i's, and that $p_M(M) = 0$. (Note that the Basis Case of the induction is a 1×1 matrix, and it is trivial to prove.)

Since for $i < k$, $\text{tr}(A^i) = \text{tr}(M^i)$, it follows from (3) and the induction hypothesis that for $i < k$, $(-1)^i s_i = c_i$ (note that $s_0 = c_0 = 1$).

Next we show that $(-1)^k s_k = c_k$. By definition (i.e., by (3)) we have that s_k is equal to:

$$\frac{1}{k}(s_{k-1}\text{tr}(A) - s_{k-2}\text{tr}(A^2) + \cdots + (-1)^{k-2}s_1\text{tr}(A^{k-1}) + (-1)^{k-1}s_0\text{tr}(A^k))$$

and by the induction hypothesis and the fact that for $i < k$ $\text{tr}(A^i) = \text{tr}(M^i)$ we have:

$$= \frac{1}{k}(-1)^{k-1}(c_{k-1}\text{tr}(M) + c_{k-2}\text{tr}(M^2) + \cdots + c_1\text{tr}(M^{k-1}) + c_0\text{tr}(A^k)).$$

Note that $\text{tr}(A^k) = -kc_k + \text{tr}(M^k)$, so:

$$= \frac{1}{k}(-1)^{k-1}\left[c_{k-1}\text{tr}(M) + c_{k-2}\text{tr}(M^2) + \cdots + c_1\text{tr}(M^{k-1}) + c_0\text{tr}(M^k)\right]$$
$$+ (-1)^k c_k$$

Observe that

$$\text{tr}(c_{k-1}M + c_{k-2}M^2 + \cdots + c_1 M^{k-1} + c_0 M^k) = \text{tr}(p_M(M)M) = \text{tr}(0) = 0$$

since $p_M(M) = 0$ by the induction hypothesis. Therefore, $s_k = (-1)^k c_k$.

It remains to prove that $p_A(A) = \sum_{i=0}^k c_i A^{k-i} = 0$. First, show that for $1 \le i \le (k-1)$:

$$A^{i+1} = \left(\begin{array}{c|c} 0 & RM^i \\ \hline M^i S & \sum_{j=0}^{i-1} M^j SRM^{(i-1)-j} + M^{i+1} \end{array} \right) \tag{14}$$

(For A of the form given by (12), and R, S, M defined as in the first paragraph of the proof.) Define w_i, X_i, Y_i, Z_i as follows:

$$
\begin{aligned}
A^{i+1} &= \begin{pmatrix} w_{i+1} & X_{i+1} \\ Y_{i+1} & Z_{i+1} \end{pmatrix} = \begin{pmatrix} w_i & X_i \\ Y_i & Z_i \end{pmatrix} \begin{pmatrix} 0 & R \\ S & M \end{pmatrix} \\
&= \begin{pmatrix} X_i S & w_i R + X_i M \\ Z_i S & Y_i R + Z_i M \end{pmatrix}
\end{aligned} \tag{15}
$$

We want to show that the right-most matrix of (15) is equal to the right-hand side of (14). First note that:

$$X_{i+1} = \sum_{j=0}^{i} w_{i-j} RM^j \qquad w_{i+1} = \sum_{j=0}^{i-1} (RM^j S) w_{i-1-j} \tag{16}$$

With the convention that $w_0 = 1$. See [8, lemma 5.1] for an **LAP**-proof of (16). Since $w_1 = 0$, a straight-forward induction shows that $w_{i+1} = 0$. Therefore, at this point the right-most matrix of (15) can be simplified to:

$$\begin{pmatrix} 0 & RM^i \\ Z_i S & Y_i R + Z_i M \end{pmatrix}$$

Again by [8, lemma 5.1] we have:

$$Y_{i+1} = M^i S + \sum_{j=0}^{i-2} (RM^j S) Y_{i-1-j} \qquad Z_{i+1} = M^{i+1} + \sum_{j=0}^{i-1} Y_{i-1-j} RM^j$$

By the same reasoning as above, $\sum_{j=0}^{i-2} (RM^j S) Y_{i-1-j} = 0$, so putting it all together we obtain the right-hand side of (14).

Using the induction hypothesis $(p_M(M) = 0)$ it is easy to show that the first row and column of $p_A(A)$ are zero. Also, by the induction hypothesis, the term M^{i+1} in the principal submatrix of $p_A(A)$ disappears but leaves $c_k I$. Therefore, it will follow that $p_A(A) = 0$ if we show that

$$\sum_{i=2}^{k} c_{k-i} \sum_{j=0}^{i-2} M^j SRM^{(i-2)-j} \tag{17}$$

is equal to $-c_k I$.

Some observations about (17): for $0 \le j \le i - 2 \le k - 2$, the first column of M^j is just e_{j+1}. And SR is a matrix of zeros, with $-c_k$ in the upper-right corner. Thus $M^j SR$ is a matrix of zeros except for the last column which is $-c_k e_{j+1}$. Thus, $M^j SRM^{(i-2)-j}$ is a matrix with zeros everywhere, except in row $(j+1)$

where it has the bottom row of $M^{(i-2)-j}$ multiplied by $-c_k$. Let $\mathbf{m}^{(i-2)-j}$ denote the $1 \times (k-1)$ row vector consisting of the bottom row of $M^{(i-2)-j}$. Therefore, (17) is equal to:

$$
- c_k \cdot \left(
\begin{array}{c}
\sum_{i=2}^{k} c_{k-i}\mathbf{m}^{(i-2)} \\
\hline
\sum_{i=3}^{k} c_{k-i}\mathbf{m}^{(i-3)} \\
\hline
\vdots \\
\hline
\sum_{i=k}^{k} c_{k-i}\mathbf{m}^{(i-k)}
\end{array}
\right)
\tag{18}
$$

We want to show that (18) is equal to $-c_k I$ to finish the proof of $p_A(A) = 0$. To accomplish this, let l denote the l-th row of the matrix in (18) starting with the bottom row. We want to show, by induction on l, that the l-th row is equal to e_{k-l}.

The Basis Case is $l = 0$:

$$
\sum_{i=k}^{k} c_{k-i}\mathbf{m}^{(i-k)} = c_0\mathbf{m}^0 = e_k,
$$

and we are done.

For the induction step, note that $\mathbf{m}^{l+1}$ is equal to $\mathbf{m}^l$ shifted to the left by one position, and with

$$
\mathbf{m}^l \cdot (-c_{k-1} \; -c_{k-2} \; \cdots \; -c_1)^t
\tag{19}
$$

in the last position. We introduce some more notation: let $\mathbf{r}_l$ denote the $k-l$ row of (18). Thus $\mathbf{r}_l$ is $1 \times (k-1)$ row vector. Let $\overleftarrow{\mathbf{r}}_l$ denote $\mathbf{r}_l$ shifted by one position to the left, and with a zero in the last position. This can be stated succinctly in **LAP** as follows:

$$
\overleftarrow{\mathbf{r}}_l \overset{\text{def}}{=} \lambda ij\langle 1, (k-1), e(\mathbf{r}_l, 1, i+1)\rangle.
$$

Based on (18) and (19) we can see that:

$$
\mathbf{r}_{l+1} = \overleftarrow{\mathbf{r}}_l + [\mathbf{r}_l \cdot (-c_{k-1} \; -c_{k-2} \; \cdots \; -c_1)^t]e_k + c_l\mathbf{m}^0.
$$

(Here the "·" in the square brackets denotes the dot product of the two vectors.) Using the induction hypothesis: $\overleftarrow{\mathbf{r}}_l = e_{k-(l+1)}$, and

$$
\mathbf{r}_l \cdot (-c_{k-1} \; -c_{k-2} \; \cdots \; -c_1)^t = e_{k-l} \cdot (-c_{k-1} \; -c_{k-2} \; \cdots \; -c_1)^t = -c_l
$$

so $\mathbf{r}_{l+1} = e_{k-l} - c_l e_k + c_l e_k = e_{k-(l+1)}$ as desired. This finishes the proof of the fact that the matrix in (18) is the identity matrix, which in turn proves that (17) is equal to $-c_k I$, and this ends the proof of $p_A(A) = 0$, which finally finishes the main induction argument, and proves the lemma.

A Propositional Proof System for Log Space

Steven Perron

Department of Computer Science, University of Toronto
sperron@cs.toronto.edu

Abstract. The proof system G_0^* of the quantified propositional calculus corresponds to NC^1, and G_1^* corresponds to P, but no formula-based proof system that corresponds log space reasoning has ever been developed. This paper does this by developing GL^*.

We begin by defining a class $\Sigma CNF(2)$ of quantified formulas that can be evaluated in log space. Then GL^* is defined as G_1^* with cuts restricted to $\Sigma CNF(2)$ formulas and no cut formula that is not quantifier free contains a non-parameter free variable.

To show that GL^* is strong enough to capture log space reasoning, we translate theorems of Σ_0^B-rec into a family of tautologies that have polynomial size GL^* proofs. Σ_0^B-rec is a theory of bounded arithmetic that is known to correspond to log space. To do the translation, we find an appropriate axiomatization of Σ_0^B-rec, and put Σ_0^B-rec proofs into a new normal form.

To show that GL^* is not too strong, we prove the soundness of GL^* in such a way that it can be formalized in Σ_0^B-rec. This is done by giving a log space algorithm that witnesses GL^* proofs.

1 Introduction

Recently there has been lots of research looking into the connection between computational complexity, bounded arithmetic, and propositional proof complexity. A recent survey on this topic can be found at [1]. In this paper, we give a method of restricting the proof system G^* to get a proof system GL^*, which corresponds to log space. The proof system G^* is a tree-like proof system for the quantified propositional calculus based on Gentzen's LK [2]. This affirms the belief of Cook that there exists a formula-based proof system that corresponds to log space [1]. Before this the only proof system for log space was based on liar games [3], and it has never been well developed.

The definition of GL^* is similar to the definition of G_i^*: it is obtain from G^* by restricting cuts. We restrict cuts to the class of formulas $\Sigma CNF(2)$. This class of formulas is closely related to $CNF(2)$ formulas. These are CNF formulas where no variable appears more than twice.

Another attempt to capture reasoning between NC^1 and P resorted to putting a bound on the depth of the proof and the number of cuts along a branch [4], but it is not obvious how to capture other complexity classes using this type of restriction. In contrast to that, it seems plausible that a proof system for NL could be defined in the same spirit as GL^*.

L. Ong (Ed.): CSL 2005, LNCS 3634, pp. 509–524, 2005.

The rest of the paper is organized as follows. In the next section, we define the $\Sigma CNF(2)$ formulas and the proof system GL^*, and we give the reason for the restrictions on the cut formulas. Section 3 is devoted to translating theorems of Σ_0^B-rec into tautologies with polynomial size GL^* proofs. Σ_0^B-rec is a theory of bounded arithmetic introduced by Zambella [12] that is known to correspond to log space [1]. This proves GL^* is strong enough to capture log space reasoning. In section 4, we prove in Σ_0^B-rec that GL^* is sound. This is sometimes called the reflection principle. This tells us that GL^* does not capture reasoning for a higher complexity class.

This paper is based on my Masters thesis [9], where more details can be found.

2   The Proof System

The proof system PK is the Gentzen-style sequent calculus for propositional logic [5, 6]. The initial sequents are $\bot \to$, $\to \top$, and $A \to A$, for any propositional formula A. The rules of inference include structural rules, which are weakening, contraction, and exchange; propositional rules, which are used to add propositional connective to formulas in the sequents; and the cut rule, which infers $\Gamma \to \Delta$ from $A, \Gamma \to \Delta$ and $\Gamma \to \Delta, A$. We can then extend PK to the proof system G by adding rules for quantifiers over propositional variables [7]. Anytime you deal with quantifiers, you will have problems with the substituting terms or, in this case, formulas for variables. To help avoid these problems, we use the following convention.

Notation 1 *In propositional formulas, all bound variables will be z, $z_1, z_2, \ldots$ and will be called z-variables. The free variables will be x, $x_1, x_2 \ldots$ and will be called x-variables.*

This way we know, x-variables are never quantified. The quantifier rules are

$$\exists\text{-left} \ \frac{A(x), \Gamma \to \Delta}{\exists z A(z), \Gamma \to \Delta} \qquad \exists\text{-right} \ \frac{\Gamma \to \Delta, A(B)}{\Gamma \to \Delta, \exists z A(z)}$$

$$\forall\text{-left} \ \frac{\Gamma \to \Delta, A(x)}{\Gamma \to \Delta, \forall z A(z)} \qquad \forall\text{-right} \ \frac{A(B), \Gamma \to \Delta}{\forall z A(z), \Gamma \to \Delta}$$

where x does not appear in the bottom sequent of the $\exists$-left and $\forall$-right rules, and B is a Σ_0^q formula that does not mention any z-variable. The Σ_0^q formulas are propositional formulas that do not contain any quantifiers. In general, the Σ_i^q formulas are quantified propositional formulas in prenex form with at most $i - 1$ quantifier alternations, beginning with $\exists$ on the outside. In these rules, x is called the *eigenvariable* and B is called the *target formula*. The proof system G^* is G restricted to tree-like proofs. The fragment G_i is G with cuts restricted to Σ_i^q formulas, and G_i^* is G_i restricted to tree-like proofs [7].

These proof systems have been extensively studied. In [2], Krajicek and Pudlak showed a connection between G_i and Buss's theories T_2^i for $i > 0$. This result was shown for a different definition of G_i, but the result still holds [7]. Later a connection between G_i^* and S_2^i [8] was found. With the connection between these theories and the polynomial-time hierarchy, we get an indirect connection between the polynomial-time hierarchy and the proof systems. In particular, G_1^* is connected to P. Later the proof system G_0^* was shown to be directly connected to NC^1, and connections to the theory VNC^1 were also considered. Since our proof system is going to correspond to log space, it must be between G_0^* and G_1^*; the cut formulas must be some subset of Σ_1^q formulas. This subset is the $\Sigma CNF(2)$ formulas.

Definition 1 *The set of formulas $\Sigma CNF(2)$ is the smallest set*

1. *containing Σ_0^q,*
2. *containing every formula $\exists z, \phi(z, x)$ where (1) ϕ is a quantifier-free CNF formula $\bigwedge_{i=1}^{m} C_i$ and (2) existence of a z-literal l in C_i and C_j, $i \neq j$, implies existence of an x-variable x such that $x \in C_i$ and $\neg x \in C_j$ or vice versa, and*
3. *closed under substitution of Σ_0^q formulas that contain only x-variables for x-variables.*

Then, GL^* is defined as follows.

Definition 2 *Given a proof of a sequent $\Gamma \to \Delta$, a variable is a parameter variable if it appears free in $\Gamma \to \Delta$.*

Definition 3 *GL^* is the propositional proof system G^* with cuts restricted to $\Sigma CNF(2)$ formulas in which no cut formula that is not Σ_0^q contains a nonparameter free variable.*

We chose this class of formulas because we are able to evaluate $\Sigma CNF(2)$ formulas, given an assignment to the free variables. Moreover, this problem is log space complete.

Lemma 1 ([9], Lemma 4.2.2). *Evaluating $\Sigma CNF(2)$ formulas is log space complete.*

This is an easy corollary to a theorem in [10], where Johannsen proved that the satisfiability problem for $CNF(2)$ formulas, $SAT(2)$, is log space complete. A $CNF(2)$ formula is a CNF formula where no variable appears more than twice. In Section 4, we will show how to find the assignment to the quantified variables in log space that satisfies the formula whenever possible.

The restriction on the free variables in cut formulas might seem unnatural, but it is necessary. Let H^* be the proof system G^* with cuts restricted to $\Sigma CNF(2)$ formulas and no restriction on the free variables. At first, we tried to show that H^* captured log space reasoning, but we found that it p-simulates G_1^* proofs of Σ_1^q formulas.

At a high level, H^* proves there exists an output to a circuit by proving that there exists an output to gate i, given the values for the inputs and the output of the first $i - 1$ gates. This can be expressed as a $\Sigma CNF(2)$ formula. Then, with repeated cutting, we can prove there exists an output to the ciruit given the values of the inputs. Since H^* allows non-parameter variables in the cut formulas, the cute can be done. This is a problem since determining the output of a circuit is P-complete. Using this idea, we are able to prove the following theorem.

Theorem 1. H^* *p-simulates* G_1^* *for* Σ_1^q *formulas.*

A complete proof can be found in [9].

3   Propositional Translations

This section is motivated by results in [11]. In that paper, Cook showed that theorems of the equational theory PV can be translated into a family of tautologies that have polynomial size extended Frege proofs. We will show that theorems of Σ_0^B-rec can be translated into a family of tautologies that have polynomial size GL^* proofs. This tells us that GL^* captures log space reasoning in the same way extended Frege captures polynomial time reasoning.

In this paper, the theories will be two sorted. Numbers are the first sort. We use a, b, x, y, z as number variables, and they are intended to range over the natural numbers. Binary strings (or finite sets of numbers) are the second sort. We use X, Y, Z as string variables, and they are intended to be strings of 0's and 1's, with leading 0's removed.

The standard language is $\mathcal{L}_A^2 = [0, 1, +, \cdot, ||; =_1, =_2, ()]$. The constants, 0 and 1, and the binary functions, $+$ and $\cdot$, have their usual meaning. The final function $|X|$ returns to size of the string X (or the least upper bound of the finite set X). The predicates $=_1$ and $=_2$ are equality between numbers and string, respectively. The predicate $\leq$ is the usual inequality between two numbers. The final predicate is used to access the bits of X. So, $X(b)$ is true if the bth bit of X is 1, and false otherwise.

In two sorted bounded arithmetic, strings are the important sort. This is why we define classes of formula based on string quantifiers. The class Σ_0^B of formulas is the set of $\mathcal{L}_A^2$ formulas with no string quantifiers and all number quantifiers are bounded. That is, the number quantifiers are of the form $\exists x \leq b$. The class Σ_1^B of formulas is the class of formulas with an initial block of bounded existential string quantifiers followed by a Σ_0^B formula. A bounded string quantifier is of the form $\exists Z \leq b\phi$, which means $\exists Z, |Z| \leq b \wedge \phi$. The class Σ_1^1 is the same as Σ_1^B except the string quantifiers do not have to be bounded.

Given a Σ_1^B formula $\phi(x, X)$ over the language $\mathcal{L}_A^2$, we want to translate it into a family of propositional formulas $||\phi(x, X)||[n]$, where the sizes of the formulas are bounded by a polynomial in n and the values assigned to x. We use the translation described in [1, 7]. It is a modification of the Paris-Wilkie translation (see [1]).

The first step is to substitute constant values for all the number variables x. For now we assume ϕ has no free number variables. The formula $||\phi(\boldsymbol{X})||[\boldsymbol{n}]$ is meant to be a formula that says $\phi(\boldsymbol{X})$ is true whenever $|X_i| = n_i$ for every string in $\boldsymbol{X}$ and the number variables are equal to the constants that replaced them. Then if $\phi(\boldsymbol{X})$ is true for all $\boldsymbol{X}$, then $||\phi(\boldsymbol{X})||[\boldsymbol{n}]$ is a tautology for all values for $\boldsymbol{n}$. Note that any term t that appears in ϕ can be evaluated immediately. This is because there are no number variables and the size of each string variable is known. So we will let $val(t(\boldsymbol{n}))$ be the value of the term. The variables $\boldsymbol{n}$ will often be omitted since they are understood. The free variables in the propositional formula will be $p_j^{X_i}$ for $j < n_i - 1$. The variable $p_j^{X_i}$ is meant to represent the value of the jth bit of X_i; we know that the n_ith bit is 1, and for $j > n_i$, we know the jth bit is 0. The definition proceeds by structural induction on ϕ.

Suppose ϕ is an atomic formula. Then it has one of the following forms: $s = t$, $s < t$, $X_i(t)$, or one of the trivial formulas $\bot$ and $\top$, for terms s and t. In the first case, we define $||\phi(\boldsymbol{X})||[\boldsymbol{n}]$ as the formula $\top$, if $val(s) = val(t)$, and $\bot$, otherwise. A similar construction is done for $s < t$. If ϕ is one of the trivial formulas, then $||\phi(\boldsymbol{X})||[\boldsymbol{n}]$ is the same trivial formula. So now suppose $\phi(\boldsymbol{X}) =_{syn} X_i(t)$. Let $j = val(t)$. Then the translation is defined as follows:

$$||\phi(\boldsymbol{X})||[\boldsymbol{n}] =_{syn} \begin{cases} p_j^{X_i} & \text{if } j < n_i - 1 \\ \top & \text{if } j = n_i - 1 \\ \bot & \text{if } j > n_i - 1 \end{cases}$$

Now for the inductive part of the definition. Suppose $\phi =_{syn} \alpha \wedge \beta$. Then

$$||\phi(\boldsymbol{X})||[\boldsymbol{n}] =_{syn} ||\alpha(\boldsymbol{X})||[\boldsymbol{n}] \wedge ||\beta(\boldsymbol{X})||[\boldsymbol{n}].$$

When the connective is $\vee$ or $\neg$, the definition is similar. Let $j = val(t)$. If the outer most connective is a quantifier, then the translation is defined as

$$||\exists y \le t, \alpha(y, \boldsymbol{X})||[\boldsymbol{n}] =_{syn} \bigvee_{i=0}^{j} ||\alpha(i, \boldsymbol{X})||[\boldsymbol{n}]$$

$$||\forall y \le t, \alpha(y, \boldsymbol{X})||[\boldsymbol{n}] =_{syn} \bigwedge_{i=0}^{j} ||\alpha(i, \boldsymbol{X})||[\boldsymbol{n}]$$

$$||\exists Y \le t, \alpha(Y, \boldsymbol{X})||[\boldsymbol{n}] =_{syn} \exists p_0^Y, \dots, \exists p_{j-2}^Y, \bigvee_{i=0}^{j} ||\alpha(Y, \boldsymbol{X})||[i, \boldsymbol{n}]$$

This completes the definition. Note that $\exists Y \le b$ means there exists a string with size at most b.

3.1   The Theory VL'

There have been a number of theories that capture log space reasoning. We will use Σ_0^B-rec [12]. The theory Σ_0^B-rec is axiomatized by the axioms of V^0

(explained below) plus the X-rec axiom below. This formula says that, if every vertex in the graph with nodes $\{0, \ldots, a\}$ and edge relation $X(i, j)$ has out-degree at least 1, then there exists a path of length b.

$$[\forall x \leq a \exists y \leq a X(x, y)] \supset \exists Z, \forall w \leq b \exists! x \leq a Z(w, x)$$
$$\wedge \forall w < b \forall x \leq a \forall y \leq a[Z(w, x) \wedge Z(w + 1, y) \supset X(x, y)]. \quad (X\text{-rec})$$

In this formula, $\exists! x \phi$ means "there exists a unique x such that ϕ is true." The theory V^0 has axioms that define addition, multiplication, the inequalities, and the size of a string. In addition, V^0 has the Σ_0^B-COMP axiom:

$$\exists Z \leq a \forall i < a, Z(i) \iff \phi(i),$$

where ϕ is a Σ_0^B formula that does not mention Z. V^0 is known to correspond to AC^0. See [1, 6] for details on V^0.

We know Σ_0^B-rec corresponds to log space because of the Σ_1^1-definability theorem below.

Theorem 2. *A function is Σ_1^1-definable in Σ_0^B-rec if and only if it is a log space function.*

See [9, 12] for a proof.

We want to reformulate the axioms of Σ_0^B-rec so they translate into $\Sigma CNF(2)$ formulas. With the exception of Σ_0^B-COMP, the axioms of V^0 are Σ_0^B, so they translate into Σ^q formulas, which are $\Sigma CNF(2)$. This means we only need to consider Σ_0^B-COMP and X-rec. We are not going to worry about Σ_0^B-COMP; we will handle this axiom the same way Cook and Morioka did in [7]. That is, if the proof system is asked to cut the translation of an instance of the Σ_0^B-COMP axiom, then the propositional proof is changed so that the cut becomes $\bigwedge_{i=0}^t |||\phi(i)|| \iff ||\phi(i)|||$, which is $\Sigma CNF(2)$. To take care of X-rec, we will define a new theory that is equivalent to Σ_0^B-rec by replacing the X-rec axiom.

Let Σ_0^B-edge-rec be the axiom scheme

$$\exists Z \leq 1 + \langle b, a, a \rangle [\rho_1 \wedge \rho_2 \wedge \rho_3 \wedge \rho_4 \wedge \rho_5 \wedge \rho_6 \wedge \rho_7 \wedge \rho_8],$$

where

$$\rho_1 =_{syn} \forall j < a, \neg Z(0, 0, j) \vee \phi(0, j) \vee \exists l < j \phi(0, l))$$
$$\rho_2 =_{syn} \forall j \leq a \forall k < j, \neg Z(0, 0, j) \vee \neg \phi(0, k) \vee \exists l < k \phi(0, l))$$
$$\rho_3 =_{syn} \forall i \leq a \forall j \leq a, i = 0 \vee \neg Z(0, i, j)$$
$$\rho_4 =_{syn} \forall w < b \forall i \leq a \forall j \leq a, \neg Z(w + 1, i, j)$$
$$\vee \exists h \leq a Z(w, h, i) \vee \neg \phi(i, j) \vee \exists l < j \phi(i, l)$$
$$\rho_5 =_{syn} \forall w < b \forall i \leq a \forall j < a, \neg Z(w + 1, i, j) \vee \phi(i, j) \vee \exists l < j \phi(i, l)$$
$$\rho_6 =_{syn} \forall w < b \forall i \leq a \forall j \leq a \forall k < j, \neg Z(w + 1, i, j) \vee \neg \phi(i, k) \vee \exists l < k \phi(i, l)$$
$$\rho_7 =_{syn} \exists i \leq a \exists j \leq a, Z(b, i, j)$$
$$\rho_8 =_{syn} \forall \langle w, i, j \rangle \leq \langle b, a, a \rangle, [w > b \vee i > a \vee j > a] \supset \neg Z(w, i, j)$$

where $\phi(i, j)$ is a Σ_0^B formula that does not mention Z, but may have other free variables. Informally this axiom says there exists a string Z that gives a pseudo-path of length b in the graph with a nodes and edge relation $\phi(i, j)$. This path starts at node 0. If (i, j) is an edge in this path, then j is the smallest number with an edge from i to j, or $j = a$ when there are no outgoing edges. Note that the edge may not exists in the original graph when $j = a$. This is why we call is a pseudo-path. If (i, j) is the wth edge in the path, then $Z(w, i, j)$ is true, and $Z(w, i', j')$ is false for every other pair. It is not immediately obvious the axiom says this, so we will look at it closer.

Let Z be a string that witnesses the axiom. We want to make sure Z is the path described above. Looking at ρ_3, we see the path starts at 0. Suppose $Z(0, 0, j)$ is true. We must show that j is the first node adjacent to 0. This follows from ρ_1, which guarantees $\phi(i, j)$ is true when $j < a$, and ρ_2, which guarantees $\phi(i, k)$ is false when $k < j$. A similar argument can be made with ρ_5 and ρ_6 to show that every node is the smallest node adjacent to its predecessor. To make sure the path is long enough, we have ρ_7, which says there is a bth edge, and ρ_4, which says if there is a $(w + 1)$th edge there is a wth. As you may have noticed, there are parts of this formula that semantically are not needed. For example, the $\exists l < j \phi(0, l)$ in ρ_1 is not needed. It is used to make sure the axiom translates into a $\Sigma CNF(2)$ formula. We add ρ_8 to make sure there is a unique Z that witnesses this axiom.

Notation 2 *For simplicity, ψ_ϕ is the Σ_0^B part of the Σ_0^B-edge-rec axiom instantiated with ϕ. Note this includes the bound on the size of Z. So the axiom can be written as $\exists Z \psi_\phi$.*

So now we define the theory VL'.

Definition 4 *VL' is the theory axiomatized by the axioms of V^0, the Σ_0^B-edge-rec axioms, and Axiom (1). The language of VL' is the language of V^0 plus a string constant C with defining axiom*

$$|C| = 0 \tag{1}$$

We add the string constant to the language so we can put VL' proofs in free variable normal form (below). We do not use the constant for any other reason. Also, in the translation, we can treat C a string variable with $n = 0$.

In [9], the author proved that VL' and Σ_0^B-rec are equivalent. Since the X-rec and Σ_0^B-edge-rec axioms are semantically similar this is not too difficult, but to prove the Σ_0^B-edge-rec axiom from X-rec does require a trick used in [12] to get the path to start at 0. This gives us the following lemma.

Lemma 2. *The theory Σ_0^B-rec is equivalent to VL'.*

So now we know that VL' captures log space reasoning, and it does not capture reasoning for a larger complexity class.

The next step is to be sure the translation of this axiom is a $\Sigma CNF(2)$ formula. This is done by a careful inspection of the formula and is left to the reader.

Lemma 3. *The formula* $||\exists Z \psi_\phi(a,b,Z)||$ *is a* $\Sigma CNF(2)$ *formula.*

3.2   Cut Variable Normal Form

In this section, we want to find a normal form for VL' proofs that makes sure the translation of VL' proofs satisfy the variable restriction for GL^*. The normal form we want is *cut variable normal form* (CVNF) and is defined as follows.

Definition 5 *A formula* $\phi(Y)$ *is bit-dependent on* Y *if there is an atomic subformula of* ϕ *of the form* $Y(t)$, *for some term t.*

Definition 6 *A proof is in free variable normal form if every non-parameter free variable y or Y that appears in the proof is used as an eigenvariable of an inference exactly once, and every formula that contains y or Y appears before that inference.*

Definition 7 *A cut in a proof is anchored if the cut formula is an instance of an axiom.*

Definition 8 *A* VL' *proof* π *is in* cut variable normal form *if* π *is (1) in free variable normal form, (2) every cut with a non-Σ_0^B cut formula is anchored, and (3) no cut formula that is an instance of the Σ_0^B-edge-rec axiom is bit-dependent on a non-parameter free string variable.*

It is known how to find a proof with the first two properties [5, 6], but, to our knowledge, no property similar to the third has ever been considered.

The main theorem of this section is

Theorem 3. *Suppose* $VL' \vdash \exists Z \le t\phi(Z)$ *for some* Σ_0^B *formula* ϕ. *Then there exists a* VL'-*proof* π *of* $\exists Z \le t\phi(Z)$ *such that* π *is in CVNF.*

The proof of this theorem is the most technical in this paper. At a high level, it amounts to showing Σ_0^B-edge-rec is closed under substitution of strings defined by Σ_0^B-edge-rec and Σ_0^B-COMP. We begin with an anchored proof that is in free variable normal. We want to change every cut that violates condition (3) in the definition of CVNF. Consider the proof given in Figure 1. This is a simple example of what can go wrong. The general case is handled in the same way, so we will only consider this case.

Since all Σ_1^B cut formulas are anchored and the $\exists Y \gamma(Y)$ must eventually be cut, it is be an instance of Σ_0^B-COMP or Σ_0^B-edge-rec. The hard part is to prove the following lemma.

$$P$$

$$\frac{\exists Z \psi_{\phi(Y)}(Z), \gamma(Y), \Gamma \to \Delta \qquad \gamma(Y), \Gamma \to \Delta, \exists Z \psi_{\phi(Y)}(Z)}{\dfrac{\gamma(Y), \Gamma \to \Delta}{\exists Y \gamma(Y), \Gamma \to \Delta}}$$

Fig. 1. Example of a proof that is not in CVNF

Lemma 4. *For any Σ_0^B formula $\phi(Y)$, there exist Σ_0^B formulas ϕ_1 and ϕ_2 such that ϕ_1 is not bit-dependent on Y and V^0 proves the sequent*

$$\gamma(Y), \psi_{\phi_1}(Z), \forall i < t[Z'(i) \iff \phi_2(Z)] \to \psi_{\phi(Y)}(Z').$$

Proof sketch. This proof is divided into two cases. In the first case, we assume

$$\gamma(Y) =_{syn} Y \leq t \forall i < t[Y(i) \iff \phi'(i)].$$

That is, it is an instance of Σ_0^B-COMP. We know Y must appear in that position because it gets quantified. In this case, ϕ_1 is ϕ with every atomic formula of the form $Y(s)$ replaced by $s < t \wedge \phi'(s)$, and ϕ_2 is not needed.

For the second case, we assume $\gamma(Y) =_{syn} \psi_{\phi'}(Y)$. In this case we use branching programs. We give a Σ_0^B description of a branching program BP_1 that computes Z' and a branching program BP_2 that computes Y. BP_1 at some point branches on Y. So we construct BP_3 by composing BP_1 and BP_2. Anytime BP_1 needs to branch on Y it runs BP_2 to see what it should do. The formula ϕ_1 is the Σ_0^B description of BP_3. The formula ϕ_2 represents the AC^0 function that extracts Z' from the run Z of BP_3. We use branching programs because log space can be defined in terms of branching programs.

Using this lemma, we are able to change the proof in Figure 1 into the proof in Figure 2. In that proof, P' is the proof P with the rules that introduced $\exists Z$ ignored, and Q is an anchored V^0 proof, which we know exists by the lemma above. This gives us a new proof of the same formula that still satisfies properties (1) and (2) in Definition 8 and it contains one less cut that is bit-dependent on Y, but it might be bit-dependent on different non-parameter variables. However, if we do things in the correct order, we can repeat the transformation and, eventually, we will get a proof that is in $CVNF$.

Using this manipulation, we prove Theorem 3.

Proof (Proof of Theorem 3). It would be nice to be able to simply say we can repeatedly apply the manipulations above and eventually the proof will be in CVNF, but this is not obvious. In the manipulation, if $\gamma(Y)$ is bit-dependent on a string variable other than Y, then the new Σ_0^B-edge-rec cut formula is bit-dependent on that variable. This includes non-parameter string variables. So we need to state our induction hypothesis more carefully.

Let $Y_1, \ldots, Y_n$ be all the non-parameter free string variables that appear in π ordered such that the variable Y_i is used as a eigenvariable before Y_j for $i < j$. This implies Y_i does not appear in $\gamma(Y_j)$ in the manipulations above. So now suppose no Σ_0^B-edge-rec cut formula is bit-dependent on the variables $Y_1, \ldots, Y_k$, for some $k < n$. Then we can manipulate π such that the same holds for the variables $Y_1, \ldots, Y_{k+1}$. To accomplish this, we simply manipulate every Σ_0^B-edge-rec cut formula that is bit-dependent on Y_{k+1} as described above. Since $Y_1, \ldots, Y_k$ cannot appear in $\gamma(Y_{k+1})$, those variables will not violate the condition. So by induction, we can get a proof that is in CVNF.

$$P' \qquad\qquad\qquad Q$$

$$\cdots \vdots \cdots \qquad\qquad\qquad \cdots \vdots \cdots$$

$$\cfrac{\cfrac{\psi_{\phi(Y)}(Z), \gamma(Y), \Gamma \to \Delta \qquad \gamma(Y), \psi_{\phi_1}(Z), \tau(Z') \to \Delta, \psi_{\phi(Y)}(Z)}{\psi_{\phi_1}(Z), \tau(Z'), \gamma(Y), \Gamma \to \Delta}}{\psi_{\phi_1}(Z), \exists Z' \tau(Z'), \gamma(Y), \Gamma \to \Delta}$$

$$\cfrac{\cfrac{\psi_{\phi_1}(Z), \exists Z' \tau(Z'), \gamma(Y), \Gamma \to \Delta \qquad \to \exists Z' \tau(Z')}{\psi_{\phi_1}(Z), \gamma(Y), \Gamma \to \Delta}}{\exists Z \psi_{\phi_1}(Z), \gamma(Y), \Gamma \to \Delta}$$

$$\cfrac{\cfrac{\exists Z \psi_{\phi_1}(Z), \gamma(Y), \Gamma \to \Delta \qquad \to \exists Z \psi_{\phi_1}(Z)}{\gamma(Y), \Gamma \to \Delta}}{\exists Y \gamma(Y), \Gamma \to \Delta}$$

Fig. 2. Modification of the proof in Figure 1. The formula $\tau(Z')$ is used to replace $\forall i < t[Z'(i) \iff \phi_2(Z)]$

3.3   Translating Theorems of VL'

We are now prepared to prove the translation theorem. The proof is done by induction on the length of the proof. For the base case, we need to prove the translation of the axioms of VL'. Since every axiom except Σ_0^B-edge-rec is an axiom of VNC^1, we know those axioms have polynomial size G_0^* proofs [7] and, therefore, polynomial size GL^* proofs as well. Axiom (1) is easy to prove since it translates to $\to \top$. We still need to show how to prove the Σ_0^B-edge-rec axiom in GL^*. Recall that we write the axiom as $\exists Z \psi_\phi(a, b, Z)$. Note that the axiom does have a bound on Z, but it has been omitted since the specific bound is not important.

Lemma 5. *The formula* $\|\exists Z \psi_\phi(a, b, Z)\|$ *has a GL^* proof of size $p(a, b)$ for some polynomial p.*

Proof sketch. The proof is done by a brute force induction. We prove, in GL^*, that, if there exists a pseudo-path of length b, then there exists a pseudo-path of length $b + 1$. It is easy to prove there exists a pseudo-path of length 0. With repeated cutting we get our final result. The entire path is quantified, so we do not cut non-parameter free variables.

We now prove the main theorem of this section.

Theorem 4. *Suppose VL' proves $\exists Z < t\phi(\boldsymbol{x}, \boldsymbol{X}, Z)$. Then there are polynomial size GL^* proofs of $\|\exists Z < t\phi(\boldsymbol{x}, \boldsymbol{X}, Z)\|[\boldsymbol{n}]$.*

Proof. By Theorem 3, there exists a VL' proof π of $\exists Z < t\phi(\boldsymbol{x}, \boldsymbol{X}, Z)$ that is in CVNF.

We proceed by induction on the depth of π. The base case follows from Lemma 5 and the comments that precede it. The inductive step is divided into cases: one for each rule. With the exception of cut, every rule can be handled the same way it is handled in the $V^1 - G_1^*$ Translation Theorem [6], and will not be repeated here.

When looking at the cut rule, there are three cases. If the cut formula is Σ_0^B, then we simply cut the corresponding Σ_0^q formula in the GL^* proof. If the cut formula is not Σ_0^B, then it must be anchored since the proof is in CVNF. This means the cut formula is an instance of Σ_0^B-edge-rec or an instance of Σ_0^B-COMP. First suppose it is an instance of Σ_0^B-edge-rec. Then we are able to cut the corresponding formula in the GL^* proof. This is because the axiom translates into a $\Sigma CNF(2)$ formula, and the free variables in the translation are parameter variables since the formula is not bit-dependent on non-parameter string variables,

When the cut formula is an instance of Σ_0^B-COMP, we apply the same transformation as in the proof of the $VNC^1 - G_1^*$ translation theorem [7]. That is, we remove the quantifiers by replacing the variables with Σ_0^q formulas that witness the quantifiers. This change does not effect other cuts since their free variables are parameter variables or they are Σ_0^q formulas and remain Σ_0^q after the substitution. The current cut formula becomes a Σ_0^q formula, which can be cut.

4 Proving GL^* Is Sound in Σ_0^B-rec

In this section, we show that GL^* does not capture reasoning for a higher complexity class. This is done by proving, in Σ_0^B-rec, that GL^* is sound. This idea comes from [11] where Cook showed that PV proves extended Frege is sound and [2] where Krajicek and Pudlak showed T_2^i proves G_i is sound for $i > 0$.

We will actually show that $\overline{VL}$ proves GL^* is sound. This theory is a universal conservative extension of Σ_0^B-rec. In [9], the author proved that $\overline{VL}$ proves induction on Σ_0^B formulas that contain log space functions. Formally, this is $\Sigma_0^B(\mathcal{L}_{FL})$-IND. In general, we will give informal proofs, but we will be sure induction hypotheses do not use functions that are not log space. Once the proof in $\overline{VL}$ is done, we will know that it can be done in Σ_0^B-rec as well since $\overline{VL}$ is a conservative extension of Σ_0^B-rec [9].

We start the proof by giving a log space algorithm that witnesses $\Sigma CNF(2)$ formulas when the formula is true. This algorithm is the algorithm given in [10] with a few additions to find the satisfying assignment. This algorithm can be formalized in $\overline{VL}$ since it is a log space algorithm, and $\overline{VL}$ proves that it is correct. This means we can use this function in induction hypotheses. We then use this function to define a log space function that witnesses GL^* proofs. We finish by proving in $\overline{VL}$ that the algorithm is correct.

4.1 Witnessing $\Sigma CNF(2)$ Formulas

We want to define a function W that takes as input a $\Sigma CNF(2)$ formula and an assignment to the free variables, and returns an assignment to the quantified variables that satisfies the quantifier free portion of the formula whenever possible. Let $\exists z F(z, x)$ be a $\Sigma CNF(2)$ formula, where F is quantifier free, and let v be the values assigned to x. We begin by using v to simplify F. By the form of $\Sigma CNF(2)$ formulas, the simplified formula is $CNF(2)$. We now need to find a satisfying assignment to the simplified formula, if one exists. The simplified formula will also be called F.

To find the satisfying assignment, we construct an undirected tagged graph (G, T) based on F. This is done as in [10]. The graph G has a vertex v_i for every clause C_i in F. There is an edge between two vertices v_i and v_j if there is a literal l such that l is in C_i and $\bar{l}$ is in C_j. A vertex is tagged if the corresponding clause contains a pure literal.

Based on this construction, Johannsen proved the following lemma [10].

Lemma 6. *F is satisfiable if and only if it is possible to direct the edges of G such that there are no untagged sinks.*

The proof of this lemma can be done in $\overline{VL}$ since the direction for the edges can be easily construction from the satisfying assignment and vice versa. Johannsen noted that the edges can be properly directed if and only if G does not contain an untagged tree. The only if direction is easy to prove using a counting argument. A tree has fewer edges than nodes, so directing the edges cannot make every node a non-sink. This can be formalized in $\overline{VL}$ since $\overline{VL}$ extends VTC^0, and, therefore, proves the pigeon hole principle for $\Sigma_0^B(\mathcal{L}_{FL})$ formulas [9]. To prove the other direction in $\overline{VL}$, we give a log space algorithm that directs the edges appropriately, and prove the correctness of the algorithm in $\overline{VL}$. This also gives us the function W we are looking for.

We use a trick first used in [13] to find a cycle in a graph. For each edge $e = (u, v)$ in G, we call its two end points e^u and e^v, with the obvious meaning. Given an end point p, we the edge can be obtain by $e(p)$ and the vertex by $v(p)$. Then we can define the permutations σ_G, which is the product of the transposition $(e^u e^v)$ for every edge in G, and ρ_G, which is the product of the cycles $(e_1^v \dots e_n^v)$ for each vertex v, where $e_1, \dots, e_n$ are the edge incident to v. Then the permutation $\pi_G = \rho_G \circ \sigma_G$. We will often talk about the graph of π_G, which will also be called π_G. This permutation is useful because of the following nice property first proved in [13].

Definition 9 *An end point e^u of a vertex $e = (u, v)$ is trivially traversed if the there are no end points e_1^u on the path from e^u to e^v in π_G. If the path from e^u to e^v does not exists, then e^u is not trivially traversed[1].*

Lemma 7. *The connected component of G that contains edge $e = (u, v)$ is a tree if and only if every end point in the cycle of π_G that contains e^u is trivially traversed. Moreover, this is provable in $\overline{VL}$.*

This gives us a method of finding cycles.

The algorithm to find appropriate directions for the edges is done in stages. In stage i, we consider the ith vertex v_i. If the vertex is tagged, we continue.

---

[1] This definition comes from a personal communication from Mark Braverman.

Otherwise, we search G, using π_G, for a cycle or tagged vertex that can be reached from v_i. If no cycle or tagged vertex is found, then we found an untagged tree and we can stop. The edges on the path from v_i to the cycle or tagged vertex are directed away from v_i. The edges in the cycle are directed to form a directed cycle, which direction does not matter. If one of these edges was given a direction in an earlier stage, that direction is overwritten with the new value.

If we let $FindTagOrCycle(G, p)$ be a function that returns the closest end point q to p in π_G such that $v(q)$ is tagged or q is not trivially traversed. (This can be done using the algorithm described in [10].) Then the algorithm described above can be implemented as in Algorithm 1.

```
for i = 1...n do
    if v_i is not tagged then
        w = FindTagOrCycle(G, e^{v_i}) for some edge e adjacent to v_i.
        If w = null, stop.
        Else, w' = e^{v_i}
        while w' ≠ w do
            Direct e(w') away from v(w')
            w' = π_G(w')
        end while
    end if
end for
```

Algorithm 1: Algorithm to direct edges on a tagged graph

The correctness of the algorithm follows from the following invariant: After stage i, the vertices $v_1, \ldots, v_i$ are tagged or are not sinks. It is easy to check that this holds.

We claim this algorithm can be done in log space. Note that an edge may be directed multiple times, but the values are never used. So, to determine the direction of an edge e, we can run the algorithm, but only keep track of e's direction. The cycles and tagged vertex are found by searching π_G, which has out-degree 1. Therefore the search can be done in log space. Note that we are using that log space is closed under composition.

The final thing to do is prove the correctness of the algorithm in $\overline{VL}$. This can be done by induction on the invariant above. Since the algorithm is log space, the invariant can be stated as a $\Sigma_0^B(\mathcal{L}_{FL})$ formula, so the induction can be done in $\overline{VL}$.

The function W can be defined by applying the reduction to the input and output of this algorithm. We add that pure literals are assigned $\top$ and variables that are still not assigned a value are assigned $\bot$.

4.2   Witnessing GL^* Proofs

Let π be a GL^* proof of a Σ_1^q formula $\exists z P(\boldsymbol{x}, \boldsymbol{z})$, and let A be an assignment to the parameter variables (Definition 2). We will assume π is in free variable normal form (Definition 6). If it is not, we can rename variables to put it in free variable normal form. The renaming can be done in log space since all that is really required is to traverse the proof, which is a tree, to determine an appropriate name.

Let $\Gamma_i \to \Delta_i$ be the ith sequent in π. To prove the soundness of GL^*, we define a function $Wit(i, \pi, A)$ that will find a formula in Γ_i that is false or a formula in Δ_i that is true. We will prove by induction that for any assignment to all of the free variables if Γ_i and Δ_i, $Wit(i, \pi)$ will find at least one formula that satisfies the sequent.

There are two things to note. Every formula in Γ_i is $\Sigma CNF(2)$, which means it can be evaluated. Also, we need an assignment that gives appropriate values to the non-parameter free variables that could appear. To take care of this second point, we extend A to an assignment A' as follows:

1: Given a non-parameter free variable y, find the $\exists$-left inference in π that uses y as an eigenvariable. Let z be the new bound variable and let F be the principal formula.

2: Find the descendant of F that is used as a cut formula. Let F' be the cut formula. Note that F is a subformula of F', and, because of the variable restriction on cut formulas, every free variable in F' is a parameter variable.

3: Assign y the value that $W(F', A)$ assigns z.

The reason for this particular assignment will become evident in the proof of Lemma 8.

We can now define $Wit(i, \pi, A')$, which witnesses $\Gamma_i \to \Delta_i$. Wit will go through each formula in the sequent to find a formula that satisfies the sequent. $\Sigma CNF(2)$ formula are evaluated using the algorithm described in the previous section. We will not focus our attention on other Σ_1^q formulas, which must appear in Δ_i. Each Σ_1^q formula $F =_{syn} \exists z F^*(z)$ in Δ is evaluated by finding a witness to the quantifiers as follows:

1: Find a formula F' in π that is a ancestor of F, is satisfied by A', and is a Σ_0^q formula of the form $F^*(z_1/B_1, \ldots, z_n/B_n)$, where each B_i is Σ_0^q

2: z_i is assigned $\top$ if A' satisfies B_i, otherwise it is assigned $\bot$

3: if no such F' exists, then every bound variable is assigned $\bot$.

Lemma 8. *For every sequent $\Gamma_i \to \Delta_i$ in π, $Wit(i, \pi, A')$ finds a false formula in Γ_i or a true formula in Δ_i.*

Proof. We prove the theorem by induction on the depth of the sequent. For the base case, the sequent is an axiom, and the theorem obviously holds. For the inductive step, we need to look at each rule. We can ignore $\forall$-left and $\forall$-right since universal quantifiers do not appear in π.

We will not assume all formulas in Γ_i are true and all $\Sigma CNF(2)$ formulas in Δ_i as false. So we need to find a Σ_1^q formula in Δ_i that is true.

Consider cut. Suppose the inference is

$$\frac{F, \Gamma \to \Delta \qquad \Gamma \to \Delta, F}{\Gamma \to \Delta}$$

First suppose F is true. By induction, with the upper left sequent, Wit witnesses one of the formulas in Δ. Then the corresponding formula in the bottom sequent is witnessed by Wit. This is because the ancestor of the formula in the upper sequent that gives the witness is also an ancestor of the corresponding formula in the lower sequent. If F is false, it cannot be the formula that was witnessed in the upper right sequent, and a similar argument can be made.

Consider $\exists$-right. Suppose the inference is

$$\frac{\Gamma \to \Delta, F(B)}{\Gamma \to \Delta, \exists z F(z)}$$

First suppose $F(B)$ is Σ_0^q. If it is false, we can apply the inductive hypothesis, and, by an argument similar to the previous case, prove one of the formulas in Δ must be witnessed. If $F(B)$ is true, then Wit will witness $\exists z F(z)$ since $F(B)$ is the ancestor that gives the witness. If $F(B)$ is not Σ_0^q, then we can apply the inductive hypothesis, and, by the same argument, find a formula that is witnessed.

The last rule we will look at is $\exists$-left. Suppose the inference is

$$\frac{F(y), \Gamma \to \Delta}{\exists z F(z), \Gamma \to \Delta}$$

To be able to apply the inductive hypothesis, we need to be sure that $F(y)$ is satisfied. If $\exists z F(z)$ it true, then we know $F(y)$ is satisfied by the construction of A': the value assigned to y is chosen to satisfy $F(y)$ if it is possible. Otherwise, $\exists z F(z)$ is false, and we do not need induction.

For the other rules the inductive hypothesis can be applied directly and the witness found as in the previous cases.

Theorem 5. $\overline{VL}$ *proves* GL^* *is sound for proofs of* Σ_1^q *formulas.*

Proof. The functions W and Wit are log space functions and can be formalized in $\overline{VL}$. A function that finds A', given A, can also be formalized since it is log space. The final thing to note is that the proof of Lemma 8 can be formalized in $\overline{VL}$ since the induction hypothesis can be express as a $\Sigma_0^B(\mathcal{L}_{FL})$ formula and the induction carried out.

The reason this proof does not work for a larger proof system, say G_1^*, is because W cannot be formalized for the larger class of cut formulas. Also, if the variable restriction was not present, we would not be able to find A' in log space, and the proof would, once again, break down.

5   Concluding Remarks

To summarize, we have a proof system that corresponds to log space reasoning. This is a formula-based proof system, and it corresponds to Σ_0^B-rec in the usual way. This treatment of the proof system suggests a proof system for NL, which would be based on $2 - SAT$ instead of $SAT(2)$.

One drawback with GL^* is the variable restriction. It forced us to prove a normal form for VL' proofs. This normal form is specific to VL', and this proof would have to be completely redone for other theories. On the other hand, the proof that GL^* is sound can be easily changed to work for other theories. All that really needs to be changed for the case of NL is the definition of the function W.

I would like to thank my supervisor Stephen Cook for providing many useful comments.

References

1. Cook, S.: Theories for complexity classes and their propositional translations. In Krajicek, J., ed.: Complexity of computations and proofs. Quaderni di Matematica (2003) 175–227. Also available at *http://www.cs.toronto.edu/~sacook/*
2. Krajicek, J., Pudlak, P.: Quantified propositional calculi and fragments of bounded arithmetic. Zeitschrift f. Mathematikal Logik u. Grundlagen d. Mathematik **36** (1990) 29–46
3. Cook, S.: A survey of complexity classes and their associated propositional proof systems and theories, and a proof system for log space. Available at *http://www.cs.toronto.edu/~sacook/* (2001)
4. Pollett, C.: A propositional proof system for R_2^i. In Beame, P.W., Buss, S.R., eds.: Proof Complexity and Feasible Arithmetics. Volume 39 of DIMACS: Series in Discrete Mathematics and Theoretical Computer Science. AMS (1997) 253–278
5. Buss, S.R.: Introduction to proof theory. In Buss, S.R., ed.: Handbook of Proof Theory. Elsevier Science Publishers, Amsterdam (1998) 1–78
6. Cook, S.: CSC 2429s: Proof complexity and bounded arithmetic. Course notes. Available at *http://www.cs.toronto.edu/~sacook/csc2429h.02* (2002)
7. Cook, S., Morioka, T.: Quantified propositional calculus and a second-order theory for NC^1. (Accepted for Archive for Math. Logic)
8. Krajicek, J.: Bounded Arithmetic, Propositional Logic, and Complexity Theory. Cambridge University Press (1995)
9. Perron, S.: GL^*: A propositional proof system for logspace. Master's thesis, University Of Toronto (2005). Available at *http://www.cs.toronto.edu/~sperron/*
10. Johannsen, J.: Satisfiability problem complete for deterministic logarithmic space. In: STACS 2004, 21st Annual Symposium on Theoretical Aspects of Computer Science, Proceedings, Springer (2004) 317–325
11. Cook, S.A.: Feasibly constructive proofs and the propositional calculus. In: Proceedings of the 7-th ACM Symposium on the Theory of computation. (1975) 83–97
12. Zambella, D.: End extensions of models of linearly bounded arithmetic. Annals of Pure and Applied Logic **88** (1997) 263–277
13. Cook, S.A., McKenzie, P.: Problems complete for deterministic logarithmic space. Journal of Algorithms **8** (1987) 385–394

Identifying Polynomial-Time Recursive Functions

Carsten Schürmann and Jatin Shah

Department of Computer Science, Yale University, New Haven, CT 06511

Abstract. We present a sound and sufficient criterion for identifying polynomial-time recursive functions over higher-order data structures generalizing the seminal results by Bellantoni and Cook [1] and Leivant [2] to complex structural data-types. The criterion, presented as a deductive system, always terminates with a response that is either *yes* or *don't know*. The criterion is complete in the sense that every polynomial-time recursive function over binary strings has at least one implementation that is identified by our criterion; whether this is also true for arbitrary higher-order data structures remains an open problem. Logic programming serves as the underlying model of computation and our results apply to the Horn fragment as well to the fragment of hereditary Harrop formulas.

1 Introduction

The task of deciding if a function over binary numbers is computable in polynomial time is well-understood and based on a series of results that date back to Cobham [3]. However, deciding if a general recursive function over arbitrary, possibly higher-order data-types is computable in polynomial time remains difficult, and requires in general a reformulation into a previously established formalism, e.g. as a function in bounded recursion on notation [3], a function in Bellantoni and Cook's algebra [1] or Leivant's algebra [2, 4], or a function typeable in Bellantoni et al. [5] or Hofmann's type systems [6, 7]. In this paper we reconcile the simplicity of characterizing polynomial-time functions over *binary numbers* with the expressiveness of general recursive functions over arbitrary domains.

Our underlying model is backward-chaining logic programming, where functions are declared as relations. This idea is fundamentally different from Ganzinger and McAllester [8] and Givan and McAllester [9] who have given various criteria for identifying polynomial time predicates for forward-chaining logic programming. Our measure of complexity is captured as the size of the execution derivation, a logical deduction (should it exist), in terms of the size of the input arguments. We consider only the class of logic programs that implement functions, and we show that our notion of complexity is compatible with the usual one. Furthermore, we give a *sufficient* criterion that decides if a logic program runs in polynomial time.

The criterion is based on the observation that the sum of the size of arguments passed to the recursive calls must not exceed the size of the input arguments of the function. In addition, all calls to auxiliary (non-recursive) functions that take recursively computed arguments as inputs must be shown to be non-size increasing. Aehlig, *et al.* [10] and Hofmann [11] have also used the latter condition to extend Hofmann's polynomial-time type system to include a larger class of functions. If the criterion is

L. Ong (Ed.): CSL 2005, LNCS 3634, pp. 525–540, 2005.

satisfied, we show that the logic programming engine will terminate in a number of steps that is bounded by a polynomial in the size of the input.

The methodology of using logic programming as the model of computation applies to several logical formalisms including the Horn and hereditary Harrop fragment of first-order logic for a higher-order, simply-typed term algebra, which are discussed here. We suspect, that it can also be used to classify logic programs written in other extensions and for other term algebras, but leave further investigations to future work.

The paper is organized as follows. First, we discuss logic programming as a model of computation in Section 2. Next, we develop the criterion that classifies polynomial time computable recursive functions in Section 3. In Section 4, we then extend our results to higher-order hereditary Harrop formulas and illustrate the expressiveness of our results with some examples before we conclude and assess results in Section 5.

2   Functions as Logic Programs

We are interested in studying general recursive functions and classifying their running time into complexity classes using syntactic criteria. We think of a recursive function $(y_1, \ldots, y_n) = f(x_1, \ldots, x_m)$ as a predicate $P_f(x_1, \ldots, x_m; y_1, \ldots, y_n)$ that relates input arguments x_i with output arguments y_i. These relations fall into a subclass of well-moded logic programs that compute ground output terms from ground input terms. A *ground term* is a term not containing any free logic variables. In logic programming the underlying model of computation is proof search; and thus a complete computation trace corresponds to a closed proof derivation, which determines ground terms in all output positions. Such logic programs are considered to have a well-defined *mode* behavior. The reader may refer Rohwedder and Pfenning [12] for more information on algorithms for identifying *mode correct* logic programs.

2.1   Term Algebra

We choose the simply-typed λ-calculus as logical framework.

$$
\begin{array}{lll}
Types & A, B & ::= a \mid A \to B \\
Canonical\ Terms & M, N & ::= \lambda x : A.N \mid R \\
Atomic\ Terms & R & ::= c \mid x \mid R\,N
\end{array}
$$

where a and c are type and object level constants declared a priori. For studying run-time complexity, it is convenient to consider only canonical terms, i.e. terms without β-redexes. However, in Section 4, we extend the results to non-canonical terms.

2.2   Logic Programming as Model of Computation

Logic programming can serve as a model of computation where traces are captured by proof derivations. For simplicity we consider only the Horn fragment in this section, but extend the results to the fragment of hereditary Harrop formulas in Section 4. Predicates are given by

$$P(M_1, \ldots, M_m; N_1, \ldots, N_n)$$

$$\frac{}{\mathcal{F} \to \top} \text{ g\_True} \qquad \frac{D \in \mathcal{F} \quad \mathcal{F} \to D \gg P}{\mathcal{F} \to P} \text{ g\_Atom}$$

$$\frac{}{\mathcal{F} \to P \gg P} \text{ c\_Atom} \qquad \frac{\mathcal{F} \to [M/x]D \gg P}{\mathcal{F} \to \forall x : A.D \gg P} \text{ c\_ForAll} \qquad \frac{\mathcal{F} \to D \gg P \quad \mathcal{F} \to G}{\mathcal{F} \to G \supset D \gg P} \text{ c\_Imp}$$

Fig. 1. Proof search semantics for the Horn fragment

where inputs M_i are separated from outputs N_i by a semicolon. The formulation of Horn-logic in terms of goals G and definite clauses D is standard.

Goals	G	$::= \top \mid P$
Clauses	D	$::= G \supset D \mid \forall x : A.D \mid P$
Programs	$\mathcal{F}$	$::= \bullet \mid \mathcal{F}, D$

A logic program $\mathcal{F}$ is simply a collection of clauses. Often we find it convenient to reverse the direction of $G \supset D$ and use $D \subset G$ instead. $\supset$ is right-associative. In addition, we always omit the leading $\bullet$ from programs.

Definition 1 (Predicate symbol, head of a clause). *For a clause D or a goal G, we define predicate symbol of D or G and head of a clause D as given below:*

$$\text{symbol}(P(\cdot;\cdot)) = P \qquad\qquad \text{head}(P) = P$$
$$\text{symbol}(\forall x : A.D) = \text{symbol}(D) \qquad \text{head}(\forall x : A.D) = \text{head}(D)$$
$$\text{symbol}(G \supset D) = \text{symbol}(D) \qquad \text{head}(G \supset D) = \text{head}(D)$$

2.3 Function Computation Through Proof Search

The proof search semantics of Horn logic is given in Figure 1. Given a program $\mathcal{F}$ and a goal G with ground terms in its input positions, the interpreter constructs a derivation of the judgment $\mathcal{F} \to G$. In the rule g_Atom an appropriate clause D corresponding to the goal G is selected. It is possible to construct a derivation for the judgment $\mathcal{F} \to D \gg P$ only if head of D can be made equal to P. For the sake of our analysis, we assume that an oracle predicts the correct instantiations of the universally quantified formulas (c_Forall, c_Atom). In an actual implementation, however, one would postpone non-deterministic choice by employing logic variables that are eventually instantiated by unification, as all logic programs considered here are mode-correct.

This logic program implements the Fibonacci function on natural numbers: $\mathcal{F} = +(\mathsf{z}, Y; Y), +(X, Y; Z) \supset +(\mathsf{s}\ X, Y; \mathsf{s}\ Z), \mathsf{fib}(\mathsf{z}; \mathsf{s}\ \mathsf{z}), \mathsf{fib}(\mathsf{s}\ \mathsf{z}; \mathsf{s}\ \mathsf{z}), +(X, Y; Z) \supset \mathsf{fib}(N; X) \supset \mathsf{fib}(\mathsf{s}\ N; Y) \supset \mathsf{fib}(\mathsf{s}\ (\mathsf{s}\ N); Z)$, where the constants z, s, and fib are appropriately defined, and all uppercase variables are of type nat and implicitly universally quantified at the beginning of the respective clause.

For a logic program $\mathcal{F}$, we denote a proof search derivation for a goal G by $\mathcal{D} :: \mathcal{F} \to G$ and measure the size of this derivation as the number of inference rules in the derivation. In the Section 2.5, we show that every rule can be implemented on a random access machine (RAM) in a constant number of steps.

Definition 2 (Size of proof search derivation). *Given a logic program $\mathcal{F}$ and a derivation $\mathcal{D} :: \mathcal{F} \to G$, we define the size of $\mathcal{D}$, $\mathsf{sz}(\mathcal{D})$ as the number of rules in $\mathcal{D}$.*

$$\#(x) = \#(c) = 1 \qquad\qquad sz_u(\top) = 0 \qquad\qquad sz_i(P(M_1,\ldots,M_m;\cdot)) = \sum_{i=1}^{m} \#(M_i)$$
$$\#(R\ N) = \#(R) + \#(N) \qquad sz_u(G \supset D) = sz_u(D) \qquad sz_o(P(\cdot;N_1,\ldots,N_n)) = \sum_{i=1}^{n} \#(N_i)$$
$$\#(\lambda x.N) = \#(N) \qquad\qquad sz_u(\forall x : A.D) = sz_u(D)$$

Fig. 2. Size function for goals G and clauses D ($u = i$ or $u = o$)

2.4   Size of Terms, Goals and Clauses

The relevant size functions are defined in Figure 2. $\#$ counts the number of variables and constants in a term. The size of a goal G or a clause D is defined using $sz_i(\cdot)$ and $sz_o(\cdot)$ depending on whether we wish to compute the size of *input* or *output* arguments. $sz_i(G)$ computes the sum of $\#$-sizes of all the *input* arguments in the goal G and $sz_i(D)$ computes the sum of $\#$-sizes of all the *input* arguments in predicate P in the clause D.

2.5   Translation to a Random Access Machine (RAM)

The following two conditions are sufficient to show that the proof search algorithm from Figure 1 can be implemented on a RAM in time proportional to the number of proof search rules in a proof search derivation. The mode-correct logic program must be

1. deterministic and non-backtracking,
2. and the time required to solve the individual unification problem is independent of input or output arguments.

The first condition is satisfied if the cases have non-overlapping patterns and all output positions of recursive calls contain only variables. The second is satisfied if all variables that occur in input arguments in the head of a clause are *linear* (i.e. variables occur exactly once) and form higher-order patterns in the sense of Miller [13]. Linearity guarantees that logical variables are only instantiated once and hence limit the complexity of unification by the size of the pattern. This may sound as a prohibitive restriction as clauses such as $P(x,x;x) \subset \top$ are disallowed. However, those clauses may be *linearized* by providing explicitly an equality predicate, such as $P(x,y;x) \subset$ equal$(x,y;) \subset \top$. *Higher-order patterns* are simply-typed β-normal λ-terms whose universally bound variables X are applied exclusively to a sequence of distinct bound variables. Qian [14] has also given a linear time and space unification algorithm for higher-order patterns. Under those two conditions the logic programming engine can be implemented on a RAM without increase in its asymptotic complexity.

Theorem 1. *Given a logic program $\mathcal{F}$ and a goal G satisfying the conditions given above. If there exists a derivation $\mathcal{D} :: \mathcal{F} \to G$, then*

1. *The goal G can be represented on a RAM in size proportional to $sz_i(G)$.*
2. *The corresponding proof search can be implemented in time proportional to $sz(\mathcal{D})$.*

3   Conditions for Polynomial-Time Functions

In this section, we describe criteria for classifying recursive functions into the polynomial complexity class, FP. These criteria are decidable and can be checked in time depending only on the size of the logic program corresponding to the function. Our criteria are sound. We prove completeness for functions over binary strings; whether these criteria are also complete for arbitrary higher-order data structures is an open problem. Thus, a checker implementing these criteria can only have two responses *yes* and *don't know*.

First, we present a general theorem on integer valued recursive functions given by

$$T(x) = \sum_{i=1}^{m} T(x_i) + f(x) \quad \text{if } x > K$$
$$T(x) = b \quad\quad\quad\quad\quad\quad \text{if } 1 \leq x \leq K \tag{1}$$

where $x, x_i \in \mathbb{Z}^+$ and there exists functions $g_i(\cdot)$ (not defined using $T(\cdot)$) such that $x_i = g_i(x)$ for all $i = 1, \ldots, m$ such that $x_i < x$, each $f(x)$ is an integer valued function defined on $\mathbb{Z}^+$ (not defined using $T(\cdot)$), b and K are positive integers; and m is an positive integer constant.

Theorem 2 ([15]). *Given a recursive function $T(x)$ defined in equation 1. If $f(x)$ is a monotonically increasing function such that $f(x) > 0$ for all $1 \leq x \leq K$, and $x \geq \sum_{i=1}^{m} x_i$, then there exists a constant $c \geq 1$ such that $T(x) \leq cx^2 f(x)$ for all $x \geq 1$.*

For example, if $T(x) = T(\lfloor x/3 \rfloor) + T(\lfloor x/4 \rfloor) + x$, then $T(x) = O(x^3)$ as $x \geq \lfloor x/3 \rfloor + \lfloor x/4 \rfloor$. On the other hand, we know that $T(x) = T(x-1) + T(x-2) + 1$ when $x \geq 2$ and $T(0) = T(1) = 1$ is not a polynomial. In this case, $x \not\geq (x-1) + (x-2)$.

Theorem 2 can be generalized to a set of functions $\mathcal{T} = \{T_1(\cdot), T_2(\cdot), \ldots, T_k(\cdot)\}$ where each $T_i(\cdot)$ is defined as

$$T_i(x) = \sum_{j=1}^{m_i} T_{l_j}(x_{ij}) + f_i(x) \quad \text{if } x > K_i$$
$$T_i(x) = b_i \quad\quad\quad\quad\quad\quad\quad \text{if } 1 \leq x \leq K_i \tag{2}$$

where m_i, K_i and b_i are positive integer constants, each $l_j \in \{1, \ldots, k\}$, every $f_i(x)$ is an integer-valued function defined on $\mathbb{Z}^+$ (not defined using $T(\cdot)$), $x, x_{ij} \in \mathbb{Z}^+$ and there exists functions $g_{ij}(\cdot)$ (not defined using $T(\cdot)$) such that $x_{ij} = g_{ij}(x)$.

Theorem 3. *Given a set of recursive functions $\mathcal{T} = \{T_1(\cdot), T_2(\cdot), \ldots, T_k(\cdot)\}$ such that each function is given by equation 2. If for all $i = 1, \ldots, k$:*

1. *$f_i(\cdot)$ are monotonically increasing functions such that $f_i(x) > 0$ for all $1 \leq x \leq K_i$.*
2. *$x \geq \sum_{j=1}^{m_i} x_{ij}$*

then there exists a constant $c \geq 1$ such that $T_i(x) \leq cx^2 F(x)$ for all $x \geq 1$ where $F(x) = \max(f_1(x), f_2(x), \ldots, f_k(x))$.

We present our result in two stages. In Section 3.1 we present the *basic criterion* which captures the essence of our solution. The functions identified by this criterion satisfy the following condition: the sum of the sizes of the recursive input arguments to the recursive calls is less than the original recursive input arguments. Thus, we are generalizing the results of Theorem 2 and 3 to higher-order data structures by defining an appropriate size function. In Section 3.3, we first show that Cobham's function class [3] is a special case of this criteria. Later in Section 3.4, we will also extend our criteria to identify functions where size of the output is bounded by a polynomial.

3.1 Basic Criteria

We generalize Theorems 2 and 3 to mutually recursive functions on arbitrary simply-typed λ-terms.

Definition 3 (goals). *Given a clause D, we define the set* goals(D) *as given below.*

$$\begin{aligned} \text{goals}(P) &= \{\} \\ \text{goals}(G \supset D) &= \{G\} \cup \text{goals}(D) \\ \text{goals}(\forall x : A.D) &= \text{goals}(D) \end{aligned}$$

Definition 4 (Mutually recursive predicate symbols). *Given a logic program $\mathcal{F}$, a set S of predicate symbols is said to be mutually recursive if and only if for any predicate symbols $P_f, P_g \in S$ there exists clauses $D_1, D_2 \in \mathcal{F}$ such that* symbol$(D_1) = P_f$, symbol$(D_2) = P_g$ *and there exist goals $G_1 \in$ goals(D_1) and $G_2 \in$ goals(D_2) such that* symbol$(G_1) = P_g$ *and* symbol$(G_2) = P_f$.

Figure 3 shows a deductive system for identifying logic programs corresponding to polynomial time functions. We say that a logic program $\mathcal{F}$ and a corresponding set S of mutually recursive predicate symbols computes a polynomial-time function, if we can construct a proof of the judgment $\vdash_S \mathcal{F}$ poly$_b$ using the rules given in Figure 3. The deductive system checks that every clause $D \in \mathcal{F}$ satisfies our polynomial time criteria and the corresponding judgment is given by $\vdash_S \Delta/D$ poly, where Δ is the list of subgoals. Initially, Δ is empty; the subgoals of D are added to Δ and they are eventually used in the base rule (rule b_Atom).

For the sake of clarity, given a program clause D and a set S of mutually recursive predicate symbols corresponding to a function, we will refer to subgoals G such that symbol$(G) \in S$ as *recursive function calls* and subgoals G such that symbol$(G) \notin S$ as *auxiliary function calls*.

Informally speaking, these conditions require that every program clause D satisfies the following properties:

1. The sum of the sizes of the inputs to all recursive function calls is no greater than the size of the input to the function. (Rule b_Atom)
2. The size of the input to a recursive function call is strictly less than the size of the input to the function. (Rule b_Imp1)
3. All auxiliary function calls are polynomial-time computable functions and the sizes of the inputs to those function calls are bounded by a polynomial in the size of the input to the function. (Rule b_Imp2)

In our deductive system, we have omitted proofs of these conditions, but they could be implemented in standard theorem provers using, say an implementation of Peano's arithmetic. The main result of this paper is shown in the theorem below.

Theorem 4 (Basic Criteria). *Given a program $\mathcal{F}$ and a set S of mutually recursive predicate symbols from $\mathcal{F}$ such that $\vdash_S \mathcal{F}$ poly$_b$, then there exists a monotonically increasing polynomial $p(\cdot)$ such that for all goals G: if* symbol$(G) \in S$ *and $\mathcal{D} :: \mathcal{F} \to G$, then* sz$(\mathcal{D}) \le p(\text{sz}_i(G))$.

Programs:

$$\frac{}{\vdash_S \bullet \; \mathsf{poly_b}} \; \mathsf{b\_empty} \qquad \frac{\mathsf{symbol}(D) \notin S \quad \vdash_S \mathcal{F} \; \mathsf{poly_b}}{\vdash_S \mathcal{F}, D \; \mathsf{poly_b}} \; \mathsf{b\_clause1}$$

$$\frac{\mathsf{symbol}(D) \in S \quad \vdash_S \bullet/D \; \mathsf{poly_b} \; \vdash_S \mathcal{F} \; \mathsf{poly_b}}{\vdash_S \mathcal{F}, D \; \mathsf{poly_b}} \; \mathsf{b\_clause2}$$

Clauses:

$$\frac{}{\vdash_S \Delta/P \; \mathsf{poly_b}} \; \mathsf{b\_Atom} \left\langle \Sigma_{\substack{G \in \Delta \\ \mathsf{symbol}(G) \in S}} \mathsf{sz}_i(G) \leq \mathsf{sz}_i(P) \right\rangle$$

$$\frac{\vdash_S \Delta, G/D \; \mathsf{poly_b} \quad \mathsf{symbol}(G) \in S}{\vdash_S \Delta/G \supset D \; \mathsf{poly_b}} \; \mathsf{b\_Imp1}\langle \mathsf{sz}_i(G) < \mathsf{sz}_i(D) \rangle$$

$$\frac{\vdash_S \Delta, G/D \; \mathsf{poly_b} \quad \mathsf{symbol}(G) \notin S \quad \vdash_T \mathcal{F} \; \mathsf{poly_b}}{\vdash_S \Delta/G \supset D \; \mathsf{poly_b}} \; \mathsf{b\_Imp2}\langle \mathsf{sz}_i(G) < f_G(\mathsf{sz}_i(D)) \rangle$$
(where T is a set of mutually recursive predicate symbols such that
$\mathsf{symbol}(G) \in T$ and $f_G(\cdot)$ is a polynomial)

$$\frac{\vdash_S \Delta/D \; \mathsf{poly_b}}{\vdash_S \Delta/\forall x : A.D \; \mathsf{poly_b}} \; \mathsf{b\_Forall}$$

Fig. 3. Basic criteria for identifying for polynomial time functions

According to Theorem 1, if the logic program $\mathcal{F}$ computes a function that satisfies the conditions given in Section 2.5, the proof derivation $\mathcal{D}$ can be implemented on a RAM in time proportional to $\mathsf{sz}(\mathcal{D})$. Since, $\mathsf{sz}(\mathcal{D})$ is bounded by a polynomial in the size of the input $\mathsf{sz}_i(G)$, we can conclude that $\mathcal{F}$ is polynomial-time computable.

Example 1 (Combinators). The combinators $c ::= \mathsf{S} \mid \mathsf{K} \mid \mathsf{MP} \; c_1 \; c_2$ that are prevalent in programming language theory are represented as constructors of type comb. We study the complexity of the bracket abstraction algorithm ba, which converts a parametric combinator M (a representation-level function of type comb → comb) into a combinator with one less parameter (of type comb) to which we refer as M'. The bracket abstraction algorithm is expressed by a predicate relating M and M'. Let $\mathcal{F}$ be defined as the following program.

$$\mathsf{ba} \; (\lambda x : \mathsf{comb}. \, x; \mathsf{MP} \; (\mathsf{MP} \; \mathsf{S} \; \mathsf{K}) \; \mathsf{K}).$$
$$\mathsf{ba} \; (\lambda x : \mathsf{comb}. \, \mathsf{K}; \mathsf{MP} \; \mathsf{K} \; \mathsf{K}).$$
$$\mathsf{ba} \; (\lambda x : \mathsf{comb}. \, \mathsf{S}; \mathsf{MP} \; \mathsf{K} \; \mathsf{S}).$$
$$\mathsf{ba} \; (\lambda x : \mathsf{comb}. \, \mathsf{MP} \; (C_1 \; x) \; (C_2 \; x); \mathsf{MP} \; (\mathsf{MP} \; \mathsf{S} \; D_1) \; D_2)$$
$$\subset \mathsf{ba} \; (\lambda x : \mathsf{comb}. \, C_1 \; x; D_1)$$
$$\subset \mathsf{ba} \; (\lambda x : \mathsf{comb}. \, C_2 \; x; D_2).$$

It is easy to see that $\Sigma_{i=1}^{2} \#(\lambda x : \mathsf{comb}. \, C_i \; x) < \#(\lambda x : \mathsf{comb}. \, \mathsf{MP} \; (C_1 \; x) \; (C_2 \; x))$, and hence $\vdash_{\mathsf{ba}} \mathcal{F} \; \mathsf{poly_b}.\square$

3.2 Functions with Inputs from Outputs of Auxiliary Functions

When recursive function calls receive inputs from outputs of certain auxiliary functions, we may be unable to verify the first condition in our *basic criteria* directly. In such cases, we will need additional properties that relate outputs of those auxiliary functions to their inputs. Non-size increasing property described in detail later in Section 3.4 could suffice. But, in general, the user or the theorem prover could use any other property that may be known to be true regarding that auxiliary function.

3.3   Completeness on Functions over Natural Numbers

Cobham [3] gave a characterization of polynomial-time computable functions as the least class of functions containing constant, projection, successor, and the smash function $2^{|x| \cdot |y|}$ (where $|x|$ is the length of x); and closed under ordinary composition and *bounded recursion on notation* as defined below:

Definition 5 (Bounded recursion on notation). *Let g, h_0, h_1 and k be functions in the class. The function f is defined by bounded recursion on notation if*

$$
\begin{aligned}
f(0, x_1, \ldots, x_n) &= g(x_1, \ldots, x_n) \\
f(2y, x_1, \ldots, x_n) &= h_0(y, x_1, \ldots, x_n, f(y, x_1, \ldots, x_n)) \\
f(2y + 1, x_1, \ldots, x_n) &= h_1(y, x_1, \ldots, x_n, f(y, x_1, \ldots, x_n))
\end{aligned}
$$

and $f(y, x_1, \ldots, x_n) \le k(y, x_1, \ldots, x_n)$.

Of the elementary functions, constant, projection and successor can be implemented without any recursion. Consider a direct implementation of bounded recursion. In this case, we have a bound on the size of the recursive call which we can inductively assume to be a polynomial. Since polynomials are closed under composition, we can show that the total size of the inputs to h_0 and h_1 are polynomials (side condition to the rule pc_Imp2). Hence, the implementation is within our *basic criteria*. The smash function can be implemented using bounded recursion and so, it satisfies our basic criteria. The case for composition is similar – we know a polynomial bound on the size of the output of the functions being composed.

Therefore, the Cobham's functions can be implemented in our logic programming language and they always satisfy our basic criteria. It is possible to show a similar result for Cobham's functions when defined over binary strings.

3.4   Polynomial Time Functions with Bounded Recursion

It is clear from the discussion in Section 3.3 that our *basic criteria* are unable to identify functions that have function calls which use as inputs, outputs of other recursive functions unless we know an *apriori* bound on the size of those outputs. Based on the ideas first introduced by Caseiro [16], Aehlig, *et al.* [10] and Hofmann [11] have developed type systems for identifying functions which recurse on their *safe* inputs and yet remain within polynomial-time. Such functions have the property that any function that recurses on a recursively computed value is non-size increasing. Essentially, this property ensures that the size of the output of the function is bounded. In this section, we shall extend our *basic criteria* using their idea to identify functions which have bounded output.

Non-size Increasing Functions. We say that a function f is non-size increasing if and only if, the sum of the sizes of the output arguments is never greater than the sizes of its input arguments within an additive constant, i.e. $sz_o(G) = sz_i(G) + C$, where C is an integer independent of the input variables of G. The concept of multiplicity defined below will be used in building a formal deductive system to identify non-size increasing functions.

Definition 6 (Multiplicity). *Given a clause D, a goal $G \in$ goals(D) the α and β_G multiplicities of D are defined as follows.*

1. *$\alpha(D)$ is defined as the maximum number of times any input variable in* head(D) *appears in the output positions of* head(D).
2. *$\beta_G(D)$ is defined as the maximum number of times any output variable in G appears in the output positions of* head(D).

For example $\alpha(\forall N_1 N_2 M. + (N_1, M; N_2) \supset +(sN_1, M; sN_2))$ and $\beta_{+(N_1, M; N_2)}(\forall N_1 N_2 M. + (N_1, M; N_2) \supset +(sN_1, M; sN_2))$ corresponding to the second declaration of addition $+$ operation are given by 0 and 1 respectively. Similarly, for a clause of the form $P(N; cNN)$, $\alpha(P(N; cNN))$ is given by 2.

The following lemma relates the size of output of a logic program in terms of its input. We would like to note that goals $G \in$ goals(D) may have free variables, while goals $G \in$ GOALS(G) have no free variables.

Definition 7 (GOALS). *Given a clause D and a predicate P such that* symbol$(D) = $ symbol(G) *and a derivation $\mathcal{D} :: \mathcal{F} \to D \gg P$, we define the set* GOALS$(\mathcal{D})$ *as given below.*

$$\text{GOALS}\left(\overline{\mathcal{F} \to P \gg P}\right) = \{\}$$

$$\text{GOALS}\left(\frac{\overset{\mathcal{D}_1}{\mathcal{F} \to D \gg P} \quad \overset{\mathcal{D}_2}{\mathcal{F} \to G}}{\mathcal{F} \to G \supset D \gg P}\right) = \{\mathcal{D}_2\} \cup \text{GOALS}(\mathcal{D}_1)$$

$$\text{GOALS}\left(\frac{\overset{\mathcal{D}'}{\mathcal{F} \to [M/x]D \gg P}}{\mathcal{F} \to \forall x : A.D \gg P}\right) = \text{GOALS}(\mathcal{D}')$$

Lemma 1. *Given a program $\mathcal{F}$ and a set S of mutually recursive predicate symbols from $\mathcal{F}$. Given a predicate P and a clause $D \in \mathcal{F}$ such that* symbol$(P) = $ symbol$(D) \in S$. *If $\mathcal{D} :: \mathcal{F} \to D \gg P$, then*

$$sz_o(P) = \alpha(D)sz_i(P) + \sum_{\mathcal{D}_H :: \mathcal{F} \to H \in \text{GOALS}(D)} \beta_G(D)sz_o(H) + \gamma(D)$$

where $\gamma(D)$ is a constant depending only on the structure of D and not its ground input terms.

The judgment corresponding to the non-size increasing property is written as $\vdash_S \mathcal{F}$ nsi and the corresponding deductive system for functions which make recursive function calls is given in Figure 4. The case when a function has no recursive function calls is easier to analyze and we will omit it from our discussion here. The deductive system ensures that the following conditions hold for all program clauses D:

Programs:

$$\frac{}{\vdash_S \bullet \text{nsi}} \text{ nsi\_empty} \qquad \frac{\text{symbol}(D) \notin S \quad \vdash_S \mathcal{F} \text{ nsi}}{\vdash_S \mathcal{F}, D \text{ nsi}} \text{ nsi\_clause1} \qquad \frac{\text{symbol}(D) \in S \quad \vdash_S \bullet/D \text{ nsi} \quad \vdash_S \mathcal{F} \text{ nsi}}{\vdash_S \mathcal{F}, D \text{ nsi}} \text{ nsi\_clause2}$$

Clauses:

$$\frac{\gamma(P)=0 \wedge \sum_{\substack{G \in \Delta \\ \text{symbol}(G) \in S}} \beta_G(P)=1}{\vdash_S \Delta/P \text{ nsi}} \text{ nsi\_Atom}\left(\sum_{\substack{G \in \Delta \\ \text{symbol}(G) \in S}} \beta_G(P)sz_i(G) + \sum_{\substack{G \in \Delta \\ \text{symbol}(G) \notin S}} \beta_G(P)sz_o(G) \le (1 - \alpha(P))sz_i(P)\right)$$

$$\frac{\vdash_S \Delta, G/D \text{ nsi} \quad \text{symbol}(G) \in S}{\vdash_S \Delta/G \supset D \text{ nsi}} \text{ nsi\_Imp1}\langle sz_i(G) < sz_i(D)\rangle$$

$$\frac{\vdash_S \Delta, G/D \text{ nsi} \quad \text{symbol}(G) \notin S \quad \vdash_T \mathcal{F} \text{ nsi}}{\vdash_S \Delta/G \supset D \text{ nsi}} \text{ nsi\_Imp2}$$

(where T is a set of mutually recursive predicate symbols such that $\text{symbol}(G) \in T$)

$$\frac{\vdash_S \Delta/D \text{ nsi}}{\vdash_S \Delta/\forall x : A.D \text{ nsi}} \text{ nsi\_Forall}$$

Fig. 4. *Sufficient* conditions for non-size increasing functions

1. The sum of the contribution to the output of the function due to the original inputs given by $\alpha(D)sz_i(D)$ and due to outputs from the subgoal calls given by $\sum_{\substack{G \in \Delta \\ \text{symbol}(G) \in S}} \beta_G(P)sz_i(G) + \sum_{\substack{G \in \Delta \\ \text{symbol}(G) \notin S}} \beta_G(P)sz_o(G)$ is equal to the input to the function $sz_i(D)$. $\gamma(D) = 0$ and $\sum_{\substack{G \in \Delta \\ \text{symbol}(G) \in S}} \beta_G(P) = 1$. In addition, (Rule nsi_Atom)
2. The sum of all input sizes of recursive calls is less than the input to the function. (Rule nsi_Imp1)
3. All auxiliary function calls are non-size increasing. (Rule nsi_Imp2)

This condition is sufficient to ensure that the predicate corresponding to the clause D is non-size increasing.

Theorem 5 (Non-size increasing functions). *Given a logic program $\mathcal{F}$ and a set S of mutually recursive predicate symbols from $\mathcal{F}$ such that $\vdash_S \mathcal{F}$ nsi. For all goals G, if $\mathcal{D} :: \mathcal{F} \to G$, then $sz_o(G) = sz_i(G) + C$ where C is a constant integer depending on the logic program $\mathcal{F}$.*

Dependence Paths. The definitions of *dependence paths* given below assist us in keeping track of outputs of function calls when they are used as inputs to other function calls.

Definition 8. *Given a clause D, and goals G and H in the clause, $H \leftsquigarrow_m G$ iff variables of G in output positions appear in input positions of H and no variable of G appears more than m times in H.*

Definition 9 (Dependence Path). *Given a clause D and goals $H = G_0, G_1, \ldots, G_n = G \in \text{goals}(D)$, a dependence path from G to H of length n denoted by $H \leftsquigarrow G$ is a sequence of goal and positive integer pairs $(G_1, m_1), \ldots, (G_n = G, m_n)$ such that for each pair of goals G_i, G_{i+1} for $i = 0, \ldots, n - 1$, $G_i \leftsquigarrow_{m_{i+1}} G_{i+1}$. The width of this dependence path is defined as $\Pi_{i=1}^n m_i$.*

$$\frac{\vdash_S H \triangleleft D}{\vdash_S H \triangleleft \forall x : A.D} \; \text{dp\_Forall} \qquad\qquad \frac{\vdash_S H \triangleleft D}{\vdash_S H \triangleleft G \supset D} \; \text{dp\_Imp1} \langle H \not\leftsquigarrow G \rangle$$

$$\frac{\text{symbol}(G) \in S}{\vdash_S H \triangleleft G \supset D} \; \text{dp\_Imp2} \langle H \leftsquigarrow G \rangle \qquad \frac{\text{symbol}(G) \notin S \quad \vdash_S G \triangleleft D}{\vdash_S H \triangleleft G \supset D} \; \text{dp\_Imp3/1} \langle H \leftsquigarrow G \rangle$$

$$\frac{\text{symbol}(G) \notin S \quad \vdash_S H \triangleleft D}{\vdash_S H \triangleleft G \supset D} \; \text{dp\_Imp3/2} \langle H \leftsquigarrow G \rangle$$

$$\frac{\vdash_S H \ntriangleleft D}{\vdash_S H \ntriangleleft \forall x : A.D} \; \text{ndp\_Forall} \qquad\qquad \frac{\vdash_S H \ntriangleleft D}{\vdash_S H \ntriangleleft G \supset D} \; \text{ndp\_Imp1} \langle H \not\leftsquigarrow G \rangle$$

$$\frac{\text{symbol}(G) \notin S \quad \vdash_S G \ntriangleleft D \quad \vdash_S H \ntriangleleft D}{\vdash_S H \ntriangleleft G \supset D} \; \text{ndp\_Imp2} \langle H \leftsquigarrow G \rangle$$

Fig. 5. Proving existence and non-existence of dependence paths

For example, consider the example of Fibonacci numbers from Section 2.3. In this case, there are two dependence paths each of length 1 from fib($N;X$) to +($X,Y;Z$) and from fib(s $N; Y$) to +($X,Y;Z$).

It is worth noting that dependence paths are a structural property of a logic program and hence identifying dependence paths is independent of any of the inputs to the program.

Definition 10 (Set of Dependence Paths). *Given a clause D and two goals $G, H \in$* goals(D), $H \triangleleft^* G$ *is the set of all dependence paths from G to H*

For a clause D and a goal H, we define a judgment $\vdash_S H \triangleleft D$ which is provable if and only if there exists a goal $G \in$ goals(D) such that symbol(G) $\in S$ and there is a dependence path from G to H. Similarly, we define the judgment $\vdash_S H \ntriangleleft D$. Figure 5 gives the deductive systems corresponding to these judgments.

Criteria for Functions with Bounded Recursion. Now we can define an extended version of the conditions given in Figure 3; the corresponding judgment is given by $\vdash_S \mathcal{F}$ poly$_{br}$. In this case, $\vdash_S \mathcal{F}$ poly$_{\{b,br\}}$ means that either $\vdash_S \mathcal{F}$ poly$_b$ or $\vdash_S \mathcal{F}$ poly$_{br}$ is true.

These conditions are given in Figure 6 below and they generalize the conditions given earlier. In this case, we distinguish between functions that have function calls that use output of a recursive function call and functions that do not. We require that in the former case, the function calls which use output of a recursive call are non-size increasing in addition to being polynomial-time computable (compare rules br_Imp2/1 and br_Imp2/2). The conditions ensure that the size of the output of the logic programs which satisfy these criteria is polynomially bounded in their input. In the rule pc_Atom we require that the sum of all the inputs to the recursive calls is not larger than the original input. We require that we count the inputs to those recursive calls whose outputs have been used either as input to other function calls or in the final output (with corre-

Programs:

$$\frac{}{\vdash_S \bullet \text{ poly}_{br}} \text{ br\_empty} \quad \frac{\text{symbol}(D) \in S \quad \vdash_S \bullet/D \text{ poly}_{br} \quad \vdash_S \mathcal{F} \text{ poly}_{br}}{\vdash_S \mathcal{F}, c : D \text{ poly}_{br}} \text{ br\_clause1} \quad \frac{\text{symbol}(D) \notin S \quad \vdash_S \mathcal{F} \text{ poly}_{br}}{\vdash_S \mathcal{F}, c : D \text{ poly}_{br}} \text{ br\_clause2}$$

Clauses:

$$\frac{}{\vdash_S \Delta/P \text{ poly}_{br}} \text{ br\_Atom} \left\langle \sum_{\substack{G \in \Delta \\ \text{symbol}(G) \in S}} \beta_G(P)\text{sz}_i(G) + \sum_{\substack{H \in \Delta \\ \text{symbol}(H) \notin S}} \sum_{\substack{G \in \Delta \\ \text{symbol}(G) \in S \\ p \in H \triangleleft^* G}} \beta_H(P)\text{sz}_i(G)\text{width}(p) \leq \text{sz}_i(P) \right\rangle$$

$$\frac{\vdash_S \Delta/D \text{ poly}_{br}}{\vdash_S \Delta/\forall x : A.D \text{ poly}_{br}} \text{ br\_Forall} \quad \frac{\vdash_S \Delta, G/D \text{ poly}_{br} \quad \text{symbol}(G) \in S}{\vdash_S \Delta/G \supset D \text{ poly}_{br}} \text{ br\_Imp1} \langle \text{sz}_i(G) < \text{sz}_i(D) \rangle$$

$$\frac{\vdash_S \Delta, G/D \text{ poly}_{br} \quad \text{symbol}(G) \notin S \quad \vdash_S G \triangleleft D \quad \vdash_T \mathcal{F} \text{ nsi} \quad \vdash_T \mathcal{F} \text{ poly}_{(br,u)}}{\vdash_S \Delta/G \supset D \text{ poly}_{br}} \text{ br\_Imp2/1}$$
(where T is a set of mutually recursive predicate symbols such that $\text{symbol}(G) \in T$)

$$\frac{\vdash_S \Delta, G/D \text{ poly}_{br} \quad \text{symbol}(G) \notin S \quad \vdash_S G \ntriangleleft D \quad \vdash_T \mathcal{F} \text{ poly}_{(br,u)}}{\vdash_S \Delta/G \supset D \text{ poly}_{br}} \text{ br\_Imp2/2}$$
(where T is a set of mutually recursive predicate symbols such that $\text{symbol}(G) \in T$)

Fig. 6. Criteria for identifying polynomial-time functions with bounded recursion

```
mergesort(nil; nil)                          merge(nil, w; w)
mergesort(cons x xs; w)                       merge(w, nil; w)
    ⊂ split(cons x xs; y, z)                  merge(cons x xs, cons y ys; cons u z)
    ⊂ mergesort(y; y₁)                            ⊂ compare(x, y; t)
    ⊂ mergesort(z; z₁)                            ⊂ merge'(t, cons x xs, cons y ys; u, v, w)
    ⊂ merge(y₁, z₁; w)                            ⊂ merge(v, w; z)

split(nil; nil, nil)                          merge'(true, cons x xs, cons y ys; x, xs,
split(cons x nil; cons x nil, nil)                cons y ys)
split(cons x (cons y xs); cons x x₁, cons y y₁)  merge'(false, cons x xs, cons y ys; y,
    ⊂ split(xs; x₁, y₁)                           cons x xs, ys)
```

Fig. 7. Merge Sort

sponding multiplicities). Thus, the sum $\sum_{\substack{G \in \Delta \\ \text{symbol}(G) \in S}} \beta_G(P)\text{sz}_i(G)$ accounts for the first

case and $\displaystyle\sum_{\substack{H \in \Delta \\ \text{symbol}(H) \notin S}} \sum_{\substack{G \in \Delta \\ \text{symbol}(G) \in S \\ p \in H \triangleleft^* G}} \beta_H(P)\text{sz}_i(G)\text{width}(p)$ for the second.

This ensures that the input arguments to goal H are polynomial in the original input arguments of the clause D. Hence, the third condition of our basic criteria (rule br_Imp2 in Figure 3) is satisfied.

Theorem 6 (Bounded Recursion). *Given a program $\mathcal{F}$ and a set S of mutually recursive predicate symbols from $\mathcal{F}$ such that $\vdash_S \mathcal{F}$ poly$_{br}$, then there exists monotonically increasing polynomials $p(\cdot)$ and $p'(\cdot)$ such that for all goals G: if $\text{symbol}(G) \in S$ and $\mathcal{D} :: \mathcal{F} \rightarrow G$, then $\text{sz}_0(G) \leq p(\text{sz}_i(G))$ and $\text{sz}(\mathcal{D}) \leq p'(\text{sz}_i(G))$.*

Example 2 (Merge Sort). Consider a representation of a list using the constants nil and cons. The logic program $\mathcal{F}$ corresponding to merge sort is given in Figure 7.

In this example compare($x, y; t$), t is true if $x < y$ and t is false otherwise (clauses omitted for brevity). It is not hard to see that $\vdash_{compare} \mathcal{F}$ poly$_b$

It is also clear that $\vdash_{split} \mathcal{F}$ poly$_b$ as #(xs) $\leq$ #(cons (cons y xs)) for the third declaration of split. The predicate merge' is also in polynomial time as it is not recursive. We can also check that $\vdash_{merge'} \mathcal{F}$ nsi. In this case, the side condition of nsi_Atom is satisfied because $\alpha(\cdot) = 1$ and $\beta_G(\cdot) = 0$ for both declarations of merge'. In fact, we can show that $sz_o(merge'(G)) = sz_i(merge'(G)) - 2$ when given some input through a goal G

We can also show that $\vdash_{merge} \mathcal{F}$ poly$_b$. For this we need to show that #(v) + #(w) $\leq$ #(cons x xs) + #(cons y ys). It is true because merge' is non-size increasing and we know that $1 + $#(cons x xs) + #(cons y ys) $- 2 = $#($u$) + #($v$) + #($w$). We can also show that merge is non-size increasing. Here $\alpha(merge'(\cdot)) = \alpha(merge(\cdot)) = 1$ and we need to show that #(cons) + #(u) + #(v) + #(w) $\leq$ #(cons x xs) + #(cons y ys). This follows from the fact that merge' is non-size increasing.

Finally, it needs to be shown that $\vdash_{mergesort} \mathcal{F}$ poly$_{br}$ as the outputs y_1 and z_1 of mergesort are given as inputs to the predicate merge. In this case, $\beta_{mergesort}(\cdot) = 0$ for both the mergesort subgoals and $\beta_{merge}(\cdot) = 1$ for the second declaration of mergesort. There are also two dependence paths of length $= 1$ from mergesort to merge. Thus, this conditions in Figure 6 require that merge is non-size increasing and #(y) + #(z) $\leq$ #(cons x xs). This follows from split being non-size increasing.$\square$

3.5   Decidability

The formal deductive systems presented in Figures 3, 4 and 6 are terminating if the side conditions can be proved or disproved. These side conditions are simply linear multi-variable inequalities which depend only on the input variables of the function and output variables of the function calls. We have commented in Section 3.2 on some techniques to eliminate output variables in the conditions. After we have removed all the output variables, we simply need to check that the resulting inequality holds over all positive integer values of its input variables[1]. Therefore, these deductive systems are decidable.

4   Extending to Hereditary Harrop Formulas

The results presented so far are quite general and even apply to logic programming languages with dependent types, higher-order terms, and embedded implication. Let us consider hereditary Harrop formulas [17, 18] which allow embedded implications by extending Horn goals G as shown below.

$$\begin{aligned} \textit{Goals} \quad & G ::= \top \mid P \mid \forall x : A.G \mid D \supset G \\ \textit{Clauses} \quad & D ::= G \supset D \mid \forall x : A.D \mid P \end{aligned}$$

The proof search semantics are extended as shown in Figure 8. The embedded implication is operationally interpreted as extending the logic program dynamically during proof-search. Thus, a logic program with hereditary Harrop formulas is polynomial time if we can ensure that all embedded implications satisfy the polynomial time conditions that we have presented so far.

---

[1] The inequality $\sum_{i=0}^{n} a_i x_i + b > 0$ is true for all $x_i > 0$ if $a_i > 0$ and $b > 0$.

$$\frac{}{\mathcal{F} \to \top} \text{ g\_True} \qquad \frac{D \in \mathcal{F} \quad \mathcal{F} \to D \gg P}{\mathcal{F} \to P} \text{ g\_Atom} \qquad \frac{\mathcal{F}, D \to G}{\mathcal{F} \to D \supset G} \text{ g\_Imp}$$

$$\frac{c \text{ new} \quad \mathcal{F} \to [c/x]G}{\mathcal{F} \to \forall x : A.G} \text{ g\_Forall}$$

$$\frac{}{\mathcal{F} \to P \gg P} \text{ c\_Atom} \qquad \frac{\mathcal{F} \to [M/x]D \gg P}{\mathcal{F} \to \forall x : A.D \gg P} \text{ c\_ForAll} \qquad \frac{\mathcal{F} \to D \gg P \quad \mathcal{F} \to G}{\mathcal{F} \to G \supset D \gg P} \text{ c\_Imp}$$

Fig. 8. Proof search semantics for the hereditary Harrop formulas

Example 3 (β-redexes). Since the arguments to predicates P have to be in canonical form, it is not possible to represent functions such as eval which simplify a term in lambda-calculus to its β-normal form.

eval (lam E) (lam E) $\subset \top$,
eval (app E_1 E_2) $V \subset$ eval E_1 (lam E_1') $\subset$ eval E_2 $V_2 \subset$ eval $(E_1'$ $V_2)$ V

However, such predicates can be represented by defining a predicate subst^{A,B} : $(A \to B) \to A \to B$ which performs the substitution explicitly and computes the canonical form. For example, if $A = B = $ exp then subst$^{\text{exp,exp}}$ (written as subst^1 for clarity) is given by

subst$^1(\lambda x.x, V; V) \subset \top$,
subst$^1(\lambda x.\text{app } (E_1 x) (E_2 x), V; (\text{app } (E_1') (E_2')))$
    $\subset$ subst$^1(\lambda x.(E_1 x); E_1') \subset$ subst$^1(\lambda x.(E_2 x); E_2')$,
subst$^1(\lambda x.\text{lam } (\lambda y.(E \ x \ y))), V; \text{lam } (\lambda y.(E'y)))$
    $\subset (\forall y : \text{exp.subst}^1(\lambda x.y, V; y) \supset \text{subst}^1(\lambda x.(E \ x \ y), V; (E' \ y)))$

In this case, we observe that for logic program $\mathcal{F}$ corresponding to subst$^{\text{exp,exp}}$, $\vdash_{\text{subst}^{\text{exp,exp}}} \mathcal{F}$ poly$_b$ because the first declaration is non-recursive, $\sum_{i=1}^2 \#(\lambda x.(E_i x)) < \#(\lambda x.\text{app } (E_1 x) (E_2 x))$ in the second declaration, and the embedded implication in the third declaration in non-recursive.

On the other hand, when $A = \text{exp} \to \text{exp}$ and $B = \text{exp}$ then subst$^{\text{exp}\to\text{exp,exp}}$ (written as subst^2 for clarity) is given by

subst$^2(\lambda f.f, V; V) \subset \top$,
subst$^2(\lambda f.(\text{app } (E_1 \ f) (E_2 \ f)), V; \text{app } E_1' \ E_2')$
    $\subset$ subst$^2(\lambda f.(E_1 f), V; E_1') \subset$ subst$^2(\lambda f.(E_2 f), V; E_2')$,
subst$^2(\lambda f.\text{lam } \lambda y.(E \ f \ y), V; \text{lam } \lambda y.(E' \ y))$
    $\subset (\forall y : \text{exp.subst}^2(\lambda f.y, V; y) \supset \text{subst}^2(\lambda f.(E \ f \ y), V; (E' \ y)))$,
subst$^2(\lambda f.f \ (Ef), V; E'')$
    $\subset$ subst$^2(\lambda f.E \ f, V; E') \subset$ subst$^2(\lambda x.Vx, E'; E'')$

In this case, the first three declarations satisfy the polynomial time conditions we have described so far. In the fourth declaration, output term E' from the recursive call subst$^{\text{exp}\to\text{exp,exp}}$ is provided as input to subst$^{\text{exp,exp}}$. It is easy to see that Stage 1 conditions do not hold for this case because, it is not possible to determine the run time of subst$^{\text{exp,exp}}$ as we do not know the size of its input E'. Stage 2 conditions do not hold either because, subst$^{\text{exp,exp}}$ is a size-increasing function. Now the eval (app E_1 E_2) V is changed to

eval (app E_1 E_2) V
   $\subset$ eval E_1 (lam E_1') $\subset$ eval E_2 V_2 $\subset$ subst$^{A,\exp}(E_1', V_2; E_1'') \subset$ eval $(E_1''$ $V)$

where an appropriate subst$^{A,\exp}$ is chosen.

Therefore, when $A = \exp$ we know that β-reduction is a polynomial time operation, but when A is a higher-order type, our conditions can no longer guarantee that β-reduction is in polynomial time.□

Example 4 (Combinators cont'd). Recall the bracket abstraction algorithm from Example 1 that is used in the conversion from λ-expressions into combinators. We follow standard practice and define a new type exp together with the two constructors app of type exp $\to$ exp $\to$ exp and lam of type (exp $\to$ exp) $\to$ exp. Using our syntax, extend the program $\mathcal{F}$ from Example 1 to a program $\mathcal{F}'$ by the following new declarations.

   convert(app E_1 E_2; MP C_1 C_2) $\subset$ convert($E_1; C_1$) $\subset$ convert($E_2; C_2$),
   convert(lam E); D)
      $\subset$ ($\forall x : \exp. \forall y : \text{comb. ba}(\lambda z : \text{comb. } y; \text{MP K } y)$
         $\supset$ convert$(y; z) \supset$ convert $(E$ $x; C$ $y))$
      $\subset$ ba $(\lambda y : \text{comb. } C$ $y; D)$

We observe that $\vdash_{\text{convert}}$ $\mathcal{F}'$ poly$_{\text{br}}$ because the first declaration satisfies $\sum_{i=1}^{2} \#(E_i) <$ $\#(\text{app } E_1$ $E_2)$, and each embedded implication in the second is non-recursive. Furthermore $\#(E$ $x) < \#(\text{lam } E)$ because E is applied to a parameter x (and not an arbitrary term). In addition, $\vdash_{\text{ba}}$ $\mathcal{F}'$ nsi by rule nsi_Atom where we choose $\alpha(\cdot) = 0$ and $\beta_{\text{ba}}(\cdot) = 1$ for the two recursive calls, and hence the dynamic extension of the bracket abstraction algorithm ba is non-size increasing.□

5   Conclusions

In this paper, we have given criteria for identifying general recursive functions over a simply-typed higher-order term algebra that can be executed in polynomial time. The criteria are informally intuitive and have been rigorously proven sound. Moreover, they are complete for functions over binary numbers as Cobham's functions fall within the criteria. [3].

In future, we wish to use the results presented here to determine if reductions between two NP complete problems represented as logic programs (such as [19]) are poly-time computable. The mathematical foundations of this work presented in Section 3 allow us to identify super-polynomial complexity classes as well; investigating how this observation can be translated into an appropriate decision procedures for super-polynomial checkers is also left to future work.

Acknowledgements

We would like to thank Dana Angluin, Harald Ganzinger, Arvind Krishnamurthy, Adam Poswolsky, Valery Trifonov, and Rakesh Verma for their comments and suggestions during various stages of this research.

References

1. Bellantoni, S., Cook, S.: A new recursion-theoretic characterization of the polytime functions. In: Twenty-fourth Annual ACM Symposium on Theory of Computing. (1992)
2. Leivant, D.: A foundational delineation of computational feasibility. In: Sixth Annual IEEE Symposium on Logic in Computer Science, IEEE (1991) 2–11
3. Cobham, A.: The intrinsic computational complexity of functions. In Bar-Hellel, Y., ed.: Proceedings of the 1964 International Congress for Logic, Methodology, and the Philosophy of Science. (1965) 24–30
4. Leivant, D.: Subrecursion and lambda representation over free algebras. In Buss, S., Scott, P., eds.: Feasible Mathematics. Birkhauser (1990) 281–292
5. Bellantoni, S., Niggl, K.H., Schwichtenberg, H.: Higher type recursion, ramification and polynomial time. Annals of Pure and Applied Logic **104** (2000)
6. Hofmann, M.: Typed lambda calculi for polynomial-time computation. Habilitation thesis, TU Darmstadt (1998)
7. Hofmann, M.: Safe recursion with higher types and BCK-algebra. Annals of Pure Applied Logic **104** (2000) 113–166
8. Ganzinger, H., McAllester, D.: A new meta-complexity theorem for bottom-up logic programs. In: Proc. International Joint Conference on Automated Reasoning. Volume 2083 of Lecture Notes in Computer Science., Springer-Verlag (2001) 5114–5128
9. Givan, R., McAllester, D.: Polynomial-time computation via local inference relations. ACM Transactions on Computational Logic **3** (2002) 521–541
10. Aehlig, K., Schwichtenberg, H.: A syntactical analysis of non-size-increasing polynomial time computation. In: Fifteenth Annual IEEE Symposium on Logic in Computer Science. (2000)
11. Hofmann, M.: Linear types and non-size-increasing polynomial time computation. Information and Computation **183** (2003) 57–85
12. Rohwedder, E., Pfenning, F.: Mode and termination checking for higher-order logic programs. In Nielson, H.R., ed.: Proceedings of the European Symposium on Programming, Linköping, Sweden, Springer-Verlag (1996) 296–310
13. Miller, D.: A logic programming language with lambda-abstraction, function variables, and simple unification. Journal of Logic and Computation **1** (1991) 497–536
14. Qian, Z.: Unification of higher-order patterns in linear time and space. J. Log. Comput. **6** (1996) 315–341
15. Verma, R.M.: General techniques for analyzing recursive algorithms with applications. SIAM Journal of Computing **26** (1997) 568–581
16. Caseiro, V.H.: Equations for defining poly-time functions. PhD thesis, University of Oslo (1997)
17. Harrop, R.: Concerning formulas of the types $A \to B \vee C, A \to (Ex)(Bx)$. Journal of Symbolic Logic **25** (1960) 27–32
18. Miller, D.: Hereditary harrop formulas and logic programming. In: Proceedings of the VIII International Congress of Logic, Methodology, and Philosophy of Science, Moscow (1987)
19. Schürmann, C., Shah, J.: Representing reductions of NP-complete problems in logical frameworks: A case study. In: Eighth ACM SIGPLAN International Conference on Functional Programming, Workshop on Mechanized reasoning about languages with variable binding, MERLIN, ACM (2003)

Confluence of Shallow Right-Linear Rewrite Systems[*]

Guillem Godoy[1] and Ashish Tiwari[2]

[1] Technical University of Catalonia
Jordi Girona 1, Barcelona, Spain
ggodoy@lsi.upc.es
[2] SRI International, Menlo Park, CA 94025
tiwari@csl.sri.com

Abstract. We show that confluence of shallow and right-linear term rewriting systems is decidable. This class of rewriting system is expressive enough to include nontrivial nonground rules such as commutativity, identity, and idempotence. Our proof uses the fact that this class of rewrite systems is known to be regularity-preserving, which implies that its reachability and joinability problems are decidable. The new decidability result is obtained by building upon our prior work for the class of ground term rewriting systems and shallow linear term rewriting systems. The proof relies on the concept of extracting more general rewrite derivations from a given rewrite derivation.

1 Introduction

Term rewriting systems provide a Turing-complete formalism for modeling computation. Terms over a signature encode the state of a system and the rewriting rules specify the dynamics. Rewriting systems have been used this way to model and verify discrete state transition systems, see for instance [2, 6, 11]. Under a slightly different interpretation, rewriting rules can be viewed as defining an equational theory over terms. The direction of the rule, in this case, generally indicates which equivalent form is simpler. This viewpoint has been successfully used for equational reasoning in theorem proving, see for instance [1].

Confluence is a central property of rewrite systems. It guarantees that the order of application of rewrite rules is not significant. When viewed as a model of computation, confluence provides a more general definition of *determinism*. For purposes of verification, confluence generalizes the condition required for partial-order reduction. In the context of equational reasoning, confluence and termination of a computable rewrite relation imply decidability of the word problem for the induced equational theory.

The expressive power of a rewrite system can be limited by imposing additional constraints on the form of terms. For instance, if variables are not allowed,

---

[*] The first author was partially supported by Spanish Min. of Educ. and Science by the LogicTools project (TIN2004-03382). The second author was supported in part by the National Science Foundation under grants ITR-CCR-0326540 and CCR-0311348.

L. Ong (Ed.): CSL 2005, LNCS 3634, pp. 541–556, 2005.

we get *ground* term rewrite system, which have been extensively studied, mainly via mapping them to tree automata [3]. Richer classes of rewrite systems are obtained by allowing restricted variable occurrences in the term rewrite system (or the tree automata transitions). In going from special to more general classes of rewrite systems, the complexity of deciding various fundamental problems, like termination and confluence, increases until all these problems become undecidable. It is, therefore, fruitful to study these properties for some intermediate classes, especially if they are expressive enough to capture interesting rules.

In this context, we consider shallow right-linear term rewrite systems, where every rule $l \rightarrow r$ is such that every variable occurs at most once in r, and all variables in l, r occur at depth 0 or 1. Some examples of shallow right-linear rules are $0 \wedge x \rightarrow 0$, $x \wedge x \rightarrow x$, $1 \wedge x \rightarrow x$, $x \vee x \rightarrow x$ and $x \vee y \rightarrow y \vee x$.

The class of shallow right-linear rewrite systems is very close to the frontier of classes for which confluence is undecidable. A (generally) simpler problem like reachability is known to be undecidable for linear TRS's, and also for shallow TRS's [10]. On the positive side, Takai, Kaji, and Seki [13] showed that right-linear finite-path-overlapping systems effectively preserve recognizability. Since shallow right-linear systems are right-linear and finite-path-overlapping, it follows that the reachability and joinability problems for these systems are decidable. The exact location of the barrier for decidability of termination and confluence inside the class of right-linear finite-path-overlapping systems is still open.

We prove the decidability of confluence for shallow right-linear TRS's. This result uses the decidability of reachability and joinability for this class as a black box. We extend and simplify the ideas presented in [8] where decidability of confluence of shallow linear TRS's was proved. Here, we eliminate the necessity of constructing a rewrite closure for the original TRS (which is difficult for shallow right-linear TRS's, if possible), and the notion of rewriting with marked (sub-)terms. Non-linearity forces us to use extended counterexample witnesses to confluence: pairs $\{s, t\}$ were used in [8], but now larger sets $\{s_1, \ldots, s_n\}$ are needed. Moreover, as in [8], the computation of top-stabilizable constants (constants equivalent to some term that cannot be reduced to a constant) is crucial, but much more difficult here. In fact, we can compute all such constants only when the system is confluent, and the final proof shows that when not all of them are computed, a non-confluence witness is detected.

The procedure to decide confluence of R has two steps. First we add new rules to R and obtain $\overline{R} \supseteq R$ in Section 3.1. Then, in Section 3.3, we present a simple decidable characterization of confluence of R in terms of R- and $\overline{R}$-joinability of certain flat terms.

2   Preliminaries

We use standard notation from the term rewriting literature. A signature Σ is a (finite) set of function symbols, which is partitioned as $\cup_i \Sigma_i$ such that $f \in \Sigma_n$ if arity of f is n. Symbols in Σ_0, called *constants*, are denoted by

a, b, c, d, with possible subscripts. The elements of a set $\mathcal{V}$ of variable symbols are denoted by x, y, z with possible subscripts. The set $\mathcal{T}(\Sigma, \mathcal{V})$ of *terms* over Σ and $\mathcal{V}$, *position* p in a term, *subterm* $t|_p$ of term t at position p, and the term $t[s]_p$ obtained by replacing $t|_p$ by s are defined in the standard way. For example, if t is $f(a, g(b, h(c)), d)$, then $t|_{2.2.1} = c$, and $t[d]_{2.2} = f(a, g(b, d), d)$. The empty sequence, denoted by λ, corresponds to the root position. We denote $t[s_1]_{p_1}[s_2]_{p_2} \cdots [s_n]_{p_n}$ by either $t[s_1, s_2, \ldots, s_n]_{p_1, p_2, \ldots, p_n}$, or $t[s_1, \ldots, s_n]_P$, where $P = \{p_1, \ldots, p_n\}$. By $t|_P$ we denote the set $\{t|_{p_i} : p_i \in P\}$. By $Pos(t)$ we denote the set of all positions p such that $t|_p$ is defined. We write $p_1 \succ p_2$ (equivalently, $p_2 \prec p_1$) and say p_1 is below p_2 (equivalently, p_2 is above p_1) if p_2 is a proper prefix of p_1, that is, $p_1 = p_2.p_2'$ for some nonempty p_2'. Positions p and q are *disjoint* if $p \not\succeq q$ and $q \not\succeq p$.

We will often denote a term $f(t_1, \ldots, t_n)$ by the simplified form $ft_1 \ldots t_n$, and $t[s]_p$ by $t[s]$ when p is clear by the context or not important. By $Vars(t)$ we denote the set of all variables occurring in t. The *height* of a term s is 0 if s is a variable or a constant, and $1 + max_i height(s_i)$ if $s = f(s_1, \ldots, s_m)$. The *depth* of a position p is the length of p. The *size* of a term $fs_1 \ldots s_m$ is $1 + \Sigma_{i=1}^m size(s_i)$.

A substitution σ is sometimes presented explicitly as $\{x_1 \mapsto t_1, \ldots, x_n \mapsto t_n\}$. We assume standard definition for a *rewrite rule* $l \to r$, a *rewrite system* R, the *one step rewrite relation at position p induced by R* $\to_{R,p}$, and the *one step rewrite relation induced by R* (at any position) $\to_R$. The notations $\leftrightarrow$, $\to^+$, and $\to^*$, are standard [5].

A rewrite system R is *confluent* if the relation $\leftarrow_R^* \circ \to_R^*$ is contained in $\to^* \circ \leftarrow^*$, which is equivalent to the relation $\leftrightarrow_R^*$ being contained in $\to^* \circ \leftarrow^*$ (called the Church-Rosser property). A term t is *reachable* from s by R (or, R-reachable) if $s \to_R^* t$. A set S of terms is said to be *equivalent* by R (or, R-equivalent) if $s \leftrightarrow_R^* t$ for all $s, t \in S$. A set S of terms is R-*joinable* if there is a term that is R-reachable from all of them. A *(rewrite) derivation or proof* (from s) is a sequence of rewrite steps (starting from s), that is, a sequence $s \to_R s_1 \to_R s_2 \to_R \cdots$.

A term t is called *ground* if t contains no variables. It is called *shallow* if all variable positions in t are at depth 0 or 1. It is called *linear* if every variable occurs at most once in t. It is *flat* if its height is at most 1. A rule $l \to r \in R$ is called *shallow right-linear* if the term r is linear, and both l, r are shallow; it is *flat* if both l, r are flat terms. A flat rule $l \to r$ is called a *permutation rule* if $height(l) = height(r) = 1$; it is called a *decreasing rule* if $height(l) = 1$ and $height(r) = 0$, and an *increasing rule* if $height(l) = 0$ and $height(r) = 1$.

3   Confluence

Let R be such that every rule $l \to r \in R$ is shallow and right-linear. Using standard transformation rules, the rewrite system R can be transformed into a rewrite system R' such that every rule $l \to r \in R'$ is flat and right-linear. This transformation is achieved by introducing new constants and adding new rewrite rules [7, 9]. Additionally, we can also assume that $\Sigma = \Sigma_0 \cup \{f\}$, where f is of arity m. All these transformations preserve confluence. We assume henceforth that R is a flat and right-linear TRS defined over a signature $\Sigma = \Sigma_0 \cup \{f\}$. We want to decide if R is confluent.

We use the fact that R-equivalence is decidable [4, 12] repeatedly below. Since shallow and right-linear systems are finite-path overlapping and right-linear, the R-reachability and R-joinability relations are also decidable [13]. We assume that the theory of the rewrite system R is not trivial, that is, it is not the case that $x \leftrightarrow_R^* y$. Confluence of such systems can be decided by simply checking if x and y are R-joinable. We also assume that R contains no rule of the form $x \to t$ where $x \notin Vars(t)$. Any R that contains such a rule is trivially confluent.

3.1   Top-Stabilizable Constants

A term $t \in \mathcal{T}(\Sigma, \mathcal{V})$ is called *top-stable* if it cannot be rewritten to a constant in Σ_0 by R, that is, there is no constant $c \in \Sigma_0$ s.t $t \to_R^* c$. A constant c is *top-stabilizable* if it is R-equivalent to a top-stable term. Our intention is, for every set $\{c_1, \ldots, c_k\}$ of equivalent constants that are top-stabilizable, to choose a new *marked* representative constant, say $\overline{c_1}$, and add the k rules $c_1 \to \overline{c_1}$, $c_2 \to \overline{c_1}, \ldots, c_k \to \overline{c_1}$ to R. The intuitive idea for adding these rules is that they allow the rewrite system to replace a top-stabilizable constant by an equivalent constant which can not be *used* by the rewrite system (note that these new constants do not appear in R).

Let $\overline{\Sigma}_0$ be a new set of constants obtained by picking a representative constant from each set of R-equivalent constants and marking it.

$$\overline{\Sigma}_0 = \{\overline{c} : c \in \Sigma_0, \ c \text{ is a chosen representative for its } R\text{-equivalence class}\}$$

We next define $\overline{R}$ (over the new signature $\Sigma \cup \overline{\Sigma}_0$) by adding *certain* rewrite rules of the form $c \to \overline{d}$, where c and d are R-equivalent. The construction of $\overline{R}$ is achieved through a fixpoint computation.

$$R_0 = R$$
$$R_{i+1} = R_i \cup \{c \to \overline{d} : c, d \in \Sigma_0, \exists \text{ flat term } t \in \mathcal{T}(\Sigma \cup \overline{\Sigma}_0, \mathcal{V}) : t \leftrightarrow_{R_i}^* c \leftrightarrow_R^* d,$$
$$\overline{d} \in \overline{\Sigma}_0, \ t \text{ does not rewrite to a constant in } \Sigma_0 \text{ by } R_i\}$$

When we add $c \to \overline{d}$, we also add all rewrite rules $c' \to \overline{d}$, where $c' \leftrightarrow_R^* c$, in the same iteration, and hence, the fixpoint iterations terminate in at most $|\overline{\Sigma}_0|$ steps. Let $\overline{R}$ be the final result.

Example 1. If $R_0 = \{fa \to b, a \to a', fb \to c\}$, then $R_1 = R_0 \cup \{b \to \overline{b}\}$ (due to the witness fa') and subsequently $R_2 = R_1 \cup \{c \to \overline{c}\}$ (due to the witness $f\overline{b}$).

The next lemma states that the addition of the new constants does not change the congruence relation (over the original signature).

Lemma 1. *If $s, t \in \mathcal{T}(\Sigma, \mathcal{V})$, then for all i, $s \leftrightarrow_{R_i}^* t$ iff $s \leftrightarrow_R^* t$.*

We would want a rule $c \to \bar{d}$ to be added if and only if the constant c is top-stabilizable. This is not always the case, but the left-to-right implication is stated in Lemma 5. Its proof uses the following definitions and lemmas. These lemmas are not conceptually difficult but they are of technical nature, and they are used again later, in the last part of Lemma 12.

Definition 1. *Let $s \to_{l\to r,q} t$ be a one-step rewrite derivation with a flat right-linear rule and let $p \in Pos(s)$. We say that p goes to a certain position p' in this derivation, denoted by $Post(s \to t)(p) = p'$, whenever:*

- *$p' = p$, and either (i) q is disjoint from p, or (ii) $p \prec q$, or (iii) $p = q$ and l is not a variable, or*
- *$p = q.q_1.p''$ and $p' = q.q_2.p''$, for some q_1, q_2 such that $l|_{q_1}$ and $r|_{q_2}$ are the same variable.*

If there is no such p', then we say that p does not go anywhere, denoted by $Post(s \to t)(p) = \bot$.

We extend this definition to derivations of arbitrary length. If $s = s_1 \to s_2 \to \cdots \to s_n$, we say that p_1 goes to p_n, or $Post(s \to^* t)(p_1) = p_n$, if there exists $p_2, \ldots, p_{n-1}$ such that for all i, $Post(s_i \to s_{i+1})(p_i) = p_{i+1}$. If no such sequence of positions exists, we say that p does not go anywhere.

For example, in the derivation $f(faab)bc \to_{fxxy \to fxbc,1} f(fabc)bc$, position 1 goes to position 1, 1.1 and 1.2 go to 1.1, and 1.3 does not go anywhere.

Lemma 2. *Let R' be any flat right-linear rewrite system and $Post(s \to^*_{R'} t)(\lambda) = p$. Then $s \to^*_{R'} t|_{p'}$ for any $p' \preceq p$.*

The above lemma depends on the right-shallowness of insertion rules in R.

Lemma 3. *Let R_i be one of the rewrite systems appearing in the construction of $\bar{R}$. Let $s \to^*_{R_i} t$ be any derivation. Let $p_1, \ldots, p_k$ be disjoint positions, that are also disjoint with $Post(s \to^* t)(\lambda)$ whenever it is not $\bot$, and such that every $t|_{p_j}$ is R_i-equivalent to some bar constant $\bar{c}_j$.*

Then $s \to^*_{R_i} t[\overline{c_1}]_{p_1} \cdots [\overline{c_k}]_{p_k}$, where λ goes to the same position as before.

Alternatively, if none of the terms $t|_{p_j}$ is R_i-equivalent to any constant, then there is a derivation $s \to^*_{R_i} t[z_1]_{p_1} \cdots [z_k]_{p_k}$, where λ goes to the same position as before and $z_1, \ldots, z_k$ are new variables.

Proof. We use induction on the length of the derivation $s \to^*_{R_i} t$. For length 0 the result is trivial since there are no positions disjoint with λ. Hence, let the derivation be of the form $s \to^*_{R_i} t' \to_{l\to r,q} t$.

If $q \succeq p_j$ for some j, then it is the case that $t'[\overline{c_1}]_{p_1} \cdots [\overline{c_k}]_{p_k} \equiv t[\overline{c_1}]_{p_1} \cdots [\overline{c_k}]_{p_k}$. Moreover, the p_j's are disjoint with $Post(s \to^*_{R_i} t')(\lambda)$. By induction hypothesis, $s \to^*_{R_i} t'[\overline{c_1}]_{p_1} \cdots [\overline{c_k}]_{p_k}$ with λ going to the same position as before.

Now suppose that $q \not\succeq p_j$ for any j. For every p_j, define a set of maximal disjoint positions $P_j = Pre(t' \to t)(p_j) \subseteq Pos(t')$ as follows:

$$Pre(t' \to t)(p_j) = Maximal(\{p \in Pos(t') : Post(t' \to t)(p) = p_j\}),$$

where $Maximal(P)$ denotes the set of maximal positions in P (wrt $\succ$). Now, note that every P_j is a set of disjoint positions, but moreover, all of them are disjoint with $Post(s \rightarrow^*_{R_i} t')(\lambda)$, and $\cup_{j \in \{1...k\}} P_j$ is a set of disjoint positions. Hence, by induction hypothesis, $s \rightarrow^*_{R_i} t'[\overline{c_1}, \ldots, \overline{c_1}]_{P_1} \ldots [\overline{c_k}, \ldots, \overline{c_k}]_{P_k}$ with λ going to the same place, and $t'[\overline{c_1}, \ldots, \overline{c_1}]_{P_1} \ldots [\overline{c_k}, \ldots, \overline{c_k}]_{P_k} \rightarrow t''$, where t'' may differ from $t[\overline{c_1}]_{p_1} \ldots [\overline{c_k}]_{p_k}$ in the positions p_j of the form $q.l$ for some $l \in \{1, \ldots, m\}$ such that $r|_l$ is a constant. For such a position p_j, $t''|_{p_j} \equiv t|_{p_j} \equiv r|_l$, and hence, this constant is R_i-equivalent to the corresponding $\overline{c_j}$. Therefore, by applying rules of the form $c \rightarrow \overline{c_j}$ on t'', the term $t[\overline{c_1}]_{p_1} \ldots [\overline{c_k}]_{p_k}$ is reached, and hence, this term is also R_i-reachable from the initial s, and with λ going to the same position. The proof for the alternate claim follows the same pattern.   ∎

Example 2. Let $S = \{x \rightarrow gxc, c \rightarrow fc\}$, $R = R_0 \cup S$ and $\overline{R} = R_2 \cup S$, where R_0 and R_2 are as in Example 1. In the $\overline{R}$-derivation $b \rightarrow gbc \rightarrow g(gbc)c \rightarrow g(gbc)fc$, the position λ goes to position 1.1. We can replace c and fc in the disjoint positions 1.2 and 2 by $\overline{c}$ and get a new derivation $b \rightarrow gbc \rightarrow g(gbc)c \rightarrow g(gb\overline{c})\overline{c}$.

We can replace top-stable subterms by equivalent bar-constants in certain R_i-derivations (Lemma 4). Recall that a rewrite step using $l \rightarrow r$ is called decreasing if $height(l) = 1$ and $height(r) = 0$.

Lemma 4. *Let R_i be one of the rewrite systems appearing in the construction of $\overline{R}$ and let $fs_1 \ldots s_m$ be a flat term over $\Sigma \cup \overline{\Sigma_0}$. Let $fs'_1 \ldots s'_m$ be a term obtained from $fs_1 \ldots s_m$ by replacing every bar constant by an R_i-equivalent and R-top-stable term in $\mathcal{T}(\Sigma, \mathcal{V})$. Let $fs'_1 \ldots s'_m \rightarrow^*_R t[ft'_1 \ldots t'_m]_p$ be a derivation in which λ goes to p, and that does not have any decreasing steps applied at the positions where λ goes to.*

Then, there exists a derivation $fs_1 \ldots s_m \rightarrow^*_{R_i} t[ft_1 \ldots t_m]_p$, where λ goes to p, and $ft_1 \ldots t_m$ is obtained from $ft'_1 \ldots t'_m$ by replacing every R-top-stable t'_j R_i-equivalent to a bar constant t_j by this corresponding t_j, and leaving the other t'_j unchanged, that is, $t_j = t'_j$.

Proof. We induct on the length of the derivation $fs'_1 \ldots s'_m \rightarrow^*_R t[ft'_1 \ldots t'_m]_p$. For length 0 the result is trivial since in this case t is the empty context and every t'_j coincides with the corresponding s'_j. Hence, let this derivation be of the form $fs'_1 \ldots s'_m \rightarrow^*_R t' \rightarrow_{l \rightarrow r \in R, q} t[ft'_1 \ldots t'_m]_p$. Now, we distinguish several cases depending on the relationship between q and p.

q and p are disjoint. In this case $Post(fs'_1 \ldots s'_m \rightarrow^*_{R_i} t')(\lambda) = p$ and t' is actually $t'[ft'_1 \ldots t'_m]_p$. By induction hypothesis, $fs_1 \ldots s_m \rightarrow^*_{R_i} t'[ft_1 \ldots t_m]_p$ in which λ goes to p. Moreover, $t'[ft_1 \ldots t_m]_p \rightarrow_{l \rightarrow r, q} t[ft_1 \ldots t_m]_p$, and hence $t[ft_1 \ldots t_m]_p$ is R_i-reachable from $fs_1 \ldots s_m$, with λ going to p.

$q \prec p$, or $q = p$ and $Post(fs'_1 \ldots s'_m \rightarrow^*_{R_i} t')(\lambda) \neq q$. (In the latter case, the step $t' \rightarrow_{l \rightarrow r \in R_i, q} t[ft'_1 \ldots t'_m]_p$ is necessarily a decreasing step.) Let $P = Pre(t' \rightarrow t[\ldots]_p)(p)$. Note that there is a position $p' \in P$ such that $Post(fs'_1 \ldots s'_m \rightarrow^* t')(\lambda) = p'$. By induction hypothesis, $fs_1 \ldots s_m \rightarrow^*_{R_i} t'[ft_1 \ldots t_m]_{p'}$. Using Lemma 3 on this derivation at positions below the other positions in P, we get $fs_1 \ldots s_m \rightarrow^*_{R_i} t''$, where

t'' is $t'[ft_1 \ldots t_m, \ldots, ft_1 \ldots t_m]_P$. We have set up t'' so that $t'' \rightarrow_{l \rightarrow r, q}$ $t[ft_1 \ldots t_m]_p$, and hence $t[ft_1 \ldots t_m]_p$ is R_i-reachable from $fs_1 \ldots s_m$, with λ going to p.

$q = p$ **and** $Post(fs'_1 \ldots s'_m \rightarrow^*_{R_i} t')(\lambda) = q$. In this case, $l \rightarrow r$ has to be a permutation rule: by assumption, no decreasing rules occur at the positions where λ goes to, and an increasing step would imply $Post(fs'_1 \ldots s'_m \rightarrow^*_{R_i}$ $t[ft'_1 \ldots t'_m]_p)(\lambda) \neq p$. Therefore, t' is of the form $t[fr'_1 \ldots r'_m]_p$ and $fr'_1 \ldots r'_m$ rewrites to $ft'_1 \ldots t'_m$ by $l \rightarrow r$. By induction hypothesis, $fs_1 \ldots s_m \rightarrow^*_{R_i}$ $t[fr_1 \ldots r_m]_p$, where λ goes to p. Note that the same permutation rule $l \rightarrow r$ is applicable at λ in $fr_1 \ldots r_m$, since $r'_j \equiv r'_k$ implies $r_j \equiv r_k$ for all j, k. Hence, $fs_1 \ldots s_m \rightarrow^*_{R_i} t[ft_1 \ldots t_m]_p$, where λ goes to p in this derivation.

$p \prec q$. In this case, t' is of the form $t[ft'_1 \ldots t'_{j-1} r'_j t'_{j+1} \ldots t'_m]_p$, where $r'_j \rightarrow_{l \rightarrow r} t'_j$. Note that either $t_j \equiv t'_j$ or t_j is a bar-constant R_i-equivalent to t'_j. In either case, t_j is R_i-equivalent to r'_j.

If $t_j \equiv t'_j$, then r'_j is not an R-top-stable term that is R_i-equivalent to a bar constant occurring in R_i. By induction hypothesis there is a derivation $fs_1 \ldots s_m \rightarrow^*_{R_i} t[ft_1 \ldots t_{j-1} r'_j t_{j+1} \ldots t_m]_p$, where λ goes to p. Since $r'_j \rightarrow t'_j$, it follows that $fs_1 \ldots s_m \rightarrow^*_{R_i} t[ft_1 \ldots t_m]_p$ with λ going to p.

If t_j is a bar-constant R_i-equivalent to t'_j, then t'_j is R-top-stable. If r'_j is also R-top-stable, then induction hypothesis gives exactly what we wanted to prove. Otherwise, if r'_j is not R-top-stable, then, induction hypothesis gives a derivation $fs_1 \ldots s_m \rightarrow^*_{R_i} t[ft_1 \ldots t_{j-1} r'_j t_{j+1} \ldots t_m]_p$, where λ goes to p. Moreover, since r'_j is not R-top-stable, we have $r'_j \rightarrow^*_R c$ for some constant $c \in \Sigma$, which is R_i-equivalent to the bar-constant t_j, and hence, the rule $c \rightarrow t_j$ occurs in R_i. Therefore, $t[ft_1 \ldots t_m]_p$ is R_i-reachable from $fs_1 \ldots s_m$, with λ going to p.   ∎

Example 3. Using $\overline{R}$ from Example 2, in the derivation $g(fa')c \rightarrow g(fa')fc \rightarrow g(gfa'c)fc$, λ goes to λ, and by Lemma 4 we would have a derivation $g\overline{bc} \rightarrow^* g\overline{bc}$.

Lemma 5. *If there is a rule $c \rightarrow \overline{d}$ in $\overline{R}$, then d is R-top-stabilizable.*

Proof. We prove by induction on i that if $c \rightarrow \overline{d}$ is introduced in R_i, then d is R-top-stabilizable. Suppose this is not true for certain i and $c \rightarrow \overline{d}$; i.e., d is not top-stabilizable but this rule has been introduced in the i'th step. By the construction of R_i, there exists a flat term $s = fs_1 \ldots s_m$ in $\mathcal{T}(\Sigma \cup \overline{\Sigma}_0, \mathcal{V})$ R_{i-1}-equivalent to d and such that s does not rewrite to any constant in Σ_0 by R_{i-1}. We can assume that any s_j that is a bar constant already occurs in R_{i-1}: a bar constant s_j not occurring in R_{i-1} can be replaced by a new variable, preserving all properties and the proof. We construct a term $s' = fs'_1 \ldots s'_m$ as follows: if s_j is some bar constant $\overline{e}$ then s'_j is chosen as an R-top-stable term in $\mathcal{T}(\Sigma, \mathcal{V})$ R-equivalent to e that exists by induction hypothesis, and if s_j is not a bar constant then we make s'_j equal to s_j. By Lemma 1, s' and d are R-equivalent. Since d is not R-top-stabilizable, $s' \rightarrow^*_R c'$ for some c' equivalent to d. Wlog, assume that all terms that occur in the derivation $s' \rightarrow^*_R c'$ are in $\mathcal{T}(\Sigma, \mathcal{V})$.

First, suppose that there is a decreasing step at some position where λ goes to in the derivation $s' \rightarrow^*_R c'$. Hence, this derivation can be written as

$fs'_1 \ldots s'_m \to^*_R t[ft'_1 \ldots t'_m]_p \to_{R,p} t' \to^*_R c'$, where the step $t[ft'_1 \ldots t'_m]_p \to_{R,p} t'$ is the first decreasing step at the positions where λ goes to. By Lemma 4, there exists a derivation $fs_1 \ldots s_m \to^*_{R_{i-1}} t[ft_1 \ldots t_m]_p$ where λ goes to p, and $ft_1 \ldots t_m$ is obtained from $ft'_1 \ldots t'_m$ by replacing every R-top-stable t'_j that is R_{i-1}-equivalent to a bar constant t_j by this t_j. By Lemma 2, there exists a derivation $fs_1 \ldots s_m \to^*_{R_{i-1}} ft_1 \ldots t_m$. Since R is flat, the decreasing step applied on the subterm $ft'_1 \ldots t'_m$ can be also applied on $ft_1 \ldots t_m$ and the result is either a constant or one of the t_j's. The first case is not possible since then $fs_1 \ldots s_m$ would R_{i-1}-reach a constant, which is a contradiction. In the second case, we have $fs_1 \ldots s_m \to^*_{R_{i-1}} t_j$. Note that t_j can not be one of the introduced bar constants, since they are not equivalent to d. Hence, $t_j \in \mathcal{T}(\Sigma, \mathcal{V})$ and t_j is R-equivalent to d. Since d is not R-top-stabilizable, there is a derivation $t_j \to^*_R c''$ for some constant equivalent to d. But this derivation is also an R_{i-1}-derivation since $R \subseteq R_{i-1}$, and hence, $fs_1 \ldots s_m$ R_{i-1}-reaches a constant, which is a contradiction.

We can now assume that there are no decreasing steps at the positions where λ goes to in $s' \to^*_R c'$. This implies, in particular, that λ does not go anywhere in this derivation. Hence, the derivation is of the form $fs'_1 \ldots s'_m \to^*_R t[ft'_1 \ldots t'_m]_p \to_{l \to r \in R,p'} t' \to^*_R c'$, where λ goes to p, and there exists $p'' \in Pos(l)$ such that $l|_{p''}$ is a variable not occurring in r, and $p'.p'' \preceq p$. As before, by Lemma 4, there exists a derivation $fs_1 \ldots s_m \to^*_{R_{i-1}} t[ft_1 \ldots t_m]_p$ where λ goes to p. From $t = t[ft_1 \ldots t_m]_p$ we construct a new term $t'' = t[t|_{p'.p''}]_{p'.p_1} \ldots [t|_{p'.p''}]_{p'.p_k}$, where $p_1 \ldots p_k$ are all the positions in l where the variable $l|_{p''}$ occurs. By Lemma 3, $fs_1 \ldots s_m \to^*_{R_{i-1}} t''$. By construction of t'', $t'' \to_{l \to r \in R,p'} t' \to^*_R c'$, and since $R \subseteq R_{i-1}$, it follows that $fs_1 \ldots s_m$ R_{i-1}-reaches a constant, which contradicts the assumption on $fs_1 \ldots s_m$. ∎

3.2   Detection of Top-Stabilizable Constants

Not all the top-stabilizable constants are necessarily detected by the fix-point computation. But under certain confluence assumptions, we can guarantee that some of them will be detected.

Lemma 6. *Let R be confluent upto height h, i.e., any set of equivalent terms with height smaller than or equal to h is joinable.*

Then, if t is a top-stable term with height smaller than or equal to $h + 1$ and equivalent to some constant c, then $c \to \overline{d} \in \overline{R}$ for some $\overline{d} \in \overline{\Sigma_0}$.

For proving the previous lemma, we first need some properties about the congruence relation induced by R. Since R is a shallow TRS, R-equivalence can be decided using a paramodulation-based completion procedure [4, 12]. The resulting saturated TRS can be used as the set $Congr(R)$, or alternatively, we can just use the following:

$$Congr(R) = \{l \to r : l, r \text{ are flat}, l \leftrightarrow^*_R r, height(l) \geq height(r)\}.$$

This set could have nonlinear terms on the right-hand sides. In this section, we study some properties of rewriting with a flat TRS R.

Definition 2. *Let t be a term. A set of disjoint positions $P \neq \{\lambda\}$ of t is called maximally equivalent if all $t|_{p \in P}$ are equivalent, and for any $t|_q$ equivalent to them and different from t there exists some $p \in P$ with $p \preceq q$. If a term s (or a set of terms S) is equivalent to $t|_P$, we say that P is the maximally equivalent set in t equivalent to s (S).*

The following lemma shows that, in some cases, the relation $\to_R^*$ is preserved after replacing some subterms by variables.

Lemma 7. *Let R be a flat TRS. Let $s \to_{Congr(R)}^* t$, and let P be maximally equivalent in s such that the terms $s|_{p \in P}$ are not equivalent to a constant.*
Then, either $s[z, \ldots, z]_P \to_{Congr(R)}^* z$ and s is equivalent $s|_P$, or $s[z, \ldots, z]_P \to_{Congr(R)}^* t[z, \ldots, z]_{P'}$, where P' is the maximally equivalent set in t equivalent to the terms $s|_{p \in P}$.

Proof. We prove it for one step derivations, since the proof inductively extends to any length. Hence, assume that $s \to_{Congr(R), q} t$ and P is as above.

If $p \preceq q$ for some $p \in P$, then $s|_{p'}$ and $t|_{p'}$ are equivalent for any $p' \preceq p$ and for any p' disjoint with p, and the result trivially follows. Hence, assume that $p \npreceq q$ for any $p \in P$. Note that $s|_{p'}$ and $t|_{p'}$ are equivalent for any $p' \preceq q$ and for any p' disjoint with q.

If the applied rule is of the form $f(\ldots, x, \ldots) \to x$ with the x of the left-hand side occurring in a position i such that $q.i \in P$, then $t|_q$ is equivalent to terms $s|_P$. In such a case, $s|_q$ is also equivalent to $s|_P$, and since q was not in P, it can only be that $q = \lambda$ and $s[z, \ldots, z]_P \to_{Congr(R)} z$ trivially.

In any other case, the maximally equivalent set in t equivalent to $s|_P$ is the set $\{p'|p' \in P$ disjoint with $q\} \cup \{q.p'|p'$ is in the maximally equivalent set of $t|_q$ equivalent to $s|_P\}$. Let $l \to r$ and σ be the applied rule and substitution. Let σ' be as σ except for the variables $x \in Vars(r) - Vars(l)$, for which we define $x\sigma' = x\sigma[z, \ldots, z]_{P_x}$, where $P_x = \{\lambda\}$ if $x\sigma$ is equivalent to $s|_P$, and P_x is the maximally equivalent set in $x\sigma$ equivalent to $s|_P$, otherwise. Then, $s[z, \ldots, z]_P$ rewrites into $t[z, \ldots, z]_{P'}$ applying $l \to r$ with σ' at position q (note that positions with constants in r are not equivalent to $s|_P$ since, by the assumptions of the lemma, $s|_P$ is not equivalent to a constant). ∎

Corollary 1. *Let $s_1, \ldots, s_n$ be terms that reach a term t by $\to_{Congr(R)}^*$. Let $P_1, \ldots, P_n$ be maximally equivalent positions in $s_1, \ldots, s_n$, respectively, such that all terms in $s_1|_{P_1}, \ldots, s_n|_{P_n}$ are equivalent, but not equivalent to a constant.*
Then, either some $s_i[z, \ldots, z]_{P_i}$ reaches z by $\to_{Congr(R)}^*$ and the s_j's are all equivalent to $s_i|_{P_i}$, or all $s_i[z, \ldots, z]_{P_i}$'s reach the same term $t[z \ldots z]_{P'}$ by $Congr(R)$, where P' is the maximal set in t equivalent to $s_i|_{P_i}$; and hence, they are all equivalent.

Let R and $\overline{R}$ be as in the previous subsection. The following lemma shows that, in some cases, an $\overline{R}$-derivation can be transformed into an R-one.

Lemma 8. *For every bar constant $\bar{c}$, let $t_{\bar{c}}$ be a term R-reachable from all constants equivalent to c. Let s and t be two terms satisfying $s \to_{\overline{R}}^* t$.*
Then, $s\{\ldots \bar{c} \mapsto t_{\bar{c}} \ldots\} \to_R^* t\{\ldots \bar{c} \mapsto t_{\bar{c}} \ldots\}$.

Proof. It is enough to prove it for one step derivations, since then it inductively extends to any length. Hence, assume $s \to_{\overline{R}} t$. If this step uses a rule in R, then the result is trivial. Otherwise, it uses a rule of the form $d \to \overline{c}$, and the result trivially follows from the fact that $t_{\overline{c}}$ is reachable from d. ∎

Now, we are ready to prove Lemma 6

Proof. (of Lemma 6) The proof is by contradiction. We consider a term t as a counterexample witness to the goal, if its height is smaller than or equal to $h+1$, t is top-stable, and it is equivalent to a constant that has not been detected as top-stabilizable. We compare witnesses by the size ordering.

Assume that the minimal counterexample witness is a certain term t equivalent to some constant c.

First, we show that *all terms occurring in t at depth 1 and with non-zero height are equivalent to constants.* Suppose not. Let s be a height non-zero subterm of t at depth 1 that is not equivalent to a constant. Let P be the maximally equivalent set of positions of t equivalent to s. Since $t \to^*_{Congr(R)} c$, by Lemma 7, either $t[z, \ldots, z]_P \to^*_{Congr(R)} z$ and t is equivalent to s, or $t[z, \ldots, z]_P \to^*_{Congr(R)} c[z, \ldots, z]_{P'}$, where P' is the maximally equivalent set in c equivalent to s. The first case is not possible, since s is not equivalent to a constant and t is equivalent to c. In the second case, $c[z, \ldots, z]_{P'} = c$, and hence $t[z, \ldots, z]_P$ is a term equivalent to c and smaller than t. Therefore, by the minimality assumptions on t, the term $t[z, \ldots, z]_P$ is not stable, i.e. $t[z, \ldots, z]_P \to^*_R d$ for some constant d equivalent to c. On the other hand, the set $t|_P$ is equivalent and all its terms have height smaller than $h+1$, and hence it is joinable to some term r. Consequently t reaches $t[r, \ldots, r]_P$, and the derivation $(t[z, \ldots, z]_P \to^*_R d)\{z \mapsto r\}$ shows that $t[r, \ldots, r]_P$ reaches d, which contradicts the fact that t is top-stable.

Furthermore, by minimality of t it follows that every such height non-zero subterm s at depth 1 in t is top-stable, and moreover, its corresponding equivalent constant has been detected as top-stabilizable, since otherwise, we would have a smaller counterexample witness $\{s\}$.

Now, let t' be like t but where every height non-zero depth 1 subterm is replaced by its corresponding $\overline{R}$-equivalent bar-constant. Clearly, t' is flat and $t' \leftrightarrow^*_{\overline{R}} c$. Since c was not detected as top-stabilizable, t' $\overline{R}$-rewrites to some constant d equivalent to c. For every bar-constant $\overline{e}$, let $S_{\overline{e}}$ be the set of all constants R-equivalent to e, plus all the terms equivalent to e occurring at depth 1 in t. Each such $S_{\overline{e}}$ is joinable, since the height of its terms is smaller than $h+1$, and hence, we can choose a term $t_{\overline{e}}$ reachable from $S_{\overline{e}}$. By Lemma 8, $t'\{\ldots \overline{e} \mapsto t_{\overline{e}} \ldots\} \to^*_R d\{\ldots \overline{e} \mapsto t_{\overline{e}} \ldots\}$. But $d\{\ldots \overline{e} \mapsto t_{\overline{e}} \ldots\}$ is d and $t'\{\ldots \overline{e} \mapsto t_{\overline{e}} \ldots\}$ is reachable from t, and this contradicts the fact that t is top-stable. ∎

3.3 Deciding Confluence for Shallow Right-Linear Systems

Before proving the decidability of confluence, we need some additional lemmas that show that rewrite derivations using shallow and right-linear rules can be generalized to yield "more-general" rewrite derivations.

Lemma 9. *Let s be a flat term such that every constant in s also occurs in $\overline{R}$. Let $s \to^*_{\overline{R}} t[r]_p$ be a derivation where λ goes to p, and without decreasing steps applied at the positions where λ goes to. Let r' be $r[z]_i$ where $i \in \{1, \ldots, m\}$ and either $r|_i$ is equivalent to the variable z, or $r|_i$ is not equivalent to any height 0 term and z is a new variable.*

Then, there exists a derivation $s \to^*_{\overline{R}} t[r']_p$, where λ goes to p.

The proof of the previous lemma, generalized to several positions $i_1, \ldots, i_k$, is completely analogous to the one of Lemma 4.

Lemma 10. *Suppose $s \to^*_{\overline{R}} t$ and $Post(s \to^* t)(\lambda) = p \neq \perp$. Let α be a bar-constant $\overline{R}$-equivalent to s if such a bar-constant exists, and let α be a variable if s is not $\overline{R}$-equivalent to any constant.*

Then, for all $p' \preceq p$, there is a derivation $\alpha \to^*_{\overline{R}} t[\alpha]_{p'}$ in which λ goes to the position p'.

Proof. We prove by induction on the length of the derivation $s \to^*_{\overline{R}} t$. The base case is trivial. Assume the above derivation is of the form $s \to^*_{\overline{R}} t[l\sigma]_q \to t[r\sigma]_q$, and take $p' \preceq p$. We analyze the following cases:
(a) If p and q are disjoint, then λ goes to p in $s \to^* t[l\sigma]_q$. Applying induction hypothesis on this derivation we have $s \to^* t[l\sigma]_q[\alpha]_{p'}$ with λ going to p', and the same rule $l \to r$ applied at position q finishes the proof.
(b) If $p' \preceq q$ then λ goes to a position below or at p' in $s \to^*_{\overline{R}} t[l\sigma]$, and induction hypothesis on this derivation establishes the claim.
(c) Suppose $p = q.i.q'$ and $p' = q.i.q''$. We again distinguish two cases. If l is a variable, then λ goes to $q.q'$ in $s \to^* t[l\sigma]$. By induction hypothesis, we get $\alpha \to^* t[\alpha]_{q.q''}$. Now, apply the rule $l \to r$ at position q to rewrite $t[\alpha]_{q.q''}$ into $t[\alpha]_{q.i.q''}$. This concludes the first case. In the second case, we assume that l is not a variable. Apply induction hypothesis on the subderivation $s \to^* t[l\sigma]_q$ at the appropriate position in $P = Pre(t[l\sigma] \to t[r\sigma])(p')$. We replace the terms at other positions in P by α using Lemma 3 to finally get the derivation $\alpha \to^*_{\overline{R}} (t[l\sigma]_q)[\alpha, \ldots, \alpha]_P$. Note that this is required because l may be nonlinear. Now, we can use the rule $l \to r$ to rewrite $(t[l\sigma]_q)[\alpha, \ldots, \alpha]_P$ into $(t[r\sigma]_q)[\alpha]_{p'}$. ∎

Example 4. Applying Lemma 10 to the derivation $b \to^* g(gbc)fc$ of Example 2, we infer that there will be derivations $\overline{b} \to^* g(g\overline{b}c)fc$, $\overline{b} \to^* g\overline{b}fc$, and $\overline{b} \to^* \overline{b}$.

The following lemma is an addendum to Lemma 4.

Lemma 11. *Let s be a flat term with constants occurring in $\overline{R}$, and let s' be obtained by replacing in s every bar-constant by an $\overline{R}$-equivalent R-top-stable term in $T(\Sigma, V)$. Let $s' \to^*_R t$ be a derivation where λ does not go anywhere, and without decreasing steps applied at the positions where λ goes to.*

Then, there exists a derivation $s \to^*_{\overline{R}} t$, where λ does not go anywhere. Moreover, if s is not $\overline{R}$-equivalent to a constant, then there exists a derivation $z \to^*_{\overline{R}} t$ for any variable z, and where λ does not go anywhere.

We are ready to give a characterization for confluence of R.

Lemma 12. *R is confluent iff the following two conditions hold:*

(i) Every R-equivalent set of constants of Σ is R-joinable.

(ii) Let $\{\alpha_1, \ldots, \alpha_k, t_1, \ldots, t_n\}$ be an $\overline{R}$-equivalent set of terms, where $n \geq 1$, $k+n \geq 2$, α_i's are constants in Σ_0 or variables (in this case there is at most one variable, i.e. $k \in \{0, 1\}$), and t_i's are flat terms over $\Sigma \cup \overline{\Sigma}_0$ such that no t_i can reach a constant in Σ or a variable by $\rightarrow^*_{\overline{R}}$.
Then, there exist $t'_1, \ldots, t'_n$ such that every t'_i is either t_i or $\overline{c}$ or x, some t'_i coincides with t_i, and the set $\{\alpha_1, \ldots, \alpha_k, t'_1, \ldots, t'_n\}$ is $\overline{R}$-joinable. Here $\overline{c}$ is the (possible) bar-constant in the equivalence class of the set and x is the (possible) variable in the equivalence class of the set.

Proof. $\Leftarrow$: For the right-to-left direction, we will prove a more general statement: R is confluent, and all top-stabilizable constants have been detected (during the fixpoint computation that constructs $\overline{R}$), i.e., $\overline{d}$ occurs in $\overline{R}$ if $\overline{d} \in \overline{\Sigma}_0$ and d is top-stabilizable. The proof is by contradiction, and we consider two kinds of counterexamples to the goal: a multiset $\{t_1, \ldots, t_n\}$ with $n \geq 2$ is a counterexample witness to confluence if it is equivalent but not joinable. A single set $\{t\}$ is a witness to the top-stabilizability detection if t is top-stable but it is equivalent to a constant that has not been detected as top-stabilizable. We compare witnesses $\{t_1, \ldots, t_n\}$ using the multiset extension of the size ordering.

Witness to Top-Stabilizability Detection: Assume that the minimal counterexample witness is $\{t\}$, where t is top-stable and equivalent to some constant c that has not been detected as top-stabilizable.

By minimality of $\{t\}$, all equivalent sets of terms with height smaller than the height of t are joinable. Therefore, by Lemma 6, we conclude that c has been detected as top-stabilizable, which is a contradiction.

Witness to Confluence: Assume that the minimal counterexample witness is a witness to confluence. The witness can not contain only constants, due to condition (i). Let $\{\alpha_1, \ldots, \alpha_k, t_1, \ldots, t_n\}$ be the minimal counterexample witness to confluence, where the t_i's are not constants, $n \geq 1$ and $n + k \geq 2$.

We first prove that all terms occurring in the t_i's at depth 1 and with height non-zero are equivalent to constants. Suppose not. Let s be a height nonzero subterm at depth 1 of some term t_i such that s is not equivalent to a constant. We pick a term t reachable from all $\{t_1, \ldots, t_n\}$ by $\rightarrow_{Congr(R)}$. If there are constants in the witness ($k \geq 1$), we choose t to be a constant. Let $P_1, \ldots, P_n$ be the maximal positions equivalent to s in $t_1, \ldots, t_n$, respectively. By Corollary 1, either for some t_i, say t_1, it happens that $t_1[z \ldots z]_{P_1} \rightarrow_{Congr(R)} z$ and t_1 is equivalent to $t_1|_{P_1}$ and to s (in this case no constant appears in the witness, i.e., $k = 0$), or all the $t_i[z, \ldots, z]_{P_i}$'s reach the same term $t[z \ldots z]_{P'}$ by $Congr(R)$, where P' is maximal in t, and hence they are all equivalent (in this case, if there are constants in the witness, P' is empty and $t[z \ldots z]_{P'}$ is exactly the constant t).

In the first case, $\{t_1[z, \ldots, z]_{P_1}, z\}$ and $t_1|_{P_1} \cup \{t_2, \ldots, t_n\}$ are both equivalent sets smaller than the original witness. Therefore, each of these two sets is separately joinable, say to terms u and v respectively. Instantiating z by v in

$z \rightarrow^* u$, we get $v \rightarrow^* u\{z \mapsto v\}$. As a result we infer that $u\{z \mapsto v\}$ is reachable from every term in $\{t_1, \ldots, t_n\}$, contradicting that this set is a counterexample to confluence.

In the second case, $\{\alpha_1, \ldots, \alpha_k, t_1[z \ldots z]_{P_1}, \ldots, t_n[z \ldots z]_{P_n}\}$ and $t_1|_{P_1} \cup \ldots \cup t_n|_{P_n}$ are both R-equivalent sets that are smaller than the original witness. Therefore, both these sets are separately R-joinable, say to terms u and v respectively. It is again easy to see that any term in $\{\alpha_1, \ldots, \alpha_k, t_1, \ldots, t_n\}$ reaches $u\{z \mapsto v\}$, contradicting the fact this set is a counterexample for confluence.

We now know that in the minimal counterexample $\{\alpha_1, \ldots, \alpha_k, t_1, \ldots, t_n\}$, any height non-zero term s occurring at depth 1 is equivalent to a constant c. By the minimality of the counterexample, s is top-stable, and hence the constant c is top-stabilizable. Moreover, this has been detected, since $\{s\}$ is smaller than the above witness, and therefore, a bar-constant for the class of c exists. Let $t'_1, \ldots, t'_n$ be as $t_1, \ldots, t_n$, but where every nonzero term occurring at depth 1 has been replaced by its $\overline{R}$-equivalent bar-constant.

First, we show that no t'_i can reach a constant of Σ. Suppose that some t'_i, say t'_1, reaches a constant $c \in \Sigma$. For every bar-constant $\overline{d}$, let $S_{\overline{d}}$ be the set of all constants R-equivalent to d, plus all the terms equivalent to d occurring at depth 1 in t_1. Each such $S_{\overline{d}}$ is joinable, since it is a smaller set compared to the original witness. Hence, we can choose a term $t_{\overline{d}}$ reachable from $S_{\overline{d}}$. By Lemma 8, $t'_1\{\ldots \overline{d} \mapsto t_{\overline{d}} \ldots\} \rightarrow^*_R c$, and since t_1 reaches $t'_1\{\ldots \overline{d} \mapsto t_{\overline{d}} \ldots\}$ by $\rightarrow^*_R$, it follows that t_1 reaches a constant, contradicting that t_1 is top-stable (which follows from minimality of the counterexample).

Now, by Condition (ii), there exist terms $t''_1, \ldots, t''_n$ such that every t''_i is either t'_i or the corresponding bar-constant of its class (if it is equivalent to a constant), some t''_i coincides with t'_i, and the set $\{\alpha_1, \ldots, \alpha_k, t''_1, \ldots, t''_n\}$ is $\overline{R}$-joinable to a certain term r. For every bar-constant $\overline{c}$, let $S_{\overline{c}}$ be now the set of all constants R-equivalent to c, plus all the terms equivalent to c occurring at depth 1 in the t_i's such that t''_i is not a bar-constant, plus all the terms t_i equivalent to c such that t''_i is a bar-constant. Each such $S_{\overline{c}}$ is joinable, since it is a smaller set than the initial witness, and hence, we can choose a term $t_{\overline{c}}$ reachable from $S_{\overline{c}}$. By Lemma 8, every term in $\{\alpha_1, \ldots, \alpha_k, t''_1, \ldots, t''_n\}\{\ldots \overline{c} \mapsto t_{\overline{c}} \ldots\}$ reaches $r\{\ldots \overline{c} \mapsto t_{\overline{c}} \ldots\}$. Since every t_i reaches $t''_i\{\ldots \overline{c} \mapsto t_{\overline{c}} \ldots\}$, this proves that $\{\alpha_1, \ldots, \alpha_k, t_1, \ldots, t_n\}$ is joinable, which contradicts that it is a witness to confluence.

$\Rightarrow$: Assume that R is confluent. This immediately implies Condition (i). Moreover, by Lemma 6, all the top-stabilizable constants have been detected. To show that Condition (ii) is also true, let $\{\alpha_1, \ldots, \alpha_k, t_1, \ldots, t_n\}$ be a set as in Condition (ii).

We choose terms $t'_1, \ldots, t'_n$ such that every t'_i can be obtained by replacing in t_i every bar-constant at depth 1 by an $\overline{R}$-equivalent R-top-stable term of the original signature. The set $\{\alpha_1, \ldots, \alpha_k, t'_1, \ldots, t'_n\}$ is $\overline{R}$-equivalent and contains terms of the original signature, and by Lemma 1 it is also R-equivalent. By confluence, it is R-joinable to a certain term t, i.e. there exist derivations of the form $\alpha_i \rightarrow^*_R t$ and $t'_i \rightarrow^*_R t$. We choose such a t to be a minimal one in size satisfying such condition. Note that, due to this minimality, every height nonzero subterm of t is top-stable.

Let t_i, say t_1, be such that a decreasing step occurs in the derivation $t_1' \to_R^* t$ at some position where λ goes to. Let $t_1' \to_R^* s[r_1']_p \to s[r_2']_p$ be the initial subderivation where a decreasing step appears for the first time. Note that λ goes to p in this derivation, no decreasing step occurs at the positions where λ goes to in $t_1' \to_R^* s[r_1']_p$, and r_1' rewrites to r_2' with a decreasing step at root position. By Lemmas 4 and 9, $t_1 \to_{\overline{R}}^* s[r_1]_p$ where λ goes to p, and r_1 is as r_1' but where, fixing a new variable z, every height nonzero subterm at depth 1 has been replaced by, either z if it is not equivalent to any height 0 term, or by an equivalent bar-constant if it is equivalent to some constant, or by an equivalent variable if it is equivalent to a variable. The same decreasing rule used in $r_1' \to_R r_2'$ can be applied to r_1 obtaining a certain term r_2 that is either a bar-constant, or a variable. Therefore, $t_1 \to_{\overline{R}}^* s[r_2]_p$ where λ goes to p, and by Lemma 2 $t_1 \to_{\overline{R}}^* r_2$. Therefore, $\{\alpha_1, \ldots, \alpha_k, t_1, r_2, \ldots, r_2\}$ is $\overline{R}$-joinable and we are done.

At this point, we can assume that no derivation $t_i' \to_R^* t$ contains a decreasing step at some position where λ goes to, and we distinguish two cases.

The t_i''s Are Not Equivalent to Any Height 0 Term. In this case $k = 0$.

First we show that *in every derivation $t_i' \to_R^* t$, λ goes to somewhere*. Suppose not, i.e. wlog. assume that λ does not go anywhere in $t_1' \to_R^* t$. By Lemma 11 there exists a derivation $z \to_{\overline{R}}^* t$ for any variable z. We choose z to be a new variable not occurring in t. This shows that the theory induced by R is trivial, which contradicts the initial assumptions of this section.

Next we show that *the positions where λ goes to in the derivations $t_i' \to_R^* t$ are not disjoint*. Suppose not, i.e. wlog. assume that λ goes to disjoint p_1 and p_2 in $t_1' \to_R^* t$ and $t_2' \to_R^* t$, respectively. Using Lemma 10 on these two derivations, we get derivations $x \to_{\overline{R}}^* t[x]_{p_1}$ and $y \to_{\overline{R}}^* t[y]_{p_2}$. Now, using Lemma 3 on these, we get derivations $x \to_{\overline{R}}^* t[x]_{p_1}[y]_{p_2}$ and $y \to_{\overline{R}}^* t[y]_{p_2}[x]_{p_1}$. This shows that x and y are $\overline{R}$-equivalent, and by Lemma 1 R-equivalent, which contradicts again the initial assumption of the non-triviality of the theory induced by R.

Finally we show that *the positions where λ goes to in the derivations $t_i' \to_R^* t$ coincide*. Suppose not, i.e. wlog. assume that λ goes to p_1 and $p_1.i.p_2$ in $t_1' \to_R^* t$ and $t_2' \to_R^* t$, respectively. By Lemma 4 applied to $t_1' \to_R^* t$, we get $t_1 \to_{\overline{R}}^* t[r]_{p_1}$. Note that, by Lemma 2 applied to $t_2' \to_R^* t$, t_2 is R-equivalent to $t|_{p_1.i}$, and hence $t|_{p_1.i}$ is not R-equivalent to a height 0 term, and $t|_{p_1.i}$ and $r|_i$ coincide. Therefore, using Lemma 9 we get $t_1 \to_{\overline{R}}^* t[r[y]_i]_{p_1}$. On the other hand, by Lemma 3, applied to $t_2' \to_R^* t$ we get $t_2 \to_{\overline{R}}^* t[r]_{p_1}$. Using Lemma 10 on this derivation, we get $y \to_{\overline{R}}^* t[r[y]_i]_{p_1}$. This shows that an arbitrary variable y is $\overline{R}$-equivalent to t_1, contradicting the non-triviality of the theory induced by R.

Once we know that there exists a position p where λ goes to in all of the derivations $t_i' \to_R^* t$, it is easy to conclude: by Lemma 4 any t_i $\overline{R}$-reaches $t[r]_p$, where r is obtained from $t|_p$ by replacing every height nonzero subterm at depth 1 R-equivalent to a constant by a bar-constant, and hence $\{t_1, \ldots, t_n\}$ is $\overline{R}$-joinable.

The Set $\{\alpha_1, \ldots, \alpha_k, t_1, \ldots, t_n\}$ Is $\overline{R}$-Equivalent to a Height 0 Term. If λ does not go anywhere in a certain derivation $t_i' \to_R^* t$, by Lemma 11,

$t_i \rightarrow^*_{\overline{R}} t$. Therefore, if λ does not go anywhere in any derivation $t'_i \rightarrow^*_{\overline{R}} t$, then $\{\alpha_1, \ldots, \alpha_k, t_1, \ldots, t_n\}$ is $\overline{R}$-joinable to t and we are done. Hence, from now on we assume that λ goes to somewhere in some of the derivations $t'_i \rightarrow^*_{\overline{R}} t$, say in $t'_1 \rightarrow^*_{\overline{R}} t, \ldots, t'_l \rightarrow^*_{\overline{R}} t$, for some $l \geq 1$ and $l \leq n$, and let $p_1, \ldots, p_l$ the positions where λ goes to in each of these derivations, respectively. From the list of positions $p_1, \ldots, p_l$ we are interested in the ones that are minimal. Wlog. assume that, for some o, $p_1, \ldots, p_o$ are the minimal ones, i.e., for every i in $\{o+1, \ldots, l\}$ there exists a j in $\{1, \ldots, o\}$ such that $p_j \prec p_i$. Now, we define the term t' by fixing a new variable z, and replacing in t every height nonzero subterm at position $p_i.j$, for i in $\{1, \ldots, o\}$ and j in $\{1, \ldots, m\}$, by either an equivalent bar-constant if it is R-equivalent to a constant, or by an equivalent variable if it is R-equivalent to a variable, or by z if it is not R-equivalent to a height 0 term.

Let α be the height 0 term $\overline{R}$-equivalent to $\{\alpha_1, \ldots, \alpha_k, t_1, \ldots, t_n\}$. The term α can be a bar-constant or variable. We finish the proof by showing that t' is $\overline{R}$-reachable from all terms in $\{\alpha_1, \ldots, \alpha_k, t_1, \ldots t_o, \alpha, \ldots, \alpha, t_{l+1}, \ldots, t_n\}$ (the t_i's for i in $\{o+1, \ldots, l\}$ are replaced by α).

- For a term t_i with i in $\{1, \ldots, o\}$, this follows from Lemmas 4, 9 and 3.
- For a term t_i with i in $\{l+1, \ldots, n\}$, this follows from Lemmas 11 and 3.
- For a term t_i with i in $\{o+1, \ldots, l\}$, λ goes to a certain $p_j.j'.p'_i$ in the derivation $t_i \rightarrow^*_R t$ for some j in $\{1, \ldots, o\}$ and j' in $\{1, \ldots, m\}$. By Lemma 10, there is a derivation $\alpha \rightarrow^*_{\overline{R}} t[\alpha]_{p_j.j'}$, where λ goes to $p_j.j'$, and now, the fact that $\alpha \rightarrow^*_{\overline{R}} t'$ follows from Lemma 3.
- For a term α_i such that λ does not go anywhere in $\alpha_i \rightarrow^*_R t$ or it goes to a position p disjoint with $p_1, \ldots, p_o$, this follows from Lemma 3.
- For a term α_i such that λ goes to a position $p_j.j'.p$ for some i in $\{1, \ldots, o\}$ and j' in $\{1, \ldots, m\}$, from Lemma 10 it follows that $\alpha \rightarrow^*_{\overline{R}} t[\alpha]_{p_j.j'}$, where λ goes to $p_j.j'$. If α is a variable then it coincides with α_i, and if α is a bar constant then $\alpha_i \rightarrow \alpha$ is a rule in $\overline{R}$. In either case there is a derivation $\alpha_i \rightarrow^*_{\overline{R}} t[\alpha]_{p_j.j'}$ where λ goes to $p_j.j'$. Now, the fact that $\alpha_i \rightarrow^*_{\overline{R}} t'$ follows from Lemma 3. ∎

Theorem 1. *Confluence of shallow right-linear rewrite systems is decidable.*

Proof. Since a shallow and right-linear system R is finite-path overlapping and right-linear, R-reachability and R-joinability are decidable [13]. R-equivalence is decidable for shallow rewrite systems [4, 12]. As a result, the set $\overline{R}$ can be constructed and the conditions of Lemma 12 can be tested. ∎

4 Conclusion

In this paper, we showed that confluence is decidable for shallow right-linear rewrite systems, thus generalizing the result for shallow linear rewrite systems [8]. The new proof uses the decidability results for reachability and joinability [13] and the word problem [4, 12]. We also prove many properties about rewriting using shallow TRSs and also shallow right-linear TRSs, which are used to prove the main results of this paper. The decidability of termination and confluence for other classes of finite-path overlapping systems is left for future investigation.

References

1. L. Bachmair and H. Ganzinger. Rewrite-based equational theorem proving with selection and simplification. *J. of Logic and Computation*, 4:217–247, 1994.
2. A. Bouajjani. Languages, rewriting systems, and verification of infinite-state systems. In *ICALP*, volume 2076 of *LNCS*, pages 24–39. Springer, 2001.
3. H. Comon, M. Dauchet, R. Gilleron, F. Jacquemard, D. Lugiez, S. Tison, and M. Tommasi. Tree automata techniques and applications. Available on: http://www.grappa.univ-lille3.fr/tata, 1997.
4. H. Comon, M. Haberstrau, and J.-P. Jouannaud. Syntacticness, cycle-syntacticness, and shallow theories. *Information and Computation*, 111(1):154–191, 1994.
5. N. Dershowitz and J. P. Jouannaud. Rewrite systems. In J. van Leeuwen, editor, *Handbook of Theoretical Computer Science (Vol. B: Formal Models and Semantics)*, pages 243–320, Amsterdam, 1990. North-Holland.
6. J. Giesl and H. Zantema. Liveness in rewriting. In *RTA*, volume 2706 of *LNCS*, pages 321–336. Springer, 2003.
7. G. Godoy and A. Tiwari. Deciding fundamental properties of right-(ground or variable) rewrite systems by rewrite closure. In *Intl. Joint Conf. on Automated Deduction, IJCAR*, volume 3097 of *LNAI*, pages 91–106, 2004.
8. G. Godoy, A. Tiwari, and R. Verma. On the confluence of linear shallow term rewrite systems. In *20th Intl. Symp. on Theor. Aspects of Comp. Sci. STACS 2003*, volume 2607 of *LNCS*, pages 85–96. Springer, 2003.
9. G. Godoy, A. Tiwari, and R. Verma. Deciding confluence of certain term rewriting systems in polynomial time. *Annals of Pure and Applied Logic*, 130(1-3):33–59, Dec 2004.
10. F. Jacquemard. Reachability and confluence are undecidable for flat term rewriting systems. *Inf. Process. Lett.*, 87(5):265–270, 2003.
11. N. Martí-Oliet and J. Meseguer. Rewriting logic: roadmap and bibliography. *Theor. Comput. Sci.*, 285(2):121–154, 2002.
12. R. Nieuwenhuis. Basic paramodulation and decidable theories. In *Proc. 11th IEEE Symp. on Logic In Comp. Sc. LICS*, pages 473–483. IEEE Computer Society, 1996.
13. T. Takai, Y. Kaji, and H. Seki. Right-linear finite path overlapping term rewriting systems effectively preserve recognizability. In *Rewriting Techniques and Applications, RTA*, volume 1833 of *LNCS*, pages 246–260, 2000.

The Ackermann Award 2005

Erich Grädel, Janos Makowsky, and Alexander Razborov

Members of EACSL Jury for the Ackermann Award

The Ackermann Award

At the annual conference of the EACSL, CSL'04, it was suggested to the newly elected president of EACSL that steps be taken to make the Annual Conference of EACSL even more attractive for young researchers in Logic and Computer Science. In response to this suggestion, the EACSL Board decided in November 2004 to launch the

Ackermann Award,
the EACSL Outstanding Dissertation Award
for Logic in Computer Science.

The **Ackermann Award** is presented to the recipients at the annual conference of the EACSL. The jury is entitled to give more than one award per year. The award consists of a diploma, an invitation to present the thesis at the CSL conference, the publication of the abstract of the thesis and the citation in the CSL proceedings, and travel support to attend the conference.

The first **Ackermann Award** is presented at this CSL'05. Eligible for the 2005 **Ackermann Award** were PhD dissertations in topics specified by the EACSL and LICS conferences, which were formally accepted as PhD theses at a university or equivalent institution between 1 January 2003 and 31 December 2004.

The jury for the **Ackermann Award** consists of seven members, three of them ex officio, namely the president and the vice-president of EACSL, and one member of the LICS organizing committee.

The current jury consists of

- S. Abramsky (Oxford, LICS Organizing Committee)
- B. Courcelle (Bordeaux)
- E. Grädel (Aachen)
- M. Hyland (Cambridge)
- J.A. Makowsky (Haifa, President of EACSL)
- D. Niwinski (Warsaw, Vice President of EACSL)
- A. Razborov (Moscow and Princeton)

L. Ong (Ed.): CSL 2005, LNCS 3634, pp. 557–565, 2005.

Wilhelm Ackermann 1896-1962

Wilhelm Ackermann[1] was born on March 29, 1896, in Schoönebeck (Kreis Altena) in Westphalia, Germany, at the time belonging to Prussia. His first studies at the University of Göttingen, which were interrupted during the First World War, were devoted to mathematics, physics and philosophy. He obtained his PhD in 1924 under the guidance of David Hilbert. From 1927 until 1961 he taught in secondary schools (Gymnasium). In 1953 he became a corresponding member of the Göttingen Academy of Sciences, and in the same year the university of Münster made him a honorary professor at the Faculty of Mathematics and Exact Sciences. He lectured until three days before his death on December 24, 1962.

His logic textbook, *Grundzuege der Theoretischen Logik* written together with David Hilbert and first published in 1928, was the most influential textbook in the formative years of mathematical logic. Its fourth edition was published in 1959. The book was translated into several languages.

Although W. Ackermann did not pursue an academic career, he nevertheless continued his research work and helped to shape mathematical logic as a tool of scientific investigations. He also played an important role in establishing mathematical logic as a discipline in post-war Germany. His work includes major contributions on

- consistency of arithmetic, set theory and other comprehensive mathematical systems;
- various strengthenings of strict implication;
- complexity and the rate of growth of recursive functions; and
- decision problems in predicate logic.

Every Computer Science student knows the Ackermann function, a recursive function (given by a simple recursive definition) which is provably not primitive recursive. But computer scientists are less aware of his other contributions. Gödel's completeness theorem proves the completeness of the system presented and proved sound by Hilbert and Ackermann. Ackermann was also the main contributor to the logical system known as the epsilon calculus, originally due to Hilbert. Finally, Ackermann solved the decision problem for $\exists^*\forall\exists^*$-formulas positively. As one of the pioneers of logic, he left his mark in shaping logic and the theory of computation. Several of his publications discussed topics which were later further developed in papers presented at the LICS and EACSL conferences.

---

[1] The following sources are used and sometimes quoted verbatim: An obituary, written in English by H. Hermes, appeared in the Notre Dame Journal of Formal Logic, Vol. uVIII, No 1-2, April 1967.
Links on the WEB: Biographical note written by W. Felscher
http://www-gap.dcs.st-and.ac.uk/~history/Mathematicians/Ackermann.html
and Biography base
http://www.biographybase.com/biography/Ackermann_Wilhelm.html

Report of the Jury

The jury received 15 nominations for the **Ackermann Award 2005**. The candidates came from 9 different nationalities and received their PhDs in 7 diferent countries in Europe and North America.

The topics covered the full range of Logic and Computer Science as represented by the LICS and CSL Conferences. All the submissions were of very high standard and contained outstanding results in their particular domain. The jury decided finally, to give for the year 2005 three awards, one for work on tree-automata and their applications to model checking and verification, one for work on logical and algorithmic properties of Knuth-Bendix orderings and their application to automated reasoning, and one for work on lower bounds for the complexity of propositional proof systems and their applications to complexity theory.

The jury considers the awarding of three awards truly exceptional, due to the outstanding quality of all three winners, and plans in the future to give at most two awards per year.

The 2005 **Ackermann Award** winners are, in alphabetical order,

– Mikołaj Bojańczyk from Poland, for his thesis
  Decidable Properties of Tree Languages,
  awarded by Warsaw University, Poland in 2004
  (Supervisor: I. Walukiewicz, Bordeaux)

– Konstantin Korovin from Russia, for his thesis
  Knuth-Bendix orders in automated deduction and term rewriting,
  awarded by the University of Manchester, England in 2003
  (Supervisor: A. Voronkov)

– Nathan Segerlind from the USA, for his thesis
  New Separations in Propositional Proof Complexity,
  awarded by the University of California at San Diego, USA in 2003
  (Supervisors: S. Buss and R. Impagliazzo)

The 2005 Ackermann Award Winners

Mikołaj Bojańczyk

Citation. Mikołaj Bojańczyk receives the 2005 Ackermann Award of the European Association of Computer Science Logic (EACSL) for his thesis

Decidable Properties of Tree Languages

in which he substantially advanced our understanding of the definability and decidability properties of theories of finite and infinite trees.

Background of the Thesis. The automata theoretic and logical study of tree properties has two branches, depending on whether one considers finite or infinite trees. Despite their common origin in the analysis of monadic second-order logic (MSO) over trees in the fundamental papers by Doner, Thatcher/Wright, Rabin, and others, the two research directions have developed in different ways and with different domains of application.

A motivating challenge for the the first part of Bojańcyk's thesis is to find an analogue of Schützenberger's Theorem for trees. More generally, the goal is to decide whether a given regular tree language belongs to some fixed definability class of trees. This is a very ambitious goal. Indeed, while this kind of definability problems are well-understood for words, and despite the fact that the theory of tree languages and tree automata is a bustling field, with close ties to verification and model checking, the question of deciding, or even characterising first-order definability of tree languages has not seen any progress for many years – not for lack of interest, but because previous efforts by doctoral students and experienced researchers alike in the 1990s had been inconclusive. A possible explanation for this is that in the case of words, these questions are tackled through an algebraic approach. In the case of trees, the algebraic approach is much more difficult and much less developed.

The motivating challenges for the second part of Bojańcyk's thesis are extensions of Rabin's Theorem on the decidability of the monadic theory of the infinite binary tree and their applications to program logics and games.

Bojańczyk's Thesis. The thesis *Decidable Properties of Tree Languages* by M. Bojańczyk represents an important advance in the classification theory of regular sets of finite trees, and secondly it introduces an intriguing extension of MSO-logic over infinite trees with interesting applications in programming logics.

The first part deepens our understanding of various logics over finite trees, in particular first-order logic (including the partial tree ordering $<$ in the signature) and chain logic (where second-order variables only range over sets that are linearly ordered by $<$). Other systems are temporal logics like CTL* or fragments thereof. After about twenty years of little progress, the results of the thesis provide an important step forward in the aim of fixing syntactic invariants that allow to decide for a given MSO-property whether it is expressible in such a fragment. Bojańczyk develops a new machinery to study these questions:

- A new class of "path-oriented" tree automata is introduced, called word sum automata, and shown to capture the Boolean closure of deterministic tree languages (accepted by deterministic top-down tree automata).
- A cascade product construction is developed for word sum automata and shown to capture chain logic over trees (while the aperiodic version captures first-order logic).
- The cascade length is shown to capture the quantifier nesting of CTL*-formulae, giving an elegant new proof of the strictness of the corresponding hierarchy.

– A very interesting effective property of tree languages (and tree automata), called confusion, is worked out which has the potential to characterize those regular tree properties that cannot be defined in chain logic.

These concepts and results are beautiful and represent real innovations. The results are then applied to characterize effectively the tree languages definable by three different fragments of CTL*, in which the temporal operators are restricted to EX, EF, respectively the use of both (but excluding the "until" construct). The given characterizations are very attractive for the effective criteria they provide, and they are technical master-pieces for the corresponding completeness proofs.

The second part of the thesis addresses the search for proper extensions of Rabin's Theorem that the monadic theory of the infinite binary tree (S2S) is decidable. There are two approaches: to change the model (e.g., by considering certain infinite graphs instead of the binary tree), or to modify the means of quantification. The thesis follows the second path, rather neglected up to now in the literature.

Mikołaj Bojańczyk introduces an interesting new quantifier B which allows to sharpen the existence of finite sets by the requirement that their size has to be bounded. Two decidability results on the satisfiability question are shown: first for the closure of MSO-logic by B, the existential quantifier plus the connectives "or", "and", secondly for the MSO-formulas preceded by the dual U-quantifier. The proof offers the interesting idea of "quasi-regular" tree languages which are shown to be the appropriate basis for B-quantification. The central contribution of the chapter is the set-up of a subtle balance between expressiveness and decidability. This is very convincingly documented in three different applications: the finite satisfiability problem for the two-way μ-calculus, the bounded tree-width problem for graphs defined inside the full binary tree, and solving pushdown games with stack unboundedness conditions.

The thesis is written in a very concise and fresh style and conveys a spirit of original thought and intuitive clarity.

Biographical Sketch. Mikołaj Bojańczyk was born on 8 June 1977. He studied computer science in Warsaw where he graduated in 2000 with a MSc thesis on two-way alternating automata. For this thesis he obtained the second prize in a national competition of the Polish Informatics Society for best MSc awards in Computer Science. Between 2000 and 2004 he was a PhD student, also at Warsaw, under the supervision of Igor Walukiewicz. He is currently a post-doc researcher at University Paris 7.

His brilliant dissertation is not the only point of excellence in Bojańczyk's work. His results on tree-walking automata, jointly with Colcombet and completely independent of the dissertation, solved a long-standing open problem and got a best paper award at ICALP 2004.

Konstantin Korovin

Citation. Konstantion Korovin recieves the **2005 Ackermann Award** of the European Association of Computer Science Logic (EACSL) for his thesis

Knuth-Bendix orders in automated deduction and term rewriting.

In this thesis he has advanced single-handedly the theoretical and algorithmic foundations of Knuth-Bendix orderings, separating effectively the feasible applicabilty of Knuth-Bendix orderings from its less feasible aspects.

Background of the Thesis. Automated deduction is an important branch of Computer science, which has applications in various areas including specification and verfication of software and hardware, synthesis of safe programs, database systems, computer algebra and others. One of the most popular methods used in automated deduction are resolution based-theorem proving, which can be implemented efficiently, yet is powerful enough for many applications. Incorporated in the resolution method are various unification algorithms and term rewrite techniques. Because of the practical importance of resolution, unification and term rewriting, intensive research has been devoted both to theoretical improvements as well as implementation issues. Among the main tools developed for termination proofs and improved implementation strategies are orderings on term algebras, and the use of ordering restrictions, which allow to cut down the search space.

There are two classes of orderings that are widely used in automated deduction: In the seminal 1970 paper by D. Knuth and P. Bendix *Simple word problems in universal algebra*, Knuth-Bendix Orderings (KBOs), were introduced. Knuth-Bendix orders have a hybrid nature. They are defined on weighted terms of term algebras, relying both on syntactic precedence and a numeric weight function, hence introducing a (non-trivial) combination of integers and terms. In 1979, N. Dershowitz introduced recursive path orderings (RPOs) for the same purpose. Recursive path orders are defined on term algebras, relying on syntactic precendence only, without weights. The literature is rich in variations of the concept of RPOs. A popular variant are the lexicographical path orderings (LPOs) introduced by Levy and Kamin in 1980. Both types of term orderings are are widely used in almost all currently implemented and widely used automated theorem provers. Knuth-Bendix orderings (KBOs) is the main family of orderings used in the theorem provers VAMPIR, E, Waldmeister, and SPASS.

The first order theory of RPOs was proven undecidable in 1992, by R. Treinen, and for LPOs in 1997. by H. Comon and R. Treinen. In 2000 P. Narendran and M. Rusinowitch showed that the first order theory of unary RPOs is decidable. They also showed that solving RPOs constraints is in NP in 1998. It was known to be NP-hard since 1984. There exists an extensive literature on RPOs and LPOs. For a survey and historic details we refer to the handbook article on Rewriting by N. Dershowitz and D.A. Plaisted in *Handbook of Automated Reasoning*, edited by A. Robinson and A. Voronkov, MIT Press and Elsevier, 2001.

Although there have been many results on the properties of all variants of recursive path orderings, virtually nothing was known about the KBOs, before the work of K. Korovin, which was published jointly with his supervisor A. Voronkov. It seems the main problem with proving results about KBOs is the (non-trivial) combination of integers and term algebras, as compared to pure combinatorics on term algebras in the case of RPOs and KBOs.

Korovin's Thesis. Konstantin Korovins thesis deals with the algorithmic properties of Knuth-Bendix orders. In his thesis, he has constructed polynomial time algorithms for the fundamental problems of solving ordering constraints of single inequalities, of the orientabilty of systems of equalities and rewrite rules by KBOs, and for term comparison. He has given lower bounds for the complexity of these problems showing that orientability is P-time complete, and the problem of solving ordering constraints is NP-complete. The general first order decision problem for KBOs is widely believed to be solvable, but no proof of this fact has been found so far. Korovin has shown the decidability of first ordering constraints for unary signatures. The proofs of the main results display a high level of interdisciplinarity, with a blend of optimization theory, complexity theory, and a mastery of a multitude of techniques for establishing effective decision procedures.

Biographical Sketch. Konstantin Korovin was born on 4 April 1976 in Sarapul, Russia (Soviet Union). He received his secondary education at the Specialized Scientific Study Center for Physics, Mathematics, Chemistry and Biology in Novosibirsk in the period from 1992-93. At the age of 17 he entered Novosibirsk University and received both his undergraduate and graduate education there. In 1998 he completed his M.Sc. studies under the supervision of Prof. Andrei Morozov. The title of his M.Sc. thesis is *Compositions of permutations and algorithmic reducibilities*.

From 1999-2002 he was a PhD student under the supervision of Prof. Andrei Voronkov, and received his PhD in 2003 for his thesis *Knuth-Bendix orders in automated deduction and term rewriting*. For this thesis he already received the best PhD Thesis Award of the University of Manchester.

He spent the years 2003 and 2004 as a researcher at the Max Planck Institute in Saarbrücken, Germany, working with Professor Harald Ganzinger, in the difficult period when Ganzinger was already very ill and until his untimely death. He wrote several important papers with Harald Ganzinger, but it was his sole responsibility to elaborate, develope and complete Ganzinger's ideas. After Ganzinger's death he returned to Manchester University where he works as a research associate.

Nathan Segerlind

Citation. Nathan Segerlind recieves the **2005 Ackermann Award** of the European Association of Computer Science Logic (EACSL) for his thesis

New Separations in Propositional Proof Complexity.

His thesis extends switching lemmas, one of the most primary tools in the area, in a very unexpected way. This has allowed Segerlind to solve a host of difficult open problems in propositional proof complexity and, in particular, to take in a single step the proof system $Res(k)$ from an almost complete mystery to being almost completely understood.

Background of the Thesis. The central question of propositional proof complexity can be formulated in a deceivingly simple way. Given a (sound and complete) proof system P for propositional logic and a tautology ϕ, what is the "simplest" (in most cases meaning the shortest) P-proof of ϕ? Partly due to the universal nature of the notion of a propositional tautology, this is an interdisciplinary area on the border between (and with motivations from) mathematical logic, theory of computing, automated theorem proving, cryptography, algebra and combinatorics among other.

Largely influenced by the automated theorem proving (Davis, Putnam (1960), Davis, Longemann, Loveland (1962), Robinson (1965)), the most widely studied proof system in the area is (propositional) Resolution. After considerable efforts by many researchers beginning with the seminal paper by Tseitin (1968), the resolution proof system is by now fairly well understood. We have rather general, industrial methods to analyze the complexity of resolution proofs (like the width-size relation by Ben-Sasson, Wigderson (1999)), as well as concrete results concerning virtually all combinatorial principles normally used as "benchmarks" in the whole area (Haken (1985), Urquhart (1987), Chvátal, Szemerédi (1988), Raz (2001), Razborov (2001)).

Until the work represented by Segerlind's thesis, however, everything looked very different (meaning much more obscure) beyond Resolution. As a typical example, take one of the most influential results in the whole area, exponential lower bounds on the complexity of the *pigeon-hole principle* in the *constant-depth Frege proof system* (Beame, Impagliazzo, Krajíček, Pitassi, Pudlák, Woods (1992)) based on one of the most powerful tools in the area, *switching lemmas for random restrictions*. We do not know how to apply this method to many tautologies where it should have been applicable, and even when successful (like e.g. Beame, Riis (1998)), the switching lemmas and other techniques, already extremely complicated, have to be re-done almost from the scratch. The situation is very similar for the intermediate proof system $Res(k)$ that operates like Resolution but allows in clauses conjunctions of size $\leq k$ rather than just literals. This system is of great potential interest for automated proving (perhaps, of better interest than constant-depth Frege) since it is structured almost as well as Resolution but surprisingly can do more interesting things than the latter (Maciel, Pitassi, Woods (2000)).

Just to give an idea of the state of the art in the area, *random 3-CNF* is one of the most popular benchmark models both in theoretical and practical communities. They had been shown to be hard for Resolution by Chvátal and Szemerédi in 1988, but the same question was widely open for *any* reasonable extension of Resolution, including $Res(2)$.

Another important subject addressed in the thesis is *algebraic proof systems* (like Nullstellensatz and Polynomial Calculus) and hybrid systems combining logical and algebraic reasoning (constant-depth Frege with counting axioms or modular gates). By the time of Segerlind's work, purely algebraic proof systems and relations between them were in general understood much better than purely logical systems, but several important questions remained open.

Segerlind's Thesis. In his thesis Segerlind proved important and nice results about the relative power of algebraic and mixed proof systems, both positive and negative, that had been open for a while. Among other things he showed that counting gates are more powerful than counting axioms (Chapter VI), but the counting axioms can efficiently simulate any Nullstellensatz proof (Chapter V). The most striking part of the thesis, however, concerns the systems $Res(k)$ for small values of k (Chapters III and IV).

The upshot of these latter results is very simple: we now understand the systems $Res(k)$ almost as well as Resolution itself. In his thesis Segerlind analyzed the complexity of the weak pigeon-hole principle and random w-CNFs (both are standard "tests for maturity") in these proof systems, and he also gave separation results between the systems $Res(k)$ and $Res(w)$ for some $w > k$. His techniques were already used in different situations by Razborov and Alekhnovich. So, it really looks like what he has found is a general, powerful and flexible method rather than ad hoc argument. All in all, Segerlind's work changed the situation with these prominent systems from a few scattered and weak results to a few important problems left open. And two features of this development look particularly striking.

The first is its speed. For Resolution it took decades to have reached the level of understanding that was reached here in a single step. Of course, this comparison is not quite fair since the general methodology gathered during these decades also played very essential role in Segerlind's work. But even with this disclaimer the speed with which it all happened was quite remarkable and totally unexpected.

The second surprise came in the form of proof methods. The novelty essentially consists in a new version of switching lemmas called in the thesis *switching with small restrictions*. And, in order to appreciate this, one should be fully aware to *which* extent this tool is basic in both computational complexity and proof complexity. Many researchers have been looking at these lemmas since the seminal work by Håstad (1986). They have been scrutinized and analyzed from every possible perspective and in all fine and technical details. The fact that Segerlind was able to say a substantially new word in an area so often re-visited by strongest researchers is also quite remarkable.

Biographical Sketch. Nathan Segerlind was born on 31 December 1973 in Marlette, Michigan. In 1992-1998 he studied Computer Science and Mathematics at Carnegie Mellon University, Pittsburg. Between 1998 and 2003 he was a PhD student at the University of California at San Diego, under joint supervision by Samuel Bass and Russell Impagliazzo. He spent the next year 2003-2004 as a postdoctoral member at the Institute for Advanced Study, Princeton. Currently he is continuing his post-doc research at the University of Washington, Seattle.

Clemens Lautemann: 1951-2005
An Obituary

The treasurer and board member of the European Association for Computer Science Logic (EACSL), Prof. Dr. Clemens Lautemann passed away on 29 April 2005 at the age of only 53 years after a short, serious illness. He is survived by his wife Rose and his two children Anne (21) and Christopher (16).

Clemens Lautemann joined the board of the EACSL as treasurer in 1997. We, the EACSL community, and all who knew him, will miss him and remember him warmly.

After studying Mathematics, Political Sciences, Philosophy and Computer Sciences at the Universities of Marburg, Bielefeld and Berlin, Clemens Lautemann earned a doctorate in Natural Sciences in June 1983. He dedicated himself to scientific research in Berlin, Osnabrück, Bremen, Edinburgh and Mainz. Theoretical Computer Science was his field of interest, especially Logic and Computational Complexity, and he obtained important results in the fields of Graph Theory, Structural Complexity and Finite Model Theory. His scientific work gained worldwide recognition.

Clemens Lautemann joined the Institute for Computer Science at the Johannes Gutenberg University as Professor for Theoretical Computer Science in 1989. He was an inspiring teacher who motivated his students in his special field of expertise and he was actively engaged in encouraging young people. He supported the *Bundeswettbewerb Informatik* by accompanying pupils interested in the Natural Sciences. Several of his students and PhD candidates were awarded scientific prizes.

Clemens Lautemann was deeply concerned with the further development of the Institute for Computer Science. He cared very much for the Institute's future and played a major part in establishing the new Bachelor of Computer Science program. He was also interested in interdisciplinary cooperation and was one of the founders of the research group *Bioinformatik* at his university.

Clemens Lautemann distinguished himself in many ways, as a warm and forthcoming person, as an original scientist, and as a committed teacher and supervisor with a vision beyond the shortlived fashions of the day. We all liked him very much and will miss his friendly, open-hearted personality. We are very sad that he will not be with us any longer. Clemens Lautemann's family, friends, students, and colleagues will keep pleasant memories of him with them for a long time.

J.A. Makowsky
President of EACSL

L. Ong (Ed.): CSL 2005, LNCS 3634, p. 566, 2005.
© Springer-Verlag Berlin Heidelberg 2005

Author Index

Lecture Notes in Computer Science

For information about Vols. 1–3517

please contact your bookseller or Springer